MERCURY/MARINER Outboards
1990-00 REPAIR MANUAL
2.5-275 HORSEPOWER, 1-6 CYLINDER

Managing Partners	Dean F. Morgantini, S.A.E.
	Barry L. Beck
Executive Editor	Kevin M. G. Maher, A.S.E.
Manager-Marine/Recreation	James R. Marotta, A.S.E.
Production Specialists	Melinda Possinger
	Ronald Webb
Author	Scott Freeman

DISCARD

Manufactured in USA
© 2000 Seloc Publications
104 Willowbrook Lane
West Chester, PA 19382
ISBN 0-89330-051-9
Library of Congress Catalog Card No. 99-076327
4567890123 3210987654

www.selocmarine.com
1-866-SELOC55

Contents

1 GENERAL INFORMATION AND BOATING SAFETY

- **1-2** HOW TO USE THIS MANUAL
- **1-3** BOATING SAFETY
- **1-10** SAFETY IN SERVICE

2 TOOLS AND EQUIPMENT

- **2-2** TOOLS AND EQUIPMENT
- **2-4** TOOLS
- **2-11** FASTENERS, MEASUREMENTS AND CONVERSIONS

3 MAINTENANCE

- **3-2** ENGINE MAINTENANCE
- **3-8** BOAT MAINTENANCE
- **3-10** TUNE-UP
- **3-38** WINTER STORAGE CHECKLIST
- **3-39** SPRING COMMISSIONING CHECKLIST

4 FUEL SYSTEM

- **4-2** FUEL AND COMBUSTION
- **4-3** CARBURETED FUEL SYSTEM
- **4-50** ELECTRONIC FUEL INJECTION (EFI)
- **4-66** OPTIMAX DIRECT FUEL INJECTION (DFI)

5 IGNITION AND ELECTRICAL SYSTEMS

- **5-2** UNDERSTANDING AND TROUBLESHOOTING ELECTRICAL SYSTEMS
- **5-8** BREAKER POINTS IGNITION (MAGNETO IGNITION)
- **5-14** CAPACITOR DISCHARGE IGNITION (CDI) SYSTEM
- **5-30** CHARGING CIRCUIT
- **5-40** STARTER CIRCUIT
- **5-42** IGNITION AND ELECTRICAL WIRING DIAGRAMS

6 LUBRICATION AND COOLING

- **6-2** OIL INJECTION SYSTEM
- **6-8** COOLING SYSTEM
- **6-22** WARNING SYSTEMS
- **6-27** OPTIMAX WARNING SYSTEMS

Contents

7-2	ENGINE MECHANICAL	**7-19**	POWERHEAD RECONDITIONING	**POWERHEAD 7**

8-2	LOWER UNIT	**8-11**	JET DRIVE	**LOWER UNIT 8**

9-2	MANUAL TILT	**9-7**	SINGLE RAM INTEGRAL POWER TILT/TRIM	**TRIM AND TILT 9**
9-5	GAS ASSIST TILT SYSTEM			
9-6	POWER TRIM/TILT	**9-12**	THREE RAM INTEGRAL POWER TILT/TRIM SYSTEM	

10-2	REMOTE CONTROL BOX	**10-6**	TILLER HANDLE	**REMOTE CONTROL 10**

11-2	HAND REWIND STARTER	**11-2**	OVERHEAD TYPE STARTER	**HAND REWIND STARTER 11**

11-25	GLOSSARY	**GLOSSARY**

11-27	INDEX	**MASTER INDEX**

See last page for information on additional titles

SAFETY NOTICE

Proper service and repair procedures are vital to the safe, reliable operation of all marine engines, as well as the personal safety of those performing repairs. This manual outlines procedures for servicing and repairing stern drives using safe, effective methods. The procedures contain many NOTES, CAUTIONS and WARNINGS which should be followed, along with standard procedures, to eliminate the possibility of personal injury or improper service which could damage the vessel or compromise its safety.

It is important to note that repair procedures and techniques, tools and parts for servicing marine engines, as well as the skill and experience of the individual performing the work, vary widely. It is not possible to anticipate all of the conceivable ways or conditions under which these engines may be serviced, or to provide cautions as to all possible hazards that may result. Standard and accepted safety precautions and equipment should be used during cutting, grinding, chiseling, prying, or any other process that can cause material removal or projectiles.

Some procedures require the use of tools specially designed for a specific purpose. Before substituting another tool or procedure, you must be completely satisfied that neither your personal safety, nor the performance of the marine engine, will be compromised.

Although information in this manual is based on industry sources and is complete as possible at the time of publication, the possibility exists that some vehicle manufacturers made later changes which could not be included here. While striving for total accuracy, Nichols Publishing cannot assume responsibility for any errors, changes or omissions that may occur in the compilation of this data.

PART NUMBERS

Part numbers listed in this reference are not recommendations by Nichols Publishing for any product brand name. They are references that can be used with interchange manuals and aftermarket supplier catalogs to locate each brand supplier's discrete part number.

SPECIAL TOOLS

Special tools are recommended by the marine manufacturer to perform a specific task. Use has been kept to a minimum, but, where absolutely necessary, they are referred to in the text by the part number of the tool manufacturer. These tools can be purchased, under the appropriate part number, from your local dealer or regional distributor, or an equivalent tool can be purchased locally from a tool supplier or parts outlet. Before substituting any tool for the one recommended, read the SAFETY NOTICE at the top of this page.

ALL RIGHTS RESERVED

No part of this publication may be reproduced, transmitted or stored in any form or by any means, electronic or mechanical, including photocopy, recording, or by information storage or retrieval system, without prior written permission from the publisher.

ACKNOWLEDGMENTS

Nichols Publishing expresses appreciation to the following companies who supported the production of this manual:

- Mercury Marine—Fond du Lac, WI
- Belk's Marine—Holmes, PA
- Marine Mechanics Institute—Orlando, FL
- Rapair/CDI Electronics—Madison, AL

Thanks to Steve Ehle and the Marine Mechanics Institute for providing the outboard engines and miscellaneous parts to get this project going, to John Hartung and Judy Belk of Belk's Marine for allowing us full access to their dealership for a portion of our photoshoot and to Clark Beard of CDI Electronics for teaching us everything he knows about troubleshooting outboards.

Nichols Publishing would like to express thanks to all of the fine companies who participate in the production of our books:
- Hand tools supplied by Craftsman are used during all phases of our vehicle teardown and photography.
- Many of the fine specialty tools used in our procedures were provided courtesy of Lisle Corporation.
- Lincoln Automotive Products (1 Lincoln Way, St. Louis, MO 63120) has provided their industrial shop equipment, including jacks (engine, transmission and floor), engine stands, fluid and lubrication tools, as well as shop presses.
- Rotary Lifts (1-800-640-5438 or www.Rotary-Lift.com), the largest automobile lift manufacturer in the world, offering the biggest variety of surface and in-ground lifts available, has fulfilled our shop's lift needs.
- Much of our shop's electronic testing equipment was supplied by Universal Enterprises Inc. (UEI).
- Safety-Kleen Systems Inc. has provided parts cleaning stations and assistance with environmentally sound disposal of residual wastes.
- United Gilsonite Laboratories (UGL), manufacturer of Drylok® concrete floor paint, has provided materials and expertise for the coating and protection of our shop floor.

MARINE TECHNICIAN TRAINING

INDUSTRY SUPPORTED PROGRAMS
OUTBOARD, STERNDRIVE & PERSONAL WATERCRAFT

- Dyno Testing • Boat & Trailer Rigging • Electrical & Fuel System Diagnostics
- Powerhead, Lower Unit & Drive Rebuilds • Powertrim & Tilt Rebuilds
- Instrument & Accessories Installation

TRAIN IN SUNNY FLORIDA!

For information regarding housing, financial aid and employment opportunities in the marine industry, contact us today:

CALL TOLL FREE
1-800-528-7995

An Accredited Institution

SM

Name
Address
City   State   Zip
Phone

MEMBER NMMA

MARINE MECHANICS INSTITUTE
A Division of CTI

9751 Delegates Drive • Orlando, Florida 32837
2844 W. Deer Valley Rd. • Phoenix, AZ 85027

FINANCIAL ASSISTANCE AVAILABLE FOR THOSE WHO QUALIFY!

HOW TO USE THIS MANUAL 1-2
CAN YOU DO IT? 1-2
WHERE TO BEGIN 1-2
AVOIDING TROUBLE 1-2
MAINTENANCE OR REPAIR? 1-2
DIRECTIONS AND LOCATIONS 1-2
PROFESSIONAL HELP 1-2
PURCHASING PARTS 1-3
AVOIDING THE MOST COMMON
  MISTAKES 1-3
BOATING SAFETY 1-3
REGULATIONS FOR YOUR BOAT 1-3
  DOCUMENTING OF VESSELS 1-4
  REGISTRATION OF BOATS 1-4
  NUMBERING OF VESSELS 1-4
  SALES AND TRANSFERS 1-4
  HULL IDENTIFICATION NUMBER 1-4
  LENGTH OF BOATS 1-4
  CAPACITY INFORMATION 1-4
  CERTIFICATE OF COMPLIANCE 1-4
  VENTILATION 1-4
  VENTILATION SYSTEMS 1-5
REQUIRED SAFETY EQUIPMENT 1-5
  TYPES OF FIRES 1-5
  FIRE EXTINGUISHERS 1-5
  WARNING SYSTEM 1-6
  PERSONAL FLOTATION DEVICES 1-6
  SOUND PRODUCING DEVICES 1-8
  VISUAL DISTRESS SIGNALS 1-8
EQUIPMENT NOT REQUIRED BUT
 RECOMMENDED 1-9
  SECOND MEANS OF
    PROPULSION 1-9
  BAILING DEVICES 1-9
  FIRST AID KIT 1-9
  ANCHORS 1-9
  VHF-FM RADIO 1-10
  TOOLS AND SPARE PARTS 1-10
COURTESY MARINE
 EXAMINATIONS 1-10
SAFETY IN SERVICE 1-10
DO'S 1-10
DON'TS 1-10

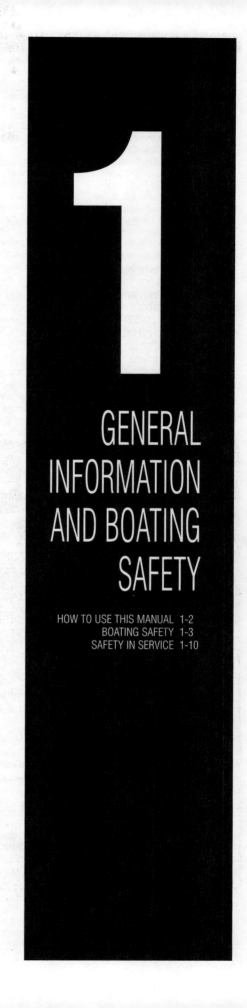

1

GENERAL INFORMATION AND BOATING SAFETY

HOW TO USE THIS MANUAL 1-2
BOATING SAFETY 1-3
SAFETY IN SERVICE 1-10

1-2 GENERAL INFORMATION AND BOATING SAFETY

HOW TO USE THIS MANUAL

This manual is designed to be a handy reference guide to maintaining and repairing your Mercury or Mariner Outboard. We strongly believe that regardless of how many or how few years experience you may have, there is something new waiting here for you.

This manual covers the topics that a factory service manual (designed for factory trained mechanics) and a manufacturer owner's manual (designed more by lawyers these days) covers. It will take you through the basics of maintaining and repairing your outboard, step-by-step, to help you understand what the factory trained mechanics already know by heart. By using the information in this manual, any boat owner should be able to make better informed decisions about what they need to do to maintain and enjoy their outboard.

Even if you never plan on touching a wrench (and if so, we hope that you will change your mind), this manual will still help you understand what a mechanic needs to do in order to maintain your engine.

Can You Do It?

If you are not the type who is prone to taking a wrench to something, NEVER FEAR. The procedures in this manual cover topics at a level virtually anyone will be able to handle. And just the fact that you purchased this manual shows your interest in better understanding your outboard.

You may find that maintaining your outboard yourself is preferable in most cases. From a monetary standpoint, it could also be beneficial. The money spent on hauling your boat to a marina and paying a tech to service the engine could buy you fuel for a whole weekend's boating. If you are unsure of your own mechanical abilities, at the very least you should fully understand what a marine mechanic does to your boat. You may decide that anything other than maintenance and adjustments should be performed by a mechanic (and that's your call), but know that every time you board your boat, you are placing faith in the mechanic's work and trusting him or her with your well-being, and maybe your life.

It should also be noted that in most areas a factory trained mechanic will command a hefty hourly rate for off site service. This hourly rate is charged from the time they leave their shop to the time they return home. The cost savings in doing the job yourself should be readily apparent at this point.

Where to Begin

Before spending any money on parts, and before removing any nuts or bolts, read through the entire procedure or topic. This will give you the overall view of what tools and supplies will be required to perform the procedure or what questions need to be answered before purchasing parts. So read ahead and plan ahead. Each operation should be approached logically and all procedures thoroughly understood before attempting any work.

Avoiding Trouble

Some procedures in this manual may require you to "label and disconnect . . ." a group of lines, hoses or wires. Don't be lulled into thinking you can remember where everything goes — you won't. If you reconnect or install a part incorrectly, things may operate poorly, if at all. If you hook up electrical wiring incorrectly, you may instantly learn a very, very expensive lesson.

A piece of masking tape, for example, placed on a hose and another on its fitting will allow you to assign your own label such as the letter "A", or a short name. As long as you remember your own code, the lines can be reconnected by matching letters or names. Do remember that tape will dissolve when saturated in fluids. If a component is to be washed or cleaned, use another method of identification. A permanent felt-tipped marker can be very handy for marking metal parts; but remember that fluids will remove permanent marker.

SAFETY is the most important thing to remember when performing maintenance or repairs. Be sure to read the information on safety in this manual.

Maintenance or Repair?

Proper maintenance is the key to long and trouble-free engine life, and the work can yield its own rewards. A properly maintained engine performs better than one that is neglected. As a conscientious boat owner, set aside a Saturday morning, at least once a month, to perform a thorough check of items which could cause problems. Keep your own personal log to jot down which services you performed, how much the parts cost you, the date, and the amount of hours on the engine at the time. Keep all receipts for parts purchased, so that they may be referred to in case of related problems or to determine operating expenses. As a do-it-yourselfer, these receipts are the only proof you have that the required maintenance was performed. In the event of a warranty problem, these receipts will be invaluable.

It's necessary to mention the difference between maintenance and repair. Maintenance includes routine inspections, adjustments, and replacement of parts that show signs of normal wear. Maintenance compensates for wear or deterioration. Repair implies that something has broken or is not working. A need for repair is often caused by lack of maintenance.

For example: draining and refilling the engine oil is maintenance recommended by all manufacturers at specific intervals. Failure to do this can allow internal corrosion or damage and impair the operation of the engine, requiring expensive repairs. While no maintenance program can prevent items from breaking or wearing out, a general rule can be stated: MAINTENANCE IS CHEAPER THAN REPAIR.

Directions and Locations

♦ See Figure 1

Two basic rules should be mentioned here. First, whenever the Port side of the engine (or boat) is referred to, it is meant to specify the left side of the engine when you are sitting at the helm. Conversely, the Starboard means your right side. The Bow is the front of the boat and the Stern is the rear.

Most screws and bolts are removed by turning counterclockwise, and tightened by turning clockwise. An easy way to remember this is: righty-tighty; lefty-loosey. Corny, but effective. And if you are really dense (and we have all been so at one time or another), buy a ratchet that is marked ON and OFF, or mark your own.

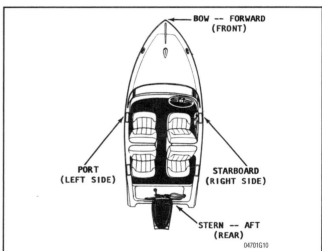

Fig. 1 Common terminology used for reference designation on boats of all size. These terms are used though out the manual

Professional Help

Occasionally, there are some things when working on an outboard that are beyond the capabilities or tools of the average Do-It-Yourselfer (DIYer). This shouldn't include most of the topics of this manual, but you will have to be the judge. Some engines require special tools or a selection of special parts, even for basic maintenance.

Talk to other boaters who use the same model of engine and speak with a trusted marina to find if there is a particular system or component on your engine that is difficult to maintain. For example, although the technique of valve adjustment on some engines may be easily understood and even performed by a DIYer, it might require a handy assortment of shims in various sizes and a few hours of disassembly to get to that point. Not having the assortment of shims handy might mean multiple trips back and forth to the parts store, and this might not be worth your time.

You will have to decide for yourself where basic maintenance ends and where professional service should begin. Take your time and do your research first (starting with the information in this manual) and then make your own decision. If you really don't feel comfortable with attempting a procedure, DON'T DO IT. If you've gotten into something that may be over your head, don't panic. Tuck your tail between your legs and call a marine mechanic. Marinas and independent shops will be able to finish a

GENERAL INFORMATION AND BOATING SAFETY

job for you. Your ego may be damaged, but your boat will be properly restored to its full running order. So, as long as you approach jobs slowly and carefully, you really have nothing to lose and everything to gain by doing it yourself.

Purchasing Parts

♦ See Figures 2 and 3

When purchasing parts there are two things to consider. The first is quality and the second is to be sure to get the correct part for your engine. To get quality parts, always deal directly with a reputable retailer. To get the proper parts always refer to the information tag on your engine prior to calling the parts counter. An incorrect part can adversely affect your engine performance and fuel economy, and will cost you more money and aggravation in the end.

Just remember, a tow back to shore will cost plenty. That charge is per hour from the time the towboat leaves their home port, to the time they return to their home port. Get the picture.....$$$?

So who should you call for parts? Well, there are many sources for the parts you will need. Where you shop for parts will be determined by what kind of parts you need, how much you want to pay, and the types of stores in your neighborhood.

Your marina can supply you with many of the common parts you require. Using a marina for as your parts supplier may be hand because of location (just walk right down the dock) or because the marina specializes in your particular brand of engine. In addition, it is always a good idea to get to know the marina staff (especially the marine mechanic).

The marine parts jobber, who is usually listed in the yellow pages or whose name can be obtained from the marina, is another excellent source for parts. In addition to supplying local marinas, they also do a sizeable business in over-the-counter parts sales for the do-it-yourselfer.

Almost every community has one or more convenient marine chain stores. These stores often offer the best retail prices and the convenience of one-stop shopping for all your needs. Since they cater to the do-it-yourselfer, these stores are almost always open weeknights, Saturdays, and Sundays, when the jobbers are usually closed.

The lowest prices for parts are most often found in discount stores or the auto department of mass merchandisers. Parts sold here are name and private brand parts bought in huge quantities, so they can offer a competitive price. Private brand parts are made by major manufacturers and sold to large chains under a store label.

Fig. 2 By far the most important asset in purchasing parts is a knowledgeable and enthusiastic parts person

Avoiding the Most Common Mistakes

There are 3 common mistakes in mechanical work:
1. Incorrect order of assembly, disassembly or adjustment. When taking something apart or putting it together, performing steps in the wrong order usually just costs you extra time; however, it CAN break something. Read the entire procedure before beginning disassembly. Perform everything in the order in which the instructions say you should, even if you can't immediately see a reason for it. When you're taking apart something that is very intricate, you might want to draw a picture of how it looks when assembled at one point in order to make sure you get everything back in its proper position. When making adjustments, perform them in the proper order; often, one adjustment affects another, and you cannot expect satisfactory results unless each adjustment is made only when it cannot be changed by another.
2. Overtorquing (or undertorquing). While it is more common for overtorquing to cause damage, undertorquing may allow a fastener to vibrate loose causing serious damage. Especially when dealing with aluminum parts, pay attention to torque specifications and utilize a torque wrench in assembly. If a torque figure is not available, remember that if you are using the right tool to perform the job, you will probably not have to strain yourself to get a fastener tight enough. The pitch of most threads is so slight that the tension you put on the wrench will be multiplied many times in actual force on what you are tightening.
3. Crossthreading. This occurs when a part such as a bolt is screwed into a nut or casting at the wrong angle and forced. Crossthreading is more likely to occur if access is difficult. It helps to clean and lubricate fasteners, then to start threading with the part to be installed positioned straight in. Always start a fastener, etc. with your fingers. If you encounter resistance, unscrew the part and start over again at a different angle until it can be inserted and turned several times without much effort. Keep in mind that some parts may have tapered threads, so that gentle turning will automatically bring the part you're threading to the proper angle, but only if you don't force it or resist a change in angle. Don't put a wrench on the part until it has been tightened a couple of turns by hand. If you suddenly encounter resistance, and the part has not seated fully, don't force it. Pull it back out to make sure it's clean and threading properly.

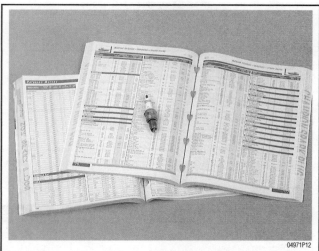

Fig. 3 Parts catalogs, giving application and part number information, are provided by manufacturers for most replacement parts

BOATING SAFETY

In 1971 Congress ordered the U.S. Coast Guard to improve recreational boating safety. In response, the Coast Guard drew up a set of regulations.

Beside these federal regulations, there are state and local laws you must follow. These sometimes exceed the Coast Guard requirements. This section discusses only the federal laws. State and local laws are available from your local Coast Guard. As with other laws, "Ignorance of the boating laws is no excuse." The rules fall into two groups: regulations for your boat and required safety equipment on your boat.

Regulations For Your Boat

Most boats on waters within Federal jurisdiction must be registered or documented. These waters are those that provide a means of transportation between two or more states or to the sea. They also include the territorial waters of the United States.

1-4 GENERAL INFORMATION AND BOATING SAFETY

DOCUMENTING OF VESSELS

A vessel of five or more net tons may be documented as a yacht. In this process, papers are issued by the U.S. Coast Guard as they are for large ships. Documentation is a form of national registration. The boat must be used solely for pleasure. Its owner must be a U.S. citizen, a partnership of U.S. citizens, or a corporation controlled by U.S. citizens. The captain and other officers must also be U.S. citizens. The crew need not be.

If you document your yacht, you have the legal authority to fly the yacht ensign. You also may record bills of sale, mortgages, and other papers of title with federal authorities. Doing so gives legal notice that such instruments exist. Documentation also permits preferred status for mortgages. This gives you additional security and aids financing and transfer of title. You must carry the original documentation papers aboard your vessel. Copies will not suffice.

REGISTRATION OF BOATS

If your boat is not documented, registration in the state of its principal use is probably required. If you use it mainly on an ocean, a gulf, or other similar water, register it in the state where you moor it.

If you use your boat solely for racing, it may be exempt from the requirement in your state. States may also exclude dinghies. Some require registration of documented vessels and non-power driven boats.

All states, except Alaska, register boats. In Alaska, the U.S. Coast Guard issues the registration numbers. If you move your vessel to a new state of principal use, a valid registration certificate is good for 60 days. You must have the registration certificate (certificate of number) aboard your vessel when it is in use. A copy will not suffice. You may be cited if you do not have the original on board.

NUMBERING OF VESSELS

A registration number is on your registration certificate. You must paint or permanently attach this number to both sides of the forward half of your boat. Do not display any other number there.

The registration number must be clearly visible. It must not be placed on the obscured underside of a flared bow. If you can't place the number on the bow, place it on the forward half of the hull. If that doesn't work, put it on the superstructure. Put the number for an inflatable boat on a bracket or fixture. Then, firmly attach it to the forward half of the boat. The letters and numbers must be plain block characters and must read from left to right. Use a space or a hyphen to separate the prefix and suffix letters from the numerals. The color of the characters must contrast with that of the background, and they must be at least three inches high.

In some states your registration is good for only one year. In others, it is good for as long as three years. Renew your registration before it expires. At that time you will receive a new decal or decals. Place them as required by state law. You should remove old decals before putting on the new ones. Some states require that you show only the current decal or decals. If your vessel is moored, it must have a current decal even if it is not in use.

If your vessel is lost, destroyed, abandoned, stolen, or transferred, you must inform the issuing authority. If you lose your certificate of number or your address changes, notify the issuing authority as soon as possible.

SALES AND TRANSFERS

Your registration number is not transferable to another boat. The number stays with the boat unless its state of principal use is changed.

HULL IDENTIFICATION NUMBER

A Hull Identification Number (HIN) is like the Vehicle Identification Number (VIN) on your car. Boats built between November 1, 1972 and July 31, 1984 have old format HINs. Since August 1, 1984 a new format has been used.

Your boat's HIN must appear in two places. If it has a transom, the primary number is on its starboard side within two inches of its top. If it does not have a transom or if it was not practical to use the transom, the number is on the starboard side. In this case, it must be within one foot of the stern and within two inches of the top of the hull side. On pontoon boats, it is on the aft crossbeam within one foot of the starboard hull attachment. Your boat also has a duplicate number in an unexposed location. This is on the boat's interior or under a fitting or item of hardware.

LENGTH OF BOATS

For some purposes, boats are classed by length. Required equipment, for example, differs with boat size. Manufacturers may measure a boat's length in several ways. Officially, though, your boat is measured along a straight line from its bow to its stern. This line is parallel to its keel.

The length does not include bowsprits, boomkins, or pulpits. Nor does it include rudders, brackets, outboard motors, outdrives, diving platforms, or other attachments.

CAPACITY INFORMATION

▶ See Figure 4

Manufacturers must put capacity plates on most recreational boats less than 20 feet long. Sailboats, canoes, kayaks, and inflatable boats are usually exempt. Outboard boats must display the maximum permitted horsepower of their engines. The plates must also show the allowable maximum weights of the people on board. And they must show the allowable maximum combined weights of people, engines, and gear. Inboards and stern drives need not show the weight of their engines on their capacity plates. The capacity plate must appear where it is clearly visible to the operator when underway. This information serves to remind you of the capacity of your boat under normal circumstances. You should ask yourself, "Is my boat loaded above its recommended capacity" and, "Is my boat overloaded for the present sea and wind conditions?" If you are stopped by a legal authority, you may be cited if you are overloaded.

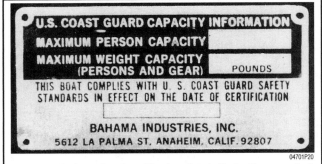

Fig. 4 A U.S. Coast Guard certification plate indicates the amount of occupants and gear appropriate for safe operation of the vessel

CERTIFICATE OF COMPLIANCE

Manufacturers are required to put compliance plates on motorboats greater than 20 feet in length. The plates must say, "This boat," or "This equipment complies with the U. S. Coast Guard Safety Standards in effect on the date of certification." Letters and numbers can be no less than one-eighth of an inch high. At the manufacturer's option, the capacity and compliance plates may be combined.

VENTILATION

A cup of gasoline spilled in the bilge has the potential explosive power of 15 sticks of dynamite. This statement, commonly quoted over 20 years ago, may be an exaggeration, however, it illustrates a fact. Gasoline fumes in the bilge of a boat are highly explosive and a serious danger. They are heavier than air and will stay in the bilge until they are vented out.

Because of this danger, Coast Guard regulations require ventilation on many power boats. There are several ways to supply fresh air to engine and gasoline tank compartments and to remove dangerous vapors. Whatever the choice, it must meet Coast Guard standards.

➡ **The following is not intended to be a complete discussion of the regulations. It is limited to the majority of recreational vessels. Contact your local Coast Guard office for further information.**

General Precautions

Ventilation systems will not remove raw gasoline that leaks from tanks or fuel lines. If you smell gasoline fumes, you need immediate repairs. The best device for sensing gasoline fumes is your nose. Use it! If you smell gasoline in an engine compartment or elsewhere, don't start your engine. The smaller the compartment, the less gasoline it takes to make an explosive mixture.

Ventilation for Open Boats

In open boats, gasoline vapors are dispersed by the air that moves through them. So they are exempt from ventilation requirements.

GENERAL INFORMATION AND BOATING SAFETY

To be "open," a boat must meet certain conditions. Engine and fuel tank compartments and long narrow compartments that join them must be open to the atmosphere." This means they must have at least 15 square inches of open area for each cubic foot of net compartment volume. The open area must be in direct contact with the atmosphere. There must also be no long, unventilated spaces open to engine and fuel tank compartments into which flames could extend.

Ventilation for All Other Boats

Powered and natural ventilation are required in an enclosed compartment with a permanently installed gasoline engine that has a cranking motor. A compartment is exempt if its engine is open to the atmosphere. Diesel powered boats are also exempt.

VENTILATION SYSTEMS

There are two types of ventilation systems. One is "natural ventilation." In it, air circulates through closed spaces due to the boat's motion. The other type is "powered ventilation." In it, air is circulated by a motor driven fan or fans.

Natural Ventilation System Requirements

A natural ventilation system has an air supply from outside the boat. The air supply may also be from a ventilated compartment or a compartment open to the atmosphere. Intake openings are required. In addition, intake ducts may be required to direct the air to appropriate compartments.

The system must also have an exhaust duct that starts in the lower third of the compartment. The exhaust opening must be into another ventilated compartment or into the atmosphere. Each supply opening and supply duct, if there is one, must be above the usual level of water in the bilge. Exhaust openings and ducts must also be above the bilge water. Openings and ducts must be at least three square inches in area or two inches in diameter. Openings should be placed so exhaust gasses do not enter the fresh air intake. Exhaust fumes must not enter cabins or other enclosed, non-ventilated spaces. The carbon monoxide gas in them is deadly.

Intake and exhaust openings must be covered by cowls or similar devices. These registers keep out rain water and water from breaking seas. Most often, in-take registers face forward and exhaust openings aft. This aids the flow of air when the boat is moving or at anchor since most boats face into the wind when anchored.

Power Ventilation System Requirements

♦ See Figure 5

Powered ventilation systems must meet the standards of a natural system. They must also have one or more exhaust blowers. The blower duct can serve as the exhaust duct for natural ventilation if fan blades do not obstruct the air flow when not powered. Openings in engine compartment, for carburetion are in addition to ventilation system requirements.

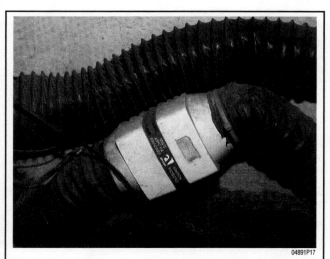

Fig. 5 Typical blower and duct system to vent fumes from the engine compartment

Required Safety Equipment

Coast Guard regulations require that your boat have certain equipment aboard. These requirements are minimums. Exceed them whenever you can.

TYPES OF FIRES

There are four common classes of fires:
- Class A—fires are in ordinary combustible materials such as paper or wood.
- Class B—fires involve gasoline, oil and grease.
- Class C—fires are electrical.
- Class D—fires involve ferrous metals

One of the greatest risks to boaters is fire. This is why it is so important to carry the correct number and type of extinguishers onboard.

The best fire extinguisher for most boats is a Class B extinguisher. Never use water on Class B or Class C fires, as water spreads these types of fires. You should never use water on a Class C fire as it may cause you to be electrocuted.

FIRE EXTINGUISHERS

♦ See Figure 6

If your boat meets one or more of the following conditions, you must have at least one fire extinguisher aboard. The conditions are:
- Inboard or stern drive engines
- Closed compartments under seats where portable fuel tanks can be stored
- Double bottoms not sealed together or not completely filled with flotation materials
- Closed living spaces
- Closed stowage compartments in which combustible or flammable materials are stored
- Permanently installed fuel tanks
- Boat is 26 feet or more in length.

Fig. 6 An approved fire extinguisher should be mounted close to the operator for emergency use

1-6 GENERAL INFORMATION AND BOATING SAFETY

Contents of Extinguishers

Fire extinguishers use a variety of materials. Those used on boats usually contain dry chemicals, Halon, or Carbon Dioxide (CO3). Dry chemical extinguishers contain chemical powders such as Sodium Bicarbonate—baking soda.

Carbon dioxide is a colorless and odorless gas when released from an extinguisher. It is not poisonous but caution must be used in entering compartments filled with it. It will not support life and keeps oxygen from reaching your lungs. A fire-killing concentration of Carbon Dioxide is lethal. If you are in a compartment with a high concentration of CO3, you will have no difficulty breathing. But the air does not contain enough oxygen to support life. Unconsciousness or death can result.

HALON EXTINGUISHERS

Some fire extinguishers and `built-in' or `fixed' automatic fire extinguishing systems contain a gas called Halon. Like carbon dioxide it is colorless and odorless and will not support life. Some Halons may be toxic if inhaled.

To be accepted to the Coast Guard, a fixed Halon system must have an indicator light at the vessel's helm. A green light shows the system is ready. Red means it is being discharged or has been discharged. Warning horns are available to let you know the system has been activated. If your fixed Halon system discharges, ventilate the space thoroughly before you enter it. There are no residues from Halon but it will not support life.

Although Halon has excellent fire fighting properties, it is thought to deplete the earth's ozone layer and has not been manufactured since January 1, 1994. Halon extinguishers can be refilled from existing stocks of the gas until they are used up, but high federal excise taxes are being charged for the service. If you discontinue using your Halon extinguisher, take it to a recovery station rather than releasing the gas into the atmosphere. Compounds such as FE 241, designed to replace Halon, are now available.

Fire Extinguisher Approval

Fire extinguishers must be Coast Guard approved. Look for the approval number on the nameplate. Approved extinguishers have the following on their labels: "Marine Type USCG Approved, Size . . . , Type . . . , 162.208/," etc. In addition, to be acceptable by the Coast Guard, an extinguisher must be in serviceable condition and mounted in its bracket. An extinguisher not properly mounted in its bracket will not be considered serviceable during a Coast Guard inspection.

Care and Treatment

Make certain your extinguishers are in their stowage brackets and are not damaged. Replace cracked or broken hoses. Nozzles should be free of obstructions. Sometimes, wasps and other insects nest inside nozzles and make them inoperable. Check your extinguishers frequently. If they have pressure gauges, is the pressure within acceptable limits? Do the locking pins and sealing wires show they have not been used since recharging?

Don't try an extinguisher to test it. Its valves will not reseat properly and the remaining gas will leak out. When this happens, the extinguisher is useless.

Weigh and tag carbon dioxide and Halon extinguishers twice a year. If their weight loss exceeds 10 percent of the weight of the charge, recharge them. Check to see that they have not been used. They should have been inspected by a qualified person within the past six months, and they should have tags showing all inspection and service dates. The problem is that they can be partially discharged while appearing to be fully charged.

Some Halon extinguishers have pressure gauges the same as dry chemical extinguishers. Don't rely too heavily on the gauge. The extinguisher can be partially discharged and still show a good gauge reading. Weighing a Halon extinguisher is the only accurate way to assess its contents.

If your dry chemical extinguisher has a pressure indicator, check it frequently. Check the nozzle to see if there is powder in it. If there is, recharge it. Occasionally invert your dry chemical extinguisher and hit the base with the palm of your hand. The chemical in these extinguishers packs and cakes due to the boat's vibration and pounding. There is a difference of opinion about whether hitting the base helps, but it can't hurt. It is known that caking of the chemical powder is a major cause of failure of dry chemical extinguishers. Carry spares in excess of the minimum requirement. If you have guests aboard, make certain they know where the extinguishers are and how to use them.

Using a Fire Extinguisher

A fire extinguisher usually has a device to keep it from being discharged accidentally. This is a metal or plastic pin or loop. If you need to use your extinguisher, take it from its bracket. Remove the pin or the loop and point the nozzle at the base of the flames. Now, squeeze the handle, and discharge the extinguisher's contents while sweeping from side to side. Recharge a used extinguisher as soon as possible.

If you are using a Halon or carbon dioxide extinguisher, keep your hands away from the discharge. The rapidly expanding gas will freeze them. If your fire extinguisher has a horn, hold it by its handle.

Legal Requirements for Extinguishers

You must carry fire extinguishers as defined by Coast Guard regulations. They must be firmly mounted in their brackets and immediately accessible.

A motorboat less than 26 feet long must have at least one approved hand-portable, Type B-1 extinguisher. If the boat has an approved fixed fire extinguishing system, you are not required to have the Type B-1 extinguisher. Also, if your boat is less than 26 feet long, is propelled by an outboard motor, or motors, and does not have any of the first six conditions described at the beginning of this section, it is not required to have an extinguisher. Even so, it's a good idea to have one, especially if a nearby boat catches fire, or if a fire occurs at a fuel dock.

A motorboat 26 feet to under 40 feet long, must have at least two Type B-1 approved hand-portable extinguishers. It can, instead, have at least one Coast Guard approved Type B-2. If you have an approved fire extinguishing system, only one Type B-1 is required.

A motorboat 40 to 65 feet long must have at least three Type B-1 approved portable extinguishers . It may have, instead, at least one Type B-1 plus a Type B-2. If there is an approved fixed fire extinguishing system, two Type B-1 or one Type B-2 is required.

WARNING SYSTEM

Various devices are available to alert you to danger. These include fire, smoke, gasoline fumes, and carbon monoxide detectors. If your boat has a galley, it should have a smoke detector. Where possible, use wired detectors. Household batteries often corrode rapidly on a boat.

You can't see, smell, nor taste carbon monoxide gas, but it is lethal. As little as one part in 10,000 parts of air can bring on a headache. The symptoms of carbon monoxide poisoning—headaches, dizziness, and nausea—are like sea sickness. By the time you realize what is happening to you, it may be too late to take action. If you have enclosed living spaces on your boat, protect yourself with a detector. There are many ways in which carbon monoxide can enter your boat.

PERSONAL FLOTATION DEVICES

Personal Flotation Devices (PFDs) are commonly called life preservers or life jackets. You can get them in a variety of types and sizes. They vary with their intended uses. To be acceptable, they must be Coast Guard approved.

Type I PFDs

A Type I life jacket is also called an offshore life jacket. Type I life jackets will turn most unconscious people from facedown to a vertical or slightly backward position. The adult size gives a minimum of 22 pounds of buoyancy. The child size has at least 11 pounds. Type I jackets provide more protection to their wearers than any other type of life jacket. Type I life jackets are bulkier and less comfortable than other types. Furthermore, there are only two sizes, one for children and one for adults.

Type I life jackets will keep their wearers afloat for extended periods in rough water. They are recommended for offshore cruising where a delayed rescue is probable.

Type II PFDs

♦ See Figure 7

A Type II life jacket is also called a near-shore buoyant vest. It is an approved, wearable device. Type II life jackets will turn some unconscious people from facedown to vertical or slightly backward positions. The adult size gives at least 15.5 pounds of buoyancy. The medium child size has a minimum of 11 pounds. And the small child and infant sizes give seven pounds. A Type II life jacket is more comfortable than a Type I but it does not have as much buoyancy. It is not recommended for long hours in rough water. Because of this, Type IIs are recommended for inshore and inland cruising on calm water. Use them where there is a good chance of fast rescue.

Type III PFDs

Type III life jackets or marine buoyant devices are also known as flotation aids. Like Type IIs, they are designed for calm inland or close offshore water where there is a good chance of fast rescue. Their minimum buoyancy is 15.5 pounds. They will not turn their wearers face up.

GENERAL INFORMATION AND BOATING SAFETY

Fig. 7 Type II approved flotation devices are recommended for inshore and inland cruising on calm water. Use them where there is a good chance of fast rescue

Fig. 8 Type IV buoyant cushions are made to be thrown to people in the water. If you can squeeze air out of the cushion, it is faulty and should be replaced

Fig. 9 Type IV throwables, such as this ring life buoy, are not designed as personal flotation devices for unconscious people, non-swimmers, or children

Type III devices are usually worn where freedom of movement is necessary. Thus, they are used for water skiing, small boat sailing, and fishing among other activities. They are available as vests and flotation coats. Flotation coats are useful in cold weather. Type IIIs come in many sizes from small child through large adult.

Life jackets come in a variety of colors and patterns—red, blue, green, camouflage, and cartoon characters. From a safety standpoint, the best color is bright orange. It is easier to see in the water, especially if the water is rough.

Type IV PFDs

♦ See Figures 8 and 9

Type IV ring life buoys, buoyant cushions and horseshoe buoys are Coast Guard approved devices called throwables. They are made to be thrown to people in the water, and should not be worn. Type IV cushions are often used as seat cushions. Cushions are hard to hold onto in the water. Thus, they do not afford as much protection as wearable life jackets.

The straps on buoyant cushions are for you to hold onto either in the water or when throwing them. A cushion should never be worn on your back. It will turn you face down in the water.

Type IV throwables are not designed as personal flotation devices for unconscious people, non-swimmers, or children. Use them only in emergencies. They should not be used for, long periods in rough water.

Ring life buoys come in 18, 20, 24, and 30 inch diameter sizes. They have grab lines. You should attach about 60 feet of polypropylene line to the grab rope to aid in retrieving someone in the water. If you throw a ring, be careful not to hit the person. Ring buoys can knock people unconscious

Type V PFDs

Type V PFDs are of two kinds, special use devices and hybrids. Special use devices include boardsailing vests, deck suits, work vests, and others. They are approved only for the special uses or conditions indicated on their labels. Each is designed and intended for the particular application shown on its label. They do not meet legal requirements for general use aboard recreational boats.

Hybrid life jackets are inflatable devices with some built-in buoyancy provided by plastic foam or kapok. They can be inflated orally or by cylinders of compressed gas to give additional buoyancy. In some hybrids the gas is released manually. In others it is released automatically when the life jacket is immersed in water.

The inherent buoyancy of a hybrid may be insufficient to float a person unless it is inflated. The only way to find this out is for the user to try it in the water. Because of its limited buoyancy when deflated, a hybrid is recommended for use by a non-swimmer only if it is worn with enough inflation to float the wearer.

If they are to count against the legal requirement for the number of life jackets you must carry on your vessel, hybrids manufactured before February 8, 1995 must be worn whenever a boat is underway and the wearer is not below decks or in an enclosed space. To find out if your Type V hybrid must be worn to satisfy the legal requirement, read its label. If its use is restricted it will say, "REQUIRED TO BE WORN" in capital letters.

Hybrids cost more than other life jackets, but this factor must be weighed against the fact that they are more comfortable than Type I, II, or III life jackets. Because of their greater comfort, their owners are more likely to wear them than are the owners of Type I, II, or III life jackets.

The Coast Guard has determined that improved, less costly hybrids can save lives since they will be bought and used more frequently. For these reasons a new federal regulation was adopted effective February 8, 1995. The regulation increases both the deflated and inflated buoyancys of hybrids, makes them available in a greater variety of sizes and types, and reduces their costs by reducing production costs.

Even though it may not be required, the wearing of a hybrid or a life jacket is encouraged whenever a vessel is underway. Like life jackets, hybrids are now available in three types. To meet legal requirements, a Type I hybrid can be substituted for a Type I life jacket. Similarly Type II and III hybrids can be substituted for Type II and Type III life jackets. A Type I hybrid, when inflated, will turn most unconscious people from facedown to vertical or slightly backward positions just like a Type I life jacket. Type I and III hybrids function like Type II and III life jackets. If you purchase a new hybrid, it should have an owner's manual attached which describes its life jacket type and its deflated and inflated buoyancys. It warns you that it may have to be inflated to float you. The manual also tells you how to don the life jacket and how to inflate it. It also tells you how to change its inflation mechanism, recommended testing exercises, and inspection and maintenance procedures. The manual also tells you why you need a life jacket and why you should wear it. A new hybrid must be packaged with at least three gas cartridges. One of these may already be loaded into the inflation mechanism. Likewise, if it has an automatic inflation mechanism, it must be packaged with at least three of these water sensitive elements. One of these elements may be installed.

Legal Requirements

A Coast Guard approved life jacket must show the manufacturer's name and approval number. Most are marked as Type I, II, III, IV, or V. All of the newer hybrids are marked for type.

You are required to carry at least one wearable life jacket or hybrid for each person on board your recreational vessel. If your vessel is 16 feet or more in length and is not a canoe or a kayak, you must also have at least one Type IV on board. These requirements apply to all recreational vessels that are propelled or controlled by machinery, sails, oars, paddles, poles, or another vessel. Sailboards are not required to carry life jackets.

You can substitute an older Type V hybrid for any required Type I, II, or III life jacket provided that its approval label shows it is approved for the activity the vessel is engaged in, approved as a substitute for a life jacket of the type required on the vessel, used as required on the labels, and used in accordance with any requirements in its owner's manual, if the approval label makes reference to such a manual.

A water skier being towed is considered to be on board the vessel when judging compliance with legal requirements.

You are required to keep your Type I, II, or III life jackets or equivalent hybrids readily accessible, which means you must be able to reach out and get them when needed. All life jackets must be in good, serviceable condition.

General Considerations

The proper use of a life jacket requires the wearer to know how it will perform. You can gain this knowledge only through experience. Each person on your boat should be assigned a life jacket. Next, it should be fitted to the person who will wear it. Only then can you be sure that it will be ready for use in an emergency.

Boats can sink fast. There may be no time to look around for a life jacket. Fitting one on you in the water is almost impossible. This advice is good even if the water is calm, and you intend to boat near shore. Most drownings occur in inland waters within a few feet of safety. Most victims had life jackets, but they weren't wearing them.

Keeping life jackets in the plastic covers they came wrapped in and in a cabin assures that they will stay clean and unfaded. But this is no way to keep them when

1-8 GENERAL INFORMATION AND BOATING SAFETY

you are on the water. When you need a life jacket it must be readily accessible and adjusted to fit you. You can't spend time hunting for it or learning how to fit it.

There is no substitute for the experience of entering the water while wearing a life jacket. Children, especially, need practice. If possible, give your guests this experience. Tell them they should keep their arms to their sides when jumping in to keep the life jacket from riding up. Let them jump in and see how the life jacket responds. Is it adjusted so it does not ride up? Is it the proper size? Are all straps snug? Are children's life jackets the right sizes for them? Are they adjusted properly? If a child's life jacket fits correctly, you can lift the child by the jacket's shoulder straps and the child's chin and ears will not slip through. Non-swimmers, children, handicapped persons, elderly persons and even pets should always wear life jackets when they are aboard. Many states require that everyone aboard wear them in hazardous waters.

Inspect your lifesaving equipment from time to time. Leave any questionable or unsatisfactory equipment on shore. An emergency is no time for you to conduct an inspection.

Indelibly mark your life jackets with your vessel's name, number, and calling port. This can be important in a search and rescue effort. It could help concentrate effort where it will do the most good.

Care of Life Jackets

Given reasonable care, life jackets last many years. Thoroughly dry them before putting them away. Stow them in dry, well ventilated places. Avoid the bottoms of lockers and deck storage boxes where moisture may collect. Air and dry them frequently.

Life jackets should not be tossed about or used as fenders or cushions. Many contain kapok or fibrous glass material enclosed in plastic bags. The bags can rupture and are then unserviceable. Squeeze your life jacket gently. Does air leak out? If so, water can leak in and it will no longer be safe to use. Cut it up so no one will use it, and throw it away. The covers of some life jackets are made of nylon or polyester. These materials are plastics. Like many plastics, they break down after extended exposure to the ultraviolet light in sunlight. This process may be more rapid when the materials are dyed with bright dyes such as "neon" shades.

Ripped and badly faded fabric are clues that the covering of your life jacket is deteriorating. A simple test is to pinch the fabric between your thumbs and forefingers. Now try to tear the fabric. If it can be torn, it should definitely be destroyed and discarded. Compare the colors in protected places to those exposed to the sun. If the colors have faded, the materials have been weakened. A fabric covered life jacket should ordinarily last several boating seasons with normal use. A life jacket used every day in direct sunlight should probably be replaced more often.

SOUND PRODUCING DEVICES

All boats are required to carry some means of making an efficient sound signal. Devices for making the whistle or horn noises required by the Navigation Rules must be capable of a four second blast. The blast should be audible for at least one-half mile. Athletic whistles are not acceptable on boats 12 meters or longer. Use caution with athletic whistles. When wet, some of them come apart and loose their "pea." When this happens, they are useless.

If your vessel is 12 meters long and less than 20 meters, you must have a power whistle (or power horn) and a bell on board. The bell must be in operating condition and have a minimum diameter of at least 200 mm (7.9 inches) at its mouth.

VISUAL DISTRESS SIGNALS

♦ See Figure 10

Visual Distress Signals (VDS) attract attention to your vessel if you need help. They also help to guide searchers in search and rescue situations. Be sure you have the right types, and learn how to use them properly.

It is illegal to fire flares improperly. In addition, they cost the Coast Guard and its Auxiliary many wasted hours in fruitless searches. If you signal a distress with flares and then someone helps you, please let the Coast Guard or the appropriate Search And Rescue Agency (SAR) know so the distress report will be canceled.

Recreational boats less than 16 feet long must carry visual distress signals on coastal waters at night. Coastal waters are:
• The ocean (territorial sea)
• The Great Lakes
• Bays or sounds that empty into oceans
• Rivers over two miles across at their mouths upstream to where they narrow to two miles.

Recreational boats 16 feet or longer must carry VDS at all times on coastal waters. The same requirement applies to boats carrying six or fewer passengers for hire. Open sailboats less than 26 feet long without engines are exempt in the daytime as are manually propelled boats. Also exempt are boats in organized races, regattas,

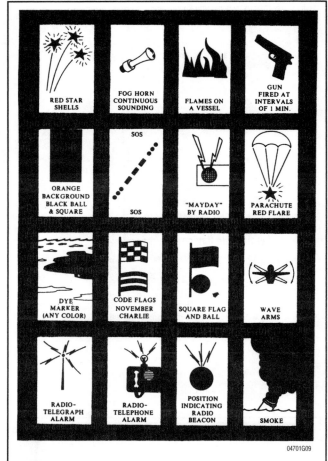

Fig. 10 Internationally accepted distress signals

parades, etc. Boats owned in the United States and operating on the high seas must be equipped with VDS.

A wide variety of signaling devices meet Coast Guard regulations. For pyrotechnic devices, a minimum of three must be carried. Any combination can be carried as long as it adds up to at least three signals for day use and at least three signals for night use. Three day/night signals meet both requirements. If possible, carry more than the legal requirement.

➥**The American flag flying upside down is a commonly recognized distress signal. It is not recognized in the Coast Guard regulations, though. In an emergency, your efforts would probably be better used in more effective signaling methods.**

Types of VDS

VDS are divided into two groups; daytime and nighttime use. Each of these groups is subdivided into pyrotechnic and non-pyrotechnic devices.

DAYTIME NON-PYROTECHNIC SIGNALS

A bright orange flag with a black square over a black circle is the simplest VDS. It is usable, of course, only in daylight. It has the advantage of being a continuous signal. A mirror can be used to good advantage on sunny days. It can attract the attention of other boaters and of aircraft from great distances. Mirrors are available with holes in their centers to aid in "aiming." In the absence of a mirror, any shiny object can be used. When another boat is in sight, an effective VDS is to extend your arms from your sides and move them up and down. Do it slowly. If you do it too fast the other people may think you are just being friendly. This simple gesture is seldom misunderstood, and requires no equipment.

DAYTIME PYROTECHNIC DEVICES

Orange smoke is a useful daytime signal. Hand-held or floating smoke flares are very effective in attracting attention from aircraft. Smoke flares don't last long, and are not very effective in high wind or poor visibility. As with other pyrotechnic

GENERAL INFORMATION AND BOATING SAFETY

devices, use them only when you know there is a possibility that someone will see the display.

To be usable, smoke flares must be kept dry. Keep them in airtight containers and store them in dry places. If the "striker" is damp, dry it out before trying to ignite the device. Some pyrotechnic devices require a forceful "strike" to ignite them.

All hand-held pyrotechnic devices may produce hot ashes or slag when burning. Hold them over the side of your boat in such a way that they do not burn your hand or drip into your boat.

Nighttime Non-Pyrotechnic Signals

An electric distress light is available. This light automatically flashes the international morse code SOS distress signal (••• --- •••). Flashed four to six times a minute, it is an unmistakable distress signal. It must show that it is approved by the Coast Guard. Be sure the batteries are fresh. Dated batteries give assurance that they are current.

Under the Inland Navigation Rules, a high intensity white light flashing 50-70 times per minute is a distress signal. Therefore, use strobe lights on inland waters only for distress signals.

Nighttime Pyrotechnic Devices

♦ See Figure 11

Aerial and hand-held flares can be used at night or in the daytime. Obviously, they are more effective at night.

Currently, the serviceable life of a pyrotechnic device is rated at 42 months from its date of manufacture. Pyrotechnic devices are expensive. Look at their dates before you buy them. Buy them with as much time remaining as possible.

Like smoke flares, aerial and hand-held flares may fail to work if they have been damaged or abused. They will not function if they are or have been wet. Store them in dry, airtight containers in dry places. But store them where they are readily accessible.

Aerial VDSs, depending on their type and the conditions they are used in, may not go very high. Again, use them only when there is a good chance they will be seen.

A serious disadvantage of aerial flares is that they burn for only a short time.

Fig. 11 Moisture protected flares should be carried onboard any vessel for use as a distress signal

Most burn for less than 10 seconds. Most parachute flares burn for less than 45 seconds. If you use a VDS in an emergency, do so carefully. Hold hand-held flares over the side of the boat when in use. Never use a road hazard flare on a boat, it can easily start a fire. Marine type flares are carefully designed to lessen risk, but they still must be used carefully.

Aerial flares should be given the same respect as firearms since they are firearms! Never point them at another person. Don't allow children to play with them or around them. When you fire one, face away from the wind. Aim it downwind and upward at an angle of about 60 degrees to the horizon. If there is a strong wind, aim it somewhat more vertically. Never fire it straight up. Before you discharge a flare pistol, check for overhead obstructions. These might be damaged by the flare. They might deflect the flare to where it will cause damage.

Disposal of VDS

Keep outdated flares when you get new ones. They do not meet legal requirements, but you might need them sometime, and they may work. It is illegal to fire a VDS on federal navigable waters unless an emergency exists. Many states have similar laws.

Emergency Position Indicating Radio Beacon (EPIRB)

There is no requirement for recreational boats to have EPIRBs. Some commercial and fishing vessels, though, must have them if they operate beyond the three mile limit. Vessels carrying six or fewer passengers for hire must have EPIRBs under some circumstances when operating beyond the three mile limit. If you boat in a remote area or offshore, you should have an EPIRB. An EPIRB is a small (about 6 to 20 inches high), battery-powered, radio transmitting buoy-like device. It is a radio transmitter and requires a license or an endorsement on your radio station license by the Federal Communications Commission (FCC). EPIRBs are activated by being immersed in water or by a manual switch.

Equipment Not Required But Recommended

Although not required by law, there are other pieces of equipment that are good to have onboard.

SECOND MEANS OF PROPULSION

All boats less than 16 feet long should carry a second means of propulsion. A paddle or oar can come in handy at times. For most small boats, a spare trolling or outboard motor is an excellent idea. If you carry a spare motor, it should have its own fuel tank and starting power. If you use an electric trolling motor, it should have its own battery.

BAILING DEVICES

All boats should carry at least one effective manual bailing device in addition to any installed electric bilge pump. This can be a bucket, can, scoop, hand operated pump, etc. If your battery "goes dead" it will not operate your electric pump.

FIRST AID KIT

♦ See Figure 12

All boats should carry a first aid kit. It should contain adhesive bandages, gauze, adhesive tape, antiseptic, aspirin, etc. Check your first aid kit from time to time. Replace anything that is outdated. It is to your advantage to know how to use your first aid kit. Another good idea would be to take a Red Cross first aid course.

ANCHORS

♦ See Figure 13

All boats should have anchors. Choose one of suitable size for your boat. Better still, have two anchors of different sizes. Use the smaller one in calm water or when anchoring for a short time to fish or eat. Use the larger one when the water is rougher or for overnight anchoring.

Carry enough anchor line of suitable size for your boat and the waters in which you will operate. If your engine fails you, the first thing you usually should do is lower your anchor. This is good advice in shallow water where you may be driven aground by the wind or water. It is also good advice in windy weather or rough water. The anchor will usually hold your bow into the waves.

1-10 GENERAL INFORMATION AND BOATING SAFETY

Fig. 12 Always carry an adequately stocked first aid kit on board for the safety

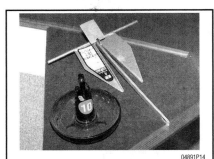

Fig. 13 Choose and anchor of sufficient weight to secure the boat without dragging. In some cases separate anchors may be needed for different situations

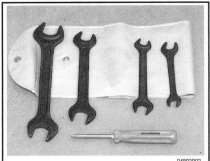

Fig. 14 A few wrenches, a screwdriver and maybe a pair of pliers can be very helpful to make emergency repairs

VHF-FM RADIO

Your best means of summoning help in an emergency or in case of a breakdown is a VHF-FM radio. You can use it to get advice or assistance from the Coast Guard. In the event of a serious illness or injury aboard your boat, the Coast Guard can have emergency medical equipment meet you ashore.

TOOLS AND SPARE PARTS

▶ See Figures 14

Carry a few tools and some spare parts, and learn how to make minor repairs. Many search and rescue cases are caused by minor breakdowns that boat operators could have repaired. If your engine is an inboard or stern drive, carry spare belts and water pump impellers and the tools to change them.

Courtesy Marine Examinations

One of the roles of the Coast Guard Auxiliary is to promote recreational boating safety. This is why they conduct thousands of Courtesy Marine Examinations each year. The auxiliarists who do these examinations are well-trained and knowledgeable in the field.

These examinations are free and done only at the consent of boat owners. To pass the examination, a vessel must satisfy federal equipment requirements and certain additional requirements of the coast guard auxiliary. If your vessel does not pass the Courtesy Marine Examination, no report of the failure is made. Instead, you will be told what you need to correct the deficiencies. The examiner will return at your convenience to redo the examination.

If your vessel qualifies, you will be awarded a safety decal. The decal does not carry any special privileges, it simply attests to your interest in safe boating.

SAFETY IN SERVICE

It is virtually impossible to anticipate all of the hazards involved with maintenance and service, but care and common sense will prevent most accidents.

The rules of safety for mechanics range from "don't smoke around gasoline," to "use the proper tool(s) for the job." The trick to avoiding injuries is to develop safe work habits and to take every possible precaution. Whenever you are working on your boat, pay attention to what you are doing. The more you pay attention to details and what is going on around you, the less likely you will be to hurt yourself or damage your boat.

Do's

- Do keep a fire extinguisher and first aid kit handy.
- Do wear safety glasses or goggles when cutting, drilling, grinding or prying, even if you have 20–20 vision. If you wear glasses for the sake of vision, wear safety goggles over your regular glasses.
- Do shield your eyes whenever you work around the battery. Batteries contain sulfuric acid. In case of contact with the eyes or skin, flush the area with water or a mixture of water and baking soda, then seek immediate medical attention.
- Do use adequate ventilation when working with any chemicals or hazardous materials.
- Do disconnect the negative battery cable when working on the electrical system. The secondary ignition system contains EXTREMELY HIGH VOLTAGE. In some cases it can even exceed 50,000 volts.
- Do follow manufacturer's directions whenever working with potentially hazardous materials. Most chemicals and fluids are poisonous if taken internally.
- Do properly maintain your tools. Loose hammerheads, mushroomed punches and chisels, frayed or poorly grounded electrical cords, excessively worn screwdrivers, spread wrenches (open end), cracked sockets, or slipping ratchets can cause accidents.
- Likewise, keep your tools clean; a greasy wrench can slip off a bolt head, ruining the bolt and often harming your knuckles in the process.
- Do use the proper size and type of tool for the job at hand. Do select a wrench or socket that fits the nut or bolt. The wrench or socket should sit straight, not cocked.
- Do, when possible, pull on a wrench handle rather than push on it, and adjust your stance to prevent a fall.
- Do be sure that adjustable wrenches are tightly closed on the nut or bolt and pulled so that the force is on the side of the fixed jaw. Better yet, avoid the use of an adjustable if you have a fixed wrench that will fit.
- Do strike squarely with a hammer; avoid glancing blows. But, we REALLY hope you won't be using a hammer much in basic maintenance.
- Do use common sense whenever you work on your boat or motor. If a situation arises that doesn't seem right, sit back and have a second look. It may save an embarrassing moment or potential damage to your beloved boat.

Don'ts

- Don't run the engine in an enclosed area or anywhere else without proper ventilation—EVER! Carbon monoxide is poisonous; it takes a long time to leave the human body and you can build up a deadly supply of it in your system by simply breathing in a little every day. You may not realize you are slowly poisoning yourself.
- Don't work around moving parts while wearing loose clothing. Short sleeves are much safer than long, loose sleeves. Hard-toed shoes with neoprene soles protect your toes and give a better grip on slippery surfaces. Jewelry, watches, large belt buckles, or body adornment of any kind is not safe working around any vehicle. Long hair should be tied back under a hat.
- Don't use pockets for toolboxes. A fall or bump can drive a screwdriver deep into your body. Even a rag hanging from your back pocket can wrap around a spinning shaft.
- Don't smoke when working around gasoline, cleaning solvent or other flammable material.
- Don't smoke when working around the battery. When the battery is being charged, it gives off explosive hydrogen gas. Actually, you shouldn't smoke anyway. Save the cigarette money and put it into your boat!
- Don't use gasoline to wash your hands; there are excellent soaps available. Gasoline contains dangerous additives which can enter the body through a cut or through your pores. Gasoline also removes all the natural oils from the skin so that bone dry hands will suck up oil and grease.
- Don't use screwdrivers for anything other than driving screws! A screwdriver used as an prying tool can snap when you least expect it, causing injuries. At the very least, you'll ruin a good screwdriver.

TOOLS AND EQUIPMENT 2-2
SAFETY TOOLS 2-2
    WORK GLOVES 2-2
    EYE & EAR PROTECTION 2-2
    WORK CLOTHES 2-2
CHEMICALS 2-2
    LUBRICANTS & PENETRANTS 2-2
    SEALANTS 2-3
    CLEANERS 2-3
TOOLS 2-4
HAND TOOLS 2-4
    SOCKET SETS 2-4
    WRENCHES 2-6
    PLIERS 2-7
    SCREWDRIVERS 2-7
    HAMMERS 2-7
OTHER COMMON TOOLS 2-7
SPECIAL TOOLS 2-8
ELECTRONIC TOOLS 2-8
GAUGES 2-9
MEASURING TOOLS 2-9
    MICROMETERS & CALIPERS 2-9
    DIAL INDICATORS 2-10
    TELESCOPING GAUGES 2-10
    DEPTH GAUGES 2-10
FASTENERS, MEASUREMENTS AND CONVERSIONS 2-11
BOLTS, NUTS AND OTHER THREADED RETAINERS 2-11
TORQUE 2-11
STANDARD AND METRIC MEASUREMENTS 2-11
SPECIFICATIONS CHARTS
CONVERSION FACTORS 2-12

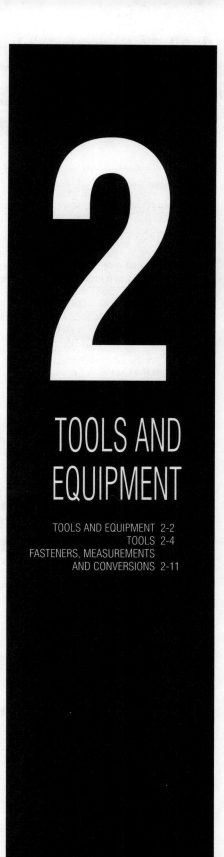

2
TOOLS AND EQUIPMENT

TOOLS AND EQUIPMENT 2-2
TOOLS 2-4
FASTENERS, MEASUREMENTS AND CONVERSIONS 2-11

2-2 TOOLS AND EQUIPMENT

TOOLS AND EQUIPMENT

Safety Tools

WORK GLOVES

♦ See Figures 1 and 2

Unless you think scars on your hands are cool, enjoy pain and like wearing bandages, get a good pair of work gloves. Canvas or leather are the best. And yes, we realize that there are some jobs involving small parts that can't be done while wearing work gloves. These jobs are not the ones usually associated with hand injuries.

A good pair of rubber gloves (such as those usually associated with dish washing) or vinyl gloves is also a great idea. There are some liquids such as solvents and penetrants that don't belong on your skin. Avoid burns and rashes. Wear these gloves.

And lastly, an option. If you're tired of being greasy and dirty all the time, go to the drug store and buy a box of disposable latex gloves like medical professionals wear. You can handle greasy parts, perform small tasks, wash parts, etc. all without getting dirty! These gloves take a surprising amount of abuse without tearing and aren't expensive. Note however, that it has been reported that some people are allergic to the latex or the powder used inside some gloves, so pay attention to what you buy.

EYE & EAR PROTECTION

♦ See Figures 3 and 4

Don't begin any job without a good pair of work goggles or impact resistant glasses! When doing any kind of work, it's all too easy to avoid eye injury through this simple precaution. And don't just buy eye protection and leave it on the shelf. Wear it all the time! Things have a habit of breaking, chipping, splashing, spraying, splintering and flying around. And, for some reason, your eye is always in the way!

If you wear vision correcting glasses as a matter of routine, get a pair made with polycarbonate lenses. These lenses are impact resistant and are available at any optometrist.

Often overlooked is hearing protection. Power equipment is noisy! Loud noises damage your ears. It's as simple as that! The simplest and cheapest form of ear protection is a pair of noise-reducing ear plugs. Cheap insurance for your ears. And, they may even come with their own, cute little carrying case.

More substantial, more protection and more money is a good pair of noise reducing earmuffs. They protect from all but the loudest sounds. Hopefully those are sounds that you'll never encounter since they're usually associated with disasters.

WORK CLOTHES

Everyone has "work clothes." Usually these consist of old jeans and a shirt that has seen better days. That's fine. In addition, a denim work apron is a nice accessory. It's rugged, can hold some spare bolts, and you don't feel bad wiping your hands or tools on it. That's what it's for.

When working in cold weather, a one-piece, thermal work outfit is invaluable. Most are rated to below zero (Fahrenheit) temperatures and are ruggedly constructed. Just look at what the marine mechanics are wearing and that should give you a clue as to what type of clothing is good.

Chemicals

There is a whole range of chemicals that you'll find handy for maintenance work. The most common types are, lubricants, penetrants and sealers. Keep these handy onboard. There are also many chemicals that are used for detailing or cleaning.

When a particular chemical is not being used, keep it capped, upright and in a safe place. These substances may be flammable, may be irritants or might even be caustic and should always be stored properly, used properly and handled with care. Always read and follow all label directions and be sure to wear hand and eye protection!

LUBRICANTS & PENETRANTS

♦ See Figure 5

Anti-seize is used to coat certain fasteners prior to installation. This can be especially helpful when two dissimilar metals are in contact (to help prevent corrosion that might lock the fastener in place). This is a good practice on a lot of different fasteners, BUT, NOT on any fastener which might vibrate loose causing a problem. If anti-seize is used on a fastener, it should be checked periodically for proper tightness.

Lithium grease, chassis lube, silicone grease or a synthetic brake caliper grease can all be used pretty much interchangeably. All can be used for coating rust-prone fasteners and for facilitating the assembly of parts that are a tight fit. Silicone and synthetic greases are the most versatile.

➥**Silicone dielectric grease is a non-conductor that is often used to coat the terminals of wiring connectors before fastening them. It may sound**

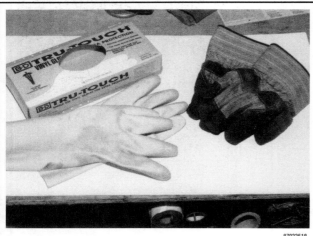

Fig. 1 Three different types of work gloves. The box contains latex gloves

Fig. 2 Latex gloves come in handy when you are doing those messy jobs, like handling filters

Fig. 3 Don't begin any job without a good pair of work goggles or impact resistant glasses. Also good noise reducing earmuffs are cheap insurance to protect your hearing

Fig. 4 Things have a habit of breaking, chipping, splashing, spraying, splintering and flying around. And, for some reason, your eye is always in the way

TOOLS AND EQUIPMENT

Fig. 5 Antiseize, penetrating oil, lithium grease, electronic cleaner and silicone spray. These products have hundreds of uses and should be a part of your chemical tool collection

Fig. 6 Sealants are essential for preventing leaks

Fig. 7 On some engines, RTV is used instead of gasket material to seal components

odd to coat metal portions of a terminal with something that won't conduct electricity, but here is it how it works. When the connector is fastened the metal-to-metal contact between the terminals will displace the grease (allowing the circuit to be completed). The grease that is displaced will then coat the non-contacted surface and the cavity around the terminals, SEALING them from atmospheric moisture that could cause corrosion.

Silicone spray is a good lubricant for hard-to-reach places and parts that shouldn't be gooped up with grease.

Penetrating oil may turn out to be one of your best friends when taking something apart that has corroded fasteners. Not only can they make a job easier, they can really help to avoid broken and stripped fasteners. The most familiar penetrating oils are Liquid Wrench® and WD-40®. A newer penetrant, PB Blaster® also works well. These products have hundreds of uses. For your purposes, they are vital!

Before disassembling any part (especially on an exhaust system), check the fasteners. If any appear rusted, soak them thoroughly with the penetrant and let them stand while you do something else (for particularly rusted or frozen parts you may need to soak them a few days in advance). This simple act can save you hours of tedious work trying to extract a broken bolt or stud.

SEALANTS

▶ See Figures 6 and 7

Sealants are an indispensable part for certain tasks, especially if you are trying to avoid leaks. The purpose of sealants is to establish a leak-proof bond between or around assembled parts. Most sealers are used in conjunction with gaskets, but some are used instead of conventional gasket material.

The most common sealers are the non-hardening types such as Permatex®No.2 or its equivalents. These sealers are applied to the mating surfaces of each part to be joined, then a gasket is put in place and the parts are assembled.

➡A sometimes overlooked use for sealants like RTV is on the threads of vibration prone fasteners.

One very helpful type of non-hardening sealer is the "high tack" type. This type is a very sticky material that holds the gasket in place while the parts are being assembled. This stuff is really a good idea when you don't have enough hands or fingers to keep everything where it should be.

The stand-alone sealers are the Room Temperature Vulcanizing (RTV) silicone gasket makers. On some engines, this material is used instead of a gasket. In those instances, a gasket may not be available or, because of the shape of the mating surfaces, a gasket shouldn't be used. This stuff, when used in conjunction with a conventional gasket, produces the surest bonds.

RTV does have its limitations though. When using this material, you will have a time limit. It starts to set-up within 15 minutes or so, so you have to assemble the parts without delay. In addition, when squeezing the material out of the tube, don't drop any glops into the engine. The stuff will form and set and travel around the oil gallery, possibly plugging up a passage. Also, most types are not fuel-proof. Check the tube for all cautions.

CLEANERS

▶ See Figures 8 and 9

There are two types of cleaners on the market today: parts cleaners and hand cleaners. The parts cleaners are for the parts; the hand cleaners are for you. They are not interchangeable.

Fig. 8 The new citrus hand cleaners not only work well, but they smell pretty good too. Choose one with pumice for added cleaning power

Fig. 9 The use of hand lotion seals your hands and keeps dirt and grease from sticking to your skin

2-4 TOOLS AND EQUIPMENT

There are many good, non-flammable, biodegradable parts cleaners on the market. These cleaning agents are safe for you, the parts and the environment. Therefore, there is no reason to use flammable, caustic or toxic substances to clean your parts or tools.

As far as hand cleaners go, the waterless types are the best. They have always been efficient at cleaning, but leave a pretty smelly odor. Recently though, just about all of them have eliminated the odor and added stuff that actually smells good. Make sure that you pick one that contains lanolin or some other moisture-replenishing additive. Cleaners not only remove grease and oil but also skin oil.

➡ Most women will tell you to use a hand lotion when you're all cleaned up. It's okay. Real men DO use hand lotion! Believe it or not, using hand lotion before your hands are dirty will actually make them easier to clean when you're finished with a dirty job. Lotion seals your hands, and keeps dirt and grease from sticking to your skin.

TOOLS

♦ See Figure 10

Tools; this subject could fill a completely separate manual. The first thing you will need to ask yourself, is just how involved do you plan to get. If you are serious about your maintenance you will want to gather a quality set of tools to make the job easier, and more enjoyable. BESIDES, TOOLS ARE FUN!!!

Almost every do-it-yourselfer loves to accumulate tools. Though most find a way to perform jobs with only a few common tools, they tend to buy more over time, as money allows. So gathering the tools necessary for maintenance does not have to be an expensive, overnight proposition.

When buying tools, the saying "You get what you pay for . . ." is absolutely true! Don't go cheap! Any hand tool that you buy should be drop forged and/or chrome vanadium. These two qualities tell you that the tool is strong enough for the job. With any tool, go with a name that you've heard of before, or, that is recommended buy your local professional retailer. Let's go over a list of tools that you'll need.

Most of the world uses the metric system. However, some American-built engines and aftermarket accessories use standard fasteners. So, accumulate your tools accordingly. Any good DIYer should have a decent set of both U.S. and metric measure tools.

➡ Don't be confused by terminology. Most advertising refers to "SAE and metric", or "standard and metric." Both are misnomers. The Society of Automotive Engineers (SAE) did not invent the English system of measurement; the English did. The SAE likes metrics just fine. Both English (U.S.) and metric measurements are SAE approved. Also, the current "standard" measurement IS metric. So, if it's not metric, it's U.S. measurement.

Hand Tools

SOCKET SETS

♦ See Figures 11 thru 17

Socket sets are the most basic hand tools necessary for repair and maintenance work. For our purposes, socket sets come in three drive sizes: ¼ inch, ⅜ inch and ½ inch. Drive size refers to the size of the drive lug on the ratchet, breaker bar or speed handle.

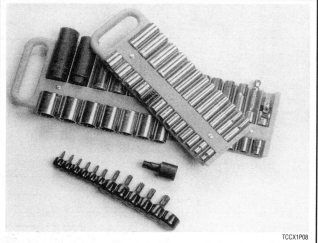

Fig. 10 Socket holders, especially the magnetic type, are handy items to keep tools in order

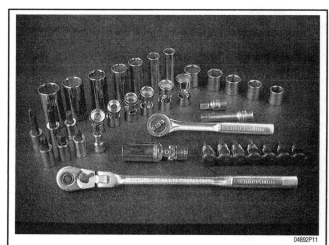

Fig. 11 A ⅜ inch socket set is probably the most versatile tool in any mechanic's tool box

Fig. 12 A swivel (U-joint) adapter (left), a ¼-inch-to-⅜inch adapter (center) and a ⅜inch-to-¼inch adapter (right)

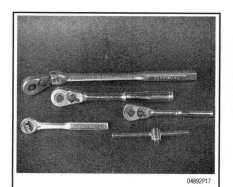

Fig. 13 Ratchets come in all sizes and configurations from rigid to swivel-headed

Fig. 14 Standard length sockets (top) are good for just about all jobs. However, some bolts may require deep sockets (bottom)

TOOLS AND EQUIPMENT

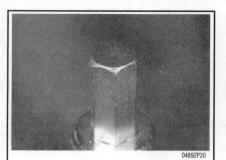

Fig. 15 Hex-head fasteners retain many components on modern powerheads. These fasteners require a socket with a hex shaped driver

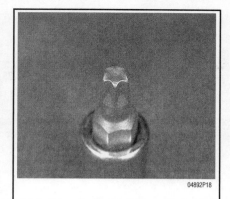

Fig. 16 Torx® drivers . . .

Fig. 17 . . . and tamper resistant drivers are required to remove special fasteners installed by the manufacturers

A ⅜ inch set is probably the most versatile set in any mechanic's tool box. It allows you to get into tight places that the larger drive ratchets can't and gives you a range of larger sockets that are still strong enough for heavy duty work. The socket set that you'll need should range in sizes from ⅜ inch through 1 inch for standard fasteners, and a 6mm through 19mm for metric fasteners.

You'll need a good ½ inch set since this size drive lug assures that you won't break a ratchet or socket on large or heavy fasteners. Also, torque wrenches with a torque scale high enough for larger fasteners are usually ½ inch drive.

¼ inch drive sets can be very handy in tight places. Though they usually duplicate functions of the ⅜ inch set, ¼ inch drive sets are easier to use for smaller bolts and nuts.

As for the sockets themselves, they come in standard and deep lengths as well as 6 or 12 point. 6 and 12 points refers to how many sides are in the socket itself. Each has advantages. The 6 point socket is stronger and less prone to slipping which would strip a bolt head or nut. 12 point sockets are more common, usually less expensive and can operate better in tight places where the ratchet handle can't swing far.

Standard length sockets are good for just about all jobs, however, some stud-head bolts, hard-to-reach bolts, nuts on long studs, etc., require the deep sockets.

Most manufacturers use recessed hex-head fasteners to retain many of the engine parts. These fasteners require a socket with a hex shaped driver or a large sturdy hex key. To help prevent torn knuckles, we would recommend that you stick to the sockets on any tight fastener and leave the hex keys for lighter applications. Hex driver sockets are available individually or in sets just like conventional sockets.

More and more, manufacturers use Torx® head fasteners, which were once known as tamper resistant fasteners (because many people did not have tools with the necessary odd driver shape). They are still used where the manufacturer would prefer only knowledgeable mechanics or advanced Do-It-Yourselfers (DIYers) to work.

Torque Wrenches

♦ See Figure 18

In most applications, a torque wrench can be used to assure proper installation of a fastener. Torque wrenches come in various designs and most stores will carry a variety to suit your needs. A torque wrench should be used any time you have a specific torque value for a fastener. Keep in mind that because there is no worldwide standardization of fasteners, the charts at the end of this section are a general guideline and should be used with caution. If you are using the right tool for the job, you should not have to strain to tighten a fastener.

BEAM TYPE

♦ See Figures 19 and 20

The beam type torque wrench is one of the most popular styles in use. If used properly, it can be the most accurate also. It consists of a pointer attached to the head that runs the length of the flexible beam (shaft) to a scale located near the handle. As the wrench is pulled, the beam bends and the pointer indicates the torque using the scale.

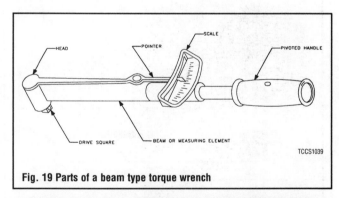

Fig. 19 Parts of a beam type torque wrench

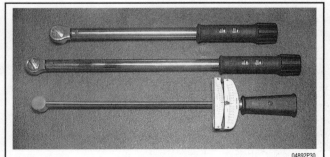

Fig. 18 Three types of torque wrenches. Top to bottom: a ⅜ inch drive beam type that reads in inch lbs., a ½ inch drive clicker type and a ½ inch drive beam type

Fig. 20 A beam type torque wrench consists of a pointer attached to the head that runs the length of the flexible beam (shaft) to a scale located near the handle

2-6 TOOLS AND EQUIPMENT

CLICK (BREAKAWAY) TYPE

▶ See Figures 21 and 22

Another popular torque wrench design is the click type. The clicking mechanism makes achieving the proper torque easy and most use ratcheting head for ease of bolt installation. To use the click type wrench you pre-adjust it to a torque setting. Once the torque is reached, the wrench has a reflex signaling feature that causes a momentary breakaway of the torque wrench body, sending an impulse to the operator's hand.

Breaker Bars

▶ See Figure 23

Breaker bars are long handles with a drive lug. Their main purpose is to provide extra turning force when breaking loose tight bolts or nuts. They come in all drive sizes and lengths. Always take extra precautions and use proper technique when using a breaker bar.

WRENCHES

▶ See Figures 24, 25, 26, 27 and 28

Basically, there are 3 kinds of fixed wrenches: open end, box end, and combination.

Open end wrenches have 2-jawed openings at each end of the wrench. These wrenches are able to fit onto just about any nut or bolt. They are extremely versatile but have one major drawback. They can slip on a worn or rounded bolt head or nut, causing bleeding knuckles and a useless fastener.

Box-end wrenches have a 360° circular jaw at each end of the wrench. They come

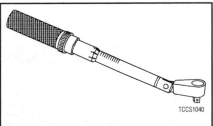

Fig. 21 A click type or breakaway torque wrench—note this one has a pivoting head

Fig. 22 Setting the proper torque on a click type torque wrench involves turning the handle until the proper torque specification appears on the dial

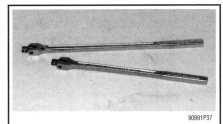

Fig. 23 Breaker bars are great for loosening large or stuck fasteners

INCHES	DECIMAL	DECIMAL	MILLIMETERS
1/8"	.125	.118	3mm
3/16"	.187	.157	4mm
1/4"	.250	.236	6mm
5/16"	.312	.354	9mm
3/8"	.375	.394	10mm
7/16"	.437	.472	12mm
1/2"	.500	.512	13mm
9/16"	.562	.590	15mm
5/8"	.625	.630	16mm
11/16"	.687	.709	18mm
3/4"	.750	.748	19mm
13/16"	.812	.787	20mm
7/8"	.875	.866	22mm
15/16"	.937	.945	24mm
1"	1.00	.984	25mm

Fig. 24 Comparison of U.S. measure and metric wrench sizes

TOOLS AND EQUIPMENT

Fig. 25 Always use a backup wrench to prevent rounding flare nut fittings

Fig. 26 Note how the flare wrench sides are extended to grip the fitting tighter and prevent rounding

Fig. 27 Several types and sizes of adjustable wrenches

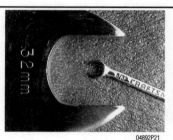

Fig. 28 Occasionally you will find a nut which requires a particularly large or particularly small wrench. Rest assured that the proper wrench to fit is available at your local tool store

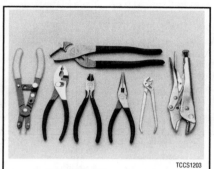

Fig. 29 Pliers and cutters come in many shapes and sizes. You should have an assortment on hand

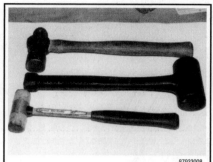

Fig. 30 Three types of hammers. Top to bottom: ball peen, rubber dead-blow, and plastic

in both 6 and 12 point versions just like sockets and each type has the same advantages and disadvantages as sockets.

Combination wrenches have the best of both. They have a 2-jawed open end and a box end. These wrenches are probably the most versatile.

As for sizes, you'll probably need a range similar to that of the sockets, about ¼ inch through 1 inch for standard fasteners, or 6mm through 19mm for metric fasteners. As for numbers, you'll need 2 of each size, since, in many instances, one wrench holds the nut while the other turns the bolt. On most fasteners, the nut and bolt are the same size so having two wrenches of the same size comes in handy.

➡Although you will typically just need the sizes we specified, there are some exceptions. Occasionally you will find a nut which is larger. For these, you will need to buy ONE expensive wrench or a very large adjustable. Or you can always just convince the spouse that we are talking about safety here and buy a whole (read expensive) large wrench set.

One extremely valuable type of wrench is the adjustable wrench. An adjustable wrench has a fixed upper jaw and a moveable lower jaw. The lower jaw is moved by turning a threaded drum. The advantage of an adjustable wrench is its ability to be adjusted to just about any size fastener.

The main drawback of an adjustable wrench is the lower jaw's tendency to move slightly under heavy pressure. This can cause the wrench to slip if it is not facing the right way. Pulling on an adjustable wrench in the proper direction will cause the jaws to lock in place. Adjustable wrenches come in a large range of sizes, measured by the wrench length.

PLIERS

▶ See Figure 29

Pliers are simply mechanical fingers. They are, more than anything, an extension of your hand. At least 3 pair of pliers are an absolute necessity—standard, needle nose and channel lock.

In addition to standard pliers there are the slip-joint, multi-position pliers such as ChannelLock® pliers and locking pliers, such as Vise Grips®.

Slip joint pliers are extremely valuable in grasping oddly sized parts and fasteners. Just make sure that you don't use them instead of a wrench too often since they can easily round off a bolt head or nut.

Locking pliers are usually used for gripping bolts or studs that can't be removed conventionally. You can get locking pliers in square jawed, needle-nosed and pipe-jawed. Locking pliers can rank right up behind duct tape as the handy-man's best friend.

SCREWDRIVERS

You can't have too many screwdrivers. They come in 2 basic flavors, either standard or Phillips. Standard blades come in various sizes and thicknesses for all types of slotted fasteners. Phillips screwdrivers come in sizes with number designations from 1 on up, with the lower number designating the smaller size. Screwdrivers can be purchased separately or in sets.

HAMMERS

▶ See Figure 30

You always need a hammer for just about any kind of work. You need a ball-peen hammer for most metal work when using drivers and other like tools. A plastic hammer comes in handy for hitting things safely. A soft-faced dead-blow hammer is used for hitting things safely and hard. Hammers are also VERY useful with non air-powered impact drivers.

Other Common Tools

There are a lot of other tools that every DIYer will eventually need (though not all for basic maintenance). They include:
- Funnels (for adding fluid)
- Chisels
- Punches
- Files
- Hacksaw

2-8 TOOLS AND EQUIPMENT

- Portable Bench Vise
- Tap and Die Set
- Flashlight
- Magnetic Bolt Retriever
- Gasket scraper
- Putty Knife
- Screw/Bolt Extractors
- Prybar

Hacksaws have just one use—cutting things off. You may wonder why you'd need one for something as simple as maintenance, but you never know. Among other things, guide studs to ease parts installation can be made from old bolts with their heads cut off.

A tap and die set might be something you've never needed, but you will eventually. It's a good rule, when everything is apart, to clean-up all threads, on bolts, screws and threaded holes. Also, you'll likely run across a situation in which stripped threads will be encountered. The tap and die set will handle that for you.

Gasket scrapers are just what you'd think, tools made for scraping old gasket material off of parts. You don't absolutely need one. Old gasket material can be removed with a putty knife or single edge razor blade. However, putty knives may not be sharp enough for some really stubborn gaskets and razor blades have a knack of breaking just when you don't want them to, inevitably slicing the nearest body part! As the old saying goes, "always use the proper tool for the job". If you're going to use a razor to scrape a gasket, be sure to always use a blade holder.

Putty knives really do have a use in a repair shop. Just because you remove all the bolts from a component sealed with a gasket doesn't mean it's going to come off. Most of the time, the gasket and sealer will hold it tightly. Lightly driving a putty knife at various points between the two parts will break the seal without damage to the parts.

A small — 8-10 inches (20–25 centimeters) long — prybar is extremely useful for removing stuck parts.

➡**Never use a screwdriver as a prybar! Screwdrivers are not meant for prying. Screwdrivers, used for prying, can break, sending the broken shaft flying!**

Screw/bolt extractors are used for removing broken bolts or studs that have broke off flush with the surface of the part.

Special Tools

◆ See Figure 31

Almost every marine engine around today requires at least one special tool to perform a certain task. In most cases, these tools are specially designed to overcome some unique problem or to fit on some oddly sized component.

When manufacturers go through the trouble of making a special tool, it is usually necessary to use it to assure that the job will be done right. A special tool might be designed to make a job easier, or it might be used to keep you from damaging or breaking a part.

Don't worry, MOST basic maintenance procedures can either be performed without any special tools OR, because the tools must be used for such basic things, they are commonly available for a reasonable price. It is usually just the low production, highly specialized tools (like a super thin 7-point star-shaped socket capable of 150 ft. lbs. (203 Nm) of torque that is used only on the crankshaft nut of the limited production what-dya-callit engine) that tend to be outrageously expensive and hard to find. Luckily, you will probably never need such a tool.

Special tools can be as inexpensive and simple as an adjustable strap wrench or as complicated as an ignition tester. A few common specialty tools are listed here, but check with your dealer or with other boaters for help in determining if there are any special tools for YOUR particular engine. There is an added advantage in seeking advice from others, chances are they may have already found the special tool you will need, and know how to get it cheaper.

Electronic Tools

Battery Testers

The best way to test a non-sealed battery is using a hydrometer to check the specific gravity of the acid. Luckily, these are usually inexpensive and are available at most parts stores. Just be careful because the larger testers are usually designed for larger batteries and may require more acid than you will be able to draw from the battery cell. Smaller testers (usually a short, squeeze bulb type) will require less acid and should work on most batteries.

Electronic testers are available and are often necessary to tell if a sealed battery is usable. Luckily, many parts stores have them on hand and are willing to test your battery for you.

Battery Chargers

◆ See Figure 32

If you are a weekend boater and take your boat out every week, then you will most likely want to buy a battery charger to keep your battery fresh. There are many types available, from low amperage trickle chargers to electronically controlled battery maintenance tools which monitor the battery voltage to prevent over or undercharging. This last type is especially useful if you store your boat for any length of time (such as during the severe winter months found in many Northern climates).

Even if you use your boat on a regular basis, you will eventually need a battery charger. Remember that most batteries are shipped dry and in a partial charged state. Before a new battery can be put into service it must be filled and properly charged. Failure to properly charge a battery (which was shipped dry) before it is put into service will prevent it from ever reaching a fully charged state.

Digital Volt/Ohm Meter (DVOM)

◆ See Figure 33

Multimeters are an extremely useful tool for troubleshooting electrical problems. They can be purchased in either analog or digital form and have a price range to suit any budget. A multimeter is a voltmeter, ammeter and ohmmeter (along with other features) combined into one instrument. It is often used when testing solid state circuits because of its high input impedance (usually 10 megaohms or more). A brief description of the multimeter main test functions follows:

- Voltmeter—the voltmeter is used to measure voltage at any point in a circuit, or to measure the voltage drop across any part of a circuit. Voltmeters usually have various scales and a selector switch to allow the reading of different voltage ranges. The voltmeter has a positive and a negative lead. To avoid damage to the meter, always connect the negative lead to the negative (-) side of the circuit (to ground or nearest the ground side of the circuit) and connect the positive lead to the positive (+) side of the circuit (to the power source or the nearest power source). Note that the negative voltmeter lead will always be black and

Fig. 31 Almost every outboard requires at least one special tool to perform certain tasks

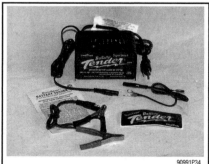

Fig. 32 The Battery Tender® is more than just a battery charger, when left connected, it keeps your battery fully charged

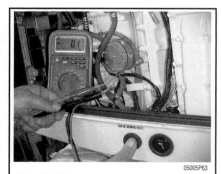

Fig. 33 Multimeters are an extremely useful tool for troubleshooting electrical problems

that the positive voltmeter will always be some color other than black (usually red).

• Ohmmeter—the ohmmeter is designed to read resistance (measured in ohms) in a circuit or component. Most ohmmeters will have a selector switch which permits the measurement of different ranges of resistance (usually the selector switch allows the multiplication of the meter reading by 10, 100, 1,000 and 10,000). Some ohmmeters are "auto-ranging" which means the meter itself will determine which scale to use. Since the meters are powered by an internal battery, the ohmmeter can be used like a self-powered test light. When the ohmmeter is connected, current from the ohmmeter flows through the circuit or component being tested. Since the ohmmeter's internal resistance and voltage are known values, the amount of current flow through the meter depends on the resistance of the circuit or component being tested. The ohmmeter can also be used to perform a continuity test for suspected open circuits. In using the meter for making continuity checks, do not be concerned with the actual resistance readings. Zero resistance, or any ohm reading, indicates continuity in the circuit. Infinite resistance indicates an opening in the circuit. A high resistance reading where there should be none indicates a problem in the circuit. Checks for short circuits are made in the same manner as checks for open circuits, except that the circuit must be isolated from both power and normal ground. Infinite resistance indicates no continuity, while zero resistance indicates a dead short.

※※ WARNING

Never use an ohmmeter to check the resistance of a component or wire while there is voltage applied to the circuit.

• Ammeter—an ammeter measures the amount of current flowing through a circuit in units called amperes or amps. At normal operating voltage, most circuits have a characteristic amount of amperes, called "current draw" which can be measured using an ammeter. By referring to a specified current draw rating, then measuring the amperes and comparing the two values, one can determine what is happening within the circuit to aid in diagnosis. An open circuit, for example, will not allow any current to flow, so the ammeter reading will be zero. A damaged component or circuit will have an increased current draw, so the reading will be high. The ammeter is always connected in series with the circuit being tested. All of the current that normally flows through the circuit must also flow through the ammeter; if there is any other path for the current to follow, the ammeter reading will not be accurate. The ammeter itself has very little resistance to current flow and, therefore, will not affect the circuit, but it will measure current draw only when the circuit is closed and electricity is flowing. Excessive current draw can blow fuses and drain the battery, while a reduced current draw can cause motors to run slowly, lights to dim and other components to not operate properly.

Gauges

Compression Gauge

▸ See Figure 34

An important element in checking the overall condition of your engine is to check compression. This becomes increasingly more important on outboards with high hours. Compression gauges are available as screw-in types and hold-in types. The screw-in type is slower to use, but eliminates the possibility of a faulty reading due to escaping pressure. A compression reading will uncover many problems that can cause rough running. Normally, these are not the sort of problems that can be cured by a tune-up.

Vacuum Gauge

▸ See Figures 35 and 36

Vacuum gauges are handy for discovering air leaks, late ignition or valve timing, and a number of other problems.

Measuring Tools

Eventually, you are going to have to measure something. To do this, you will need at least a few precision tools in addition to the special tools mentioned earlier.

MICROMETERS & CALIPERS

Micrometers and calipers are devices used to make extremely precise measurements. The simple truth is that you really won't have the need for many of these items just for simple maintenance. You will probably want to have at least one precision tool such as an outside caliper to measure rotors or brake pads, but that should be sufficient to most basic maintenance procedures.

Should you decide on becoming more involved in boat engine mechanics, such as repair or rebuilding, then these tools will become very important. The success of any rebuild is dependent, to a great extent on the ability to check the size and fit of components as specified by the manufacturer. These measurements are made in thousandths and ten-thousandths of an inch.

Micrometers

▸ See Figure 37

A micrometer is an instrument made up of a precisely machined spindle which is rotated in a fixed nut, opening and closing the distance between the end of the spindle and a fixed anvil.

Outside micrometers can be used to check the thickness parts such shims or the outside diameter of components like the crankshaft journals. They are also used during many rebuild and repair procedures to measure the diameter of components such as the pistons. The most common type of micrometer reads in 1/1000 of an inch. Micrometers that use a vernier scale can estimate to 1/10 of an inch.

Inside micrometers are used to measure the distance between two parallel surfaces. For example, in powerhead rebuilding work, the inside mike measures cylinder bore wear and taper. Inside mikes are graduated the same way as outside mikes and are read the same way as well.

Remember that an inside mike must be absolutely perpendicular to the work being measured. When you measure with an inside mike, rock the mike gently from side to side and tip it back and forth slightly so that you span the widest part of the bore. Just to be on the safe side, take several readings. It takes a certain amount of experience to work any mike with confidence.

Metric micrometers are read in the same way as inch micrometers, except that the measurements are in millimeters. Each line on the main scale equals 1 mm. Each fifth line is stamped 5, 10, 15, and so on. Each line on the thimble scale equals 0.01

Fig. 34 Cylinder compression test results are extremely valuable indicators of internal engine condition

Fig. 35 Vacuum gauges are useful for many diagnostic tasks including testing of some fuel pumps

Fig. 36 In a pinch, you can also use the vacuum gauge on a hand operated vacuum pump

2-10 TOOLS AND EQUIPMENT

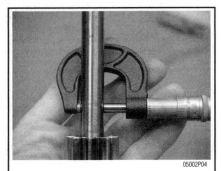

Fig. 37 Outside micrometers can be used to measure the thickness of shims or the outside diameter of a shaft

Fig. 38 Calipers, such as this dial caliper, are the fast and easy way to make precise measurements

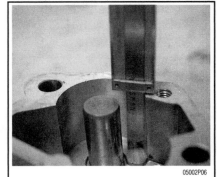

Fig. 39 Calipers can also be used to measure depth . . .

Fig. 40 . . . and inside diameter measurements, usually to 0.001 inch accuracy

Fig. 41 Here, a dial indicator is used to measure the axial clearance (end play) of a crankshaft during a powerhead rebuilding procedure

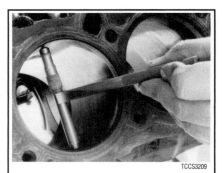

Fig. 42 Telescoping gauges are used during powerhead rebuilding procedures to measure the inside diameter of bores

mm. It will take a little practice, but if you can read an inch mike, you can read a metric mike.

Calipers

♦ See Figures 38, 39 and 40

Inside and outside calipers are useful devices to have if you need to measure something quickly and precise measurement is not necessary. Simply take the reading and then hold the calipers on an accurate steel rule.

DIAL INDICATORS

♦ See Figure 41

A dial indicator is a gauge that utilizes a dial face and a needle to register measurements. There is a movable contact arm on the dial indicator. When the arms moves, the needle rotates on the dial. Dial indicators are calibrated to show readings in thousandths of an inch and typically, are used to measure end-play and runout on various parts.

Dial indicators are quite easy to use, although they are relatively expensive. A variety of mounting devices are available so that the indicator can be used in a number of situations. Make certain that the contact arm is always parallel to the movement of the work being measured.

TELESCOPING GAUGES

♦ See Figure 42

A telescope gauge is used during rebuilding procedures (NOT usually basic maintenance) to measure the inside of bores. It can take the place of an inside mike for some of these jobs. Simply insert the gauge in the hole to be measured and lock the plungers after they have contacted the walls. Remove the tool and measure across the plungers with an outside micrometer.

DEPTH GAUGES

♦ See Figure 43

A depth gauge can be inserted into a bore or other small hole to determine exactly how deep it is. One common use for a depth gauge is measuring the distance the piston sits below the deck of the block at top dead center. Some outside calipers contain a built-in depth gauge so money can be saved by just buying one tool.

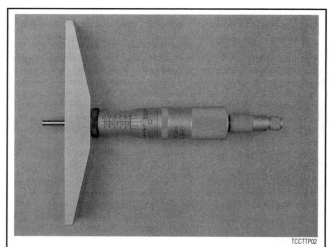

Fig. 43 Depth gauges are used to measure the depth of bore or other small holes

TOOLS AND EQUIPMENT 2-11

FASTENERS, MEASUREMENTS AND CONVERSIONS

Bolts, Nuts and Other Threaded Retainers

♦ See Figures 44 and 45

Although there are a great variety of fasteners found in the modern boat engine, the most commonly used retainer is the threaded fastener (nuts, bolts, screws, studs, etc). Most threaded retainers may be reused, provided that they are not damaged in use or during the repair.

➜Some retainers (such as stretch bolts or torque prevailing nuts) are designed to deform when tightened or in use and should not be reused.

Whenever possible, we will note any special retainers which should be replaced during a procedure. But you should always inspect the condition of a retainer when it is removed and you should replace any that show signs of damage. Check all threads for rust or corrosion which can increase the torque necessary to achieve the desired clamp load for which that fastener was originally selected. Additionally, be sure that the driver surface of the fastener has not been compromised by rounding or other damage. In some cases a driver surface may become only partially rounded, allowing the driver to catch in only one direction. In many of these occurrences, a fastener may be installed and tightened, but the driver would not be able to grip and loosen the fastener again. (This could lead to frustration down the line should that component ever need to be disassembled again).

If you must replace a fastener, whether due to design or damage, you must always be sure to use the proper replacement. In all cases, a retainer of the same design, material and strength should be used. Markings on the heads of most bolts will help determine the proper strength of the fastener. The same material, thread and pitch must be selected to assure proper installation and safe operation of the vehicle afterwards.

Thread gauges are available to help measure a bolt or stud's thread. Most part or hardware stores keep gauges available to help you select the proper size. In a pinch, you can use another nut or bolt for a thread gauge. If the bolt you are replacing is not too badly damaged, you can select a match by finding another bolt which will thread in its place. If you find a nut which threads properly onto the damaged bolt, then use that nut to help select the replacement bolt. If however, the bolt you are replacing is so badly damaged (broken or drilled out) that its threads cannot be used as a gauge, you might start by looking for another bolt (from the same assembly or a similar location) which will thread into the damaged bolt's mounting. If so, the other bolt can be used to select a nut; the nut can then be used to select the replacement bolt.

In all cases, be absolutely sure you have selected the proper replacement. Don't be shy, you can always ask the store clerk for help.

✱✱ WARNING

Be aware that when you find a bolt with damaged threads, you may also find the nut or drilled hole it was threaded into has also been damaged. If this is the case, you may have to drill and tap the hole, replace the nut or otherwise repair the threads. NEVER try to force a replacement bolt to fit into the damaged threads.

Torque

Torque is defined as the measurement of resistance to turning or rotating. It tends to twist a body about an axis of rotation. A common example of this would be tightening a threaded retainer such as a nut, bolt or screw. Measuring torque is one of the most common ways to help assure that a threaded retainer has been properly fastened.

When tightening a threaded fastener, torque is applied in three distinct areas, the head, the bearing surface and the clamp load. About 50 percent of the measured torque is used in overcoming bearing friction. This is the friction between the bearing surface of the bolt head, screw head or nut face and the base material or washer (the surface on which the fastener is rotating). Approximately 40 percent of the applied torque is used in overcoming thread friction. This leaves only about 10 percent of the applied torque to develop a useful clamp load (the force which holds a joint together). This means that friction can account for as much as 90 percent of the applied torque on a fastener.

Standard and Metric Measurements

Specifications are often used to help you determine the condition of various components, or to assist you in their installation. Some of the most common measurements include length (in. or cm/mm), torque (ft. lbs., inch lbs. or Nm) and pressure (psi, in. Hg, kPa or mm Hg).

In some cases, that value may not be conveniently measured with what is available in your toolbox. Luckily, many of the measuring devices which are available today will have two scales so Standard or Metric measurements may easily be taken. If any of the various measuring tools which are available to you do not contain the same scale as listed in your specifications, use the accompanying conversion factors to determine the proper value.

The conversion factor chart is used by taking the given specification and multiplying it by the necessary conversion factor. For instance, looking at the first line, if you have a measurement in inches such as "free-play should be 2 in." but your ruler reads only in millimeters, multiply 2 in. by the conversion factor of 25.4 to get the metric equivalent of 50.8mm. Likewise, if the specification was given only in a Metric measurement, for example in Newton Meters (Nm), then look at the center column first. If the measurement is 100 Nm, multiply it by the conversion factor of 0.738 to get 73.8 ft. lbs.

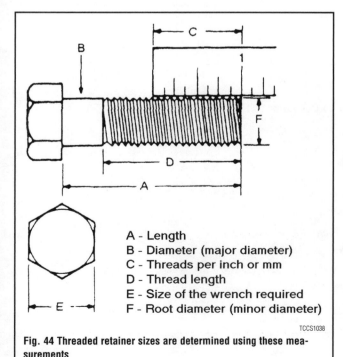

A - Length
B - Diameter (major diameter)
C - Threads per inch or mm
D - Thread length
E - Size of the wrench required
F - Root diameter (minor diameter)

Fig. 44 Threaded retainer sizes are determined using these measurements

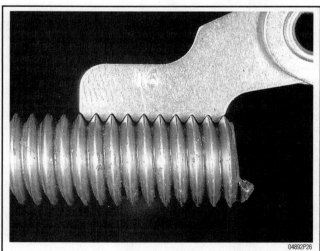

Fig. 45 Thread gauges measure the threads-per-inch and the pitch of a bolt or stud's threads

2-12 TOOLS AND EQUIPMENT

CONVERSION FACTORS

LENGTH–DISTANCE

Inches (in.)	x 25.4	= Millimeters (mm)	x .0394	= Inches
Feet (ft.)	x .305	= Meters (m)	x 3.281	= Feet
Miles	x 1.609	= Kilometers (km)	x .0621	= Miles

VOLUME

Cubic Inches (in3)	x 16.387	= Cubic Centimeters	x .061	= in3
IMP Pints (IMP pt.)	x .568	= Liters (L)	x 1.76	= IMP pt.
IMP Quarts (IMP qt.)	x 1.137	= Liters (L)	x .88	= IMP qt.
IMP Gallons (IMP gal.)	x 4.546	= Liters (L)	x .22	= IMP gal.
IMP Quarts (IMP qt.)	x 1.201	= US Quarts (US qt.)	x .833	= IMP qt.
IMP Gallons (IMP gal.)	x 1.201	= US Gallons (US gal.)	x .833	= IMP gal.
Fl. Ounces	x 29.573	= Milliliters	x .034	= Ounces
US Pints (US pt.)	x .473	= Liters (L)	x 2.113	= Pints
US Quarts (US qt.)	x .946	= Liters (L)	x 1.057	= Quarts
US Gallons (US gal.)	x 3.785	= Liters (L)	x .264	= Gallons

MASS–WEIGHT

Ounces (oz.)	x 28.35	= Grams (g)	x .035	= Ounces
Pounds (lb.)	x .454	= Kilograms (kg)	x 2.205	= Pounds

PRESSURE

Pounds Per Sq. In. (psi)	x 6.895	= Kilopascals (kPa)	x .145	= psi
Inches of Mercury (Hg)	x .4912	= psi	x 2.036	= Hg
Inches of Mercury (Hg)	x 3.377	= Kilopascals (kPa)	x .2961	= Hg
Inches of Water (H_2O)	x .07355	= Inches of Mercury	x 13.783	= H_2O
Inches of Water (H_2O)	x .03613	= psi	x 27.684	= H_2O
Inches of Water (H_2O)	x .248	= Kilopascals (kPa)	x 4.026	= H_2O

TORQUE

Pounds–Force Inches (in-lb)	x .113	= Newton Meters (N·m)	x 8.85	= in–lb
Pounds–Force Feet (ft-lb)	x 1.356	= Newton Meters (N·m)	x .738	= ft–lb

VELOCITY

Miles Per Hour (MPH)	x 1.609	= Kilometers Per Hour (KPH)	x .621	= MPH

POWER

Horsepower (Hp)	x .745	= Kilowatts	x 1.34	= Horsepower

FUEL CONSUMPTION*

Miles Per Gallon IMP (MPG)	x .354	= Kilometers Per Liter (Km/L)
Kilometers Per Liter (Km/L)	x 2.352	= IMP MPG
Miles Per Gallon US (MPG)	x .425	= Kilometers Per Liter (Km/L)
Kilometers Per Liter (Km/L)	x 2.352	= US MPG

*It is common to covert from miles per gallon (mpg) to liters/100 kilometers (1/100 km), where mpg (IMP) x 1/100 km = 282 and mpg (US) x 1/100 km = 235.

TEMPERATURE

Degree Fahrenheit (°F) = (°C x 1.8) + 32
Degree Celsius (°C) = (°F − 32) x .56

ENGINE MAINTENANCE 3-2
SERIAL NUMBER IDENTIFICATION 3-2
2-STROKE OIL 3-2
  OIL RECOMMENDATIONS 3-2
  FILLING 3-2
LOWER UNIT 3-3
  OIL RECOMMENDATIONS 3-3
  DRAINING AND FILLING 3-3
FUEL FILTER 3-4
  RELIEVING FUEL SYSTEM PRESSURE 3-4
  REMOVAL & INSTALLATION 3-5
  CLEANING & INSPECTION 3-5
FUEL/WATER SEPARATOR 3-5
  SERVICE 3-5
LUBRICATION POINTS 3-6
  INSPECTION & LUBRICATION 3-6
PROPELLER 3-6
ANODES (ZINCS) 3-6
  SERVICING 3-7
  INSPECTION 3-7
BOAT MAINTENANCE 3-8
BATTERIES 3-8
  MAINTENANCE 3-8
  CLEANING 3-8
  TESTING 3-8
  STORAGE 3-9
FIBERGLASS HULL 3-9
TUNE-UP 3-10
INTRODUCTION 3-10
COMPRESSION TEST 3-10
  PRIMARY COMPRESSION TEST 3-10
  SECONDARY COMPRESSION TEST 3-10
SPARK PLUGS 3-11
  SPARK PLUG HEAT RANGE 3-11
  SPARK PLUG SERVICE 3-11
  REMOVAL & INSTALLATION 3-11
  READING SPARK PLUGS 3-12
  INSPECTION & GAPPING 3-13
SPARK PLUG WIRES 3-14
  TESTING 3-14
  REMOVAL & INSTALLATION 3-14
IGNITION SYSTEM 3-14
TIMING AND SYNCHRONIZATION 3-14
2.5, 3 AND 3.3 HP 3-15
  IGNITION TIMING 3-15
  IDLE SPEED 3-15
  IDLE MIXTURE 3-15
  THROTTLE JET NEEDLE 3-16
4 AND 5 HP 3-16
  PRELIMINARY ADJUSTMENTS 3-16
  IGNITION TIMING 3-16
  IDLE SPEED & MIXTURE 3-17
6, 8, 9.9 AND 15 HP 3-17
  PRELIMINARY ADJUSTMENTS 3-17
  IGNITION TIMING 3-18
  FAST IDLE SPEED 3-18
  IDLE SPEED & MIXTURE 3-18
20 AND 25 HP 3-19
  PRELIMINARY ADJUSTMENTS 3-19
  FULL THROTTLE STOP 3-19
  IGNITION TIMING 3-19
  IDLE SPEED & MIXTURE 3-20
  FAST IDLE SPEED 3-21
  STARTER INTERLOCK ADJUSTMENT 3-21
  CARBURETOR THROTTLE CAM 3-21
  DASHPOT ADJUSTMENT 3-21
30 AND 40 HP (2-CYLINDER) 3-22
  PRELIMINARY ADJUSTMENTS 3-22
  CAM FOLLOWER 3-22
  IGNITION TIMING 3-22
  IDLE SPEED & MIXTURE 3-22
  OIL PUMP 3-23
40 HP (4-CYLINDER) 3-23
  PRELIMINARY ADJUSTMENTS 3-23
  CARBURETOR SYNCHRONIZATION 3-23
  IGNITION TIMING 3-23
  FULL THROTTLE STOP 3-24
  IDLE SPEED & MIXTURE 3-24
1990–93 50–60 HP 3-25
  PRELIMINARY ADJUSTMENTS 3-25
  CARBURETOR SYNCHRONIZATION 3-25
  IGNITION TIMING 3-25
  IDLE SPEED & MIXTURE 3-25
40 (3-CYLINDER), 1994–00 50 AND 60 HP 3-26
  PRELIMINARY ADJUSTMENTS 3-26
  CARBURETOR SYNCHRONIZATION 3-26
  IGNITION TIMING 3-26
  IDLE SPEED & MIXTURE 3-26
  THROTTLE CAM 3-26
  OIL PUMP 3-27
75 AND 90 HP 3-27
  PRELIMINARY ADJUSTMENTS 3-27
  CARBURETOR SYNCHRONIZATION 3-27
  IGNITION TIMING 3-27
  IDLE SPEED & MIXTURE 3-28
  THROTTLE CAM 3-28
  OIL PUMP 3-28
100–125 HP 3-28
  PRELIMINARY ADJUSTMENTS 3-28
  CARBURETOR SYNCHRONIZATION 3-28
  THROTTLE CAM 3-28
  ACCELERATOR PUMP 3-29
  IGNITION TIMING 3-29
  IDLE SPEED & MIXTURE 3-29
  OIL PUMP 3-29
  THROTTLE CABLE PRELOAD 3-29
135–225 HP CARBURETED 3-30
  PRELIMINARY ADJUSTMENTS 3-30
  THROTTLE CAM 3-30
  CARBURETOR SYNCHRONIZATION 3-30
  FULL THROTTLE STOP 3-30
  OIL PUMP 3-30
  IGNITION TIMING 3-30
  IDLE SPEED & MIXTURE 3-31
  THROTTLE CABLE PRELOAD 3-31
225 HP (3.0L) CARBURETED 3-31
  IGNITION TIMING 3-32
  THROTTLE CAM 3-32
  CARBURETOR SYNCHRONIZATION 3-32
  OIL PUMP 3-32
  THROTTLE POSITION SENSOR (TPS) 3-33
  CRANKSHAFT POSITION SENSOR 3-33
  IDLE SPEED 3-33
250 AND 275 HP 3-33
  IGNITION TIMING 3-33
  CARBURETOR SYNCHRONIZATION 3-34
  FULL THROTTLE STOP 3-34
  OIL PUMP 3-34
  IDLE SPEED 3-34
150–200 HP ELECTRONIC FUEL INJECTION (EFI) 3-35
  IGNITION TIMING 3-35
  THROTTLE CAM 3-36
  FULL THROTTLE STOP 3-36
  OIL PUMP 3-36
  IDLE SPEED 3-36
  THROTTLE POSITION SENSOR (TPS) 3-37
115–225 DIRECT FUEL INJECTION (DFI) OPTIMAX 3-37
  CRANK POSITION SENSOR 3-37
  THROTTLE CAM 3-37
  FULL THROTTLE STOP 3-37
  THROTTLE PLATE SCREW 3-38
WINTER STORAGE CHECKLIST 3-38
SPRING COMMISSIONING CHECKLIST 3-39
SPECIFICATIONS CHARTS
  GENERAL ENGINE SPECIFICATIONS 3-40
  TUNEUP SPECIFICATIONS CHART 3-41
  MAINTENANCE INTERVAL CHART 3-42
  CAPACITIES 3-42

3

MAINTENANCE

ENGINE MAINTENANCE 3-2
BOAT MAINTENANCE 3-8
TUNE-UP 3-10
WINTER STORAGE CHECKLIST 3-38
SPRING COMMISSIONING CHECKLIST 3-39

3-2 MAINTENANCE

ENGINE MAINTENANCE

Serial Number Identification

▶ See Figures 1 thru 4

The engine serial numbers are the manufactures key to engine changes. These numbers identify the year of manufacture, the horsepower rating and the parts book identification. If any correspondence or parts are required, the engine serial number must be used or proper identification is not possible.

The serial number establishes the year in which the engine was produced and not necessarily the year of first installation. Two serial numbers are used on each of the outboard units covered in this manual. The most accessible location is on the serial/instruction plate on the swivel bracket. The other location is on the engine block.

For more information, refer to the "General Engine Specifications" charts at the end of this section.

2-Stroke Oil

OIL RECOMMENDATIONS

▶ See Figures 5 and 6

Use only Quicksilver Premium Plus TC-W3 two-cycle oil or NMMA (National Marine Manufacturers Association) certified 2-stroke lubricants. These oils are proprietary lubricants designed to ensure optimal engine performance and to minimize combustion chamber deposits, avoid detonation and prolong spark plug life. If certified lubricant is unavailable, use only 2-stroke type outboard oil. Never use automotive motor oil.

→Remember, it is this oil, mixed with the gasoline, that lubricates the internal parts of the engine. Lack of lubrication due to the wrong mix or improper type of oil can cause catastrophic powerhead failure.

FILLING

There are two methods of adding 2-stroke oil to an outboard. The first is the pre-mix method used on most low horsepower and on some commercial outboards. The second is the automatic oil injection method which automatically injects the correct quantity of oil into the engine based on throttle position and operating conditions.

Pre-Mix

Mixing the engine lubricant with gasoline before pouring it into the tank is by far the simplest method of lubrication for 2-stroke outboards. However, this method is the most messy and causes the most amount of harm to our environment.

The most important part of filling a pre-mix system is to determine the proper fuel/oil ratio. Most manufacturers use a 50:1 ratio (that is 50 parts of fuel to 1 part of oil) or a 100:1 ratio. Consult your owners manual to determine what the appropriate ratio should be for your engine.

The procedure itself is uncomplicated. Simply add the correct amount of lubricant to your fuel tank and then fill the tank with gasoline. The order in which you do this is important because as the gasoline is poured into the fuel tank it will mix with and agitate the oil for a complete blending.

If you are attempting to top off your tank, here is a general guideline to determine how much oil to add. For three gallons of fuel you would add 4 ounces of oil to obtain a 100:1 ratio; 8 ounces of oil to obtain a 50:1 ratio and 16 ounces of oil to obtain a 25:1 ratio.

Fig. 1 The serial number "1" for the 2.5–3.3hp outboards is located on top of the swivel bracket and the engine serial number "2" is stamped on the starboard side of the cylinder block

Fig. 2 Serial number for the 3.3hp outboard is shown on the top of the swivel bracket

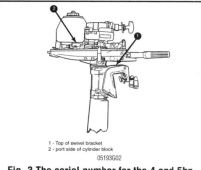

Fig. 3 The serial number for the 4 and 5hp outboards is located on the top surface of the swivel bracket "1" and the engine serial number is stamped on the port side of the cylinder block "2"

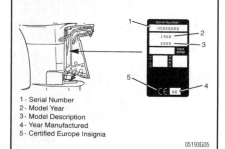

Fig. 4 The serial number for the 6–275hp is located on the lower starboard side of the engine block. Another serial number is located on the starboard side of the swivel bracket

Fig. 5 2-Stroke outboard oils are proprietary lubricants designed to ensure optimal engine performance and to minimize combustion chamber deposits, avoid detonation and prolong spark plug life

Fig. 6 This scuffed piston is an example of what can happen when the proper 2-stroke oil is not used. The outboard required a complete overhaul

MAINTENANCE 3-3

Oil Injection

Most outboard manufacturers use a mechanically driven oil pump mounted on the engine block that is connected to the throttle by way of a linkage arm. The system is powered by the crankshaft, which drives a gear in the pump, creating oil pressure. As the throttle lever is advanced to increase engine speed, the linkage arm also moves, opening a valve that allows more oil to flow into the oil pump.

Most mechanical-injection systems incorporate low-oil warning alarms that are also connected to an engine-overheating sensor. Also, these systems may have a built-in speed limiter. This sub-system is designed to reduce engine speed automatically when oil problems occur. This important feature goes a long way toward preventing severe engine damage in the event of an oil injection problem.

The procedure for filling these systems is simple. On each powerhead there is an auxiliary oil reservoir which holds the 2-stroke oil. Simply fill the oil take to the proper capacity.

➡ It is highly advisable to carry several spare bottles of 2-stroke oil with you onboard.

For more information on the oil injection system refer to the "Lubrication and Cooling" section of this manual.

Lower Unit

♦ See Figures 7, 8 and 9

Regular maintenance and inspection of the lower unit is critical for proper operation and reliability. A lower unit can quickly fail if it becomes heavily contaminated with water or excessively low on oil. The most common cause of a lower unit failure is water contamination.

Water in the lower unit is usually caused by fishing line or other foreign material, becoming entangled around the propeller shaft and damaging the seal. If the line is not removed, it will eventually cut the propeller shaft seal and allow water to enter the lower unit. Fishing line has also been known to cut a groove in the propeller shaft if left neglected over time. This area should be checked frequently.

OIL RECOMMENDATIONS

♦ See Figure 10

Use only Quicksilver Gear Lube or and equivalent high quality SAE 85-90 weight hypoid gear oil. These oils are proprietary lubricants designed to ensure optimal performance and to minimize corrosion in the lower unit.

➡ Remember, it is this lower unit lubricant that prevents corrosion and lubricates the internal parts of the drive gears. Lack of lubrication due to water contamination or the improper type of oil can cause catastrophic lower unit failure.

DRAINING AND FILLING

♦ See Figures 11 and 12

⁂ CAUTION

The EPA warns that prolonged contact with used engine oil may cause a number of skin disorders, including cancer! You should make every effort to minimize your exposure to used engine oil. Protective gloves should be worn when changing the oil. Wash your hands and any other exposed skin areas as soon as possible after exposure to used engine oil. Soap and water or waterless hand cleaner should be used.

1. Place a suitable container under the lower unit.
2. Loosen the oil level plug on the lower unit. This step is important! If the oil level plug cannot be loosened or removed, the complete lower unit lubricant service cannot be performed.

➡ Never remove the vent or filler plugs when the lower unit is hot. Expanded lubricant will be released through the plug hole.

3. Remove the fill plug from the lower end of the gear housing followed by the oil level plug.
4. Allow the lubricant to completely drain from the lower unit.
5. If applicable, check the magnet end of the drain screw for metal particles. Some amount of metal is considered normal wear is to be expected but if there are signs of metal chips or excessive metal particles, the gear case needs to be disassembled and inspected.
6. Inspect the lubricant for the presence of a milky white substance, water or metallic particles. If any of these conditions are present, the lower unit should be serviced immediately.
7. Place the outboard in the proper position for filling the lower unit. The lower unit should not list to either port or starboard and should be completely vertical.
8. Insert the lubricant tube into the oil drain hole at the bottom of the lower unit and inject lubricant until the excess begins to come out the oil level hole.

➡ The lubricant must be filled from the bottom to prevent air from being trapped in the lower unit. Air displaces lubricant and can cause a lack of lubrication or a false lubricant level in the lower unit.

9. Oil should be squeezed in using a tube or with the larger quantities, by using a pump kit to fill the gear case through the drain plug.
10. Using new gaskets, (washers) install the oil level and vent plugs first, then install the oil fill plug.
11. Wipe the excess oil from the lower unit and inspect the unit for leaks.
12. Place the used lubricant in a suitable container for transportation to an authorized recycling facility.

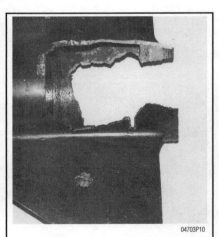

Fig. 7 This lower unit was destroyed because the bearing carrier was frozen due to lack of lubrication

Fig. 8 Excellent view of rope and fishing line entangled behind the propeller. Entangled fishing line can actually cut through the seal, allowing water to enter the lower unit and lubricant to escape

Fig. 9 This lower unit has seen better days. It had been left in the water far too long with very little if any maintenance done to it

3-4 MAINTENANCE

Fig. 10 Approved lower unit lubricant. To protect the investment made in the outboard purchase, only the very best products should be used. The small added cost is worth the increased expense many times over

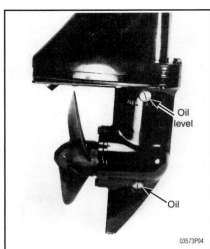

Fig. 11 The lower unit is filled through the oil plug and allowed to rise through the case to the oil level plug. This is done to allow any trapped air to escape

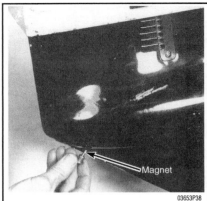

Fig. 12 Most drain plugs are magnetic to trap any metal floating around in the oil. Some amount of metal is considered normal wear is to be expected but if there are signs of metal chips or excessive metal particles, the gear case needs to be disassembled and inspected

Fuel Filter

♦ See Figure 13

The fuel filter is designed to keep particles of dirt and debris from entering the carburetor(s) and clogging the internal passages. A small speck of dirt or sand can drastically affect the ability of the carburetor(s) to deliver the proper amount of air and fuel to enter the engine. If a filter becomes clogged, the flow of gasoline will be impeded. This could cause lean fuel mixtures, hesitation and stumbling and idle problems.

Regular replacement of the fuel filter will decrease the risk of blocking the flow of fuel to the engine, which could leave you stranded on the water. Fuel filters are usually inexpensive and replacement is a simple task. Change your fuel filter on a regular basis to avoid fuel delivery problems to the carburetor.

In addition to the fuel filter mounted on the engine, a filter is usually found inside or near the fuel tank. Because of the large variety of differences in both portable and fixed fuel tanks, it is impossible to give a detailed procedure for removal and installation. Most in-tank filters are simply a screen on the pick-up line inside the fuel tank. Filters of this type usually only need to be cleaned and returned to service. Fuel filters on the outside of the tank are typically of the inline type and are replaced by simply removing the clamps, disconnecting the hoses and installing a new filter. When installing the new filter, make sure the arrow on the filter points in the direction of fuel flow.

RELIEVING FUEL SYSTEM PRESSURE

♦ See Figure 14

On fuel injected engines, always relieve system pressure prior to disconnecting any fuel system component, fitting or fuel line.

✱✱ CAUTION

Exercise extreme caution whenever relieving fuel system pressure to avoid fuel spray and potential serious bodily injury. Please be advised that fuel under pressure may penetrate the skin or any part of the body it contacts.

To avoid the possibility of fire and personal injury, always disconnect the negative battery cable.

Always place a shop towel or cloth around the fitting or connection prior to loosening to absorb any excess fuel due to spillage. Ensure that all fuel spillage is removed from engine surfaces. Ensure that all fuel soaked clothes or towels in suitable waste container.

1. Remove the plastic cap over the pressure port located portside at the bottom of the intake manifold.
2. A screwdriver is need to depress the schraeder valve and release the pressure in the system and allow the fuel to drain from the valve.
3. Place the screwdriver tip lightly over the valve tip and wrap a clean shop towel around the valve and screwdriver. This operation is similar to letting air out of a tire

Fig. 13 Typical in-line fuel filter installation

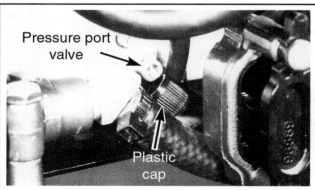

Fig. 14 Fuel pressure can be depressurized at the pressure port valve. Always use extreme caution while working around fuel under pressure

MAINTENANCE 3-5

REMOVAL & INSTALLATION

✻✻ CAUTION

Observe all applicable safety precautions when working around fuel. Whenever servicing the fuel system, always work in a well-ventilated area. Do not allow fuel spray or vapors to come in contact with a spark or open flame. Do not smoke while working around gasoline. Keep a dry chemical fire extinguisher near the work area. Always keep fuel in a container specifically designed for fuel storage; also, always properly seal fuel containers to avoid the possibility of fire or explosion.

Externally Mounted Primary Fuel Filter

A fuel filter is installed at the base of the pick-up tube in the fuel tank. Another fuel filter is installed at the inlet fitting of the crankcase vacuum operated fuel pump. In some cases, an in-line filter is installed between the fuel tank and the primary pump.
1. Disconnect the fuel hoses each side of the fuel filter.
2. Remove the fuel filter and discard.

To install:

3. Install the replacement filter with the embossed arrow pointing towards the fuel pump.
4. Inspect the fuel hoses for swelling, softness or other deterioration. Replace the hoses as necessary.

Engine Mounted Secondary Fuel Filter

♦ See Figures 15 and 16

➡ Before the secondary fuel filter cover may be removed, the EFI system must be depressurized.

1. Remove the plastic cap over the pressure port located portside at the bottom of the intake manifold. A screwdriver will be used to depress the valve tip to release the pressure and allow fuel to drain from the valve. Place the screwdriver tip lightly over the valve tip and wrap a clean shop towel around the valve and screwdriver. This operation is similar to letting air out of tires! Once all the pressure has been released and the fuel flow ceases, carefully proceed with the next step.
2. Loosen but do not remove the center bolt on the filter cover. It is not necessary to remove the inlet fuel hose from the cover. Pry the cover from the filter. Remove the two O-rings. One large O-ring is located around the circumference of the cover and the other smaller O-ring will be found either inside the cover around the securing bolt or on the filter screen.

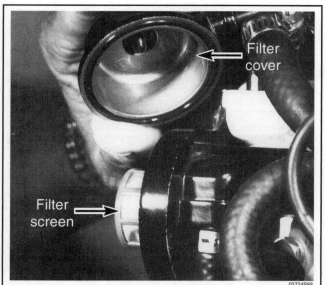

Fig. 15 Pry the cover from the filter and remove the two O-rings. One large O-ring is located around the circumference of the cover and the other smaller O-ring will be found either inside the cover around the securing bolt or on the filter screen

Fig. 16 Remove the filter screen and the large O-ring located around the inner circumference

3. Remove the filter screen and the large O-ring located around the inner circumference.

To install:

4. Position a new O-ring around the inner circumference of the filter screen and then place the screen in the filter housing.
5. Position a new O-ring around the circumference of the filter cover and another O-ring inside the cover around the center bolt. Install the filter cover over the filter screen. Tighten the center bolt to a torque valve of 25 inch lbs. (3 Nm).
6. Start the powerhead and check for fuel leaks.

✻✻ WARNING

The system will be pressurized almost instantly as the powerhead is cranked. Watch for leaks around the cover and the center screw of the secondary fuel filter. Because the system is pressurized, a leak will not appear as drips. Instead, fuel will be sprayed all over the powerhead. Should this occur, shut down the powerhead immediately. do not forget to depressurize the system at the pressure port on the intake manifold before making necessary repairs.

CLEANING & INSPECTION

All three O-rings should be replaced. If they must be used again, they must be subjected to close inspection. If any O-ring appears to have the slightest damage or distortion, the O-ring must be removed, discarded and replaced. These O-rings provide a seal for fuel under pressure.

The screen can be rinsed in solvent and blown dry with compressed air. If the screen is damaged or distorted, the filter must be replaced.

Fuel/Water Separator

SERVICE

Externally Mounted

♦ See Figure 17

In addition to the engine and inline fuel filters, there is usually another filter located in the fuel supply line. The fuel/water separator is used to remove water particles from the fuel prior to entering the engine or inline filter. Water can enter the fuel supply from a variety of sources and can lead to poor engine performance and ultimately, serious engine damage. The presence of water in the fuel will alter the proportion of air/fuel mixture to the "lean" side, resulting in a higher powerhead operating temperature and possible damage to pistons if not corrected.

A water presence sensor is located at the base of the water separating filter. A Tan lead connects the sensor to a water sensor warning module, with an LED indicator warning light to inform the operator of an unacceptable quantity/level of water in the

3-6 MAINTENANCE

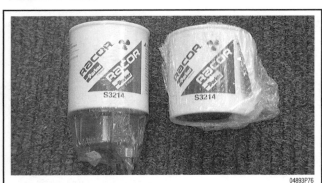

Fig. 17 Typical fuel filter (right) and a water separating fuel filter assembly (left)

fuel. A signal is also transmitted to the Lube Alert Oil Warning Module which houses a horn, sounded at the signal of excess water in the filter.

Because of the large variety of differences in both portable and fixed fuel tanks, it is impossible to give a single procedure to cover all applications. Check with the boat manufacturer or dealership who rigged the boat to get specifics on your particular fuel filtration system.

The filter is drained by opening the drain plug on the bottom and allowing the accumulated moisture and sediment into a suitable container.

Engine Mounted

♦ See Figure 18

This filter removes moisture and also debris from the fuel. If the filter becomes filled with water, the water can be removed. If the filter becomes clogged with debris, the filter must be replaced with a new element.

➡ **On some outboards, a warning system will turn on when water in the filter reaches the full level.**

1. Turn the ignition key to the **OFF** position.
2. If equipped with a sensor, disconnect the wire at the bottom of the filter.

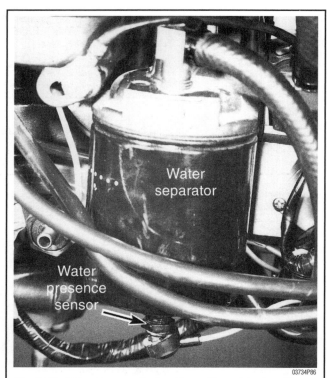

Fig. 18 The EFI water separator is a disposable canister-type filter with a sensor mounted on the filter base

3. Remove the filter by turning the filter clockwise. Tip the filter to drain the fluid into a suitable container.

To install:

4. Lubricate the sealing ring on the filter with a small amount of motor oil. Thread on the filter and tighten the filter securely by hand.
5. If equipped with a sensor, connect the wire to the filter.

✷✷ CAUTION

Visually inspect for fuel leakage from the filter by squeezing the primer bulb until it is firm, forcing fuel into the filter.

Lubrication Points

INSPECTION & LUBRICATION

♦ See Figures 19 and 20

As with every type mechanical invention with moving parts, lubrication plays a prominent role in operation, enjoyment and longevity of the unit.

If an outboard unit is operated in salt water the frequency of applying lubricant to fittings is usually cut in half for the same fitting if the unit is used in fresh water. The few minutes involved in moving around the outboard applying lubricant and at the same time making a visual inspection of its general condition will pay in rich rewards with years of continued service.

It is not uncommon to see outboard units well over 20-years of age moving a boat through the water as if the unit had recently been purchased from the current line of models. An inquiry with the proud owner will undoubtedly reveal his main credit for its performance to be regular periodic maintenance.

The steering head and other pivot points on the outboard need periodic lubrication with marine grade grease to provide smooth operation and prevent corrosion. Usually, these pivot points are easily lubricated by simply attaching a grease gun to the fittings.

If the engine is used in salt water, the frequency of applying lubricant is usually doubled in comparison to operation in fresh water. Due to the very corrosive nature of salt water, an anti-seize thread compound should be used on all exposed fasteners outside of the cowling to reduce the chance of them seizing in place and breaking off when you try to remove them.

➡ **Rinsing off the engine after each use is a very good habit to get into, not only does it help preserve the appearance of the engine, it virtually eliminates the corrosive effects of operating in salt water.**

Always wipe off the grease fitting prior to pumping in the grease and wipe it off again after you are finished. If the fitting is equipped with rubber caps, always use them to cover the fitting.

When pumping in the grease, only pump until the old grease escapes from the component being greased and the new begins to flow. Then wipe off all the excess grease. This grease never fails to get picked up on clothes or shoes and ends up being tracked all over the boat making a mess.

Propeller

♦ See Figure 21

The propeller should be inspected regularly to be sure the blades are in good condition. If any of the blades become bent or nicked, this condition will set up vibrations in the motor. Remove and inspect the propeller. Use a file to trim nicks and burrs. Take care not to remove any more material than is absolutely necessary.

Also, check the rubber and splines inside the propeller hub for damage. If there is damage to either of these, take the propeller to your local marine dealer or a "prop shop". They can evaluate the damaged propeller and determine if it can be saved by rehubbing.

Additionally, the propeller should be removed each time the boat is hauled from the water at the end of the season. Any material entangled behind the propeller should be removed before any damage to the shaft and seals can occur. This may seem like a waste of time but the small amount of time involved in removing the propeller is returned many times by reduced maintenance and repair, including the replacement of expensive parts.

Anodes (Zincs)

The idea behind anodes is simple: When dissimilar metals are dunked in water and a small current is leaked between or amongst them, the less-noble metal (galvanically speaking) is sacrificed.

MAINTENANCE 3-7

Fig. 19 Use only a good quality marine grade grease for lubrication

Fig. 20 The steering cable core must be completely retracted into the housing before applying lubricant at the fitting. Failure to retract, could cause hydraulic lock of the cable

Fig. 21 Damage was caused to this unit when the propeller struck an underwater object. If the propeller should suffer this much abuse, the propeller shaft should be carefully checked

Fig. 22 What a trim tab should look like when it's in good condition

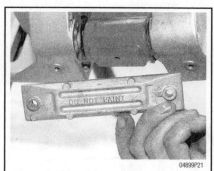

Fig. 23 Other types of anodes are also used throughout the outboard, like this one mounted on a stern bracket

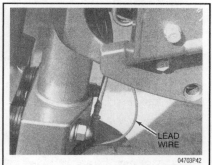

Fig. 24 One of the many lead wires used to connect bracketed parts. Lead wires are used to assist corrosion resistance

The zinc alloy that the anodes are made of is designed to be less noble than the aluminum alloy your outboard is made from. If there's any electrolysis and there almost always is, the inexpensive zinc anodes are consumed in lieu of the expensive outboard motor.

These zincs need a little attention in order to do their job. Make sure they're there, solidly attached to a clean mounting site and not covered with any kind of paint or wax.

Periodically inspect them to make sure they haven't eroded too much. At a certain point in the erosion process, the mounting holes start to enlarge, which is when the zinc might fall off. Obviously, once that happens your engine no longer has any protection.

SERVICING

▶ See Figures 22, 23 and 24

Depending on what kind of power your boat has, you might have anywhere from one to six (or more) zincs. Regardless of the number, there are some fundamental rules to follow that will give your boat's sacrificial anodes the ability to do the best job protecting your boat's underwater hardware that they can.

The first thing to remember is that zincs are electrical components and like all electrical components, they require good clean connections. So after you've undone the mounting hardware and removed last year's zincs, you want to get the zinc mounting sites clean and shiny.

Get a piece of coarse emery cloth or some 80-grit sandpaper. Thoroughly rough up the areas where the zincs attach (there's often a bit of corrosion residue in these spots). Make sure to remove every trace of corrosion.

Zincs are attached with stainless steel machine screws that thread into the mounting for the zincs. Over the course of a season, this mounting hardware is inclined to loosen. Mount the zincs and tighten the mounting hardware securely. Tap the zincs with a hammer hitting the mounting screws squarely. This process tightens the zincs and allows the mounting hardware to become a bit loose ill the process. Now, do the final tightening. This will insure your zincs stay put for the entire season.

INSPECTION

▶ See Figure 25

If you use your outboard in salt water, an your zincs never wear, inspect them carefully. Paint or wax on zincs prevents them from working properly. They must be left bare. If the zincs are installed properly and not painted or waxed, inspect around them for sings of corrosion. If corrosion is found, strip it off immediately and repaint with a rust inhibiting paint. If in doubt, replace the zinc's.

On the other hand, if your zinc seems to erode in no time at all, this may be a symptom of the zinc's themselves. Each manufacturer uses a specific blend of metals in their zincs. If you are using zincs with the wrong blend of metals, they may erode more quickly or leave you with diminished protection.

Fig. 25 Such extensive erosion of a trim tab compared with a new tab suggests an electrolysis problem or complete disregard for periodic maintenance

3-8 MAINTENANCE

BOAT MAINTENANCE

Batteries

MAINTENANCE

▶ See Figure 26

Batteries require periodic servicing and a definite maintenance program will ensure extended life. If the battery should test satisfactorily but still fails to perform properly, one of four problems could be the cause.

1. An accessory might have accidentally been left on overnight or for a long period during the day. Such an oversight would result in a discharged battery.
2. Using more electrical power than the stator assembly or lighting coil can replace would result in an undercharged condition.
3. A defect in the charging system. A faulty stator assembly or lighting coil, defective rectifier or high resistance somewhere in the system could cause the battery to become undercharged.
4. Failure to maintain the battery in good order. This might include a low level of electrolyte in the cells, loose or dirty cable connections at the battery terminals or possibly an excessively dirty battery top.

The most common procedure for maintaining a battery is to check the electrolyte level. This is done by removing the cell caps and visually observing the level in the cells. The bottom of each cell has a split vent which will cause the surface of the electrolyte to appear distorted when it makes contact. When the distortion first appears at the bottom of the split vent, the electrolyte level is correct.

During hot weather and periods of heavy use, the electrolyte level should be checked more often than during normal operation. Add distilled water to bring the level of electrolyte in each cell to the proper level. Take care not to overfill, because adding an excessive amount of water will cause loss of electrolyte and any loss will result in poor performance, short battery life and will contribute quickly to corrosion.

➡ Never add electrolyte from another battery. Use only distilled water.

CLEANING

▶ See Figures 27 and 28

Dirt and corrosion should be cleaned from the battery as soon as it is discovered. Any accumulation of acid film or dirt will permit current to flow between the terminals. Such a current flow will drain the battery over a period of time.

Clean the exterior of the battery with a solution of diluted ammonia or a paste made from baking soda and water. This neutralize any acid which may be present. Flush the cleaning solution off with plenty of clean water.

➡ Take care to prevent any of the neutralizing solution from entering the cells.

A poor contact at the terminals will add resistance to the charging circuit. This resistance will cause the voltage regulator to register a fully charged battery and thus cut down on the stator assembly or lighting coil output adding to the low battery charge problem.

At least once a season, the battery terminals and cable clamps should be cleaned. Loosen the clamps and remove the cables, negative cable first. On batteries with top mounted posts, the use of a puller specially made for this purpose is recommended. These are inexpensive and available in most parts stores.

Clean the cable clamps and the battery terminal with a wire brush, until all corrosion, grease, etc., is removed and the metal is shiny. It is especially important to clean the inside of the clamp thoroughly (a wire brush is useful here), since a small deposit of foreign material or oxidation there will prevent a sound electrical connection and inhibit either starting or charging. It is also a good idea to apply some dielectric grease to the terminal, as this will aid in the prevention of corrosion.

After the clamps and terminals are clean, reinstall the cables, negative cable last, do not hammer the clamps onto battery posts. Tighten the clamps securely but do not distort them. Give the clamps and terminals a thin external coating of grease after installation, to retard corrosion.

Check the cables at the same time that the terminals are cleaned. If the insulation is cracked or broken or if its end is frayed, that cable should be replaced with a new one of the same length and gauge.

Fig. 26 Explosive hydrogen gas is normally released from the cells under a wide range of circumstances. This battery exploded when the gas ignited from someone smoking in the area when the caps were removed. Such an explosion could also be caused by a spark from the battery terminals

TESTING

▶ See Figure 29

A hydrometer is a device to measure the percentage of sulfuric acid in the battery electrolyte in terms of specific gravity. When the condition of the battery drops from

Fig. 27 A battery post cleaner is used to clean the battery posts . . .

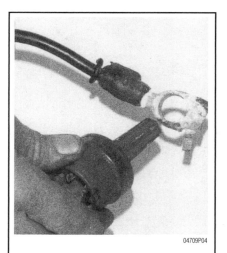

Fig. 28 . . . and the battery terminals

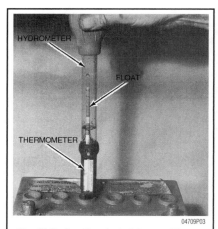

Fig. 29 Testing the electrolyte specific gravity using a temperature corrected hydrometer

fully charged to discharged, the acid leaves the solution and enters the plates, causing the specific gravity of the electrolyte to drop.

It may not be common knowledge but hydrometer floats are calibrated for use at 80°F (27°C). If the hydrometer is used at any other temperature, hotter or colder, a correction factor must be applied.

➡ **Remember, a liquid will expand if it is heated and will contract if cooled. Such expansion and contraction will cause a definite change in the specific gravity of the liquid, in this case the electrolyte.**

A quality hydrometer will have a thermometer/temperature correction table in the lower portion, as illustrated in the accompanying illustration. By knowing the air temperature around the battery and from the table, a correction factor may be applied to the specific gravity reading of the hydrometer float. In this manner, an accurate determination may be made as to the condition of the battery.

When using a hydrometer, pay careful attention to the following points:

1. Never attempt to take a reading immediately after adding water to the battery. Allow at least ¼ hour of charging at a high rate to thoroughly mix the electrolyte with the new water. This time will also allow for the necessary gases to be created.
2. Always be sure the hydrometer is clean inside and out as a precaution against contaminating the electrolyte.
3. If a thermometer is an integral part of the hydrometer, draw liquid into it several times to ensure the correct temperature before taking a reading.
4. Be sure to hold the hydrometer vertically and suck up liquid only until the float is free and floating.
5. Always hold the hydrometer at eye level and take the reading at the surface of the liquid with the float free and floating.
6. Disregard the slight curvature appearing where the liquid rises against the float stem. This phenomenon is due to surface tension.
7. Do not drop any of the battery fluid on the boat or on your clothing, because it is extremely caustic. Use water and baking soda to neutralize any battery liquid that does accidentally drop.
8. After drawing electrolyte from the battery cell until the float is barely free, note the level of the liquid inside the hydrometer. If the level is within the Green band range for all cells, the condition of the battery is satisfactory. If the level is within the white band for all cells, the battery is in fair condition.
9. If the level is within the Green or white band for all cells except one, which registers in the red, the cell is shorted internally. No amount of charging will bring the battery back to satisfactory condition.
10. If the level in all cells is about the same, even if it falls in the Red band, the battery may be recharged and returned to service. If the level fails to rise above the Red band after charging, the only solution is to replace the battery.

A check of the electrolyte in the battery should be on the maintenance schedule of any boat. A hydrometer reading of 1.300 or in the green band, indicates the battery is in satisfactory condition. If the reading is 1.150 or in the red band, the battery must be charged.

STORAGE

If the boat is to be laid up for the winter or for more than a few weeks, special attention must be given to the battery to prevent complete discharge or possible damage to the terminals and wiring. Before putting the boat in storage, disconnect and remove the batteries. Clean them thoroughly of any dirt or corrosion and then charge them to full specific gravity reading. After they are fully charged, store them in a clean cool dry place where they will not be damaged or knocked over, preferably on a couple blocks of wood. Storing the battery up off the deck, will permit air to circulate freely around and under the battery and will help to prevent condensation.

Never store the battery with anything on top of it or cover the battery in such a manner as to prevent air from circulating around the filler caps. All batteries, both new and old, will discharge during periods of storage, more so if they are hot than if they remain cool. Therefore, the electrolyte level and the specific gravity should be checked at regular intervals. A drop in the specific gravity reading is cause to charge them back to a full reading.

In cold climates, care should be exercised in selecting the battery storage area. A fully-charged battery will freeze at about 60° below zero. A discharged battery, almost dead, will have ice forming at about 19° above zero.

Fiberglass Hull

♦ **See Figures 30, 31 and 32**

Fiberglass reinforced plastic hulls are tough, durable and highly resistant to impact. However, like any other material they can be damaged. One of the advantages of this type of construction is the relative ease with which it may be repaired.

A fiberglass hull has almost no internal stresses. Therefore, when the hull is broken or stove-in, it retains its true form. It will not dent to take an out-of-shape set. When the hull sustains a severe blow, the impact will be either absorbed by deflection of the laminated panel or the blow will result in a definite, localized break. In addition to hull damage, bulkheads, stringers and other stiffening structures attached to the hull may also be affected and therefore, should be checked. Repairs are usually confined to the general area of the rupture.

➡ **The best way to care for a fiberglass hull is to wash it thoroughly, immediately after hauling the boat while the hull is still wet.**

A foul bottom can seriously affect boat performance. This is one reason why racers, large and small, both powerboat and sail, are constantly giving attention to the condition of the hull below the waterline.

In areas where marine growth is prevalent, a coating of vinyl, anti-fouling bottom paint should be applied. If growth has developed on the bottom, it can be removed with a solution of Muriatic acid applied with a brush or swab and then rinsed with clear water. Always use rubber gloves when working with Muriatic acid and take extra care to keep it away from your face and hands. The fumes are toxic. Therefore, work in a well-ventilated area or if outside, keep your face on the windward side of the work.

Barnacles have a nasty habit of making their home on the bottom of boats which have not been treated with anti-fouling paint. Actually they will not harm the fiberglass hull but can develop into a major nuisance.

If barnacles or other crustaceans have attached themselves to the hull, extra work will be required to bring the bottom back to a satisfactory condition. First, if practical, put the boat into a body of fresh water and allow it to remain for a few days. A large percentage of the growth can be removed in this manner. If this remedy is not possible, wash the bottom thoroughly with a high-pressure fresh water source and use a scraper. Small particles of hard shell may still hold fast. These can be removed with sandpaper.

Fig. 30 In areas where marine growth is a problem, a coating of anti-foul bottom paint should be applied

Fig. 31 The best way to care for a fiberglass hull is to wash it thoroughly

Fig. 32 Fiberglass, vinyl and rubber care products, such as those available from Meguiar's are available to protect every part of your boat

3-10 MAINTENANCE

TUNE-UP

Introduction

A proper tune-up is the key to long and trouble-free outboard life and the work can yield its own rewards. Studies have shown that a properly tuned and maintained outboard can achieve better fuel economy than an out-of-tune engine. As a conscientious boater, set aside a Saturday morning, say once a month, to check or replace items which could cause major problems later. Keep your own personal log to jot down which services you performed, how much the parts cost you, the date and the number of hours on the engine at the time. Keep all receipts for such items as oil and filters, so that they may be referred to in case of related problems or to determine operating expenses. As a do-it-yourselfer, these receipts are the only proof you have that the required maintenance was performed. In the event of a warranty problem, these receipts will be invaluable.

The efficiency, reliability, fuel economy and enjoyment available from boating are all directly dependent on having your outboard tuned properly. The importance of performing service work in the proper sequence cannot be over emphasized. Before making any adjustments, check the specifications. Never rely on memory when making critical adjustments.

Before tuning any outboard, insure it has satisfactory compression. An outboard with worn or broken piston rings, burned pistons or scored cylinder walls, will not perform properly no matter how much time and expense is spent on the tune-up. Poor compression must be corrected or the tune-up will not give the desired results.

The extent of the engine tune-up is usually dependent on the time lapse since the last service. In this section, a logical sequence of tune-up steps will be presented in general terms. If additional information or detailed service work is required, refer to the section of this manual containing the appropriate instructions.

Compression Test

PRIMARY COMPRESSION TEST

Because the 2-stroke powerhead is a pump, the crankcase must be sealed against pressure created on the down stroke of the piston and vacuum created when the piston moves toward top dead center. If there are air leaks into the crankcase, insufficient fuel will be brought into the crankcase and into the cylinder for normal combustion.

➡ **If it is a very small leak, the powerhead will run poorly, because the fuel mixture will be lean and cylinder temperatures will be hotter than normal.**

Air leaks are possible around any seal, O-ring, cylinder block mating surface or gasket. Always replace O-rings, gaskets and seals when service work is performed.

If the powerhead is running, soapy water can be sprayed onto the suspected sealing areas. If bubbles develop, there is a leak at that point. Oil around sealing points and on ignition parts under the flywheel indicates a crankcase leak.

The base of the powerhead and lower crankshaft seal is impossible to check on an installed powerhead. When every test and system have been checked out and the bottom cylinder seems to be effecting performance, then the lower seal should be tested.

Adapter plates available from tool manufacturers to seal the inlet, exhaust and base of the powerhead. Adapter plates can also be manufactured by cutting metal block off plates from pieces of plate steel or aluminum. A pattern made from the gaskets can be used for an accurate shape. Seal these plates using rubber or silicone gasket making compound.

1. Install adapter plates over the intake ports and the exhaust ports to completely seal the powerhead.

➡ **When installing the adapter plates, make sure to leave the water jacket holes open.**

2. Into one adapter, place an air fitting which will accept a hand air pump.
3. Using the hand pump (or another regulated air source), pressurize the crankcase to five pounds of pressure.
4. Spray soapy water around the lower seal area and other sealed areas watching for bubbles which indicate a leaking point.
5. Turn the powerhead upside down and fill the water jacket with water. If bubbles show up in the in the water when a positive pressure is applied to the crankcase, there may be cracks or corrosion holes in the cooling system passages. These holes can cause a loss of cooling system effectiveness and lead to overheating.
6. After the pressure test is completed, pull a vacuum to stress the seals in the opposite direction and watch for a pressure drop.
7. Note the leaking areas and replace the seals or gaskets.

SECONDARY COMPRESSION TEST

♦ **See Figure 33**

The actual pressure measured during a secondary compression test is not as important as the variation from cylinder to cylinder. On multi-cylinder powerheads, a variation of 15 psi or more is considered questionable. On single cylinder powerheads, a drop of 15 psi from the normal compression pressure you established when it was new is cause for concern (you did do a compression test on it when it was new, didn't you?).

➡ **If the powerhead been in storage for an extended period, the piston rings may have relaxed. This will often lead to initially low and misleading readings. Always run an engine to operating temperature to ensure that the reading you get is accurate.**

1. Disable the ignition system by removing the lanyard clip. If you do not have a lanyard, take a wire jumper lead and connect one end to a good engine ground and the other end to the metal connector inside the spark plug boot, using one jumper for each plug wire. Never simply disconnect all the plug wires.

✱✱ CAUTION

Removing all the spark plugs and cranking over the powerhead can lead to an explosion if raw fuel/oil sprays out of the plug holes. A plug wire could spark and ignite this mix outside of the combustion chamber if it isn't grounded to the engine.

2. Remove all the spark plugs and be sure to keep them in order. Carefully inspect the plugs, looking for any inconsistency in coloration and for any sign of water or rust near the tip.
3. Thread the compression gauge into the No. 1 spark-plug hole, taking care to not crossthread the fitting.
4. Open the throttle to the wide open throttle position and hold it there.

➡ **Some engines allow only minimal opening if the gearshift is in neutral, to guard against over-revving.**

5. Crank over the engine an equal number of times for each cylinder you test, zeroing the gauge for each cylinder.
6. If you have electric start, count the number of seconds you count. On manual start, pull the starter rope four to five times for each cylinder you are testing.
7. Record your readings from each cylinder. When all cylinders are tested, compare the readings and determine if pressures are within the 15 psi criterion.
8. If compression readings are lower than normal for any cylinders, try a "wet" compression test, which will temporarily seal the piston rings and determine if they are the cause of the low reading.
9. Using a can of fogging oil, fog the cylinder with a circular motion to distribute oil spray all around the perimeter of the piston. Retest the cylinder.
   a. If the compression rises noticeably, the piston rings are sticking. You may be able to cure the problem by de-carbonizing the powerhead.
   b. If the dry compression was really low and no change is evident during the wet test, the cylinder is dead. The piston and/or are worn beyond specification and a powerhead overhaul or replacement is necessary.

Fig. 33 Typical two-stroke powerhead secondary compression test

MAINTENANCE 3-11

10. If two adjacent cylinders on a multi-cylinder engine give a similarly low reading then the problem may be a faulty head gasket. This should be suspected if there was evidence of water or rust on the spark plugs from these cylinders.

Spark Plugs

The spark plug performs four main functions:
- It fills a hole in the cylinder head.
- It acts as a dielectric insulator for the ignition system.
- It provides spark for the combustion process to occur.
- It removes heat from the combustion chamber.

It is important to remember that spark plugs do not create heat, they help remove it. Anything that prevents a spark plug from removing the proper amount of heat can lead to pre-ignition, detonation, premature spark plug failure and even internal engine damage, especially in two stroke engine.

In the simplest of terms, the spark plug acts as the thermometer of the engine. Much like a doctor examining a patient, this "thermometer" can be used to effectively diagnose the amount of heat present in each combustion chamber.

Spark plugs are valuable tuning tools, when interpreted correctly. They will show symptoms of other problems and can reveal a great deal about the engine's overall condition. By evaluating the appearance of the spark plug's firing tip, visual cues can be seen to accurately determine the engine's overall operating condition, get a feel for air/fuel ratios and even diagnose driveability problems.

As spark plugs grow older, they lose their sharp edges and material from the center and ground electrodes is slowly eroded away. As the gap between these two points grows, the voltage required to bridge this gap increases proportionately. The ignition system must work harder to compensate for this higher voltage requirement and hence there are a greater rate of misfires or incomplete combustion cycles. Each misfire means lost horsepower, reduced fuel economy and higher emissions. Replacing worn out spark plugs with new ones (with sharp new edges) effectively restores the ignition system's efficiency and reduces the percentage of misfires, restoring power, economy and reducing emissions.

How long spark plugs last will depend on a variety of factors, including engine compression, fuel used, gap, center/ground electrode material and the conditions in which the outboard is operated.

SPARK PLUG HEAT RANGE

▶ See Figure 34

Spark plug heat range is the ability of the plug to dissipate heat from the combustion chamber. The longer the insulator (or the farther it extends into the engine), the hotter the plug will operate; the shorter the insulator (the closer the electrode is to the block's cooling passages) the cooler it will operate.

Selecting a spark plug with the proper heat range will ensure that the tip will maintain a temperature high enough to prevent fouling, yet be cool enough to prevent pre-ignition. A plug that absorbs little heat and remains too cool will quickly accumulate deposits of oil and carbon since it is not hot enough to burn them off. This leads to plug fouling and consequently to misfiring. A plug that absorbs too much heat will have no deposits but, due to the excessive heat, the electrodes will burn away quickly and might possibly lead to pre-ignition or other ignition problems.

Pre-ignition takes place when plug tips get so hot that they glow sufficiently to ignite the air/fuel mixture before the actual spark occurs. This early ignition will usually cause a pinging during heavy loads and if not corrected will result in severe engine damage. While there are many other things that can cause pre-ignition,

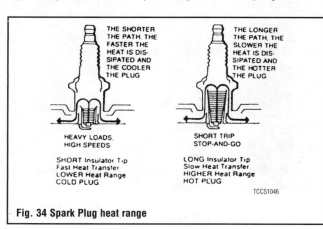

Fig. 34 Spark Plug heat range

selecting the proper heat range spark plug will ensure that the spark plug itself is not a hot-spot source.

SPARK PLUG SERVICE

➡ New technologies in spark plug and ignition system design have pushed the recommended replacement interval to every 100 hours of operation (6 months). However, this depends on usage and conditions.

Spark plugs should only require replacement once a season. The electrode on a new spark plug has a sharp edge but with use, this edge becomes rounded by wear, causing the plug gap to increase. As the gap increases, the plug's voltage requirement also increases. It requires a greater voltage to jump the wider gap and about two to three times as much voltage to fire a plug at high speeds than at idle.

Tools needed for spark plug replacement include: a ratchet, short extension, spark plug socket (there are two types; either $13/16$ inch or $5/8$ inch, depending upon the type of plug), a combination spark plug gauge and gapping tool and a can anti-seize type compound.

REMOVAL & INSTALLATION

▶ See Figures 35, 36 and 37

1. When removing spark plugs, work on one at a time. Don't start by removing the plug wires all at once, because unless you number them, they may become mixed up. Take a minute before you begin and number the wires with tape.
2. Disconnect the negative battery cable or turn the battery switch **OFF**.
3. If the engine has been run recently, allow the engine to thoroughly cool. Attempting to remove plugs from a hot cylinder head could cause the plugs to seize and damage the threads in the cylinder head, especially on aluminum heads!
4. Carefully twist the spark plug wire boot to loosen it, then pull the boot using a twisting motion to remove it from the plug. Be sure to pull on the boot and not on the wire, otherwise the connector located inside the boot may become separated from the high-tension wire.

➡ A spark plug wire removal tool is recommended as it will make removal easier and help prevent damage to the boot and wire assembly.

5. Using compressed air (and safety glasses), blow debris from the spark plug area to assure that no harmful contaminants are allowed to enter the combustion chamber when the spark plug is removed. If compressed air is not available, use a rag or a brush to clean the area. Compressed air is available from both an air compressor or from compressed air in cans available at photography stores.

➡ Remove the spark plugs when the engine is cold, if possible, to prevent damage to the threads. If plug removal is difficult, apply a few drops of penetrating oil to the area around the base of the plug and allow it a few minutes to work.

6. Using a spark plug socket that is equipped with a rubber insert to properly hold the plug, turn the spark plug counterclockwise to loosen and remove the spark plug from the bore.

✱✱ WARNING

Avoid the use of a flexible extension on the socket. Use of a flexible extension may allow a shear force to be applied to the plug. A shear force could break the plug off in the cylinder head, leading to costly and frustrating repairs. In addition, be sure to support the ratchet with your other hand—this will also help prevent the socket from damaging the plug.

7. Evaluate each cylinder's performance by comparing the spark condition. Check each spark plug to be sure they are all of the same manufacturer and have the same heat range rating. Inspect the threads in the spark plug opening of the block and clean the threads before installing the plug.
8. When purchasing new spark plugs, always ask the dealer if there has been a spark plug change for the engine being serviced. Many times manufacturers will update the type of spark plug used in an engine to offer better efficiency or performance.
9. Crank the engine through several revolutions to blow out any material which might have become dislodged during cleaning. Always use a new gasket (if applicable). The gasket must be fully compressed on clean seats to complete the heat transfer process and to provide a gas tight seal in the cylinder.
10. Inspect the spark plug boot for tears or damage. If a damaged boot is found, the spark plug boot and possible the entire wire will need replacement.

3-12 MAINTENANCE

Fig. 35 First remove the spark plug cap using a twisting motion

Fig. 36 Using a spark plug socket equipped with a rubber insert to properly hold the plug insulator, turn the spark plug counterclockwise to loosen it . .

Fig. 37 . . . and remove the spark plug from the cylinder head

11. Check the spark plug gap prior to installing the plug. Most spark plugs do not come gapped to the proper specification.

12. Apply a thin coating of anti-seize on the thread of the plug. This is extremely important on aluminum head engines.

13. Carefully thread the plug into the bore by hand. If resistance is felt before the plug completely bottomed, back the plug out and begin threading again.

✲✲ WARNING

Do not use the spark plug socket to thread the plugs. Always carefully thread the plug by hand or using an old plug wire to prevent the possibility of crossthreading and damaging the cylinder head bore.

14. Carefully tighten the spark plug. If the plug you are installing is equipped with a crush washer, tighten the plug until the washer seats, then turn it ¼ turn to crush the washer. Whenever possible, spark plugs should be tightened to the factory torque specification.

15. Apply a small amount of silicone dielectric grease to the end of the spark plug lead or inside the spark plug boot to prevent sticking, then install the boot to the spark plug and push until it clicks into place. The click may be felt or heard. Gently pull back on the boot to assure proper contact.

16. Connect the negative battery cable or turn the battery switch **ON** .
17. Start the outboard and insure proper operation.

READING SPARK PLUGS

♦ See Figures 38 thru 43

Reading spark plugs can be a valuable tuning aid. By examining the insulator firing nose color, an you can determine much about the engine's overall operating condition.

In general, a light tan/gray color tells you that the spark plug is at the optimum temperature and that the engine is in good operating condition.

Dark coloring, such as heavy black wet or dry deposits usually indicate a fouling problem. Heavy, dry deposits can indicate an overly rich condition, too cold a heat range spark plug, possible vacuum leak, low compression, overly retarded timing or too large a plug gap.

If the deposits are wet, it can be an indication of a breached head gasket, oil control from ring problems or an extremely rich condition, depending on what liquid is present at the firing tip.

Also look for signs of detonation, such as silver specs, black specs or melting or breakage at the firing tip.

Fig. 38 A normally worn spark plug should have light tan or gray deposits on the firing tip (electrode)

Fig. 39 A carbon-fouled plug, identified by soft, sooty black deposits, may indicate an improperly tuned powerhead

Fig. 40 A physically damaged spark plug may be evidence of severe detonation in that cylinder. Watch the cylinder carefully between services, as a continued detonation will not only damage the plug but will most likely damage the powerhead

MAINTENANCE 3-13

Fig. 41 An oil-fouled spark plug indicates a powerhead with worn piston rings or a malfunctioning oil injection system which allows excessive oil to enter the combustion chamber

Fig. 42 This spark plug has been left in the powerhead too long, as evidenced by the extreme gap. Plugs with such an extreme gap can cause misfiring and stumbling accompanied by a noticeable lack of power

Fig. 43 A bridged or almost bridged spark plug, identified by the build-up between the electrodes caused by excessive carbon or oil build up on the plug

Compare your plugs to the illustrations shown to identify the most common plug conditions.

Fouled Spark Plugs

A spark plug is said to be fouled when the insulator nose at the firing tip becomes coated with a foreign substance, such as fuel, oil or carbon. This coating makes it easier for the voltage to follow along the insulator nose and leach back down into the metal shell, grounding out, rather than bridging the gap normally.

Fuel, oil and carbon fouling can all be caused by different things but in any case, once a spark plug is fouled, it will not provide voltage to the firing tip and that cylinder will not fire properly. In many cases, the spark plug cannot be cleaned sufficiently to restore normal operation. It is therefore recommended that fouled plugs be replaced.

Signs of fouling or excessive heat must be traced quickly to prevent further deterioration of performance and to prevent possible engine damage.

Overheated Spark Plugs

When a spark plug tip shows signs of melting or is broken, it usually means that excessive heat and/or detonation was present in that particular combustion chamber or that the spark plug was suffering from thermal shock.

Since spark plugs do not create heat by themselves, one must use this visual clue to track down the root cause of the problem. In any case, damaged firing tips most often indicate that cylinder pressures or temperatures were too high. Left unresolved, this condition usually results in more serious engine damage.

Detonation refers to a type of abnormal combustion that is usually preceded by pre-ignition. It is most often caused by a hot spot formed in the combustion chamber.

As air and fuel is drawn into the combustion chamber during the intake stroke, this hot spot will "pre-ignite" the air fuel mixture without any spark from the spark plugs.

Detonation

Detonation exerts a great deal of downward force on the pistons as they are being forced upward by the mechanical action of the connecting rods. When this occurs, the resulting concussion, shock waves and heat can be severe. Spark plug tips can be broken or melted and other internal engine components such as the pistons or connecting rods themselves can be damaged.

Left unresolved, engine damage is almost certain to occur, with the spark plug usually suffering the first signs of damage.

➡When signs of detonation or pre-ignition are observed, they are symptom of another problem. You must determine and correct the situation that caused the hot spot to form in the first place.

INSPECTION & GAPPING

▸ See Figures 44 and 45

A particular spark plug might fit hundreds of powerheads and although the factory will typically set the gap to a pre-selected setting, this gap may not be the right one for your particular powerhead.

Insufficient spark plug gap can cause pre-ignition, detonation, even engine damage. Too much gap can result in a higher rate of misfires, noticeable loss of power, plug fouling and poor economy.

Check spark plug gap before installation. The ground electrode (the L-shaped one connected to the body of the plug) must be parallel to the center electrode and the specified size wire gauge must pass between the electrodes with a slight drag.

Do not use a flat feeler gauge when measuring the gap on a used plug, because the reading may be inaccurate. A round-wire type gapping tool is the best way to check the gap. The correct gauge should pass through the electrode gap with a slight drag. If you're in doubt, try a wire that is one size smaller and one larger. The smaller gauge should go through easily, while the larger one shouldn't go through at all.

Wire gapping tools usually have a bending tool attached. Use this tool to adjust the side electrode until the proper distance is obtained. Never attempt to bend the center electrode. Also, be careful not to bend the side electrode too far or too often as it may weaken and break off within the engine, requiring removal of the cylinder head to retrieve it.

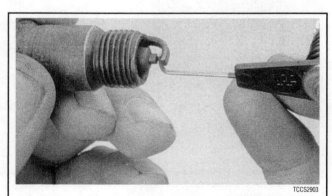

Fig. 44 Using a wire-type spark plug gapping tool to check the distance between center and ground electrodes

3-14 MAINTENANCE

Spark Plug Wires

TESTING

Each time you remove the engine cover, visually inspect the spark plug wires for burns, cuts or breaks in the insulation. Check the boots on the coil and at the spark plug end. Replace any wire that is damaged.

Once a year, usually when you change your spark plugs, check the resistance of the spark plug wires with an ohmmeter. Wires with excessive resistance will cause misfiring and may make the engine difficult to start. In addition worn wires will allow arcing and misfiring in humid conditions.

Remove the spark plug wire from the engine. Test the wires by connecting one lead of an ohmmeter to the coil end of the wire and the other lead to the spark plug end of the wire. Resistance should measure approximately 7000 ohms per foot of wire. If a spark plug wire is found to have excessive (high) resistance, the entire set should be replaced.

REMOVAL & INSTALLATION

When installing a new set of spark plug wires, replace the wires one at a time so there will be no confusion. Coat the inside of the boots with dielectric grease to prevent sticking. Install the boot firmly over the spark plug until it clicks into place. The click may be felt or heard. Gently pull back on the boot to assure proper contact. Repeat the process for each wire.

➡ **It is important to route the new spark plug wire the same as the original and install it in a similar manner on the powerhead. Improper routing of spark plug wires may cause powerhead performance problems.**

Ignition System

♦ See Figure 46

The flywheel magneto system was installed on the 2.5 and 3.0hp powerheads from 1990–92. It consists of a breaker point set, condenser and primary coil. These components are all located on the stator plate under and protected by the flywheel.

The breaker points on an outboard engine are an extremely important part of the ignition system. A set of points may appear to be in good condition but they may be the source of hard starting, misfiring or poor engine performance. The rules and knowledge gained from association with 4-cycle engines do not necessarily apply to a 2-cycle engine. The points should be replaced every 100 hundred hours of operation or at least once a year. Remember, the less an outboard engine is operated, the more care it needs. Allowing an outboard engine to remain idle will do more harm than if it is used regularly.

A breaker point set consists of two points. One is attached to a stationary bracket and does not move. The other point is attached to a moveable mount. A spring is used to keep the points in contact with each other, except when they are separated by the action of a cam built into the crankshaft. Both points are constructed with a steel base and tungsten cap fused to the base.

Take a close look at the points themselves and make sure that both points are correctly aligned and gapped and that there is no corrosion or dirt on them.

Magnetos installed on outboard engines will usually operate over extremely long periods of time without requiring adjustment or repair. However, if ignition system problems are encountered and the usual corrective actions such as replacement of the spark plugs does not correct the problem, the magneto output should be checked to determine if the unit is functioning properly.

Modern electronic CDI ignition systems have become one of the most reliable components on an outboard. There is very little maintenance involved in the operation of these ignition systems and even less to repair if they fail. Most systems are sealed and there is no option other than to replace failed components.

It is very important to narrow down the problem to the ignition system and replace the correct components rather than just replace parts hoping to solve the problem. Electronic components can be very expensive and are usually not returnable.

Refer to the "Ignition and Electrical" section for more information on troubleshooting and repairing ignition systems.

Timing and Synchronization

♦ See Figures 47 and 48

Timing and synchronization on an outboard engine is extremely important to obtain maximum efficiency. The powerhead cannot perform properly and produce its designed horsepower output if the fuel and ignition systems have not been precisely adjusted.

Outboards equipped with a mechanical advance type Capacitor Discharge Ignition (CDI) system use a series of link rods between the carburetor and the ignition base plate assembly. At the time the throttle is opened, the ignition base plate assembly is rotated by means of the link rod, thus advancing the timing.

On electronically controlled ignition models, a microcomputer decides when to advance or retard the timing based on input from various sensors. Therefore, there is no link rod between the magneto control lever and the stator assembly.

Many models have timing marks on the flywheel and CDI base. A timing light is normally used to check the ignition timing with the powerhead operating (dynamically). An alternate method is to check the timing with the powerhead not operating (static timing). This second method requires the use of a dial indicator gauge.

Various models have unique methods of checking ignition timing. As appropriate, these differences will be explained in detail in the text.

In simple terms, synchronization is timing the fuel system to the ignition. As the throttle is advanced to increase powerhead rpm, the fuel and the ignition systems are both advanced equally and at the same rate.

Any time the fuel system or the ignition system on a powerhead is serviced to replace a faulty part or any adjustments are made for any reason, powerhead timing and synchronization must be carefully checked and verified.

On the breaker point ignitions, synchronization is automatic once the point gap and the piston travel or timing mark alignments are correct.

Models equipped with electronic ignitions are statically timed by aligning the timing marks on the throttle cam or throttle stopper with timing marks on the flywheel. Initial timing and timing advance are both set this way before using a timing light to check the timing.

➡ **Before making any adjustments to the ignition timing or synchronizing the ignition to the fuel system, both systems should be verified to be in good working order.**

Timing and synchronizing the ignition and fuel systems on an outboard motor are critical adjustments. The following equipment is essential and is called out repeat-

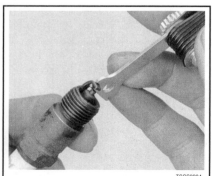

Fig. 45 Most spark plug gapping tools have an adjusting tool used to bend the ground electrode. USE IT! This tool greatly reduces the chance of breaking off the electrode and is much more accurate

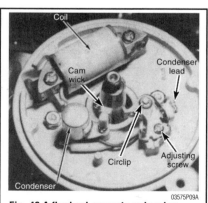

Fig. 46 A flywheel magneto or breaker points type ignition system was installed on the 2.5 and 3.0hp powerheads from 1990–92

Fig. 47 Timing marks on the flywheel with the flywheel cover removed . . .

MAINTENANCE 3-15

Fig. 48 . . . and with the cover in place showing the timing pointer in the viewing area of the cover

edly in this section. This equipment must be used as described, unless otherwise instructed by the equipment manufacturer. Naturally, the equipment is removed following completion of the adjustments.

Some manufacturers recommend the use of a test wheel instead of a normal propeller in order to put a load on the engine and propeller shaft. The use of the test wheel prevents the engine from excessive rpm.

• Dial Indicator—Top dead center (TDC) of the No. 1 (top) piston must be precisely known before the timing adjustment can be made. TDC can only be determined through installation of a dial indicator into the No. 1 spark plug opening

• Timing Light—During many procedures in this section, the timing mark on the flywheel must be aligned with a stationary timing mark on the engine while the powerhead is being cranked or is running. Only through use of a timing light connected to the No. 1 spark plug lead, can the timing mark on the flywheel be observed while the engine is operating

• Tachometer—A tachometer connected to the powerhead must be used to accurately determine engine speed during idle and high-speed adjustment. Engine speed readings range from 0–6,000 RPM in increments of 100 rpm. Choose a tachometers with solid state electronic circuits which eliminates the need for relays or batteries and contribute to their accuracy. For maximum performance, the idle RPM should be adjusted under actual operating conditions. Under such conditions it might be necessary to attach a tachometer closer to the powerhead than the one installed on the control panel.

• Flywheel Rotation—The instructions may call for rotating the flywheel until certain marks are aligned with the timing pointer. When the flywheel must be rotated, always move the flywheel in the indicated direction. If the flywheel should be rotated in the opposite direction, the water pump impeller vanes would be twisted. Should the powerhead be started with the pump tangs bent back in the wrong direction, the tangs may not have time to bend in the correct direction before they are damaged. The least amount of damage to the water pump will affect cooling of the powerhead

• Test Tank—Since the engine must be operated at various times and engine speeds during some procedures, a test tank or moving the boat into a body of water, is necessary. If installing the engine in a test tank, outfit the engine with an appropriate test propeller

➡Remember the powerhead will not start without the emergency tether in place behind the kill switch knob.

※※ CAUTION

Never operate the powerhead above a fast idle with a flush attachment connected to the lower unit. Operating the powerhead at a high RPM with no load on the propeller shaft could cause the powerhead to runaway causing extensive damage to the unit.

2.5, 3 and 3.3 HP

Perform synchronization of the 2.5, 3 and 3.3hp powerheads in the following order: set the idle speed, check idle mixture, adjust the throttle jet needle and finally, test drive the boat to insure proper full throttle speed.

※※ CAUTION

Water must circulate through the powerhead any time it is running to prevent damage to the water pump and possible powerhead overheating. Never run the engine over 3000 RPM without an adequate load applied to the propeller.

IGNITION TIMING

The ignition timing on these models is not adjustable.

IDLE SPEED

1. Connect a tachometer to the powerhead.
2. Connect a flushing attachment to the lower unit if the outboard is not in a test tank or body of water.
3. Start the engine and allow it to warm up to operating temperature.
4. Adjust the throttle lever to the lowest speed and then adjust the idle speed screw until the powerhead idles at the speed specified in the "Tune-Up Specifications" chart.

IDLE MIXTURE

♦ See Figure 49

1. Mount the outboard unit in a test tank, on the boat in a body of water or connect a flush attachment and hose to the lower unit.
2. Connect a tachometer to the powerhead.

➡While making the following adjustment, make every effort to prevent a powerhead stalling.

3. Start the powerhead and allow it to warm to normal operating temperature.
4. Slowly rotate the idle mixture screw counterclockwise a little at a time from the initial setting.
5. The powerhead must be allowed to idle for at least 10–15 seconds to permit the powerhead to stabilize, between each adjustment.
6. Continue slowly rotating the screw until the RPM decreases to a point where the powerhead is just about to die from an excessively rich mixture.
7. Slowly rotate the screw clockwise and keep track of how much the screw is rotated until the RPM increases and powerhead operation is smooth. Continue slowly rotating the screw and keeping track of how much the screw is rotated until the RPM decreases again and the powerhead is just about to die from an excessively lean mixture.

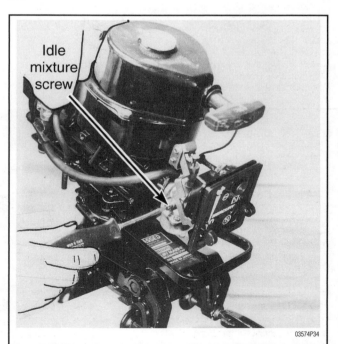

Fig. 49 Slowly rotate the idle mixture screw from the initial setting

3-16 MAINTENANCE

8. Divide the amount of clockwise rotation by two. Rotate the idle mixture screw counterclockwise from the too lean position by the amount of movement calculated. This procedure will set the idle mixture screw midway between the too lean position and the too rich operating limits.

THROTTLE JET NEEDLE

♦ See Figure 50

1. This carburetor is equipped with a throttle valve using a tapered jet needle to change the fuel flow as the carburetor is opened. Normally the E-clip is located in the second groove.
2. Lowering the E-clip on the needle gives a richer fuel mixture. Raising the E-clip will result in a leaner fuel mixture.

➥The manufacturer recommends the adjustment be made a little on the rich side rather than on the lean side.

3. Disconnect and remove the carburetor from the powerhead. Remove and disassemble the throttle valve assembly.
4. Change the E-clip to the next higher or lower groove to obtain the desired fuel mixture.
5. Assemble the throttle valve and install the carburetor onto the powerhead.

4 and 5 HP

Perform synchronization of the 4 and 5hp powerheads in the following order: perform preliminary adjustments, set the ignition timing, set the idle speed and mixture and finally, test drive the boat to insure proper full throttle speed.

※ CAUTION

Water must circulate through the powerhead any time it is running to prevent damage to the water pump and possible powerhead overheating. Never run the engine over 3000 RPM without an adequate load applied to the propeller.

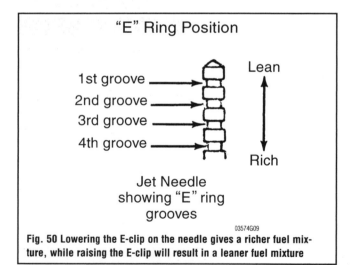

Fig. 50 Lowering the E-clip on the needle gives a richer fuel mixture, while raising the E-clip will result in a leaner fuel mixture

PRELIMINARY ADJUSTMENTS

1. Rotate the throttle twist grip to the idle position (on tiller handle models), or place the remote control handle in the neutral (idle) position (on remote control equipped models).
2. Loosen the set screw to allow the throttle arm to move freely on the wire.
3. Back the idle screw off the throttle arm.
4. Turn the idle speed screw inward (clockwise) until if just touches the throttle arm, then inward tow additional turns to slightly open the throttle plate in the carburetor.

5. With the throttle arm of the carburetor against the idle speed screw, pull the slack from the throttle wire and secure the wire into the retainer by tightening the setscrew.
6. Turn the idle mixture screw 1½turns outward (counterclockwise) from a lightly seated position.

IGNITION TIMING

Static Timing

♦ See accompanying illustrations

1. Remove the spark plug and install a dial indicator in the spark plug hole. Slowly rotate the flywheel clockwise and determine TDC for the cylinder using the dial indicator.
2. After TDC has been determined, observe the two vertical lines embossed on the port side of the flywheel. The aft line should align with split line of the cylinder block and the crankcase cover.
   - If the aft line is centered as explained, the flywheel has been installed correctly and the Woodruff key is in place.
   - If the aft line is not centered, the hand rewind starter and flywheel must be removed and the Woodruff key and keyway checked for damage and if it is installed correctly.

Step 2

Dynamic Timing

♦ See accompanying illustrations

1. Remove the dial indicator, install the spark plug and connect a timing light to the powerhead.
2. Mount the outboard in a test tank or on a boat in a body of water with the boat well secured to the dock or slip.
3. Obtain and connect a timing light to the high tension spark plug lead as per the manufacturers instructions.
4. Start the powerhead and allow it to reach the correct operating temperature.
5. Allow it run at idle speed in **FORWARD**. Aim the timing light at the port side of the powerhead.

MAINTENANCE 3-17

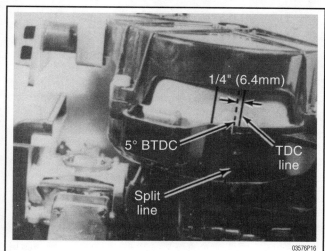

Step 6

charging coil, the trigger and the ignition coil are outlined in "Ignition and Electrical" section.

  8. Twist the throttle grip to the **FAST** position with the outboard still in **FORWARD**. Aim the timing light at the port side of the powerhead.

  9. The forward line embossed on the flywheel should align with the split line between the cylinder block and the crankcase cover. This position corresponds to 30°BTDC advance at the flywheel.

  10. If the timing is not as indicated and the powerhead operates roughly or misfires, the problem is most likely electrical. Detailed procedures for testing the capacitor charging coil, trigger and ignition coil are covered in the "Ignition and Electrical" section.

IDLE SPEED & MIXTURE

▶ See Figure 51

1. Connect a tachometer to the powerhead.
2. Start the engine and allow it to warm up to normal operating temperature.
3. Throttle the engine back to the idle position and shift into forward gear.
4. Adjust the idle to the speed specified in the "Tune-Up Specifications" chart.
5. Slowly turn the idle mixture screw counterclockwise until the engine starts to "load up" or fire unevenly because of an over-rich mixture.
6. Slowly turn the idle mixture screw clockwise until the engine picks-up speed and fires evenly.
7. Do not adjust leaner than necessary for reasonably smooth idling. It is preferable to set it a little rich than too lean.
8. If the engine hesitates during acceleration after adjusting the idle mixture, the mixture is too lean. Richen the idle mixture slightly until the engine accelerates smoothly.
9. Return the engine to idle and recheck the idle speed.
10. Readjust the idle speed screw if necessary.
11. Shift the engine to neutral and shut down the engine.

6, 8, 9.9 and 15 HP

Perform synchronization of the 6, 8, 9.9 and 15hp powerheads in the following order: perform preliminary adjustments, set the ignition timing, set the idle speed and mixture and finally, test drive the boat to insure proper full throttle speed.

✴✴ CAUTION

Water must circulate through the powerhead any time it is running to prevent damage to the water pump and possible powerhead overheating. Never run the engine over 3000 RPM without an adequate load applied to the propeller.

PRELIMINARY ADJUSTMENTS

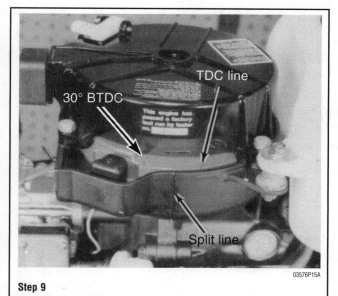

Step 9

▶ See Figures 52 and 53

1. On tiller handle equipped models, adjust the throttle cables so there is no slack and equal travel in each direction.
2. Push the fast idle knob inward and at the same time rotate the knob counterclockwise as far as possible.
3. On 6hp models, loosen the throttle cam locking screw and push the cam follower down until it contacts the throttle cam. Tighten the locking screw.

  6. The split line between the cylinder block and the crankcase cover should be misaligned by approximately ¼ in. (6.4mm) toward the forward side of the aft line embossed on the flywheel. This position corresponds to 5°BTDC advance at the flywheel.

  7. If the timing is not as indicated and the powerhead operates roughly or misfires, the problem is most likely electrical. Detailed procedures to test the capacitor

Fig. 51 Idle speed screw and pilot screw location—4 and 5hp

Fig. 52 Rotate the idle speed screw counterclockwise until the screw is off the cam follower—8-15hp models

Fig. 53 If the model being serviced has an air intake cover, remove the idle mixture screw access plug

MAINTENANCE

4. On 8–15hp models, turn the idle speed screw counterclockwise until it does not touch the cam follower. Turn it back in until it just touches the cam follower and then an additional ¼–½ turn as an initial adjustment.

5. Rotate the idle mixture screw slowly clockwise (inward) until it barely seats and then back it out counterclockwise the number of turns specified in the "Tune-Up Specifications" chart.

IGNITION TIMING

▶ See accompanying illustrations

➡ This procedure must be performed with the powerhead running at full speed. Mount the outboard in a test tank or put the boat in a body of water.

1. Connect a tachometer and timing light to the powerhead.
2. Start the powerhead and allow it to reach operating temperature.
3. Move the shift lever to the **FORWARD** position. Advance the throttle to the wide open throttle position.
4. Check the maximum advance timing mark on the flywheel. The mark should be aligned with the pointer.
5. Adjust the timing by loosening the jam-nut on the maximum advance screw. Rotate the advance spark screw clockwise until the maximum advance timing mark on the flywheel is aligned with the pointer.
6. When the timing marks are aligned, tighten the jam-nut to hold the adjustment.
7. Return the throttle to idle and leave the unit in the **FORWARD** position.

➡ It may be necessary to adjust the idle speed screw in order to obtain a satisfactory idle rpm.

8. Adjust the idle timing screw until the pointer aligns with the specified idle timing mark as indicated in the "Tune-Up Specifications" chart.
9. After the timing set, shut down the powerhead and disconnect the tachometer and timing light.

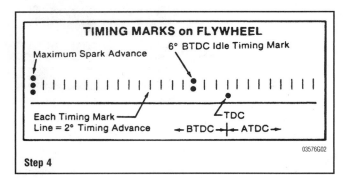

Step 4

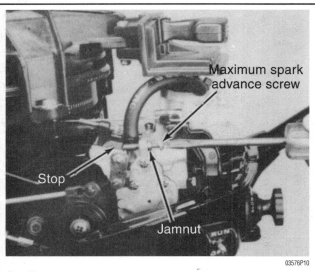

Step 5

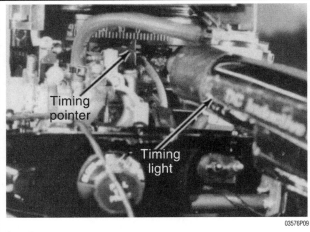

Step 6

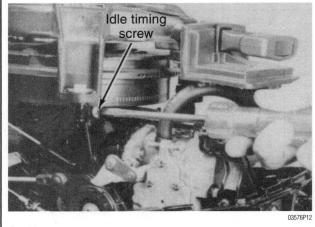

Step 8

FAST IDLE SPEED

▶ See Figure 54

1. Push the fast idle knob inward and then turn the knob counterclockwise as far as possible.
2. Shift the unit into **NEUTRAL**, and adjust the upper screw on the ratchet to remove any clearance between the idle wire and the trigger.
3. The wire should barely make contact with the trigger.

IDLE SPEED & MIXTURE

▶ See Figure 55 and 56

1. Connect a tachometer to the powerhead.
2. Start the powerhead and allow it to reach operating temperature.
3. With the engine running at idle in **FORWARD** gear adjust the idle speed screw to obtain the recommended idle speed found in the "Tune-Up Specifications" chart.

➡ Make an effort to prevent the powerhead from shutting down during the following adjustments.

4. Move the throttle to the idle speed position and allow the RPM to stabilize. Push the primer/fast idle knob inward and then rotate it to the full counterclockwise position. Shift the unit into **FORWARD**. With the powerhead operating at idle rpm, rotate the idle mixture screw counterclockwise until the powerhead begins to misfire because of an over-rich fuel mixture.
5. Slowly rotate the idle mixture screw clockwise and count the turns until the powerhead is operating evenly and the RPM increases. Continue rotating the mixture screw clockwise until the powerhead begins to slow down and eventually begins to misfire because of a too lean air/fuel mixture.

MAINTENANCE 3-19

Fig. 54 Adjust the idle wire to set the fast idle speed

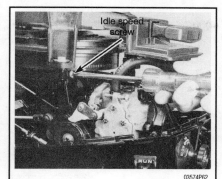

Fig. 55 Idle speed screw location—6, 8, 9.9, 10 and 15hp

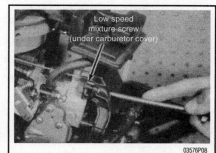

Fig. 56 Insert a screwdriver through the carburetor cover opening and adjust the idle mixture screw for proper air/fuel ratio at idle speed

6. Count and move the idle mixture screw back to the halfway point between the too rich mixture and the too lean mixture.

➡ Do not adjust the mixture any leaner than necessary to obtain a smooth idle. It is better for the mixture to be on the "too rich" side rather than the "too lean" side.

20 and 25 HP

Perform synchronization of the 1990-94 20 and 25hp powerheads in the following order: perform preliminary adjustments, set the ignition timing (this includes idle timing, pick-up timing and maximum advance timing), set the throttle cam, set the idle speed and mixture, set the fast idle speed and finally, test drive the boat to insure proper full throttle speed.

Perform synchronization of the 1994-00 20 and 25hp powerheads in the following order: perform preliminary adjustments, set full throttle stop, set the ignition timing (this includes idle timing and maximum advance timing), set the idle speed and mixture, set the fast idle speed, check starter interlock, shift and throttle cables and finally, test drive the boat to insure proper full throttle speed.

✴ CAUTION

Water must circulate through the powerhead any time it is running to prevent damage to the water pump and possible powerhead overheating. Never run the engine over 3000 RPM without an adequate load applied to the propeller.

PRELIMINARY ADJUSTMENTS

♦ See Figure 57

1. On tiller handle equipped models, adjust the throttle cables so there is no slack and equal travel in each direction.
2. Push the fast idle knob inward and at the same time rotate the knob counterclockwise as far as possible.
3. On 1994-00 models, turn the idle speed screw counterclockwise until it does not touch the cam follower. Turn it back in until it just touches the cam follower and then an additional turn as an initial adjustment.
4. Rotate the idle mixture screw slowly clockwise (inward) until it barely seats and then back it out counterclockwise the number of turns specified in the "Tune-Up Specifications" chart.

FULL THROTTLE STOP

1. Shift the unit into **FORWARD** and twist the throttle grip to the **FULL THROTTLE** position.
2. On tiller shift and remote control models, the throttle return spring on the cam follower plate should just contact the fuel pump housing at wide open throttle position. If not, adjust the throttle link rod. Do not allow the throttle spring to act as a throttle stop.
3. On side shift models, adjust the throttle cable jam nuts so that the throttle plate opens fully throttle return spring on the cam follower plate just contacts the fuel pump housing. Ensure all slack is removed from the throttle cables.

IGNITION TIMING

Maximum Advance Timing

♦ See accompanying illustrations

1. Connect a tachometer and timing light to the powerhead.
2. Start the powerhead and allow it to reach operating temperature.
3. Move the shift lever to the **FORWARD** position.
4. On mechanical advance models, advance the throttle to the wide open throttle position and direct the timing light to the flywheel and pointer mark on the starboard

Fig. 57 Turn the idle speed screw in until the cam follower just makes contact with the throttle cam, then turn in one additional turn to slightly open the throttle plate

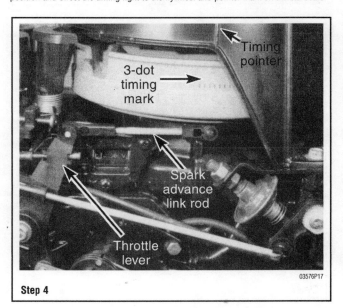

Step 4

3-20 MAINTENANCE

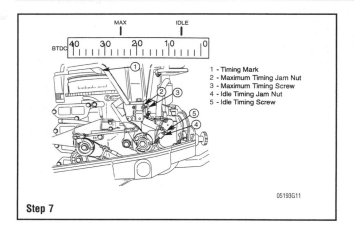

1 - Timing Mark
2 - Maximum Timing Jam Nut
3 - Maximum Timing Screw
4 - Idle Timing Jam Nut
5 - Idle Timing Screw

Step 7

side of the powerhead. The full throttle (3 dots) timing mark on the flywheel should align with the timing pointer. note the timing.

5. On electronic advance models, with the engine running at approximately 3000 rpm, the timing should be as specified in the "Tune-Up Specifications" chart.

➡ **At 5500 rpm, the timing will be automatically retarded.**

6. On 1990–1994 mechanical advance models and electronic advance models, if timing adjustment is necessary, shut the powerhead down. Pry the end of the spark advance link rod free of the throttle lever with a screwdriver. Adjust the length of the rod.

- Lengthening the link rod will advance the timing. It will move the timing mark to the right while the powerhead is operating at wide open throttle.
- Shortening the link rod will retard the timing. It will move the timing mark to the left while the powerhead is operating at wide open throttle.

7. On 1997–00 mechanical advance models, if timing adjustment is necessary, shut the powerhead down. Loosen the maximum advance screw and align the timing mark specified in the "Tune-Up Specifications" chart with the timing pointer. Stop the engine and tighten the jam nut.

Idle (Pick-Up) Timing

▶ **See accompanying illustrations**

➡ **Before the throttle pick-up timing adjustment can be properly made, the throttle cam must be correctly adjusted.**

1. Connect a tachometer and timing light to the powerhead.
2. Start the powerhead and allow it to reach operating temperature.
3. Reduce the engine speed to idle and shift the outboard into **FORWARD**.
4. Remove the throttle link rod from the throttle lever.

Step 4

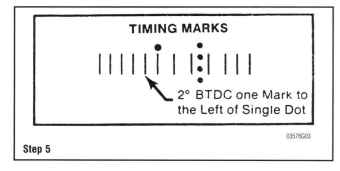

Step 5

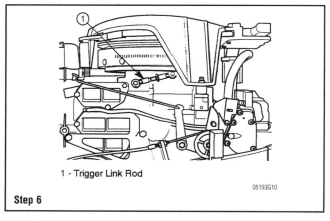

1 - Trigger Link Rod

Step 6

✲✲ CAUTION

Do not attempt to pry the throttle link rod free of the throttle cam. Such action could very well damage the throttle cam.

5. With the idle speed screw against the stop, adjust the idle speed screw until the powerhead has reached normal operating temperature and is operating with the timing at the setting specified in the "Tune-Up Specifications" chart.
6. Hold the throttle cam in position with the throttle roller just touching the cam and with the center of the roller aligned with the scribe mark on the cam. Continue to hold this position and at the same time, adjust the length of the throttle link rod until it will just slip over the throttle lever anchor ball.
7. Rotate the idle speed screw slightly to obtain the proper idle speed with the unit in **FORWARD**.

IDLE SPEED & MIXTURE

▶ **See Figure 58**

1. Connect a tachometer to the powerhead.
2. Start the powerhead and allow it to reach operating temperature.
3. With the engine running at idle in **FORWARD** gear adjust the idle speed

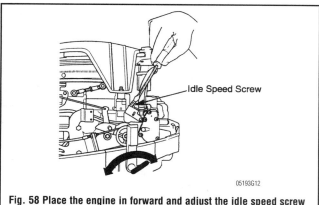

Fig. 58 Place the engine in forward and adjust the idle speed screw

MAINTENANCE 3-21

screw to obtain the recommended idle speed found in the "Tune-Up Specifications" chart.

→ **Make an effort to prevent the powerhead from shutting down during the following adjustments.**

4. Move the throttle to the idle speed position and allow the RPM to stabilize. Push the primer/fast idle knob inward and then rotate it to the full counterclockwise position.
5. With the powerhead operating at idle rpm, rotate the idle mixture screw counterclockwise until the powerhead begins to misfire because of an over-rich fuel mixture.
6. Slowly rotate the idle mixture screw clockwise and count the turns until the powerhead is operating evenly and the RPM increases. Continue rotating the mixture screw clockwise until the powerhead begins to slow down and eventually begins to misfire because of a too lean air/fuel mixture.
7. Count and move the idle mixture screw back to the halfway point between the too rich mixture and the too lean mixture.

→ **Do not adjust the mixture any leaner than necessary to obtain a smooth idle. It is better for the mixture to be on the "too rich" side rather than the "too lean" side.**

8. Check for a too lean mixture on acceleration (engine will hesitate or stall on acceleration). Readjust the mixture if necessary.
9. Readjust the idle speed, as necessary and install the access plug.

FAST IDLE SPEED

♦ See Figure 59

1. Connect a tachometer and timing light to the powerhead.
2. Start the powerhead and allow it to reach operating temperature.
3. Push the primer/fast idle knob in fully and turn clockwise.
4. Place the lower unit in **FORWARD** gear and adjust the fast idle speed screw to obtain the recommended idle speed found in the "Tune-Up Specifications" chart.

STARTER INTERLOCK ADJUSTMENT

1990–93

♦ See Figure 60

1. Shift the engine into the **NEUTRAL** position and shut the engine off.
2. Pull the fast idle knob to the full out position.
3. Adjust the interlock screw to permit the interlock lever to clear the ratchet on the starter sheave.
4. Once the adjustment is made, pull the starter cord to verify that the powerhead can be manually started.

1994–00

1. Shift the engine into the **NEUTRAL** position and shut the engine off.
2. Push the primer/fast idle knob in fully and turn clockwise.
3. Make sure the throttle is in the idle position
4. Adjust the interlock screw to permit the interlock lever to clear the ratchet on the starter sheave.
5. Once the adjustment is made, pull the starter cord to verify that the powerhead can be manually started.

CARBURETOR THROTTLE CAM

♦ See Figure 61

1. The throttle cam adjustment cannot be made unless the carburetor throttle shutter is completely closed. Therefore, the idle speed screw may have to be backed out; the dashpot adjusted away from the throttle cam; and/or the neutral RPM screw rotated counterclockwise before a proper throttle cam/roller alignment can be made.
2. With the engine shut down, move the throttle cam to check the cam adjustment. The bottom scribe line on the cam should be aligned with the center of the throttle roller and the roller just barely touches the cam.
3. If an adjustment is necessary, first loosen the hex bolt and with the punch mark on the eccentric facing down and forward of the hex bolt axis, rotate the eccentric counterclockwise to reposition the cam until the throttle roller is aligned with the scribe line.

DASHPOT ADJUSTMENT

♦ See Figure 62

The stem of the dashpot should be fully depressed when the powerhead is operating at the specified idle rpm. Adjust the dashpot to obtain this condition.

Fig. 62 Adjust the stem on the dashpot so that it is fully depressed when at idle speed

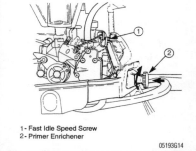

Fig. 59 With the engine in neutral and the primer enricher pushed to the full in and full clockwise position, adjust the fast idle speed screw to specification

Fig. 60 Turning the interlock screw will adjust the interlock lever

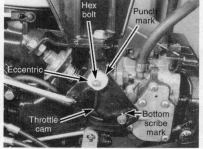

Fig. 61 Location of the components necessary to adjust the throttle cam—20 and 25hp

3-22 MAINTENANCE

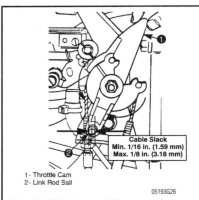

Fig. 63 With the cam follower resting on the throttle cam, tighten the cam follower screw

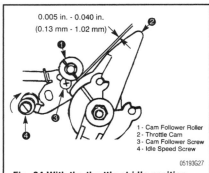

Fig. 64 With the throttle at idle position, turn the idle speed screw clockwise in until a gap of 0.005–0.040 in. (0.13–1.02 mm) is achieved between the throttle cam and cam follower

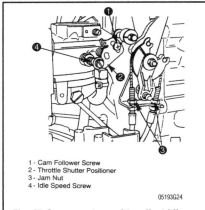

Fig. 65 Components used to adjust idle speed

30 and 40 HP (2-Cylinder)

Perform synchronization of the 30 and 40hp (2-Cylinder) powerheads in the following order: perform preliminary adjustments, set cam follower, set the ignition timing, set the idle speed and mixture, set oil pump linkage and finally, test drive the boat to insure proper full throttle speed.

✱✱ CAUTION

Water must circulate through the powerhead any time it is running to prevent damage to the water pump and possible powerhead overheating. Never run the engine over 3000 RPM without an adequate load applied to the propeller.

PRELIMINARY ADJUSTMENTS

1. Loosen the cam follower screw and turn the idle speed screw until it no longer touches the throttle arm and the throttle plates are fully closed.
2. Rotate the idle mixture screw slowly clockwise (inward) until it barely seats and then back it out counterclockwise the number of turns specified in the "Tune-Up Specifications" chart.

CAM FOLLOWER

◆ See Figures 63 and 64

1. Loosen the throttle cable jam nuts.
2. With the throttle at the idle position, place the cam follower roller against the throttle cam. Center the roller with raised mark on the throttle cam by adjusting the position of the throttle cable sleeves in the mounting bracket.

➡**When positioning throttle cables, a minimum of 1/16 in. (1.59 mm) to a maximum of 1/8 in. (3.18 mm) slack must be allowed to prevent throttle cables from binding. (Rock throttle cam side to side and measure the amount of throttle cam travel at link rod ball).**

3. Tighten the throttle cable jam nuts.
4. With the cam follower resting on the throttle cam, tighten the cam follower screw.
5. With the throttle at idle position, turn the idle speed screw clockwise in until a gap of 0.005–0.040 in. (0.13–1.02 mm) is achieved between the throttle cam and cam follower.

IGNITION TIMING

➡**Due to the electronic ignition system design, there are no timing adjustments required for the 30 or 40hp (2-Cylinder) powerheads. However, the timing can be verified using the following procedure.**

1. Connect a tachometer and timing light to the powerhead.
2. Start the powerhead and allow it to reach operating temperature.
3. Move the shift lever to the **FORWARD** position and insure the throttle is a idle position.

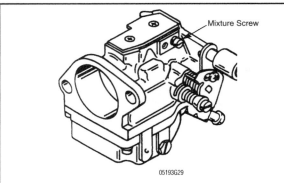

Fig. 66 With the engine running at idle speed and the gear box in the FORWARD position, turn the mixture screw in (clockwise) until the engine starts to bog down and misfire. Back out the screw 1/4 turn or more

4. Check the maximum advance timing mark on the flywheel. The mark should be aligned with the pointer. Timing specifications can be found in the "Tune-Up Specifications" chart.

➡**The idle stabilization feature of these engines will cause the timing to fluctuate 2–3° at idle.**

5. Move the throttle to the wide open throttle position and recheck the timing.
6. If the ignition timing is not within specification, see the "Ignition and Electrical" section for troubleshooting procedures.
7. After the timing set, shut down the powerhead and disconnect the tachometer and timing light.

IDLE SPEED & MIXTURE

◆ See Figures 65 and 66

1. Connect a tachometer to the powerhead.
2. Start the powerhead and allow it to reach operating temperature.
3. With the engine running at idle in **FORWARD** gear adjust the idle speed screw to obtain the recommended idle speed found in the "Tune-Up Specifications" chart.

➡**Make an effort to prevent the powerhead from shutting down during the following adjustments.**

4. Move the throttle to the idle speed position and allow the RPM to stabilize. Push the primer/fast idle knob inward and then rotate it to the full counterclockwise position.
5. With the powerhead operating at idle rpm, rotate the idle mixture screw counterclockwise until the powerhead begins to misfire because of an over-rich fuel mixture.
6. Slowly rotate the idle mixture screw clockwise and count the turns until the powerhead is operating evenly and the RPM increases. Continue rotating the mixture

MAINTENANCE 3-23

screw clockwise until the powerhead begins to slow down and eventually begins to misfire because of a too lean air/fuel mixture.

7. Count and move the idle mixture screw back to the halfway point between the too rich mixture and the too lean mixture.

→ Do not adjust the mixture any leaner than necessary to obtain a smooth idle. It is better for the mixture to be on the "too rich" side rather than the "too lean" side.

8. Check for a too lean mixture on acceleration (engine will hesitate or stall on acceleration). Readjust the mixture if necessary.
9. Readjust the idle speed, as necessary and install the access plug.

OIL PUMP

♦ See Figure 67

→ The oil pump linkage must be synchronized to the throttle linkage any time the throttle linkage is adjusted

While holding the throttle arm at idle position, adjust the length of the link rod so the stamped mark of the oil pump body aligns with the stamped mark of the oil pump lever.

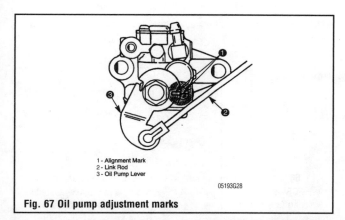

Fig. 67 Oil pump adjustment marks

40 HP (4-Cylinder)

Perform synchronization of the 40hp (4-Cylinder) powerheads in the following order: perform preliminary adjustments, set the throttle plate, set the ignition timing, set the full throttle stop, set the idle speed and mixture, and finally, test drive the boat to insure proper full throttle speed.

✱✱✱ CAUTION

Water must circulate through the powerhead any time it is running to prevent damage to the water pump and possible powerhead overheating. Never run the engine over 3000 RPM without an adequate load applied to the propeller.

PRELIMINARY ADJUSTMENTS

1. Disconnect the throttle cables from the throttle lever arm.
2. Rotate the idle mixture screw slowly clockwise (inward) until it barely seats and then back it out counterclockwise the number of turns specified in the "Tune-Up Specifications" chart.

CARBURETOR SYNCHRONIZATION

1. Turn the idle speed screw counterclockwise until the throttle cam is not touching the throttle arm and the throttle plates on the upper carburetor are fully closed.
2. Loosen the adjustment screw on the tie bar connecting the carburetors.
3. Open both carburetor throttle plates and allow them to close against spring pressure. Verify that both carburetor throttle plates are fully closed.
4. Hold the tie bar in position and tighten the adjustment screw.
5. Once again, open the carburetor throttle plates (this time they should work in unison) and allow them to close against spring pressure. Verify that both carburetor throttle plates are working in unison.

IGNITION TIMING

Pick-Up Timing

→ After all timing adjustments have been made, the idle RPM screw will have to be readjusted to provide the proper idle RPM for the engine.

1990–93

♦ See accompanying illustrations

1. Connect a tachometer and timing light to the powerhead.
2. Crank the engine while pointing the timing light at the flywheel.
3. Slowly advance the throttle spark arm until the timing mark is aligned with the "V".
4. Hold the throttle spark arm stationary and at the same time, adjust the idle RPM screw until the screw contacts the stop.
5. Verify the primary timing mark is still aligned with the "V".
6. Loosen the two throttle actuator plate retaining screws.
7. Hold the throttle spark arm against the idle RPM stop screw and at the same time rotate the actuator plate until the primary throttle cam just contacts the pick-up arm on the carburetor cluster. Tighten the actuator plate retaining screws to hold this adjustment.

1994–97

1. Connect a tachometer and timing light to the powerhead.
2. Start the powerhead and allow it to reach operating temperature.
3. Move the shift lever to the **FORWARD** position and insure the throttle is a idle position.
4. Use a timing light to read the current timing with the powerhead at idle speed.
5. Hold the throttle spark arm stationary and at the same time, adjust the idle RPM screw until the proper timing is achieved. Timing specifications can be found in the "Tune-Up Specifications" chart
6. Stop the powerhead and loosen the two throttle actuator plate retaining screws.

Step 3

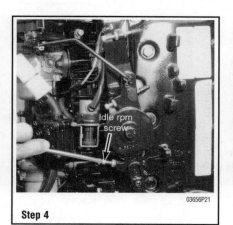

Step 4

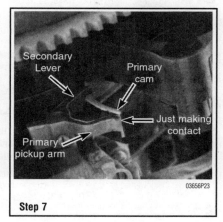

Step 7

7. Hold the throttle spark arm against the idle RPM stop screw and at the same time rotate the actuator plate until the primary throttle cam just contacts the pick-up arm on the carburetor cluster. Tighten the actuator plate retaining screws to hold this adjustment.

Secondary Pick-Up

▶ See Figure 68

1. Loosen the locknut on the secondary throttle pick-up screw.
2. Advance the throttle lever until the maximum advance screw is against the stop on the exhaust cover.
3. Hold the throttle lever against the stop and at the same time, adjust the secondary throttle pick-up screw until the end of the screw contacts the secondary lever of the carburetor cluster.
4. Tighten the locknut to hold this adjustment.

Maximum Advance Timing

▶ See Figure 69

1. Loosen the locknut on the maximum advance screw.
2. Move throttle lever until the maximum advance screw is against the stop and hold the throttle in this position.
3. Adjust the secondary throttle pick-up screw so is just barely contacts the secondary lever.
4. Tighten the secondary throttle pick-up screw locknut to hold this position.

FULL THROTTLE STOP

▶ See Figures 70 and 71

1. Loosen the locknut on the maximum advance screw.
2. Move throttle lever until full throttle stop screw is against the stop and hold the throttle in this position.
3. Adjust the full throttle stop screw so the carburetor throttle valves are in the wide open throttle position.
4. Ensure that the carburetor throttle valves are not serving as the throttle stop. There should be 0.010–0.015 in. (0.25–0.38mm) free play between the secondary pick-up screw and the secondary lever. If there is insufficient free-play, readjust the full throttle stop screw.
5. Tighten the full throttle stop screw locknut to hold this position.

IDLE SPEED & MIXTURE

▶ See Figure 72 and 73

1. Connect a tachometer to the powerhead.
2. Start the powerhead and allow it to reach operating temperature.
3. With the engine running at idle in **FORWARD** gear adjust the idle speed screw to obtain the recommended idle speed found in the "Tune-Up Specifications" chart.

➡ Make an effort to prevent the powerhead from shutting down during the following adjustments.

Fig. 68 Hold the throttle lever against the stop and at the same time, adjust the secondary throttle pick-up screw until the end of the screw contacts the secondary lever of the carburetor cluster

Fig. 69 While holding the spark arm against the stop, adjust the screw until the timing marks align correctly

Fig. 70 Advance the throttle lever to the full throttle position and the throttle stop screw is against the stop on the exhaust cover

Fig. 71 A .010in.–.015in. (0.254mm–0.381mm) free "play" should exist between the secondary pick-up screw and carburetor cluster secondary lever

Fig. 72 Idle speed screw location— 40hp (4-Cylinder)

Fig. 73 Idle mixture screw location—40hp (4-Cylinder)

MAINTENANCE 3-25

4. Move the throttle to the idle speed position and allow the RPM to stabilize.

5. With the powerhead operating at idle rpm, rotate the idle mixture screw counterclockwise until the powerhead begins to misfire because of an over-rich fuel mixture.

6. Slowly rotate the idle mixture screw clockwise and count the turns until the powerhead is operating evenly and the RPM increases. Continue rotating the mixture screw clockwise until the powerhead begins to slow down and eventually begins to misfire because of a too lean air/fuel mixture.

7. Count and move the idle mixture screw back to the halfway point between the too rich mixture and the too lean mixture.

➡ Do not adjust the mixture any leaner than necessary to obtain a smooth idle. It is better for the mixture to be on the "too rich" side rather than the "too lean" side.

8. Check for a too lean mixture on acceleration (engine will hesitate or stall on acceleration). Readjust the mixture if necessary.

9. Readjust the idle speed, as necessary and install the access plug.

1990-93 50-60 HP

Perform synchronization of the 50–60 HP powerheads in the following order: perform preliminary adjustments, synchronize the carburetors, set the timing pointer, set the ignition timing, set the idle speed and mixture and finally, test drive the boat to insure proper full throttle speed.

✻✻ CAUTION

Water must circulate through the powerhead any time it is running to prevent damage to the water pump and possible powerhead overheating. Never run the engine over 3000 RPM without an adequate load applied to the propeller.

PRELIMINARY ADJUSTMENTS

1. Disconnect the throttle cables from the throttle lever arm.
2. Rotate the idle mixture screw slowly clockwise (inward) until it barely seats and then back it out counterclockwise the number of turns specified in the "Tune-Up Specifications" chart.

CARBURETOR SYNCHRONIZATION

1. Turn the idle speed screw counterclockwise until the throttle cam is not touching the throttle arm and the throttle plates on the upper carburetor are fully closed.
2. Loosen the adjustment screw on the tie bar connecting the carburetors.
3. Open the carburetor throttle plates and allow them to close against spring pressure. Verify that the carburetor throttle plates are fully closed.
4. Hold the tie bar in position and tighten the adjustment screw.
5. Once again, open the carburetor throttle plates (this time they should work in unison) and allow them to close against spring pressure. Verify that the carburetor throttle plates are working in unison.

IGNITION TIMING

Timing Pointer

♦ See Figure 74

1. Remove all spark plugs.
2. Install a dial indicator in the No. 1 (top) spark plug opening.
3. Rotate the flywheel clockwise until the No. 1 piston is at top dead center (TDC). At this point, set the dial indicator a 0°.
4. Rotate the flywheel counterclockwise until the dial indicator reads as follows:
   - .464 in. for powerheads up to Serial Number D000750
   - .459 in. for powerheads after Serial Number D000750.
5. Loosen the attaching screws and shift the timing pointer until the correct mark on the timing decal is aligned with the groove in the timing pointer.
6. Tighten the pointer attaching screws to hold the adjustment.
7. Remove the dial indicator and install the spark plugs.

Idle Timing

♦ See Figure 75

1. Disconnect the spark plug leads from the plugs and ground each lead to the powerhead.
2. Connect a timing light to the powerhead.
3. Place the remote shift lever in the **NEUTRAL** position.
4. Loosen the locknut on the idle timing screw and hold the throttle lever against the idle stop.
5. Crank the powerhead with the starter motor and at the same time, aim the timing light at the flywheel cover timing pointer.
6. Adjust the idle timing screw until the timing pointer aligns with the timing mark specified in the "Tune-Up Specifications" chart.
7. Tighten the locknut to hold this position.

Maximum Advance Timing

♦ See Figure 76

1. Disconnect the spark plug leads from the plugs and ground each lead to the powerhead.
2. Connect a timing light to the powerhead.
3. Place the remote shift lever in the **NEUTRAL** position.
4. Loosen the locknut on the maximum advance screw. Hold the throttle lever in the aft position until the maximum advance screw contacts the stop.
5. Crank the powerhead with the starter motor and at the same time adjust the maximum advance screw until the timing pointer is aligned in the timing mark specified in the "Tune-Up Specifications" chart.
6. Tighten the locknut on the maximum advance screw to hold the adjustment.
7. Remove the timing light.

IDLE SPEED & MIXTURE

1. Connect a tachometer to the powerhead.
2. Start the powerhead and allow it to reach operating temperature.
3. With the engine running at idle in **FORWARD** gear adjust the idle speed

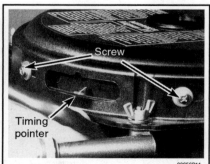

Fig. 74 Shift the timing pointer until the correct mark on the timing decal is aligned with the groove in the timing pointer

Fig. 75 Idle timing screw location—1990-93 50-60hp

Fig. 76 Maximum advance screw location—1990-93 50-60hp

3-26 MAINTENANCE

screw to obtain the recommended idle speed found in the "Tune-Up Specifications" chart.

➡ **Make an effort to prevent the powerhead from shutting down during the following adjustments.**

4. Move the throttle to the idle speed position and allow the RPM to stabilize.
5. With the powerhead operating at idle rpm, rotate the idle mixture screw counterclockwise until the powerhead begins to misfire because of an over-rich fuel mixture.
6. Slowly rotate the idle mixture screw clockwise and count the turns until the powerhead is operating evenly and the RPM increases. Continue rotating the mixture screw clockwise until the powerhead begins to slow down and eventually begins to misfire because of a too lean air/fuel mixture.
7. Count and move the idle mixture screw back to the halfway point between the too rich mixture and the too lean mixture.

➡ **Do not adjust the mixture any leaner than necessary to obtain a smooth idle. It is better for the mixture to be on the "too rich" side rather than the "too lean" side.**

8. Check for a too lean mixture on acceleration (engine will hesitate or stall on acceleration). Readjust the mixture if necessary.
9. Readjust the idle speed, as necessary and install the access plug.

40 (3-Cylinder), 1994–00 50 and 60 HP

Perform synchronization of the 40 (3-Cylinder), 50 and 60hp (1994–00) powerheads in the following order: perform preliminary adjustments, set the timing pointer, synchronize the carburetors, adjust the throttle cam, set the ignition timing, set the idle speed and mixture and finally, test drive the boat to insure proper full throttle speed.

✱✱ CAUTION

Water must circulate through the powerhead any time it is running to prevent damage to the water pump and possible powerhead overheating. Never run the engine over 3000 RPM without an adequate load applied to the propeller.

PRELIMINARY ADJUSTMENTS

1. Disconnect the throttle cables from the throttle lever arm.
2. Rotate the idle mixture screw slowly clockwise (inward) until it barely seats and then back it out counterclockwise the number of turns specified in the "Tune-Up Specifications" chart.

CARBURETOR SYNCHRONIZATION

1. Turn the idle speed screw counterclockwise until the throttle cam is not touching the throttle arm and the throttle plates on the upper carburetor are fully closed.
2. Loosen the adjustment screw on the tie bar connecting the carburetors.
3. Open the carburetor throttle plates and allow them to close against spring pressure. Verify that the carburetor throttle plates are fully closed.
4. Hold the tie bar in position and tighten the adjustment screw.
5. Once again, open the carburetor throttle plates (this time they should work in unison) and allow them to close against spring pressure. Verify that the carburetor throttle plates are working in unison.

IGNITION TIMING

Timing Pointer

1. Remove all spark plugs.
2. Install a dial indicator in the No. 1 (top) spark plug opening.
3. Rotate the flywheel clockwise until the No. 1 piston is at top dead center (TDC). At this point, set the dial indicator a 0°.
4. Rotate the flywheel counterclockwise until the dial indicator reads .459 in.
5. Loosen the attaching screws and shift the timing pointer until the correct mark on the timing decal is aligned with the groove in the timing pointer.
6. Tighten the pointer attaching screws to hold the adjustment.
7. Remove the dial indicator and install the spark plugs.

Idle Timing

1. Disconnect the spark plug leads from the plugs and ground each lead to the powerhead.
2. Connect a timing light to the powerhead.
3. Place the remote shift lever in the **NEUTRAL** position.
4. Loosen the locknut on the idle timing screw and hold the throttle lever against the idle stop.
5. Crank the powerhead with the starter motor and at the same time, aim the timing light at the flywheel cover timing pointer.
6. Adjust the idle timing screw until the timing pointer aligns with the timing mark specified in the "Tune-Up Specifications" chart.
7. Tighten the locknut to hold this position.

Maximum Advance Timing

1. Disconnect the spark plug leads from the plugs and ground each lead to the powerhead.
2. Connect a timing light to the powerhead.
3. Place the remote shift lever in the **NEUTRAL** position.
4. Loosen the locknut on the maximum advance screw. Hold the throttle lever in the aft position until the maximum advance screw contacts the stop.
5. Crank the powerhead with the starter motor and at the same time adjust the maximum advance screw until the timing pointer is aligned in the timing mark specified in the "Tune-Up Specifications" chart.
6. Tighten the locknut on the maximum advance screw to hold the adjustment.
7. Remove the timing light.

IDLE SPEED & MIXTURE

1. Connect a tachometer and timing light to the powerhead.
2. Start the powerhead and allow it to reach operating temperature.
3. With the engine running at idle in **FORWARD** gear adjust the idle speed screw to obtain the recommended idle speed found in the "Tune-Up Specifications" chart.

➡ **Make an effort to prevent the powerhead from shutting down during the following adjustments.**

4. Move the throttle to the idle speed position and allow the RPM to stabilize.
5. With the powerhead operating at idle rpm, rotate the idle mixture screw counterclockwise until the powerhead begins to misfire because of an over-rich fuel mixture.
6. Slowly rotate the idle mixture screw clockwise and count the turns until the powerhead is operating evenly and the RPM increases. Continue rotating the mixture screw clockwise until the powerhead begins to slow down and eventually begins to misfire because of a too lean air/fuel mixture.
7. Count and move the idle mixture screw back to the halfway point between the too rich mixture and the too lean mixture.

➡ **Do not adjust the mixture any leaner than necessary to obtain a smooth idle. It is better for the mixture to be on the "too rich" side rather than the "too lean" side.**

8. Check for a too lean mixture on acceleration (engine will hesitate or stall on acceleration). Readjust the mixture if necessary.
9. Readjust the idle speed, as necessary and install the access plug.

THROTTLE CAM

♦ **See accompanying illustrations**

1. Loosen the locknut on the idle stop screw. Push the throttle lever until the idle stop screw rests against the stop.
2. Position the throttle roller against the throttle cam.
3. Adjust the idle stop screw until the raised mark on the cam is aligned with the center of the throttle roller. Tighten the locknut on the idle stop screw.
4. Pull back and hold the throttle lever against the idle stop. Adjust the throttle roller until a clearance of 0.005–0.040 in. (0.13–1.02mm).
5. Ensure the throttle cam mark is aligned with the center of the roller and then tighten the screw.
6. Loosen the locknut on the full throttle stop screw and adjust the full throttle stop screw while observe all three throttle shutter valves.

MAINTENANCE 3-27

Step 2

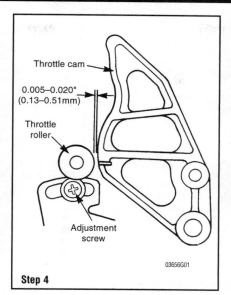
Step 4

Step 6

➡ The shutter valves must be in the full open position. Check to be sure that the valves do not act as a throttle stop and approximately 0.015 in. (0.40 mm) free play exists in the throttle linkage at wide open throttle.

7. Tighten the locknut on the full throttle stop screw to hold the adjustment.

OIL PUMP

♦ See Figure 77

1. The oil pump may have two markings on the body of the pump. The second mark should be disregarded for the alignment and adjustment procedures as shown in the illustration below.
2. Hold the throttle arm against the idle stop.
3. Visually check the alignment marks on the oil pump body and the oil pump synchronization lever.
4. If the marks are not aligned, disconnect and adjust the length of the carburetor/oil pump link rod.
5. Attach the link rod and check the marks for correct alignment.

Fig. 77 Visually check the alignment marks on the oil pump body and the oil pump synchronization lever

75 and 90 HP

Perform synchronization of the 75 and 90hp powerheads in the following order: perform preliminary adjustments, synchronize the carburetors, adjust the throttle cam, set the full throttle stop, set the ignition timing, set the idle speed and mixture, adjust the oil pump linkage and finally, test drive the boat to insure proper full throttle speed.

✱✱ CAUTION

Water must circulate through the powerhead any time it is running to prevent damage to the water pump and possible powerhead overheating. Never run the engine over 3000 RPM without an adequate load applied to the propeller.

PRELIMINARY ADJUSTMENTS

1. Disconnect the throttle cables from the throttle lever arm.
2. Rotate the idle mixture screw slowly clockwise (inward) until it barely seats and then back it out counterclockwise the number of turns specified in the "Tune-Up Specifications" chart.

CARBURETOR SYNCHRONIZATION

1. Turn the idle speed screw counterclockwise until the throttle cam is not touching the throttle arm and the throttle plates on the upper carburetor are fully closed.
2. Loosen the adjustment screw on the tie bar connecting the carburetors.
3. Open the carburetor throttle plates and allow them to close against spring pressure. Verify that the carburetor throttle plates are fully closed.
4. Hold the tie bar in position and tighten the adjustment screw.
5. Once again, open the carburetor throttle plates (this time they should work in unison) and allow them to close against spring pressure. Verify that the carburetor throttle plates are working in unison.

IGNITION TIMING

Idle Timing

1. Disconnect the spark plug leads from the plugs and ground each lead to the powerhead.
2. Connect a timing light to the powerhead.
3. Place the remote shift lever in the **NEUTRAL** position.
4. Loosen the locknut on the idle timing screw and hold the throttle lever against the idle stop.
5. Crank the powerhead with the starter motor and at the same time, aim the timing light at the flywheel cover timing pointer.
6. Adjust the idle timing screw until the timing pointer aligns with the timing mark specified in the "Tune-Up Specifications" chart.
7. Tighten the locknut to hold this position.

Maximum Advance Timing

1. Disconnect the spark plug leads from the plugs and ground each lead to the powerhead.

3-28 MAINTENANCE

2. Connect a timing light to the powerhead.
3. Place the remote shift lever in the **NEUTRAL** position.
4. Loosen the locknut on the maximum advance screw. Hold the throttle lever in the aft position until the maximum advance screw contacts the stop.
5. Crank the powerhead with the starter motor and at the same time adjust the maximum advance screw until the timing pointer is aligned in the timing mark specified in the "Tune-Up Specifications" chart.
6. Tighten the locknut on the maximum advance screw to hold the adjustment.
7. Remove the timing light.

IDLE SPEED & MIXTURE

1. Connect a tachometer and timing light to the powerhead.
2. Start the powerhead and allow it to reach operating temperature.
3. With the engine running at idle in **FORWARD** gear adjust the idle speed screw to obtain the recommended idle speed found in the "Tune-Up Specifications" chart.

➡ **Make an effort to prevent the powerhead from shutting down during the following adjustments.**

4. Move the throttle to the idle speed position and allow the RPM to stabilize.
5. With the powerhead operating at idle rpm, rotate the idle mixture screw counterclockwise until the powerhead begins to misfire because of an over-rich fuel mixture.
6. Slowly rotate the idle mixture screw clockwise and count the turns until the powerhead is operating evenly and the RPM increases. Continue rotating the mixture screw clockwise until the powerhead begins to slow down and eventually begins to misfire because of a too lean air/fuel mixture.
7. Count and move the idle mixture screw back to the halfway point between the too rich mixture and the too lean mixture.

➡ **Do not adjust the mixture any leaner than necessary to obtain a smooth idle. It is better for the mixture to be on the "too rich" side rather than the "too lean" side.**

8. Check for a too lean mixture on acceleration (engine will hesitate or stall on acceleration). Readjust the mixture if necessary.
9. Readjust the idle speed, as necessary and install the access plug.

THROTTLE CAM

1. Loosen the locknut on the idle stop screw. Push the throttle lever until the idle stop screw rests against the stop.
2. Position the throttle roller against the throttle cam.
3. Adjust the idle stop screw until the raised mark on the cam is aligned with the center of the throttle roller. Tighten the locknut on the idle stop screw.
4. Pull back and hold the throttle lever against the idle stop.
5. Adjust the throttle roller to 0.005–0.020 in. (0.13–0.51 mm).
6. Ensure the throttle cam mark is aligned with the center of the roller and then tighten the screw.
7. Loosen the locknut on the full throttle stop screw and adjust the full throttle stop screw while observe all three throttle shutter valves.

➡ **The shutter valves must be in the full open position. Check to be sure that the valves do not act as a throttle stop and approximately 0.015 in. (0.40 mm) free play exists in the throttle linkage at wide open throttle.**

8. Tighten the locknut on the full throttle stop screw to hold the adjustment.

OIL PUMP

1. The oil pump may have two markings on the body of the pump. The second mark should be disregarded for the alignment and adjustment procedures as shown in the illustration below.
2. Hold the throttle arm against the idle stop.
3. Visually check the alignment marks on the oil pump body and the oil pump synchronization lever.
4. If the marks are not aligned, disconnect and adjust the length of the carburetor/oil pump link rod.
5. Attach the link rod and check the marks for correct alignment.

100–125 HP

Perform synchronization of the 100–125hp powerheads in the following order: perform preliminary adjustments, synchronize the carburetors, adjust the throttle cam, adjust the accelerator pump clearance, set the ignition timing, set the idle speed and mixture, adjust the oil pump linkage, adjust the throttle cable preload and finally, test drive the boat to insure proper full throttle speed.

✸✸ CAUTION

Water must circulate through the powerhead any time it is running to prevent damage to the water pump and possible powerhead overheating. Never run the engine over 3000 RPM without an adequate load applied to the propeller.

PRELIMINARY ADJUSTMENTS

1. Disconnect the throttle cables from the throttle lever arm.
2. Rotate the idle mixture screw slowly clockwise (inward) until it barely seats and then back it out counterclockwise the number of turns specified in the "Tune-Up Specifications" chart.

CARBURETOR SYNCHRONIZATION

1. Turn the idle speed screw counterclockwise until the throttle cam is not touching the throttle arm and the throttle plates on the upper carburetor are fully closed.
2. Loosen the adjustment screw on the tie bar connecting the carburetors.
3. Open the carburetor throttle plates and allow them to close against spring pressure. Verify that the carburetor throttle plates are fully closed.
4. Hold the tie bar in position and tighten the adjustment screw.
5. Once again, open the carburetor throttle plates (this time they should work in unison) and allow them to close against spring pressure. Verify that the carburetor throttle plates are working in unison.

THROTTLE CAM

♦ See accompanying illustrations

1. Loosen the locknut on the idle stop screw. Hold the throttle lever aft until the idle stop screw rests against the stop on the outer exhaust cover.
2. Position the throttle roller against the throttle cam as far as possible. Adjust the idle stop screw until the raised mark on the throttle cam is centered with the throttle roller. Tighten the idle stop screw locknut.

Step 1

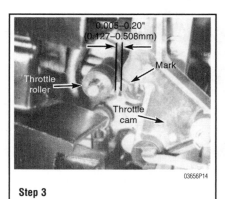

Step 3

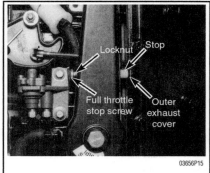

Step 4

MAINTENANCE 3-29

3. Hold the throttle arm against the idle stop and at the same time, adjust the throttle roller to a clearance of 0.005–0.020 in. (.127mm–.508mm) between the throttle roller and the throttle cam. Tighten the adjustment screw below the throttle roller to hold this adjustment.

4. Move the throttle lever aft until the full throttle stop screw contacts the stop on the outer exhaust cover.

5. Loosen the locknut on the full throttle stop screw. Adjust the full throttle stop screw and at the same time observe all four throttle shutter valves. The shutter valves must be in the full open position.

6. Check to be sure the throttle valves do not act as a throttle stop and approximately 0.015 in. (0.40mm) free "play" exists in the throttle linkage at wide open throttle. Tighten the locknut on the full throttle stop screw to hold the adjustment.

ACCELERATOR PUMP

♦ See Figure 78

1. Move the throttle lever to the full wide open throttle position. This action will cause the throttle cam to fully depress the plunger on top of the accelerator pump.

2. With the plunger depressed, measure the distance between the contact point of the cam and the plunger to the upper surface of the accelerator pump. This measurement should be exactly 0.030in. (0.76mm).

3. If an adjustment is necessary, loosen the two accelerator pump mounting bolts and reposition the pump body until the required measurement is obtained.

4. Tighten the two bolts securely to hold the new adjustment.

IGNITION TIMING

♦ See Figures 79 and 80

Idle Timing

1. Disconnect the spark plug leads from the plugs and ground each lead to the powerhead.
2. Connect a timing light to the powerhead.
3. Place the remote shift lever in the **NEUTRAL** position.
4. Loosen the locknut on the idle timing screw and hold the throttle lever against the idle stop.
5. Crank the powerhead with the starter motor and at the same time, aim the timing light at the flywheel cover timing pointer.
6. Adjust the idle timing screw until the timing pointer aligns with the timing mark specified in the "Tune-Up Specifications" chart.
7. Tighten the locknut to hold this position.

Maximum Advance Timing

♦ See Figure 81

1. Disconnect the spark plug leads from the plugs and ground each lead to the powerhead.
2. Connect a timing light to the powerhead.
3. Place the remote shift lever in the **NEUTRAL** position.
4. Loosen the locknut on the maximum advance screw. Hold the throttle lever in the aft position until the maximum advance screw contacts the stop.
5. Crank the powerhead with the starter motor and at the same time adjust the maximum advance screw until the timing pointer is aligned in the timing mark specified in the "Tune-Up Specifications" chart.

6. Tighten the locknut on the maximum advance screw to hold the adjustment.
7. Remove the timing light.

IDLE SPEED & MIXTURE

1. Connect a tachometer to the powerhead.
2. Start the powerhead and allow it to reach operating temperature.
3. With the engine running at idle in **FORWARD** gear adjust the idle speed screw to obtain the recommended idle speed found in the "Tune-Up Specifications" chart.

➥Make an effort to prevent the powerhead from shutting down during the following adjustments.

4. Move the throttle to the idle speed position and allow the RPM to stabilize.
5. With the powerhead operating at idle rpm, rotate the idle mixture screw counterclockwise until the powerhead begins to misfire because of an over-rich fuel mixture.
6. Slowly rotate the idle mixture screw clockwise and count the turns until the powerhead is operating evenly and the RPM increases. Continue rotating the mixture screw clockwise until the powerhead begins to slow down and eventually begins to misfire because of a too lean air/fuel mixture.
7. Count and move the idle mixture screw back to the halfway point between the too rich mixture and the too lean mixture.

➥Do not adjust the mixture any leaner than necessary to obtain a smooth idle. It is better for the mixture to be on the "too rich" side rather than the "too lean" side.

8. Check for a too lean mixture on acceleration (engine will hesitate or stall on acceleration). Readjust the mixture if necessary.
9. Readjust the idle speed, as necessary and install the access plug.

OIL PUMP

♦ See Figure 82

➥On some model powerheads the oil pump may have two markings on the body of the pump. The second mark should be disregarded for the alignment and adjustment procedures.

1. Hold the throttle arm against the idle stop.
2. Visually check the alignment marks on the oil pump body and the oil pump synchronization lever.
3. If the marks are not aligned, disconnect and adjust the length of the carburetor/oil pump link rod.
4. Attach the link rod and check the alignment marks for correct alignment.

THROTTLE CABLE PRELOAD

♦ See Figure 83

1. With the end of the throttle cable connected to the throttle lever anchor pin, hold the throttle lever against the idle stop. Adjust the throttle cable barrel until it will slip into the barrel retainer on the cable anchor bracket. If necessary, adjust the throttle cable barrel until there is a very light preload on the throttle lever against the idle stop. Lock the barrel in place.

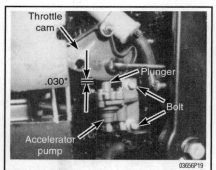

Fig. 78 Make sure that the gap is correct when the accelerator pump plunger is depressed

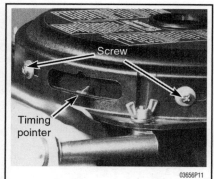

Fig. 79 Timing mark window opening and timing mark pointer

Fig. 80 Adjust the idle timing screw until the timing pointer aligns with the specified mark on the decal

3-30 MAINTENANCE

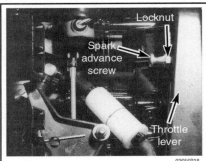

Fig. 81 Adjust the maximum advance screw until the timing pointer aligns with the specified mark on the timing decal

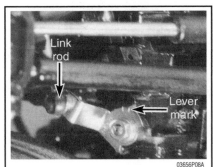

Fig. 82 Visually check the alignment marks on the oil pump body and the oil pump synchronization lever

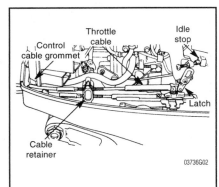

Fig. 83 Throttle cable assembly with major parts identified

➥An excessive preload on the throttle cable will cause difficulty when shifting from FORWARD gear to NEUTRAL. The preload may be easily checked by placing a piece of paper between the idle stop screw and the idle stop and then withdrawing it. If the paper does not tear but drag can be felt, the preload is correct. Adjust the cable barrel, if necessary, to obtain the proper preload just described. Install the powerhead cowling.

2. If sufficient throttle cable barrel adjustment is not available, a check must be made for correct installation of the link rod (located between the throttle lever and the throttle cam). Each end of the link rod must be threaded into its plastic barrel until it bottoms against the throttle lever or the throttle cam casting and then backed out from this position only far enough to obtain correct orientation of the link rod. The link rod must be backed out less than one full turn. All timing adjustments must be reset after this procedure has been completed. Disconnect the tachometer and install the powerhead cowling.

135–225 HP Carbureted

Perform synchronization of the 135–225hp powerheads in the following order: perform preliminary adjustments, set the timing pointer, adjust the throttle cam, synchronize the carburetors, set the full throttle stop, adjust the oil pump linkage, set the ignition timing, set the idle speed and mixture, and finally, test drive the boat to insure proper full throttle speed.

✳✳ CAUTION

Water must circulate through the powerhead any time it is running to prevent damage to the water pump and possible powerhead overheating. Never run the engine over 3000 RPM without an adequate load applied to the propeller.

PRELIMINARY ADJUSTMENTS

1. Disconnect the throttle cables from the throttle lever arm.
2. Rotate the idle mixture screw slowly clockwise (inward) until it barely seats and then back it out counterclockwise the number of turns specified in the "Tune-Up Specifications" chart.

THROTTLE CAM

1. Place the outboard into **FORWARD** gear and ensure the throttle lever is in the idle position.
2. Loosen the cam follower screw to allow the cam to move freely.
3. Hold the throttle lever against the idle stop and check the alignment mark on throttle cam. The alignment mark must be centered with the cam roller.
4. If the mark is not aligned, loosen the locknut on the idle stop screw and adjust the idle stop screw until the mark on the cam is centered in the cam roller. Do not tighten the cam follower screw at this time.
5. Tighten the locknut on the idle stop screw to hold the adjustment.

CARBURETOR SYNCHRONIZATION

1. Turn the idle speed screw counterclockwise until the throttle cam is not touching the throttle arm and the throttle plates on the upper carburetor are fully closed.

2. Loosen the adjustment screw on the tie bar connecting the carburetors.
3. Open the carburetor throttle plates and allow them to close against spring pressure. Verify that the carburetor throttle plates are fully closed.
4. Hold the tie bar in position and tighten the adjustment screw.
5. Once again, open the carburetor throttle plates (this time they should work in unison) and allow them to close against spring pressure. Verify that the carburetor throttle plates are working in unison.

FULL THROTTLE STOP

1. Move the throttle lever to the wide open throttle position.
2. Loosen the locknut and adjust the full throttle stop screw until the carburetor shutters are fully open at wide open throttle. Check to be sure the throttle shutters do not act as throttle stops.
3. Use a feeler gauge and check for 0.010–0.015 in. (.25mm–.38mm) clearance between the cam follower roller and the throttle cam.
4. Tighten the locknut on the full throttle stop screw to hold this adjustment.

OIL PUMP

1. Move the powerhead throttle lever so the idle stop screw is against the idle stop.
2. Check the alignment mark on the oil pump lever. The mark should be aligned with the casting mark on the oil pump body.
3. If the mark is not aligned, disconnect the linkage rod from the oil pump lever. Adjust the rod end to align the mark on the oil pump body with the mark on the lever.
4. Connect the rod end onto the oil pump lever.

IGNITION TIMING

Timer Pointer

1. Remove all spark plugs.
2. Install a dial indicator into the No.1 (top cylinder starboard bank) spark plug opening. Slowly rotate the flywheel clockwise until the No. 1 piston is at top dead center (TDC). Set the dial indicator to "0". Slowly rotate the flywheel counterclockwise until the dial indicator needle is ¼ turn beyond the .462 in. (11.73 mm) mark.
3. On some models with a one piece flywheel, a ⅛ (45°) mark on the flywheel is equal to the .462 mark on flexplate type flywheels.
4. Slowly rotate the flywheel back clockwise until the dial indicator is exactly at the .462 in. (11.73 mm). Observe the timing pointer on the flywheel cover and the .462 in. (11.73 mm) mark on the flywheel.
5. If the flywheel pointer is not exactly on the mark, loosen the pointer adjustment screws and align the pointer with the .462 in. (11.73 mm) mark. Tighten the pointer adjustment screws.
6. Remove the dial indicator from the No. 1 spark plug opening.

Preliminary Timing Adjustments

▶ See Figure 84

1. Measure the length of the trigger link rod from the center line of the 90°bend to the edge of the locknut.
2. The dimension must be ¹¹⁄₁₆ in. (17.5mm) for proper timing adjustment.
3. If the measurement is not as specified, disconnect the link rod and adjust.

MAINTENANCE 3-31

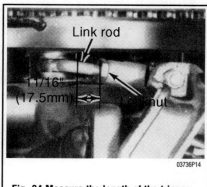

Fig. 84 Measure the length of the trigger rod

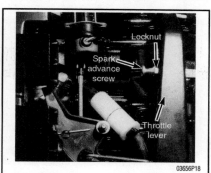

Fig. 85 Adjust the maximum advance screw until the timing pointer aligns with the specified mark on the timing decal

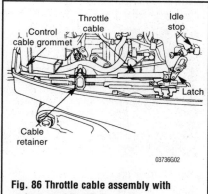

Fig. 86 Throttle cable assembly with major parts identified

4. Disconnect the idle stabilizer module at the White/Black wire bullet connector. Tape the connector to prevent accidental grounding.

➡All models except XR4, XR6, Magnum II, Magnum III and 175hp equipped with the idle stabilizer shift kit require the maximum advance timing be retarded 3°. The idle stabilizer shift kit is standard equipment on the above listed models. Therefore, the timing specification in the "Tune-Up Specifications" chart has been adjusted for these models.

Pick-Up Timing

1. Disconnect the spark plug leads from the plugs and ground each lead to the powerhead.
2. Connect a timing light to the powerhead.
3. Place the remote shift lever in the **NEUTRAL** position.
4. Hold the throttle lever against the idle stop.
5. Crank the powerhead with the starter motor and at the same time, aim the timing light at the flywheel cover timing pointer.
6. Adjust the primary pick-up screw until the timing pointer aligns with the timing mark specified in the "Tune-Up Specifications" chart.
7. Tighten the locknut to hold this position.

Maximum Advance Timing

♦ See Figure 85

1. Disconnect the spark plug leads from the plugs and ground each lead to the powerhead.
2. Connect a timing light to the powerhead.
3. Place the remote shift lever in the **NEUTRAL** position.
4. Loosen the locknut on the maximum advance screw. Hold the throttle lever in the aft position until the maximum advance screw contacts the stop.
5. Crank the powerhead with the starter motor and at the same time adjust the maximum advance screw until the timing pointer is aligned in the timing mark specified in the "Tune-Up Specifications" chart.
6. Tighten the locknut on the maximum advance screw to hold the adjustment.
7. Remove the timing light.

IDLE SPEED & MIXTURE

1. Connect a tachometer to the powerhead.
2. Start the powerhead and allow it to reach operating temperature.
3. With the engine running at idle in **FORWARD** gear adjust the idle timing screw to obtain the recommended idle speed found in the "Tune-Up Specifications" chart.

➡Make an effort to prevent the powerhead from shutting down during the following adjustments.

4. Move the throttle to the idle speed position and allow the RPM to stabilize.
5. With the powerhead operating at idle rpm, rotate the idle mixture screw counterclockwise until the powerhead begins to misfire because of an over-rich fuel mixture.
6. Slowly rotate the idle mixture screw clockwise and count the turns until the powerhead is operating evenly and the RPM increases. Continue rotating the mixture screw clockwise until the powerhead begins to slow down and eventually begins to misfire because of a too lean air/fuel mixture.
7. Count and move the idle mixture screw back to the halfway point between the too rich mixture and the too lean mixture.

➡Do not adjust the mixture any leaner than necessary to obtain a smooth idle. It is better for the mixture to be on the "too rich" side rather than the "too lean" side.

8. Check for a too lean mixture on acceleration (engine will hesitate or stall on acceleration). Readjust the mixture if necessary.
9. Remove the throttle cable barrel from the retainer on the bracket.
10. Readjust the idle speed using the idle timing screw, as necessary.
11. Perform the throttle cable preload adjustment and reconnect the throttle cable.

THROTTLE CABLE PRELOAD

♦ See Figure 86

1. With the end of the throttle cable connected to the throttle lever anchor pin, hold the throttle lever against the idle stop. Adjust the throttle cable barrel until it will slip into the barrel retainer on the cable anchor bracket. If necessary, adjust the throttle cable barrel until there is a very light preload on the throttle lever against the idle stop. Lock the barrel in place.

➡An excessive preload on the throttle cable will cause difficulty when shifting from FORWARD to NEUTRAL.

2. The preload may be easily checked by placing a piece of paper between the idle stop screw and the idle stop and then withdrawing it. If the paper does not tear but drag can be felt, the preload is correct. Adjust the cable barrel, if necessary, to obtain the proper preload just described. Install the powerhead cowling.
3. If sufficient throttle cable barrel adjustment is not available, a check must be made for correct installation of the link rod (located between the throttle lever and the throttle cam). Each end of the link rod must be threaded into its plastic barrel until it bottoms against the throttle lever or the throttle cam casting and then backed out from this position only far enough to obtain correct orientation of the link rod. The link rod must be backed out less than one full turn.

225 HP (3.0L) Carbureted

Perform synchronization of the 225hp (3.0L) powerheads in the following order: set the timing pointer, adjust the throttle cam, synchronize the carburetors, adjust the oil pump linkage, set the ignition timing, set the idle speed and finally, test drive the boat to insure proper full throttle speed.

✷✷✷ CAUTION

Water must circulate through the powerhead any time it is running to prevent damage to the water pump and possible powerhead overheating. Never run the engine over 3000 RPM without an adequate load applied to the propeller.

3-32 MAINTENANCE

IGNITION TIMING

Timing Pointer

1. Remove all spark plugs.
2. Install a dial indicator into the No.1 (top cylinder starboard bank), spark plug opening. Slowly rotate the flywheel counterclockwise until the dial indicator needle is ¼ turn beyond the .526 in. mark.
3. Slowly rotate the flywheel back clockwise until the dial indicator is exactly at the .526 in. (11.7mm) mark. Observe the timing pointer on the powerhead and the .526 in.(11.7 mm) mark on the flywheel.
4. If the timing pointer is not exactly on the mark, loosen the pointer adjustment bolt and align the pointer with the .526 in. (11.7 mm) mark on the flywheel.
5. Tighten the pointer adjustment bolt.
6. Remove the dial indicator from the No. 1 spark plug opening.

Timing Adjustment

Maximum spark timing is controlled by the electronic control unit (ECU) and is not adjustable. As long as the electronic control unit, crankcase position and throttle position sensors are functioning properly, the maximum timing will be correct.

THROTTLE CAM

♦ See Figure 87

1. Check to be sure the remote control throttle lever is in the **NEUTRAL** and idle position.
2. Loosen the cam follower screw to allow the cam to move freely.
3. Hold the throttle lever against the idle stop and check the alignment mark on throttle cam. The mark should be centered with the cam roller. If the mark is not aligned, loosen the locknut on the idle stop screw and adjust the idle stop screw until the mark on the cam is centered in the cam roller.
4. Tighten the locknut on the idle stop screw to hold the adjustment.
5. Do not tighten the cam follower screw at this time.

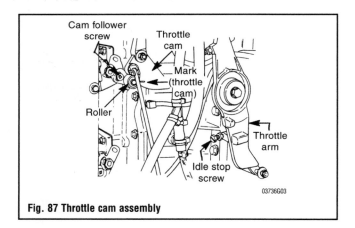

Fig. 87 Throttle cam assembly

CARBURETOR SYNCHRONIZATION

1994–95

♦ See Figures 88 and 89

1. Loosen the two carburetor synchronization screws on the two upper carburetor linkage arms.
2. Check to be sure the shutter plates in the carburetors are completely closed.
3. Hold the throttle lever with the idle stop screws against the idle stop.
4. Move the cam follower roller next to the throttle cam until the roller barely makes contact with the throttle cam.
5. Carefully, tighten the carburetor synchronization screws and the cam follower roller screw without disturbing any of the adjustments.
6. Move the throttle lever from idle to half throttle while checking the shutter plates on all three carburetors. The shutters should open and close simultaneously. Repeat this step if any shutter plate lags.
7. Move the throttle lever to the wide open throttle position. Loosen the locknut and adjust the full throttle stop screw until the carburetor shutters are fully open at wide open throttle. Check to be sure the throttle shutters do not act as throttle stops. Use a feeler gauge and check for .010–.015" (.25mm–.38mm) clearance between the cam follower roller and the throttle cam. Tighten the locknut on the full throttle stop screw to hold this adjustment.

1996–97

1. Adjust the idle stop screw until the bend in the linkage is ¼ in. (6.3mm) from the crankcase casting. The threaded idle stop boss on the back side of the throttle arm should be close to the middle of the idle stop screw.
2. Loosen the carburetor synchronization screws to allow the shutter plates to close completely.
3. Position the throttle lever so that the idle stop screw is against the idle stop. Without moving the linkage, retighten the center screw first and then the upper and lower screws.
4. Move the throttle lever from idle to half throttle while checking the shutter plates on all three carburetors. The shutters should open and close simultaneously. Repeat this step if any shutter plate lags.
5. Move the throttle lever to the wide open throttle position. Loosen the locknut and adjust the full throttle stop screw until the carburetor shutters are fully open at wide open throttle. Check to be sure the throttle shutters do not act as throttle stops. Use a feeler gauge and check for .010–.015" (.25mm–.38mm) clearance between the cam follower roller and the throttle cam. Tighten the locknut on the full throttle stop screw to hold this adjustment.

OIL PUMP

1994–95

Although the 225hp powerhead is equipped with an oil injection pump, the 1994–95 carbureted models do not have an adjustable oil link rod.

1996–97

♦ See Figure 90

1. Move the throttle lever so the idle stop screw is against the idle stop.

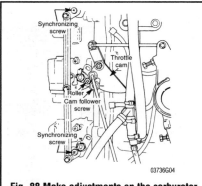

Fig. 88 Make adjustments on the carburetor linkage arms to synchronize the carburetors

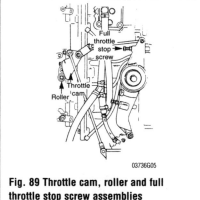

Fig. 89 Throttle cam, roller and full throttle stop screw assemblies

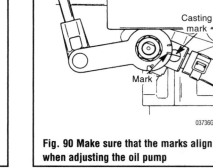

Fig. 90 Make sure that the marks align when adjusting the oil pump

MAINTENANCE 3-33

2. Check the alignment mark on the oil pump lever. The mark should be aligned with the casting mark on the oil pump body.
3. If the marks are not aligned, disconnect the linkage rod from the oil pump lever.
4. Adjust the rod end to align the mark on the oil pump body with the mark on the lever.
5. Connect the rod end onto the oil pump lever.

THROTTLE POSITION SENSOR (TPS)

♦ See Figure 91

The TPS transmits throttle position information to the engine controller in the form of a low voltage signal. This low voltage signal will range from .950 VDC at the idle position to 3.91VDC at wide open throttle setting.

A lower voltage setting will cause the powerhead timing to retard and a higher voltage will advance the timing. A 0.050 volt change in the TPI setting, results in a 1° change in the powerhead timing.

Alignment and adjustment of the TPS is a critical step in the timing and synchronizing of the powerhead.

1. Disconnect the TPS from the ignition harness.
2. Connect a multimeter to the harness using a Mercury Test Lead Assembly (84-825207A1) or equivalent.
3. On 1994–95 models, connect the Black meter lead to the Black test lead and the Red meter lead to the White test lead.
4. On 1996–97 models, connect the Black meter lead to the White test lead and the Red meter lead to the Red test lead.
5. Loosen the screws on the TPS bracket and turn the ignition switch **ON**.
6. Rotate the TPS to obtain a voltage of 0.9–1.00, then tighten the screws.
7. Disconnect the throttle control cable and slowly rotate the throttle lever to the full open position while monitoring the TPS voltage. The voltage reading should increase and decrease smoothly.
8. Maximum voltage at wide open throttle should be 3.7–3.9 volts.

➡ If maximum voltage is not correct, verify that the full throttle stop is set correctly.

9. Remove the test leads and connect the TPS to the harness.

CRANKSHAFT POSITION SENSOR

♦ See Figure 92

1. Remove the flywheel cover, if it is still in place.
2. Rotate the flywheel clockwise by hand and align one of the flywheel sensor teeth directly in line (perpendicular), with the crankshaft position sensor.
3. Using a feeler gauge, measure the gap between the sensor and flywheel tooth. The allowable gap measurement is 0.020–0.060 in. (0.51–1.53 mm). If the gap is incorrect, loosen the two sensor bracket bolts.
4. Set the gap to the correct measurement and tighten the two sensor bracket bolts.
5. Install the flywheel cover.

IDLE SPEED

➡ The following procedures must be performed with the outboard in a test tank or the boat and outboard in a body of water. The idle speed adjustment procedures can only be preformed with the outboard running with the lower unit in FORWARD gear and the propeller under an actual load condition.

1. Start the powerhead and allow it to warm to operating temperature.
2. Place the outboard into **FORWARD** gear and monitor the powerhead rpm. If the powerhead RPM is not correct, verify carburetor synchronization and/or carburetor mixture settings. If synchronization and fuel mixture are correct, the idle speed may be adjusted by repositioning the TPS sensor. See "TPS Adjustment" this section.
3. With the end of the throttle cable connected to the throttle lever anchor pin hold the throttle lever against the idle stop. Adjust the throttle cable barrel until it will slip into the barrel retainer on the cable anchor bracket. If necessary, adjust the throttle cable barrel until there is a very light preload on the throttle lever against the idle stop.

➡ An excessive preload on the throttle cable will cause difficulty when shifting between gears.

4. The preload may be easily checked by placing a piece of paper between the idle stop screw and the idle stop and then withdrawing it. If the paper does not tear but drag can be felt, the preload is correct. Adjust the cable barrel, if necessary, to obtain the proper preload just described. Install the powerhead cowling.

250 and 275 HP

Perform synchronization of the 250 and 275hp powerheads in the following order: set the timing pointer, synchronize the carburetors, set the ignition timing, set the full throttle stop, adjust the oil pump linkage, set the idle speed and finally, test drive the boat to insure proper full throttle speed.

※ CAUTION

Water must circulate through the powerhead any time it is running to prevent damage to the water pump and possible powerhead overheating. Never run the engine over 3000 RPM without an adequate load applied to the propeller.

IGNITION TIMING

Timing Pointer

1. Remove all spark plugs.
2. Install a dial indicator into the No.1 (top cylinder starboard bank), spark plug opening. Slowly rotate the flywheel counterclockwise until the dial indicator needle is ¼ turn beyond the .526 in. mark.
3. Slowly rotate the flywheel back clockwise until the dial indicator is exactly at the .557 in. (14.14 mm) mark. Observe the timing pointer on the powerhead and the .557 in. (14.14 mm) mark on the flywheel.
4. If the timing pointer is not exactly on the mark, loosen the pointer adjustment bolt and align the pointer with the .557 in. (14.14 mm)mark on the flywheel.
5. Tighten the pointer adjustment bolt.

Preliminary Timing Adjustments

♦ See Figure 93

1. Measure the length of the trigger link rod from the center line of the 90°bend to the edge of the locknut.

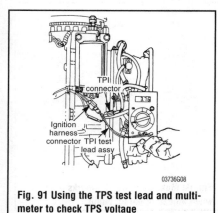

Fig. 91 Using the TPS test lead and multimeter to check TPS voltage

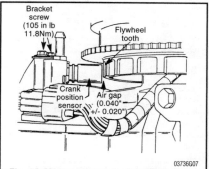

Fig. 92 After removing the flywheel cover it is possible to check the gap on the crankshaft position sensor

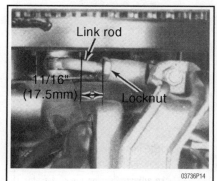

Fig. 93 Measure the length of the trigger rod

MAINTENANCE

2. The dimension must be 11/16 in. (17.5mm) for proper timing adjustment.
3. If the measurement is not as specified, disconnect the link rod and adjust.
4. Disconnect the idle stabilizer module at the White/Black wire bullet connector. Tape the connector to prevent accidental grounding.

Pick-Up Timing

♦ See Figure 94

1. Disconnect the spark plug leads from the plugs and ground each lead to the powerhead.
2. Connect a timing light to the powerhead.
3. Place the remote shift lever in the **NEUTRAL** position.
4. Hold the throttle lever against the idle stop.
5. Crank the powerhead with the starter motor and at the same time, aim the timing light at the flywheel cover timing pointer.
6. Adjust the primary pick-up screw until the timing pointer aligns with the timing mark specified in the "Tune-Up Specifications" chart.
7. Tighten the locknut to hold this position.

Maximum Advance Timing

♦ See Figure 95

1. Disconnect the spark plug leads from the plugs and ground each lead to the powerhead.
2. Connect a timing light to the powerhead.
3. Place the remote shift lever in the **NEUTRAL** position.
4. Loosen the locknut on the maximum advance screw. Hold the throttle lever in the aft position until the maximum advance screw contacts the stop.
5. Crank the powerhead with the starter motor and at the same time adjust the maximum advance screw until the timing pointer is aligned in the timing mark specified in the "Tune-Up Specifications" chart.
6. Tighten the locknut on the maximum advance screw to hold the adjustment.
7. Remove the timing light.

CARBURETOR SYNCHRONIZATION

♦ See Figure 96

1. Position the throttle lever idle stop screw against the idle stop. Measure the distance between the throttle arm barrel and the cam barrel. The measurement from center to center should be 5-13/32 in. (137.32mm). If necessary, adjust the link rod for the correct measurement.
2. Loosen the six carburetor synchronizing screws. Position the throttle lever against the idle stop. Position the throttle roller to just make contact with the throttle cam. Adjust the idle stop screw to align the slash mark on the throttle cam with the center of the throttle roller. Tighten the idle stop screw locknut. Hold the throttle roller arm steady and tighten the six carburetor screws to hold the adjustment.
3. Move the throttle lever from idle to half throttle while looking at the carburetor shutters. Be sure all carburetor shutters open and close simultaneously. Repeat this step if any carburetor shutter does not fully close or move at the same time. Install the sound box cover and secure with screws.

FULL THROTTLE STOP

♦ See Figure 97

1. Move the throttle lever to the wide open throttle position.
2. Loosen the locknut and adjust the full throttle stop screw to permit full throttle shutter opening at wide open throttle. Check to be sure the throttle shutter does not act as a throttle stop.
3. Use a feeler gauge and check the clearance between the roller and the throttle cam at wide open throttle. This clearance should be 0.010–0.015in. (0.25–0.38mm).
4. Tighten the locknut to hold the adjustment.
5. Install and tighten the spark plugs.

OIL PUMP

♦ See Figure 98

1. Move the powerhead throttle lever until the idle stop screw makes contact with the idle stop.
2. Check the alignment mark on the oil pump lever (first shortest mark) is aligned with the casting mark on the oil pump body.
3. If the alignment mark is not properly aligned, disconnect the linkage rod from the oil pump lever.
4. Adjust the rod end to align the mark on the oil pump body with the mark on the lever.
5. Connect the rod end onto the oil pump lever.

IDLE SPEED

1. Place the outboard in a test tank or the boat in a body of water. Shift the lower unit into **FORWARD** gear.
2. Connect a tachometer to the powerhead.
3. Start the powerhead and allow it to warm to operating temperature.
4. Remove the cable barrel from the barrel retainer and loosen the locknut on the idle adjustment screw.
5. Adjust the idle to the proper speed, as listed in the "Tune-Up Specifications" chart. Tighten the locknut to hold the adjustment.
6. With the end of the throttle cable connected to the throttle lever, hold the throttle lever against the idle stop.
7. Adjust the throttle cable barrel to slip into the barrel retainer on the cable anchor bracket with a very light preload of throttle lever against the idle stop. Lock the barrel in place.

➡An excessive preload on the throttle cable will cause difficulty when shifting gears.

8. The preload may be easily checked by placing a piece of paper between the idle stop screw and the idle stop and then withdrawing it. If the paper does not tear but drag is felt, the preload is correct. Adjust the cable barrel, if necessary, to obtain the proper preload just described.

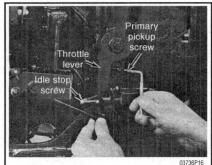

Fig. 94 Hold the throttle arm against the throttle stop while cranking the engine with the starter motor

Fig. 95 Make the advance timing adjustment by turning the maximum advance screw until the timing mark is aligned

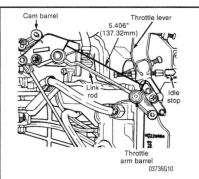

Fig. 96 Position the throttle lever idle stop screw against the idle stop

MAINTENANCE 3-35

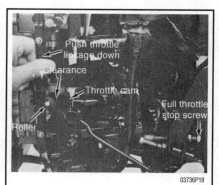

Fig. 97 Throttle linkage assembly with major components identified

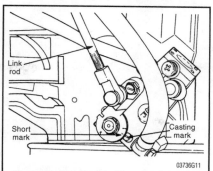

Fig. 98 Make adjustments on the link rod to align the mark on the oil pump lever and casting mark on the oil pump body

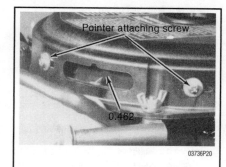

Fig. 99 Slowly rotate the flywheel back—clockwise—until the dial indicator is exactly at the .462 in. (11.73 mm) mark

150–200 HP Electronic Fuel Injection (EFI)

Perform synchronization of the 150–200hp EFI powerheads in the following order: set the timing pointer, set the throttle cam, set the full throttle stop, adjust the oil pump linkage, set the ignition timing, set the idle speed, adjust the throttle position sensor and finally, test drive the boat to insure proper full throttle speed.

✽✽ CAUTION

Water must circulate through the powerhead any time it is running to prevent damage to the water pump and possible powerhead overheating. Never run the engine over 3000 RPM without an adequate load applied to the propeller.

IGNITION TIMING

Timing Pointer

▶ See Figure 99

1. Remove all spark plugs.
2. Install a dial indicator into the No. 1 (top cylinder starboard bank), spark plug opening.
3. Slowly rotate the flywheel clockwise until the No. 1 piston is at top dead center (TDC). Set the dial indicator to "0". Slowly rotate the flywheel counterclockwise until the dial indicator needle is ¼ turn beyond the .462 in. (11.73 mm) mark.
4. Slowly rotate the flywheel back—clockwise—until the dial indicator is exactly at the .462 in. (11.73 mm) mark. Observe the timing pointer on the flywheel cover and the .462 in. (11.73 mm) mark on the flywheel.
5. If the flywheel pointer is not exactly on the mark, loosen the pointer adjustment screws and align the pointer with the .462 in. (11.73 mm) mark.
6. Tighten the pointer adjustment screws and remove the dial indicator from the No. 1 spark plug opening.

Preliminary Timing Adjustments

▶ See Figures 100 and 101

1. Measure the length of the trigger link rod from the center line of the 90°bend to the edge of the locknut.

Fig. 100 Disconnect the Black/White wire from the idle stabilization module

2. A dimension of 11/16 in. (17.5mm) is required for proper timing adjustment.
3. If the measurement is not as specified, disconnect the link rod and adjust.
4. Disconnect the Black/White wire lead from the idle stabilization module at the bullet connector next to the module. Wrap tape around the wire end to prevent it from shorting out.
5. Disconnect the control unit harness connector for the ignition timing procedures.

Idle Timing

▶ See Figures 102 and 103

1. Disconnect the spark plug leads from the plugs and ground each lead to the powerhead.
2. Connect a timing light to the powerhead.
3. Place the remote shift lever in the **NEUTRAL** position.
4. Loosen the locknut on the idle timing screw.
5. Move the throttle lever until the idle stop screw contacts the idle stop.
6. Crank the powerhead with the starter motor and at the same time rotate the idle timing screw until the marks on the timing decal align with the timing mark specified in the "Tune-Up Specifications" chart.
7. Tighten the locknut on the idle timing screw.

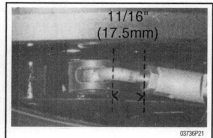

Fig. 101 A dimension of 11/16 in. (17.5mm) is required for proper timing adjustment

Fig. 102 Move the throttle until the idle stop screw just contacts the idle stop

Fig. 103 Location of the idle timing screw

Maintenance

Maximum Advance Timing

♦ See Figures 104 and 105

1. Disconnect the spark plug leads from the plugs and ground each lead to the powerhead.
2. Connect a timing light to the powerhead.
3. Place the remote shift lever in the **NEUTRAL** position.
4. Loosen the locknut on the maximum advance screw. Hold the throttle lever in the aft position until the maximum advance screw contacts the stop.
5. Crank the powerhead with the starter motor and at the same time adjust the maximum advance screw until the timing pointer is aligned in the timing mark specified in the "Tune-Up Specifications" chart.
6. Tighten the locknut on the maximum advance screw to hold the adjustment.
7. Remove the timing light.

Detonation Control System

♦ See Figure 106

This system is used on the 200hp model only.
1. Connect a timing and tachometer to the powerhead.
2. Start the powerhead and allow it to warm to operating temperature.
3. Shift the unit into **FORWARD** gear.
4. Move the throttle lever forward until the powerhead indicates 3500 RPM on the tachometer.
5. Using a multimeter, verify the timing has been electronically advanced to 26°BTDC. The advancement indicates the detonation (knock), control circuit is functioning properly.
6. Return the throttle lever to the idle position and shift the outboard into **NEUTRAL**. Shut down the powerhead and then disconnect the timing light and tachometer.

THROTTLE CAM

♦ See Figures 107 and 108

1. Loosen the locknut on the idle stop screw and move the throttle lever until the idle stop screw makes contact with the idle stop. Hold the throttle lever in this position.
2. Loosen the cam follower roller screw behind the cam.
3. Rotate the idle stop screw until the cam follower roller contacts the throttle cam and the embossed line on the throttle cam is centered on the cam follower roller.
4. Tighten the idle stop screw locknut and the cam follower screw with the cam follower roller barely making contact with the cam.
5. Release the throttle lever back to the idle position.

FULL THROTTLE STOP

1. Loosen the locknut on the full throttle stop screw.
2. Move the throttle lever forward until the lever contacts the full throttle stop screw.
3. Rotate the screw until all the throttle shutters are fully open. A slight clearance should be maintained between the throttle shaft arm and the stop on the manifold.
4. Check for free play at the cam follower roller and the throttle cam. This indicates there is no binding of the throttle linkage.
5. Readjust the full throttle stop screw, if required.

OIL PUMP

♦ See Figure 109

1. Position and hold the throttle lever against the idle stop. Verify the alignment mark on the oil pump lever is aligned with the casting mark on the oil pump body. If not, disconnect the linkage rod from the oil pump lever.
2. Screw the rod end, in or out, to align the oil pump lever mark with the mark on the body casting.
3. Connect the rod end onto the oil pump lever ball.

IDLE SPEED

♦ See Figure 110

1. Mount the outboard unit in a test tank or the craft and outboard in a body of water.

Fig. 104 Maximum advance screw

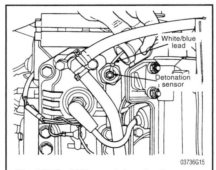

Fig. 105 On 200hp models only, disconnect the White/Blue lead from the detonation sensor on the port side cylinder head

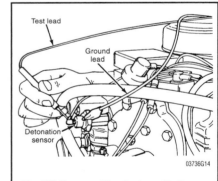

Fig. 106 Use a multimeter to verify the detonation sensor is functioning properly

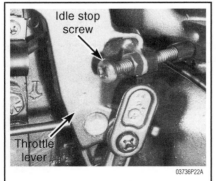

Fig. 107 Loosen the locknut on the idle stop screw

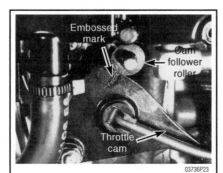

Fig. 108 Make sure the mark on the throttle cam is centered on the cam follower roller

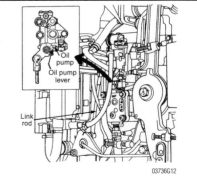

Fig. 109 Align the marks on the oil pump lever and oil pump body

MAINTENANCE 3-37

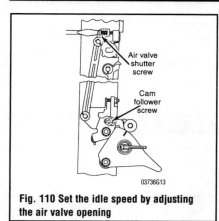

Fig. 110 Set the idle speed by adjusting the air valve opening

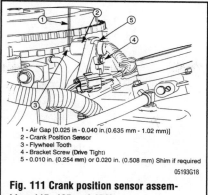

Fig. 111 Crank position sensor assembly—115, 135 and 150hp

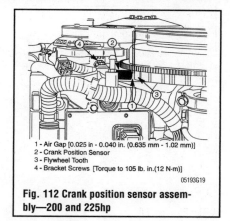

Fig. 112 Crank position sensor assembly—200 and 225hp

2. Connect a tachometer to the rectifier Gray wire or the terminal block Yellow wire.
3. Start the powerhead and allow it to warm to operating temperature.
4. Loosen the screw on the cam follower and hold the throttle lever against the idle stop screw.
5. Adjust the idle speed to specification by increasing or decreasing the air valve opening.
6. Tighten the screw on the cam follower.
7. Connect the end of the throttle cable to the throttle lever. Move the throttle lever forward against the idle stop screw. Adjust the throttle cable barrel until it will slip into the barrel recess in the control cable anchor bracket with only a light preload against the idle stop.
8. Lock the throttle cable barrel into the barrel recess on the control cable anchor bracket.

➥An excessive preload on the throttle cable will cause difficulty when shifting from FORWARD gear into NEUTRAL.

9. The preload may be easily checked by placing a piece of paper between the idle stop screw and the idle stop and then withdrawing it. If the paper does not tear but drag can be felt, the preload is correct. Adjust the cable barrel, if necessary, to obtain the proper preload just described. Install the powerhead cowling.

THROTTLE POSITION SENSOR (TPS)

The TPS transmits throttle position information to the engine controller in the form of a low voltage signal. This low voltage signal will range from .950 VDC at the idle position to 3.91VDC at wide open throttle setting.

A lower voltage setting will cause the powerhead timing to retard and a higher voltage will advance the timing. A 0.050 volt change in the TPI setting, results in a 1° change in the powerhead timing.

Alignment and adjustment of the TPS is a critical step in the timing and synchronizing of the powerhead.

1. Disconnect the TPS from the ignition harness.
2. Disconnect the cylinder head temperature sender wiring harness.
3. Connect a multimeter to the harness using a Mercury Test Lead Assembly (84-825207A1) or equivalent.
4. Loosen the screws on the TPS bracket and turn the ignition switch **ON**.
5. On 1990–93 models using ECU number 14632A13 and lower, rotate the TPS to maintain a voltage of 0.125–0.145 volts.
6. On 1990–93 models using ECU number 14632A15 and higher, rotate the TPS to maintain a voltage of 0.240–0.260 volts.
7. Disconnect the throttle control cable and slowly rotate the throttle lever to the full open position while monitoring the TPS voltage. The voltage reading should increase and decrease smoothly.
8. Maximum voltage at wide open throttle should not exceed 7.46 volts.

➥If maximum voltage is not correct, verify that the full throttle stop is set correctly.

9. Remove the test leads and connect the TPS and cylinder head temperature sender wiring harnesses.

115–225 Direct Fuel Injection (DFI) OPTIMAX

The ignition and fuel systems on these models are controlled by the Engine Control Module (ECM) which performs the following functions:
- Calculates the precise fuel and ignition timing requirements based on engine speed, throttle position, manifold pressure and coolant temperature
- Controls the fuel injectors for each cylinder
- Controls ignition timing for each cylinder
- Controls all alarm horn and warning lamp functions
- Controls rpm and rpm limiter function
- Supplies tachometer signal to the gauge
- Monitors the shift interrupt switch
- Records engine running information

CRANK POSITION SENSOR

▶ See Figures 111 and 112

1. Remove the flywheel cover.
2. Using a feeler gauge, measure the air gap between the crank position sensor and a tooth on the flywheel. The gap should measure 0.025–0.040 in. (0.635–1.02 mm).
3. As necessary, loosen both screws and set gap to specification. 0.010 in. (0.254 mm) and 0.020 in. (0.508 mm) shims are available for the 115, 135 and 150hp models to properly position the sensor. Retighten the screws.

✸✸ WARNING

The crank position sensor must be perpendicular to the flywheel tooth.

4. Reinstall the flywheel cover.

THROTTLE CAM

1. Loosen the roller arm screw allowing the roller to move freely.
2. Allow the roller to rest on the throttle cam. Adjust the idle stop screw on the throttle arm to align the cam roller in the pocket of the throttle cam.
3. Tighten the roller arm screw to provide clearance of 0.00–0.10 in. (0.0–0.254 mm) between the roller and cam.

FULL THROTTLE STOP

▶ See Figures 113 and 114

1. Hold the throttle arm against the full throttle stop.
2. Adjust the full throttle stop screw (located behind the electric fuel pump) to allow full throttle valve opening while maintaining a clearance of 0.010 in. (0.254 mm) on the 200 and 225hp models and a clearance of 0.020 in. (0.508 mm) on the 115/135/150 HP models, between the arm of the throttle shaft and the stop on the attenuator box.

3-38 MAINTENANCE

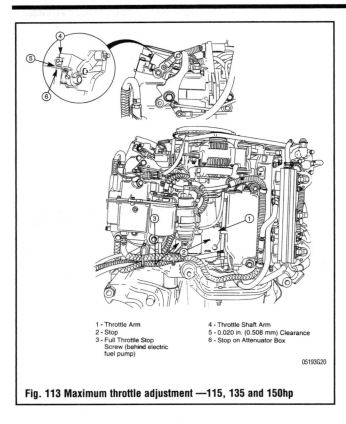

1 - Throttle Arm
2 - Stop
3 - Full Throttle Stop Screw (behind electric fuel pump)
4 - Throttle Shaft Arm
5 - 0.020 in. (0.508 mm) Clearance
6 - Stop on Attenuator Box

Fig. 113 Maximum throttle adjustment —115, 135 and 150hp

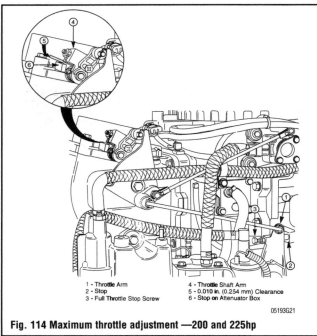

1 - Throttle Arm
2 - Stop
3 - Full Throttle Stop Screw
4 - Throttle Shaft Arm
5 - 0.010 in. (0.254 mm) Clearance
6 - Stop on Attenuator Box

Fig. 114 Maximum throttle adjustment —200 and 225hp

3. Tighten the jam nut on the full throttle stop screw.
4. Check for free play (roller lifts from cam) between the roller and cam at full throttle to prevent the linkage from binding. Readjust the full throttle stop screw if necessary.

THROTTLE PLATE SCREW

♦ See Figures 115 and 116

※※ WARNING

It is not recommended that the throttle plate screw be adjusted from the factory setting. However, should the throttle plate require adjustment, use the throttle plate stop screw to set the total throttle plate clearance as follows.

1. On the 200 and 225hp models, using suitable drills (so that the combined air gap fore and aft on the throttle plate equals a total of 0.149 in. (3.78 mm) the correct clearance is 0.149 in. (3.78 mm).
2. On the 135 and 150hp models, the clearance is 0.131 in. (0.7937 mm) using a #68 drill.
3. On the 115hp model close the throttle plate and turn the throttle plate screw in one turn.

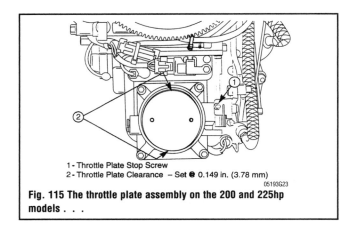

1 - Throttle Plate Stop Screw
2 - Throttle Plate Clearance – Set @ 0.149 in. (3.78 mm)

Fig. 115 The throttle plate assembly on the 200 and 225hp models . . .

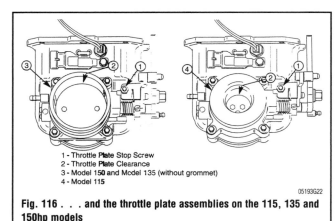

1 - Throttle Plate Stop Screw
2 - Throttle Plate Clearance
3 - Model 150 and Model 135 (without grommet)
4 - Model 115

Fig. 116 . . . and the throttle plate assemblies on the 115, 135 and 150hp models

WINTER STORAGE CHECKLIST

Taking extra time to store the boat properly at the end of each season will increase the chances of satisfactory service at the next season. Remember, storage is the greatest enemy of an outboard motor. The unit should be run on a monthly basis. The boat steering and shifting mechanism should also be worked through complete cycles several times each month. If a small amount of time is spent in such maintenance, the reward will be satisfactory performance, increased longevity and greatly reduced maintenance expenses.

For many years there has been the widespread belief simply shutting off the fuel at the tank and then running the powerhead until it stops is the proper procedure before storing the engine for any length of time. Right? WRONG!

First, it is not possible to remove all fuel in the carburetor by operating the powerhead until it stops. Considerable fuel is trapped in the float chamber and other passages and in the line leading to the carburetor. The only guaranteed method of removing all fuel is to take the time to remove the carburetors and drain the fuel.

MAINTENANCE 3-39

Proper storage involves adequate protection of the unit from physical damage, rust, corrosion and dirt. The following steps provide an adequate maintenance program for storing the unit at the end of a season.

1. On four-stroke outboards it is important drain the engine oil and replace with new oil and filter.
2. Squirt a small quantity of engine oil into each spark plug hole and crank the engine over to distribute the oil around the engine internals. Reinstall the old spark plugs (you will install new spark plugs in the spring).
3. Drain all fuel from the carburetor float bowls. On fuel injected models, drain the fuel from the vapor separator.
4. Drain the fuel tank and the fuel lines Store the fuel tank in a cool dry area with the vent OPEN to allow air to circulate through the tank. Do not store the fuel tank on bare concrete. Place the tank to allow air to circulate around it.
5. Change the fuel filter.
6. Drain and then fill the lower unit with new lower unit gear oil.
7. Lubricate the throttle and shift linkage and the steering pivot shaft.
8. Clean the outboard unit thoroughly. Coat the powerhead with a commercial corrosion and rust preventative spray. Install the cowling and then apply a thin film of fresh engine oil to all painted surfaces.
9. Remove the propeller. Apply Perfect Seal® or a waterproof sealer to the propeller shaft splines and then install the propeller back in position.
10. Be sure all drain holes in the gear housing are open and free of obstructions. Check to be sure the flush plug has been removed to allow all the water to drain. Trapped water could freeze, expand and cause expensive castings to crack.
11. Always store the outboard unit off the boat with the lower unit below the powerhead to prevent any water from being trapped inside.
12. Be sure to consult your owners manual for any particular storage procedures applicable to your specific model.

SPRING COMMISSIONING CHECKLIST

A spring tune-up is essential to getting the most out of your engine. If the engine has been properly winterized, it is usually no problem to get it in top running condition again in the springtime. If the engine has just been put in the garage and forgotten for the winter, then it is doubly important to do a complete tune up before putting the engine back into service. If you have ever been stranded out on the water because your engine has died and you had to suffer the embarrassment of having to be towed back to the marina, now is the time to prevent that from occurring.

Satisfactory performance and maximum enjoyment can be realized if a little time is spent in preparing the outboard unit for service at the beginning of the season. Assuming the unit has been properly stored, a minimum amount of work is required to prepare the unit for use. The following steps outline an adequate and logical sequence of tasks to be performed before using the outboard the first time in a new season.

1. Lubricate the outboard according to the manufacturer's recommendations.
2. Perform a tune-up on the engine. This should include replacing the spark plugs and making a thorough check of the ignition system. The ignition system check should include the ignition coils, stator assembly, condition of the wiring and the battery.
3. If a built-in fuel tank is installed, take time to check the gasoline tank and all fuel lines, fittings, couplings, valves, including the flexible tank fill and vent. Turn on the fuel supply valve at the tank. If the fuel was not drained at the end of the previous season, make a careful inspection for gum formation. If a six-gallon fuel tank is used, take the same action. When gasoline is allowed to stand for long periods of time, particularly in the presence of copper, gummy deposits form. This gum can clog the filters, lines and passageways in the carburetor.
4. Replace the oil in the lower unit.
5. Replace the fuel filter.
6. Replace the engine oil and filter. Make sure to use only a quality four stroke engine oil and NEVER use two stroke oil in a four stroke engine.
7. Close all water drains. Check and replace any defective water hoses. Check to be sure the connections do not leak. Replace any spring-type hose clamps with band-type clamps, if they have lost their tension or if they have distorted the water hose.
8. The engine can be run with the lower unit in water to flush it. If this is not practical, a flush attachment may be used. This unit is attached to the water pick-up in the lower unit. Attach a garden hose, turn on the water, allow the water to flow into the engine for awhile and then run the engine.
9. Check the exhaust outlet for water discharge. Check for leaks. Check operation of the thermostat.
10. Check the electrolyte level in the battery and the voltage for a full charge. Clean and inspect the battery terminals and cable connections. Take time to check the polarity, if a new battery is being installed. Cover the cable connections with grease or special protective compound as a prevention to corrosion formation. Check all electrical wiring and grounding circuits.
11. Check all electrical parts on the engine and lower portions of the hull. Rubber boots help keep electrical connections clean and reduce the possibility of arcing.

➡ **Starter motors and high tension wiring harnesses should be of a marine type that cannot cause an explosive mixture to ignite.**

12. If a water separating filter is installed between the fuel tank and the powerhead fuel filter, replace the element at least once each season. This filter removes water and fuel system contaminants such as dirt, rust and other solids, thus reducing potential problems.
13. As a last step in spring commissioning, perform a full engine tune-up.

✲✲ CAUTION

Before putting the boat in the water, take time to verify the drain plugs are installed. Countless number of boating excursions have had a very sad beginning because the boat was eased into the water only to have the boat begin to fill with the water.

General Engine Specifications

Year	Model (HP)	Engine Type	Displace cu.in. (cc)	Bore and Stroke	Oil Injection System	Ignition System	Starting System	Cooling System
1993-00	2.5	1-cyl	4.6 (74.6)	1.85" x 1.69"	Pre-Mix	CDI	Manual	Water-cooled
1990-92	3	1-cyl	4.6 (74.6)	1.85" x 1.69"	Pre-Mix	Magneto w/ points	Manual	Water-cooled
1993-00	3.3	1-cyl	4.6 (74.6)	1.85" x 1.69"	Pre-Mix	CDI	Manual	Water-cooled
1990-00	4	1-cyl	6.2 (102)	2.16" x 1.69"	Pre-Mix	CDI	Manual	Water-cooled
1990-00	5	1-cyl	6.2 (102)	2.16" x 1.69"	Pre-Mix	CDI	Manual	Water-cooled
1993-00	6	2-cyl	12.8 (210)	2.13" x 1.77"	Pre-Mix	CDI	Manual (electric optional)	Water-cooled w/ thermostat
1990-00	8	2-cyl	12.8 (210)	2.13" x 1.77"	Pre-Mix	CDI	Manual (electric optional)	Water-cooled w/ thermostat
1990-00	9.9	2-cyl	16 (262)	2.38" x 1.80"	Pre-Mix	CDI	Manual or Electric	Water-cooled w/ thermostat
1990-00	15	2-cyl	16 (262)	2.38" x 1.80"	Pre-Mix	CDI	Manual or Electric	Water-cooled w/ thermostat
1990-00	20	2-cyl	24.4 (400)	2.56" x 2.38"	Pre-Mix	CDI	Manual or Electric	Water-cooled w/ thermostat
1990-00	25	2-cyl	24.4 (400)	2.56" x 2.38"	Pre-Mix	CDI	Manual or Electric	Water-cooled w/ thermostat
1992-00	30	2-cyl	39.3 (645)	2.99" x 2.80"	Standard	Modular CDI or CDI	Electric	Water-cooled w/ thermostat & pressure control
1998-00	40	3-cyl	59 (967)	2.99" x 2.80"	Standard	Modular CDI	Electric	Water-cooled w/ thermostat & pressure control
1995	40	2-cyl	39.3 (644)	2.99" x 2.80"	Standard	CDI	Electric	Water-cooled w/ thermostat & pressure control
1997-90	40	4-cyl	43.9 (719)	2.56" x 2.13"	Standard	CDI	Manual or Electric	Water-cooled w/ thermostat & pressure control
1990-00	50	3-cyl	59 (967)	2.99" x 2.80"	Standard	Modular CDI or CDI	Electric	Water-cooled w/ thermostat & pressure control
1990-00	60	3-cyl	59 (967)	2.99" x 2.80"	Standard	Modular CDI or CDI	Electric	Water-cooled w/ thermostat & pressure control
1990-00	75	3-cyl	84.6 (1386)	3.50" x 2.93"	Standard	Modular CDI or CDI	Electric	Water-cooled w/ thermostat & pressure control
1990-00	90	3-cyl	84.6 (1386)	3.50" x 2.93"	Standard	Modular CDI or CDI	Electric	Water-cooled w/ thermostat & pressure control
1990-00	100	4-cyl	113 (1848)	3.50" x 2.93"	Standard	Modular CDI or CDI	Electric	Water-cooled w/ thermostat & pressure control
1991-00	115	4-cyl	113 (1848)	3.50" x 2.93"	Standard	Modular CDI or CDI	Electric	Water-cooled w/ thermostat & pressure control
1993-00	125	4-cyl	113 (1848)	3.50" x 2.93"	Standard	Modular CDI or CDI	Electric	Water-cooled w/ thermostat & pressure control
1990-00	135	V-6 (60°)	121.9 (1998)	3.13" x 2.65"	Standard	CDI	Electric (turnkey)	Water-cooled w/ thermostat & pressure control
1997-00	135 Optimax	V-6 (60°)	153 (2507)	3.50" x 2.65"	Standard	Digital Inductive	Electric (turnkey)	Water-cooled w/ thermostat & pressure control
1990-96	150	V-6 (60°)	121.9 (1998)	3.13" x 2.65"	Standard	CDI	Electric (turnkey)	Water-cooled w/ thermostat & pressure control
1993-00	150 EFI	V-6 (60°)	153 (2507)	3.50" x 2.65"	Standard	CDI or CDI w/ electronic control (ECM)	Electric (turnkey)	Water-cooled w/ thermostat & pressure control
1997-00	150 Optimax	V-6 (60°)	153 (2507)	3.50" x 2.65"	Standard	Digital Inductive	Electric (turnkey)	Water-cooled w/ thermostat & pressure control
1990-00	175	V-6 (60°)	153 (2507)	3.50" x 2.65"	Standard	CDI	Electric (turnkey)	Water-cooled w/ thermostat & pressure control
1991-00	175 EFI	V-6 (60°)	153 (2507)	3.50" x 2.65"	Standard	CDI or CDI w/ electronic control (ECM)	Electric (turnkey)	Water-cooled w/ thermostat & pressure control
1990-00	200	V-6 (60°)	153 (2507)	3.50" x 2.65"	Standard	CDI	Electric (turnkey)	Water-cooled w/ thermostat & pressure control
1990-00	200 EFI	V-6 (60°)	153 (2507)	3.50" x 2.65"	Standard	CDI or CDI w/ electronic control (ECM)	Electric (turnkey)	Water-cooled w/ thermostat & pressure control
1997-00	200 Optimax	V-6 (60°)	185 (3032)	3.63" x 3.00"	Standard	Digital Inductive	Electric (turnkey)	Water-cooled w/ thermostat & pressure control
1997-00	225	V-6 (60°)	185 (3032)	3.63" x 3.00"	Standard	Modular CDI w/ computer control	Electric (turnkey)	Water-cooled w/ thermostat & pressure control
1995-00	225 EFI	V-6 (60°)	185 (3032)	3.63" x 3.00"	Standard	Modular CDI w/ computer control	Electric (turnkey)	Water-cooled w/ thermostat & pressure control
1997-00	225 Optimax	V-6 (60°)	185 (3032)	3.63" x 3.00"	Standard	Digital Inductive	Electric (turnkey)	Water-cooled w/ thermostat & pressure control
1995-00	250 EFI	V-6 (60°)	185 (3032)	3.63" x 3.00"	Standard	Modular CDI w/ computer control	Electric (turnkey)	Water-cooled w/ thermostat & pressure control
1990-94	275	V-6 (60°)	207 (3392)	3.74" x 3.15"	Standard	CDI	Electric	Water-cooled w/ thermostat & pressure control
1999-00	XR6 Classic (150)	V-6 (60°)	153 (2507)	3.50" x 2.65"	Standard	CDI	Electric	Water-cooled w/ thermostat & pressure control

Tuneup Specifications Chart

Model	Spark Plug NGK	Spark Plug Champion	Spark Plug Gap Inch(mm)	Idle Timing ° BTDC	Pick-Up	Max BDTC Cranking	WOT Timing ° BTDC	Idle Speed RPM (In Gear)	WOT RPM	Idle Mixture Screw Turns out
2.5	BPR6HS-10	RL87YC	0.040 (1.0)	-	-	-	-	900-1000	900-1000	E-clip in second groove
3.0	BPR6HS-10	RL87YC	0.040 (1.0)	-	-	-	Point Gap .012-.016 in.	900-1000	900-1000	E-clip in second groove
3.3	BPR6HS-10	RL87YC	0.040 (1.0)	-	-	-	-	900-1000	4500-5500	E-clip in second groove
4	BP7HS-10	L82YC; L81Y	0.040 (1.0)	5° BTDC	-	-	30° BTDC	800-900	4500-5500	1-1/2
5	BP7HS-10	L82YC; L81Y	0.040 (1.0)	5° BTDC	-	-	30° BTDC	800-900	4500-5500	1-1/2
8	BP8HS-15	-	0.040 (1.0)	8° BTDC	-	-	36° BTDC	675-775	5000-6000	1 to 2
9.9	BP8HS-15	-	0.060 (1.5)	7°-9° BTDC	-	-	36° BTDC	675-775	5000-6000	1-1/2
15	BP8HS-15	-	0.040 (1.0)	8° BTDC	-	-	36° BTDC	675-775	5000-6000	1 to 2
20	BP8H-N-10	-	0.040 (1.0)	-	TDC to 2° BTDC	-	25° BTDC	700-800	4500-5500	1 to 2
25	BP8H-N-10	-	0.040 (1.0)	-	TDC to 2° BTDC	-	2° BTDC	700-800	5000-6000	1 to 2
30	BP8H-N-10	-	0.040 (1.0)	0°- 6° BTDC	-	32° BTDC	22°- 28° BTDC	700-750	4500-5500	1-3/4
40 (2-cyl)	BP8H-N-10	-	0.040 (1.0)	0°- 6° BTDC	-	-	22°- 28° BTDC	700-750	5000-5500	1-1/4
40 (3-cyl)	BP8H-N-10	-	0.040 (1.0)	2°- 6° ATDC	-	24° BTDC	22° BTDC	650-750	5000-5500	1-1/4
40 (4-cyl)	BUHW-2	L78V	Surface Effect (No gap)	3°-10° ATDC	2° ATDC to 2° BTDC	32° BTDC	30° BTDC	600-700	5000-5500	1-1/4 to 1-3/4
50	BP8H-N-10	-	0.040 (1.0)	2°- 6° ATDC	-	24° BTDC	22° BTDC	600-750	5000-5500	1-1/4 to 1-3/4
60	BP8H-N-10	-	0.040 (1.0)	2°- 6° ATDC	-	24° BTDC	22°BTDC	650-750	5000-5500	1-1/4 to 1-3/4
75	BUHW-2	L78V	Surface Effect (No gap)	2° ATDC to 6° BTDC	-	20° BTDC	18° BTDC	600-750	4750-5250	1-1/4 to 1-3/4
90	BP8H-N-10	-	0.040 (1.0)	2° ATDC to 6° BTDC	-	22° BTDC	20° BTDC	625-725	5000-5500	1-1/2
100	BP8H-N-10	-	0.040 (1.0)	4° ATDC to 2° BTDC	-	24° BTDC	20° BTDC	625-725	4750-5250	1-1/2
115	BP8H-N-10	-	0.040 (1.0)	4° ATDC to 2° BTDC	-	22° BTDC	20° BTDC	625-725	4750-5250	1-3/4
125	BP8H-N-10	-	0.040 (1.0)	4° ATDC to 2° BTDC	-	22° BTDC	20° BTDC	625-725	4750-5250	1-1/2
135	BU8H	-	Surface Effect (No gap)	2°-9° ATDC	2°-9° ATDC	21° BTDC	19° BTDC	600-700	4750-5250	1-1/8 to 1-3/8
150	BU8H	-	Surface Effect (No gap)	2°-9° ATDC	2°-9° ATDC	21° BTDC	19° BTDC	600-700	5000-5500	1-1/4 to 1-3/4
150 XR6 / EFI	BU8H	-	Surface Effect (No gap)	0°-9° ATDC	0°-9° ATDC	20° BTDC	19° BTDC	625-725	5000-5600	1-1/4 to 1-3/4
150 Pro Max EFI	BU8H	-	Surface Effect (No gap)	0°-9° ATDC	0°-9° ATDC	20° BTDC	19° BTDC	600-700	5000-5500	1-1/4 to 1-3/4
175	BU8H	-	Surface Effect (No gap)	0°-9° ATDC	0°-9° ATDC	20° BTDC	19° BTDC	600-700	6200-6500	1-1/4 to 1-3/4
175 EFI	BU8H	-	Surface Effect (No gap)	0°-9° ATDC	0°-9° ATDC	20° BTDC	19° BTDC	600-700	5000-5500	1-1/4 to 1-3/4
200	BU8H	-	Surface Effect (No gap)	0°-9° ATDC	0°-9° ATDC	20° BTDC	20° BTDC	600-700	5000-5500	1-1/4 to 1-3/4
200 EFI	BU8H	-	Surface Effect (No gap)	0°-9° ATDC	0°-9° ATDC	22° BTDC	22° BTDC	600-700	5000-5600	1-1/4 to 1-3/4
200 Pro Max EFI	BU8H	-	Surface Effect (No gap)	0°-9° ATDC	0°-9° ATDC	16° BTDC	25° BTDC	600-700	6200-6500	1-1/4 to 1-3/4
225	BP28H-N-10	QL77CC	.040 in. (NGK) .035 in. (Champion)	-	-	20° BTDC	19° BTDC @ 5000 rpm 24° BTDC @ 5500 rpm	600-700	5000-5500	1-1/4 to 1-3/4
225 EFI	BP28H-N-10	QL77CC	.040 in. (NGK) .035 in. (Champion)	4°- 9° ATDC	-	-	24° BTDC @ 5000 rpm 24° BTDC @ 5500 rpm	600-700	5000-5600	1-1/4 to 1-3/4
250 EFI	BP28H-N-10	QL77CC	.040 in. (NGK) .035 in. (Champion)	4°- 9° ATDC	-	-	24° BTDC @ 5000 rpm 28° BTDC @ 5500 rpm	600-700	6200-6500	1-1/4 to 1-3/4
275	BU8H	-	Surface Effect (No gap)	5° ATDC	-	22° BTDC	20° BTDC	650-750	5000-5500	N/A (fixed jet)
135 DFI Optimax	PZFR5F-11	RC12MC4	0.040 (1.0)	Timing controlled by the ECM and is not adjustable				525-575	5000-5600	N/A
150 DFI Optimax	PZFR5F-11	RC12MC4	0.040 (1.0)	Timing controlled by the ECM and is not adjustable				525-575	5000-5600	N/A
200 DFI Optimax	PZFR5F-11	QC12GMC	0.040 (1.0)	Timing controlled by the ECM and is not adjustable				525-575	5000-5750	N/A
225 DFI Optimax	PZFR5F-11	QC12GMC	0.040 (1.0)	Timing controlled by the ECM and is not adjustable				525-575	5000-5750	N/A

Maintenance Interval Chart

Component	First 1mth/10hrs	Every 3mths/50hrs	Every 6mths/100hrs	Off Season
Bolts and nuts	T	T		T
Spark plugs		C&A		C&A
Wire harness	I	I		
Starter Motor Brush Length			I	
Ignition timing			C&A	C&A
Carburetor	C&A	C&A		C&A
Gear oil	R	R		R
Pistons, cylinder and head		De-Carbon		
Propeller	I	I		
Choke	I	I		
Fuel tank				I
Fuel strainer		I		I
Fuel hoses	I	I		I
Water pump		I		I
Handle	L	L & A		L
Clutch lever		L		L
Starter rope		I		
Tilt		A		
Neutral start interlock switch		I		
Zinc Anodes		I		I

A-Adjust
C-Clean
I-Inspect and Clean, Adjust, Lubricate or Replace
L-Lubricate
R-Replace
T-Tighten

Capacities

Model	Injection Oil Quart (Liter)	Lower Unit Oz. (ml)	Fuel Tank Gal. (Liter)
2.5	PreMix	3.0 (90)	0.375 (1.4)
3	PreMix	3.0 (90)	0.375 (1.4)
3.3	PreMix	3.0 (90)	0.375 (1.4)
4	PreMix	6.6 (195)	0.66 (2.5)
5	PreMix	6.6 (195)	0.66 (2.5)
6	PreMix	6.5 (200)	3.2 (12)
8	PreMix	6.5 (200)	3.2 (12)
9.9	PreMix	6.5 (200)	3.2 (12)
15	PreMix	6.5 (200)	6.6 (25)
20	PreMix	7.6 (225)	6.6 (25)
25	PreMix	8.8 (260)	6.6 (25)
30, 40 (2-cyl.)	1.5 (1.5)	14.9 (440)	6.6 (25)
40 (4-cyl.)	3.74 (3.54)	12.5 (370)	6.6 (25)
40, 50 (3-cyl.)	3.0 (2.8)	14.9 (440)	6.6 (25)
55, 60 (3-cyl.)	3.0 (2.8)	11.5 (340)	6.6 (25)
70, 75, 80, 90	4.0 (3.78)	22.5 (665.3)	6.6 (25)
100, 115, 125	5.6 (5.3)	22.5 (665.3)	6.6 (25)
135, 150	12 (11.4)	24.25 (717)	6.6 (25)
XR4, Magnum II	12 (11.4)	21 (625)	6.6 (25)
175, 200, 225	12 (11.4)	24.25 (717)	6.6 (25)
115, 135, 150, 200, 225 DFI	12 (11.4)	22.5 (665.3)	6.6 (25)

FUEL AND COMBUSTION 4-2
  FUEL 4-2
    OCTANE RATING 4-2
    VAPOR PRESSURE 4-2
    ALCOHOL-BLENDED FUELS 4-2
    HIGH ALTITUDE OPERATION 4-2
    RECOMMENDATIONS 4-2
  COMBUSTION 4-2
CARBURETED FUEL SYSTEM 4-3
  CARBURETION 4-3
    BASIC FUNCTIONS 4-3
    FUEL & AIR METERING 4-4
    CARBURETOR CIRCUITS 4-4
    FUEL PUMP 4-5
  TROUBLESHOOTING THE FUEL SYSTEM 4-6
    COMMON PROBLEMS 4-6
    COMBUSTION RELATED PISTON FAILURES 4-7
  2.5, 3 AND 3.3 HP 4-8
    REMOVAL & INSTALLATION 4-8
    DISASSEMBLY 4-8
    CLEANING & INSPECTION 4-10
    ASSEMBLY 4-10
  4 AND 5 HP 4-11
    REMOVAL & INSTALLATION 4-11
    DISASSEMBLY 4-12
    CLEANING & INSPECTION 4-13
    ASSEMBLY 4-13
  6, 8, 9.9, 10 AND 15 HP 4-14
    REMOVAL & INSTALLATION 4-14
    DISASSEMBLY 4-15
    CLEANING & INSPECTION 4-17
    ASSEMBLY 4-19
  20 AND 25 HP 4-20
    REMOVAL & INSTALLATION 4-20
    DISASSEMBLY 4-20
    CLEANING & INSPECTION 4-21
    ASSEMBLY 4-22
  30 AND 40 HP (2-CYLINDER) 4-23
    REMOVAL & INSTALLATION 4-23
    DISASSEMBLY 4-24
    CLEANING & INSPECTION 4-25
    ASSEMBLY 4-25
  40–125 HP 4-26
    REMOVAL & INSTALLATION 4-27
    DISASSEMBLY 4-28
    CLEANING & INSPECTION 4-30
    ASSEMBLY 4-31
  1990 135–200 HP 4-33
    REMOVAL & INSTALLATION 4-33
    DISASSEMBLY 4-35
    CLEANING & INSPECTION 4-35
    ASSEMBLY 4-36
  1991–00 135–200 HP 4-37
    REMOVAL & INSTALLATION 4-37
    DISASSEMBLY 4-38
    CLEANING & INSPECTION 4-40
    ASSEMBLY 4-40
  1991–00 275 HP 4-41
    REMOVAL & INSTALLATION 4-42
    DISASSEMBLY 4-43
    CLEANING & INSPECTION 4-44
    ASSEMBLY 4-44
  FUEL PUMP 4-45
    TESTING 4-45
    REMOVAL & INSTALLATION 4-47
    OVERHAUL 4-48
  FUEL LINES 4-49
ELECTRONIC FUEL INJECTION (EFI) 4-50
  DESCRIPTION AND OPERATION 4-50
    FUEL INJECTION BASICS 4-50
    MERCURY ELECTRONIC FUEL INJECTION 4-50
  TROUBLESHOOTING ELECTRONIC FUEL
    INJECTION 4-53
  FUEL INJECTORS 4-54
    TESTING 4-54
    REMOVAL & INSTALLATION 4-56
    CLEANING & INSPECTION 4-60
  COOLANT TEMPERATURE SENSOR 4-60
    TESTING 4-60
    REMOVAL & INSTALLATION 4-60
  AIR TEMPERATURE SENSOR 4-60
    TESTING 4-60
    REMOVAL & INSTALLATION 4-60
  MECHANICAL FUEL PUMP 4-61
    TESTING 4-61
    REMOVAL & INSTALLATION 4-61
  ELECTRIC FUEL PUMP 4-61
    TESTING 4-61
    REMOVAL & INSTALLATION 4-62
  VAPOR SEPARATOR 4-62
    REMOVAL & INSTALLATION 4-62
    CLEANING & INSPECTION 4-63
  FUEL PRESSURE REGULATOR 4-64
    REMOVAL & INSTALLATION 4-64
  THROTTLE POSITION SENSOR 4-65
    TESTING 4-65
    REMOVAL & INSTALLATION 4-65
    ADJUSTMENT 4-65
  DETONATION SENSOR AND MODULE 4-66
    TESTING 4-66
OPTIMAX DIRECT FUEL INJECTION (DFI) 4-66
  DESCRIPTION AND OPERATION 4-66
    OPTIMAX COMPONENTS 4-68
  TROUBLESHOOTING THE OPTIMAX FUEL INJECTION
    SYSTEM 4-69
    WITHOUT THE DIGITAL DIAGNOSTIC TERMINAL
      (DDT) 4-69
    WITH THE DIGITAL DIAGNOSTIC
      TERMINAL (DDT) 4-70
  AIR COMPRESSOR 4-70
    TESTING 4-70
    REMOVAL & INSTALLATION 4-70
  AIR PRESSURE REGULATOR 4-72
    REMOVAL & INSTALLATION 4-72
  AIR TEMPERATURE SENSOR 4-72
    TESTING 4-72
    REMOVAL & INSTALLATION 4-72
  CRANKSHAFT POSITION SENSOR (CPS) 4-72
    TESTING 4-72
    REMOVAL & INSTALLATION 4-72
  THROTTLE POSITION SENSOR (TPS) 4-73
    TESTING 4-73
    REMOVAL & INSTALLATION 4-73
  DIRECT INJECTOR 4-73
    TESTING 4-73
    REMOVAL & INSTALLATION 4-73
    CLEANING & INSPECTION 4-74
  ELECTRONIC CONTROL MODULE (ECM) 4-74
    REMOVAL & INSTALLATION 4-74
  FUEL INJECTOR 4-74
    TESTING 4-74
    REMOVAL & INSTALLATION 4-74
    CLEANING & INSPECTION 4-74
  FUEL PRESSURE REGULATOR 4-75
    REMOVAL & INSTALLATION 4-75
    CLEANING & INSPECTION 4-75
  FUEL PUMP 4-75
    TESTING 4-75
  FUEL RAIL 4-76
    REMOVAL & INSTALLATION 4-76
    CLEANING & INSPECTION 4-76
  MANIFOLD ABSOLUTE PRESSURE (MAP)
    SENSOR 4-76
    TESTING 4-76
    REMOVAL & INSTALLATION 4-76
  SHIFT INTERRUPT SWITCH 4-76
    TESTING 4-76
    REMOVAL & INSTALLATION 4-76
  THROTTLE PLATE ASSEMBLY 4-77
    REMOVAL & INSTALLATION 4-77
  TRACKER VALVE 4-77
    REMOVAL & INSTALLATION 4-77
  VAPOR SEPARATOR 4-77
    REMOVAL & INSTALLATION 4-77
    DISASSEMBLY 4-77
    CLEANING & INSPECTION 4-77
    ASSEMBLY 4-78
  WATER TEMPERATURE SENSOR 4-78
    TESTING 4-78
    REMOVAL & INSTALLATION 4-78
SPECIFICATIONS CHART
  WME CARBURETOR SPECIFICATIONS 4-26

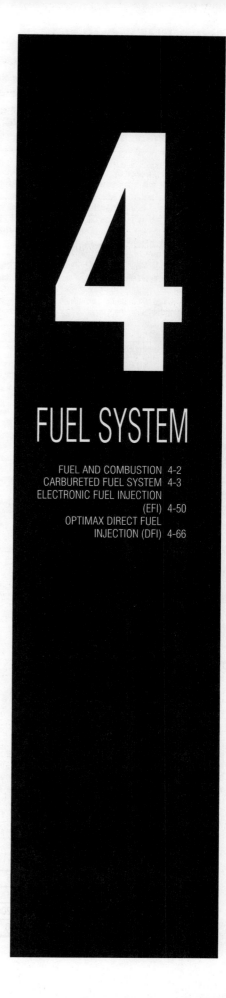

4

FUEL SYSTEM

FUEL AND COMBUSTION 4-2
CARBURETED FUEL SYSTEM 4-3
ELECTRONIC FUEL INJECTION
(EFI) 4-50
OPTIMAX DIRECT FUEL
INJECTION (DFI) 4-66

4-2 FUEL SYSTEM

FUEL AND COMBUSTION

Fuel

Fuel recommendations have become more complex as the chemistry of modern gasoline changes. The major driving force behind the changes in gasoline chemistry is the search for additives to replace lead as an octane booster and lubricant. These new additives are governed by the types of emissions they produce in the combustion process. Also, the replacement additives do not always provide the same level of combustion stability, making a fuel's octane rating less meaningful.

In the 1960's and 1970's, leaded fuel was common. The lead served two functions. First, it served as an octane booster (combustion stabilizer) and second, in 4-stroke engines, it served as a valve seat lubricant. For 2-stroke engines, the primary benefit of lead was to serve as a combustion stabilizer. Lead served very well for this purpose, even in high heat applications.

Today, all lead has been removed from the refining process. This means that the benefit of lead as an octane booster has been eliminated. Several substitute octane boosters have been introduced in the place of lead. While many are adequate in an automobile engines, most do not perform nearly as well as lead did, even though the octane rating of the fuel is the same.

OCTANE RATING

A fuel's octane rating is a measurement of how stable the fuel is when heat is introduced. Octane rating is a major consideration when deciding whether a fuel is suitable for a particular application. For example, in an engine, we want the fuel to ignite when the spark plug fires and not before, even under high pressure and temperatures. Once the fuel is ignited, it must burn slowly and smoothly, even though heat and pressure are building up while the burn occurs. The unburned fuel should be ignited by the traveling flame front, not by some other source of ignition, such as carbon deposits or the heat from the expanding gasses. A fuel's octane rating is known as a measurement of the fuel's anti-knock properties (ability to burn without exploding).

Usually a fuel with a higher octane rating can be subjected to a more severe combustion environment before spontaneous or abnormal combustion occurs. To understand how two gasoline samples can be different, even though they have the same octane rating, we need to know how octane rating is determined.

The American Society of Testing and Materials (ASTM) has developed a universal method of determining the octane rating of a fuel sample. The octane rating you see on the pump at a gasoline station is known as the pump octane number. Look at the small print on the pump. The rating has a formula. The rating is determined by the R+M/2 method. This number is the average of the research octane reading and the motor octane rating.

- The Research Octane Rating is a measure of a fuel's anti-knock properties under a light load or part throttle conditions. During this test, combustion heat is easily dissipated.
- The Motor Octane Rating is a measure of a fuel's anti-knock properties under a heavy load or full throttle conditions, when heat buildup is at maximum.

Because a 2-stroke engine has a power stroke every revolution, with heat buildup every revolution, it tends to respond more to the motor octane rating of the fuel than the research octane rating. Therefore, in an outboard motor, the motor octane rating of the fuel is the best indication of how it will perform.

VAPOR PRESSURE

Fuel vapor pressure is a measure of how easily a fuel sample evaporates. Many additives used in gasoline contain aromatics. Aromatics are light hydrocarbons distilled off the top of a crude oil sample. They are effective at increasing the research octane of a fuel sample but can cause vapor lock (bubbles in the fuel line) on a very hot day. If you have an inconsistent running engine and you suspect vapor lock, use a piece of clear fuel line to look for bubbles, indicating that the fuel is vaporizing.

One negative side effect of aromatics is that they create additional combustion products such as carbon and varnish. If your engine requires high octane fuel to prevent detonation, de-carbon the engine more frequently with an internal engine cleaner to prevent ring sticking due to excessive varnish buildup.

ALCOHOL-BLENDED FUELS

When the Environmental Protection Agency mandated a phase-out of the leaded fuels in January of 1986, fuel suppliers needed an additive to improve the octane rating of their fuels. Although there are multiple methods currently employed, the addition of alcohol to gasoline seems to be favored because of its favorable results and low cost. Two types of alcohol are used in fuel today as octane boosters, methanol (wood alcohol) or ethanol (grain alcohol).

When used as a fuel additive, alcohol tends to raise the research octane of the fuel, so these additives will have limited benefit in an outboard motor. There are, however, some special considerations due to the effects of alcohol in fuel.

- Since alcohol contains oxygen, it replaces gasoline without oxygen content and tends to cause the air/fuel mixture to become leaner.
- On older outboards, the leaching affect of alcohol will, in time, cause fuel lines and plastic components to become brittle to the point of cracking. Unless replaced, these cracked lines could leak fuel, increasing the potential for hazardous situations.
- When alcohol blended fuels become contaminated with water, the water combines with the alcohol then settles to the bottom of the tank. This leaves the gasoline (and the oil for models using premix) on a top layer.

➡**Modern outboard fuel lines and plastic fuel system components have been specially formulated to resist alcohol leaching effects.**

HIGH ALTITUDE OPERATION

At elevated altitudes there is less oxygen in the atmosphere than at sea level. Less oxygen means lower combustion efficiency and less power output. As a general rule, power output is reduced three percent for every thousand feet above sea level.

On carbureted engines, re-jetting for high altitude does not restore lost power, it simply corrects the air-fuel ratio for the reduced air density and makes the most of the remaining available power. The most important thing to remember when re-jetting for high altitude is to reverse the jetting when return to sea level. If the jetting is left lean when you return to sea level conditions, the correct air/fuel ratio will not be achieved and possible powerhead damage may occur.

RECOMMENDATIONS

According to the fuel recommendations that come with your outboard, there is no engine in the product line that requires more than 87 octane. Most Mercury/Mariner engines need only 87 octane or less. An 89 octane rating generally means middle grade unleaded. Premium unleaded is more stable under severe conditions but also produces more combustion products. Therefore, when using premium unleaded, more frequent de-carboning is necessary.

Combustion

▶ See Figure 1

Unlike a 4-stroke engine, a 2-stroke engine has a power stroke every revolution of the crankshaft. Therefore, the 2-stroke engine has twice as many power strokes for any given RPM. If the displacement of the two types of engines is identical, then the 2-stroke engine has to dissipate twice as much heat as the 4-stroke engine.

In such a high heat environment, the fuel must be very stable to avoid detonation. If any parameters affecting combustion change suddenly (the engine runs lean for example), uncontrolled heat buildup will occurs very rapidly.

The combustion process is affected by several interrelated factors. This means that when one factor is changed, the other factors also must be changed to maintain the same controlled burn and level of combustion stability.

- Compression—determines the level of heat buildup in the cylinder when the air-fuel mixture is compressed. As compression increases, so does the potential for heat buildup
- Ignition Timing—determines when the gasses will start to expand in relation to the motion of the piston. If the ignition timing is too advanced, gasses will be ignited and begin to expand too soon, such as they would during preignition. The motion of the piston opposes the expansion of the gasses, resulting in extremely high combustion chamber pressures and heat. If the ignition timing is retarded, the gases are ignited later in relation to piston position. This means that the piston has already traveled back down the bore toward the bottom of the cylinder, resulting in less usable power.
- Fuel Mixture—determines how efficient the burn will be. A rich mixture burns slower than a lean one. If the mixture is too lean, it can't become explosive. The slower the burn, the cooler the combustion chamber, because pressure buildup is gradual.
- Fuel Quality (Octane Rating)—determines how much heat is necessary to ignite the mixture. Once the burn is in progress, heat is on the rise. The unburned poor quality fuel is ignited all at once by the rising heat instead of burning gradually as a flame front of the burn passing by. This action results in detonation (pinging).

There are two types of abnormal combustion—preignition and detonation.

- Preignition—occurs when the air-fuel mixture is ignited by some other incandescent source other than the correctly timed spark from the spark plug.

FUEL SYSTEM 4-3

- Detonation—occurs when excessive heat and or pressure ignites the air/fuel mixture rather than the spark plug. The burn becomes explosive.

In general, anything that can cause abnormal heat buildup can be enough to push an engine over the edge to abnormal combustion, if any of the four basic factors previously discussed are already near the danger point, for example, excessive carbon buildup raises the compression and retains heat as glowing embers.

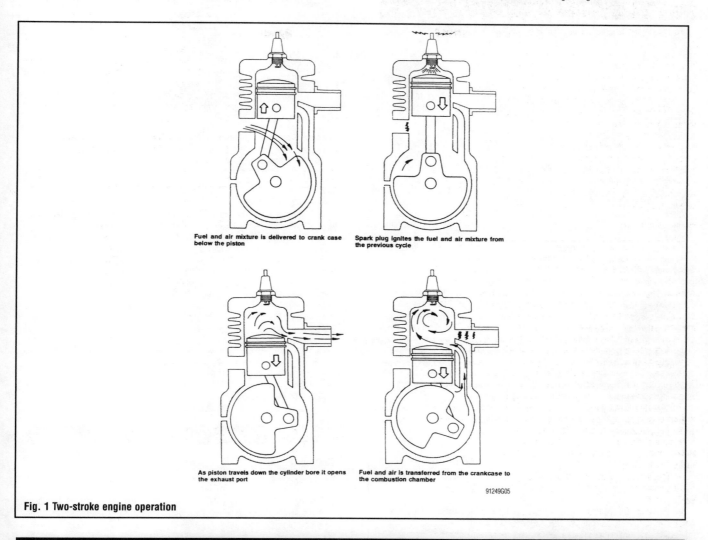

Fig. 1 Two-stroke engine operation

CARBURETED FUEL SYSTEM

Carburetion

BASIC FUNCTIONS

♦ See Figure 2

Traditional carburetor theory often involves a number of laws and principles. The diagram illustrates several carburetor basics. If you blow air across a straw inserted into a container of liquid, a pressure drop is created in the straw column. As the liquid in the column is expelled, an atomized mixture (air and fuel droplets) is created. In a carburetor this is mostly air and a little fuel.

The actual ratio of air to fuel differs with engine conditions but is usually from 15 parts air to one part fuel at optimum cruise to as little as 7 parts air to one part fuel at full choke.

Using our example, what if the top of the container is covered and sealed around the straw, what will happen? No flow. This is typical of a clogged carburetor bowl vent. If the base of the straw is clogged or restricted what will happen? No flow or low flow. This represents a clogged main jet. If the liquid in the glass is lowered and you blow through the straw with the same force what will happen? Not as much fuel will flow. A lean condition occurs. If the fuel level is raised and you blow again at the same velocity what happens? The result is a richer mixture.

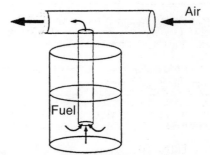

Fig. 2 If you blow air across a straw inserted into a container of liquid, a pressure drop is created in the straw column. As the liquid in the column is expelled, an atomized mixture (air and liquid droplets) is created

4-4 FUEL SYSTEM

FUEL & AIR METERING

The carburetor is merely a metering device for mixing fuel and air in the proper proportions for efficient engine operation. At idle speed, an outboard engine requires a mixture of about 8 parts air to 1 part fuel. At high speed or under heavy duty service, the mixture may change to as much as 12 parts air to 1 part fuel.

Float Systems

♦ See Figure 3

A small chamber in the carburetor serves as a fuel reservoir. A float valve admits fuel into the reservoir to replace the fuel consumed by the engine. If the carburetor has more than one reservoir, the fuel level in each reservoir (chamber) is controlled by identical float systems.

Fuel level in each chamber is extremely critical and must be maintained accurately. Accuracy is obtained through proper adjustment of the floats. This adjustment will provide a balanced metering of fuel to each cylinder at all speeds.

Following the fuel through its course, from the fuel tank to the combustion chamber of the cylinder, will provide an appreciation of exactly what is taking place. In order to start the engine, the fuel must be moved from the tank to the carburetor by a squeeze bulb installed in the fuel line. This action is necessary because the fuel pump does not have sufficient pressure to draw fuel from the tank during cranking before the engine starts.

The fuel for some small horsepower units is gravity fed from a tank mounted at the rear of the powerhead. Even with the gravity feed method, a small fuel pump may be an integral part of the carburetor.

After the engine starts, the fuel passes through the pump to the carburetor. All systems have some type of filter installed somewhere in the line between the tank and the carburetor. Many units have a filter as an integral part of the carburetor.

At the carburetor, the fuel passes through the inlet passage to the needle and seat and then into the float chamber (reservoir). A float in the chamber rides up and down on the surface of the fuel. After fuel enters the chamber and the level rises to a predetermined point, a tang on the float closes the inlet needle and the flow entering the chamber is cut off. When fuel leaves the chamber as the engine operates, the fuel level drops and the float tang allows the inlet needle to move off its seat and fuel once again enters the chamber. In this manner, a constant reservoir of fuel is maintained in the chamber to satisfy the demands of the engine at all speeds.

A fuel chamber vent hole is located near the top of the carburetor body to permit atmospheric pressure to act against the fuel in each chamber. This pressure assures an adequate fuel supply to the various operating systems of the powerhead.

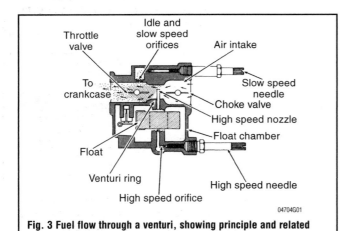

Fig. 3 Fuel flow through a venturi, showing principle and related parts controlling intake and outflow

Air/Fuel Mixture

♦ See Figure 4

A suction effect is created each time the piston moves upward in the cylinder. This suction draws air through the throat of the carburetor. A restriction in the throat, called a venturi, controls air velocity and has the effect of reducing air pressure at this point.

The difference in air pressures at the throat and in the fuel chamber, causes the fuel to be pushed out of metering jets extending down into the fuel chamber. When the fuel leaves the jets, it mixes with the air passing through the venturi. This fuel/air

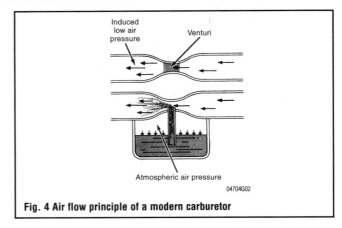

Fig. 4 Air flow principle of a modern carburetor

mixture should then be in the proper proportion for burning in the cylinders for maximum engine performance.

In order to obtain the proper air/fuel mixture for all engine speeds, some models have high and low speed jets. These jets have adjustable needle valves which are used to compensate for changing atmospheric conditions. In almost all cases, the high-speed circuit has fixed high-speed jets and are not adjustable.

A throttle valve controls the flow of air/fuel mixture drawn into the combustion chambers. A cold powerhead requires a richer fuel mixture to start and during the brief period it is warming to normal operating temperature. A choke valve is placed ahead of the metering jets and venturi. As this valve begins to close, the volume of air intake is reduced, thus enriching the mixture entering the cylinders.

When this choke valve is fully closed, a very rich fuel mixture is drawn into the cylinders.

The throat of the carburetor is usually referred to as the barrel. Carburetors with single, double or four barrels have individual metering jets, needle valves, throttle and choke plates for each barrel. Single and two barrel carburetors are fed by a single float and chamber.

CARBURETOR CIRCUITS

The following section illustrates the circuit functions and locations of a typical marine carburetor.

Starting Circuit

♦ See Figure 5

The choke plate is closed, creating a partial vacuum in the venturi. As the piston rises, negative pressure in the crankcase draws the rich air-fuel mixture from the float bowl into the venturi and on into the engine.

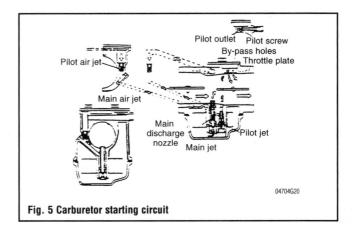

Fig. 5 Carburetor starting circuit

Low Speed Circuit

♦ See Figure 6

Zero–one-eighth throttle, when the pressure in the crankcase is lowered, the air-fuel mixture is discharged into the venturi through the pilot outlet because the throttle

FUEL SYSTEM   4-5

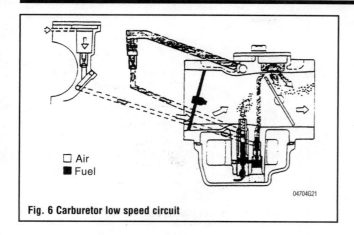

Fig. 6 Carburetor low speed circuit

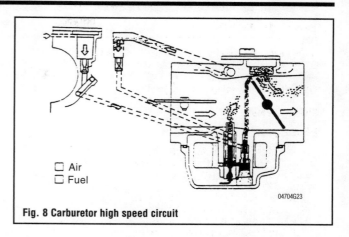

Fig. 8 Carburetor high speed circuit

plate is closed. No other outlets are exposed to low venturi pressure. The fuel is metered by the pilot jet. The air is metered by the pilot air jet. The combined air-fuel mixture is regulated by the pilot air screw.

Mid-Range Circuit

♦ See Figure 7

One-eighth–three-eighths throttle, as the throttle plate continues to open, the air-fuel mixture is discharged into the venturi through the bypass holes. As the throttle plate uncovers more bypass holes, increased fuel flow results because of the low pressure in the venturi. Depending on the model, there could be two, three or four bypass holes.

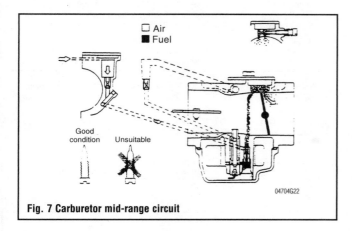

Fig. 7 Carburetor mid-range circuit

High Speed Circuit

♦ See Figure 8

Three-eighths–wide-open throttle, as the throttle plate moves toward wide open, we have maximum air flow and very low pressure. The fuel is metered through the main jet and is drawn into the main discharge nozzle. Air is metered by the main air jet and enters the discharge nozzle, where it combines with fuel. The mixture atomizes, enters the venturi and is drawn into the engine.

FUEL PUMP

Basic Functions

♦ See Figures 9, 10 and 11

A fuel pump is a basic mechanical device that utilizes crankcase positive and negative pressures to pump fuel from the fuel tank to the carburetors.

This device contains a flexible diaphragm and two check valves (flappers or fingers) that control flow. As the piston goes up, crankcase pressure drops (negative pressure) and the inlet valve opens, pulling fuel from the tank. As the piston nears TDC, pressure in the pump area is neutral (atmospheric pressure). At this point both valves are closed. As the piston comes down, pressure goes up (positive pressure)

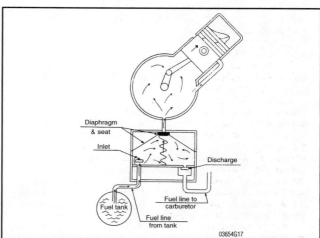

Fig. 9 Simplified drawing of the fuel pump with the powerhead piston on the upward stroke. Notice the position of the diaphragm, the inlet disc is open, and the discharge disc is closed. The springs to preload the discs are not shown for clarity

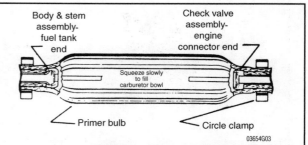

Fig. 10 Major parts of a typical fuel line squeeze bulb, used to prime the system and deliver fuel to the carburetor until the engine is operating and the pump/s can deliver fuel on their own

Fig. 11 A squeeze bulb is used on most carbureted applications to prime the fuel pump prior to initial start

and the fuel is pushed toward the carburetor bowl by the diaphragm through the now open outlet valve.

This is a reliable method to move fuel but can have several problems. Sometimes an engine backfire can rupture the diaphragm. The diaphragm and valves are moving parts subject to wear. The flexibility of the diaphragm material can go away, reducing or stopping flow. Rust or dirt can hang a valve open and reduce or stop fuel flow.

Most pumps consist of a diaphragm, two similar spring loaded disc valves, one for inlet (suction) and the other for outlet (discharge) and a small opening leading directly into the crankcase bypass. The suction and compression created, as the piston travels up and down in the cylinder, causes the diaphragm to flex.

As the piston moves upward, the diaphragm will flex inward displacing volume on its opposite side to create suction. This suction will draw liquid fuel in through the inlet disc valve.

When the piston moves downward, compression is created in the crankcase. This compression causes the diaphragm to flex in the opposite direction. This action causes the discharge valve disc to lift off its seat. Fuel is then forced through the discharge valve into the carburetor.

The pump has the capacity to lift fuel two feet and deliver approximately five gallons per hour at 4 psi pressure.

Problems with the fuel pump are limited to possible leaks in the flexible neoprene suction lines, a punctured diaphragm, air leaks between sections of the pump assembly or possibly from the disc valves not seating properly.

The pump is activated by one cylinder. If this cylinder indicates a wet fouled condition, as evidenced by a wet fouled spark plug, be sure to check the fuel pump diaphragm for possible puncture or leakage.

Mercury Fuel Pumps

The fuel pump used on the Mikuni round bowl side-draft carburetor cannot be serviced and is considered a "disposable" pump.

The integral fuel pumps on the "Keihin" single float carburetor and the Walbro "WM" double float carburetor are rebuildable and replacement parts are available.

The fuel pump used on most of the other powerheads is a diaphragm displacement type. This pump is attached to the cylinder bypass and is operated by crankcase impulses. A hand-operated squeeze bulb is installed in the fuel line to fill the fuel pump and carburetor with fuel before the engine starts. After engine start, the pump is able to supply an adequate supply of fuel to the carburetor to meet engine demands under all speeds and conditions.

Troubleshooting the Fuel System

♦ See Figures 12 and 13

Troubleshooting fuel systems requires the same techniques used in other areas. A thorough, systematic approach to troubleshooting will pay big rewards. Build your troubleshooting checklist, with the most likely offenders at the top. Use your experience to adjust your list for local conditions. Everyone has been tempted to jump into the carburetor on a vague hunch. Pause a moment and review the facts when this urge occurs.

In order to accurately troubleshoot a carburetor or fuel system problem, you must first verify that the problem is fuel related. Many symptoms can have several different possible causes. Be sure to eliminate mechanical and electrical systems as the potential fault. Carburetion is the number one cause of most engine problems but there are other possibilities.

One of the toughest tasks with a fuel system is the actual troubleshooting. Several tools are at your disposal for making this process very simple. A timing light works well for observing carburetor spray patterns. Look for the proper amount of fuel and for proper atomization in the two fuel outlet areas (main nozzle and bypass holes). The strobe effect of the lights helps you see in detail the fuel being drawn through the throat of the carburetor. On multiple carburetor engines, always attach the timing light to the cylinder you are observing so the strobe doesn't change the appearance of the patterns. If you need to compare two cylinders, change the timing light hookup each time you observe a different cylinder.

Pressure testing fuel pump output can determine whether the fuel spray is adequate and if the fuel pump diaphragms are functioning correctly. A pressure gauge placed between the fuel pumps and the carburetors will test the entire fuel delivery system. Normally a fuel system problem will show up at high speed where the fuel demand is the greatest. A common symptom of a fuel pump output problem is surging at wide open throttle but normal operation at slower speeds. To check the fuel pump output, install the pressure gauge and accelerate the engine to wide open throttle. Observe the pressure gauge needle. It should always swing up to some value between 5–6 psi and remain steady. This reading would indicate a system that is functioning properly.

If the needle gradually swings down toward zero, fuel demand is greater than the fuel system can supply. This reading isolates the problem to the fuel delivery system (fuel tank or line). To confirm this, an auxiliary tank should be installed and the engine re-tested. Be aware that a bad anti-siphon valve on a built-in tank can create enough restriction to cause a lean condition and serious engine damage.

If the needle movement becomes erratic, suspect a ruptured diaphragm in the fuel pump.

A quick way to check for a ruptured fuel pump diaphragm is while the engine is at idle speed, to squeeze the primer bulb and hold steady firm pressure on it. If the diaphragm is ruptured, this will cause a rough running condition because of the extra fuel passing through the diaphragm into the crankcase. After performing this test you should check the spark plugs for cylinders that the fuel pump supplies. If the spark plugs are OK but the fuel pumps are still suspected, you should remove the fuel pumps and completely disassemble them. Rebuild or replace the pumps as needed.

To check the boat's fuel system for a restriction, install a vacuum gauge in the line before the fuel pump. Run the engine under load at wide open throttle to get a reading. Vacuum should read no more than 4.5 in. Hg (15.2 kPa) for engines up to and including 200hp and should not exceed 6.0 in. Hg (20.2 kPa) for engines greater than 200hp.

To check for air entering the fuel system, install a clear fuel hose between the fuel screen and fuel pump. If air is in the line, check all fittings back to the boat's fuel tank.

Spark plug tip appearance is a good indication of combustion efficiency. The tip should be a light tan. A White insulator or small beads on the insulator indicate too much heat. A dark or oil fouled insulator indicates incomplete combustion. To properly read spark plug tip appearance, run the engine at the RPM you are testing for about 15 second and then immediately turn the engine OFF without changing the throttle position.

Reading spark plug tip appearance is also the proper way to test jet verifications in high altitude.

COMMON PROBLEMS

Fuel Delivery

♦ See Figure 14

Many times fuel system troubles are caused by a plugged fuel filter, a defective fuel pump or by a leak in the line from the fuel tank to the fuel pump. A defective choke may also cause problems. would you believe, a majority of starting troubles which are traced to the fuel system are the result of an empty fuel tank or aged sour fuel.

Fig. 12 One of the first things to check when troubleshooting a fuel problem is to make sure the fuel tank vent is open . . .

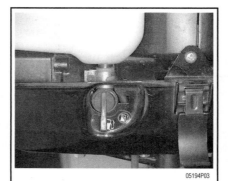

Fig. 13 . . . and that the fuel petcock is also in the open or run position

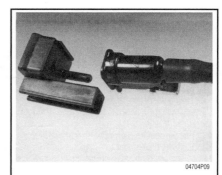

Fig. 14 An excellent way of protecting fuel hoses against contamination is an end cap filter

Sour Fuel

♦ See Figure 15

Under average conditions (temperate climates), fuel will begin to break down in about four months. A gummy substance forms in the bottom of the fuel tank and in other areas. The filter screen between the tank and the carburetor and small passages in the carburetor will become clogged. The gasoline will begin to give off an odor similar to rotten eggs. Such a condition can cause the owner much frustration, time in cleaning components and the expense of replacement or overhaul parts for the carburetor.

Even with the high price of fuel, removing gasoline that has been standing unused over a long period of time is still the easiest and least expensive preventative maintenance possible. In most cases, this old gas can be used without harmful effects in an automobile using regular gasoline.

A gasoline preservative for 2 cycle engines will keep the fuel fresh for up to twelve months. These products are available in most areas under various trade names.

Fig. 15 A gasoline stabilizer and conditioner, such as Sta-Bil may be used to prevent the fuel from "souring" for up to twelve full months

Choke Problems

When the engine is hot, the fuel system can cause starting problems. After a hot engine is shut down, the temperature inside the fuel bowl may rise to 200°F and cause the fuel to actually boil. All carburetors are vented to allow this pressure to escape to the atmosphere. However, some of the fuel may percolate over the high-speed nozzle.

If the choke should stick in the open position, the engine will be hard to start. If the choke should stick in the closed position, the engine will flood, making it very difficult to start.

In order for this raw fuel to vaporize enough to burn, considerable air must be added to lean out the mixture. Therefore, the only remedy is to remove the spark plugs, ground the leads, crank the powerhead through about ten revolutions, clean the plugs, reinstall the plugs and start the engine.

If the needle valve and seat assembly is leaking, an excessive amount of fuel may enter the reed housing in the following manner. After the powerhead is shut down, the pressure left in the fuel line will force fuel past the leaking needle valve. This extra fuel will raise the level in the fuel bowl and cause fuel to overflow into the reed housing.

A continuous overflow of fuel into the reed housing may be due to a sticking inlet needle or to a defective float, which would cause an extra high level of fuel in the bowl and overflow into the reed housing.

Rough Engine Idle

If an engine does not idle smoothly, the most reasonable approach to the problem is to perform a tune-up to eliminate such areas as:
- Defective points
- Faulty spark plugs
- Timing out of adjustment

Other problems that can prevent an engine from running smoothly include:
- An air leak in the intake manifold
- Uneven compression between the cylinders
- Sticky or broken reeds

Of course any problem in the carburetor affecting the air/fuel mixture will also prevent the engine from operating smoothly at idle speed. These problems usually include:
- Too high a fuel level in the bowl
- A heavy float
- Leaking needle valve and seat
- Defective automatic choke
- Improper adjustments for idle mixture or idle speed

Excessive Fuel Consumption

Excessive fuel consumption can be the result of any one of four conditions or a combination of all.
- Inefficient engine operation.
- Faulty condition of the hull, including excessive marine growth.
- Poor boating habits of the operator.
- Leaking or out of tune carburetor.

If the fuel consumption suddenly increases over what could be considered normal, then the cause can probably be attributed to the engine or boat and not the operator.

Marine growth on the hull can have a very marked effect on boat performance. This is why sail boats always try to have a haul-out as close to race time as possible.

While you are checking the bottom, take note of the propeller condition. A bent blade or other damage will definitely cause poor boat performance.

If the hull and propeller are in good shape, then check the fuel system for possible leaks. Check the line between the fuel pump and the carburetor while the engine is running and the line between the fuel tank and the pump when the engine is not running. A leak between the tank and the pump many times will not appear when the engine is operating, because the suction created by the pump drawing fuel will not allow the fuel to leak. Once the engine is turned off and the suction no longer exists, fuel may begin to leak.

If a minor tune-up has been performed and the spark plugs, points and timing are properly adjusted, then the problem most likely is in the carburetor and an overhaul is in order.

Check the needle valve and seat for leaking. Use extra care when making any adjustments affecting the fuel consumption, such as the float level or automatic choke.

Engine Surge

If the engine operates as if the load on the boat is being constantly increased and decreased, even though an attempt is being made to hold a constant engine speed, the problem can most likely be attributed to the fuel pump or a restriction in the fuel line between the tank and the carburetor.

COMBUSTION RELATED PISTON FAILURES

♦ See Figure 16

When an engine has a piston failure due to abnormal combustion, fixing the mechanical portion of the engine is the easiest part of the equation. The hard part is determining what caused the problem in order to prevent a repeat failure. Think back to the four basic areas that affect combustion to find the cause of the failure.

Since you probably removed the cylinder head, inspect the failed piston and look for excessive deposit buildup that could raise compression or retain heat in the combustion chamber. Statically check the wide open throttle timing. Be sure that the timing is not over advanced. It is a good idea to seal these adjustments with paint to detect tampering.

Fig. 16 This burned piston is typical of a combustion relate failure. The combustion chamber temperature got so hot that it melted the top of the piston (hole in the top of the piston)

4-8 FUEL SYSTEM

Look for a fuel restriction that could cause the engine to run lean. Don't forget to check the fuel pump, fuel tank and lines, especially if a built in tank is used. Be sure to check the anti-siphon valve on built in tanks. If everything else looks good, the final possibility is poor quality fuel.

2.5, 3 and 3.3 HP

REMOVAL & INSTALLATION

♦ See accompanying illustrations

1. Remove the screws securing the knobs at the end of the choke and throttle levers. Remove the two screws securing the rectangular intake cover to the carburetor body. Lift the cover free.

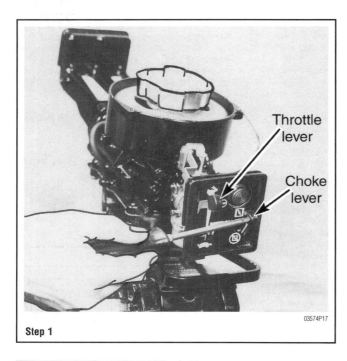
Step 1

Step 3

➡ It is not necessary to disconnect the wires attached to the cover. Simply move the cover to one side out of the way.

2. Close the fuel valve between the fuel tank and the carburetor. Squeeze the wire-type hose clamp on the fuel line enough for the hose to slip free of the inlet fuel fitting.
3. Loosen the clamp screw and remove the carburetor from the powerhead.

To install:
4. Position a new carburetor-to-power-head seal in the recess of the carburetor throat.
5. Slide the carburetor onto the crank-case cover and secure it in place with the clamp screw.
6. Connect the fuel line to the fuel inlet fitting. Open the fuel valve between the fuel tank and the carburetor.
7. Install the front cover and the knobs onto the choke and throttle levers.

DISASSEMBLY

♦ See accompanying illustrations

1. Work the carburetor-to-powerhead seal out of the recess in the carburetor throat. Discard the seal.
2. Back out the two Phillips head screws securing the float bowl to the carburetor.

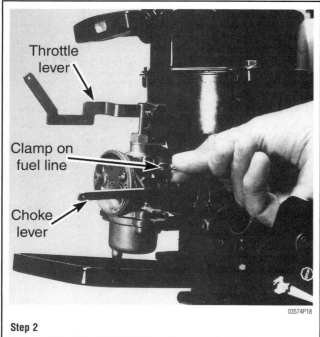
Step 2

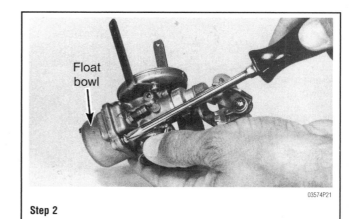
Step 2

FUEL SYSTEM   4-9

3. Remove the float from the carburetor.
4. Slide the hinge pin free and then remove the float hinge and the inlet needle.
5. Lift off and discard the float bowl gasket.
6. Unscrew the main jet from the center of the main nozzle.
7. Use the proper size wrench and remove the main nozzle.
8. Rotate the idle speed screw clockwise and count the number of turns necessary to seat it lightly. After the number of turns has been noted and recorded somewhere, back out the idle speed screw. Slide the spring free of the screw.
9. Loosen the retainer nut with the proper size wrench. Rotate the nut several turns counterclockwise but do not remove the nut.

10. Unscrew the mixing chamber cover with a pair of pliers. Once the cover is free, lift off the throttle valve assembly. Disconnect the throttle lever from the bracket.
11. Compress the spring in the throttle valve assembly to allow the throttle cable end to clear the recess in the base of the throttle valve and to slide down the slot.
12. Disassemble the throttle valve consisting of the throttle valve, spring, jet needle with an E-clip on the second groove, jet retainer and throttle cable end.

➡ **It is not necessary to remove the E-clip from the jet needle, unless replacement is required or if the powerhead is to be operated at a significantly different elevation.**

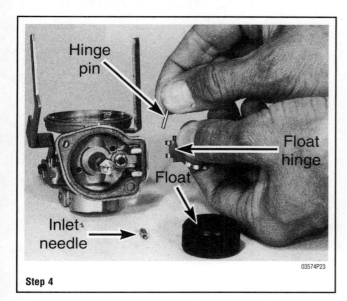

Step 4

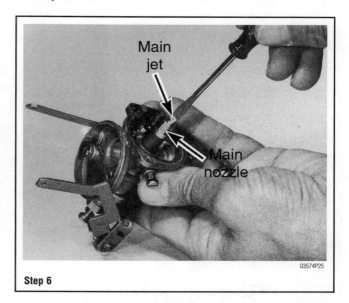

Step 6

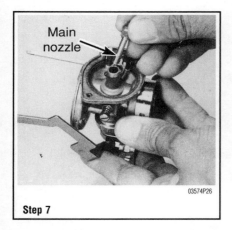

Step 7

Step 8

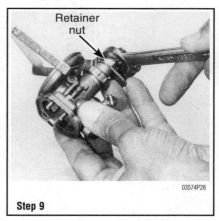

Step 9

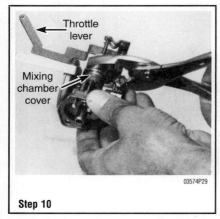

Step 10

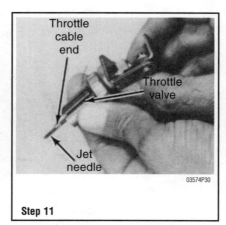

Step 11

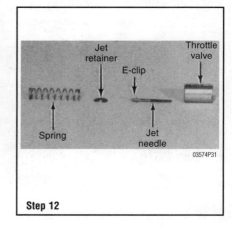

Step 12

4-10 FUEL SYSTEM

CLEANING & INSPECTION

▶ See Figure 17

Never dip rubber parts, plastic parts, diaphragms or pump plungers in carburetor cleaner, because they tend to absorb liquid and expand. These parts should be cleaned only in solvent and then blown dry with compressed air immediately.

Place all of the metal parts in a screen-type tray and dip them in carburetor cleaner until they appear completely clean—free of gum and varnish which accumulates from stale fuel. Blow the parts dry with compressed air.

Blow out all of the passageways in the carburetor with compressed air. Check all of parts and passageways to be sure they are clear and not clogged with any deposits.

Never use a piece of wire or any type of pointed instrument to clean drilled passages or calibrated holes in a carburetor.

Carefully inspect the casting for cracks, stripped threads or plugs for any sign of leakage. Inspect the float hinge in the hinge pin area for wear and the float for any sign of leakage.

Examine the inlet needle for wear. If there is any evidence of wear, the needle must be replaced.

Always replace any worn or damaged parts. A carburetor service kit is available at modest cost from the local Mercury/Mariner dealer. The kit will contain all necessary parts to perform the usual carburetor overhaul.

2. Thread the spring over the end of the throttle cable and insert the cable into the retainer end of the throttle valve. Compress the spring and at the same time guide the cable end through the slot until the end locks into place in the recess. Position the assembled throttle valve in such a manner to permit the slot to slide over the alignment pin while the throttle valve is lowered into the carburetor. Attach the throttle lever to the bracket on top of the throttle valve assembly.

3. Carefully tighten the mixing chamber cover with a pair of pliers.

4. Position the bracket to allow the throttle to just clear the front of the carburetor and then tighten the retainer nut.

5. Slide the spring onto the idle speed screw and then start to thread the screw into the carburetor body. Slowly rotate the screw until it seats lightly. From this position, back out the screw the same number of complete turns as noted during removal. If the count was lost, back the screw out two turns as a preliminary adjustment. A fine idle adjustment will be made later.

6. Thread the main nozzle into the carburetor body and tighten it just "snug" with the proper size wrench. do not overtighten the nozzle.

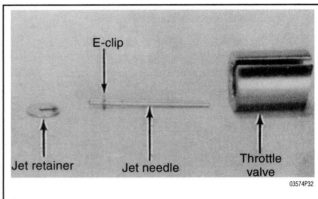

Step 1

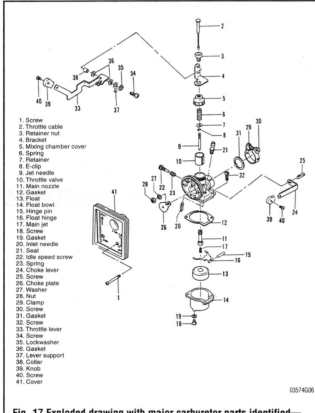

1. Screw
2. Throttle cable
3. Retainer nut
4. Bracket
5. Mixing chamber cover
6. Spring
7. Retainer
8. E-clip
9. Jet needle
10. Throttle valve
11. Main nozzle
12. Gasket
13. Float
14. Float bowl
15. Hinge pin
16. Float hinge
17. Main jet
18. Screw
19. Gasket
20. Inlet needle
21. Seat
22. Idle speed screw
23. Spring
24. Choke lever
25. Screw
26. Choke plate
27. Washer
28. Nut
29. Clamp
30. Screw
31. Gasket
32. Screw
33. Throttle lever
34. Screw
35. Lockwasher
36. Gasket
37. Lever support
38. Collar
39. Knob
40. Screw
41. Cover

Fig. 17 Exploded drawing with major carburetor parts identified—2.5, 3, 3.3 hp

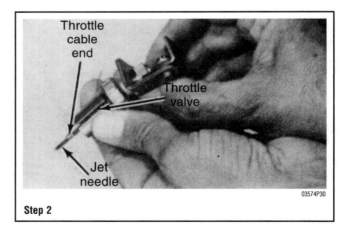

Step 2

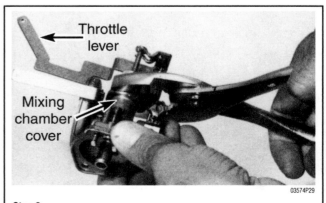

Step 3

ASSEMBLY

▶ See accompanying illustrations

➡ The E-clip must be installed into the same groove from which it was removed. If the clip is lowered, the carburetor will allow the powerhead to operate "rich". Raising the E-clip will cause the powerhead to operate "lean".

1. Assemble the throttle valve components in the following order: Insert the E-clip end of the jet needle into the throttle valve (the end with the recess for the throttle cable end). Place the needle retainer into the throttle valve over the E-clip and align the retainer slot with the slot in the throttle valve.

FUEL SYSTEM   4-11

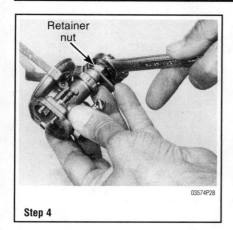
Step 4

Step 5

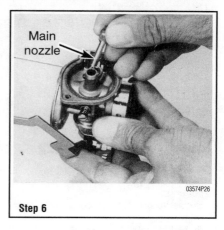
Step 6

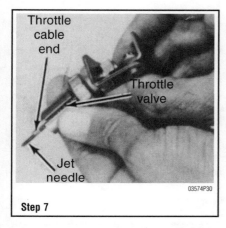
Step 7

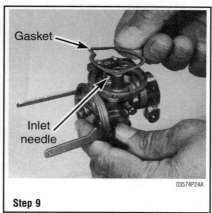
Step 9

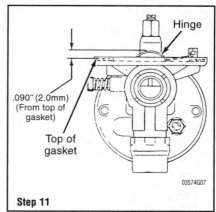
Step 11

7. Install the main jet into the main nozzle and tighten it just "snug" with a screwdriver.
8. For CDI ignition models use a No. 92 main jet and for magneto ignition models use a No. 94 main jet.
9. Position a new float bowl gasket in place and then install the inlet needle.
10. Hold the carburetor body in a perfect upright position on a firm surface. Set the hinge in position and then slide the hinge pin into place through the hinge.
11. Check the float hinge adjustment. The vertical distance between the top of the hinge and the top of the gasket should be .090 in. (2.0mm). carefully bend the hinge, as necessary, to achieve the required measurement. Install the float.
12. Place the float bowl in position on the carburetor body and then secure it with the two Phillips head screws.

4 and 5 HP

The 4 and 5hp models use a round bowl single float carburetor with "Keihin" integral fuel pump. The fuel pump is integrated into the design of this carburetor.

REMOVAL & INSTALLATION

♦ See accompanying illustrations

1. Squeeze the wire clamp around the fuel inlet line and pull the line from the pump. Plug the fuel line to prevent leakage.
2. Loosen but do not remove, the screw retaining the throttle wire in the barrel clamp. Pull the throttle wire downward, to clear the barrel clamp. Pry the choke link from the plastic retainer on the carburetor linkage.
3. Remove the two screws securing the baffle cover to the carburetor.

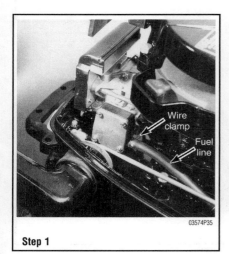
Step 1

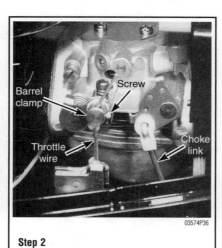
Step 2

Step 3

4-12 FUEL SYSTEM

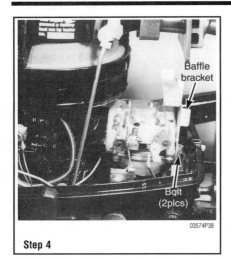
Step 4

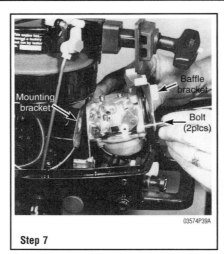
Step 7

Step 9

4. Loosen the two carburetor retaining bolts. These two long bolts extend through the baffle bracket, the body of the carburetor and into the block.
5. Lift away the baffle bracket and carburetor with integral fuel pump still attached.
6. Remove and discard the carburetor mounting gasket.

To install:

7. Slide the two long carburetor retaining bolts through the baffle bracket, carburetor body and mounting gasket.
8. Start the bolts and then tighten them, to a torque value of 5.8 ft. lbs. (8Nm).
9. Secure the baffle cover to the baffle bracket using two screws and washers. Tighten the screws securely.
10. Snap the choke link into the plastic fitting on the starboard side of the carburetor.
11. Check the action of the choke knob to be sure there is no evidence of binding.
12. Slide the throttle cable up into the barrel clamp from the bottom up but do not secure the wire until the following adjustment has been made. Back off the vertical idle speed screw until it no longer contacts the throttle arm. Rotate the same screw clockwise until it barely makes contact with the throttle arm, then continue to rotate the screw two more turns to slightly open the throttle plate. Grasp the cable end and lightly pull up to take up the slack. Tighten the screw on the clamp to secure the cable in position.
13. Slide the inlet fuel hose onto the fitting on the carburetor. Squeeze the two ends of the wire clamp and bring the clamp up over the inlet fitting.
14. Mount the outboard unit in a test tank, on the boat in a body of water or connect a flush attachment and hose to the lower unit. Connect a tachometer to the powerhead.

DISASSEMBLY

♦ See accompanying illustrations

1. Remove the pilot (low speed mixture) screw and spring. The number of turns out from a lightly seated position will be given during assembling.

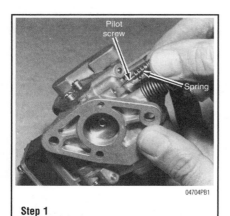
Step 1

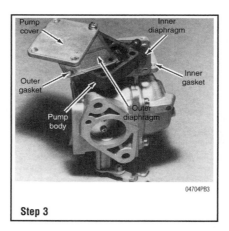
Step 3

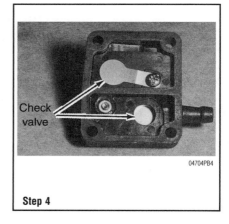
Step 4

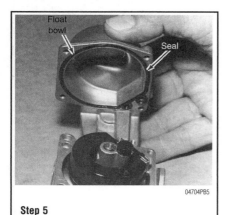
Step 5

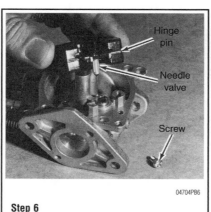
Step 6

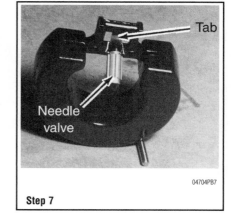
Step 7

FUEL SYSTEM 4-13

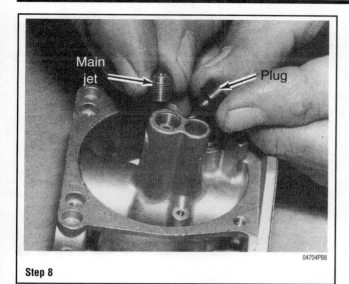

Step 8

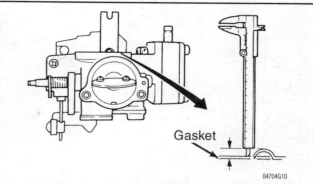

Fig. 18 Exploded drawing of the round bowl, single float carburetor, with integrated Keihin fuel pump. Major parts are identified

2. Remove the two screws securing the top cover and lift off the cover. Gently pry the two rubber plugs out with an awl. Remove the oval air jet cover and the round bypass cover.

3. Remove the four screws securing the fuel pump to the carburetor body. Disassemble the fuel pump components in the following order: the pump cover, the outer gasket, the outer diaphragm, the pump body, the inner diaphragm and finally the inner gasket.

4. Remove the screws from both sides of the fuel pump body. Remove the check valves.

5. Remove the four screws securing the float bowl cover in place. Lift off the float bowl. Remove and discard the rubber sealing ring.

6. Remove the small Phillips screw securing the float hinge to the mounting posts. Lift out the float, the hinge pin and the needle valve. Slide the hinge pin free of the float.

7. Slide the wire attaching the needle valve to the float free of the tab.

8. Use the proper size slotted screwdriver and remove the main jet, then unscrew the main nozzle from beneath the main jet. Remove the plug and then unscrew the pilot jet located beneath the plug.

CLEANING & INSPECTION

▶ See Figure 18

Never dip rubber parts, plastic parts, diaphragms or pump plungers in carburetor cleaner, because they tend to absorb liquid and expand. These parts should be cleaned only in solvent and then blown dry with compressed air immediately.

Place all of the metal parts in a screen-type tray and dip them in carburetor cleaner until they appear completely clean—free of gum and varnish which accumulates from stale fuel. Blow the parts dry with compressed air.

Blow out all of the passageways in the carburetor with compressed air. Check all of parts and passageways to be sure they are clear and not clogged with any deposits.

Never use a piece of wire or any type of pointed instrument to clean drilled passages or calibrated holes in a carburetor.

Carefully inspect the casting for cracks, stripped threads or plugs for any sign of leakage. Inspect the float hinge in the hinge pin area for wear and the float for any sign of leakage.

Examine the inlet needle for wear. If there is any evidence of wear, the needle must be replaced.

Always replace any worn or damaged parts. A carburetor service kit is available at modest cost from the local Mercury/Mariner dealer. The kit will contain all necessary parts to perform the usual carburetor overhaul.

ASSEMBLY

▶ See accompanying illustrations

1. Invert the carburetor and allow the float to rest on the needle valve. Measure the distance between the top of the float and the mixing chamber housing. This distance should be ½ in. (13 mm). This dimension, with the carburetor inverted, places the lower surface of the float parallel to the carburetor body. If the dimension is not within the limits listed, the needle valve must be replaced.

2. Carefully bend the float arm, as required, to obtain a satisfactory measurement. Install the float.

3. Slide the wire attached to the needle valve onto the float tab.

4. Guide the hinge pin through the float hinge. Lower the float and needle

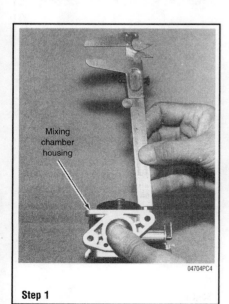

Step 1

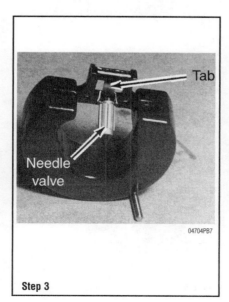

Step 3

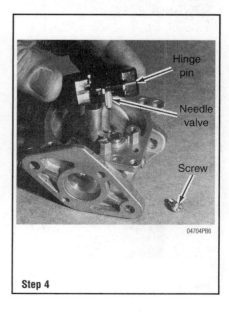

Step 4

4-14 FUEL SYSTEM

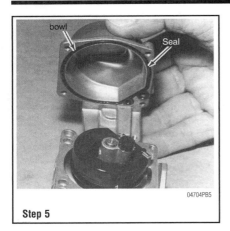
Step 5

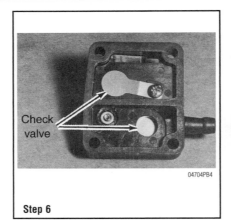
Step 6

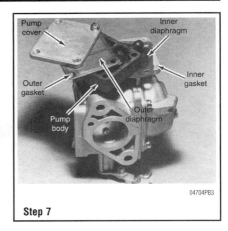
Step 7

assembly down into the float chamber and guide the needle valve into the needle seat. Check to be sure the hinge pin indexes into the mounting posts. Secure the pin in place with the small Phillips screw.

5. Insert the rubber sealing ring into the groove in the float bowl. Install the float bowl and secure it in place with the four Phillips screws.

6. Place the check valves, one at a time, in position on both sides of the fuel pump body. Secure each valve with the attaching screw.

7. Assemble the fuel pump components onto the carburetor body in the following order: the inner gasket, the inner diaphragm, the pump body, the outer diaphragm, the outer gasket and finally the pump cover. Check to be sure all the parts are properly aligned with the mounting holes. Secure it all in place with the four attaching screws.

8. Install the oval air jet cover and the round bypass cover in their proper recesses. Place the top cover over them, no gasket is used. Install and tighten the two attaching screws.

9. Slide the spring over the pilot screw and then install the screw. Tighten the screw until it barely seats. Back the pilot screw out 1½ turns.

6, 8, 9.9, 10 and 15 HP

The 6, 8, 9.9, 10 and 15hp models are equipped with a Walbro "WM" or "WMC" carburetor with integral fuel pump. The WM and WMC carburetors can be identified by the embossed letters on the mounting flange.

REMOVAL & INSTALLATION

♦ See accompanying illustrations

1. Using a pair of pliers, remove the Circlip securing the choke knob in place.
2. Remove the choke knob by pulling it straight out. Unsnap the idle wire from the primer bracket. Remove the two bolts securing the primer bracket to the carburetor. The primer bracket will come free with the bracket. Using a pair of snips, cut the Sta-strap from the fuel hose to the carburetor. Pull the fuel line free of the fitting. Remove the two carburetor mounting bolts and lift the carburetor off the powerhead.

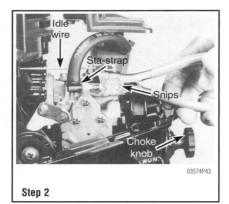
Step 2

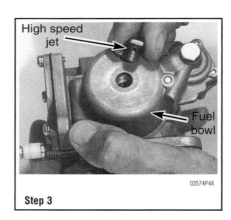
Step 3

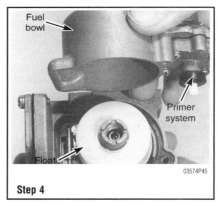
Step 4

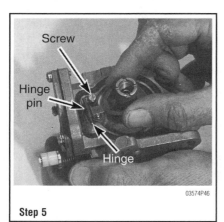
Step 5

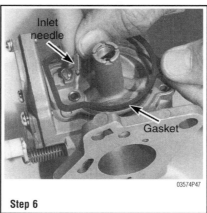
Step 6

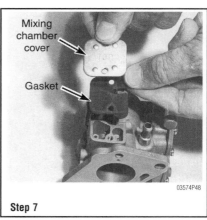
Step 7

FUEL SYSTEM 4-15

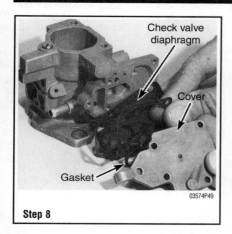
Step 8

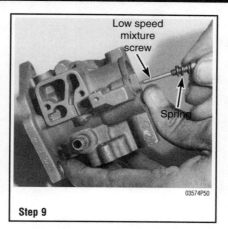
Step 9

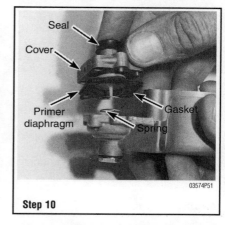
Step 10

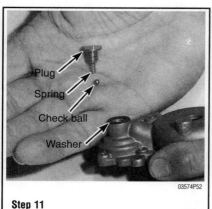
Step 11

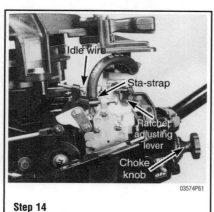
Step 14

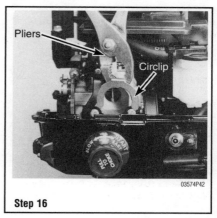
Step 16

3. Remove the high speed jet from the bottom of the carburetor. The jet also serves as a retaining bolt for the fuel bowl.

4. Lift the fuel bowl off the carburetor. The primer system is attached to the bowl and therefore, will come with it. Lift out the fuel float.

5. Back out the screw securing the hinge pin in place. Pull the hinge pin and then lift the hinge out of the carburetor body.

6. Remove the inlet needle from its position under the hinge. Remove the carburetor bowl gasket from its recess.

7. Remove the two screws securing the mixing chamber cover to the carburetor body. Remove the cover and then the gasket.

8. Remove the five screws securing the fuel pump cover in place. Remove in order, the cover, gasket and then the fuel pump check valve diaphragm.

➡ **Make sure to count the number of turns as the low speed mixture screw is removed, as an aid during assembling to making a rough bench adjustment.**

9. Back out and count the number of turns required to remove the low speed mixture screw. The spring will come with the screw.

10. Remove the four screws securing the primer system cover to the float bowl flange. Gently lift the cover and the primer diaphragm, gasket and spring will then be free, as shown. A seal is installed in the cover.

11. Remove the base plug from the underside of the float bowl flange (primer system housing). Take care not to lose the spring from the shaft of the plug or the tiny check ball.

12. Remove the washer.

To install:

13. Position a new carburetor gasket onto the powerhead. Install the carburetor to the powerhead and secure it in place with the two nuts.

14. Connect the fuel hose and secure it with a new Sta-strap. Install the small primer bracket to the side of the carburetor with the two bolts.

15. Snap the idle wire into place on the ratchet adjustment lever.

16. Position the choke knob in place through the opening in the lower cowling. Install the Circlip securing the choke knob.

17. Mount the outboard unit in a test tank, on a boat in a body of water or connect a flush attachment to the lower unit.

DISASSEMBLY

♦ **See accompanying illustrations**

1. Using a pair of pliers, remove the Circlip securing the choke knob in place.

2. Remove the choke knob by pulling it straight out. Unsnap the idle wire from the primer bracket. Remove the two bolts securing the primer bracket to the carburetor. The primer bracket will come free with the bracket.

3. Using a pair of snips, cut the Sta-strap from the fuel hose to the carburetor.

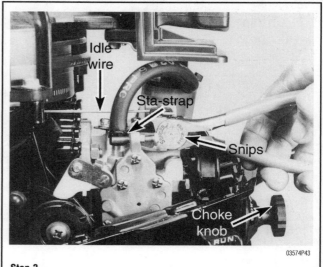
Step 3

4-16 FUEL SYSTEM

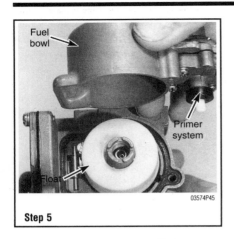
Step 5

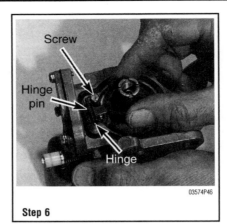
Step 6

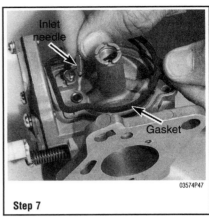
Step 7

Step 8

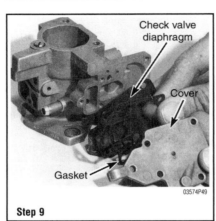
Step 9

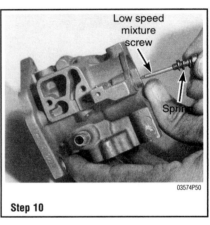
Step 10

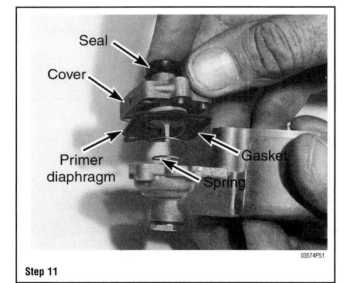
Step 11

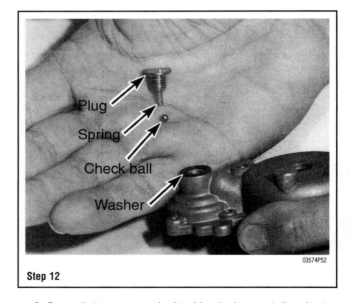
Step 12

Pull the fuel line free of the fitting. Remove the two carburetor mounting bolts and lift the carburetor off the powerhead.

4. Remove the high speed jet from the bottom of the carburetor. The jet also serves as a retaining bolt for the fuel bowl. Lift out the fuel float.

5. Lift the fuel bowl off the carburetor

➥ **The primer system is attached to the bowl and therefore, will come with it.**

6. Back out the screw securing the hinge pin in place. Pull the hinge pin, and then lift the hinge out of the carburetor body.

7. Remove the inlet needle from its position under the hinge. Remove the carburetor bowl gasket from its recess.

8. Remove the two screws securing the mixing chamber cover to the carburetor body. Remove the cover, and then the gasket.

9. Remove the five screws securing the fuel pump cover in place. Remove in order, the cover, gasket, and then the fuel pump check valve and diaphragm.

➥ **Count the number of turns as the low speed mixture screw is removed, as an aid during assembling to making a rough bench adjustment.**

10. Back out and count the number of turns required to remove the low speed mixture screw. The spring will come with the screw.

11. Remove the four screws securing the primer system cover to the float bowl

FUEL SYSTEM 4-17

Fig. 19 All rubber and plastic parts must be removed before other carburetor parts are placed in a basket to be submerged in carburetor cleaner

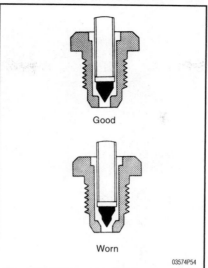

Fig. 20 Comparison of a worn and new needle and seat arrangement. The worn needle would have to be replaced for full carburetor efficiency

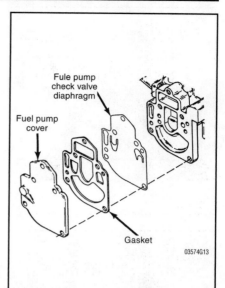

Fig. 21 Exploded drawing of the integral fuel pump to show arrangement of diaphragm and gasket

flange. Gently lift the cover and the primer diaphragm, gasket, and spring will then be free, as shown. A seal is installed in the cover.

12. Remove the base plug from the underside of the float bowl flange (primer system housing). Take care not to lose the spring from the shaft of the plug or the tiny check ball. Remove the washer.

CLEANING & INSPECTION

♦ See Figures 19 thru 24

Never dip rubber parts, plastic parts, diaphragms or pump plungers in carburetor cleaner, because they tend to absorb liquid and expand. These parts should be cleaned only in solvent and then blown dry with compressed air immediately.

Place all of the metal parts in a screen-type tray and dip them in carburetor cleaner until they appear completely clean—free of gum and varnish which accumulates from stale fuel. Blow the parts dry with compressed air.

Blow out all of the passageways in the carburetor with compressed air. Check all of parts and passageways to be sure they are clear and not clogged with any deposits.

Never use a piece of wire or any type of pointed instrument to clean drilled passages or calibrated holes in a carburetor.

Carefully inspect the casting for cracks, stripped threads or plugs for any sign of leakage. Inspect the float hinge in the hinge pin area for wear and the float for any sign of leakage.

Examine the inlet needle for wear. If there is any evidence of wear, the needle must be replaced.

Always replace any worn or damaged parts. A carburetor service kit is available at modest cost from the local Mercury/Mariner dealer. The kit will contain all necessary parts to perform the usual carburetor overhaul.

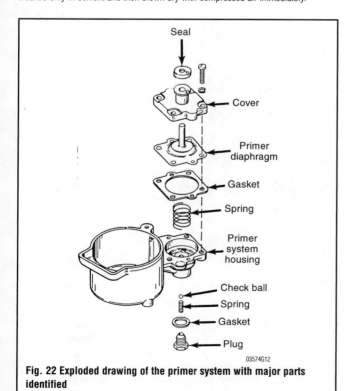

Fig. 22 Exploded drawing of the primer system with major parts identified

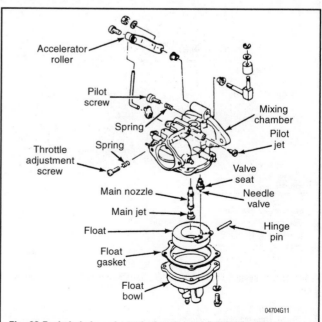

Fig. 23 Exploded view of a typical early model "WM" carburetor with integral fuel pump. Major parts are identified

4-18 FUEL SYSTEM

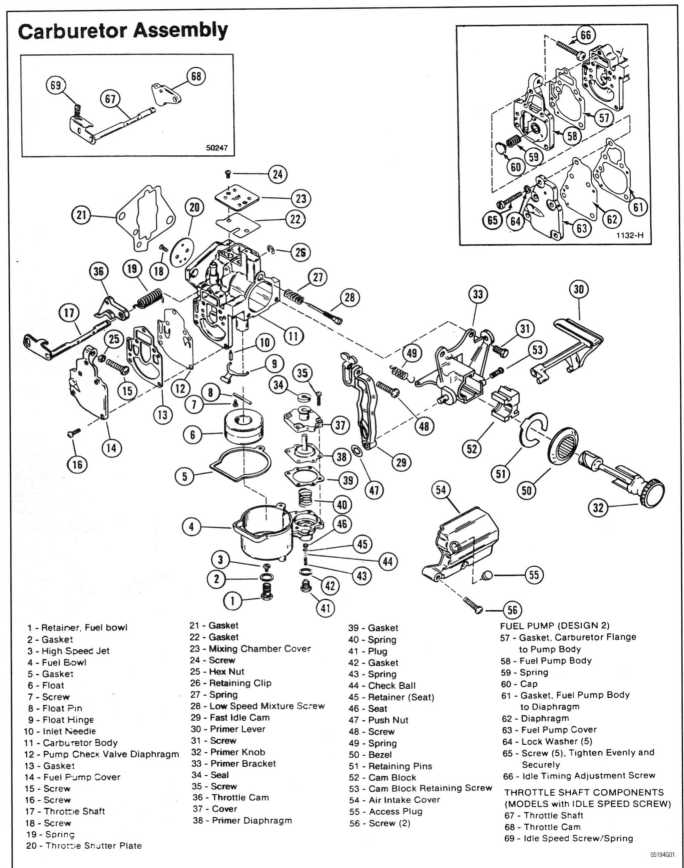

Fig. 24 Exploded view of a typical late model "WMC" carburetor with integral fuel pump. Major parts are identified

FUEL SYSTEM 4-19

ASSEMBLY

♦ See accompanying illustrations

1. Drop the tiny check ball and then the little spring into the cavity of the primer system housing. Set a new washer in place on the housing. Thread the plug into place and tighten it securely.
2. Turn the carburetor body over and set the spring in place in the primer housing.
3. Position a new gasket in place on the housing. Set the primer diaphragm on top of the spring with the holes in the diaphragm for the mounting screws roughly aligned with the holes in the primer housing.
4. Slip the seal over the lip on the cover and compress the spring with the primer system cover allowing the shaft of the primer diaphragm to pass up through the hole in the seal and the holes in the diaphragm to align perfectly with the holes in the housing.
5. Secure the cover in place with the four attaching screws and lock-washers.
6. Tighten the screws securely.
7. Slide the spring onto the shaft of the low speed mixture screw. Thread the screw into the carburetor housing and seat it lightly.

➡ Never over-tighten the screw, because such action would damage the tip.

8. From the lightly seated position, back the screw out the same number of turns recorded during disassembly. If the number of turns was lost, back the screw out 1¼ turns as a preliminary bench adjustment. This adjustment will allow the powerhead to be started.
9. Place the fuel pump check valve diaphragm onto the carburetor body with the holes for the mounting screws aligned with the holes in the body. Position a new gasket on top of the diaphragm with the mounting holes aligned.
10. Place the fuel pump cover onto the carburetor body and secure it with the five attaching screws and lockwashers. Tighten the screws securely.
11. Position a new gasket in place on the mixing chamber. Install the mixing chamber cover onto the carburetor and secure it with the two screws. Tighten the screws securely.
12. Position a new carburetor bowl gasket into place on the carburetor body. Slide the inlet needle into its hole in the body.
13. Slide the hinge pin through the hinge. Next, install the hinge and secure it in place with the screw. Tighten the screw snugly.

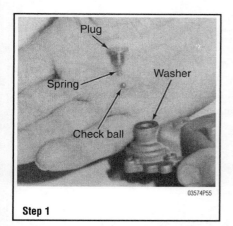
Step 1

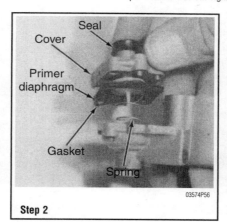
Step 2

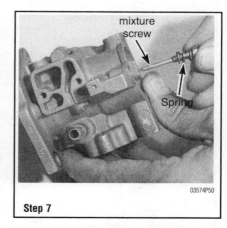
Step 7

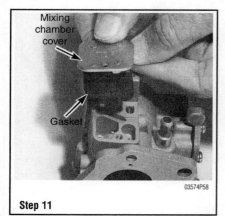
Step 11

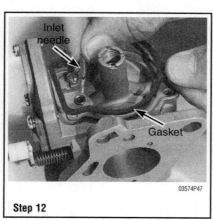
Step 12

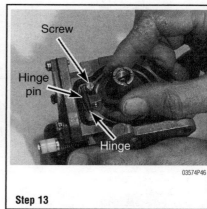
Step 13

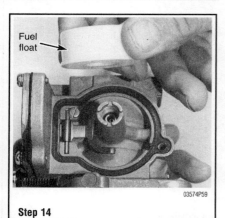
Step 14

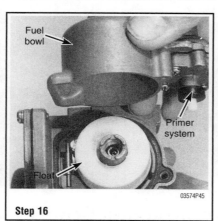
Step 16

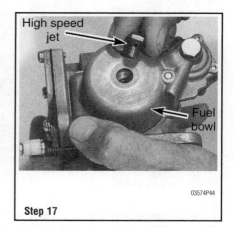
Step 17

4-20 FUEL SYSTEM

➡ Check to be sure the hinge will move without binding.

14. Slide the float down over the shaft and onto the hinge.
15. Check the float level. With the carburetor in the position shown, the distance from the lower edge of the float and the top of the retainer, should be 1 in. (25.4mm). If necessary, adjust the float level by ever so carefully bending the hinge Slightly. Just a "whisker" of change in the hinge will move the float.
16. Cover the float with the float bowl and primer system housing.
17. Slide a new gasket onto the high speed jet. Install the high speed jet. This jet also secures the bowl and primer system to the carburetor body.

20 and 25 HP

REMOVAL & INSTALLATION

♦ See accompanying illustrations

1. Using a pair of pliers, remove the circlip securing the choke knob in place.
2. Remove the choke knob by pulling it straight out. Unsnap the idle wire from the primer bracket. Remove the two bolts securing the primer bracket to the carburetor. The primer bracket will come free with the bracket. Using a pair of snips, cut the sta-strap from the fuel hose to the carburetor. Pull the fuel line free of the fitting. Remove the two carburetor mounting bolts and lift the carburetor off the powerhead.

To install:

3. Position a new carburetor gasket onto the powerhead. Install the carburetor to the powerhead and secure it in place with the two nuts.
4. Connect the fuel hose and secure it with a new sta-strap.
5. Install the small primer bracket to the side of the carburetor with the two bolts. Snap the idle wire into place on the ratchet adjustment lever.

Step 1

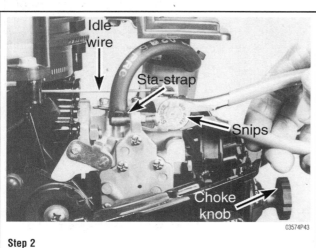
Step 2

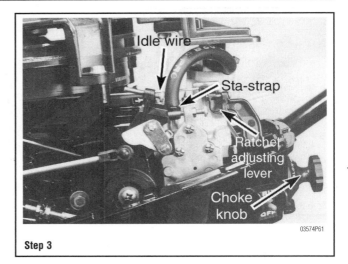
Step 3

6. Position the choke knob in place through the opening in the lower cowling. Install the Circlip securing the choke knob.

DISASSEMBLY

♦ See accompanying illustrations

1. Remove the high speed jet from the bottom of the carburetor. The jet also serves as a retaining bolt for the fuel bowl.
2. Lift the fuel bowl off the carburetor. The primer system is attached to the bowl and therefore, will come with it. Lift out the fuel float

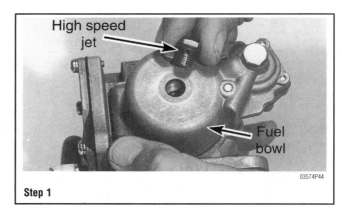
Step 1

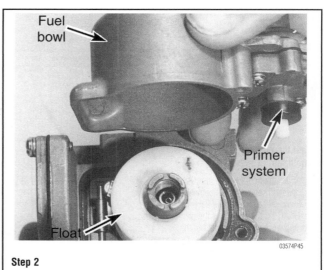
Step 2

FUEL SYSTEM   4-21

Step 3

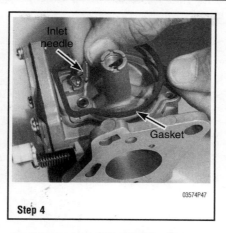
Step 4

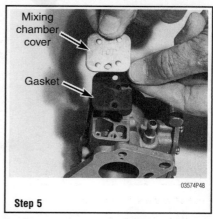
Step 5

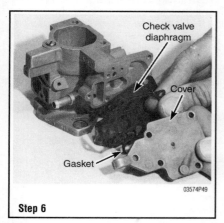
Step 6

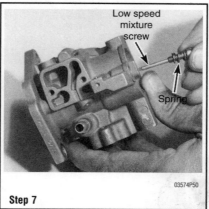
Step 7

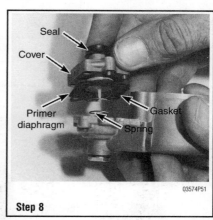
Step 8

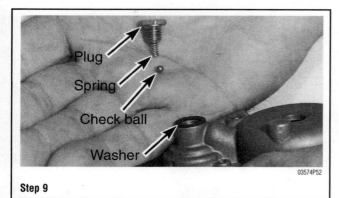
Step 9

3. Back out the screw securing the hinge pin in place. Pull the hinge pin, and then lift the hinge out of the carburetor body.

4. Remove the inlet needle from its position under the hinge. Remove the carburetor bowl gasket from its recess.

5. Remove the two screws securing the mixing chamber cover to the carburetor body. Remove the cover, and then the gasket.

6. Remove the five screws securing the fuel pump cover in place. Remove in order, the cover, gasket, and then the fuel pump check valve diaphragm.

➡ Count the number of turns as the low speed mixture screw is removed, as an aid during assembling to making a rough bench adjustment.

7. Back out and count the number of turns required to remove the low speed mixture screw. The spring will come with the screw.

8. Remove the four screws securing the primer system cover to the float bowl flange. Gently lift the cover and the primer diaphragm, gasket, and spring will then be free, as shown. A seal is installed in the cover.

9. Remove the base plug from the underside of the float bowl flange (primer system housing). Make sure not to lose the spring from the shaft of the plug or the tiny check ball. Remove the washer.

CLEANING & INSPECTION

♦ See Figures 25, 26 and 27

➡ Never dip rubber parts, plastic parts, diaphragms, or pump plungers in carburetor cleaner. These parts should be cleaned in solvent only, and then blown dry with compressed air.

Fig. 25 All rubber and plastic parts must be removed before other carburetor parts are placed in a basket to be submerged in carburetor cleaner

4-22 FUEL SYSTEM

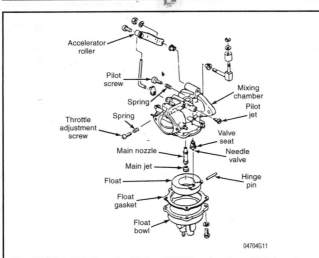

Fig. 26 Exploded view of a Walbro "WM" carburetor with integral fuel pump

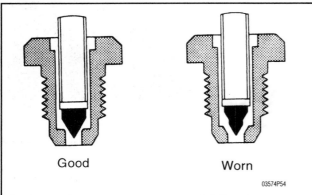

Fig. 27 Line drawing to compare a worn and new needle and seat arrangement. The worn needle would have to be replaced for full carburetor efficiency

1. Place all the metal parts in a screen-type tray and dip them in carburetor cleaner until they appear completely clean—free of gum and varnish which accumulates from stale fuel. Blow the parts dry with compressed air.

2. Blow out all passageways in the castings with compressed air. Check all of the parts and passages to be sure they are not clogged or contain any deposits. Never use a piece of wire or any type of pointed instrument to clean drilled passages or calibrated holes in a carburetor.

3. Carefully inspect the casting for cracks, stripped threads, or plugs for any sign of leakage. Inspect the float hinge in the hinge pin area for wear and the float for any sign of leakage.

4. Examine the inlet needle for wear and if there is any evidence of wear, the inlet needle must be replaced.

5. Always replace any and all worn parts.

➡A carburetor service kit is available at modest cost from the local marine dealer. The kit will contain all necessary parts to perform the usual carburetor overhaul work.

ASSEMBLY

♦ See accompanying illustrations

1. Drop the tiny check ball, and then the little spring into the cavity of the primer system housing. Set a new washer in place on the housing. Thread the plug into place and tighten it securely.

2. Turn the carburetor body over and set the spring in place in the primer housing. Next, position a new gasket in place on the housing. Set the primer diaphragm on top of the spring with the holes in the diaphragm for the mounting screws roughly aligned with the holes in the primer housing. Slip the seal over the lip on the cover

3. Now, compress the spring with the primer system cover allowing the shaft of the primer diaphragm to pass up through the hole in the seal and the holes in the diaphragm to align perfectly with the holes in the housing. Secure the cover in place with the four attaching screws and lockwashers. Tighten the screws securely.

4. Slide the spring onto the shaft of the low speed mixture screw. Thread the screw into the carburetor housing and seat it lightly. Never tighten the screw, because such action would damage the tip. From the lightly seated position, back the screw out the same number of turns recorded during disassembly. If the number of turns was lost, back the screw out 1¼ turns as a preliminary bench adjustment. This adjustment will allow the powerhead to be started.

5. Place the fuel pump check valve diaphragm onto the carburetor body with the holes for the mounting screws aligned with the holes in the body. Position a new gasket on top of the diaphragm with the mounting holes aligned.

6. Place the fuel pump cover onto the carburetor body and secure it with the five attaching screws and lockwashers. Tighten the screws securely.

7. Position a new gasket in place on the mixing chamber. Install the mixing chamber cover onto the carburetor and secure it with the two screws. Tighten the screws securely.

8. Position a new carburetor bowl gasket into place on the carburetor body. Slide the inlet needle into its hole in the body.

9. Slide the hinge pin through the hinge. Next, install the hinge and secure it in place with the screw. Tighten the screw snug.

10. Check to be sure the hinge will move without binding.

11. Slide the float down over the shaft and onto the hinge.

12. Check the float level as shown in the accompanying illustration. With the carburetor in the position shown, the distance from the lower edge of the float and the top of the retainer, should be 1 inch (25.4mm). If necessary, adjust the float level by very carefully bending the hinge slightly. The slightest movement of the hinge will move the float.

13. Cover the float with the float bowl and primer system housing.

14. Slide a new gasket onto the high speed jet. Install the high speed jet. This jet also secures the bowl and primer system to the carburetor body.

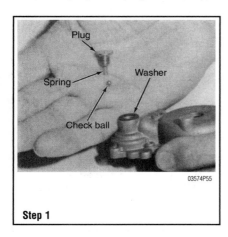

Step 1

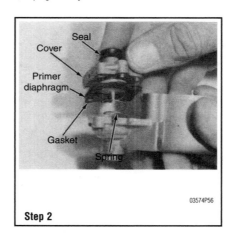

Step 2

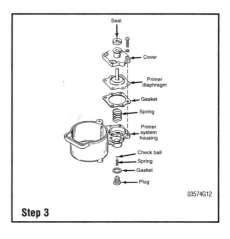

Step 3

FUEL SYSTEM 4-23

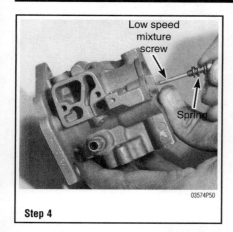
Step 4

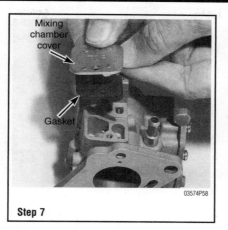
Step 7

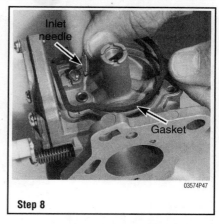
Step 8

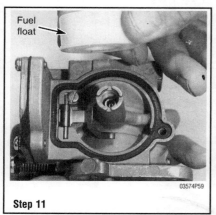
Step 11

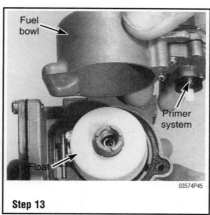
Step 13

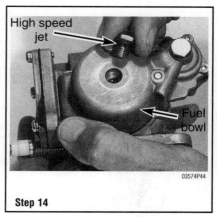
Step 14

30 and 40 HP (2-Cylinder)

Several different versions of the Walbro "WME" carburetor are used. The units are almost identical except for the main jet size

REMOVAL & INSTALLATION

▶ See accompanying illustrations

1. Remove the battery leads from the battery terminals. Pull outward on the starboard side of the front shield.
2. Remove the spring from the latch and open the cowlings.
3. Remove these screws prior to lifting off the air silencer from the carburetor.
4. Remove the air silencer.
5. Take time to identify each carburetor to ensure each will be installed back in its original position, because each carburetor is different.
6. Disconnect the throttle linkage from each carburetor.
7. Disconnect the fuel line from the engine then remove the hose clamps on each fuel line to each carburetor. Remove the fuel line from each carburetor.
8. Remove the attaching nuts securing each carburetor to the intake manifold.
9. Remove each carburetor from the powerhead.

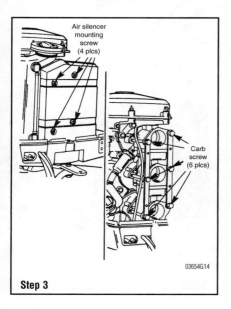
Step 3

Step 4

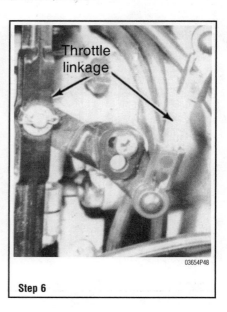
Step 6

4-24 FUEL SYSTEM

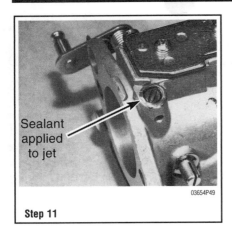
Step 11

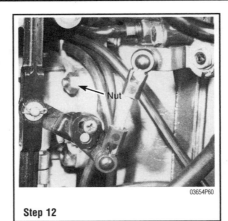
Step 12

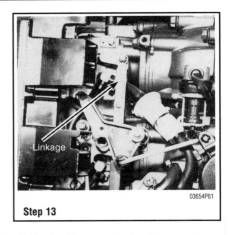
Step 13

To install:

10. Position new gaskets in place on the intake manifold. Install each carburetor onto the manifold in the same position from which it was removed. Each carburetor should have been identified as instructed during the removal procedures.

11. As appropriate, apply sealant to screws or plugs as an anti-tampering method.

12. Secure the carburetors in place with the retaining nuts. Tighten the nuts alternately and evenly to a torque value of 100 inch lbs. (11.3 Nm).

13. Connect all of the fuel lines to the carburetors and tighten the clamps securely. Connect the fuel line from the fuel tank. Activate the fuel line squeeze bulb several times and check the carburetors and fuel lines for leaks. Connect the throttle linkage to and between the carburetors. Connect the battery leads to the battery.

DISASSEMBLY

♦ See accompanying illustrations

➡ This carburetor does not have a traditional choke system. This enrichener system provides the engine with an extra fuel charge for ease of starting a cold engine.

✴✴ WARNING

Notice the application of red sealant on certain adjustment screws or plugs. This sealant was applied at the factory and is the manufacturer's method of instructing anyone servicing the carburetor not to disturb the screw or plug.

The sealant is not affected by carburetor cleaner and should remain in place on the plug or screw and the carburetor for the entire useful life of the carburetor.

1. Back out the idle mixture screw.

➡ If servicing a 3-Cylinder powerhead, remove the bowl vent jet located next to the idle mixture screw. This jet is identical on all three carburetors. The 4-Cylinder powerheads are not equipped with this jet.

2. Remove the Phillips head screws and captive lockwashers and then remove the mixing chamber cover and gasket.

3. Remove the Phillips head screws with captive lockwashers from the float bowl. Lift off the fuel bowl. Remove and discard the gasket.

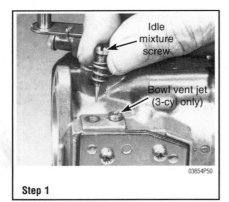
Step 1

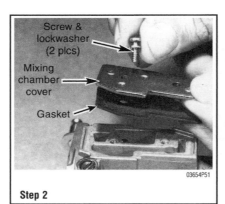
Step 2

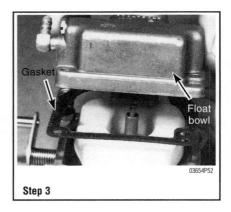
Step 3

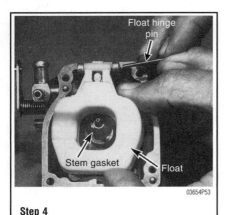
Step 4

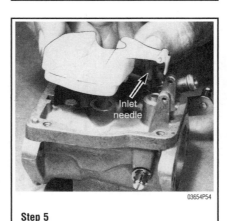
Step 5

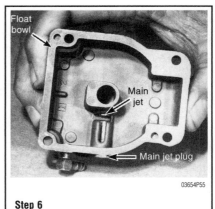
Step 6

FUEL SYSTEM 4-25

4. Remove the stem gasket from the center of the fuel bowl. Support the float and at the same time push the float hinge pin free of the mounting posts.
5. Carefully lift the float with the inlet needle attached from the needle seat. Detach the inlet needle from the float. The needle seat is not removable on this carburetor.
6. Remove the main jet plug and gasket from the exterior wall of the float bowl. This plug provides access for a screwdriver to be inserted into the opening and allows removal of the main jet from the side of the center turret. This design facilitates changing the size of the main jet, without removal of the carburetor from the powerhead. Main jet sizes must be changed when operating the powerhead at elevations higher than 2,500 feet above sea level.

➡ **Further disassembly of the carburetor is not necessary in order to clean it properly.**

CLEANING & INSPECTION

▶ See Figures 28, 29 and 30

Never dip rubber parts, plastic parts, diaphragms or pump plungers in carburetor cleaner, because they tend to absorb liquid and expand. These parts should be cleaned only in solvent and then blown dry with compressed air immediately.

Place all of the metal parts in a screen-type tray and dip them in carburetor cleaner until they appear completely clean—free of gum and varnish which accumulates from stale fuel. Blow the parts dry with compressed air.

Blow out all of the passageways in the carburetor with compressed air. Check all of parts and passageways to be sure they are clear and not clogged with any deposits.

Never use a piece of wire or any type of pointed instrument to clean drilled passages or calibrated holes in a carburetor.

Carefully inspect the casting for cracks, stripped threads or plugs for any sign of

Fig. 28 Never use a piece of wire or any type of pointed instrument to clean drilled passages or calibrated holes in a carburetor, particularly those in the mixing chamber

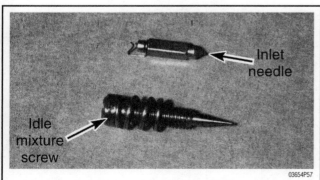

Fig. 29 Inspect the taper of the inlet needle and the idle mixture screw for evidence of a worn groove

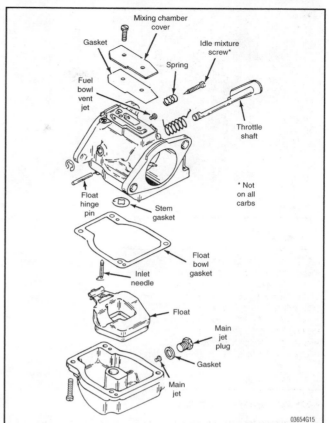

Fig. 30 Exploded drawing of Walbro "WME" carburetor, with major parts identified

leakage. Inspect the float hinge in the hinge pin area for wear and the float for any sign of leakage.

Examine the inlet needle for wear. If there is any evidence of wear, the needle must be replaced.

Always replace any worn or damaged parts. A carburetor service kit is available at modest cost from the local Mercury/Mariner dealer. The kit will contain all necessary parts to perform the usual carburetor overhaul.

ASSEMBLY

▶ See accompanying illustrations

1. Hook the inlet needle spring over the float tab and lower the needle into its seat with the float hinge between the mounting posts.

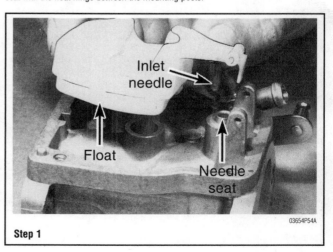

Step 1

4-26 FUEL SYSTEM

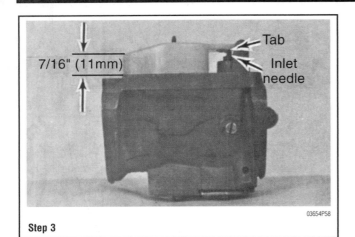
Step 3

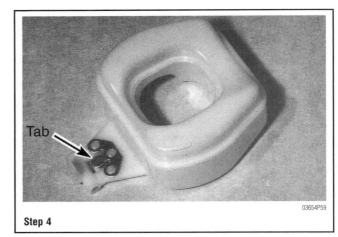
Step 4

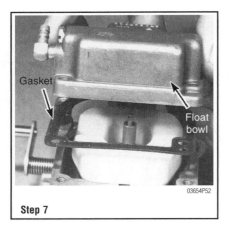
Step 7

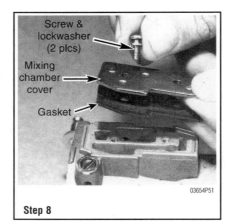
Step 8

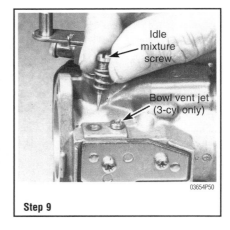
Step 9

2. Slide the float hinge pin through the posts to secure the float. Center the pin between the posts. Place the stem gasket over the center turret.

3. Hold the carburetor body in the inverted position—the same position it has been in since the start of the assembling procedures—with the float resting on the inlet needle. Measure the distance between the float bowl gasket surface and a point on the float directly opposite the hinge. Notice the surface of the float curves downward. Therefore, the measurement point is the lowest on the horizontal surface. The distance should be 9/16 in. (14.29mm).

4. If the distance is not as specified, then remove the float hinge pin to free the float. The float height adjustment may be made by bending the tab on which the inlet needle hangs.

5. Repeat the adjustments until the specified distance between the float bowl gasket surface and the lowest edge of the float is obtained.

6. Install the main jet into the center turret of the fuel bowl. Tighten the jet securely. Install the gasket and main jet plug in the exterior wall of the float bowl and tighten the plug securely.

7. With the carburetor still inverted, position a new gasket onto the body. Place the fuel bowl in position and secure it in place with the four Phillips head screws and captive lockwashers.

8. Position a new gasket over the mixing chamber. Install the cover and secure it in place with the attaching hardware.

9. Thread the idle mixture screw into the carburetor until the screw is lightly seated. From this lightly seated position, back the screw out approximately 1¼ turns as a preliminary adjustment at this time.

40–125 HP

Two different Walbro carburetors are used on the 40–125hp Mercury/Mariner powerheads.

The Walbro WMA—with integral fuel pump installed on the 40hp model and the Walbro WME—with separate fuel pump installed on all other powerheads.

Several different versions of the Walbro WME carburetor are used. The units are almost identical except for the main jet size and only the top two carburetors on a 4-carburetor installation have adjustable idle mixture screws. identification letters are stamped on top of the carburetor mounting flange.

WME CARBURETOR SPECIFICATIONS
Carburetor Number Stamped at TOP of Carburetor Mounting Flange

MODEL H.P.	CARBURETOR NUMBER	LOCATION	MAIN JET	BOWL VENT JET
50	WME-23-1 WME-23-2 WME-23-3	Top Carburetor Center Carburetor Bottom Carburetor	.052	.092
60	WME-22-1 A or B WME-22-2 A or B WME-23-3 A or B	Top Carburetor Center Carburetor Bottom Carburetor	.070	.090
60	WME-35-1 WME-35-2 WME-35-3	Top Carburetor Center Carburetor Bottom Carburetor	.068	NONE
70	WME-7 -1 2 3	Top Carburetor Center Carburetor Bottom Carburetor	.072	.094
75	WME -8 -1 2 3	Top Carburetor Center Carburetor Bottom Carburetor	.068	.094
90	WME -10 -1 2 3	Top Carburetor Center Carburetor Bottom Carburetor	.072	.094
100	WME -32 -1 2 3 4	Top Carburetor #2 Carburetor #3 Carburetor Bottom Carburetor	.046 .048 .052 .052	NONE
115	WME -33 -1 2 3 4 4A	Top Carburetor #2 Carburetor #3 Carburetor Bottom Carburetor Bottom Carburetor	.052 .056 .056 .062* .060	NONE
125	WME -34 -1 2 3 4A	Top Carburetor #2 Carburetor #3 Carburetor Bottom Carburetor	.066 .068 .070 .072	NONE

* If WME-33-4 is used, change main jet to .060.

FUEL SYSTEM 4-27

These carburetors are equipped with an integral fuel pump, which should be overhauled each time the carburetor is disassembled.

REMOVAL & INSTALLATION

WMA

♦ See accompanying illustrations

1. Remove the battery leads from the battery terminals.
2. Remove the cowling assembly.
3. Disconnect the fuel line at the fuel tank.
4. Disconnect the choke cable from the choke lever.
5. Remove the cap screw and spacer securing the choke cable to the carburetor.
6. Remove the fuel line from the inlet cover.

➥An alternative method is to remove the screw securing the inlet cover to the carburetor and leave the fuel line attached.

7. Remove the two nuts attaching the carburetor to the manifold and then remove the carburetor.

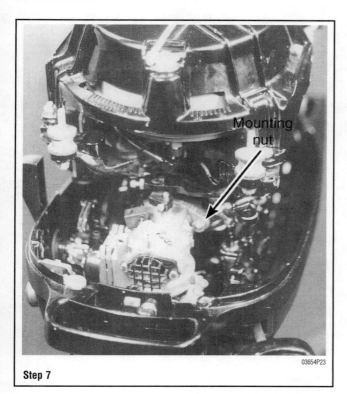
Step 7

Step 8

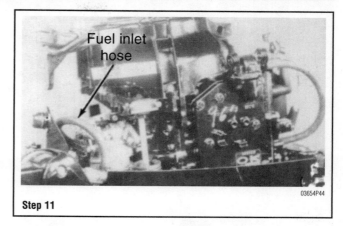
Step 11

To install:

8. Place the carburetor in position on the intake manifold.
9. Install and tighten the two carburetor retaining nuts alternately to a torque value of 100inch lbs. (11Nm).
10. Install the fuel pump inlet cover. Slide a new lockwasher onto the retaining screw and then install and tighten the screw. If the fuel line was removed from the inlet cover, install the hose and tighten the hose clamps.
11. Install the cap screw and spacer securing the choke cable to the carburetor.
12. Connect the choke cable to the lever.
13. Connect the fuel line to the tank. Activate the fuel line squeeze bulb several times. Check delivery of fuel to the carburetor and the lines and their fittings for possible leaks.
14. Connect the battery leads.

WME

♦ See accompanying illustrations

1. Remove the battery leads from the battery terminals.
2. Remove the engine cowling.
3. Remove the four mounting screws and lift the air silencer free of the carburetor.

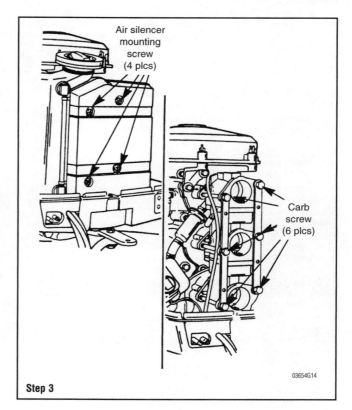
Step 3

4-28 FUEL SYSTEM

Step 4

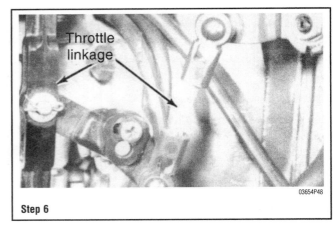
Step 6

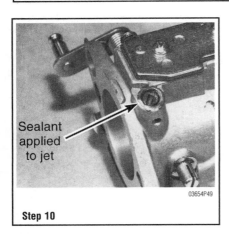
Step 10

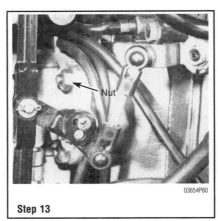
Step 13

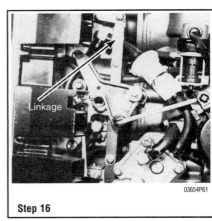
Step 16

4. Remove the air silencer.
5. Take time to identify each carburetor to ensure each will be installed back in its original position, because each carburetor is different.
- Each carburetor has slightly different linkage.
- Only the top and second carburetors installed on 100hp or 115hp powerheads have an adjustable idle screw. The third and bottom carburetors have a non-removable plug in place of the idle screw. This plug is clearly identified with a colored sealer, is preset at the factory and must not be disturbed.
- The fuel bowl of the top carburetor is unique from the others. The enrichener valve is fed by fuel from only the top carburetor float bowl.

6. Disconnect the throttle linkage from each carburetor.
7. Disconnect the fuel line from the engine.
8. Remove the hose clamps on each fuel line to each carburetor.
9. Remove the fuel line from each carburetor.
10. Apply red sealant to screws or plugs as an anti-tampering device.
11. Remove the attaching nuts securing each carburetor to the intake manifold.
12. Remove each carburetor from the powerhead. Apart from the specific differences mentioned above, which in most cases do not affect service procedures, the carburetors are identical, the following procedures are to be repeated for each carburetor. Where the differences affect service procedures, they will be clearly identified.

To install:
13. Position new gaskets in place on the intake manifold. Install each carburetor onto the manifold in the same position from which it was removed. Each carburetor should have been identified as instructed during the removal procedures. Secure the carburetors in place with the retaining nuts. Tighten the nuts alternately and evenly to a torque value of 100 inch lbs. (11.3 Nm).
14. Connect all of the fuel lines to the carburetors and tighten the clamps securely.
15. Connect the fuel line from the fuel tank. Activate the fuel line squeeze bulb several times and check the carburetors and fuel lines for leaks.
16. Connect the throttle linkage to and between the carburetors.
17. Connect the battery leads to the battery.

DISASSEMBLY

WMA

▶ See accompanying illustrations

1. Remove the four screws holding the fuel pump strainer body to the carburetor. Remove the gaskets and diaphragm. Remove the fuel pump body and gaskets.
2. Remove the bolt securing the float bowl to the carburetor casting.

➡ There is a gasket under the bolt bold and one between the float bowl and the casting.

3. Withdraw the float retaining pin and then lift off the float assembly.
4. Lift out the inlet needle valve and spring. Do not attempt to remove the needle valve seat. This seat is pressed into the carburetor body.

Step 1

FUEL SYSTEM 4-29

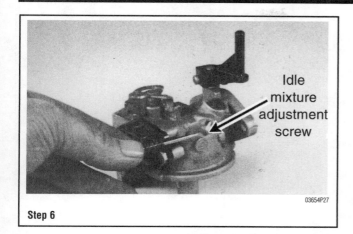
Step 6

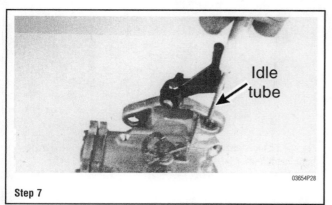
Step 7

5. Remove the main fuel (high-speed) jet. A gasket is not used under this jet. do not attempt to remove the main nozzle even though it has a screwdriver slot. The boost venturi is very difficult to install if the main nozzle has been removed.
6. Remove the idle mixture adjusting screw and spring.
7. Remove the plug screw and then unscrew the idle tube. Slide the gasket free of the idle tube.

WME

♦ See accompanying illustrations

➡ This carburetor does not have a traditional choke system. Instead, the choke function is performed by an enrichener system fed from the fuel bowl of the top carburetor. The extra fuel needed for cold powerhead startup is supplied to the intake manifold by an electrically operated device called the enrichener valve.

✲✲ WARNING

Notice the application of red sealant on certain adjustment screws or plugs. This sealant was applied at the factory and is the manufacturer's method of instructing anyone servicing the carburetor not to disturb the screw or plug.

The sealant is not affected by carburetor cleaner and should remain in place on the plug or screw and the carburetor for the entire useful life of the carburetor.
1. Back out the idle mixture screw. If servicing a 4-Cylinder powerhead, the idle mixture screw on the two lower carburetors exists but is not adjustable and must not be disturbed, as just described in the above "Very Important Words".
If servicing a 3-Cylinder powerhead, remove the bowl vent jet located next to the idle mixture screw. This jet is identical on all three carburetors. The 4-Cylinder powerheads are not equipped with this jet.
2. Remove the two Phillips head screws and captive lockwashers and then remove the mixing chamber cover and gasket.
3. Remove the four Phillips head screws with captive lockwashers from the float bowl. Lift off the fuel bowl. Remove and discard the gasket.

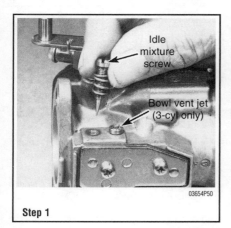
Step 1

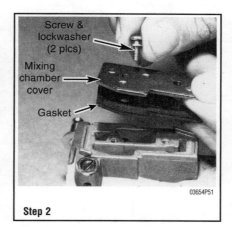
Step 2

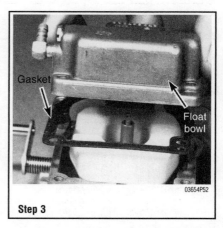
Step 3

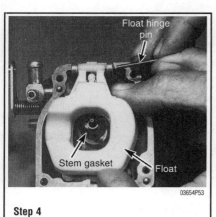
Step 4

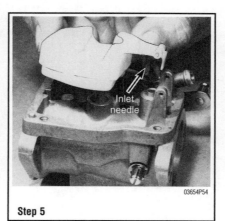
Step 5

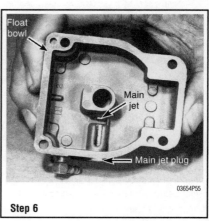
Step 6

4-30 FUEL SYSTEM

4. Remove the stem gasket from the center of the fuel bowl. Support the float and at the same time push the float hinge pin free of the mounting posts.

5. Carefully lift the float with the inlet needle attached from the needle seat. Detach the inlet needle from the float. The needle seat is not removable on this carburetor.

6. Remove the main jet plug and gasket from the exterior wall of the float bowl. This plug provides access for a screwdriver to be inserted into the opening and allows removal of the main jet from the side of the center turret. This design facilitates changing the size of the main jet, without removal of the carburetor from the powerhead. Main jet sizes must be changed when operating the powerhead at elevations higher than 2,500 feet above sea level.

→ **Further disassembly of the carburetor is not necessary in order to clean it properly.**

CLEANING & INSPECTION

♦ See Figures 31 thru 36

Never dip rubber parts, plastic parts, diaphragms or pump plungers in carburetor cleaner, because they tend to absorb liquid and expand. These parts should be cleaned only in solvent and then blown dry with compressed air immediately.

Place all of the metal parts in a screen-type tray and dip them in carburetor cleaner until they appear completely clean—free of gum and varnish which accumulates from stale fuel. Blow the parts dry with compressed air.

Blow out all of the passageways in the carburetor with compressed air. Check all of parts and passageways to be sure they are clear and not clogged with any deposits.

Never use a piece of wire or any type of pointed instrument to clean drilled passages or calibrated holes in a carburetor.

Carefully inspect the casting for cracks, stripped threads or plugs for any sign of leakage. Inspect the float hinge in the hinge pin area for wear and the float for any sign of leakage.

Examine the inlet needle for wear. If there is any evidence of wear, the needle must be replaced.

Always replace any worn or damaged parts. A carburetor service kit is available at modest cost from the local Mercury/Mariner dealer. The kit will contain all necessary parts to perform the usual carburetor overhaul.

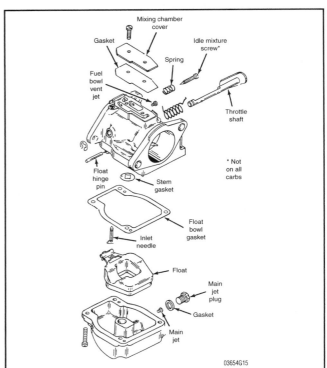

Fig. 32 Exploded drawing of Walbro "WME" carburetor, with major parts identified

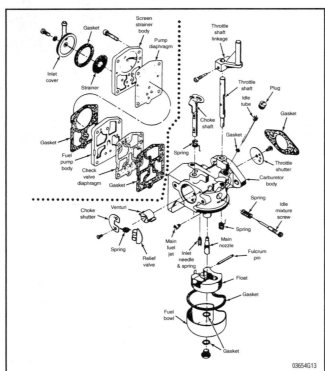

Fig. 31 Exploded drawing of Walbro "WMA" carburetor with integral fuel pump. Fuel pump parts are to the left of the dotted line. Major parts are identified

Fig. 33 NEVER use a piece of wire or any type of pointed instrument to clean drilled passages or calibrated holes in a carburetor, particularly those in the mixing chamber

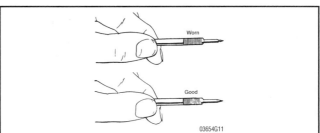

Fig. 34 Carburetor idle mixture adjustment needles. The top needle is worn and unfit for service. The bottom needle is new

FUEL SYSTEM 4-31

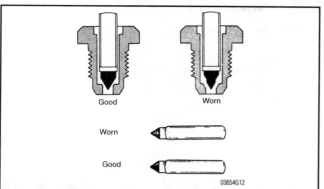

Fig. 35 Needle and seat arrangement showing a worn and new needle for comparison

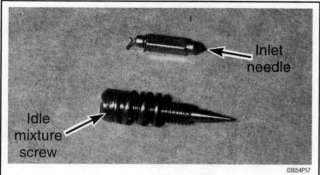

Fig. 36 Inspect the taper of the inlet needle and the idle mixture screw for evidence of a worn groove

ASSEMBLY

WMA

1. Install the main fuel (high-speed) jet. As mentioned during removal, a gasket is not used under this jet.
2. Install a new inlet valve needle and spring to reduce the chances of a leak. install a new float bowl gasket.
3. Install the float, and then insert the float retaining pin to secure the float in place.
4. Hold the carburetor as shown, and measure the distance to the bottom edge of the float. This measurement should be ¼ in. ± ¹¹⁄₆₄ in. (6.35 ± 0.40inrn). Carefully bend `the float needle actuating lever to obtain the correct measurement.
5. Hold the carburetor upside down, as shown, to allow the float to drop to its lowest point. Measure the distance from the bottom of the float to the top of the main fuel (high-speed) jet. This distance should be from ¹⁄₆₄–¹⁄₃₂ in. (0.40 to 0.80mm). Carefully bend the float tang to obtain the correct measurement.
6. Check to be sure the float bowl gasket is in place properly. Position the float bowl gasket on the carburetor casting. Install the float bowl.
7. Slide a new gasket onto the retaining bolt, and then install the bolt onto the float bowl cover.
8. Position a new fuel pump gasket onto the carburetor casting taking care to

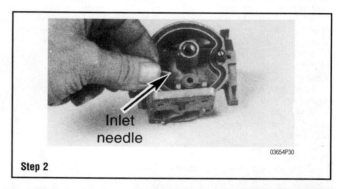

Step 2

Step 3

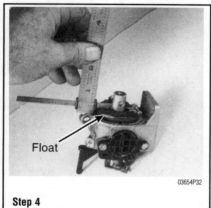

Step 4

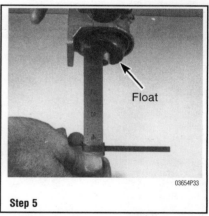

Step 5

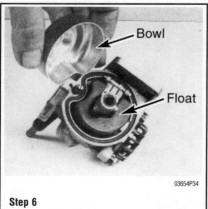

Step 6

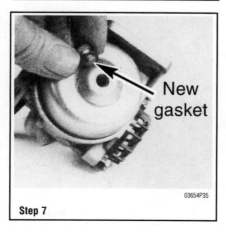

Step 7

Step 8

4-32 FUEL SYSTEM

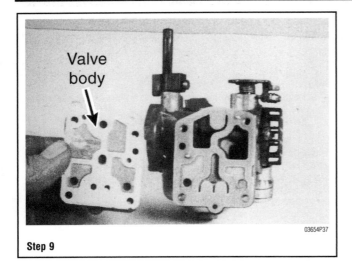
Step 9

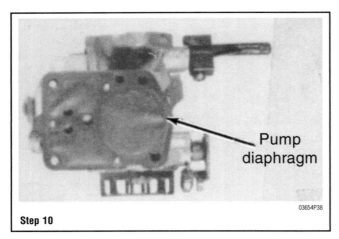
Step 10

WME

♦ See accompanying illustrations

1. Hook the inlet needle spring over the float tab and lower the needle into its seat with the float hinge between the mounting posts.
2. Slide the float hinge pin through the posts to secure the float. Center the pin between the posts. Place the stem gasket over the center turret.
3. Hold the carburetor body in the inverted position—the same position it has been in since the start of the assembling procedures—with the float resting on the inlet needle. Measure the distance between the float bowl gasket surface and a point

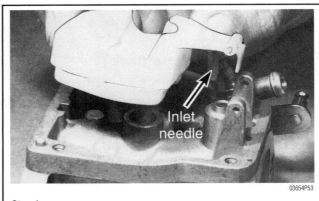
Step 1

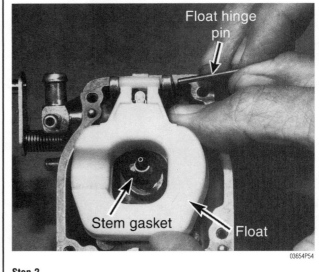
Step 2

index it over the alignment dowel. Install a new valve diaphragm. Observe the three valve flaps, two for inlet control and one for outlet.

9. Install the valve body with the alignment dowel entering the hole in the casting properly.
10. Position a new gasket on the valve body. Install a new dump diaphragm.
11. Install the fuel pump strainer body and secure it in place with the four retaining screws. Tighten the screws evenly and a little-at-a-time. Install the filter screen and a new gasket.
12. Slide a new gasket onto the idle tube and then thread it into place. Install the plug screw over the idle tube.
13. Slowly thread the idle mixture adjusting screw into the carburetor body until you can feel it seat. Do not tighten the screw or you will damage the tip. Now, as a preliminary adjustment, back it out 1¼ turns.

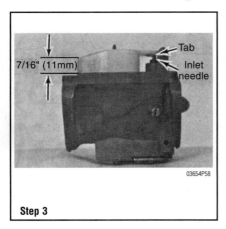
Step 3

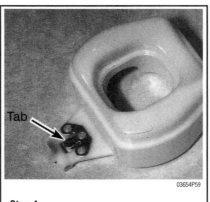
Step 4

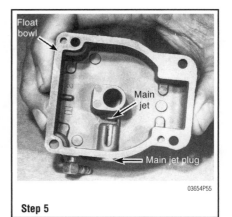
Step 5

FUEL SYSTEM 4-33

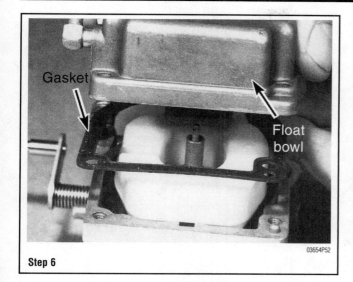

Step 6

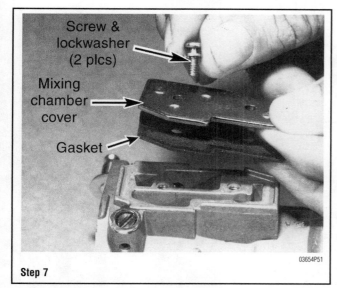

Step 7

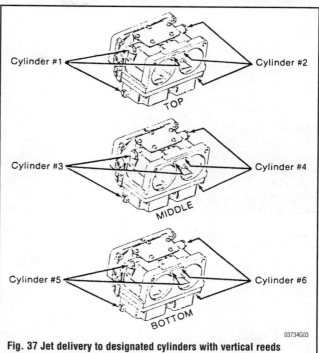

Fig. 37 Jet delivery to designated cylinders with vertical reeds installed—1990 135–175hp

on the float directly opposite the hinge. Notice the surface of the float curves downward. Therefore, the measurement point is the lowest on the horizontal surface.

4. If the distance is not as specified, then remove the float hinge pin to free the float. The float height adjustment may be made by bending the tab on which the inlet needle hangs. Repeat above steps until the specified distance between the float bowl gasket surface and the lowest edge of the float is obtained.

5. Install the main jet into the center turret of the float bowl. Tighten the jet securely. Install the gasket and main jet plug in the exterior wall of the float bowl and tighten the plug securely.

6. With the carburetor still inverted, position a new gasket onto the body. Place the fuel bowl in position and secure it in place with the four Phillips head screws and captive lockwashers.

7. Position a new gasket over the mixing chamber. Install the cover and secure it in place with the attaching hardware.

1990 135–200 HP

♦ See Figure 37

The 1990 135hp–200hp models are equipped with an early version of the Walbro WMH carburetor. These carburetors are provided with three sets of jets. The three sets of changeable jets include two main or high speed jets, two vent jets and two idle air jets.

Idle speed fuel mixture is not adjustable on this model carburetor. In order to change the fuel mixture, the size of the idle air jet must changed.

The main jets, located on each side of the carburetor fuel bowl, control the high speed air/fuel mixture. Fuel is drawn from the fuel bowl, through the main jets. A set of discharge tubes guide the fuel into the carburetor venturi. A jet with a small orifice will provide a lean air/fuel mixture. A jet with a larger orifice will provide a rich air/fuel mixture.

The vent jets, located at the carburetor throat, lower atmospheric pressure at the fuel bowl, resulting in improved fuel economy. As with the main jets, the smaller the orifice—the leaner the mixture, the larger the orifice—the richer the mixture. On some V6 models, the manufacturer calls for different size vent jets for the center and bottom carburetor from those recommended for the top carburetor. The manufacturer takes one further step by calling for different size jets on the starboard and port sides of the carburetor.

The set of idle jets, located on either side of the mixing chamber cover, meter air flow. Their operating principle is opposite to the main and vent jets. The smaller the orifice—the richer the mixture, the larger the orifice—the leaner the mixture.

The off-idle jet is located in each carburetor bore. Each venturi has one off-idle tube and one off-idle air jet. This circuit provides metered fuel during powerhead operation above idle speeds to full throttle operation. If a jet with a smaller orifice is installed, the fuel ration becomes richer. If a jet with a larger orifice is installed, the fuel/air ration becomes leaner.

The idle circuit consists of two externally adjustable fuel mixture screws, idle air bleed jet and factory set air trim screws located beneath welch plugs. The idle circuit operates independently of the progression and main jet circuits.

The two fuel mixture screws provide a maintenance capability of lean or richen the idle fuel mixture for a smoother idle speed. The screws have a limiter cap installed, from the factory, to limit the adjustments to ½ turn. When adjusting the fuel mixture screws all screws must be turned the same amount and in the same direction for the engine to operate efficiently at idle speeds. Turning the screws clockwise will lean the fuel mixture and turning them counterclockwise will richen the fuel mixture.

Do not remove the limiter caps to further richen or lean the fuel mixture. If the limiter caps and fuel mixture screws are removed for carburetor cleaning, count the number of turns clockwise (in) until the screw is lightly seated and then remove the screws.

The idle air bleed jet meters air and is located next to the back draft vent jet. If a jet with a smaller orifice is installed the idle fuel/air mixture becomes richer, if a jet with a larger orifice is installed, idle fuel/air ratio becomes leaner.

The two air trim screws are located beneath welch plugs and are preset at the factory. The manufacturer recommends these welch plugs and air trim screws not be removed.

REMOVAL & INSTALLATION

♦ See accompanying illustrations

1. Disconnect the battery leads from the battery terminals. Disconnect the fuel line at the fuel tank or at the powerhead. Remove the two bolts on the front of the

4-34 FUEL SYSTEM

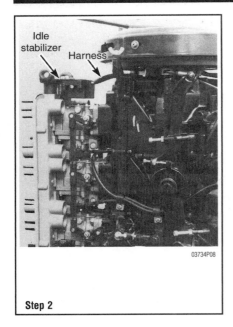
Step 2

Step 3

Step 4

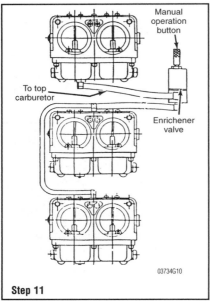
Step 11

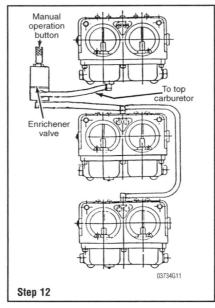
Step 12

Step 15

cover and lift the cover from the attenuator. On the rear of the attenuator plate, disconnect the bleed hose from the fitting on the port side lower corner.

2. Trace the electrical harness from the idle stabilizer, mounted on top of the attenuator plate, to the powerhead switch box. This harness will contain three multicolored wires, a White/Black, a Red/White and a Black wire. Make a note—to which terminal—at the switch box—these wires are connected, then disconnect all three.

3. Remove the six bolts securing the attenuator plate to the carburetors and then lift the plate free of the carburetors. Disconnect the fuel inlet hose to the top carburetor, middle carburetor and the bottom carburetor. Disconnect the fuel enrichener hose from each fitting on the starboard side of the carburetors.

4. Disconnect the throttle linkage by removing the three screws, one at each carburetor and the oil pump control link arm. Slide the throttle linkages off the throttle shaft. Disconnect the bleed hoses from each carburetor.

➡️ **Each carburetor must be installed back into the exact location from which it was removed. The jets on the port and starboard side may differ on an individual carburetor and the jets differ from one carburetor to another, top, center and bottom. Therefore, take time to make an identification mark on each carburetor before removing it from the powerhead. do not make a mark on the mating surface to the choke plate. A preferred location is on the float bowl.**

5. Remove the four attaching nuts securing each carburetor to the intake manifold and then remove the carburetor.

To install:

6. Place a new flange gasket for each carburetor in position on the intake manifold. Install and secure each carburetor in place with the four attaching nuts.

7. Connect all the fuel lines to the carburetors and tighten the clamps securely.

8. Slide the throttle control linkage over each of the carburetor throttle shafts. Secure each linkage to the shaft with a Phillips head screw. As the connections are made, look into each carburetor to be sure the throttle shutters are closed. If the shutters are not closed, loosen the Phillips head screw and force the shutter closed. Tighten the screw and proceed with connecting the linkage.

9. Connect the oil control lever from the oil pump to the linkage arm.

10. Connect the fuel enrichener hoses to each fitting on the starboard side of the carburetors.

11. The enrichener valve and hose routing on the port side of a V6 powerhead should look like this.

12. The enrichener valve and hose routing on the starboard side of a V6 powerhead should look like this.

13. Connect the fuel inlet hose to the top carburetor, middle carburetor and the bottom carburetor. Install the attenuator plate over the front of the carburetors. Align the holes in the plate with the carburetors. Install the six bolts through the attenuator plate, securing the plate to the carburetors.

FUEL SYSTEM   4-35

14. Install the idle stabilizer on top of the attenuator plate. Route the wire harness down to the switch box. Connect the three or four wires from the idle stabilizer to the appropriate terminals on the switch box, as noted during removal. On the rear of the attenuator plate, connect the bleed hose to the fitting on the port side lower corner.

15. Install the sound box cover over the attenuator plate and secure the plate with two bolts. Connect the fuel line at the fuel tank and at the powerhead.

16. Connect the battery leads and perform the timing and synchronizing procedures.

DISASSEMBLY

♦ See accompanying illustrations

1. Remove the two main jet plugs and gaskets, one on each side in the bottom of the carburetor bowl. After the plugs have been removed, note the size of each high speed jet inside each plug. The jets may now be removed from the plugs using the proper size screwdriver and wrench.

Main jet sizes must be changed when operating the powerhead at elevations higher than 2,500 feet above sea level.

2. Remove the fuel inlet hose fitting. Check the filter screen inside the hole. Remove the screen. The factory does not recommend using a filter in the carburetor.

Turn the carburetor upside down and remove the six screws securing the bowl to the carburetor body. Lift the bowl from the body and then remove and discard the bowl gasket and the two nozzle gaskets.

3. Remove the two screws holding the float pins to the fuel bowl.

4. Lift out both floats, pins and inlet needles. Remove both inlet needle seats and the gasket under each seat from the fuel bowl.

5. Remove the two idle jet plugs and gaskets. A plug is located on each side of the body near the top edge. Using the proper size screwdriver, remove the jets from the plugs. Take time to set the jets and other small parts aside and in order, to ensure they will be installed back in the proper sequence and location.

➡ The two idle jets are not normally disturbed during a carburetor overhaul. However, the idle jets together with the main jets and the fuel bowl vent jets, must be changed when operating the powerhead at elevations higher than 2,500 feet above sea level.

6. Remove the five screws securing the cover plate to the mixing chamber. Lift off the cover plate and the gasket.

➡ Further disassembly of the carburetor is not necessary in order to clean it properly. Normally it is not necessary to service the fuel bowl vent jets. However, if the jets are removed, they must be kept in order and installed back into the correct carburetor. Remember, the jets sizes differ between top, center and bottom carburetors.

CLEANING & INSPECTION

♦ See Figures 38 and 39

Never dip rubber parts, plastic parts, diaphragms or pump plungers in carburetor cleaner, because they tend to absorb liquid and expand. These parts should be cleaned only in solvent and then blown dry with compressed air immediately.

Place all of the metal parts in a screen-type tray and dip them in carburetor cleaner until they appear completely clean—free of gum and varnish which accumulates from stale fuel. Blow the parts dry with compressed air.

Blow out all of the passageways in the carburetor with compressed air. Check all of parts and passageways to be sure they are clear and not clogged with any deposits.

Never use a piece of wire or any type of pointed instrument to clean drilled passages or calibrated holes in a carburetor.

Carefully inspect the casting for cracks, stripped threads or plugs for any sign of leakage. Inspect the float hinge in the hinge pin area for wear and the float for any sign of leakage.

Examine the inlet needle for wear. If there is any evidence of wear, the needle must be replaced.

Always replace any worn or damaged parts. A carburetor service kit is available at modest cost from the local Mercury/Mariner dealer. The kit will contain all necessary parts to perform the usual carburetor overhaul.

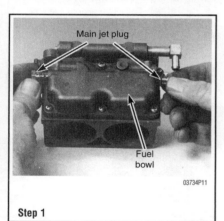
Step 1

Step 2

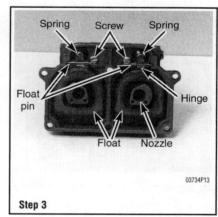
Step 3

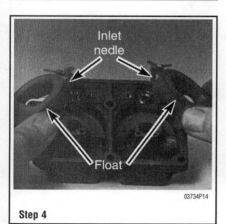
Step 4

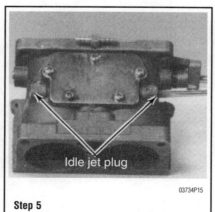
Step 5

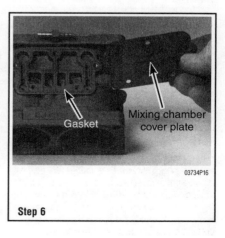
Step 6

4-36　FUEL SYSTEM

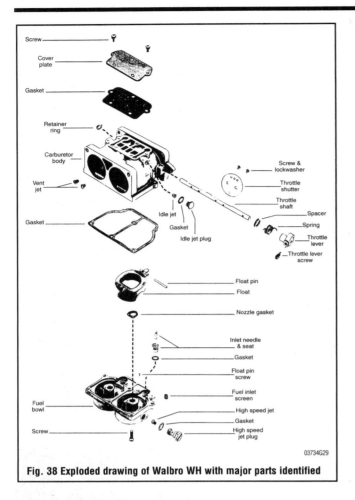

Fig. 38 Exploded drawing of Walbro WH with major parts identified

Fig. 39 Location of the fuel bowl vent jets. Changing to a jet with smaller orifice will result in a leaner air/fuel mixture at midrange. A larger orifice will result in a richer mixture at midrange

ASSEMBLY

▶ See accompanying illustrations

1. Install the two fuel bowl vent jets between the carburetor throats, if they were removed.
2. Place a new gasket over the mixing chamber. Install the mixing chamber cover and secure it with the attaching hardware.
3. Install the two idle jets into their recesses on either side of the mixing chamber cover. Position a new gasket on each idle jet plug. Thread the plugs into the same location from which they were removed, as noted during disassembling.
4. Install both inlet needle seats with new gasket into the fuel bowl. Discharge a drop of oil into each seat. Hook the float pin and inlet needle with the spring onto each float. Lower each float into the fuel bowl and at the same time guide the inlet needle into its seat. Secure the float pin to the fuel bowl with the screw tightened securely.
5. Turn the bowl upside down and allow the float to drop. Measure the distance from the base of the bowl to the bottom of the float. This measurement should be 1/16 in. (1.6mm). carefully bend the float tab to obtain the proper measurement.

Step 1

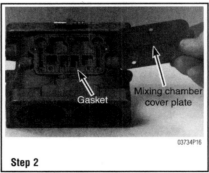

Step 2

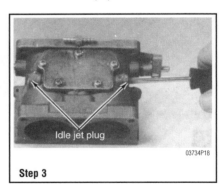

Step 3

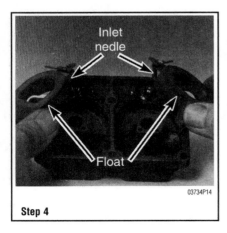

Step 4

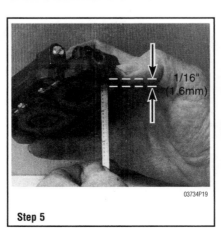

Step 5

Step 8

FUEL SYSTEM 4-37

6. Position a new fuel bowl gasket onto the carburetor body. Slide a new nozzle gasket down over each nozzle and into place on the carburetor body. With the carburetor body upside down, lower the bowl assembly onto the body.

7. Secure it in place with the six screws and lockwasher in the proper sequence. Tighten the No. 1 screw first and then tighten the others in a circular pattern working in either direction. Tighten the No. 1 screw a second time. The manufacturer does not recommend replacing the fuel filter. Install the fuel hose fitting.

8. Check to be sure each main jet plug has its jet installed. Install the main jet plugs with new gaskets into both sides of the fuel bowl.

1991–00 135–200 HP

▶ See Figures 40, 41 and 42

1991–00 135–225hp models are equipped with a later model of the Walbro WMH carburetor. These newer carburetors are provided with two off idle jets, two main (high speed) jets, one idle air bleed jet, one back draft jet and two adjustable idle mixture screws. They can be identified by having two barrels and only one float. The 1991 carburetor had an accelerator pump. After 1991, the pump circuit was no longer used and the opening for the pump was merely sealed with a plug.

The main jets, located on each side of the carburetor fuel bowl, control the high speed air/fuel mixture. Fuel is drawn from the fuel bowl, through the main jets. A set of discharge tubes guide the fuel into the carburetor venturi. A jet with a small orifice will provide a lean air/fuel mixture. A jet with a larger orifice will provide a rich air/fuel mixture.

The off-idle jet is located in each carburetor bore. Each venturi has one off-idle tube and one off-idle air jet. This circuit provides metered fuel during powerhead operation above idle speeds to full throttle operation. If a jet with a smaller orifice is installed, the fuel ration becomes richer. If a jet with a larger orifice is installed, the fuel/air ration becomes leaner.

The idle circuit consists of two externally adjustable fuel mixture screws, idle air bleed jet and factory set air trim screws located beneath welch plugs. The idle circuit operates independently of the progression and main jet circuits.

The two fuel mixture screws provide a maintenance capability of lean or richen the idle fuel mixture for a smoother idle speed. The screws have a limiter cap installed, from the factory, to limit the adjustments to ½ turn. When adjusting the fuel mixture screws all screws must be turned the same amount and in the same direction for the engine to operate efficiently at idle speeds. Turning the screws clockwise will lean the fuel mixture and turning them counterclockwise will richen the fuel mixture.

Do not remove the limiter caps to further richen or lean the fuel mixture. If the limiter caps and fuel mixture screws are removed for carburetor cleaning, count the number of turns clockwise (in) until the screw is lightly seated and then remove the screws.

The idle air bleed jet meters air and is located next to the back draft vent jet. If a jet with a smaller orifice is installed the idle fuel/air mixture becomes richer, if a jet with a larger orifice is installed, idle fuel/air ratio becomes leaner.

The two air trim screws are located beneath welch plugs and are preset at the factory. The manufacturer recommends these welch plugs and air trim screws not be removed.

The back draft circuit consists of a vent jet which supplies less than atmospheric pressure to the fuel bowl during mid-range power settings. The lower atmospheric pressure in the fuel bowl during mid-range power operation causes the fuel flow to become slightly lean and results in improved performance and fuel economy.

REMOVAL & INSTALLATION

▶ See accompanying illustrations

1. Carburetors must remain in the exact position from where they are removed. The top, center and bottom carburetors have different size main and vent jets. Identify each carburetor position prior to removal.

2. Unlatch and remove the cowling from the powerhead. Disconnect the battery leads from the battery terminals. Disconnect the main fuel line from the fuel tank and the powerhead. Remove the two screws securing the air box cover to the powerhead. Remove the six bolts securing the air box to the carburetors and lift the air box free of the powerhead.

3. Disconnect the throttle linkage from each carburetor throttle lever by inserting a flat blade screwdriver between the throttle lever and the nylon linkage arm. Pry the nylon linkage arm off the end of the throttle lever. Disconnect the oil pump link rod from the bottom carburetor throttle lever by prying the link arm off the ball stud on the throttle lever.

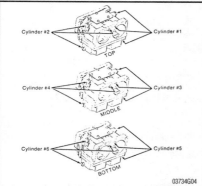

Fig. 40 Jet delivery to designated cylinders with horizontal reeds installed—1990–00 135–225hp

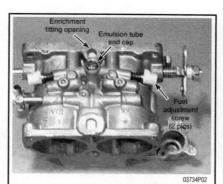

Fig. 41 Top view of the Walbro WMH carburetor showing the enrichment fitting opening, emulsion tube and cap and the fuel adjustment screw

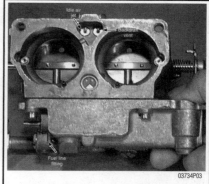

Fig. 42 Another view of the Walbro WMH, showing the idle air jet, backdraft jet and fuel line fitting

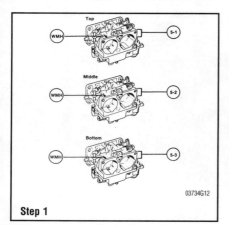

Step 1

Step 2

Step 3

4-38 FUEL SYSTEM

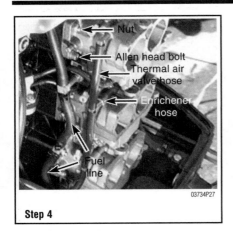

Step 4

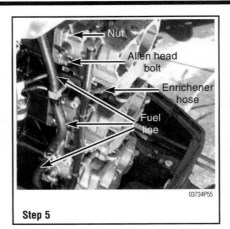

Step 5

Step 6

4. Loosen the clamps or cut the tie-straps and disconnect the fuel enrich-ment hoses, fuel lines and the thermal air valve hoses from the carburetor fit-tings. Mark each carburetor for position (Top, Center and Bottom) before removal to be sure each carburetor is installed back in its original position. Remove the two nuts and two Allen head bolts securing the carburetor to the intake manifold and lift the carburetor free of the powerhead. Discard the carburetor gasket.

To install:

5. Position a new carburetor gasket onto the intake manifold studs. Install the carburetor in the same position from which it was removed—Top, Center and Bottom—as identified during the removal procedures. Secure each carburetor to the powerhead with the two nuts and two Allen head screws. Connect the fuel enrichment hoses, fuel lines and the thermal air valve hoses to the carburetor fittings. Secure the lines and hoses with clamps or tie-straps.

6. Connect the throttle linkage to each carburetor throttle lever by pushing the nylon linkage arm onto the end of the throttle lever. Connect the oil pump link rod to the bottom carburetor throttle lever by pushing the link arm onto the ball stud on the throttle lever.

7. Install the air box over the carburetors, aligning the holes in the cover with the carburetors. Secure the air box to the carburetors with six bolts and tighten the bolts securely. Connect the main fuel line to the fuel tank and the powerhead. Connect the battery leads to the battery terminals.

➡ When operating the outboard for the first time following a carburetor overhaul, be sure to check all fuel line connections and the carburetor float bowl seams for any sign of fuel leakage. With the powerhead operating, check for fuel dribbling from the carburetor vent tube. Fuel leaks from the vent tube and float bowl seams indicate the float level is not set properly or a needle valve is sticking.

After the powerhead has been operated and verified no fuel leaks exist, install the air box cover and secure it with the two screws. Install the cowling over the powerhead and latch it closed.

DISASSEMBLY

♦ See accompanying illustrations

➡ Proper identification of the carburetor by model number is critical when purchasing replacement parts. Without these numbers the dealer has no way of ordering the correct parts. The model WMH carburetor is identified on the starboard side of the main body by a single or double digit number. If there is any doubt, take the carburetor to the dealer for his positive identification.

1. Loosen and remove the off-idle tube access plug. Discard the gasket on the plug.

2. Insert a long flat blade screwdriver down inside the counterbore and unscrew the off-idle tube. Withdraw the off-idle tube from the bore and set the tube to one side. Repeat steps to remove the other off-idle tube from the other side of the carburetor body.

3. Loosen and remove the main jet access plug. Discard the gasket on the plug.

4. Insert a long flat blade screwdriver down inside the counterbore and unscrew the main jet. Withdraw the main jet from the bore. Set the jet to one side. Repeat steps to remove the other main jet access plug and main jet from the other side of the carburetor body.

5. Use a flat blade screwdriver and remove the emulsion tube access plug on top of the carburetor. Pull the emulsion tube and plug from the carburetor. If only the plug comes out, the emulsion tube may be removed once the float bowl is removed.

➡ Limiter caps are installed on the idle air mixture screw at the factory. The caps are necessary to prevent an over rich or too lean a fuel mixture but do allow some movement for idle speed fuel mixture settings. If it is necessary to remove the idle mixture screws from the carburetor for proper cleaning, remove the limiter cap and turn the screw clockwise and count the number of turns until it is lightly seated. The mixture screw must be set back to this number of turns from the lightly seated position when the limiter caps are installed.

6. Slip the nylon limiter cap off the end of the idle mixture screw. Using a flat blade screwdriver turn the mixture screw clockwise and count the number of turns until it is lightly seated.

7. Unscrew the idle mixture screw fully from the bore. Do not loose the spring over the screw shank. Repeat for the other idle mixture screw on the other side of the carburetor.

8. The two off-idle jets, idle air jets and the vent jets are all installed in the front of the carburetor body. If the carburetor is being disassembled only for general cleaning and these jets are not to be replaced, it is recommended these jets not be

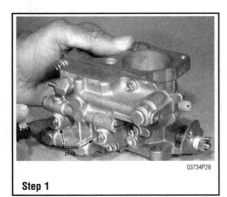

Step 1

Step 2

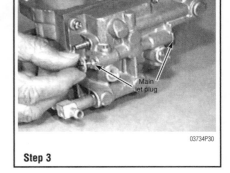

Step 3

FUEL SYSTEM 4-39

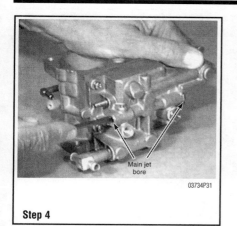
Step 4

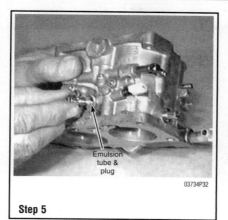
Step 5

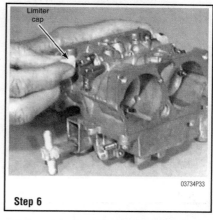
Step 6

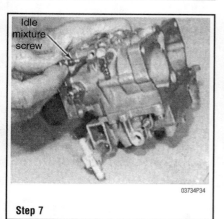
Step 7

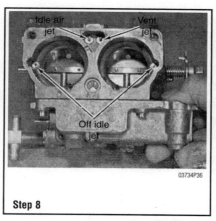
Step 8

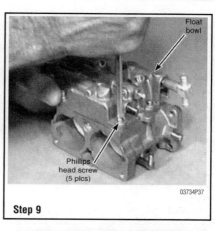
Step 9

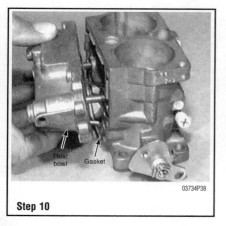
Step 10

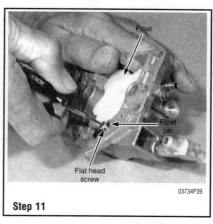
Step 11

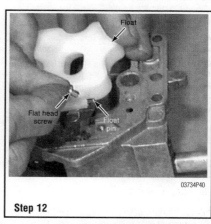
Step 12

disturbed and a good quality carburetor spray cleaner be applied. If the jets are being replaced, remove each jet using a flat blade screw driver.

9. Remove the five Phillips head screws securing the float bowl to the carburetor main body.

10. Separate the float bowl from the main body and slide the bowl assembly free of the carburetor. Discard the gasket between the float bowl and main body.

11. Hold the float bowl carefully and remove the flat head screw securing the float pin to the bowl assembly.

12. Lift out the screw, float and pin from the float bowl. Examine the float assembly for signs of cracking or saturation of fuel. Replace the float assembly if found to be defective.

13. Lift the fuel needle valve from the needle seat assembly. Examine the needle valve for signs of wear and deterioration. If the needle valve is worn, replace the needle valve and seat as an assembly. Discard the needle valve if a new assembly is to be installed.

14. Use a "wide" flat blade screwdriver and remove the needle valve seat from the bowl assembly. Discard the gasket on the valve seat. If a new needle valve and seat is to be installed, discard both the seat and gasket.

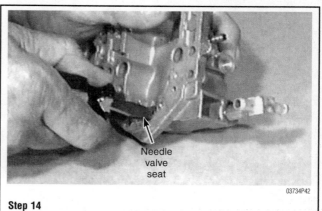
Step 14

4-40 FUEL SYSTEM

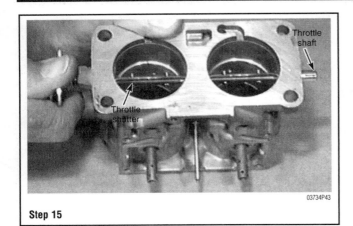

Step 15

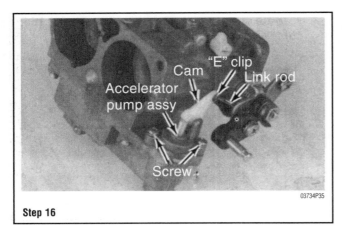

Step 16

Push and pull on the throttle shaft, if there is sign of excessive "play" in the shaft and/or the shutters are gouging the throttle bore heavily. The throttle shaft and bushings are defective or worn and must be replaced.

➥**Further disassembly of the carburetor is not necessary in order to clean it properly. Normally it is not necessary to disassemble the throttle shutters unless there is excessive "play" or the shutters are binding. Before disassembling the shutters and throttle shaft, first check with the local dealer for parts availability. The shutters on some carburetor models cannot be serviced as individual parts. In some cases the throttle body must be replaced.**

15. Operate the throttle shutters open and closed several times checking for any sticking or binding action. If the throttle shutters show any sign of binding or sticking they must be replaced.
16. The 1991 model year carburetor was equipped with an accelerator pump circuit.
17. Remove the E-clip from the throttle cam link rod.
18. Remove the Phillips head screws securing the accelerator pump assembly to the float bowl assembly.
19. Lift off the accelerator pump assembly with the plunger from the float bowl.
20. Remove the plunger return spring from the pump bore.
21. Inspect the seal around the pump cup. If the pump cup seal is hard or damaged, replace the seal.

CLEANING & INSPECTION

◆ See Figure 43

Never dip rubber parts, plastic parts, diaphragms or pump plungers in carburetor cleaner, because they tend to absorb liquid and expand. These parts should be cleaned only in solvent and then blown dry with compressed air immediately.

Place all of the metal parts in a screen-type tray and dip them in carburetor cleaner until they appear completely clean—free of gum and varnish which accumulates from stale fuel. Blow the parts dry with compressed air.

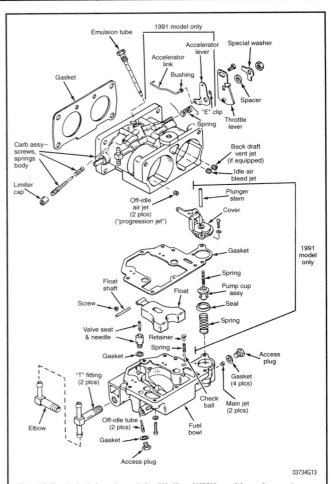

Fig. 43 Exploded drawing of the Walbro WMH—with major parts identified

Blow out all of the passageways in the carburetor with compressed air. Check all of parts and passageways to be sure they are clear and not clogged with any deposits.

Never use a piece of wire or any type of pointed instrument to clean drilled passages or calibrated holes in a carburetor.

Carefully inspect the casting for cracks, stripped threads or plugs for any sign of leakage. Inspect the float hinge in the hinge pin area for wear and the float for any sign of leakage.

Examine the inlet needle for wear. If there is any evidence of wear, the needle must be replaced.

Always replace any worn or damaged parts. A carburetor service kit is available at modest cost from the local Mercury/Mariner dealer. The kit will contain all necessary parts to perform the usual carburetor overhaul.

ASSEMBLY

◆ See accompanying illustrations

1. Slip a new gasket over the threaded end of the fuel valve seat. Install the seat into the float bowl assembly. Use a "wide" blade screw-driver to tighten the valve seat.
2. Insert the needle valve into the seat and then install the float and float pin. Be sure to center the float pin into the recess of the float bowl. Install the flat head screw to secure the float.
3. Pick up the float bowl and invert the assembly to allow the float to hang free. Place a machinist scale across the flat surface of the float bowl directly under the float. The correct float height is the float just making contact with the scale but the float does not drop below the horizontal surface of the float bowl.
4. If the float height is not correct, carefully bend the small metal tab on the tip of the float to adjust the float height.
5. Place a new gasket over the float bowl. Guide the float bowl onto the carburetor body. Secure the float bowl to the carburetor body with five Phillips head screws.

FUEL SYSTEM 4-41

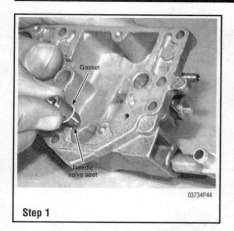
Step 1

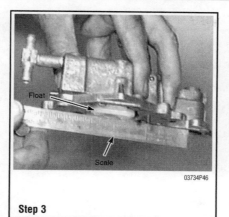
Step 3

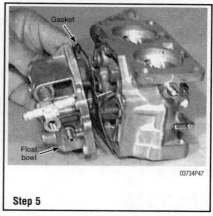
Step 5

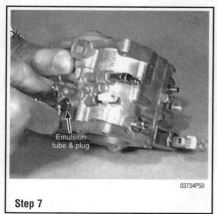
Step 7

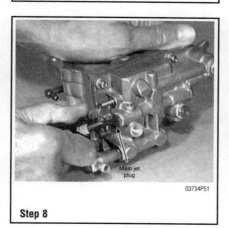
Step 8

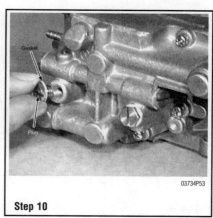
Step 10

6. Thread the idle mixture screw and spring into the bore. Turn the mixture screw all the way down until it is lightly seated. Back out the mixture screw the exact number of turns recorded during disassembly. If the count was lost or not remembered, back it out 1½ turns as a "rough" bench adjustment at this time. Fit the limiter cap over the end of the screw at approximately the 12 O'clock position. Repeat this step for the other idle mixture screw.

7. Insert the emulsion tube and cap plug into the top of the carburetor body. Tighten the emulsion tube plug securely.

8. Install the main jet into the bore using a long narrow screwdriver. Be careful not to cross thread the jet. Place a new gasket over the end of the access plug and then install the plug and tighten it securely. Repeat this step for the other main jet and access plug.

9. Insert the off-idle tube into the bore. Using a flat blade screwdriver, screw the tube into the bore until it is seated and then tighten the tube securely.

10. Place a new gasket over the end of the off-idle tube access plug. Install and tighten the access plug. Repeat for the other off-idle tube.

11. Install the off-idle jets, idle air jet and vent jet, if any were removed during the disassembly procedures. Be sure to install the proper size jet for the carburetor model.

12. On 1991 models with an accelerator pump, install a new seal onto the pump cup. Insert the plunger stem and small spring into the pump cover. The tip of the plunger must contact the tip of the lever. Insert the large spring into the accelerator pump bore. Lower the pump cover onto the fuel bowl and secure the cover with two Phillips head screws. Connect the throttle cam link rod between the pump lever and throttle cam lever, secure the link rod with the E-clip.

1991–00 275 HP

▶ See Figure 44

The Walbro "WMO" is a single barrel, dual float carburetor with a single fuel bowl. It has been used on the 275hp powerhead since 1991. Six units serve the powerhead with each unit serving one cylinder.

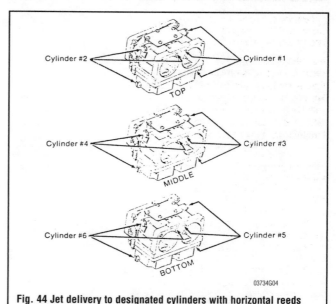

Fig. 44 Jet delivery to designated cylinders with horizontal reeds installed—1991–00 275hp

The main jets, located on each side of the carburetor fuel bowl, control the high speed air/fuel mixture. Fuel is drawn from the fuel bowl, through the main jets. A set of discharge tubes guide the fuel into the carburetor venturi. A jet with a small orifice will provide a lean air/fuel mixture. A jet with a larger orifice will provide a rich air/fuel mixture.

4-42 FUEL SYSTEM

REMOVAL & INSTALLATION

▶ See accompanying illustrations

1. Remove the front powerhead cowling. Take off the wrap around cowling.
2. Release the fuel bayonet from the bracket on the air box cover. Remove the wing nuts securing the air box cover and lift the cover and air box from the carburetors. Remove the gasket from each carburetor.
3. Disconnect the fuel line from the powerhead. Remove the hose clamps on each fuel line to each carburetor. Disconnect the fuel line from each carburetor. Disconnect the enrichener valve hose fitting on the top port carburetor.
4. Disconnect the throttle linkage from each carburetor. Take time to identify each carburetor to ensure each will be installed back in its original position. The jets are different in the top, center and bottom carburetors and the jets may differ from port and starboard carburetors.
5. Remove the attaching nuts securing each carburetor to the intake manifold. Remove each carburetor from the engine. Since the carburetors are identical, the following procedures are to be repeated for each carburetor.

To install:

6. Position a new gasket in place on the intake manifold. Install the carburetor onto the manifold in the same position from which it was removed. Each carburetor should have been identified as instructed during the removal procedures. Secure the carburetor in place with the retaining nuts.
7. Assemble and install the other carburetors in a similar manner.
8. Connect all fuel and enrichener system lines to the carburetors and secure the fuel hoses with Sta-Straps.
9. Connect the enrichener valve hose to the fitting on top port side carburetor.
10. Connect the fuel line from the fuel tank.
11. Activate the fuel line squeeze bulb several times and check the carburetors and fuel lines for leaks.
12. Connect the throttle linkage to and between the carburetors.
13. Place the air box gaskets over the studs of the carburetors. Install the air box plate with the locknuts and flat washers on each carburetor bank. Tighten the locknuts to 60 inch lbs. (6.8 Nm).
14. Install the cover over the air box and secure it in place with the attaching screws. Tighten the screw alternately and evenly to 60 inch lbs. (6.8 Nm).

Step 12

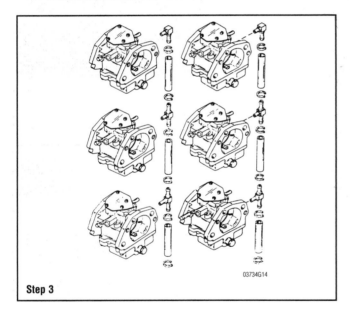
Step 3

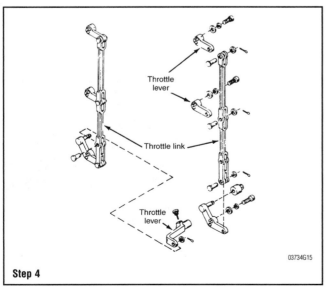
Step 4

Step 13

FUEL SYSTEM   4-43

DISASSEMBLY

♦ See accompanying illustrations

1. Remove the main jet plug located in the bottom of the carburetor bowl. Note how the main (high-speed) jet is located inside the plug. The jet may be removed, using the proper size screwdriver.
2. Remove the idle jet plug, gasket and idle jet, from the carburetor body.
3. Remove the five screws securing the mixing chamber cover. Lift off the cover and gasket.
4. Turn the carburetor upside down and remove the four screws securing the bowl to the body. Remove the bowl and then remove and discard the bowl to body gasket.
5. Observe how the float is a double unit with two hinge pins. Withdraw both hinge pins by pushing each toward the outer edge of the carburetor. After the pins are free, lift the float from the carburetor body.
6. Remove the pin securing the float lever to the carburetor body by pushing the pin toward the backside of the carburetor. Remove the lever.
7. Withdraw the inlet needle from its seat. Remove the seat and the metal gasket installed below the seat.
8. Remove the nozzle in the center of the carburetor body and at the same time observe that the venturi in the bore will now be loose. After the nozzle is out, remove the venturi from the carburetor bore.
9. Remove the throttle return spring, flat washer and rubber seal from the bottom side of the carburetor.
10. Remove the fuel bowl vent jet from the port side.
11. Make sure to note the size and location of each jet during disassembling. All three jets, the main jet, the idle jet and the vent jet, are easily accessible without disassembling the carburetor. These jets take the place of exterior carburetor adjustment screws used on earlier carburetors.

➥Further disassembly of the carburetor is not necessary in order to clean it properly.

Step 1

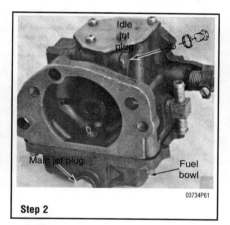
Step 2

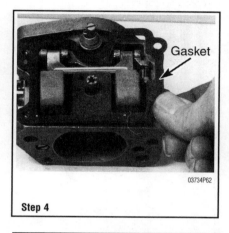
Step 4

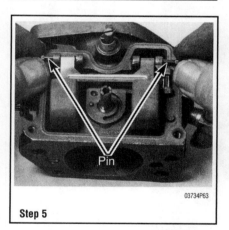
Step 5

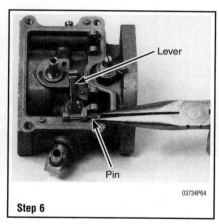
Step 6

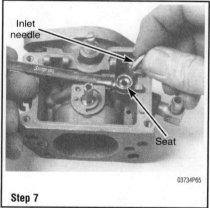
Step 7

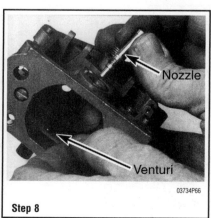
Step 8

Step 9

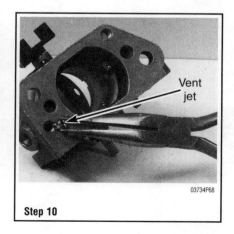
Step 10

4-44 FUEL SYSTEM

CLEANING & INSPECTION

♦ See Figure 45

Never dip rubber parts, plastic parts, diaphragms or pump plungers in carburetor cleaner, because they tend to absorb liquid and expand. These parts should be cleaned only in solvent and then blown dry with compressed air immediately.

Place all of the metal parts in a screen-type tray and dip them in carburetor cleaner until they appear completely clean—free of gum and varnish which accumulates from stale fuel. Blow the parts dry with compressed air.

Never use a piece of wire or any type of pointed instrument to clean drilled passages or calibrated holes in a carburetor.

Carefully inspect the casting for cracks, stripped threads or plugs for any sign of leakage. Inspect the float hinge in the hinge pin area for wear and the float for any sign of leakage.

Examine the inlet needle for wear. If there is any evidence of wear, the needle must be replaced.

Always replace any worn or damaged parts. A carburetor service kit is available at modest cost from the local Mercury/Mariner dealer. The kit will contain all necessary parts to perform the usual carburetor overhaul.

ASSEMBLY

♦ See accompanying illustrations

1. Install the fuel bowl vent jet into the port side of the carburetor and tighten it securely.
2. Turn the carburetor upside down. Slide the rubber seal onto the throttle shaft with the lip toward the carburetor. Observe there are two sizes of springs to be installed onto the throttle shaft of the carburetors. Always install the strongest spring, the one with the largest diameter wire, onto the top mounted carburetor on all model engines. Slide the flat washer and throttle return spring onto the shaft. Hold the shutter plate in the closed position and attach the throttle return spring.
3. Insert the venturi into the front of the carburetor and into the bore. The rounded edge must be installed into the carburetor towards the rear of the carburetor.
4. Install the nozzle through the top of the carburetor and into the venturi. The venturi is now held in place by the nozzle. Tighten the nozzle securely.
5. Position a new metal gasket onto the carburetor body at the inlet needle seat hole. Install the inlet needle seat, with rubber insert, into place and tighten the seat securely. Slide the needle into the seat.
6. Position the float lever between the posts of the carburetor and then slide the hinge pin into place from the rear of the carburetor. Use a flat end punch and seat the hinge pin until the knurled end of the pin is flush or within $\frac{1}{32}$ in. (0.80mm) from the side of the post.

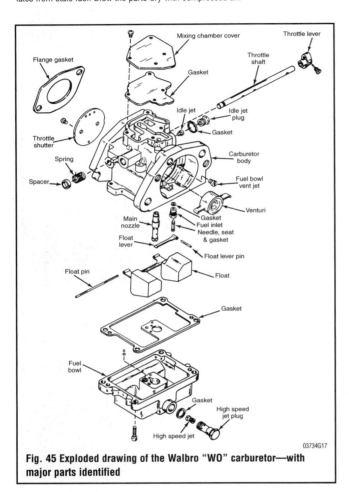

Fig. 45 Exploded drawing of the Walbro "WO" carburetor—with major parts identified

Blow out all of the passageways in the carburetor with compressed air. Check all of parts and passageways to be sure they are clear and not clogged with any deposits.

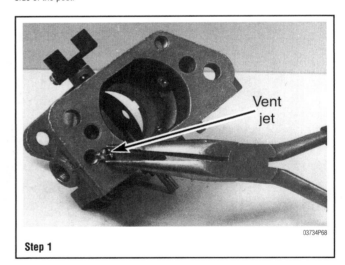

Step 1

Step 2

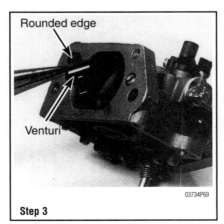

Step 3

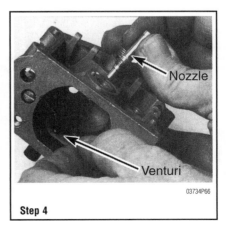

Step 4

FUEL SYSTEM 4-45

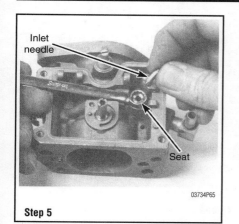
Step 5

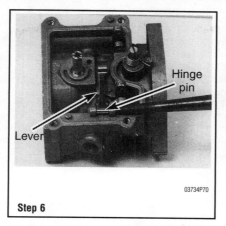
Step 6

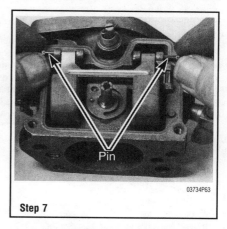
Step 7

Step 8

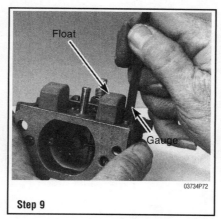
Step 9

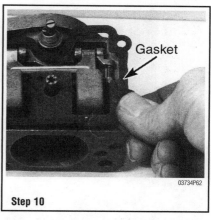
Step 10

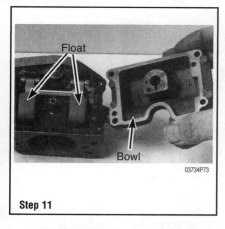
Step 11

Step 12

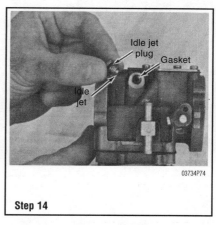
Step 14

7. Slide the float into place between the posts of the carburetor. Insert both hinge pins through the posts from the outside edge.

8. Use a flat end punch to push the pins into the posts until the knurled end of each pin is flush or with 1/32 in. (0.80mm) from the side of the post. Some models of this carburetor may have a single float pin. Installation procedures are identical to those with two pins.

9. Turn the carburetor upside down with the floats resting on the inlet needle. Measure the distance from the base of the carburetor to the bottom edge of the float. This measured distance must be 11/16 in. (17.46mm). carefully bend the float lever to obtain the correct measurement.

10. With the carburetor still upside down, position a new gasket onto the body.

11. Place the fuel bowl in position and secure it with the four attaching screws.

12. Install the main jet into the plug, if it was removed. Use a new gasket and install the plug into the carburetor bowl.

13. Position the gasket over the mixing chamber. Install the cover and secure it in place with the attaching hardware.

14. Install the idle jet and jet plug with a new gasket.

Fuel Pump

TESTING

Lack of an adequate fuel supply will cause the engine to run lean, lose rpm or cause piston scoring due to a lack of lubricant that is carried in the fuel. If an integral fuel pump carburetor is installed, the fuel pressure cannot be checked.

With a multiple carburetor installation, fuel pressure at the top carburetor should be checked whenever insufficient fuel is suspected because it is the farthest from the fuel pump.

4-46 FUEL SYSTEM

⚠️ CAUTION

Observe all applicable safety precautions when working around fuel. Whenever servicing the fuel system, always work in a well ventilated area. Do not allow fuel spray or vapors to come in contact with a spark or open flame. Keep a dry chemical fire extinguisher near the work area. Always keep fuel in a container specifically designed for fuel storage, also, always properly seal fuel containers to avoid the possibility of fire or explosion.

4–25 HP

♦ See accompanying illustrations

1. The high tension wire between the coil and the distributor can be grounded by either pulling it out of the coil and grounding it or by connecting a jumper wire from the primary (distributor) side of the ignition coil to a good ground. An alternate safety method and perhaps a better one, is to ground each spark plug lead. Disconnect the fuel line at the carburetor. Place a suitable container over the end of the fuel line to catch the fuel discharged. Now, squeeze the primer bulb and observe if there is satisfactory flow of fuel from the line.
2. If there is no fuel discharged from the line, the check valve in the squeeze bulb may be defective or there may be a break or obstruction in the fuel line.
3. Testing the fuel pickup in the fuel tank and operation of the squeeze bulb by observing fuel flow from the line disconnected at the fuel pump and discharged into a suitable container.
4. Working the squeeze bulb and observing the fuel flow from the line disconnected at the carburetor and discharged into a suitable container. This verifies fuel flow through the fuel pump.
5. If there is a good fuel flow, then crank the engine. If the fuel pump is operating properly, a healthy stream of fuel should pulse out of the line.
6. Continue cranking the engine and catching the fuel for about 15 pulses to determine if the amount of fuel decreases with each pulse or maintains a constant amount. A decrease in the discharge indicates a restriction in the line. If the fuel line is plugged, the fuel stream may stop. If there is fuel in the fuel tank but no fuel flows out of the fuel line while the engine is being cranked, the problem may be in one of four areas:
7. The line from the fuel pump to the carburetor may be plugged as already mentioned, or the fuel pump may be defective.
8. The line from the fuel tank to the fuel pump may be plugged, the line may be leaking air, or the squeeze bulb may be defective.
9. If the engine does not start even though there is adequate fuel flow from the fuel line, the fuel filter in the carburetor inlet may be plugged or the fuel inlet needle valve and the seat may be gummed together and prevent adequate fuel flow.

30–125 HP

♦ See Figures 46, 47, 48 and 49

1. Fuel pressure should be checked if a fuel tank, other than the one supplied by the outboard unit's manufacturer, is being used. When the tank is checked, be sure the fuel cap has an adequate air vent.
2. Verify that the fuel line from the tank is of sufficient size to accommodate the engine demands. An adequate size line would be one measuring from 5/16 in. –3/8 in. (7.94–9.52mm) ID (inside diameter).
3. Check the fuel filter on the end of the pickup in the fuel tank, to be sure it is not too small and that it is not clogged. Check the fuel pickup tube. The tube must be large enough to accommodate the fuel demands of the engine under all conditions. Be sure to check the filter at the carburetor. Sufficient quantities of fuel cannot pass through into the carburetor to meet engine demands if this screen becomes clogged.

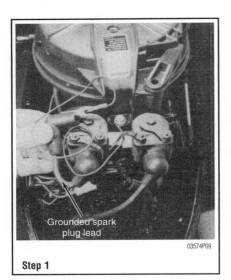

Step 1

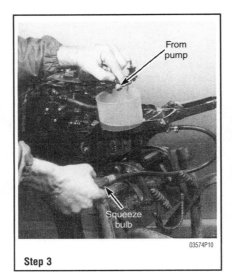

Step 3

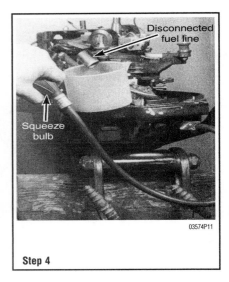

Step 4

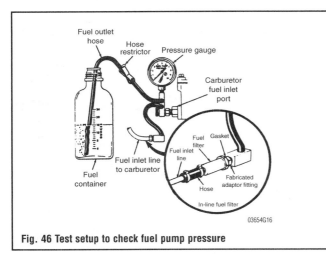

Fig. 46 Test setup to check fuel pump pressure

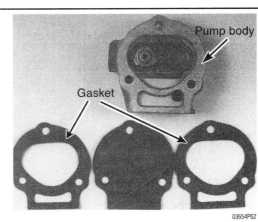

Fig. 47 Major parts of a typical separate fuel pump

FUEL SYSTEM 4-47

4. To test, install the fuel pressure gauge in the fuel line between the fuel pump and the carburetor. If multiple carburetor are installed, connect the gauge in the line to the top carburetor. Operate the engine at full throttle and check the pressure reading. The gauge should indicate at least 2 psi.

V6

1. Inspect the fuel tank and insure the fuel cap has an adequate air vent.
2. Verify that the fuel line from the tank is of sufficient size to accommodate the engine demands. An adequate size line would be one measuring from 5/16 in. – 3/8 in. (7.94–9.52mm) ID (inside diameter).
3. Check the fuel filter on the end of the pickup in the fuel tank, to be sure it is not too small and that it is not clogged. Check the fuel pickup tube. The tube must be large enough to accommodate the fuel demands of the engine under all conditions. Be sure to check the filter at the carburetor. Sufficient quantities of fuel cannot pass through into the carburetor to meet engine demands if this screen becomes clogged.
4. To test, install the fuel pressure gauge in the fuel line between the fuel pump and the carburetor. If multiple carburetor are installed, connect the gauge in the line to the top carburetor. Operate the engine at full throttle and check the pressure reading. The gauge should indicate at least 3 psi.

REMOVAL & INSTALLATION

4 and 5 HP

The fuel pump can be removed for service without removing the carburetor.
1. Disconnect the fuel hose from the fuel pump inlet.
2. Remove the fuel pump screws and separate the pump cover, body and gasket from the carburetor.

To install:
3. Install a new gasket between the fuel pump and the carburetor.
4. Install the fuel pump assembly onto the carburetor. Make sure all the components are aligned properly.
5. Install the mounting screws and tighten securely.

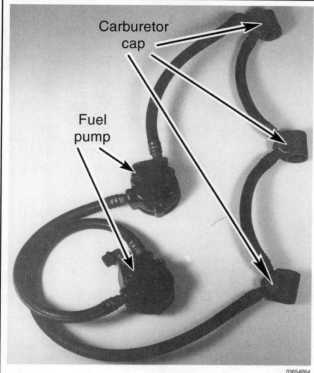

Fig. 48 Fuel pump assemblies removed from the powerhead. Notice how the fuel lines are still connected to the pumps. Further disassembling may be performed on the work bench

6–25 HP

1. Turn the fuel shut-off valve to the **OFF** position.
2. Disconnect the top and bottom fuel lines. Use a golf tee or a stubby pencil to plug the end of each disconnected hose to prevent the loss of fuel.
3. Disconnect the pulse hose from the front surface of the pump.
4. Remove the retaining screw and lift off the fuel pump from the carburetor

To install:
5. Place the fuel pump on the powerhead base using a new mounting gasket.
6. Install the two retaining screws with slotted heads through the pump and into the engine block.
7. Tighten the screws alternately to a torque value of 50–60 inch lbs. (5.5–6.8Nm).
8. Connect the fuel lines or turn the fuel valve to the **ON** position.

30–275 HP

1. On 40hp and 50–60hp models, it is necessary to remove the oil reservoir.
2. Turn the fuel shut-off valve to the OFF position.
3. Cut away the sta-straps from the three hoses at the fuel pump.
4. Disconnect the top and bottom fuel lines. Use a golf tee or a stubby pencil to plug the end of each disconnected hose to prevent the loss of fuel.
5. Disconnect the pulse hose from the front surface of the pump.
6. Observe the four bolts. Notice two of the bolts have slotted heads and two are just plain. The two bolts with slotted heads secure the pump to the block. The other two bolts will hold the pump components together while the pump is being removed. Therefore, remove the two slotted bolts and lift the pump clear of the block.

➡ **On the 250–275hp engines, two fuel pumps are used. These pumps are connected in series with the bottom pump delivering fuel to top pump and the top pump delivering fuel to the carburetors.**

7. Repeat the above steps to remove the second fuel pump if equipped.
8. Remove and discard the mounting gasket.

To install:
9. Place the fuel pump on the powerhead base using a new mounting gasket.
10. Install the two retaining screws with slotted heads through the pump and into the engine block.
11. Tighten the screws alternately to a torque value of 50–60 inch lbs. (5.5–6.8Nm).

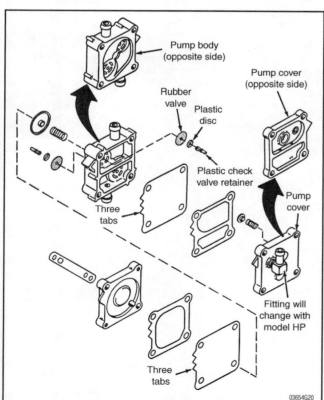

Fig. 49 Exploded drawing of a typical fuel pump. As noted on the drawing, the fitting on the pump cover will change, depending on the powerhead installation

FUEL SYSTEM

12. Connect the fuel lines.
13. Replace the oil reservoir.
14. Turn the fuel valve to the **ON** position.

OVERHAUL

4 and 5 HP

♦ See Figure 50

1. Separate the pump cover, outer gasket and diaphragm from the pump body.
2. Remove the inner gasket and diaphragm from the pump body.
3. Wash all parts thoroughly in solvent and then blow them dry with compressed air. Use care when using compressed air on the check valves. Do not hold the nozzle too close because the check valve can be damaged from an excessive blast of air.
4. Inspect each part for wear and damage. Verify that the valve seats provide a flat contact area for the valve disc. Tighten all elbows and check valve connections firmly as they are replaced.
5. Test each check valve by blowing through it with your mouth. In one direction the valve should allow air to pass through. In the other direction, air should not pass through.
6. Check the diaphragm for pin holes by holding it up to the light. If pin holes are detected or if the diaphragm is not pliable, it must be replaced.

To assemble:

7. Install the check valves and secure with the nuts and screws.
8. Make sure the check valves are centered over the valve seats.
9. Install the inner gasket and diaphragm into the pump body.
10. Attach the pump cover, outer gasket and diaphragm to the pump body.

➡**Make sure the gaskets and diaphragms are properly aligned. Do not use gasket sealing compound on any of the pump components.**

11. Tighten the pump mounting screws securely.

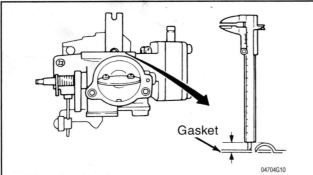

Fig. 50 Exploded view of a typical integrated fuel pump assembly and carburetor

6–25 HP

♦ See Figure 51

1. Remove the fuel pump assembly from the carburetor.
2. Separate the fuel pump cover, gasket/s and diaphragm/s.
3. Discard the used gaskets.
4. Wash all parts thoroughly in solvent and then blow them dry with compressed air. Use care when using compressed air on the check valves. Do not hold the nozzle too close because the check valve can be damaged from an excessive blast of air.
5. Inspect each part for wear and damage. Verify that the valve seats provide a flat contact area for the valve disc. Tighten all elbows and check valve connections firmly as they are replaced.
6. Test each check valve by blowing through it with your mouth. In one direction the valve should allow air to pass through. In the other direction, air should not pass through.
7. Check the diaphragm for pin holes by holding it up to the light. If pin holes are detected or if the diaphragm is not pliable, it must be replaced.

To assemble:

8. Install new gasket/s.
9. Attach the pump cover, gasket/s and diaphragm/s.

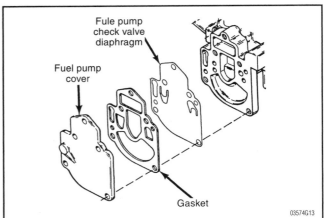

Fig. 51 Exploded view of the fuel pump assembly showing the arrangement of the diaphragm and gasket

10. Tighten the retaining screws securely.
11. Install the fuel pump assembly onto the carburetor.

30–275 HP

♦ See Figures 52 and 53

1. Remove the screws holding the fuel pump assembly together.
2. Separate the pump cover, gaskets and diaphragms from the pump body and base.
3. Always discard the used gaskets.
4. Lay the pump on a suitable work surface and remove the two remaining bolts. Now, carefully separate the parts and keep them in order as an assist in assembling.
5. Do not remove the check valves unless they are defective. Once removed, the valves cannot be used again. If the check valves are to be replaced, take time to Observe and remember how each valve faces, because it must be installed in exactly the same manner or the pump will not function.
6. To remove a check valve, grasp the retainer with a pair of needlenose pliers and pull the valve from the valve seat.
7. Wash all parts thoroughly in solvent and then blow them dry with compressed air. Use care when using compressed air on the check valves. Do not hold the nozzle too close because the check valve can be damaged from an excessive blast of air.
8. Inspect each part for wear and damage. Verify that the valve seats provide a flat contact area for the valve disc. Tighten all elbows and check valve connections firmly as they are replaced.

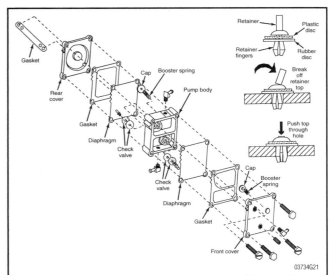

Fig. 52 Exploded view of a V6 fuel pump assembly. This is typical of a remote mounted fuel pump

FUEL SYSTEM 4-49

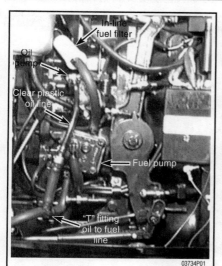

Fig. 53 Close view to show typical fuel and oil component locations on the 135–225hp powerhead

Fig. 54 To test the fuel pickup in the fuel tank, operate the squeeze bulb and observe fuel flowing from the disconnected line at the fuel pump. Discharge fuel into an approved container

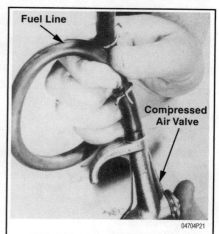

Fig. 55 Many times restrictions such as foreign material may be cleared from the fuel lines using compressed air. Ensure the open end of the hose is pointing in a clear direction to avoid personal injury

9. Test each check valve by blowing through it with your mouth. In one direction the valve should allow air to pass through. In the other direction, air should not pass through.

10. Check the diaphragm for pin holes by holding it up to the light. If pin holes are detected or if the diaphragm is not pliable, it must be replaced.

11. If servicing a Model 135hp thru 200hp powerhead—1990 and on—check the condition of the two booster springs.

To assemble:

12. The fuel pump rebuild kit will contain new gaskets, diaphragms and check valve components. Each check valve consists of a large rubber disc, a smaller plastic disc and a valve retainer.

13. Insert the fingers of the retainer into the smaller plastic disc and then the larger rubber disc.

14. Install the two discs and retainer onto the fuel pump body. Push in the retainer until the collar and both discs are tightly pressed against the pump body. Bend the end of the retainer from side to side until it breaks away from the collar.

15. Install the broken off end through the hole in the collar and through the discs. Use a hammer and tap the retainer end down into place. As this piece is forced down, it will spread the fingers of the retainer and secure the check valve within the pump body. Place the larger of the two caps and boost springs into place on the back side of the pump body.

16. All the layered components of the fuel pump have notches which must be aligned during assembling.

17. Place a new diaphragm on the pump body over the large booster spring and cap, followed by a new gasket and the rear cover. Make sure all the notches align.

18. Position a new diaphragm and gasket over the front of the pump body. Install the smaller booster spring and cap into the front cover. Mate the fuel pump front cover to the body and hold it all together.

19. Secure the pump components together with the two plain head bolts installed in the top right and bottom left bolt holes.

20. Place the fuel pump on the powerhead base using a new mounting gasket. Install the two retaining screws with slotted heads through the pump and into the engine block.

21. Tighten the screws alternately to a torque value of 50–60 inch lbs. (5.5–6.8Nm).

22. Connect the fuel lines and turn the fuel valve to the **ON** position.

Fuel Lines

▶ See Figures 54, 55 and 56

On most installations, the fuel line is provided with quick-disconnect fittings at the tank and at the engine. If there is reason to believe the problem is at the quick-disconnects, the hose ends should be replaced as an assembly. For a small additional expense, the entire fuel line can be replaced and thus eliminate this entire area as a problem source for many future seasons.

The primer squeeze bulb can be replaced in a short time. First, cut the hose line as close to the old bulb as possible. Slide a small clamp over the end of the fuel line from the tank. Next, install the small end of the check valve assembly into this side of the fuel line. The check valve always goes towards the fuel tank. Place a large clamp over the end of the check valve assembly. Use Primer Bulb Adhesive when the connections are made. Tighten the clamps. Repeat the procedure with the other side of the bulb assembly and the line leading to the engine.

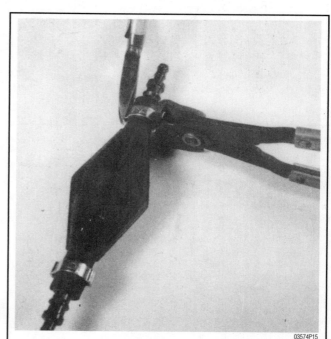

Fig. 56 Using the proper tools to install a clamp around the squeeze bulb check valve

4-50 FUEL SYSTEM

ELECTRONIC FUEL INJECTION (EFI)

Description and Operation

FUEL INJECTION BASICS

Fuel injection is not a new invention. Even as early as the 1950s, various automobile manufacturers experimented with mechanical-type injection systems. There was even a vacuum tube equipped control unit offered for one system! This might have been the first "electronic fuel injection system."

Early problems with fuel injection revolved around the control components. The electronics were not very smart or reliable. These systems have steadily improved since. Today's fuel injection technology, responding to the need for better economy and emission control, has become amazingly reliable and efficient. Computerized engine management, the brain of fuel injection, continues to get more reliable and more precise.

Components needed for a basic computer-controlled system are as follows:
• A computer-controlled engine manager, which is normally called the Electronic Control Unit (or ECU), with a set of internal maps to follow. Changes to fuel and timing are based on input information matched to the map programs.
• A set of sensors or input devices to inform the ECU of engine performance parameters.
• A set of output devices, each controlled by the ECU, to modify fuel delivery and timing.

This list gets a little more complicated when you start to look at specific components. Some fuel injection systems may have twenty or more input devices. On many systems, output control can extend beyond fuel and timing. Most modern systems provide more than just the basic functions but are still straight forward in their layout.

There are several fuel injection delivery methods. Throttle body injection is relatively inexpensive and is used widely in stern drive applications. This is usually a low pressure system running at 15 PSI or less. Often an engine with a single carburetor was selected for throttle body injection. The carburetor was recast to hold a single injector and the original manifold was retained. Throttle body injection is not as precise or efficient as port injection.

Multi-port fuel injection is defined as one or more electrically activated fuel injectors for each cylinder. Multi-port injection generally operates at higher pressures (in excess of 35.5 PSI) than throttle body systems. At present, multi-port fuel injection is the only type used on outboards.

Port injectors can be triggered two ways. One system uses simultaneous injection. All injectors are triggered at once. The fuel "hangs around" until the pressure drop in the cylinder pulls the fuel into the combustion chamber.

The second type is more precise and follows the firing order of the engine. Each cylinder gets a squirt of fuel precisely when needed.

MERCURY ELECTRONIC FUEL INJECTION

♦ See Figures 57 and 58

The type fuel injection used on some models such as the 200XRi and the 175XRi is called "Indirect Multi-Port Fuel Injection", because the fuel is injected into the intake manifold before entering the combustion chamber.

By design, the method of injection is also termed "Port Tuned Injection", as each port has minimum and equal restriction to ensure all ports pass the same amount of air into the crankcase. Pairs of injectors are pulsed sequentially and timed to the induction of air into the crankcase.

A microprocessor housed in the Electronic Control Unit (ECU) accepts data from a number of sensors and computes the new ideal air/fuel ratio and the fuel injectors deliver the correct amount of fuel. Based on the information received, the ECU signals each fuel injector to inject a precise and correct amount of fuel. The system provides the correct air/fuel ratio for all powerhead loads, rpm and temperature conditions. System input and output devices are as follows:

Air Temperature Sensor

♦ See Figure 59

An air temperature sensor is located on the starboard side of the intake manifold, the sensor is mounted under the upper trim/tilt solenoid. The sensor measures the ambient air temperature and conducts this information in the form of an electrical signal to the ECU. As the air temperature changes, the amount of oxygen per cubic foot also changes. The quantity of available oxygen has an affect on combustion and therefore must be taken into account when computing the ideal air/fuel ratio.

Fig. 57 Exterior view of a typical EFI system with some of the major parts identified

Coolant Temperature Sensor

♦ See Figure 60

The coolant temperature sensor is located in the port head. This sensor is a thermistor—an electronic device which functions in the opposite manner of a resistor.

A resistor increases resistance with temperature (decreases voltage), with an increase in temperature. A thermistor varies resistance (increases or decreases voltage), with a change in temperature. When this voltage is received by the ECU, the information is used to determine injector pulse widths and spark advance. The temperature information is also used to determine if an extra charge of fuel is necessary for a cold powerhead. Powerhead temperature information is used by the ECU to assist in starting a cold powerhead by automatic fuel enrichment. In this manner, a choke is not required. Once the powerhead has reached operating temperature, enrichment is no longer required.

Detonation Sensor and Module

♦ See Figure 61

A detonation sensor is threaded into the portside cylinder head. The sensor is able to detect the frequency of vibrations associated with pre-ignition and detonation, approximately, 8,000 Hertz. If either of these conditions is due to fuel of an insufficient octane rating (less than 87) or a sudden change in loading of the powerhead, the sensor will be activated.

An electronic signal is sent to the detonation module and the ECU. The result of these signals is ignition timing being retarded by as much as 8°.

Electronic Control Unit (ECU)

The fuel injection system is controlled by the ECU, an onboard computer mounted as far from heat and vibration as possible. The computer is sealed unit and is in no

FUEL SYSTEM 4-51

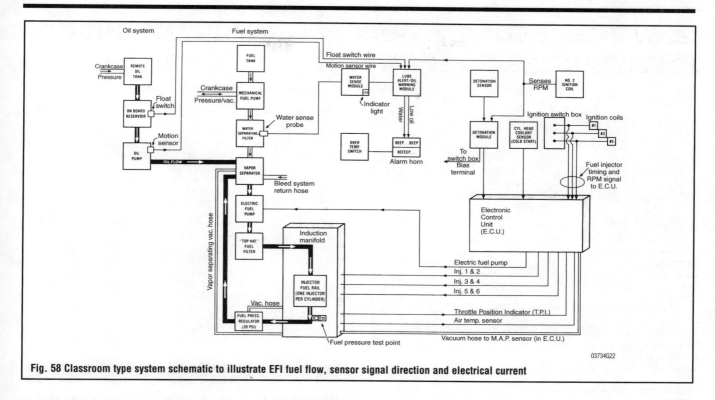

Fig. 58 Classroom type system schematic to illustrate EFI fuel flow, sensor signal direction and electrical current

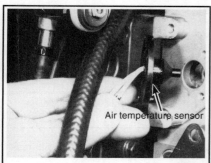

Fig. 59 The air temperature sensor is mounted under the upper trim/tilt solenoid on the starboard side of the powerhead

Fig. 60 The coolant temperature sensor is located on the port cylinder head

Fig. 61 The detonation sensor is threaded into the port cylinder head

way serviceable. The ECU receives signals from numerous sensors on the powerhead. From the signals received, the ECU determines the amount of fuel to be injected. This computer also determines the timing of the spark at the spark plugs.

The amount of fuel injected is determined by how long each injector nozzle remains open. This is commonly referred to as the "pulse width". The nozzle opens and closes in response to signals from the ECU.

The ECU receives three types of imput signals—Analog, Digital and Pressure Differential.

- Analog signals change with changing conditions. For example: the coolant temperature sensor will have more electrical resistance when the powerhead is cold, than when the powerhead is hot.
- Digital signals are a series of on and off pulses. These pulses are counted by the computer to determine a condition. For example: the signal sent from the No. 2 ignition coil will provide the computer with information on powerhead rpm.
- Pressure differential signals received through vacuum lines connecting the intake manifold to the ECU indicate powerhead loading to sensors within the computer.

Sensors located on the powerhead or inside the ECU provide information to the computer on powerhead load, rpm, temperature and other conditions affecting operation.

The computer is programmed or provided with instructions, to produce correct air/fuel mixtures and throttle openings for varying conditions.

Fuel Injectors

▶ See Figure 62

The six fuel injectors force fuel under pressure into the intake manifold. Each injector is mounted on the fuel rail.

The injectors used in this type fuel injection system are solenoid operated. Each consists of a valve body, a needle valve and valve seat. A small voltage is sent from the ECU to each injector. When this voltage is applied to the windings of the solenoid, a magnetic field is induced around the needle valve. The valve lifts off its seat and fuel is allowed to pass between the needle valve and the needle seat. Because the fuel is pressurized, a spray emerges from the injector nozzle. The nozzle spray angle of each injector remains constant and is the same for all six injectors. A small return spring seats the needle valve back onto the seat, the instant the voltage is removed.

The time interval for the injector to be open and emitting fuel is called the "pulse width". The actual "pulse width" for the injector is controlled by the ECU and must be measured in microseconds.

Two O-rings are used to secure each injector in the fuel rail. One O-ring provides a seal between the injector nozzle and the intake manifold. The other O-ring provides a seal between the injector and the fuel inlet connection. Both O-rings prevent excessive injector vibration. These O-rings are replaceable and are included in an injector overhaul kit.

4-52 FUEL SYSTEM

Fig. 62 Intake manifold removed from the powerhead to show location of the six fuel injectors

Fuel Pressure Regulator

A fuel pressure regulator is connected to the fuel rail. The pressure regulator is a mechanical device used to maintain constant fuel pressure to ensure a uniform spray from the injectors. An electric fuel pump delivers fuel under pressure to the fuel rail. Usually an excessive amount of fuel is delivered to the injectors. All the fuel passes through the injector but the excess is returned via the fuel regulator assembly in the following manner.

The fuel regulator is pre-set at the factory to operate between 36 and 39psi. When fuel pressure exceeds the regulator setting, excess fuel pushes a spring loaded diaphragm downward. This action uncovers a fuel return port and the excess fuel is routed back to the vapor separator. As the fuel pressure drops, the spring loaded diaphragm relaxes and the fuel return port is closed.

Fuel Pump

MECHANICAL

A mechanical fuel pump draws fuel from the fuel tank. Fuel from the tank passes through the primary fuel filter, a water separating filter and then a vapor separator. An electric fuel pump boosts the flow of fuel from the vacuum operated mechanical fuel pump and passes the fuel on to a secondary fuel filter. A fuel pressure regulator restricts fuel pressure from 36–39psi, as set at the factory. Fuel is passed on to the fuel rail and is distributed to each injector. Excess fuel is routed back to the vapor separator.

ELECTRIC

▶ See Figures 63 and 64

The electric fuel pump is mounted on the port side of the powerhead next to the vapor separator. The purpose of the electric fuel pump is to boost the initial fuel pressure from the crankcase vacuum operated fuel pump and deliver the fuel to the injectors under constant pressure of 36–39psi.

The electric (booster) fuel pump is a roller-cell type and is not serviceable. Once opened, the airtight seal is lost and the integrity of the seal cannot be regained. The pump must be replaced.

This booster pump is unique because the pump and electric motor are housed together in one permanently sealed case, constantly surrounded with fuel. Yes, the fuel actually flows past the electric motor brushes. If the case remains absolutely airtight, there is no danger of explosion. However, if the case develops a crack or is deformed in any way, there is the distinct possibility air may enter the case and together with the fuel form an explosive mixture. Fuel by itself will not ignite while exposed to a spark from the electric motor brushes, because there is no oxygen present for combustion. A mixture of 0.7 parts air to 1.3 parts fuel becomes explosive.

Pump action will commence the moment the ignition key switch is rotated to the **ON** position. If the key switch is held in this position for more than 30 continuous seconds, the ECU will cutoff electrical power to the pump and of course the pumping action will cease.

✴✴ WARNING

The powerhead must never be cranked without an adequate supply of fuel to the booster pump. If the electric pump draws air into the pump case, an explosive mixture may then form. On the other hand, a fuel injection system cannot be flooded. The fuel pressure regulator will activate the return circuit and excess fuel will not flood the crankcase, as in a conventional carbureted powerhead. As explained earlier, the excess fuel will be returned to the vapor separator.

Fuel Rail

▶ See Figure 65

The fuel injectors are mounted on a rail extending the length of the intake manifold. Fuel enters the rail at one end and a fuel pressure test point is provided at the other end for troubleshooting purposes. Fuel is evenly distributed from the rail to each injector. The injectors are held in place with fragile wire clips.

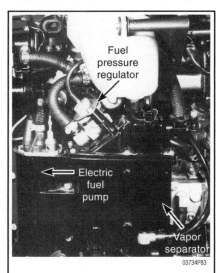

Fig. 63 The fuel pressure regulator, electric fuel pump and the vapor separator

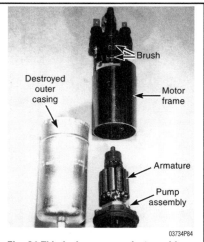

Fig. 64 This fuel pump was destroyed in order to show you the internal components. An operational pump is an air tight sealed unit and must never be opened

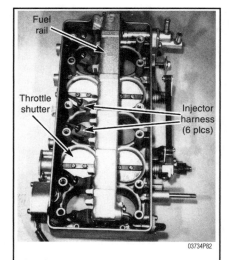

Fig. 65 Intake manifold turned over to show the fuel rail injector harness leads and the throttle shutters

FUEL SYSTEM   4-53

Manifold Absolute Pressure Sensor

A manifold absolute pressure sensor is located inside the ECU. A vacuum hose connects the intake manifold to the sensor. The sensor is a flexible type resistor. As the pressure changes, the resistor flexes and its resistance and the voltage applied across the sensor changes. This change is registered at the ECU.

Two conditions can affect the pressure in the manifold. The first and most common condition, is a reduction in manifold pressure when load on the powerhead is increased. When the operator places a power demand on the powerhead by advancing the throttle, intake manifold pressure is reduced. Conversely, a reduction in intake manifold pressure, indicates an additional load has been placed on the powerhead.

The second condition which may affect the manifold pressure is operation of the powerhead at high altitudes.

RPM Sensor

The powerhead rpm sensor is housed within the ECU. Electronic pulses from the No. 2 ignition coil through the detonation module are received by the rpm sensor. These signals indicate powerhead rpm.

Throttle Position Sensor

♦ See Figure 66

The throttle position sensor is mounted on the bottom throttle shaft. The sensor is an encased potentiometer sending a signal to the ECU indicating powerhead load for a specific rpm. Powerhead load is influenced by the load in the boat, by the propeller type and size used with the outboard unit, performance demands imposed by the operator and other operational factors.

The body of the sensor is stationary with a small shaft emerging from the center of the sensor. The shaft is connected to the throttle valve. As the throttle is advanced, movement is transferred to the sensor and the resistance changes. Therefore, the variable voltage signal sent to the ECU is directly proportional to the throttle position.

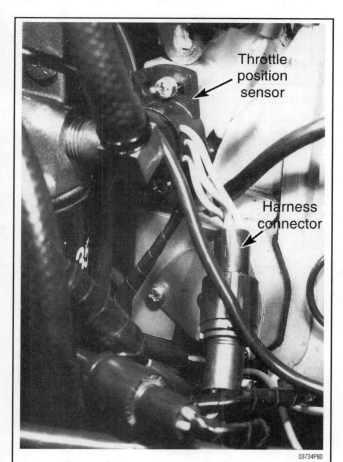

Fig. 66 The throttle position sensor is mounted on the starboard side of the powerhead on the bottom throttle shaft

Troubleshooting Electronic Fuel Injection

Diagnosis of electronic fuel injection is generally based on symptom diagnosis and stored fault codes. Diagnosis of a running problem or as it is commonly called "driveability problem" requires attention to detail and following the diagnostic procedures in the correct order. Resist the temptation to begin extensive testing before completing the preliminary diagnostic steps. A preliminary or visual inspection must be completed in detail before diagnosis begins. In many cases this will shorten diagnostic time and often cure the problem without the need for involved electronic testing.

There are two basic ways to check your fuel system for problems. They are by symptom diagnosis and by the on-board computer self-diagnostic system. The first place to start is always the preliminary inspection. Intermittent problems are the most difficult to locate. If the problem is not present at the time you are testing you may not be able to locate the fault.

The visual inspection of all components is possibly the most critical step of diagnosis. A detailed examination of connectors, wiring and vacuum hoses can often lead to a repair without further diagnosis. Also, take into consideration if the powerhead has been serviced recently. Sometimes things get reconnected in the wrong place or not at all. A careful inspector will check the undersides of hoses as well as the integrity of hard-to-reach hoses blocked by the air silencer or other components. Wiring should be checked carefully for any sign of strain, burning, crimping or terminals pulled from their connectors.

Checking connectors at components or in harnesses is required, usually, pushing them together will reveal a loose fit. Also, check electrical connectors for corroded, bent, damaged, improperly seated pins and bad wire crimps to terminals. Pay particular attention to ground circuits, making sure they are not loose or corroded. Any component or wiring in the vicinity of a fluid leak or spillage should be given extra attention during inspection. Remember how corrosive an environment it is that an outboard engine operates in. Salt spray is especially corrosive to electrical connectors.

Additionally, inspect maintenance items such as belt condition and tension and the battery charge and condition. Any of these very simple items may affect the system enough to set a fault code.

When diagnosing by symptom, the first step is to find out if the problem really exists. This may sound like a waste of time but you must be able to recreate the problem before you begin testing. This is called an "operational check". Each operational check will give either a positive or negative answer (symptom). A positive answer is found when the check gives a positive result (the instruments function when you turn the key). A negative answer is found when the check gives a negative result (the instruments do not function when you turn the key). After performing several operational checks, a pattern may develop. This pattern is used in the next step of diagnosis to determine related symptoms.

In order to determine related symptoms, perform operational checks on circuits related to the problem circuit (the instruments and the tilt and trim system do not work). These checks can be made without the use of any test equipment. Simply follow the wires in the wiring harness or, if available, obtain a copy of your engine's specific wiring diagram. If you see that the instruments and the tilt and trim system are on the same circuit, first check the instruments to see if they work. Then check the tilt and trim system. If the neither the instruments or tilt and trim system work, this tells you that there is a problem in that circuit. Perform additional operational checks on that circuit and compile a list of symptoms.

When analyzing your answers, a defect will always lie between a check which gave a positive answer and one which gave a negative answer. Look at your list of symptoms and try to determine probable areas to test. If you get negative answers on related circuits, then maybe the problem is at the common junction. After you have determined what the symptoms are and where you are going to look for defects, develop a plan for isolating the trouble. Ask a knowledgeable mechanic which components frequently fail on your vehicle. Also notice which parts or components are easiest to reach and how can you accomplish the most by doing the least amount of checks.

A common way of diagnosis is to use the split-in-half technique. Each test that is made essentially splits the trouble area in half. By performing this technique several times the area where a problem is located becomes smaller and smaller until the problem can be isolated in a single wire or component. This area is most commonly between the two closest checkpoints that produced a negative answer and a positive answer.

After the problem is located, perform the repair procedure. This may involve replacing a component, repairing a component or damaged wire or making an adjustment.

➥Never assume a component is defective until you have thoroughly tested it.

The final step is to make sure the complaint is corrected. Remember that the symptoms that you uncover may lead to several problems that require separate repairs. Repeat the diagnosis and test procedures repeatedly until all negative symptoms are corrected.

4-54 FUEL SYSTEM

Two simple tests, pressure and volume, can easily verify the mechanical integrity of the powerhead's fuel supply and return systems. Problems within the fuel supply or return system can cause many driveability problems because the control system can't compensate for a mechanical problem. Nor can the system adjust the air/fuel mixture when the system is in open loop or back-up mode. Moreover, a control system's parameters are programmed based on a set flow rate and system pressure. Therefore, even if the computer were adjusting the air/fuel mixture, the computer's calculations would be wrong whenever the system pressure or volume was incorrect.

➡When testing fuel system pressure, always use a pressure gauge that is capable of handling the system pressure. Old style vacuum/pressure gauges will not work with most fuel injection systems

The fuel system's pressure can be thoroughly verified by performing both static and dynamic tests. Dynamic tests are used to determine the system's overall operating condition while the engine is running. For example, if the pressure is low, the problem is most likely to be in the supply or high side of the system. Likewise, if the fuel pressure is high, the problem is going to be on the return or low side of the fuel system. All pressure tests should be taken on the high or supply side of the fuel system.

If a dynamic test reveals that the pressure is below manufacturer specifications, the problem is caused by one or more of the following items: weak pumps, clogged or restricted fuel filter or lines, external leaks bad check valves or a faulty fuel pressure regulator. But if a test reveals pressures above the desired specification, then the problem is caused by one or more of these items: restricted or clogged return lines, vacuum leak at the pressure regulator or a defective fuel pressure regulator.

However, if the pressure is high on a return less fuel system, the cause, most likely, is a defective pressure regulator or clogged/faulty injectors. The causes are slightly different on returnless systems because the regulator is located in or near the fuel tank and there is no fuel return line used on these systems. However, the systems which use a remotely mounted fuel pressure regulator require a short return line. Thus, there is no excess fuel beyond the injectors and the measured pressure is already the regulated pressure.

Many manufacturers are switching to returnless fuel systems because of the advantages they provide. One advantage is lower evaporative emissions which allows less frequent purge cycles and the use of smaller canisters. The evaporative emissions are lower on a return-less fuel system because the fuel doesn't absorb any heat from the engine compartment. This prevents it from turning into a vapor as it returns to the fuel tank.

Another example of how returnless fuel systems are different from conventional fuel injection systems is that they may not use a frame/engine compartment mounted fuel filter. Most return-less systems use a combination fuel filter/pressure regulator which is mounted on top of the fuel pump module. The fuel pump module is located in the tank. Other return-less systems use an in-line filter on the frame that incorporates an integral fuel pressure regulator. These systems also utilize a short return line from the fuel filter to the fuel pump module in the tank. Yet, others use a regular frame/engine mounted filter in conjunction with a fuel pressure regulator and filter assembly in the tank.

Previously we mentioned that a static pressure test should also be performed. This test can be performed before or after a dynamic test with the engine off. These tests are useful for determining the sealing capacity of the system. If the fuel system leaks down after the engine is shut off, the vehicle will experience extended crank times while the system builds up enough fuel pressure to start. If the system is prone to losing residual pressure too quickly, it also usually suffers from low fuel pressure during the dynamic test. In addition to verifying that the system is holding residual pressure, a static test can also verify that the system is receiving the initial priming pulse at startup.

However, don't assume that the correct amount of fuel is being delivered just because the system pressure is within specifications. The only way to measure fuel volume is to open the fuel system into a graduated container and measure the fuel output while cranking the engine. Some manufacturers specify that a certain amount of fuel be dispersed over a short period of time but generally the pump should be able to supply at least 1.5 oz. (45 ml.) for approximately 2 seconds. If the fuel output is below the recommended specification, there is a restriction in the fuel supply or the fuel pump is weak.

In summary, pressure and volume tests provide information about the overall condition of the fuel system. The information obtained from these tests will set the direction and path of your diagnosis.

When troubleshooting the fuel injection system, always remember to make absolutely sure that the problem is with the fuel injection system and not another component. This particular system has very sophisticated electronic controls, which in most cases, can only be diagnosed and repaired by an authorized dealer who has the expensive diagnostic equipment.

There are some trouble shooting procedures that can be done outside a professional shop. By follow the following steps:

Many times fuel system troubles are caused by a plugged fuel filter, a defective fuel pump or by a leak in the line from the fuel tank to the fuel pump. A defective choke may also cause problems. would you believe, a majority of starting troubles which are traced to the fuel system are the result of an empty fuel tank or aged sour fuel.

Under average conditions (temperate climates), fuel will begin to break down in about four months. A gummy substance forms in the bottom of the fuel tank and in other areas. The filter screen between the tank and the carburetor and small passages in the carburetor will become clogged. The gasoline will begin to give off an odor similar to rotten eggs. Such a condition can cause the owner much frustration, time in cleaning components and the expense of replacement or overhaul parts for the carburetor.

Even with the high price of fuel, removing gasoline that has been standing unused over a long period of time is still the easiest and least expensive preventative maintenance possible. In most cases, this old gas can be used without harmful effects.

The gasoline preservative will keep the fuel fresh for up to twelve months. If this particular product is not available in your area, other similar additives are produced under various trade names.

Fuel injected engines don't actually have a choke circuit in the fuel system. These engines rely on the ECU to control the air/fuel mixture and injector function according to the electronic signals sent by a host of sensors and monitors. If a fuel injected engine suffers a driveability problem and it can be traced to the fuel injection system, then the sensors and computer-controlled output components must be tested and repaired or replaced.

Fuel Injectors

TESTING

Injector Removed

♦ See Figure 67

1. Remove the fuel injectors. Once the injectors are unplugged from their wiring harness and removed from the fuel rail, a simple test with a 9V battery can be performed to determine if the solenoid inside the injector is functioning properly.
2. Obtain a 9V battery, such as those used in radios and calculators. Obtain two jumper leads with small alligator clips at each end.
3. Momentarily connect the battery across the two terminals inside the injector connector, making sure the two leads do not contact each other. For these tests the jumper cables may be connected in either way, because polarity does not affect the tests.
4. As soon as contact is made, the injector should emit a sound as the solenoid is energized and the needle is pulled off its seat. When contact is broken, the injector should again emit a sound as the needle returns under spring pressure back to its original position.
5. Perform this test on each injector in turn. If any injector fails this test, there is an internal defect in the injector. The needle is stuck or there is bad electrical connection. The injector must be replaced. A quick test to determine which of the two listed reasons prevent the injector from working properly is to perform a resistance test across the two injector terminals using a digital multimeter. Leads contacted across the two injector terminals should register a reading of 1.1 ohm.

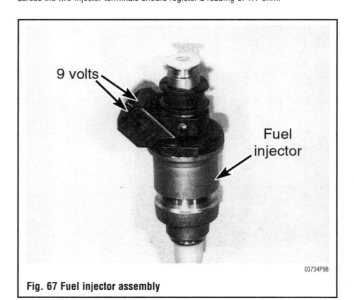

Fig. 67 Fuel injector assembly

FUEL SYSTEM 4-55

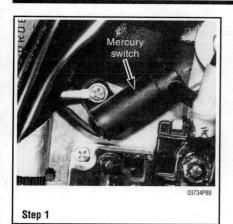
Step 1

Step 3

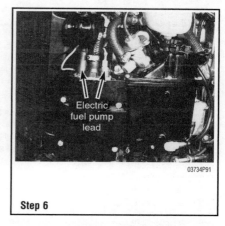
Step 6

Injector Installed

♦ See accompanying illustrations

1. Connect one end of a jumper cable to the terminal of the mercury switch and the other end to a suitable ground on the powerhead. Grounding the mercury switch is necessary to prevent the powerhead from accidentally starting while tests are being performed.

2. Keep switch grounded with the jumper cable in place while performing the test, then remove the cable.

3. Rotate the ignition key to the **START** position and crank the powerhead through several revolutions. Rotate the key switch back to the **OFF** position. Remove the spark plugs one by one and inspect the electrode end. A damp electrode will indicate the injector is functioning for the cylinder being checked.

4. If all spark plugs are damp indicating the presence of fuel, all injectors are spraying fuel into the crankcase. If some spark plugs are damp with fuel, while others are dry, the electric fuel pump is probably functioning correctly. However, either the injectors of the "dry" cylinders have a restricted fuel flow or the dry injectors have an electrical problem.

5. If all spark plugs are dry, a problem with fuel delivery exists. The problem may be a main fuel line blockage.

6. A faulty electrical component affecting operation of the entire system, such as a defective electric fuel pump, defective injector harness plug or a component within the ECU may be the cause.

Injector Wiring Harness

♦ See accompanying illustrations

1. Disconnect both the leads at the fuel pump, to prevent the fuel being pumped during this test. Position the red "hot" lead in such a manner to prevent it from making contact with any part of the powerhead.

2. To test for voltage to the injectors, the injector harness connector plug must be removed from each injector at the fuel rail.

3. Remove the ECU and the induction cover from the manifold. These steps can be performed with the unit removed from the powerhead or with the unit still in place on the powerhead. The bolts securing the cover to the powerhead, pass through the manifold. Therefore, temporarily reattach the manifold to the powerhead while performing this test.

4. Once the connector has been removed from the injector, obtain a multimeter and set the scale to 12V DC. Obtain a small jumper cable and use it to ground the manifold to the powerhead, because the three original grounding leads were removed in previously.

5. Rotate the ignition key to the **ON** position with the jumper cable still in place on the mercury switch.

6. Connect the multimeter lead to each injector in turn.

7. The meter should register at least 9V at each injector.

➡ **If voltage is present at each injector harness plug, skip the following test and proceed to testing the injector.**

8. If voltage is not present at any injector connector plug but the previous step proved the ECU is supplied with battery voltage, then the ECU may not be supplying voltage to the main injector wire harness plug.

9. Reconnect the leads at the electric fuel pump.

10. Disconnect the main injector harness at the connector plug. With the ignition key still in the **ON** position and with the jumper cable still in place on the mercury switch, check for voltage output from the ECU.

11. Make contact with the meter lead to a good ground on the powerhead. Make contact with the meter lead to the No. 6 lead terminal at the connector plug from the ECU. If the meter registers no voltage, verify voltage is being supplied to the ECU by momentarily touching the meter lead to the (+) terminal on the rectifier. If voltage is indicated, then either the ECU is defective or there is a problem in the main ECU harness plug (which probably means a new ECU is required). If no voltage is indicated, remove the ECU and take the ECU to the dealer for testing with a Quicksilver Fuel Injection Tester before purchasing a new unit.

12. If voltage is indicated at the No. 6 lead terminal at the connector plug from the ECU but no voltage is indicated at any or only some of the plug connectors at the injectors, the injector harness is defective and must be replaced with a new harness.

13. The jumper cable, temporarily grounding the mercury switch, may now be removed and the lead reconnected to the mercury switch terminal.

Step 1

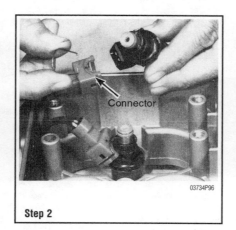
Step 2

Step 10

4-56 FUEL SYSTEM

REMOVAL & INSTALLATION

▶ See accompanying illustrations

1. Remove the two screws securing the water sensing module to the powerhead. The left screw has two small black grounding leads attached. The module will still be connected to the powerhead by other leads. Move the module aside. Remove the large bolt in the center at the base of the ECU.

➥Do not disturb the small Phillips head screws around the perimeter of the ECU. The cover of the ECU will not come away once these bolts are removed. The bolts were installed during manufacture of the unit and then the cover and the case were hermetically sealed using the Phillips head screws for alignment. Therefore, no purpose will be served, if these screws are removed.

2. Remove the two nuts on the studs at the top of the ECU. The right stud has a small Black ground wire over it.

➥It is far easier to disconnect leads and leave components still attached to the intake manifold, because some components are mounted behind others and should not be disturbed. For example: if the throttle position sensor is misaligned during installation, a digital type multimeter is required to correctly reset the sensor on its mounting bracket. Voltage in the 1/10 range must be accurately read. A misaligned sensor could send misleading signals to the ECU and consequently affect the EFI and the ignition timing.

3. Disconnect the main harness plug.
4. Disconnect the throttle position sensor harness plug.
5. Disconnect the two leads from the air temperature sensor at their quick disconnect fittings. Disconnect the leads to the up and down trim/tilt solenoids. Disconnect the fuel injector plug harness.
6. Pry the oil pump control link rod from the ball joint on the intake manifold, leaving the rod on the oil pump. Disconnect the throttle control link rod from the throttle cam.

➥Three vacuum hoses connect the intake manifold to the vapor separator, the fuel pressure regulator and the ECU. Identify these hoses and tag each one as an assist in later installation.

7. Gently ease the two vacuum hoses free. One hose leads to the fuel pressure regulator and the other hose leads to the vapor separator. Loosen the clamp from the fuel supply hose at the secondary filter and gently pull the hose free of the filter fitting.

Step 1

Step 2

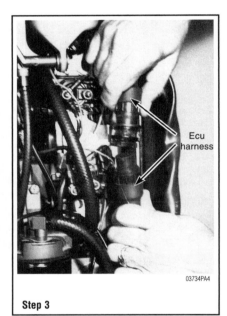
Step 3

Step 4

Step 5

Step 6

FUEL SYSTEM   4-57

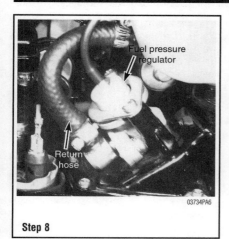
Step 8

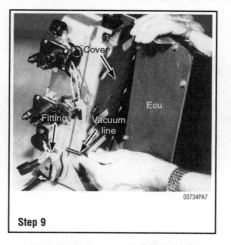
Step 9

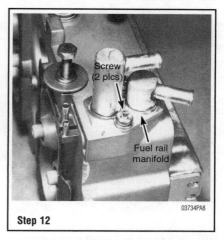
Step 12

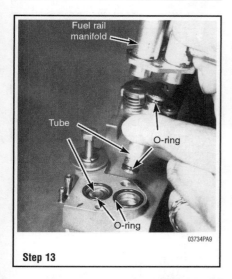
Step 13

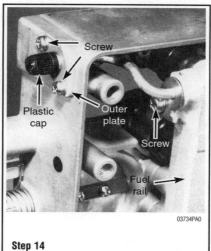
Step 14

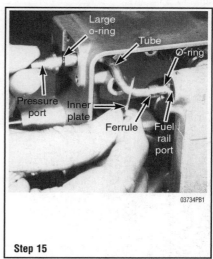
Step 15

8. Loosen the clamp and remove the return fuel hose from the fitting at the fuel pressure regulator. Lift off the cowling bracket which supported the ECU. Remove the two bolts from the lower support bracket.

9. Disconnect the vacuum line from the ECU at the fitting on the manifold next to the pressure port. Remove the ECU from the front cover.

10. Note the location of the three grounding leads on the front face of the intake manifold cover, as an aid during installation. Remove the twelve bolts securing the manifold to the powerhead. Notice all bolts are not the same length. Make a note as to length and location.

11. Lift the intake manifold from the powerhead. Remove and discard the gasket. Remove the front cover and gasket from the manifold.

12. Remove the two Phillips head screws with captive lockwashers from the fuel rail manifold.

13. Gently pull the fuel rail manifold from the intake manifold. Two tubes are installed under the fuel rail manifold which will either come away with the fuel rail manifold or stay inside the intake manifold. Remove the two tubes. Both tubes are identical and have an O-ring at each end. The O-rings need not be removed unless they are no longer fit for service. Remove the two O-rings which seal the fuel rail-to-intake manifolds.

14. Remove the plastic cap from the pressure port on the side of the intake manifold. Remove the two Phillips head screws with captive lockwashers and the outer plate securing the port to the manifold. Remove the Phillips head screw with captive lockwasher from the fuel rail.

15. Ease out the pressure port fitting. As the port is pulled just slightly from the seated position, the attached tube will come out of the fuel rail. A number of small items along the length of this tube must be accounted for and not lost. First, a small O-ring around the pressure port. Next, another small O-ring around the tube, then a ferrule and finally an inner plate. The items around the tube cannot be removed until the tube clears the port. The tube may then be guided clear of the manifold.

16. Remove the four long Allen head screws from the injector side of the manifold. Support the fuel rail and turn the manifold over.

17. Separate the fuel rail from the fuel injectors.

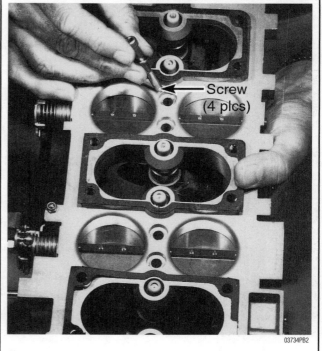
Step 16

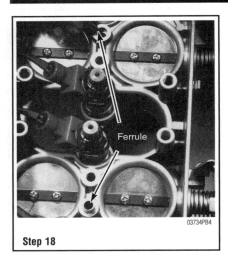
Step 18

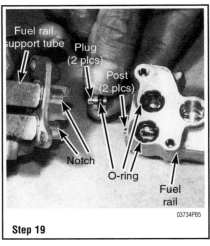
Step 19

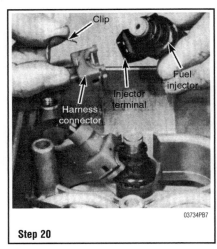
Step 20

Step 21

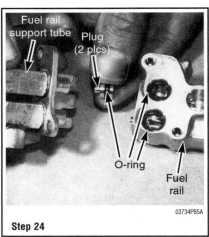
Step 24

Step 25

18. Remove the two ferrules from the manifold. These two ferrules may have remained either with the manifold or in the fuel rail.

19. Remove the two Phillips head screws with captive lockwashers from the fuel rail support tube. Lift the support tube from the rail. Lift out the two tiny plugs and a total of four O-rings from the rail.

20. Using a small slotted screwdriver, pry the retaining clip from the injector harness connector. Disconnect the harness from the injector. Disconnect the harnesses from the other injectors in a similar fashion.

21. Slowly pull each injector straight up and free of the manifold. The injectors are held in the manifold by rubber sealing rings only. These rings may harden or become gummy, depending on use, making removal of the injectors difficult.

To install:

22. Position the retaining clip around the injector wire harness. Push the injector harness connector firmly over the injector terminal. The harness can only be connected one way. Slide the clip into place to secure the harness connector to the injector.

23. Using new O-rings, push the injector into place to seat firmly into the intake manifold. Repeat Step 1 and this step for all injectors.

24. Install the two tiny plugs into the fuel rail, with the larger end of the plug going into the fuel rail first. Install the two sealing O-rings over the two fuel ports in the rail. Install fuel rail support tube over the rail with the two notches indexing over the two posts on the rail end. This ensures the support can only be installed one way over the rail. Install and tighten the two Phillips head screws with captive lockwashers.

25. Position the two ferrules over the two outer holes in the intake manifold.

26. Lower the fuel rail over the ferrules and injectors. Make sure the rail is seated properly over each injector and none are askew. Support the rail with one hand and turn the manifold over.

27. Install the four long Allen head bolts. Tighten the bolts alternately and evenly to a torque value of 25 inch lbs. (3.4 Nm).

28. Slide the large O-ring over the tube. This O-ring remains on the outside of the manifold. Insert the tube into the pressure port hole. Over the tube, first slide the inner plate, with the tang on the plate facing toward the rail upon installation, then the ferrule and finally the small O-ring. Guide the tube into the fuel rail port, with the O-ring and ferrule entering the port first. Secure the inner plate to the rail with the Phillips head screw and captive lockwasher.

29. Seat the pressure port valve and O-ring into the manifold. Slide the outer plate over the valve, with the two screw holes aligned. Install and tighten the two Phillips head screws with captive lockwashers.

30. Insert the two identical tubes, with an O-ring at each end, into the intake manifold. These tubes must seat in the fuel rail support tube. Position two O-rings over the ports. Install the fuel rail manifold over the tubes. The manifold can only be installed properly one way. The bolt holes will not align if the manifold is positioned incorrectly.

31. Install and tighten the two Phillips head screws securing the fuel rail manifold to the intake manifold.

32. Position a new gasket on the intake manifold. This gasket is purchased and installed in one piece. After installation, four sections of the gasket are removed.

33. Install the intake manifold to the powerhead. The securing bolts have different lengths. Bolts identified as No. 4, 12 and 9 have grounding wires attached. Tighten all bolts to a torque value of 90 inch lbs. (10.2Nm) in the proper sequence. Position the top cowling bracket over the two studs on top of the intake manifold.

34. Connect the vacuum hose from the MAP sensor inside the ECU to the fitting on the intake manifold.

35. Connect the fuel return hose to the fuel pressure regulator and tighten the hose clamp. Connect the vacuum hose from the intake manifold to the fitting on the pressure regulator. Connect the vacuum hose from the intake manifold to the vapor separator.

36. Connect the two leads from the air temperature sensor to the two leads from the ECU at their quick connect fittings.

Connect the leads to both trim/tilt solenoids. Consult the wiring diagram as an aid in making these connections.

37. Connect the throttle position sensor harness leads to its harness connector from the ECU. Connect the fuel injector harness to the harness connector from the ECU.

38. Connect the ECU wire harness connector to the main powerhead harness connector. The connection can only be made one way.

39. Rotate the oil pump lever clockwise and snap the oil pump control link rod on the ball joint onto the intake manifold throttle linkage. Attach the throttle control link rod to the throttle cam.

FUEL SYSTEM 4-59

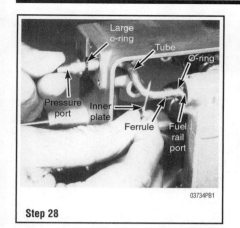
Step 28

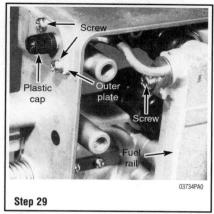
Step 29

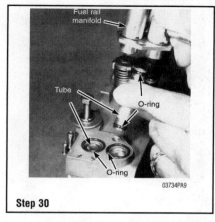
Step 30

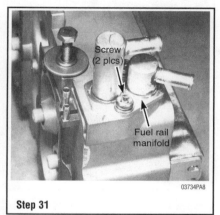
Step 31

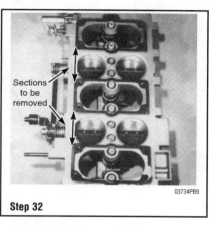
Step 32

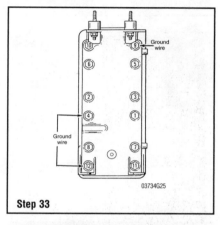
Step 33

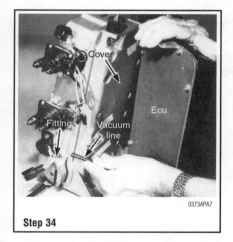
Step 34

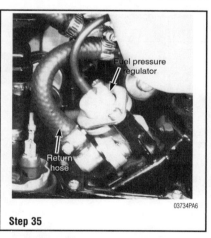
Step 35

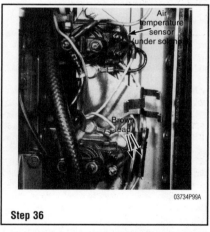
Step 36

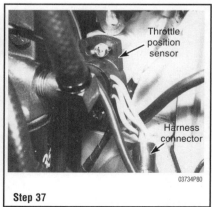
Step 37

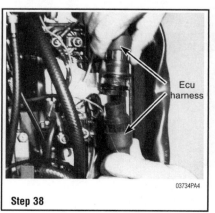
Step 38

Step 39

4-60 FUEL SYSTEM

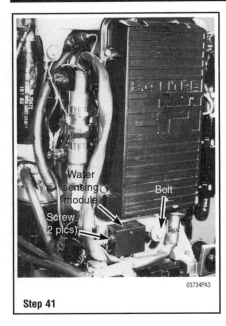

Step 41

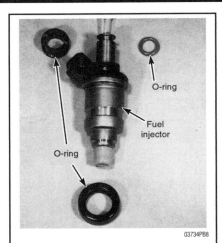

Fig. 68 An individual kit containing the necessary O-rings to seal just one injector during installation may be purchased from the local marine dealer. A new kit for each injector removed should be purchased

Fig. 69 The temperature sensor is located in the port cylinder head

40. Slide the ECU down over the two top studs. Slide the small ground lead over the right stud. Install and hand tighten the two nuts.

41. Install the single center bolt at the base of the ECU. Tighten this bolt and the two nuts installed in the previous step to a torque value of 45 inch lbs. (5 Nm). Install and tighten the two bolts on the lower support bracket. Install the water sensing module below the ECU. The left attaching Phillips head screw also acts as a grounding point for two small Black wires. Install and tighten both securing screws.

CLEANING & INSPECTION

♦ See Figure 68

The intake manifold, with sensors attached, may be carefully cleaned using a parts brush and solvent. Try not to get the solvent on the sensor or the wire harnesses.

Never dip rubber parts or plastic parts in carburetor cleaner. Blow the manifold dry with compressed air.

Move each throttle shaft back-and-forth to check for wear. If any shaft appears to be too loose, replace the manifold. Individual replacement parts are not available.

Check action of the throttle return springs. Remove and replace as necessary.

Inspect the main body of the intake manifold and cover gasket surfaces for cracks and burrs which might cause a leak.

Perform the injector operation test with a 9-volt battery.

Two O-rings are used to secure each injector in the fuel rail. One O-ring provides a seal between the injector nozzle and the intake manifold. The other O-ring provides a seal between the injector and the fuel inlet connection. Both O-rings prevent excessive injector vibration. these O-rings are replaceable and are included in an injector overhaul kit.

Inspect the sensors as far as possible without disturbing them. Check the wire harnesses for signs of chafing, cracks or corroded connections.

Coolant Temperature Sensor

TESTING

♦ See Figure 69

1. Identify the two leads from the coolant temperature sensor located in the Port cylinder head.

2. Using a multimeter set on the ohms scale, make contact with the two meter leads across the two sensor leads.

3. The meter should register approximately 1,000 ohms. If the meter registers zero or infinity, the sensor should be replaced. If the meter registers in the kilo-ohm range, the sensor is probably satisfactory.

REMOVAL & INSTALLATION

1. Remove the screw and retainer plate.
2. Disconnect the leads and remove the sensor.

To install:
3. Connect the sensor leads and install the sensor into the cylinder head.
4. Install the retainer plate and screw. Tighten the screw snugly.

Air Temperature Sensor

TESTING

♦ See Figure 70

1. The air temperature sensor need not be removed from the powerhead for testing purposes. This sensor is located on the starboard side of the intake manifold behind the upper trim/tilt solenoid.

2. Disconnect the two leads from the air temperature sensor at their quick disconnect fittings.

3. Using a multimeter set on the ohms scale, make contact with the two meter leads across the two sensor leads.

4. The meter should register approximately 8,000 ohms. If the meter registers zero or infinity, the sensor should be replaced. If the meter registers in the kilo-ohm range, the sensor is probably okay.

REMOVAL & INSTALLATION

1. Disconnect the sensor leads.
2. Remove the retaining screws and remove the sensor.
3. Remove the O-ring from the sensor.

To install:
4. Install the O-ring onto the sensor.
5. Install the sensor and retaining screws. Tighten the screws snugly.
6. Connect the sensor leads.

FUEL SYSTEM 4-61

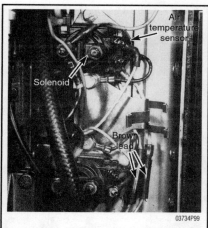

Fig. 70 Air temperature sensor location

Fig. 71 Crankcase vacuum operated pump mounted on powerhead

Fig. 72 Connect the free end of the jumper cable to the negative terminal on the electric fuel pump

Mechanical Fuel Pump

TESTING

♦ See Figure 71

1. Ground each spark plug lead.
2. Disconnect the fuel line at the top carburetor.
3. Place a suitable container beneath the container to catch the discharged fuel.
4. Squeeze the primer bulb and observe if there is satisfactory fuel being discharged.
5. Using the starter motor, crank the engine. If the fuel pump is operating correctly, a healthy stream of fuel should pulse out of the fuel line.
6. Continue cranking the engine for about 15 pulses to determine if the amount of fuel decreases with each pulse or maintains a constant amount.
7. A decrease in the discharge indicates a restriction in the line or a fault in the fuel pump. If the fuel line is clogged, clear the obstruction. If the line is clear, the problem may lie with the fuel pump.
8. Replace the fuel pump if necessary.

REMOVAL & INSTALLATION

1. Disconnect the top and bottom fuel hoses. Use a golf tee or stubby pencil to plug the fuel lines to prevent any fuel from leaking.
2. Disconnect the pulse hose from the front surface of the pump.
3. Remove the pump from the powerhead. Make sure that you remove the bolts securing the pump to the powerhead. The other bolts hold the pump assembly itself together.
4. Remove the gasket and discard.
To install:
5. Install a new mounting gasket and mount the fuel pump to the powerhead. Tighten the retaining screws to 50–60 inch lbs. (5.5–6.8 Nm).
6. Connect the pulse hose to the front for the fuel pump.
7. Remove the plug(s) and connect the top and bottom fuel hoses.

Electric Fuel Pump

TESTING

Voltage

♦ See Figure 72

1. If the fuel pump is suspected of being faulty, first perform a ground test.
2. Leave one end of the jumper cable connected to a good ground on the powerhead. Disconnect the other end of the cable from the mercury switch.
3. Connect free end of the jumper cable to the negative terminal on the electric fuel pump.

❋❋ CAUTION

Water must circulate through the lower unit to the powerhead anytime the powerhead is operating to prevent damage to the water pump in the lower unit. Just five seconds without water will damage the water pump impeller.

4. Attempt to start the powerhead. If the electric fuel pump operates and the powerhead starts, the problem is a bad ground connection to the pump. To obtain a satisfactory ground, the temporary jumper cable must be replaced with a new permanent ground lead.
5. If the powerhead did not start, remove the temporary jumper cable from the electric fuel pump terminal and connect it back to the terminal on the mercury switch.
6. Using a multimeter set on VDC, rotate the ignition key to the **ON** position, with the jumper cable connected to the mercury switch still in place, disconnect the red lead at the positive terminal on the electric fuel pump.
7. Make contact with the meter lead to the positive terminal on the pump. Make contact with the black meter lead to a suitable ground on the powerhead. The meter should register 12V.
8. If the meter fails to register 12V, keep the black meter lead in place and move the red meter lead to the positive terminal on the rectifier.
9. If the meter registers 12V at positive rectifier terminal, the ECU is defective, because the ECU is not transmitting voltage from its pickup point (here at the positive rectifier terminal) and delivering voltage to the pump.
10. If the meter fails to indicate 12V at the rectifier, the ECU is not being supplied with power and therefore cannot pass it on to the electric fuel pump. In this case the fault may lie in the battery, cables, connections, harness plug or ignition switch.
11. If the meter registered 12V at the electric fuel pump, the ECU is functioning correctly and the problem may lie in the portion of the fuel system past this connection: the electric fuel pump, the secondary filter or the injector harness plug.

Operating Pressure

♦ See Figure 73

1. Obtain a pressure gauge capable of registering 50 psi.
2. Remove the plastic cap from the pressure port located portside at the bottom of the intake manifold.
3. Connect the gauge to the pressure port. Rotate the ignition key to the **ON** position and crank the powerhead for about 15 seconds with the jumper cable in place grounding the mercury switch. Note the reading on the gauge. Normal powerhead operation requires pressure between 36 and 39psi (248.2–268.9kPa).
4. If the reading is low, the cause may be a restriction in a fuel line.
5. If there is no reading, the pump is defective and must be replaced.

Fuel Flow

♦ See Figure 74

1. Obtain a suitable container and place it under the electric fuel pump and the vapor separator. Snip the Sta-Strap around the hose connected to the bottom of the electric fuel pump. Gently pry the hose from the fitting on the pump and place the end in the container.

4-62 FUEL SYSTEM

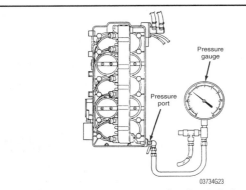

Fig. 73 Use a pressure gauge to test the electric fuel pump operating pressure

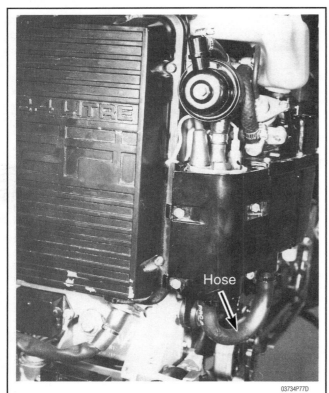

Fig. 74 After loosening one end of the fuel hose, place the hose in an appropriate container to catch any fuel being pumped

2. Rotate the ignition key to the ON position and crank the powerhead, with the jumper cable still in place on the mercury switch, for about 15 seconds.
3. If fuel flowed from the disconnected hose, then all is well with the vapor separator, crankcase vacuum operated fuel pump and water separating filter. Reconnect the hose using a new Sta-Strap. If no fuel flowed, the problem could be with one of the components just mentioned.
4. If the vapor separator is suspected, proceed to servicing the vapor separator.

REMOVAL & INSTALLATION

1. Make sure the ignition switch is in the **OFF** position. Disconnect the battery cables.
2. Remove the screws securing the fuel pump cover and then remove the cover.

✲✲ CAUTION

Make sure that the battery is disconnected when prying back the wire covers. Battery voltage is present at both fuel pump terminals with the ignition switch in the OFF position and the battery connected. Damage can occur to both the fuel pump and ECU if the pump terminals are shorted.

3. Disconnect the positive and negative fuel pump leads.
4. Remove the filter screw and separate the fuel pump cover from the filter base.
5. Disconnect the fuel pump inlet hose from the bottom of the pump assembly.
6. Remove the fuel pump mounting screws. Remove the fuel pump assembly.

To install:
7. Install the fuel pump assembly onto the vapor separator assembly.
8. Connect the fuel pump inlet hose.
9. Connect the negative and positive leads and recover the terminals with the rubber boots.
10. Install the pump mounting screws.
11. Reinstall the filter cover.

Vapor Separator

▶ See accompanying illustrations

REMOVAL & INSTALLATION

➡ If only the fuel regulator is to be serviced, the work may be just as easily performed while the regulator is still in place on the powerhead. However, if the vapor separator is to be serviced, it must be removed from the powerhead. If only the vapor separator is to be serviced and not the fuel pressure regulator, simply skip the steps which pertain to the regulator.

✲✲ CAUTION

Before the electric fuel pump, vapor separator or the fuel pressure regulator is removed, the EFI system must be depressurized.

1. Remove the plastic cap over the pressure port located portside at the bottom of the intake manifold. A screwdriver will be used to depress the valve tip to release the pressure and allow fuel to drain from the valve. Place the screwdriver tip lightly over the valve tip and wrap a clean shop towel around the valve and screwdriver. This operation is similar to letting air out of tires! Once all the pressure has been released and the fuel flow ceases, carefully proceed with the following step.
2. Remove and plug the fuel inlet hose from the water separating filter to the vapor separator cover fitting.
3. Remove and plug the oil inlet hose at the base of the vapor separator. Obtain a suitable container, then remove and drain the hose from the vapor separator to the electric fuel pump. Disconnect the leads from the electric fuel pump.
4. Remove the four cover plate mounting bolts and lift off the plate.
5. Disconnect the inlet and outlet hoses from the pump.
6. Lift the pump and rubber blanket free of the powerhead.
7. Remove and drain the fuel return hose at the forward end of the fuel pressure regulator to the fitting on the separator cover.
8. Remove the vacuum line from the intake manifold at the fitting on the fuel pressure regulator.

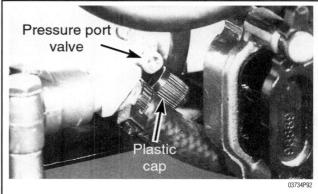

Step 1

FUEL SYSTEM 4-63

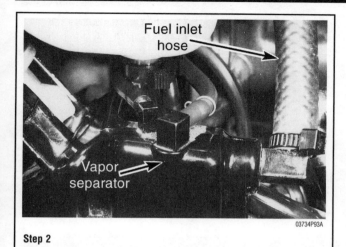

Step 2

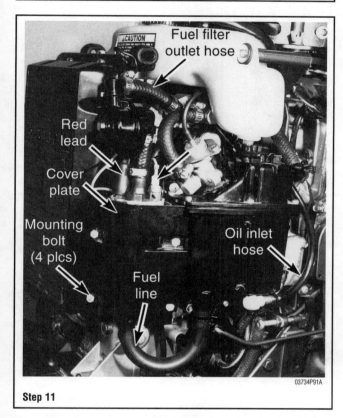

Step 11

15. Position the rubber blanket and electric fuel pump in place on the powerhead.
16. Connect the inlet and outlet hoses to the pump. Install the cover plate and secure it in place with the four mounting bolts.
17. Connect the leads to the correct terminals on the electric fuel pump.
18. Connect the fuel line between the base of the vapor separator and the electric fuel pump. Remove the plug from the oil inlet hose and connect the hose to the fitting at the base of the vapor separator.
19. Remove the plug from the fuel inlet hose and connect the hose to the fitting at the top of the vapor separator cover.
Make a quick check of all connections made in the last two steps.
Start the powerhead and check for fuel leaks.

✼✼ WARNING

The system will be pressurized as soon as the powerhead is cranked. Watch for leaks around all fuel hose connections which were disturbed. Because the system is pressurized, a leak will not appear as drips. Instead fuel will spray all over the powerhead. Should this occur, shut down the powerhead immediately. do not forget to depressurize the system at the pressure port on the intake manifold before making necessary repairs.

CLEANING & INSPECTION

▶ See Figure 75

1. The inlet needle seat is pressed into the cover, therefore, it is not removable or replaceable.

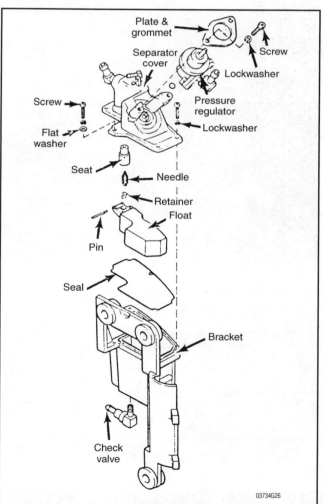

Fig. 75 Exploded drawing of the fuel pressure regulator and vapor separator, with major parts identified

9. Remove the two vacuum lines from the fittings on the vapor separator cover. One line is from the intake manifold and the other line from the crankcase regulator.
The vapor separator and fuel regulator assembly is now disconnected from the system ready for removal from the powerhead.
10. Remove the three attaching bolts and then remove the assembly from the powerhead.
To install:
11. Position the assembled vapor separator and fuel pressure regulator up against the powerhead. Install and tighten the three attaching bolts to a torque value of 90 inch lbs. (10.2 Nm).
12. Connect the two vacuum hoses, one from the crankcase bleed system and the other from the intake manifold, to the two fittings on the vapor separator cover. Observe that these hoses have two different size inner diameters to fit over the two different size fittings.
13. Connect the vacuum hose from the intake manifold to the fuel pressure regulator.
14. Connect the fuel return hose between the vapor separator and the regulator. Connect the fuel filter outlet hose between the secondary fuel filter and the fitting on the intake manifold.

4-64 FUEL SYSTEM

2. Rinse the housing in solvent and then blow it dry with compressed air.
3. Check the float for deterioration. The float is hollow, therefore, check to be sure the float does not contain any fluid.
4. Check to be sure the float tab is in good condition. If any part of the float is damaged, the unit must be replaced.
5. Check the float tab-to-needle contacting surface and replace the float if this surface has a groove worn in it.

Fuel Pressure Regulator

REMOVAL & INSTALLATION

♦ See accompanying illustrations

1. Remove the two Phillips head screws with captive lockwashers from the regulator support plate.

2. Remove the regulator from the water separator cover. Remove the O-ring from the separator cover.
3. Remove the two screws securing the regulator to the manifold plate and then remove the O-ring.
4. Remove the six Phillips head screws with captive lockwashers from the vapor separator cover. Lift the cover slowly and squarely from the housing. The cover will come away with the float assembly attached.
5. Use a pair of needle nose pliers and pull the float hinge pin from the mounting posts on the cover. Remove the float and inlet needle from the cover. The needle may be slid from the brass tab on the float. Take care not to alter the angle of the brass tab.
6. Remove the O-ring from the housing.

To install:

7. Install the O-ring around the housing.
8. Hang the inlet needle on the brass float tab. Lower the needle and float into place between the mounting posts, with the needle entering the seat. Insert the float hinge pin through the mounting posts.
9. Lower the cover assembly down into the housing. Install and tighten the six Phillips head screws with captive lockwashers.

Step 1

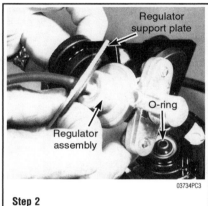
Step 2

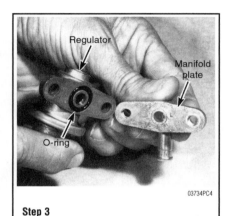
Step 3

Step 4

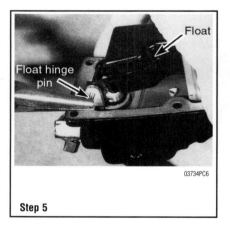
Step 5

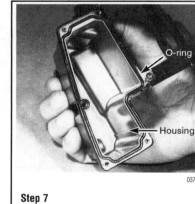
Step 7

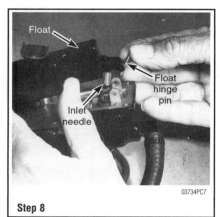
Step 8

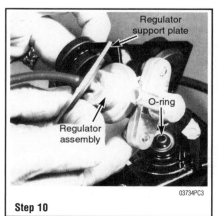
Step 10

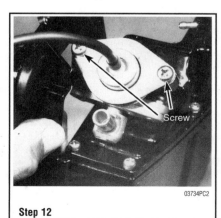
Step 12

FUEL SYSTEM 4-65

10. Position the O-ring around the regulator seat on the vapor separator cover. Install an O-ring in the regulator port and then install the manifold plate to the regulator using the two Phillips head screws with captive lockwashers.

11. Position the regulator onto the vapor separator cover and slide the regulator support plate over the vacuum hose fitting.

12. Install and tighten the two Phillips head screws with captive lockwashers through the support plate and into the vapor separator cover.

Throttle Position Sensor

TESTING

♦ See Figure 76

➡If this sensor is disturbed during testing or service procedures, a digital type multimeter is needed to correctly reset the sensor on its mounting bracket, because voltages in the $\frac{1}{10}$ range must be accurately read. A misaligned sensor could send misleading signals to the ECU and consequently affect the fuel delivery and ignition timing.

1. Start the powerhead and allow it to run at idle until warmed to operating temperature.
2. Mark the original location of the throttle position sensor before performing any tests on the sensor which may lead to disturbing its original alignment.

❊❊ CAUTION

Water must circulate through the lower unit to the powerhead anytime the powerhead is operating to prevent damage to the water pump in the lower unit. Just five seconds without water will damage the water pump impeller.

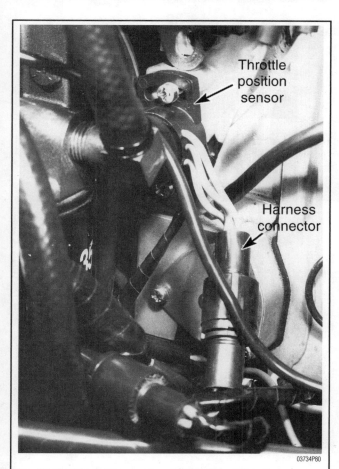

Fig. 76 Throttle position sensor and wiring harness location

➡Do not remove or change the throttle sensor position during the following test.

3. Disconnect the wire harness from the sensor.
4. Obtain a multimeter. Make contact with the Red meter lead to the Orange sensor terminal. Make contact with the Black meter lead to the Tan/Black sensor terminal.
5. The meter should register between 800 and 1200 ohms at idle speed.
6. Keeping the red meter lead in place, move the Black meter lead to the Light Blue sensor terminal. The meter should register the same reading.
7. The lower the resistance reading, the leaner the air/fuel mixture. If the powerhead is overheating at idle speeds, perhaps due to an excessively lean mixture, the problem may lie in the throttle position sensor. Conversely, if the powerhead is smoking at idle speeds, due to an over-rich mixture, again the problem may lie in a misaligned throttle position sensor.

REMOVAL & INSTALLATION

1. Mark the original location of the throttle position sensor before removal. If this sensor is misaligned during installation, a digital type multimeter is needed to correctly reset the sensor on its mounting bracket. During alignment, voltages must be accurately measured to within $\frac{1}{10}$ volt. A misaligned sensor could send misleading signals to the ECU and consequently affect the EFI and the ignition timing.
2. After marking the position of the throttle sensor, remove the two securing screws and lift the sensor free from the throttle shaft.

To install:

3. Check to be sure the throttle position sensor shaft engages the slot on the throttle shaft. Secure the sensor with the attaching hardware, in exactly the same location as before removal. The marks made during removal should be aligned. If no locating marks were made during removal, carefully examine the bracket for traces of an outline of the sensor and attempt to mount the sensor as near to the original location as possible. A difference of 0.04 in. (1mm) corresponds to 2° throttle angle.

ADJUSTMENT

➡Do not attempt this adjustment unless an instrument is available capable of indicating voltage accurately to within $\frac{1}{10}$ volt. If such an instrument is not available, it is strongly recommended qualified technicians at the local Mariner dealership install and calibrate this sensor using the necessary equipment.

1. This adjustment is performed with the powerhead not running.
2. Loosen the screw next to the cam follower to allow the follower to ride freely along the throttle cam.
3. On V6 powerheads with EFI and equipped with an idle speed screw, back off the idle speed screw, located just above the cam follower, from the stop on the intake manifold. The throttle position sensor must be adjusted with the throttle plates fully closed and with the arm of the throttle shaft against the throttle valve stop screw.
4. Disconnect the two Tan/Black leads from the powerhead coolant temperature sensor at their quick disconnect fittings. This sensor is located in the port cylinder head.
5. The meter used in this test must have either a special harness for measuring the output voltage or have probes capable of piercing through the insulation of the wires to make contact without disconnecting the leads or damaging the insulation of the leads.
6. Select the volts scale on the multimeter. Ensure the two mounting screws of the sensor are snug enough to permit the sensor to be rotated a few degrees either way and still hold its position.
7. Make contact with the red multimeter lead to the orange sensor lead. Make contact with the black multimeter lead to the tan/black sensor lead.
8. Turn the key to the **ON** position without starting the powerhead. Begin by rotating the sensor clockwise as far as possible. Note the meter reading. Now, rotate the sensor counterclockwise until the meter reading increases by 0.01–0.02 volts from the original reading. At this point, tighten both securing screws to a torque value of 23 inch lbs. (2.5Nm) to hold the adjustment.
9. A misalignment of the sensor by 0.04 in. (1mm) on the mounting bracket corresponds to 2° of throttle angle.
10. With the multimeter still connected, open and close the throttle valve a few times to be sure the output voltage fluctuates with different throttle angles. Also, check to see if the voltage reverts back to the specified level when the throttle valves are fully closed.
11. Mark the position of the new sensor in relation to the bracket as a preparation for possible future service.
12. Connect the two tan/black leads from the powerhead coolant temperature sensor at their quick connect fittings.

The operation of this sensor can be compared to the volume knob on a radio. If

4-66 FUEL SYSTEM

the listener has a favorite position for the knob, in time a "flat spot" will wear on the shaft and interference may result.

If the operator of the outboard has a favorite throttle position, in time a flat spot will wear on the sensor shaft and not provide the ECU with accurate infor-mation. This condition could lead to ignition timing which varies by as much as 20° with no change in throttle position and incorrect operation of the EFI.

> ※※ **CAUTION**
>
> **Precautions must be taken not to cause damage to expensive electronic components during these service procedures.**

Detonation Sensor and Module

TESTING

Detonation Sensor

♦ See Figure 77

1. Remove the single white/blue lead from the detonation controller to the detonation sensor at the screw terminal of the sensor. Set this black lead aside.

Obtain an multimeter. Make contact with the Red meter lead to the sensor terminal. Make contact with the black meter lead to a suitable ground on the powerhead. A reading of no continuity (infinite resistance) should register on the meter. If the reading is less than infinity, there is a short in the sensor and the sensor must be replaced.

Fig. 77 Detonation sensor and wire lead installed on the powerhead

Detonation Module

♦ See Figure 78

A digital multimeter is required to perform this test.

1. Disconnect the gray/white lead between the detonation module and the switchbox at the quick disconnect fitting.
2. With the meter set on the 10V DC scale, make contact with the red meter lead to the disconnected lead from the module and make contact with the black meter lead to a suitable ground on the powerhead.

➡**Never operate the engine at high speed with a flush device attached. The engine, operating at high speed with such a device attached, could runaway from lack of a load on the propeller, causing extensive damage.**

3. Connect a tachometer to the powerhead.
4. Start the powerhead and allow it to reach operating temperature at idle speed.
5. The meter should register only one (1) volt DC. Increase powerhead rpm to between 3500 and 4000. The meter should register zero volts.
6. If the meter registers a fluctuating voltage higher than specified, the powerhead is experiencing detonation.
7. If the meter registers a constant voltage higher than specified, the module is defective.

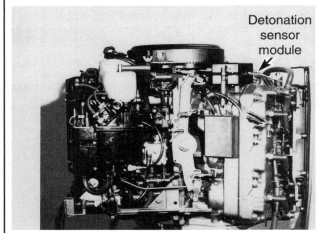

Fig. 78 Check the detonation sensor control module using a digital multimeter

OPTIMAX DIRECT FUEL INJECTION (DFI)

Description and Operation

♦ See Figures 79, 80, 81, 82 and 83

Combustion air enters the cowl through holes located in the top aft end of the cowl. The cowl liner then directs air to the bottom of the powerhead. This limits the exposure of salt air to the components inside the engine cowl.

Once inside the cowl, the air enters the plenum through the throttle shutter which is located in the plenum assembly. The air then continues through the reed valves and into the crankcase. The throttle shutter is actuated by the throttle shaft. Mounted on a separate shaft is the Throttle Position Sensor (TPS). This sensor tells the Engine Control Module (ECM) the position of the throttle.

If the TPS should fail, the dash mounted check engine light will flash and the warning horn will sound. Engine speed will be Reduced by 20%-25% and the ECM will reference the MAP sensor for fuel calibration.

Air from inside the engine cowl is drawn into the compressor through the flywheel cover. This cover acts like a muffler to quiet compressor noise and contains a filter to prevent the ingestion of debris into the compressor. The compressor is driven by a serpentine belt from a pulley mounted on the flywheel and is automatically self adjusting using a single idler pulley. This air compressor is a single cylinder unit containing a connecting rod, piston, rings, bearings, reed valves and a crankshaft. The compressor is water cooled to lower the temperature of the air charge and is lubricated by oil from the engine oil pump assembly. As the compressor piston

Fig. 79 Typical OptiMax fuel rail assembly

FUEL SYSTEM   4-67

Fig. 80 Electric fuel pump, water separating filter and vapor separator assemblies—115 hp OptiMax

Fig. 81 Engine mounted oil reservoir on the 115 hp engine . . .

Fig. 82 . . . and the remote oil tank mounted beneath the transom

moves downward inside the cylinder, air is pulled through the filter, reed valves and into the cylinder. After the compressor piston change direction, the intake reeds close and the exhaust reeds open allowing the compressed air into the hose leading to the air/fuel rails.

The air fuel rails contain two passages: one for fuel, the second is the air passage. The air passage is common between all the cylinders included in the rail. A hose connects the starboard rail air passage to the air compressor. Another hose connects the starboard air rail passage to the port air rail passage. An air pressure regulator will limit the amount of pressure developed inside the air passages to approximately 10 psi below the pressure of the fuel inside the fuel passages (i.e. 80 psi air vs. 90 psi fuel). Air exiting the pressure regulator is returned into the exhaust adaptor and exits through the propeller on the 115–150hp powerhead and returns to the air plenum on the 200–225hp powerhead.

Fuel for the engine is stored in a typical tank. A primer bulb is installed into the fuel line to allow the priming of the fuel system. A crankcase mounted pulse driven diaphragm fuel pump draws fuel through the fuel line, primer bulb, fuel pump assembly and then pushes the fuel through a water separating fuel filter. This filter removes any contaminates and water before fuel reaches the vapor separator. Fuel vapors are vented through a hose into the air compressor inlet in the front of the flywheel cover. The electric fuel pump is different than the fuel pump that is utilized on the standard EFI engine (non DFI) and is capable of developing fuel pressures in excess of 90 psi. fuel inside the rail must remain pressurized at exactly 10 psi over the air rail pressure or the ECM (map) calibrations will be incorrect. Fuel from the vapor separator is supplied to the bottom of the starboard fuel rail. A fuel line connects the bottom of the first rail to the opposite fuel rail. Fuel is stored inside the rail until an injector opens.

A fuel pressure regulator controls pressure in the fuel rails and allows excess fuel to return into the vapor separator. The fuel regulator not only regulates fuel pressure but it also regulates it at approximately 10 psi higher than whatever the air rail pressure is. The fuel regulator diaphragm is held closed with a spring that requires 10 psi to force the diaphragm off the diaphragm seat. The back side of the diaphragm is exposed to air rail pressure. As the air rail pressure increases, the fuel pressure needed to open the regulator will equally increase. For example, if there is 50 psi of air pressure on the rail side of the diaphragm, 60 psi of the pressure will be required to open the regulator. The port fuel rail is water cooled.

To equalize the pulses generated by the pumps (both air and fuel) a tracker diaphragm is installed in the starboard rail. The tracker diaphragm is positioned between the fuel and air passages. The tracker diaphragm is a rubber diaphragm which expands and retracts depending upon which side of the diaphragm senses the pressure increase (pulse).

Oil in this engine is not mixed with fuel before entering the combustion chamber. Oil is stored in a standard remote oil reservoir. Crankcase pressure will force oil from the remote oil tank into the oil reservoir on the side of the powerhead. Oil will flow from the oil reservoir into the oil pump. The oil pump is a solenoid design. It is activated by the ECM and includes 7 pistons with corresponding discharge ports. The oil pump is mounted directly onto the powerhead. Each cylinder is lubricated by one of the discharge ports. The oil is discharged into the crankcase. The seventh passage connects to the hose that leads to the air compressor for lubrication. Excess oil from the compressor returns into the plenum and is ingested through the crankcase

The ECM will change the discharge rate of the oil pump, depending upon engine demand. The ECM will also pulse the pump on initial start up to fill the oil passages eliminating the need to bleed the oil system. The ECM provides additional oil for break in, as determined by its internal clock. The oil ratio varies with engine rpm and load.

The electrical system consists of the Engine Control Module (ECM), crank posi-

Fig. 83 Electrically operated OptiMax solenoid design oil pump assembly

tion sensor (flywheel speed & crankshaft position, Throttle Position Sensor (TPS), MAP sensor, engine temperature sensor, ignition coils and injectors (fuel & direct). The engine requires a battery to start (the ignition and injection will not occur if the battery is dead). The system will run off the alternator.

The operation of the system happens in milliseconds (ms) exact timing is critical for engine performance. As the crankshaft rotates, air is drawn into the crankcase through the throttle shutter, into the plenum and through the reed valves. As the piston nears bottom dead center (BDC), air from the crankcase is forced through the transfer system into the cylinder. As the crankshaft continues to rotate, the exhaust and intake ports close. With these ports closed, fuel can be injected into the cylinder. The ECM will receive a signal from the throttle position sensor, engine temperature sensor and the crank position sensor (flywheel speed and position sensor). With this information the ECM refers to the fuel calibration (maps) to determine when to activate (open and close) the injectors and fire the ignition coils. With the piston in the correct position, the ECM opens the fuel injector, 90 psi fuel is discharged into the machined cavity inside the air chamber of the air/fuel rail. This mixes the fuel with the air charge. Next, the direct injector will open, discharging the air/fuel mixture into the combustion chamber. The direct injector directs the mixture at the bowl located in the top of the piston. The pistons bowl directs the air/fuel mixture into the center of the combustion chamber. This air/fuel mixture is then ignited by the spark plug.

To aid in starting when the air rail pressure is low and before the compressor has time to build pressure, some direct injectors are held open by the ECM. This allows the compression from inside the cylinders to pressurize the air rail faster (1 or 2 strokes or 60° of crankshaft rotation).

Idle quality is controlled by fuel volume and fuel timing. The throttle shutter will be open at idle speed. The shift cut-out switch will interrupt the fuel to 3 of the cylinders to assist in shifting.

The TPS signals the ECM to change the fuel and spark without movement of the throttle shutters. The throttle cam is manufactured to allow the TPS shaft to move before opening the throttle shutter.

OPTIMAX COMPONENTS

Air Pressure Regulator

♦ See Figure 84

The air pressure regulator is located on the port fuel rail and is designed to limit the air pressure inside the rails to approximately 80 psi.

The air regulator uses a spring (pressure) to control the air pressure. This spring (80 psi) holds the diaphragm against the diaphragm seat. The contact area blocks (closes) the air inlet passage from the excess air, return passage.

As the air pressure rises (below the diaphragm), it must reach a pressure equal to or greater than the spring pressure holding the diaphragm closed. Once this pressure is achieved, the spring collapses, allowing the diaphragm to move. The diaphragm moves away from the diaphragm seat, allowing air to exit through the diaphragm seat, into the excess air passage leading to the exhaust adaptor plate.

Fig. 84 Air compressor installation—115 hp

Air Temperature Sensor

The air temperature sensor is mounted on the intake manifold. The ECM regulates the fuel flow, in part, based on manifold air temperature. As the air temperature increases, the Air Temperature Sensor sends a signal to the ECM which then decreases fuel flow.

Charging System Alternator

♦ See Figure 85

Battery charging is contained within the belt driven alternator, including the regulator. At cranking speeds, electrical power for the engine is provided by the boat battery. The minimum recommended size is 750 or 1000 MCA, cold cranking amps. Above 550 rpm, all electrical power is provided by the alternator. Should engine rpm drop below 550 rpm, the alternator is not capable of providing sufficient output and the battery becomes the primary source of electrical power. Once the engine is running and the rpm is 650 rpm or higher, the engine will continue to run should the battery become shorted or disconnected.

Alternator output (when hot) to the battery @ 2000 rpm is approximately 33–38 amperes.

Crankshaft Position Sensor (CPS)

This is a Hall Effect-type sensor, which senses the 24-teeth located on the flywheel underneath the ring gear. It supplies the ECM with crankshaft position information and engine speed.

There is no resistance reading available for this sensor but the air gap of 0.015–0.040 in. (.381–1.016 mm) between the sensor and the flywheel must be maintained.

This is the only sensor which cannot be disconnected while the engine is running. The engine will shut down if it is disconnected or fails.

Direct Injectors

The 6 direct injectors (1 per cylinder) are used to inject an air/fuel mixture into the cylinders. The direct injector uses a tapered tip and seat. The tapered tip is ground to provide a leakproof seal.

The tapered tip is controlled by the ECM through the use of a solenoid inside the injector body. After the ECM signals the fuel injector to open, 90 psi fuel is discharged into the machined cavity inside the air chamber of the air/fuel rail. This mixes the fuel with the air charge. Next, the ECM signals the direct injector to open, discharging the mixed fuel/air mixture into the combustion chamber.

The direct injector directs the air/fuel mixture at the bowl located in the top of the piston. The piston's bowl directs the air/fuel mixture into the center of the combustion chamber. The air/fuel mixture is then ignited by the spark plug.

Electronic Control Module (ECM)

The ECM requires 8 VDC minimum to operate. If the ECM should fail, the engine will stop running. The inputs to the ECM can be monitored and tested by the Digital Diagnostic Terminal (DDT) (91–823686A2) using adapter harness (84–822560A5).

The ECM performs the following functions:
- Calculates the precise fuel and ignition timing requirements based on engine speed, throttle position, manifold pressure and coolant temperature
- Controls the fuel injectors for each cylinder, direct injectors for each cylinder and ignition for each cylinder
- Controls all alarm horn and warning lamp functions
- Supplies tachometer signal to the gauge
- Controls RPM limit function
- Montiors the shift interrupt switch

Flywheel

There are 24 teeth underneath the flywheel ring gear. As these teeth rotate past the crankshaft position sensor, they provide engine speed and crankshaft position information to the ECM.

With this information, the ECM refers to the fuel calibration (maps) to determine when to activate (open and close) the injectors and fire the ignition coils.

Fuel Injectors

The 6 fuel injectors (1 per cylinder) are used to provide fuel from the fuel rail to the direct injectors. The fuel injectors are electromagnetic devices, controlled by a solenoid inside the injector body. The solenoid is energized by an injector driver in the ECM. This driver circuit controls the "on" time of the solenoid by providing a ground.

After receiving information from the TPS, engine temperature sensor and crankshaft position sensor, the ECM refers to the fuel calibration maps and determines when to activate the fuel injectors. When the piston is in the correct position, the ECM signals the injector and 90 psi fuel is discharged into a machined cavity inside the air chamber of the air/fuel rail.

Fuel Pressure Regulator

The regulator is located in the port fuel rail.

The fuel pump is capable of delivering more fuel than the engine can consume. Excess fuel flows through the fuel pressure regulator, interconnecting passages/hoses and back to the vapor separator tank. This constant flow of fuel means that the

Fig. 85 Alternator location—115 hp OptiMax

FUEL SYSTEM 4-69

fuel system is always supplied with cool fuel, thereby preventing the formation of fuel vapor bubbles and minimizing the chances of vapor lock.

The fuel pressure regulator is calibrated to raise the fuel pressure to 10 psi above the air pressure.

The fuel regulator is mounted on the port fuel rail, near the top. This regulator relies on both air and spring pressure to control the fuel pressure. Inside the regulator assembly is a 10 lb. Spring, this spring holds the diaphragm against the diaphragm seat. The contact between the diaphragm and diaphragm seat closes the passage between the incoming fuel (from the electric fuel pump) and the fuel return passage.

When the engine is not running (no air pressure on the spring side of the diaphragm) the fuel pressure required to move the diaphragm is 10 psi.

With the engine running, air pressure from the air compressor (80 psi) is routed through the air passages, to the spring side of the fuel pressure regulator diaphragm.

The air pressure (80 psi) and spring pressure (10 psi) combine to regulate the fuel system pressure to 90 psi or 10 psi higher than the air pressure in the DFI system fuel/air rails.

Ignition Module (Coil)

Inductive type ignition coils are used on the DFI engines. DC current is from the battery or charging system is transferred through the main relay fuse to the positive terminal of all 6 ignition coil primary windings.

The negative terminal of the coil primary is connected to the engine ground through the ECM. When this circuit is closed, a magnetic field is allowed to be built up in the ignition coil. The crankshaft position sensor senses the location of the 24 teeth on the flywheel and supplies a trigger signal to the ECM.

When the ECM receives this signal, the ECM then opens the ground circuit of the coil primary. The magnetic field in the ignition coil primary then collapses, cutting across the coil secondary winding, creating a high voltage charge (50,000 volts) that is sent to the spark plug.

Three styles of ignition coils have been used on Optimax/DFI engines:
- 1997: Bosch (round)
- 1998: Marshall (square with removable plug-wire)
- 1998 ½: Marshall (square with non-removable plug wire)

Manifold Absolute Pressure (MAP) Sensor

The MAP sensor is located on the top of the air plenum. A vacuum hose connects the MAP sensor to the intake manifold. Through this vacuum hose, the MAP sensor functions through the full rpm range and is continually signaling induction manifold pressure readings to the ECM.

The ECM in turn determines fuel flow as signals are received. Drawing a vacuum on the MAP sensor will create a lean condition, altering engine operation. If no change occurs when drawing a vacuum, the MAP sensor is not functioning properly. The MAP sensor can be tested with the DDT.

Shift Interrupt Switch

The shift interrupt switch is designed to reduce the torque load on the gear case components to assist in shifting. The switch is monitored by the ECM which will interrupt the fuel flow momentarily to the cylinders when the engine speed exceeds 600 rpm in neutral.

The switch function can be monitored by the DDT. The DDT will display "ON" when the outboard is in neutral and "OFF" when in gear.

The switch is open (no continuity) when the outboard is in gear and closed (continuity) when the outboard is in neutral.

If shift operation is difficult, the shift interrupt function can be checked by the DDT or an multimeter—for open or closed operation and for a continuity check of the switch harness for shorts or open wiring.

Throttle Position Sensor (TPS)

The Throttle Position Sensor (TPS) is a potentiometer that provides a signal that is directly proportional to the throttle plate shaft. This sensor operates much like the rheostat on a light. The TPS monitors throttle plate movement and position and transmits an appropriate electrical signal to the ECM.

These signals are used by the ECM to vary the injector pulse width accordingly.

The two sensors are used in conjunction with each other. If one sensor should fail, the dash mounted CHECK ENGINE light will light up and the warning horn will sound. RPM will be limited to 3000 rpm. If both TPS sensors should fail, the engine rpm will be reduced to idle by the ECM.

➡ **Model year 2000 engines use only one TPS.**

Tracker Valve

The tracker valve is located on the starboard fuel/air rail assembly. The DFI system must maintain a constant 10 psi pressure difference between the fuel pressure and air pressure in the rails, at all times. The tracker is designed to maintain the 10 psi differential when the air or fuel pressure suddenly raises (i.e. pulses generated by the compressor's piston or by the fuel injectors opening and closing). The tracker contains a spring on the air side of the diaphragm. This spring positions the diaphragm against the diaphragm's seat (when the engine is not running).

After the engine starts and the fuel and air pressure reach normal operating range, the fuel pressure will compress the spring and the diaphragm will move slightly away from the seat (to a neutral position). At this point the pressure on both sides of the tracker diaphragm is equal (10 psi spring pressure + 80 psi air pressure = 90 psi fuel pressure).

Any air or fuel pressure "spikes" on one side of the diaphragm will transfer this pressure rise to the other system (air or fuel) on the other side of the diaphragm. Both systems will have a momentary increase in pressure so that the 10 psi difference between the air and fuel systems can be maintained.

➡ **To prevent excessive wear in the seat, the tracker is calibrated to allow the diaphragm to be slightly away from the seat during normal operation.**

Water Temperature Sensor

Coolant temperature sensors (3 on the 200–225hp models and 2 on the 115–150hp models) are used to provide cylinder head temperature information to the ECM. One sensor is mounted in each cylinder head and one in the air compressor cylinder head. The ECM uses this information to increase the injector pulse width for cold starts and to retard timing in the event of an over-heat condition.

Troubleshooting the Optimax Fuel Injection System

WITHOUT THE DIGITAL DIAGNOSTIC TERMINAL (DDT)

Troubleshooting without the Digital Diagnostic Terminal (DDT) or equalivant tester is limited to checking resistance on some of the sensors. Typical failures usually do not involve the ECM. Connectors, set-up and mechanical wear are usually most likely to fail. The OptiMax models can really only be checked using the DDT or an equalivant tester to troubleshoot the fuel injection system.

1. Make sure that all spark plug wires are securely installed all the way on the coil towers.
2. The engine may not run or may not run above idle if the wrong spark plugs are installed.
3. Swap ignition coils to see if the problem follows that particular coil or if the problem stays with original cylinder.

➡ **The ECM is capable of performing a cylinder misfire test to isolate problem cylinders. Once a suspect cylinder is located, an output load test on the ignition coil, fuel injector and direct injector may be initiated through the use of the DDT.**

4. Any sensor or connection can be disconnected and reconnected while the engine is running without damaging the ECM. Disconnecting the crank position sensor will stop the engine.

➡ **Any sensor that is disconnected while the engine is running will be recorded as a Fault in the ECM Fault History. Use the DDT to view and clear the fault history when the repair or troubleshooting is finished.**

5. If all cylinders exhibit similar symptoms, the problem is with a sensor or harness input to the ECM.
6. If the problem is speed related or intermittent, it is probably connector or contact related. Inspect the connectors for corrosion, loose wires or loose pins. Secure the connector seating by using dielectric compound.
7. Inspect the harness for obvious damage, such as, pinched wires and chaffing.
8. Secure grounds and all connections involving ring Terminals. Use Liquid Neoprene to seal the Terminals.
9. Check fuel pump connections fuel pump pressure.
10. Check air compressor pressure.

4-70 FUEL SYSTEM

WITH THE DIGITAL DIAGNOSTIC TERMINAL (DDT)

▶ See Figures 86 and 87

Attach the diagnostic cable to the ECM diagnostic connector and plug in the software cartridge. You will then be able to monitor the sensors and ECM data values including status switches.

The ECM program can help diagnose intermittent engine problems. It will record the state of the engine sensors and switches for a period of time and then can be played back to review the recorded information.

Fig. 86 Connect the DDT to the onboard ECM diagnostic cable to troubleshoot the system

Fig. 87 All OptiMax models will have a diagnostic connector which looks like this to interface with the DDT

Air Compressor

TESTING

▶ See Figures 88 and 89

1. Install the Pressure Gauge Assembly (91–852087A1/A2/A3) to the fuel rail pressure test valves. The starboard rail has the fuel test valve and the port rail has the air test valve.

➡ **After cranking the engine for 15-seconds using the starter motor, the air pressure gauge should read 77–81 psi (530.2–557.8 kPa) and the fuel gauge should read 87–91 psi (599.7–627.3 kPa).**

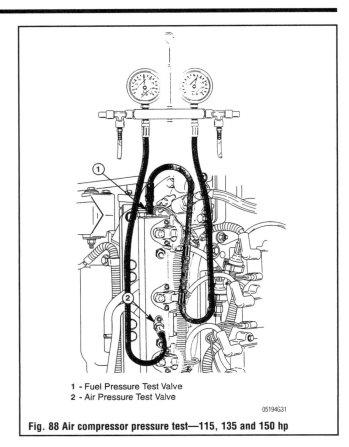

1 - Fuel Pressure Test Valve
2 - Air Pressure Test Valve

Fig. 88 Air compressor pressure test—115, 135 and 150 hp

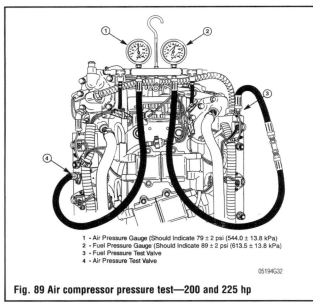

1 - Air Pressure Gauge (Should Indicate 79 ± 2 psi (544.0 ± 13.8 kPa))
2 - Fuel Pressure Gauge (Should Indicate 89 ± 2 psi (613.5 ± 13.8 kPa))
3 - Fuel Pressure Test Valve
4 - Air Pressure Test Valve

Fig. 89 Air compressor pressure test—200 and 225 hp

REMOVAL & INSTALLATION

▶ See Figures 90 and 91

1. Disconnect the battery cables from the battery terminals.
2. Remove the top cowling.

➡ **Prior to removing the flywheel cover, remove the vent hose from the fitting on the flywheel cover.**

3. Remove the flywheel cover.
4. Use a ⅜ in. (9.5mm) drive on belt tensioner arm to relieve belt tension. Then remove the belt.

FUEL SYSTEM 4-71

PROBLEM	CORRECTIVE ACTION
Fuel Pressure and Air Pressure are Both Low	1. Inspect air compressor air intake (air filter in flywheel cover) for blockage. 2. Remove air compressor cylinder head and inspect for scuffing of cylinder wall. Inspect for broken reeds and/or reed stops. 3. Tracker Valve – Remove and inspect diaphragm for cuts or tears and seat damage on diaphragm and rail. 4. Air Regulator – Remove and inspect diaphragm for cuts or tears on diaphragm and rail.
Fuel Pressure Low or Fuel Pressure Drops while Running (Air Pressure Remains Normal)	1. Each time key is turned to the RUN position, both electric pumps should operate for 2 seconds. If it they do not run, check 20 ampere fuse and wire connections. 2. If pumps run but have no fuel output, check vapor separator (remove drain plug) for fuel. 3. If no fuel present in vapor separator, check fuel/water separator for debris. Check crankcase mounted fuel pump for output. 4. Check high pressure pump amperage draw. Normal draw is 6 – 9 amperes; if draw is below 2 amperes, check fuel pump filter (base of pump) for debris. If filter is clean, replace pump. If amperage is above 9 amperes, pump is defective – replace pump. Check low pressure output – 6–9 psi. Check low pressure electric fuel pump amperage draw. Normal draw is 1 – 2 amperes; if draw is below 1 ampere, check for blockage between pump inlet fitting and vapor separator tank. If ampere draw is above 2 amperes, replace pump. 5. Fuel Regulator – Remove and inspect diaphragm for cuts or tears.
Fuel Pressure High and Air Pressure is Normal	1. Stuck check valve in fuel return hose. 2. Debris blocking fuel regulator hole. 3. Faulty pressure gauge
Fuel and Air Pressure Higher than Normal	1. Debris blocking air regulator passage. 2. Air dump hose (rail to air plenum) blocked/plugged.

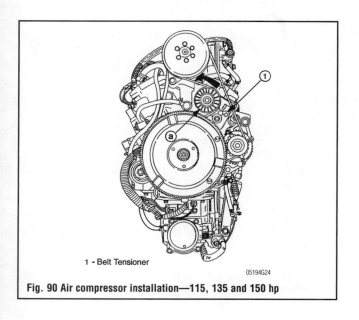

1 - Belt Tensioner

Fig. 90 Air compressor installation—115, 135 and 150 hp

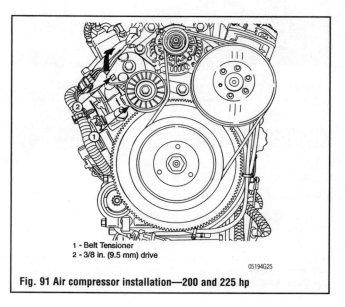

1 - Belt Tensioner
2 - 3/8 in. (9.5 mm) drive

Fig. 91 Air compressor installation—200 and 225 hp

4-72 FUEL SYSTEM

** CAUTION

If the engine has been run recently, the air pressure outlet hose fittings may be extremely hot. Allow the components to cool off before beginning disassembly.

5. Remove two screws securing retainer plate to remove air pressure outlet hose. Inspect the O-rings on the air pressure hose fitting for cuts and abrasions. Replace the O-rings as necessary.
6. Remove the air pressure outlet hose.
7. Disconnect the compressor water inlet hose.
8. Disconnect the air compressor oil inlet hose.
9. Disconnect water outlet hose (tell-tale).
10. Disconnect excess oil return hose.
11. Disconnect the bolts securing the compressor to the engine and remove the compressor.

To install:
12. Install the compressor and tighten the retaining bolts to the powerhead.
13. Connect the excess oil return hose.
14. Connect the water outlet hose (tell-tale).
15. Connect the air compressor oil inlet hose.
16. Connect the air pressure outlet hose.
17. Connect the compressor water inlet hose.
18. Connect the air pressure outlet hose.
19. Install the new O-rings and then install the air pressure outlet hose and tighten the screws securing the retainer plate.
20. Install the drive belt and then use a ⅜in. (9.5mm) drive on belt tensioner arm to adjust the belt tension.
21. Install the flywheel cover.

➥**Install the vent hose to the fitting on the flywheel cover.**

22. Install the top cowling.
23. Connect the battery cables to the battery terminals.

Air Pressure Regulator

REMOVAL & INSTALLATION

1. Remove the screws securing the regulator and remove it from the rail
2. Inspect the regulator diaphragm for cuts or tears. Replace if necessary.

To install:
3. Apply a light coat of Quicksilver 2-4-C w/Teflon to the diaphragm surface to aid in the retention of the diaphragm to the fuel rail during assembly.
4. Position the diaphragm/spring/cup onto the fuel rail in a horizontal position.
5. Apply anti-seize grease or Quicksilver 2-4-C w/Teflon to the regulator attaching screw threads.
6. Due to the stiffness of the regulator spring, it is recommended the 2 longer screws (5mm x 25mm) (10–40073 25) and 2 flat washers (12–30164) be installed through the cover first to begin compression. This will allow 2 shorter screws (5mm x 15mm) to be installed. Remove the 2 long screws and washers and install the remaining 2 short screws. Torque screws to 70 inch lbs. (8.0 Nm).

Air Temperature Sensor

TESTING

1. Disconnect the sensor electrical harness and remove the sensor.
2. Connect a digital meter set on the 20K scale to the leads on the sensor.
3. Place the sensor in a container of ice water while monitoring the meter reading.
4. The ohm readings should be 540 to 44,500 ohms. 8000 ohms with the engine at 77°F. Use the graph below for reference. Resistance does not change inversely with temperature change.
5. Replace the defective Air Temperature Sensor if the readings are not within specification.

REMOVAL & INSTALLATION

♦ **See Figures 92 and 93**

1. Disconnect the sensor wiring harness.
2. Remove the sensor from the top of the air plenum.

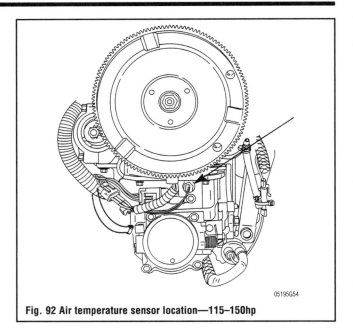

Fig. 92 Air temperature sensor location—115–150hp

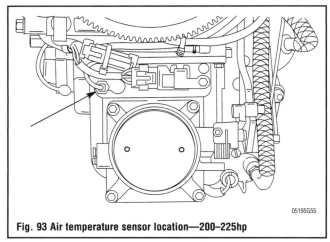

Fig. 93 Air temperature sensor location—200–225hp

To install:
3. Install the sensor into the top of the air plenum.
4. Connect the sensor wiring harness

Crankshaft Position Sensor (CPS)

TESTING

1. The crankshaft position sensor can only be tested using the DDT or equivalent tester. If the sensor should fail during operation, the engine will stop running.

REMOVAL & INSTALLATION

♦ **See Figures 94 and 95**

1. Disconnect the CPS sensor from the wiring harness.
2. Remove the bolt that secures the CPS to the bracket.
3. Remove the CPS from the powerhead.

To install:
4. Fasten the CPS to the bracket.
5. Set the air gap between the sensor and the flywheel at 0.025–0.040 in. (0.635–1.01 mm).

➥**On the 115–150hp models, there are two different size shims used to set the air gap, 0.010 in. (0.254 mm) and 0.020 in. (0.508 mm).**

FUEL SYSTEM   4-73

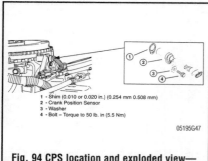

Fig. 94 CPS location and exploded view— 115–150hp

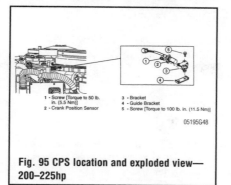

Fig. 95 CPS location and exploded view— 200–225hp

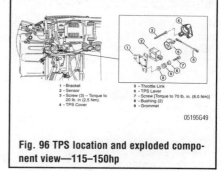

Fig. 96 TPS location and exploded component view—115–150hp

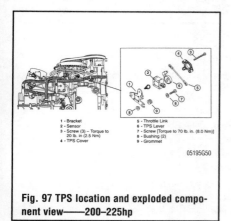

Fig. 97 TPS location and exploded component view——200–225hp

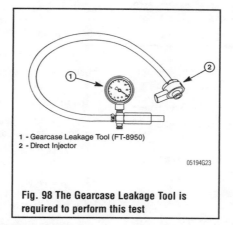

Fig. 98 The Gearcase Leakage Tool is required to perform this test

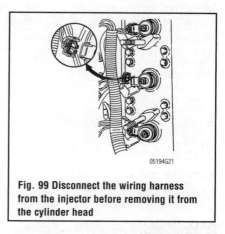

Fig. 99 Disconnect the wiring harness from the injector before removing it from the cylinder head

Throttle Position Sensor (TPS)

TESTING

1. The throttle position sensor can only be tested using the DDT or equivalent tester. The sensor position is not adjustable.

REMOVAL & INSTALLATION

♦ See Figures 96 and 97

1. Disconnect the TPS wire from the wiring harness.
2. Remove the screws securing the TPS to the bracket on the powerhead and remove the TPS.
   To install:
3. Fasten the sensor to the bracket and tighten the screws to 70 inch lbs. (8.0 Nm)

Direct Injector

TESTING

Resistance

1. With the injector lead disconnected, connect the multimeter leads between each injector terminal pin.
2. The resistance should measure 1–1.6 ohms.
3. If the injector does not within specification, it may be faulty and will need to be replaced.

Leakage

♦ See Figure 98

1. Attach a Gearcase Leakage Tool (FT-8950) to the discharge side of the injector.
2. Pump up leakage tool to indicate 25–30 psi (172.4–206.8 kPa).

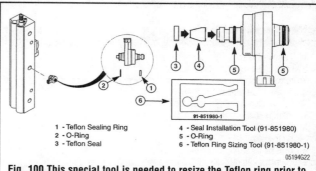

Fig. 100 This special tool is needed to resize the Teflon ring prior to installation

3. Direct injector should not leak down more than ½psi (3.5 kPa) in one minute.
4. If injector does not meet the above specifications, the injector must be replaced.

REMOVAL & INSTALLATION

♦ See Figures 99 and 100

➡ If the cylinder head is going to be replaced, removed the cup washers from each direct injector port by prying out with a flat tip screwdriver. Reinstall washers with retainers into the new cylinder head. Washers provide tension between direct injectors, cylinder head and fuel rails.

1. Remove harness connector from the direct injectors.
2. Remove the direct injector (3 per cylinder head) from the cylinder head
   To install:
3. Use Teflon Ring Sizing Tool (91-851980-1) to compress new Teflon sealing rings prior to installation of injector into the cylinder head.
4. Carefully slide the fuel rail over the mounting studs and onto the direct injectors.

4-74 FUEL SYSTEM

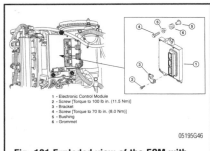

Fig. 101 Exploded view of the ECM with fasteners shown

Fig. 102 The fuel injector retaining screws must first be removed in order to remove the injector

Fig. 103 Be very careful when prying out the fuel injector to prevent damaging it

✱✱ WARNING

The air hose MUST be secured with stainless steel hose clams.

5. Secure each fuel rail with 2 nuts. Torque the nuts to 33 ft. lbs. (44 Nm).
6. Reinstall the direct injector harness connectors.

CLEANING & INSPECTION

Inspect the injector Teflon sealing ring (white) for sign of combustion blowby (Teflon ring will be streaked brownish black). If blowby is present, replace the Teflon sealing ring. If blowby is not present, the sealing ring may be reused.

Inspect the O-rings for cut or deformities. Replace any damaged components as needed.

If the Teflon seal requires replacement, use Teflon ring installation tool (91–8511980) to slide the new seal onto the injector. Following the installation of the Teflon ring sizing tool (91–851980–1) can be used to compress the Teflon seal to aid in the installation of the injector into the cylinder head.

Carbon buildup on the tip of the direct injector my be removed by using a brass wire brush.

Electronic Control Module (ECM)

REMOVAL & INSTALLATION

♦ See Figure 101

1. Remove the engine cowling.
2. Remove the flywheel cover.
3. Disconnect the ECM wiring harness connectors.
4. Remove the 3 bolts securing the ECM to the powerhead.

To install:

5. Secure the ECM to the powerhead with the 3 retaining bolts and torque to 80 inch lbs. (9 Nm).
6. Reconnect the ECM wiring harness connectors.
7. Install the flywheel cover as follows:
   - Place the cover onto the front flange
   - Push the rear of the cover down onto the rear locating pin and air intake tube for the air compressor
   - Connect the vent tube onto the flywheel cover fitting

Fuel Injector

TESTING

1. With the injector lead disconnected, connect the multimeter leads between each injector terminal pin.
2. The resistance should measure 1.7–1.9 ohms.
3. If the injector does not test within specification, it may be faulty and will need to be replaced.

REMOVAL & INSTALLATION

♦ See Figures 102 and 103

1. Remove 2 screws securing injector.

➡ Use a cotter pin extractor tool in the pry holes to remove the injectors.

To install:

➡ Apply anti-seize grease or Quicksilver 2-4-C w/Teflon to fuel injector attaching screw threads.

2. Insert fuel injector into the fuel rail with connector pins facing (inwards) towards the center of the engine.

➡ Turn injector back and forth slightly to seat the injector O-rings in the fuel rail while securing the injector with the retainer and two screws. Torque the screws to 70 inch lbs. (8.0 Nm).

CLEANING & INSPECTION

♦ See Figure 104

1. Inspect fuel injector orifices for foreign debris, O-rings for cuts or abrasions and plastic components for heat damage. Replace any components as required.
2. An ohm test of the fuel injector may be made by connecting test leads to the injector terminals. The ohm reading should be 1.7–1.9 ohms.

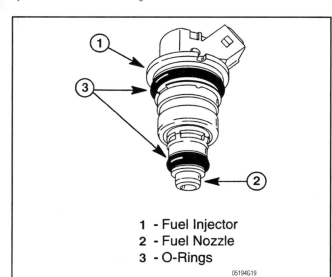

1 - Fuel Injector
2 - Fuel Nozzle
3 - O-Rings

Fig. 104 Make sure the injector O-rings are free of cuts, abrasions and debris

FUEL SYSTEM   4-75

Fuel Pressure Regulator

REMAVAL & INSTALLATION

▶ See Figure 105

1. Remove the 4 screws securing the regulator and remove the regulator from the fuel rail.

To install:

➡ Apply a light coat of Quicksilver 2-4-C w/ Teflon to the diaphragm surface and O-ring to aid in the retention of the diaphragm and O-ring on the fuel rail during reassembly.

2. Position the diaphragm on the fuel rail.
3. Position the O-ring on the fuel rail.
4. Position the spring and cup onto the diaphragm.

➡ Apply anti-seize grease (obtain locally) or Quicksilver 2-4-C w/Teflon to the regulator attaching screw threads.

5. Place the cover over the spring/cup/diaphragm assembly and secure it with 4 screws. Torque the screws to 70 inch lbs. (8.0 Nm).

CLEANING & INSPECTION

Inspect the regulator diaphragm for cuts, nicks or abrasions.
Inspect the regulator housing O-ring for cuts and abrasions. If any parts are damaged, replace components as required.

Fuel Pump

TESTING

Low Pressure Pump Output

▶ See Figures 106, 107 and 108

✱✱ WARNING

After completing fuel pressure tests, reconnect and secure fuel outlet hose to fuel pump with full circle stainless clamps.

1. Remove the outlet fuel hose from the low pressure pump. Install a short piece of hose (obtained locally) onto the pump outlet fitting. Install a Schrader valve t-fitting (22–8496060 between the outlet fuel hose (removed from the pump) and new fuel hose (installed on the pump). Secure the hose connections with clamps.

➡ **Due to the low pressure output of this pump, it is recommended that the air gauge of the Dual Fuel/Air Pressure Gauge (91–852087A1/A2) be connected to the Schrader valve. The gauge should indicate 6–9 psi (41.37–62.04 kPa)**

High Pressure Pump Output

▶ See Figures 109 and 110

1. On the 115, 135 and 150hp models, install the Dual Fuel/Air Pressure Gauge assembly to the **PORT** fuel rail pressure test valve.

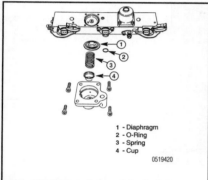

Fig. 105 Exploded view of the fuel pressure regulator assembly

Fig. 106 Remove the caps on the fuel rail Schrader valves

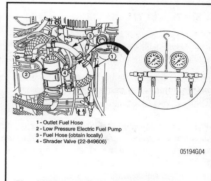

Fig. 107 Low pressure fuel pump test on the 115, 135 and 150hp models. . .

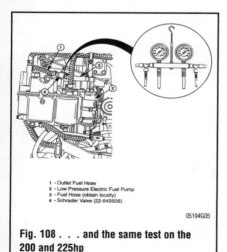

Fig. 108 . . . and the same test on the 200 and 225hp

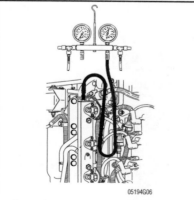

Fig. 109 Testing the high pressure fuel pump on the 115, 150 and 150 hp models . . .

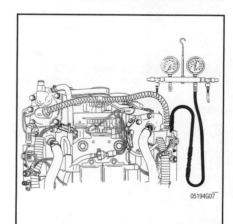

Fig. 110 . . . and the same set-up on the 200–225 hp models

4-76 FUEL SYSTEM

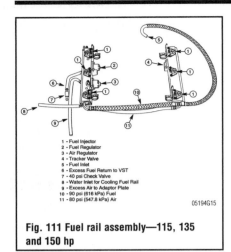

Fig. 111 Fuel rail assembly—115, 135 and 150 hp

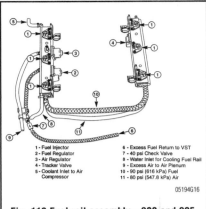

Fig. 112 Fuel rail assembly—200 and 225 hp

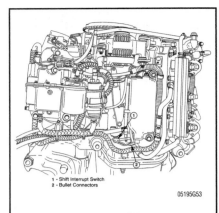

Fig. 113 Location of the Shift Interrupt Switch and the electrical connector

2. On the 200–225hp models, install the test assembly to the **STARBOARD** fuel rail pressure test valve.
3. After 15 seconds of engine cranking with the starter motor, the fuel pressure gauge should measure 87–91 psi (599.7–627.3 kPa).

Fuel Rail

REMOVAL & INSTALLATION

♦ See Figures 111 and 112

1. De-pressurize fuel system.
2. Remove the fuel injector harness from each injector by compressing the spring clip with a flat tip screw driver while pulling on the connector.

➡ Always remove the fuel/air hose and fitting together by removing fitting retainer rather than cutting the clamps.

3. Remove the fuel, water and air hoses from the fuel rail.

➡ It is recommended that direct injectors remain in the cylinder head (if they are not to be replaced) while removing the fuel rail. The direct injectors have a Teflon seal which may expand if the injector is removed from the head. This expansion may cause reinstallation difficulties or require the replacement of the seal.

4. Remove the two nuts securing the fuel rail.
5. As the fuel rail is removed, use a flat tip screwdriver to hold the direct injectors in the cylinder head.
6. The starboard fuel rail contains 3 fuel injectors and a tracker valve. The port fuel rail contains 3 fuel injectors, 1 fuel regulator and 1 air regulator.

To install:

7. Carefully slide the fuel rail over the mounting studs and onto the direct injectors.

✸✸ WARNING

The air hose must be secured with stainless steel hose clamps.

8. Secure each fuel rail with 2 nuts. Torque the nuts to 33 ft. lbs. (44 Nm).
9. Reinstall the direct injector harness connectors.

CLEANING & INSPECTION

After all fuel injectors, air regulator, tracker valve, fuel regulator, inlet hoses and outlet hoses have been removed, the fuel rails may be flushed out with a suitable parts cleaning solvent. Use compressed air to remove any remaining solvent.

Each fuel/air inlet or outlet hose adaptor has 2 O-ring seals. Theses O-rings should be inspected for cuts, deformities or abrasions and replaced as necessary when the fuel rail is disassembled for inspection and cleaning.

Manifold Absolute Pressure (Map) Sensor

TESTING

1. Connecting a hand held vacuum pump and drawing a vacuum on the MAP sensor hose will create a lean fuel condition altering engine operation. If no change occurs when drawing the vacuum, the MAP sensor is not functioning properly.
2. Using a multimeter, check the voltage change in the MAP sensor. As you vary the vacuum, the voltage reading should raise or lower. Normal voltage is 0 to 5 volts, depending on air pressure.

REMOVAL & INSTALLATION

1. Disconnect the MAP sensor wiring connector at the harness.
2. Remove the retaining bolt and remove the MAP sensor.

To install:

3. Connect the MAP sensor wiring harness at the connector.
4. Install the MAP sensor and tighten the retaining bolt.

Shift Interrupt Switch

TESTING

1. Using a multimeter, check the continuity of the switch.
2. With the outboard in neutral, the switch is closed (showing continuity) and normally open (no continuity) when in gear.
3. If the switch does not test within specification, it may be faulty and will need to be replaced.

REMOVAL & INSTALLATION

♦ See Figure 113

1. Remove the bottom cowling.
2. Disconnect the shift interrupt switch bullet connectors.
3. Remove the retaining screws and remove the switch.

To install:

4. Install the switch on the powerhead and secure with the retaining screws.
5. Connect the switch bullet connectors.
6. Install the bottom cowling.

Throttle Plate Assembly

REMOVAL & INSTALLATION

➡ The throttle plate assembly is calibrated and preset for proper running characteristics and emissions at the factory. Other than the complete assembly removal from the air plenum, no further disassembly should be made.

1. Remove the bolts securing the throttle plate to the air plenum and remove the throttle plate.

To install:

2. Install the throttle plate.
3. Install the bolts and torque to 100 inch lbs. (11.5 Nm).

Tracker Valve

REMOVAL & INSTALLATION

1. Remove the screws securing the tracker valve assembly and remove it from the fuel/air rail.

To install:

2. Apply a light coat of Quicksilver 2-4-C w/Teflon to the tracker diaphragm and cover O-ring to aid in their retention on the fuel rail while reinstalling the tracker valve to the fuel rail.

➡ Apply anti-seize grease or Quicksilver 2-4-C to the tracker valve attaching screw threads.

3. Position the diaphragm, spring and O-ring onto the fuel rail.
4. Place the cover over the diaphragm/spring/O-ring assembly and secure it with 4 screws. Torque the screws to 70 inch lbs. (8.0 Nm).

Vapor Separator

REMOVAL & INSTALLATION

1. De-pressurize the fuel system.
2. Place a suitable container underneath the vapor separator drain plug and drain the separator.
3. Disconnect the water separator sensor lead.
4. Disconnect the electric fuel pump harness connectors.
5. Remove the straps securing the fuel hoses.

➡ Upper hose is the excess fuel return from the fuel rails and the lower hose is the fuel inlet from the electric pump beside the fuel/water separator.

6. Remove the fuel inlet hose from the pulse fuel pump.
7. Remove the fuel outlet hose and the fuel return hose from the fuel rails.
8. Remove the bolts and then remove the vapor separator.

To install:

9. Secure the vapor separator to the air plenum. Torque the bolts to 140 inch lbs. (16.0 Nm).
10. Connect the fuel inlet hose from the pulse pump.
11. Connect he fuel outlet hose and fuel return hose to the vapor separator.
12. Connect the water separator sensor lead to the water separator.
13. Connect the electric fuel pump harnesses.
14. Connect the vapor separator vent hose to the air plenum.
15. Secure the hoses with tie-wraps.

DISASSEMBLY

1. Remove the screws securing the separator cover and remove the cover.
2. The fuel pump may be removed from the cover by slightly wiggling it while pulling outward.

✳✳ CAUTION

Do not twist the pump during removal. Doing so may damage the wiring harness.

3. Disconnect the harness from the pump to separate the pump from the cover
4. Loosen the screw securing the float assembly and remove the float.
5. Remove the phenolic sealing plate.

CLEANING & INSPECTION

▸ See accompanying illustrations

1. Inspect the seal in the fuel pump chamber of the separator tank for cuts, nicks or abrasions. Replace the seal if necessary. If the seal is in serviceable condition, apply a coat of Quicksilver 2-4-C w/ Teflon to the seal lips.
2. Inspect the filter screen for any debris. The screen may be pried out of the pump and cleaned if necessary.
3. Inspect the seal above the fuel pump for cuts, nicks or abrasions. Replace the seal if necessary. If the seal is in serviceable condition, apply a coat of Quicksilver 2-4-C w/ Teflon to the seal lips.
4. Inspect the float for deterioration or fuel retention. Replace the float as required.
5. Remove the phenolic sealing plate and inspect the imbedded neoprene seal on both side of plate for cuts and abrasions. Replace the plate/seal as necessary.

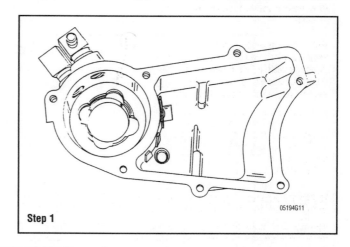
Step 1

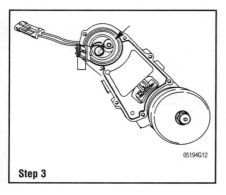
Step 3

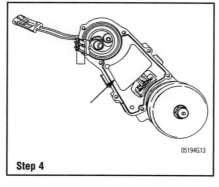
Step 4

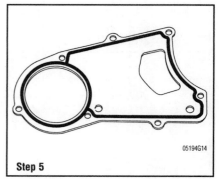
Step 5

4-78 FUEL SYSTEM

ASSEMBLY

1. Reinstall the phenolic sealing plate onto the vapor separator cover.
2. Secure the float, needle and pivot pin assembly to the separator cover with screw. Torque the screw to 10 inch lbs. (1.0 Nm).
3. Apply a marine grade grease to the lips of the seal in the separator cover.
4. Conect he electrical harness to the fuel pump. Inspect the fuel pump filter screen for debris. Remove the screen and clean as required.
5. Seat the fuel pump and harness into separator cover being careful not to pinch the harness.
6. Apply a 2-4-C w/Teflon to the seal lips of the seal in the separator tank.

Water Temperature Sensor

TESTING

♦ See Figure 114

1. Insert digital or analog test leads into both the Tan/Black sensor leads.
2. With the engine at the temperature (F°) indicated on the chart, the ohm readings should be as indicated plus or minus 10-percent.
3. Resistance measured between the Black and each Tan/Black wire should show no continuity.
4. Resistance measured between each lead and ground should read no continuity.

REMOVAL & INSTALLATION

♦ See Figures 115 and 116

1. Disconnect the sensor/s wiring harness connections.
2. Unbolt the retaining nut and remove the sensor assembly from either the air compressor or cylinder head.

   To install:
3. Install the sensor assembly and tighten the retaining bolt.
4. Connect the sensor wiring connectors to the harness.

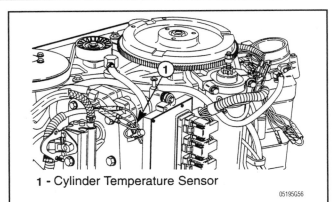

1 - Cylinder Temperature Sensor

Fig. 115 Coolant temperature sensor location—115–150hp

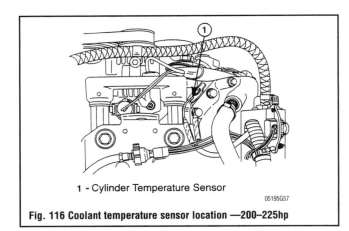

1 - Cylinder Temperature Sensor

Fig. 116 Coolant temperature sensor location —200–225hp

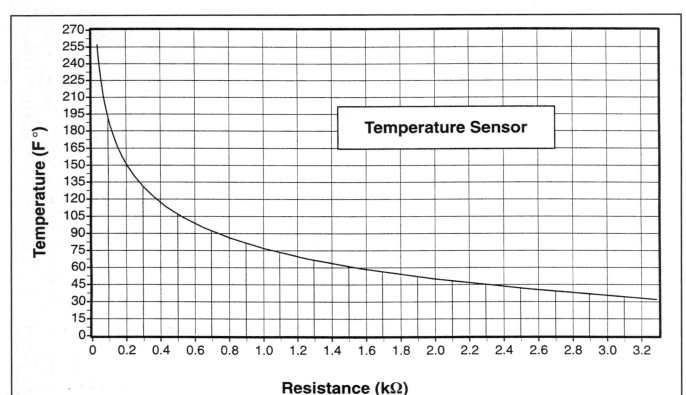

Fig. 114 Temperature sensor resistance

UNDERSTANDING AND
 TROUBLESHOOTING ELECTRICAL
 SYSTEMS 5-2
BASIC ELECTRICAL THEORY 5-2
  HOW ELECTRICITY WORKS: THE
   WATER ANALOGY 5-2
  OHM'S LAW 5-2
ELECTRICAL COMPONENTS 5-2
  POWER SOURCE 5-2
  GROUND 5-3
  PROTECTIVE DEVICES 5-3
  SWITCHES & RELAYS 5-3
  LOAD 5-4
  WIRING & HARNESSES 5-4
  CONNECTORS 5-4
TEST EQUIPMENT 5-4
  JUMPER WIRES 5-4
  TEST LIGHTS 5-5
  MULTIMETERS 5-5
TROUBLESHOOTING THE ELECTRICAL
 SYSTEM 5-6
  VOLTAGE 5-6
  VOLTAGE DROP 5-6
  RESISTANCE 5-6
  OPEN CIRCUITS 5-6
  SHORT CIRCUITS 5-7
WIRE AND CONNECTOR REPAIR 5-7
**BREAKER POINTS IGNITION
 (MAGNETO IGNITION) 5-8**
DESCRIPTION & OPERATION 5-8
  SYSTEM COMPONENTS 5-8
TROUBLESHOOTING THE BREAKER
 POINTS IGNITION SYSTEM 5-8
  SPARK PLUGS 5-9
  COMPRESSION TEST 5-9
BREAKER POINTS 5-9
  POINT GAP CHECK 5-9
  REMOVAL & INSTALLATION 5-11
CONDENSER 5-12
  TESTING 5-12
  REMOVAL & INSTALLATION 5-12
PRIMARY COIL 5-13
  TESTING 5-13
  REMOVAL & INSTALLATION 5-13
IGNITION COIL 5-13
  TESTING 5-13
  REMOVAL & INSTALLATION 5-14
**CAPACITOR DISCHARGE IGNITION
 (CDI) SYSTEM 5-14**
DESCRIPTION AND OPERATION 5-14
  2.5–3.3 HP 5-15
  4–275 HP 5-15
TROUBLESHOOTING THE CDI
 SYSTEM 5-15
  TROUBLESHOOTING WITH MINIMAL
   TEST EQUIPMENT 5-15
  TROUBLESHOOTING BATTERY CD
   IGNITIONS 5-16
  TROUBLESHOOTING ALTERNATOR
   DRIVEN IGNITIONS 5-16

TRIGGER (CHARGE) COIL 5-16
  TESTING 5-16
  REMOVAL & INSTALLATION 5-18
IGNITION COIL 5-25
  TESTING 5-25
  REMOVAL & INSTALLATION 5-26
CAPACITOR DISCHARGE MODULE
 (SWITCH BOX) 5-27
  TESTING 5-27
  REMOVAL & INSTALLATION 5-28
CHARGING CIRCUIT 5-30
DESCRIPTION AND OPERATION 5-30
TROUBLESHOOTING THE CHARGING
 SYSTEM 5-30
STATOR (ALTERNATOR) 5-30
  TESTING 5-30
  REMOVAL & INSTALLATION 5-31
RECTIFIER 5-34
  TESTING 5-34
  REMOVAL & INSTALLATION 5-34
REGULATOR/RECTIFIER 5-34
  TESTING 5-34
  REMOVAL & INSTALLATION 5-38
BATTERY 5-38
  BATTERY CONSTRUCTION 5-38
  MARINE BATTERIES 5-38
  BATTERY RATINGS 5-38
  BATTERY LOCATION 5-39
  BATTERY CHARGERS 5-39
  BATTERY CABLES 5-40
STARTER CIRCUIT 5-40
DESCRIPTION AND OPERATION 5-40
TROUBLESHOOTING THE STARTING
 SYSTEM 5-40
STARTER MOTOR 5-41
  REMOVAL & INSTALLATION 5-41
STARTER SOLENOID 5-41
  TESTING 5-41
**IGNITION AND ELECTRICAL WIRING
 DIAGRAMS 5-42**
SPECIFICATIONS CHARTS
  IGNITION TESTING
   SPECIFICATIONS 5-17
  STATOR ALTERNATOR RESISTANCE
   SPECIFICATIONS 5-30
  SMALL VOLTAGE REGULATOR TEST
   (STATIC) 5-35
TROUBLESHOOTING CHARTS
  RECTIFIER TROUBLESHOOTING - TEST
   "A" 5-37
  RECTIFIER TROUBLESHOOTING - TEST
   "B" 5-37

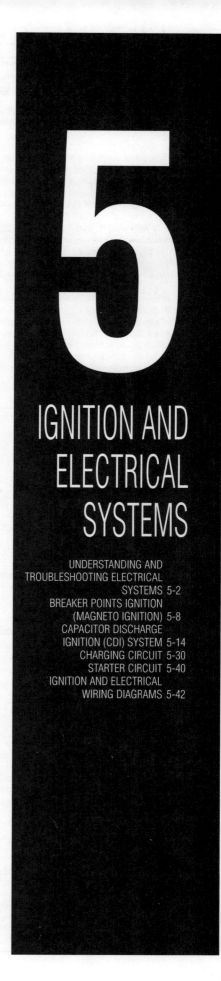

5

IGNITION AND ELECTRICAL SYSTEMS

UNDERSTANDING AND
TROUBLESHOOTING ELECTRICAL
SYSTEMS 5-2
BREAKER POINTS IGNITION
(MAGNETO IGNITION) 5-8
CAPACITOR DISCHARGE
IGNITION (CDI) SYSTEM 5-14
CHARGING CIRCUIT 5-30
STARTER CIRCUIT 5-40
IGNITION AND ELECTRICAL
WIRING DIAGRAMS 5-42

IGNITION AND ELECTRICAL SYSTEMS

UNDERSTANDING AND TROUBLESHOOTING ELECTRICAL SYSTEMS

Basic Electrical Theory

♦ See Figure 1

For any 12 volt, negative ground, electrical system to operate, the electricity must travel in a complete circuit. This simply means that current (power) from the positive terminal (+) of the battery must eventually return to the negative terminal (-) of the battery. Along the way, this current will travel through wires, fuses, switches and components. If for any reason the flow of current through the circuit is interrupted, the component(s) fed by that circuit will cease to function properly.

Perhaps the easiest way to visualize a circuit is to think of connecting a light bulb (with two wires attached to it) to the battery—one wire attached to the negative (-) terminal of the battery and the other wire to the positive (+) terminal. With the two wires touching the battery terminals, the circuit would be complete and the light bulb would illuminate. Electricity would follow a path from the battery to the bulb and back to the battery. It's easy to see that with longer wires on our light bulb, it could be mounted anywhere. Further, one wire could be fitted with a switch so that the light could be turned on and off.

The normal marine circuit differs from this simple example in two ways. First, instead of having a return wire from each bulb to the battery, the current travels through a single ground wire, which handles all the grounds for a specific circuit. Secondly, most marine circuits contain multiple components, which receive power from a single circuit. This lessens the amount of wire needed to power components.

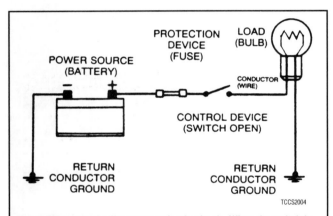

Fig. 1 This example illustrates a simple circuit. When the switch is closed, power from the positive (+) battery terminal flows through the fuse and the switch and then to the light bulb. The light illuminates and the circuit is completed through the ground wire back to the negative (-) battery terminal.

HOW ELECTRICITY WORKS: THE WATER ANALOGY

Electricity is the flow of electrons—the sub-atomic particles that constitute the outer shell of an atom. Electrons spin in an orbit around the center core of an atom. The center core is comprised of protons (positive charge) and neutrons (neutral charge). Electrons have a negative charge and balance out the positive charge of the protons. When an outside force causes the number of electrons to unbalance the charge of the protons, the electrons will split off the atom and look for another atom to balance out. If this imbalance is kept up, electrons will continue to move and an electrical flow will exist.

Many people have been taught electrical theory using an analogy with water. In a comparison with water flowing through a pipe, the electrons would be the water and the wire is the pipe.

The flow of electricity can be measured much like the flow of water through a pipe. The unit of measurement used is amps, frequently abbreviated as amps (a). You can compare amperage to the volume of water flowing through a pipe. When connected to a circuit, an ammeter will measure the actual amount of current flowing through the circuit. When relatively few electrons flow through a circuit, the amperage is low. When many electrons flow, the amperage is high.

Water pressure is measured in units such as pounds per square inch (psi), electrical pressure is measured in units called volts (v). When a voltmeter is connected to a circuit, it is measuring the electrical pressure. When electrical pressure is low, then voltage is considered to be low. When electrical pressure is high, then voltage is considered to be high.

The actual flow of electricity depends not only on voltage and amperage but also on the resistance of the circuit. The higher the resistance, the higher the force necessary to push the current through the circuit. The standard unit for measuring resistance is an ohm (Ω). Resistance in a circuit varies depending on the amount and type of components used in the circuit and the overall condition of the components and wires. If we assume that everything in our circuit is new, then, the main factors which determine resistance are:

- Material—some materials have more resistance than others. Those with high resistance are said to be insulators. Rubber materials (or rubber-like plastics) are some of the most common insulators used, as they have a very high resistance to electricity. Very low resistance materials are said to be conductors. Copper wire is among the best conductors. Silver is actually a superior conductor to copper and is used in some relay contacts but its high cost prohibits its use as common wiring. Most marine wiring is made of copper.
- Size—the larger the wire size being used, the less resistance the wire will have. This is why components which use large amounts of electricity usually have large wires supplying current to them.
- Length—for a given thickness of wire, the longer the wire, the greater the resistance. The shorter the wire, the less the resistance. When determining the proper wire for a circuit, both size and length must be considered to design a circuit that can handle the current needs of the component.
- Temperature—with many materials, the higher the temperature, the greater the resistance (positive temperature coefficient). Some materials exhibit the opposite trait of lower resistance with higher temperatures (negative temperature coefficient). These principles are used in many of the sensors on the engine.

OHM'S LAW

There is a direct relationship between current, voltage and resistance which can be summed up by a statement known as Ohm's law.

- Voltage (E) is equal to amperage (I) times resistance (R): $E = I \times R$
- Other forms of the formula are $R = E/I$ and $I = E/R$

In each of these formulas, E is the voltage in volts, I is the current in amps and R is the resistance in ohms. The basic point to remember is that as the resistance of a circuit goes up, the amount of current that flows in the circuit will go down, if voltage remains the same.

The amount of work that the electricity can perform is expressed as power. A unit of power is known as a watt (W). There is a direct relationship between power, voltage and current which can be summed up by the following formula:

- Power (W) is equal to amperage (I) times voltage (E): $W = I \times E$

➡ This formula is only true for direct current (DC) circuits, The alternating current formula is a tad different but since the electrical circuits in most boats are DC type, we need not get into AC circuit theory.

Electrical Components

POWER SOURCE

♦ See Figure 2

Power is supplied to the boat by two devices: The battery and the alternator (stator). The battery supplies electrical power during starting or during periods when the current demand of the boat's electrical system exceeds the output capacity of the alternator. The alternator supplies electrical current when the engine is running. The alternator does not just supply the current needs of the boat but it also recharges the battery.

In most modern boats, the battery is a lead/acid electrochemical device consisting of six 2 volt subsections (cells) connected in series, so that the unit is capable of producing approximately 12 volts of electrical pressure. Each subsection consists of a series of positive and negative plates held a short distance apart in a solution of sulfuric acid and water.

The two types of plates are of dissimilar metals, which sets up a chemical reaction inside the battery case. It is this reaction which produces current flow from the battery when its positive and negative terminals are connected to an electrical load. The alternator, restoring the battery to its original chemical state replaces the power removed from the battery.

IGNITION AND ELECTRICAL SYSTEMS

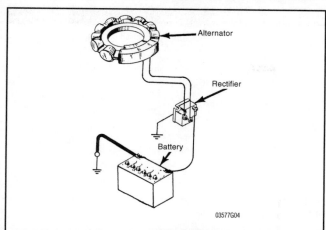

Fig. 2 Functional diagram of a typical charging circuit showing the relationship of the stator, solid state rectifier and the battery

The following is a brief description of how the system works:

The battery stores electricity and acts as a "sponge" for the whole system. It mops up generated current until it's fully charged and it releases energy on demand.

The flywheel holds the permanent magnets that create the moving magnetic field. If your engine has good spark, you can take it for granted that the magnets are in working order because the ignition and charging systems share the same magnets.

The stator windings are the stationary coils of wire the flywheel magnets rotate around. They produce the electrical charge. Simply put the more windings in your stator, the greater the potential output in amps your charging system you'll have.

The rectifier consists of a series of diodes or electrical one-way valves. The rectifier overcomes one of the disadvantages of a current-generating system using permanent magnets and stator windings, which is that the current produced within the windings is alternating current (AC). You can't use AC to charge batteries. They accept only direct current (DC). So the rectifier is designed to convert AC current to a usable form of DC current simply called "rectified AC."

On larger outboards, there may be a voltage regulator. either combined with the rectifier or standing alone. The regulator automatically reduces the output of generated current as the battery becomes fully charged.

GROUND

All boats use some sort of a ground return circuit. Direct ground components are grounded to an electrically conductive metal component through their mounting points. These electrically conductive metal components are then grounded to the battery.

All other components use some sort of ground wire which leads directly back to the battery. The electrical current runs through the ground wire and returns to the battery through the ground (-) cable. If you look, you'll see that the battery ground cable connects between the battery and a heavy gauge ground wire.

➡ It should be noted that a good percentage of electrical problems can be traced to bad grounds.

PROTECTIVE DEVICES

♦ See Figure 3

It is possible for large surges of current to pass through the electrical system of your boat. If this surge of current were to reach components in the circuit, the surge could burn them out or severely damage them. Surges can also overload the wiring, causing the harness to get hot and melt the insulation. To prevent this, fuses, circuit breakers and/or fusible links are connected into the supply wires of the electrical system. These items are nothing more than a built-in weak spot in the system. When an abnormal amount of current flows through the system, these protective devices work as follows to protect the circuit:

• Fuse—when an excessive electrical current passes through a fuse, the fuse "blows" (the conductor melts) and opens the circuit, preventing the passage of current.

• Circuit Breaker—a circuit breaker is basically a self-repairing fuse. It will open the circuit in the same fashion as a fuse but when the surge subsides, the circuit breaker can be reset and does not need replacement.

• Fusible Link—a fusible link (fuse link or main link) is a short length of spe-

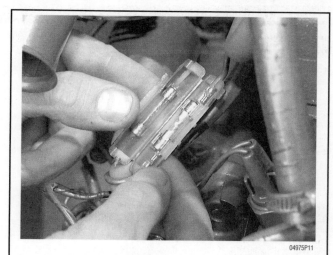

Fig. 3 Fuses protect the vessel's electrical system from abnormally high amounts of current flow

cial, high temperature insulated wire that acts as a fuse. When an excessive electrical current passes through a fusible link, the thin gauge wire inside the link melts, creating an intentional open to protect the circuit. To repair the circuit, the link must be replaced. Some newer type fusible links are housed in plug-in modules, which are simply replaced like a fuse, while older type fusible links must be cut and spliced if they melt. Since this link is very early in the electrical path, it's the first place to look if nothing on the boat works, yet the battery seems to be charged and is properly connected.

✱✱ CAUTION

Always replace fuses, circuit breakers and fusible links with identically rated components. Under no circumstances should a component of higher or lower amperage rating be substituted.

SWITCHES & RELAYS

♦ See Figure 4

Switches are used in electrical circuits to control the passage of current. The most common use is to open and close circuits between the battery and the various electric devices in the system. Switches are rated according to the amount of amperage they can handle. If a sufficient amperage rated switch is not used in a circuit, the switch could overload and cause damage.

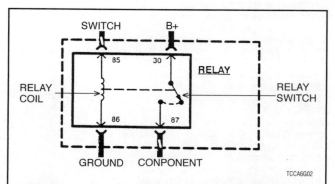

Fig. 4 Relays are composed of a coil and a switch. These two components are linked together so that when one operates, the other operates at the same time. The large wires in the circuit are connected from the battery to one side of the relay switch (B+) and from the opposite side of the relay switch to the load (component). Smaller wires are connected from the relay coil to the control switch for the circuit and from the opposite side of the relay coil to ground

5-4 IGNITION AND ELECTRICAL SYSTEMS

Some electrical components which require a large amount of current to operate use a special switch called a relay. Since these circuits carry a large amount of current, the thickness of the wire in the circuit is also greater. If this large wire were connected from the load to the control switch, the switch would have to carry the high amperage load and the space needed for wiring in the boat would be twice as big to accommodate the increased size of the wiring harness. To prevent these problems, a relay is used.

Relays are composed of a coil and a set of contacts. When the coil has a current passed though it, a magnetic field is formed and this field causes the contacts to move together, completing the circuit. Most relays are normally open, preventing current from passing through the circuit but they can take any electrical form depending on the job they are intended to do. Relays can be considered "remote control switches." They allow a smaller current to operate devices that require higher amperages. When a small current operates the coil, a larger current is allowed to pass by the contacts. Some common circuits which may use relays are horns, lights, starter, electric fuel pumps and other high draw circuits.

LOAD

Every electrical circuit must include a "load" (something to use the electricity coming from the source). Without this load, the battery would attempt to deliver its entire power supply from one pole to another. This would result in a "short circuit" of the battery. All this electricity would take a short cut to ground and cause a great amount of damage to other components in the circuit by developing a tremendous amount of heat. This condition could develop sufficient heat to melt the insulation on all the surrounding wires and reduce a multiple wire cable to a lump of plastic and copper.

WIRING & HARNESSES

The average boat contains miles of wiring, with hundreds of individual connections. To protect the many wires from damage and to keep them from becoming a confusing tangle, they are organized into bundles, enclosed in plastic or taped together and called wiring harnesses. Different harnesses serve different parts of the boat. Individual wires are color coded to help trace them through a harness where sections are hidden from view.

Marine wiring can be either single strand wire, multi-strand wire or printed circuitry. Single strand wire has a solid metal core and is usually used inside such components as alternators, motors, relays and other devices. Multi-strand wire has a core made of many small strands of wire twisted together into a single conductor. Most of the wiring in a marine electrical system is made up of multi-strand wire, either as a single conductor or grouped together in a harness. All wiring is color coded on the insulator, either as a solid color or as a colored wire with an identification stripe. A printed circuit is a thin film of copper or other conductor that is printed on an insulator backing. Occasionally, a printed circuit is sandwiched between two sheets of plastic for more protection and flexibility. A complete printed circuit, consisting of conductors, insulating material and connectors is called a printed circuit board. Printed circuitry is used in place of individual wires or harnesses in places where space is limited, such as behind instrument panels.

Since marine electrical systems are very sensitive to changes in resistance, the selection of properly sized wires is critical when systems are repaired. A loose or corroded connection or a replacement wire that is too small for the circuit will add extra resistance and an additional voltage drop to the circuit.

The wire gauge number is an expression of the cross-section area of the conductor. Boats from countries that use the metric system will typically describe the wire size as its cross-sectional area in square millimeters. In this method, the larger the wire, the greater the number. Another common system for expressing wire size is the American Wire Gauge (AWG) system. As gauge number increases, area decreases and the wire becomes smaller. An 18 gauge wire is smaller than a 4 gauge wire. A wire with a higher gauge number will carry less current than a wire with a lower gauge number. Gauge wire size refers to the size of the strands of the conductor, not the size of the complete wire with insulator. It is possible, therefore, to have two wires of the same gauge with different diameters because one may have thicker insulation than the other.

It is essential to understand how a circuit works before trying to figure out why it doesn't. An electrical schematic shows the electrical current paths when a circuit is operating properly. Schematics break the entire electrical system down into individual circuits. In a schematic, usually no attempt is made to represent wiring and components as they physically appear on the boat, switches and other components are shown as simply as possible. Face views of harness connectors show the cavity or terminal locations in all multi-pin connectors to help locate test points.

CONNECTORS

♦ See Figures 5, 6, 7 and 8

Weatherproof connectors are most commonly used where the connector is exposed to the elements. Terminals are protected against moisture and dirt by sealing rings which provide a weather tight seal. All repairs require the use of a special terminal and the tool required to service it.

Unlike standard blade type terminals, these weatherproof terminals cannot be straightened once they are bent. Make certain that the connectors are properly seated and all of the sealing rings are in place when connecting leads.

Test Equipment

Pinpointing the exact cause of trouble in an electrical circuit is most times accomplished by the use of special test equipment. The following sections describe different types of commonly used test equipment and briefly explain how to use them in diagnosis. In addition to the information covered below, the tool manufacturer's instruction manual (provided with most tools) should be read and clearly understood before attempting any test procedures.

JUMPER WIRES

♦ See Figure 9

✳✳ CAUTION

Never use jumper wires made from a thinner gauge wire than the circuit being tested. If the jumper wire is of too small a gauge, it may overheat and possibly melt. Never use jumpers to bypass high resistance loads in a circuit. Bypassing resistances, in effect, creates a short circuit. This may, in turn, cause damage and fire. Jumper wires should only be used to bypass lengths of wire or to simulate switches.

Jumper wires are simple, yet extremely valuable, pieces of test equipment. They are basically test wires which are used to bypass sections of a circuit. Although jumper wires can be purchased, they are usually fabricated from lengths of standard marine wire and whatever type of connector (alligator clip, spade connector or pin connector) that is required for the particular application being tested. In cramped,

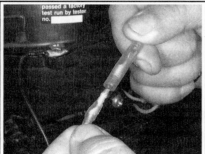

Fig. 5 Bullet connectors are some of the more common electrical connectors found on an outboard engine

Fig. 6 A typical weatherproof electrical connector

Fig. 7 Hard shell (left) and weatherproof (right) connectors have replaceable terminals

IGNITION AND ELECTRICAL SYSTEMS

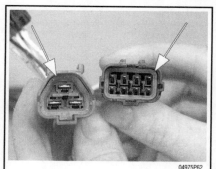

Fig. 8 The seals on weatherproof connectors must be kept in good condition to prevent the terminals from corroding

Fig. 9 Jumper wires are simple, yet extremely valuable, pieces of test equipment

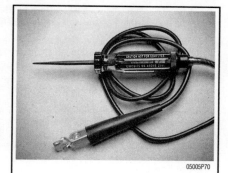

Fig. 10 A test light is used to detect the presence of voltage in a circuit

hard-to-reach areas, it is advisable to have insulated boots over the jumper wire terminals in order to prevent accidental grounding.

It is also advisable to include a standard marine fuse in any jumper wire. This is commonly referred to as a "fused jumper". By inserting an in-line fuse holder between a set of test leads, a fused jumper wire can be used for bypassing open circuits. Use a 5 amp fuse to provide protection against voltage spikes.

Jumper wires are used primarily to locate open electrical circuits. If an electrical component fails to operate, connect the jumper wire between the component and a good ground. If the component operates only with the jumper installed, the ground circuit is open.

If the ground circuit is good but the component does not operate, the circuit between the power feed and component may be open. By moving the jumper wire successively back from the component toward the power source, you can isolate the area of the circuit where the open is located. When the component stops functioning or the power is cut off, the open is in the segment of wire between the jumper and the point previously tested.

You can sometimes connect the jumper wire directly from the battery to the "hot" terminal of the component but first make sure the component uses 12 volts in operation. Some electrical components, such as fuel injectors or sensors are designed to operate on about 4 to 5 volts and running 12 volts directly to these components will cause damage.

TEST LIGHTS

▶ See Figure 10

The test light is used to check circuits and components while electrical current is flowing through them. It is used for voltage and ground tests. To use a 12 volt test light, connect the ground clip to a good ground and probe wherever necessary with the pick. The test light will illuminate when voltage is detected. This does not necessarily mean that 12 volts (or any particular amount of voltage) is present, it only means that some voltage is present.

➡ It is advisable before using the test light to touch its ground clip and probe across the battery posts or terminals to make sure the light is operating properly.

✱✱ WARNING

Do not use a test light to probe electronic ignition, spark plug or coil wires. Never use a pick-type test light to probe wiring on electronically controlled systems unless specifically instructed to do so. Any wire insulation that is pierced by the test light probe should be taped and sealed with silicone after testing.

Like the jumper wire, the 12 volt test light is used to isolate opens in circuits. But, whereas the jumper wire is used to bypass the open to operate the load, the 12 volt test light is used to locate the presence of voltage in a circuit. If the test light illuminates, there is power up to that point in the circuit, if the test light does not illuminate, there is an open circuit (no power). Move the test light in successive steps back toward the power source until the light in the handle illuminates. The open is between the probe and a point which was previously probed.

The self-powered test light is similar in design to the 12 volt test light but contains a 1.5 volt penlight battery in the handle. It is most often used in place of a multimeter to check for open or short circuits when power is isolated from the circuit (continuity test).

The battery in a self-powered test light does not provide much current. A weak battery may not provide enough power to illuminate the test light even when a complete circuit is made (especially if there is high resistance in the circuit). Always make sure that the test battery is strong. To check the battery, briefly touch the ground clip to the probe, if the light glows brightly, the battery is strong enough for testing.

✱✱ WARNING

A self-powered test light should not be used on any electronically controlled system or component. The small amount of electricity transmitted by the test light is enough to damage many electronic marine components.

MULTIMETERS

▶ See Figure 11

Multimeters are an extremely useful tool for troubleshooting electrical problems. They can be purchased in either analog or digital form and have a price range to suit any budget. A multimeter is a voltmeter, ammeter and multimeter (along with other features) combined into one instrument. It is often used when testing solid state circuits because of its high input impedance (usually 10 megaohms or more). A brief description of the multimeter main test functions follows:

• Voltmeter—the voltmeter is used to measure voltage at any point in a circuit or to measure the voltage drop across any part of a circuit. Voltmeters usually have various scales and a selector switch to allow the reading of different voltage ranges.

The voltmeter has a positive and a negative lead. To avoid damage to the meter, always connect the negative lead to the negative (-) side of the circuit (to ground or nearest the ground side of the circuit) and connect the positive lead to the positive (+) side of the circuit (to the power source or the nearest power source).

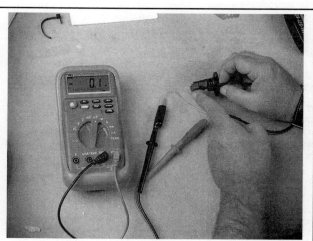

Fig. 11 Multimeters are essential for diagnosing faulty wires, switches and other electrical components

5-6 IGNITION AND ELECTRICAL SYSTEMS

➥The negative voltmeter lead will always be Black and that the positive voltmeter will always be some color other than Black (usually Red).

- **Multimeter**—the multimeter is designed to read resistance (measured in ohms) in a circuit or component. Most multimeters will have a selector switch which permits the measurement of different ranges of resistance (usually the selector switch allows the multiplication of the meter reading by 10, 100, 1,000 and 10,000). Some multimeters are "auto-ranging" which means the meter itself will determine which scale to use.

Since an internal battery powers the meters, the multimeter can be used like a self-powered test light. When the multimeter is connected, current from the multimeter flows through the circuit or component being tested. Since the multimeter's internal resistance and voltage are known values, the amount of current flow through the meter depends on the resistance of the circuit or component being tested.

The multimeter can also be used to perform a continuity test for suspected open circuits. In using the meter for making continuity checks, do not be concerned with the actual resistance readings. Zero resistance (or any ohm reading) indicates continuity in the circuit. Infinite resistance indicates an opening in the circuit. A high resistance reading where there should be none indicates a problem in the circuit.

Checks for short circuits are made in the same manner as checks for open circuits, except that the circuit must be isolated from both power and normal ground. Infinite resistance indicates no continuity, while zero resistance indicates a dead short.

✵✵ WARNING

Never use a multimeter to check the resistance of a component or wire while there is voltage applied to the circuit.

- **Ammeter**—an ammeter measures the amount of current flowing through a circuit in units called amps (amps). At normal operating voltage, most circuits have a characteristic amount of amps, called "current draw" which can be measured using an ammeter. By referring to a specified current draw rating, then measuring the amps and comparing the two values, one can determine what is happening within the circuit to aid in diagnosis.

For example, an open circuit will not allow any current to flow, so the ammeter reading will be zero. A damaged component or circuit will have an increased current draw, so the reading will be high.

The ammeter is always connected in series with the circuit being tested. All of the current that normally flows through the circuit must also flow through the ammeter, if there is any other path for the current to follow, the ammeter reading will not be accurate. The ammeter itself has very little resistance to current flow and, therefore, it will not affect the circuit but it will measure current draw only when the circuit is closed and electricity is flowing. Excessive current draw can blow fuses and drain the battery, while a Reduced current draw can cause motors to run slowly, lights to dim and other components to not operate properly.

Troubleshooting the Electrical System

When diagnosing any electrical problem organized troubleshooting is a must. The complexity of electrical systems on modern boats and their power plants demands that you approach any problem in a logical organized manner. There are certain troubleshooting techniques, which are standard:

- **Establish when the problem occurs**—Does the problem appear only under certain conditions? Were there any noises, odors or other unusual symptoms?
- **Check for obvious problems**—Problems such as broken wires and loose or dirty connections can cause major problems. Always check the obvious before assuming something complicated (or expensive) is the cause.

➥**Experience has shown that most problems tend to be the result of a fairly simple and obvious cause, such as loose or corroded connectors, bad grounds or damaged wire insulation, which causes a short. This makes careful visual inspection of components during testing essential to quick and accurate troubleshooting.**

- **Isolate the problem area**—Make some simple tests and observations, then eliminate the systems that are working properly. Test for problems systematically to determine the cause once the problem area is isolated. Are all the components functioning properly? Is there power going to electrical switches and motors. Performing careful, systematic checks will often turn up most causes on the first inspection, without wasting time checking components that have little or no relationship to the problem.
- **Verify all systems after repairs are completed**—Some causes can be traced to more than one component, so a careful verification of repair work is important in order to pick up additional malfunctions that may cause a problem to reappear

or a different problem to arise. A blown fuse, for example, is a simple problem that may require more than another fuse to repair. If you don't look for a problem that caused a fuse to blow, a shorted wire (for example) may go undetected.

VOLTAGE

♦ See Figure 12

This test determines voltage available from the battery and should be the first step in any electrical troubleshooting procedure after visual inspection. Many electrical problems, especially on electronically controlled systems, can be caused by a low state of charge in the battery. Excessive corrosion at the battery cable terminals can cause poor contact that will prevent proper charging and full battery current flow.

1. Set the voltmeter selector switch to the 10-volt position.
2. Connect the multimeter negative lead to the battery's negative (-) terminal and the positive lead to the battery's positive (+) terminal.
3. Turn the battery (ignition) switch **ON** to provide a load.
4. A well charged battery should register over 12 volts. If the meter reads below 11.5 volts, the battery power may be insufficient to operate the electrical system properly.
5. Charge the battery and retest.

VOLTAGE DROP

♦ See Figure 13

When current flows through a load, the voltage beyond the load drops. This voltage drop is due to the resistance created by the load and also by small resistances created by corrosion at the connectors and damaged insulation on the wires. The maximum allowable voltage drop under load is critical, especially if there is more than one load in the circuit, since all voltage drops are cumulative.

1. Set the voltmeter selector switch to the 10-volt position.
2. Connect the multimeter negative lead to the battery negative (-) terminal or another good ground.
3. Touch the multimeter positive lead to the battery's positive (+) terminal to determine battery voltage.
4. Operate the circuit and check the voltage prior to the first component (load).
5. There should be little or no voltage drop (from battery voltage) in the circuit prior to the first component. If a voltage drop exists, the wire or connectors in the circuit are suspect.
6. While operating the first component in the circuit, probe the ground side of the component with the positive (+) meter lead and observe the voltage readings. A small voltage drop should be noticed. The resistance of the component causes this voltage drop.
7. Repeat the test for each component (load) down the circuit.
8. If a large voltage drop is noticed, the preceding component, wire or connector is suspect.

RESISTANCE

♦ See Figures 14 and 15

✵✵ WARNING

Never use a multimeter with power applied to the circuit. The multimeter is designed to operate on its own power supply. The normal 12-volt electrical system voltage can damage the meter!

1. Isolate the circuit from the boat's power source.
2. Ensure that the battery (ignition) switch is **OFF**.
3. Isolate at least one side of the circuit to be checked, in order to avoid reading parallel resistances. Parallel circuit resistances will always give a lower reading than the actual resistance of either of the branches.
4. Connect the meter leads to both sides of the circuit (wire or component) and read the actual resistance measured in ohms on the meter scale. Make sure the selector switch is set to the proper ohm scale for the circuit being tested, to avoid misreading the multimeter test value.
5. Compare this reading to the resistance specification for the component or formulate the theoretical resistance using Ohms Law.

OPEN CIRCUITS

♦ See Figures 16 and 17

This test already assumes the existence of an open in the circuit and it is used to help locate the open portion.

IGNITION AND ELECTRICAL SYSTEMS

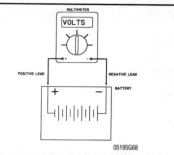

Fig. 12 The voltage test determines voltage available from the battery and should be the first step in any electrical troubleshooting procedure after visual inspection

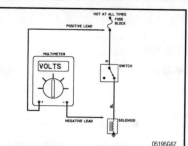

Fig. 13 Voltage drop is due to the resistance created by the load and also by small resistances created by corrosion at the connectors and damaged insulation on the wires

Fig. 14 Using a multimeter to check resistance on the secondary side of the ignition coil

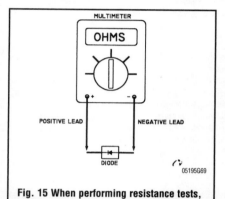

Fig. 15 When performing resistance tests, always isolate the circuit from power

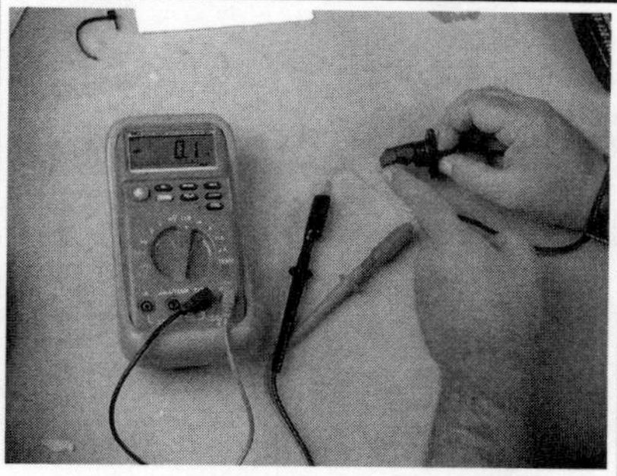

Fig. 16 The infinite display on this multimeter (1.) indicates that the circuit is open

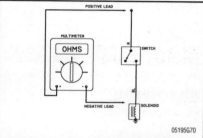

Fig. 17 The easiest way to illustrate an open circuit is to consider an example circuit with a switch. When the switch is turned OFF, power does not flow through the circuit to the load. Thus, the circuit is open

1. Isolate the circuit from power and ground.
2. Connect the self-powered test light or multimeter ground clip to the ground side of the circuit and probe sections of the circuit sequentially.
3. If the light is out or there is infinite resistance, the open is between the probe and the circuit ground.
4. If the light is on or the meter shows continuity, the open is between the probe and the end of the circuit toward the power source.

SHORT CIRCUITS

♦ See Figure 18

➡ Never use a self-powered test light to perform checks for opens or shorts when power is applied to the circuit under test. The test light can be damaged by outside power.

1. Isolate the circuit from power and ground.
2. Connect the self-powered test light or multimeter ground clip to a good ground and probe any easy-to-reach point in the circuit.
3. If the light comes on or there is continuity, there is a short somewhere in the circuit.
4. To isolate the short, probe a test point at either end of the isolated circuit (the light should be on or the meter should indicate continuity).
5. Leave the test light probe engaged and sequentially open connectors or switches, remove parts, etc. until the light goes out or continuity is broken.
6. When the light goes out, the short is between the last two circuit components, which were opened.

Wire and Connector Repair

Almost anyone can replace damaged wires, as long as the proper tools and parts are available. Wire and terminals are available to fit almost any need. Even the specialized weatherproof, molded and hard shell connectors are now available from aftermarket suppliers.

Be sure the ends of all the wires are fitted with the proper terminal hardware and connectors. Wrapping a wire around a stud is never a permanent solution and will only cause trouble later. Replace wires one at a time to avoid confusion. Always route wires in the same manner of the manufacturer.

When replacing connections, make absolutely certain that the connectors are certified for marine use. Automotive wire connectors may not meet United States Coast Guard (USCG) specifications.

➡ If connector repair is necessary, only attempt it if you have the proper tools. Weatherproof connectors require special tools to release the pins inside the connector. Attempting to repair these connectors with conventional hand tools will damage them.

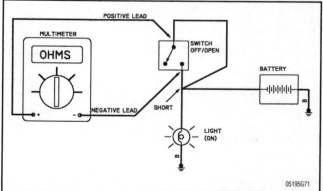

Fig. 18 In this illustration, the circuit between the battery and light should be open because the switch is turned OFF. However, battery voltage is reaching the light at the point of the short. This could possibly be caused by chaffed wires

BREAKER POINTS IGNITION (MAGNETO IGNITION)

Description & Operation

♦ See Figure 19

The breaker point ignition consists of the rotor assembly, contact point assembly, ignition coil, condenser spark plug, spark plug cap and the engine stop switch. The breaker points ignition system uses a mechanically switched, collapsing field to induce spark at the plug. A magnet moving by a coil produces current in the primary coil winding. The current in the primary winding creates a magnetic field.

At the proper time, the breaker points are separated by action of a cam designed into the collar of the crankshaft. and the primary circuit is broken. When the circuit is broken, the flow of primary current stops and causes the magnetic field about the coil to break down instantly. At this precise moment, an electrical current of extremely high voltage is induced in the fine secondary windings of the coil. This high voltage is conducted to the spark plug where it jumps the gap between the points of the plug to ignite the compressed charge of air-fuel mixture in the cylinder.

Breaker point ignition systems installed on outboard engines will usually operate over extremely long periods of time without requiring adjustment or repair. However, if ignition system problems are encountered and the usual corrective actions such as replacement of spark plugs does not correct the problem, the magneto output should be checked to determine if the unit is functioning properly.

SYSTEM COMPONENTS

Breaker Point Set

♦ See Figure 20

A breaker point set consists of two points. One is attached to a stationary bracket and does not move. The other point is attached to a movable mount. A spring is used to keep the points in contact with each other, except when they are separated by the action of a cam built into the flywheel collar which fits over the crankshaft. Both points are constructed with a steel base and a tungsten cap fused to the base.

The breaker points must be aligned accurately to provide the best contact surface. This is the only way to assure maximum contact area between the point surfaces, accurate setting of the point gap, proper synchronization and satisfactory point life. If the points are not aligned properly, the result will be premature wear or pitting. This type of damage may change the cam angle, although the actual distance will remain the same.

Condenser

♦ See Figure 21

The breaker point ignition system contains a condenser that works like a sponge in the circuit. Current that is flowing through the primary circuit tries to keep going. When the breaker point switch opens the current will arc over the widening gap. The condenser is wired in parallel with the points. The condenser absorbs some of the current flow as the points open. This reduces arc over and extends the life of the points.

In simple terms, a condenser is composed of two sheets of tin or aluminum foil laid one on top of the other but separated by a sheet of insulating material such as waxed paper, etc. The sheets are rolled into a cylinder to conserve space and then inserted into a metal case for protection and to permit easy assembling.

The purpose of the condenser is to prevent excessive arcing across the points and to extend their useful life. When the flow of primary current is brought to a sudden stop by the opening of the points, the magnetic field in the primary windings collapses instantly and is not allowed to "fade away", which would happen if the points were allowed to arc.

The condenser stores the electricity that would have arced across the points and discharges that electricity when the points open and close again. This discharge is in the opposite direction to the original flow and tends to "smooth out" the current. The more quickly the primary field collapses, the higher the voltage produced in the secondary windings and delivered to the spark plugs. In this way, the condenser (in the primary circuit), affects the voltage (in the secondary circuit) at the spark plugs.

Modern condensers seldom cause problems, therefore, it is not necessary to install a new one each time the points are replaced. However, if the points show evidence of arcing, the condenser may be at fault and should be replaced. A faulty condenser may not be detected without the use of special test equipment. Testing will reveal any defects in the condenser but will not predict the useful life left in the unit.

The modest cost of a new condenser justifies its purchase and installation to eliminate this item as a source of trouble.

Primary Coil

As the flywheel rotates, magnets attached to the edge of the flywheel create a current that will flow through the closed breaker points into the ignition coil primary windings. This flow of current through the coil primary windings builds a very strong magnetic field. When the breaker cam opens the points, the field collapses, inducing high voltage in the coil secondary winding. This voltage is sent to the spark plug. The condenser will absorb any residual current remaining in the primary windings while the points are open. This eliminates arcing at the breaker points and in turn produces a stronger spark at the spark plug. The breaker points then close and the flywheel continues to rotate.

Ignition Coil

The ignition coil is the heart of the ignition system. Essentially, it is nothing more than a transformer which takes a relatively low voltage (12 volts) available from the primary coil and increases it to the point where it will fire the spark plug with as much as 20,000 volts.

Once the voltage is discharged from the ignition coil the secondary circuit begins and only stretches from the ignition coil to the spark plugs via extremely high-tension leads. At the spark plug end, the voltage arcs in the form of a spark, across from the center electrode to the outer electrode and then to ground via spark plug threads. This completes the ignition circuit.

Troubleshooting the Breaker Points Ignition System

Always attempt to proceed with system testing in an orderly manner. The "shot in the dark" approach will only result in wasted time, incorrect diagnosis, replacement of unnecessary parts and frustration.

Begin the ignition system troubleshooting with the spark plug and continue through the system until the source of trouble is located.

To properly diagnose magneto (spark) problems, the theory of electricity flow must be understood. The flow of electricity through a wire may be compared with the flow of water through a pipe. Consider the voltage in the wire as the water pressure

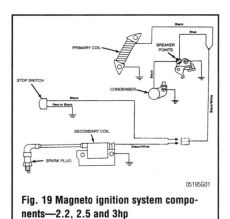

Fig. 19 Magneto ignition system components—2.2, 2.5 and 3hp

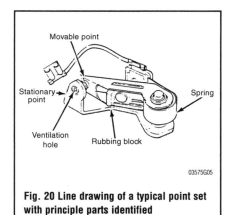

Fig. 20 Line drawing of a typical point set with principle parts identified

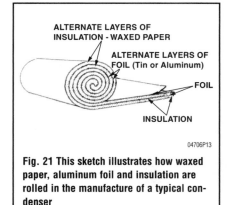

Fig. 21 This sketch illustrates how waxed paper, aluminum foil and insulation are rolled in the manufacture of a typical condenser

IGNITION AND ELECTRICAL SYSTEMS

in the pipe and the amps as the volume of water. Now, if the water pipe is broken, the water does not reach the end of the pipe. In a similar manner if the wire is broken the flow of electricity is broken. If the pipe springs a leak, the amount of water reaching the end of the pipe is reduced. Same with the wire. If the installation is defective or the wire becomes grounded, the amount of electricity (amps) reaching the end of the wire is reduced.

The breaker points in an outboard motor are an extremely important part of the ignition system. A set of points may appear to be in good condition but they may be the source of hard starting, misfiring or poor engine performance. The rules and knowledge gained from association with 4-Stroke engines does not necessarily apply to a 2-Stroke engine. The points should be replaced every 100 hours of operation or at least once a year. Remember, the less an outboard engine is operated, the more care it needs. Allowing an outboard engine to remain idle will do more harm than if it is used regularly.

SPARK PLUGS

♦ See accompanying illustrations

1. Remove the cowling. On 1990 2.5 and 3hp models, remove the fuel filler cap from the fuel tank. Press in on both sides of the upper cowling half and lift the upper cowling free.

On the 2.5hp and 3hp—1991 and on—release the two snap latches on both sides of the cowling and lift the upper half of the cowling free of the powerhead.

2. Check the plug high-tension lead to be sure it is properly connected. Check the entire length of the lead from the plug to the secondary ignition coil. If the lead is to be removed from the spark plug, always use a twisting and pulling motion as a precaution against damaging the connection.

3. Attempt to remove the spark plug by hand. This is a "rough" test to determine if the plug is tightened properly. The attempt to loosen the plug by hand should fail. The plug should be tight and require the proper size socket for removal. Remove the spark plug and evaluate its condition.

4. Use a spark tester and check for spark. Rotate the flywheel with the hand rewind starter and observe the light for the spark. If a spark tester is not available, disconnect the lead from the spark plug, insert a screwdriver into the boot to make contact with the lead "shell" inside, hold the shank of the screwdriver close to the powerhead—say about ¼ in. (6.4mm) from the powerhead, have an assistant crank the powerhead, and observe the spark jumping the gap. A strong spark over a wide gap must be observed when testing in this manner, because under compression, a strong spark is required to ignite the air/fuel mixture.

COMPRESSION TEST

♦ See Figure 22

Before spending too much time and money attempting to trace a problem to the ignition system, a compression check of the cylinder should be made. If the cylinder does not have adequate compression, troubleshooting and attempted service of the

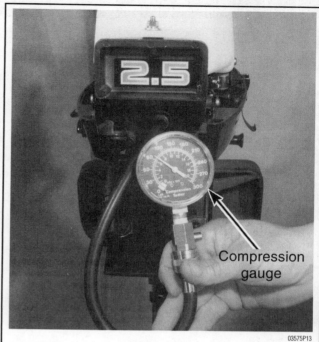

Fig. 22 Before any serious tasks are started, a compression check should be made. An acceptable pressure reading should be close to 90 psi (620.5 kPa)

ignition or fuel system will fail to give the desired results of satisfactory engine performance.

Breaker Points

POINT GAP CHECK

♦ See Figures 23 thru 28

1. If the points appear to be dirty or contaminated with oil or grease, open the points and insert a piece of notebook paper, close the points and pull the paper out. Repeat this using a clean piece of paper each time until the paper is clean between the closed points.

2. Inspect the flywheel for cracks or other damage, especially around the inside

Step 1

Step 2

Step 4

5-10 IGNITION AND ELECTRICAL SYSTEMS

of the center hub. Check to be sure that the metal parts have not become attached to the magnets. Verify each magnet has good magnetism by using a screwdriver or other similar tool.

3. Thoroughly clean the inside taper of the flywheel and the taper on the crankshaft to prevent the flywheel from attempting to "walk" on the crankshaft during operation.

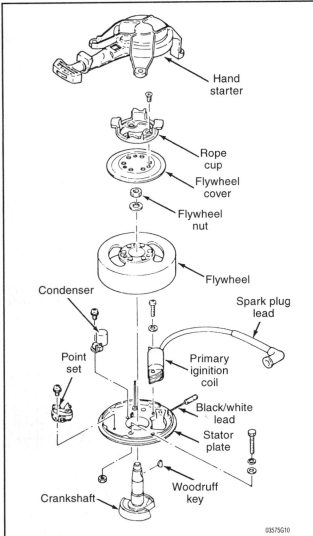

Fig. 23 Exploded drawing of a breaker point (magneto) ignition system, with major components identified

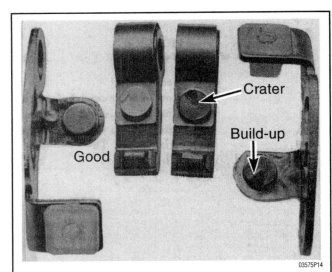

Fig. 25 A normal set of breaker points used in a magneto will show evidence of a shallow crater and build-up after a few hours of operation. The left set of points is considered normal and need not be replaced. The set on the right has been in service for more than 450 hours and should be replaced

Fig. 24 Before setting the breaker point gap, the points must be properly aligned (top). Always bend the stationary point, never the breaker lever. Attempting to adjust an old worn set of points is not practical, when compared with the modest cost of a new set, thus eliminating this area as a possible cause of trouble. If a worn set of points is to be retained for emergency use, both contact surfaces of the set should be refaced with a point file

Fig. 26 The flywheel on a single cylinder powerhead has a "window" to permit point gap adjustment using a screwdriver and feeler gauge, as shown

IGNITION AND ELECTRICAL SYSTEMS

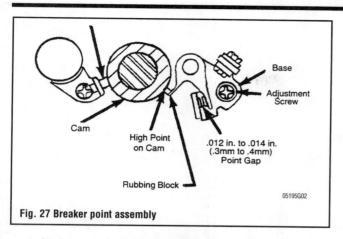

Fig. 27 Breaker point assembly

4. Check the top seal around the crankshaft to be sure no oil has been leaking onto the stator plate. If there is any evidence the seal has been leaking, it must be replaced.
5. Test the stator assembly to verify it is not loose. Attempt to lift each side of the plate. There should be little or no evidence of movement.
6. Lightly lubricate the cam wick with all-purpose lubricant. Excess lubricant will shorten the breaker point life. Inadequate lubrication will quickly wear the rubbing block from the point set and thus alter the timing. Therefore, just use a dab of lubricant.
7. Unfortunately, the breaker point set is located under the flywheel. This location requires the hand rewind starter to be removed and the flywheel to be "pulled" in order to replace the point set. However, the manufacturer made provisions for the point gap to be checked with a feeler gauge through one of two slots cut into the top of the flywheel for this purpose. Therefore, only the hand rewind starter needs to be removed to perform the task of point adjustment.
8. Rotate the flywheel to position the high point on the cam against the rubbing block of the breaker point assembly.

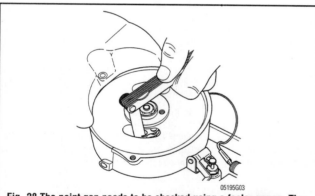

Fig. 28 The point gap needs to be checked using a feeler gauge. The point gap measurement should be 0.012–0.016 in. (0.3–0.4 mm)

9. If necessary, loosen the adjustment screw and reposition the breaker point base to obtain the correct point gap.
10. Verify that the felt pad is making light contact with the cam. The felt pad should be moistened with oil.

REMOVAL & INSTALLATION

♦ See accompanying illustrations

1. Remove the three bolts securing the hand rewind starter to the powerhead. Lift the hand rewind starter free.
2. Loosen the three bolts securing the rope cup to the flywheel. A strap wrench or large pair of channel lock pliers around the outside edge of the rope cup may be needed to hold the flywheel while loosening the rope cup bolts.

➟When a point set is new, the fixed contact point surface is a hemispherical curve. The movable contact point surface is flat to aid in adjusting the point gap. The manufacturer states it is permissible to file the curved surface flat if the points are pitted. This can be accomplished with a small flat file through the opening in the flywheel.

3. Remove the three bolts from the rope cup. Lift off the rope cup and flywheel cover (if equipped) free from the flywheel.
4. Hold the flywheel steady with a flywheel holder and remove the flywheel unit.

✳✳ WARNING

Never attempt to use a puller, which pulls on the outside edge of the flywheel.

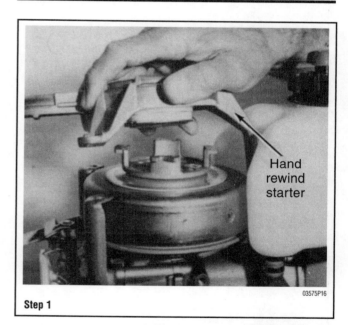

Step 1

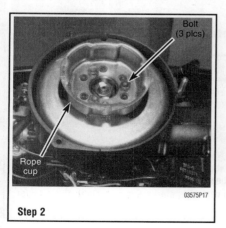

Step 2

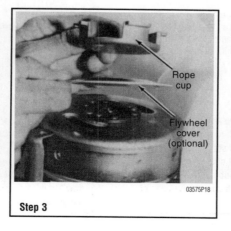

Step 3

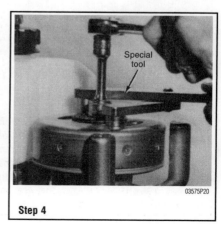

Step 4

5-12 IGNITION AND ELECTRICAL SYSTEMS

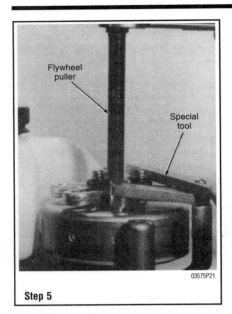
Step 5

Step 12

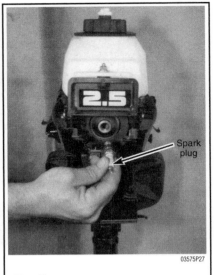
Step 18

5. Obtain a flywheel puller. Install the puller and take up the slack. Continue to tighten the puller and at the same time, shock the crankshaft with a gentle to moderate tap with a hammer on the end of the puller. This shock will assist in breaking the flywheel loose from the crankshaft.

6. Lift the flywheel free of the crankshaft. Remove and save the Woodruff key from the recess in the crankshaft.

7. Remove the nut securing the condenser and coil leads to the point set. Pry off the circlip securing the point set to the pivot. Remove the adjusting screw and lift the point set free of the stator plate. If installing a new point set, remove the screw securing the condenser and lift the condenser from the stator plate.

➡It is considered good shop practice to replace the condenser whenever a new point set is being installed.

To install:

8. Install a new condenser and secure it to the stator plate with a screw through the mounting bracket.

9. Install a new point set. Install the adjusting screw but do not tighten it at this time.

10. Align the slot in the set with the slot in the stator plate as a preliminary adjustment. The point gap will be set when the rubbing block contacts the flywheel cam.

11. Attach the coil and condenser leads to the point set. Install the circlip on the pivot post.

12. Place a tiny dab of thick lubricant on the curved surface of the Woodruff key to hold it in place while the flywheel is being installed. Press the Woodruff key into place in the crankshaft recess. Wipe away any excess lubricant to prevent the flywheel from "walking" during powerhead operation.

13. Check the flywheel magnets to ensure they are free of any metal particles. Double-check the taper in the flywheel hub and the taper on the crankshaft. to verify they are clean and contain no oil.

14. Slide the flywheel down over the crankshaft with the keyway in the flywheel aligned with the Woodruff key in place on the crankshaft. Rotate the flywheel counterclockwise to be sure it does not contact any part of the stator plate or wiring.

15. Slide the washer onto the crankshaft. and then thread the flywheel nut onto the crankshaft. Using a flywheel holder, tighten the flywheel nut to 30 ft. lbs. (40.6Nm).

16. Position the flywheel cover (if equipped) and rope cup aligning the bolt holes with the flywheel. Install the three bolts and tighten them securely.

17. Install the hand rewind starter to the powerhead. Tighten the three attaching bolts to 50 inch lbs. (5.6 Nm).

18. Install and tighten the spark plug to 20 ft. lbs. (27 Nm). Attach the high-tension lead to the spark plug and close the access-cover door.

19. Install the cowling. On 1990 2.5 and 3hp powerheads, position the cover over the latches and push down until the latch engages. On 1991 and later 2.5, 3 and 3.3hp powerheads, position and align the top cover over the bottom cover and engage the latches on both sides.

Condenser

TESTING

1. The condenser does not have to be removed to perform this test.
2. Place a multimeter test lead on the condenser output wire and another on the condenser body. The meter should read 0.22–0.28 microfarads.
3. If the condenser measurement does not meet specifications, it may be faulty and will need to be replaced.

REMOVAL & INSTALLATION

1. Remove the three bolts securing the hand rewind starter to the powerhead. Lift the hand rewind starter free.
2. Loosen the three bolts securing the rope cup to the flywheel. A strap wrench or large pair of channel lock pliers around the outside edge of the rope cup may be needed to hold the flywheel while loosening the rope cup bolts.

➡When a point set is new, the fixed contact point surface is a hemispherical curve. The movable contact point surface is flat to aid in adjusting the point gap. The manufacturer states it is permissible to file the curved surface flat if the points are pitted. This can be accomplished with a small flat file through the opening in the flywheel.

3. Remove the three bolts from the rope cup. Lift off the rope cup and flywheel cover (if equipped) free from the flywheel.
4. Install a flywheel holder and hold the flywheel steady while removing the flywheel unit.

✲✲✲ WARNING

Never attempt to use a puller, which pulls on the outside edge of the flywheel.

5. Install a flywheel puller onto the flywheel and take up the slack on the tool. Continue to tighten on the puller and at the same time, shock the crankshaft with a gentle to moderate tap with a hammer on the end of the puller. This shock will assist in "breaking" the flywheel loose from the crankshaft.

6. Lift the flywheel free of the crankshaft. Remove and save the Woodruff key from the recess in the crankshaft.

7. Remove the nut securing the condenser and coil leads to the point set and remove the condenser from the stator plate.

➡It is considered good shop practice to replace the condenser whenever a new point set is being installed.

IGNITION AND ELECTRICAL SYSTEMS

To install:

8. Install a new condenser and secure it to the stator plate with a screw through the mounting bracket.
9. Attach the coil and condenser leads to the point set. Install the circlip on the pivot post.
10. Place a tiny dab of thick lubricant on the curved surface of the Woodruff key to hold it in place while the flywheel is being installed. Press the Woodruff key into place in the crankshaft recess. Wipe away any excess lubricant to prevent the flywheel from "walking" during powerhead operation.
11. Check the flywheel magnets to ensure they are free of any metal particles. Double-check the taper in the flywheel hub and the taper on the crankshaft to verify they are clean and contain no oil.
12. Now, slide the flywheel down over the crankshaft with the keyway in the flywheel aligned with the Woodruff key in place on the crankshaft. Rotate the flywheel counterclockwise to be sure it does not contact any part of the stator plate or wiring.
13. Slide the washer onto the crankshaft and then thread the flywheel nut onto the crankshaft. Hold the flywheel with a flywheel holder and tighten the flywheel nut to 30 ft. lbs. (40.6 Nm).
14. Position the flywheel cover (if equipped) and rope cup aligning the bolt holes with the flywheel. Install the three bolts and tighten them securely.
15. Install the hand rewind starter to the powerhead. Tighten the three attaching bolts to 50 inch lbs. (5.6 Nm).
16. Install and tighten the spark plug to 20 ft. lbs. (27Nm). Attach the high-tension lead to the spark plug and close the access-cover door.
17. Install the cowling. On 1990 2.5hp and 3hp powerheads, position the cover over the latches and push down until the latch engages. On 1991 and later 2.5, 3 and 3.3hp powerheads, position and align the top cover over the bottom cover and engage the latches on both sides.

Primary Coil

TESTING

♦ See Figure 29

1. Connect the positive multimeter lead to the primary coil lead and the negative lead to the coil ground.
2. Meter should read 1.5 ohms.
3. If the reading does not meet specifications, replace the coil.

REMOVAL & INSTALLATION

1. Remove the three bolts securing the hand rewind starter to the powerhead. Lift the hand rewind starter free.
2. Loosen the three bolts securing the rope cup to the flywheel. A strap wrench or large pair of channel lock pliers around the outside edge of the rope cup may be needed to hold the flywheel while loosening the rope cup bolts.
3. Remove the three bolts from the rope cup. Lift off the rope cup and flywheel cover (if equipped) free from the flywheel.
4. Hold the flywheel steady with a flywheel holder and remove the flywheel unit.

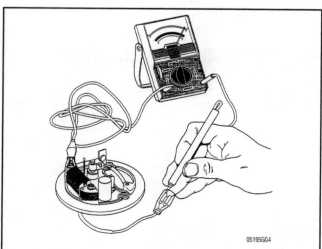

Fig. 29 Use a multimeter to test the resistance of the primary coil

➤**Never attempt to use a puller, which pulls on the outside edge of the flywheel.**

5. Install the puller onto the flywheel and then take up the slack on the puller. Continue to tighten the puller and at the same time, shock the crankshaft with a gentle to moderate tap with a hammer on the end of the puller. This shock will assist in "breaking" the flywheel loose from the crankshaft.
6. Lift the flywheel free of the crankshaft. Remove and save the Woodruff key from the recess in the crankshaft.
7. Remove the screws securing the primary coil to the stator plate and remove the primary coil.

To install:

8. Install a new primary coil and secure it to the stator plate with the fastening screws.
9. Place a tiny dab of thick lubricant on the curved surface of the Woodruff key to hold it in place while the flywheel is being installed. Press the Woodruff key into place in the crankshaft recess. Wipe away any excess lubricant to prevent the flywheel from "walking" during powerhead operation.
10. Check the flywheel magnets to ensure they are free of any metal particles. Double-check the taper in the flywheel hub and the taper on the crankshaft to verify they are clean and contain no oil.
11. Now, slide the flywheel down over the crankshaft with the keyway in the flywheel aligned with the Woodruff key in place on the crankshaft. Rotate the flywheel counterclockwise to be sure it does not contact any part of the stator plate or wiring.
12. Slide the washer onto the crankshaft. and then thread the flywheel nut onto the crankshaft. Hold the flywheel with a flywheel holder and tighten the flywheel nut to 30 ft. lbs. (40.6 Nm).
13. Position the flywheel cover (if equipped) and rope cup aligning the bolt holes with the flywheel. Install the three bolts and tighten them securely.
14. Install the hand rewind starter to the powerhead. Tighten the three attaching bolts to 50 inch lbs. (5.6 Nm).
15. Install and tighten the spark plug to 20 ft. lbs. (27 Nm). Attach the high-tension lead to the spark plug and close the access-cover door.
16. Install the cowling. On 1990 2.5 and 3hp powerheads, position the cover over the latches and push down until the latch engages. On 1991 and later 2.5, 3 and 3.3hp powerheads, position and align the top cover over the bottom cover and engage the latches on both sides.

Ignition Coil

TESTING

An multimeter test can only detect certain faults in the secondary ignition coil. Replace the ignition coil if the multimeter reading is not as specified. If the coil tests satisfactory and the coil is still suspected of being faulty, use a magneto analyzer to thoroughly check the coil. Follow the coil test procedure in the analyzer manual.

The primary and secondary coil resistance tests may be performed with the components installed on the powerhead. If chaffing or broken wiring is suspected, removal of the component is recommended to permit a complete inspection of all wiring leads.

Primary Resistance

♦ See Figure 30

1. Disconnect the primary coil electrical lead Black/White wire from the bullet connector on the side of the powerhead. Remove the two screws securing the stator plate to the powerhead.
2. Lift up on the stator plate and slowly feed the primary coil wire lead through the opening in the powerhead. Set the stator plate aside for bench testing.
3. Obtain a multimeter and set it on the Rx 1 scale. Connect the (+) meter lead to the coil Black/White lead. Connect the (-) meter lead to a good coil ground.
4. An acceptable resistance reading for the primary coil is 1.5 Ohms.
5. If the reading is not within an acceptable range, the primary coil is defective and must be replaced.

Secondary Resistance

♦ See Figure 31

➤**If not previously done, disconnect the lead on the secondary ignition coil at the bullet connector and remove the high-tension lead from the spark plug.**

1. Remove the two mounting bolts securing the coil to the powerhead and set the coil aside for bench testing.

5-14 IGNITION AND ELECTRICAL SYSTEMS

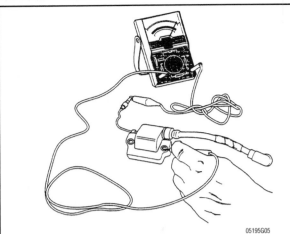

Fig. 30 Connect the positive (+) meter lead to the coil and the negative (-) meter lead to the coil ground. The meter reading should be .81–1.09 ohms

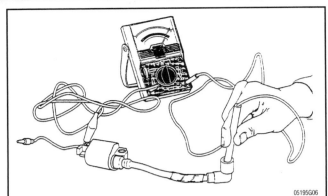

Fig. 31 Connect the (+) meter lead to the spark plug lead terminal and the negative (-) to the coil ground. The meter reading should be 4.250 – 5.750 ohms

2. Remove the two mounting bolts securing the coil to the powerhead and set the coil aside for bench testing.

To install:
3. Install the two bolts securing the coil to the power head and tighten snugly.
4. After making sure the both sides of the connector is clean, reconnect the bullet connector making sure it's tight.
5. Install the spark plug wire.

REMOVAL & INSTALLATION

1. If not previously done, disconnect the lead on the secondary ignition coil at the bullet connector and remove the high-tension lead from the spark plug.

CAPACITOR DISCHARGE IGNITION (CDI) SYSTEM

Description and Operation

♦ See Figure 32

The engine's flywheel contains magnets carefully positioned to create an electric current as they rotate past specially designed coils of wire. Current is created by magnetic induction. Simply put, that means that a magnet moving rapidly near a conductor will induce electrical flow within the conductor.

Conversely, a wire that moves rapidly through a magnetic field will also generate electrical flow. This principle governs the working of electric motors, alternators and generators.

One of the coils under your flywheel is called a charge coil. As the flywheel magnets spin past this coil, they generate in it a fairly high-voltage alternating current that travels to your system's "module" often called the power pack or CDI unit. This voltage is often in the region of 200 volts AC.

The other ignition system coil(s) found under the flywheel are called the sensor coils, pulsar coils or trigger coils. They send an electrical signal to the ignition module to tell it which cylinder to work with at the correct time.

The CDI unit is the brains of the system and serves several functions. First, it converts the alternating current (AC) from the charge coil into usable direct current (DC). Next it stores the current in a built-in capacitor. The module also interprets the timing signal from the trigger coil. This changes constantly with engine speed and moving the trigger coil's position relative to the flywheel magnets brings about the change. The coil's movement is controlled by a device called a timing plate, to which both the charge coil and trigger coils are mounted. The timing plate moves in response to changes in throttle opening, to which it is mechanically linked. The CDI unit also controls the discharge of the capacitor and sends this voltage to the primary winding of the ignition coil for the correct cylinder.

Also (depending on the system) the module may incorporate electronic circuits that limit engine speed and prevent over-revving. Some modules even have a circuit that reduces engine speed if the engine begins to run too hot for any reason. Larger engines often have an automatic ignition advance for initial start-up and for when the engine is running at temperatures less than approximately 100°F.

Manufacturers commonly use one power pack or module for each bank of cylinders on V-type powerheads. One module will control the odd numbered cylinders and the other will service the even numbered cylinders.

The voltage from the module goes to the primary winding of the ignition coil or high-tension coil. You may know this type of coil as a step-up transformer. Here, the voltage is stepped up to between 15,000 and 40,000 volts. That's the `kind of voltage needed to jump the air gap on the spark plug and ignite the air/fuel mixture in the cylinder. Your high-tension ignition coil has two sides, the primary side and the secondary side. It is really two coil assemblies combined into one neat, compact case.

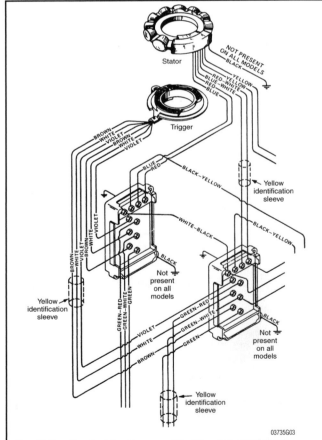

Fig. 32 Functional wiring diagram of the Thunderbolt CD ignition system. The boxes are stacked one on top of the other, so they must be separated to gain access to the inner box

IGNITION AND ELECTRICAL SYSTEMS 5-15

The internal construction of a typical ignition coil includes primary and secondary windings. It also uses the principle of magnetic induction with the magnetism generated by the primary (lower -voltage) winding creating a magnetic field around the secondary winding, which has many more windings than the primary coil. The ignition module controls the rapid turning on and off of electrical flow in the primary winding, thereby turning this magnetic field on and off. The rapid movement of this magnetic field past the secondary windings induces electrical current flow. The are greater the number of turns of wire in the secondary winding, the higher the voltage produced.

As this secondary voltage leaves the center tower of the ignition coil, it travels along the spark-plug wire, which is heavily insulated and designed to carry this high voltage.

If all is well, the high voltage will jump the gap on the spark plug between the center electrode and the ground electrode. On larger engines with surface-gap plugs, the high voltage current will jump from the center electrode to the side of the plug assembly itself, completing a circuit to ground via the engine block.

Last but certainly not least, is the stop control. You need a means to shut your engine off and a good way is to stop the spark plugs from working. Depending on your engine, this may be accomplished by a simple stop button or a key switch on larger engines. This disables the whole ignition system. On newer engines, an emergency stop button with an overboard clip and lanyard attachment is standard. This system is wired directly into the ignition module. It functions by creating a momentary short circuit inside the CDI unit, grounding the current intended for the high-tension coils and thus shutting off the ignition long enough to stop the engine. Faulty stop circuits are frequently the cause of a no spark condition.

2.5–3.3 HP

The components of this system include a flywheel and magnets, capacitor charging/trigger coil, CDI, secondary ignition coil, stop switch and spark plug. Four magnets are imbedded within the flywheel. These magnets are arranged in pairs in a North/South and South/North polarity position 180°apart. As the North/South magnets pass over the capacitor charging/trigger coil, located under the flywheel on the stator plate, a magnetic field is built up within the capacitor trigger/coil windings, producing an AC voltage. This voltage is conducted to the CDI where it is rectified and stored in a capacitor within the CDI. As the opposite polarity magnets, South/North pass over the capacitor charging/trigger coil, a triggering voltage (pulse) is generated and sent to the CDI. The bias/trigger circuit releases the stored voltage from the capacitor to the secondary ignition coil. The secondary ignition coil steps up the voltage through the internal primary and secondary windings to a voltage level where it can jump the gap at the spark plug.

The ignition coil fires the spark plug at both the top of the piston stroke and at the bottom of the stroke or every 180 degrees of crankshaft rotation. One firing results in work—the other is a free ride.

There are no mechanical devices for timing adjustments or timing advancement throughout the throttle range. When the stop switch on the tiller handle is actuated, the bias/trigger circuit within the CDI is shorted to ground. This prevents the capacitor within the CDI from discharging the stored voltage to the secondary ignition coil and firing the spark plug. The powerhead is shut down.

The manufacturer makes the following suggestion: If the powerhead will not idle to—say a good trolling speed but runs excessively fast with the throttle fully closed—the bias/trigger circuit within the CDI is defective and the CDI must be replaced.

4–275 HP

✳✳✳ CAUTION

On all powerheads equipped with a flywheel magneto, the rectifier will be damaged if the battery leads are disconnected from the battery while the powerhead is running or if the leads should accidentally be reversed. This is a safety feature designed by the manufacturer because the cost of replacing a rectifier is a fraction of the cost for a new switch box.

This CD ignition system is an alternator driven capacitor discharge system. Major components include the flywheel, stator coil, trigger coil, switch box, ignition coils (one per cylinder) and of course the spark plugs (one per cylinder).

The flywheel contains three sets of permanent magnets mounted on its inside surface. The outer rim contains two sets of magnets and the center hub has one set. The stator is mounted under the flywheel and has both a Low Speed (LS) and a High-Speed (HS) capacitor-charging coil. As the flywheel rotates, the permanent magnets on the rim surface pass over the LS and HS capacitor charging coils which produce an AC voltage. The AC voltage is conducted to the switch box where it is rectified and stored in a capacitor.

The trigger assembly is also mounted under the flywheel—in the center of the stator and contains the trigger coil. On a 4hp or 5hp powerhead, the trigger is mounted inside the "pan" under the flywheel. On all other powerheads with this type of ignition, the trigger is mounted on a stator plate around the crankshaft under the flywheel. As the inner magnets of the flywheel pass by the trigger coil, an AC pulse is produced. This pulse voltage is also conducted to the switch box where it turns on one of two electronic switches or Silicon Controlled Rectifiers (SCR) in the switch box. A positive voltage pulse turns on the SCR switch for cylinder No. 1, a negative voltage pulse turns on the SCR for cylinder No. 2.

As each SCR switch closes, the stored capacitor voltage is discharged to the primary side of the ignition coil for the respective cylinder. The ignition coil steps up the voltage level to a value high enough to jump the spark plug gap and ignite the compressed air/fuel charge in the cylinder.

Spark timing is advanced and retarded on the 6–25hp models by rotating the trigger coil stator assembly. As the trigger coil stator assembly is rotated this changes the phase relationship between the trigger magnet poles and trigger coils, which will advance or delay the opening and closing of the SCR's in the switch box, advancing or retarding the ignition timing.

When the stop switch is actuated, the voltage produced by the stator LS and HS coil is shorted directly to ground within the CDI. This short circuit action prevents the ignition coils from firing the spark plugs and causes the powerhead to shut down.

Troubleshooting the CDI System

TROUBLESHOOTING WITH MINIMAL TEST EQUIPMENT

➡ **The following conditions apply to all powerheads covered in this manual. Troubleshooting procedures are courtesy of our friends at CDI Electronics (256-772-3829).**

- **Intermittent Firing:**—This problem can be very hard to isolate. A good inductive tachometer can be used to compare the RPM on all cylinders up through wide open throttle. A big difference on one or two cylinders indicates a problem.
- **Two or more cylinders misfiring:**— It is recommended that both power packs be replaced, unless the problem is caused by a bad trigger.
- **Engine continuously blows power packs:**— When an engine starts blowing packs repeatedly, especially on the same cylinders, replace the ignition coils on those cylinders. The inductive kickback from a bad coil can destroy the packs, even if the coils check good with all known tests. A stator that tests good can also be sending spike voltages to the pack causing them to fail repeatedly.
- **Visually check the stator, trigger and flywheel:**—Cracks, burned marks and bubbling on the stator or trigger indicate a severe problem. If the stator shows bubbling around the battery charge windings, more than likely you will have to replace the rectifier/regulator in addition to the stator. Signs of rubbing on the flywheel usually indicate a bad upper or lower bearing. Check both the outer and trigger magnets for signs of cracking and to be sure that they are not loose.

The following conditions apply only to Mercury 6-Cylinder powerheads with ADI ignition:

- **No fire on 1,3,5 or 2,4,6:**— Swap Red and Red/White wires, also Blue with blue/White wires. If the problem moves to the other set of cylinders, the stator is likely at fault. Disconnect rectifier and retest. If the engine fires normally, replace the rectifier. If no change, we recommend replacement of the stator. If you replace the stator and the problem remains, try another flywheel, if possible.
- **No fire on one cylinder:**—Since this condition can be caused by the opposite switch box (pack), disconnect the White/Black jumper between the packs and retest. If the dead cylinder starts firing, replace the pack that was firing all three cylinders. As a verification, swap the trigger and spark plug wire to the cylinder closest to the dead cylinder. If the problem moves to #3, replace the opposite power pack (switch box). If #1 is still dead, swap the green and green/White coil wires. If the problem moves to #3, replace the power pack (switch box).

➡ **Remember to put the trigger wires in the correct order after you finish.**

- **Always check the bias circuit:**—Disconnect the jumper between the packs on the White/Black post. Check the resistance from the White/Black post on each pack to engine ground. Standard packs will read from 13000–15,000 ohms. Make sure that the bias reading is the same on both packs.

➡ **If the bias is out of specification on one pack you must replace both packs to prevent future damage. This circuit effects ignition timing and could cause a powerhead failure.**

5-16 IGNITION AND ELECTRICAL SYSTEMS

TROUBLESHOOTING BATTERY CD IGNITIONS

> ※※ **CAUTION**
>
> Do not use a maintenance free battery with these types of ignitions as they tend to overcharge and blow the packs.

➡ A large portion of the problems with the battery CD units are caused by low battery voltage or bad ground connections or high battery voltage. Low Voltage symptoms are weak fire or weak erratic firing of cylinders. Misfiring after a few minutes of running can be caused by excessive voltage at the pack. Battery reversal will usually destroy battery, CD units and triggers.

> ※※ **WARNING**
>
> Check the voltage on the Red (or Purple) wire at the CD unit through the RPM range. At no time should the voltage exceed 15.5 Volts DC.

The first item to check is that all battery and ground connections are solid.
- **Dead or No Fire Until You Release the Key Switch:**—Disconnect mercury switch and retest, if the engine fires, replace the mercury switch.

Check the voltage on the Red and White ignition wires at the CD unit. If the voltage is less than 9½ volts during cranking there is a problem in the battery wires or ignition switch circuit. These units require at least 9½ volts to fire properly.

Perform a peak voltage check between the White and Black wires (sometimes Black and Blue). Reading should be at least 2½ at cranking. Connect a jumper wire directly from the battery positive (+) terminal to the White ignition wire and retest. If the engine still fails to crank, recheck voltage as above. If low, replace battery and retry.

> ※※ **CAUTION**
>
> Do not connect the jumper wire to the White trigger terminal. In order to kill the engine if it cranks, the jumper wire has to be disconnected and/or choke the engine.

If there is still no fire, disconnect trigger wires and connect a CD tester. Align the rotor with a spark plug wire. Connect a spark gap tester to all spark plug wires and turn the ignition switch **ON**. If the CD unit fires to only one spark plug wire, check points wire (for breaks and shorts) or trigger. If any other spark plug wire fires besides the one the rotor is aligned with, the distributor cap and rotor should be replaced.

If the CD unit fails to fire with this hookup, it is usually bad. Check the trigger using a trigger tester to see if it is good or bad.
- **Engine Cranks and Fires As Long As The Starter Is Engaged:**—This problem usually indicates a bad trigger.
- **Check the Ignition Coil:**—An open, cracked or poorly grounded coil can burn out a battery CD.
- **Check the Peak Voltage on the Primary Input Wire to The Coil:**—The reading should be approximately 100 volts or more for OEM units and 200 volts aftermarket units.
- **Inline engines with internal exhaust plate:**—If engine speeds up when you remove one spark plug wire, the internal exhaust plate is more than likely warped.

TROUBLESHOOTING ALTERNATOR DRIVEN IGNITIONS

➡ Initial peak voltage readings should be taken with everything hooked up.

- **Disconnect the kill wire:**—Connect a multimeter between the kill wires and engine ground. Turn the ignition switch ON and OFF several times. If at any time, you see DC voltage on the kill wires, there is a problem with the harness or ignition switch. Battery voltage on the kill circuit will destroy most CD units.
- **Visually Inspect Stator for Cracks or Varnish Leakage:**—If found, replace the stator. Burned marks or discolored areas on the battery charge windings indicate a possible problem with the rectifier.
- **Unit Will Not Fire: Disconnect kill wire at the pack:**— Check for broken or bare wires on the unit, stator and trigger. Check the peak voltage of the stator, (on 3 and 6-Cylinder models read from each Red and Blue wire to engine ground. On 4-Cylinder engines, read between the two Red wires and between the two blue wires), with everything connected. The readings should be approximately 180 volts or more on the Blue wires and 30 volts or more on the Red wires. Disconnect the rectifier. If the engine fires, replace the rectifier.
- **Engine Will Not Kill:**—Check kill circuit in the pack by using a jumper wire connected to the Black/Yellow terminal or wire coming out of the pack and shorting it to ground. If this kills the engine, the kill circuit in the harness or on the boat is bad, possibly the ignition switch.
- **High Speed Miss:**—Disconnect the rectifier and retest. If miss is gone, the rectifier is usually at fault. If the miss still exists, check peak voltage (between the Red wires on 4-Cylinder, or Red wires to engine ground on 3 & 6-Cylinder) of the stator at high speed.

➡ Use caution when doing this and do not exceed the rated voltage range of your meter. The readings should show a smooth climb in voltage. If there is a sudden or fast drop in voltage right before the miss becomes apparent, the stator is usually at fault. If there is no indication of the problem, it could be a small water leak in one or two cylinders.

- **Coils Fire with Spark Plugs Out but Not In:**—Check for dragging starter or low battery causing slow cranking speed. Test peak voltage on the stator and trigger. Disconnect rectifier, regulator and retest. If the problem goes away, replace the rectifier and/or regulator.
- **Engines Runs Rough on Top or Bottom Two Cylinders (4-Cylinder Engines):**—Check peak voltage of the stator between Blue wires and ground. Readings to ground should be fairly equal. If unequal, swap stator leads (Blue with Blue/White, Red with Red/White) and see if the problem moves with the stator leads. If it does, replace the stator.

Check trigger resistance between #1 & #2, compare to resistance between #3 and #4. The readings should be approximately 850–12,5091 ohms. For test purposes only, swap trigger leads 1 & 3, and 2 & 4. If the problem moves, replace the trigger. If it does not move, swap coil primary wires, and replace the pack if the problem remains on the same terminals.
- **No Fire on One Bank (Odd or Even Cylinders on Inline 6-Cylinder Engines):**—Check peak voltage of the stator, checking from each red and blue wire to engine ground. The readings should be approximately 180 volts or more on the blue wires and 30 or more on the red wires. If a DVA meter is not available, swap both sets of the stator wires between the packs. If the problem moves, replace the stator. If the problem stays on the same bank, swap physical location and all connections of the two packs. If the problem stays with one pack, replace the pack.

➡ If the pack is bad, it is recommended that both packs be replaced. If the packs lose ground, internally or externally, the packs usually have severe damage to the bias circuit and need to be replaced as a set. Packs manufactured by CDI Electronics (256-772-3829) will withstand loss of ground connection, normally with no damage to the bias circuitry. In most cases you will just lose fire.

- **Intermittent Firing On One or More Cylinders:**—Disconnect the White/Black wire between the packs on a 6-Cylinder and retest. If all cylinders now fire, replace both packs as there is a problem in the bias circuitry. On all others, check for low voltage from the stator and trigger. Disconnect the rectifier and retest. If the problem disappears, replace the rectifier.
- **All Cylinders Fire but the Engine Will Not Crank and Run:**—On 3 and 6-Cylinder engines, disconnect White/Black wire and check the bias circuit (White/Black terminals) resistance to engine ground. Readings should be approximately 15,000 ohms. If the readings are correct on the packs, index the flywheel and check timing on all individual cylinders. If the timing varies, replace the packs. On 4-Cylinder engines the bias circuit is internal, therefore the only way to verify proper operation of the bias, circuit is to index the flywheel and check timing on each cylinder. If the timing is off, replace the packs.

Trigger (Charge) Coil

TESTING

The following tests perform a resistance check of both the capacitor charging and trigger coils. This test is performed with the components installed but disconnected at the wire lead bullet connectors. It is essential that the wires be disconnected in order to obtain an accurate reading and evaluation of the components.

If the coils should test bad, before spending the money on new replacement coils, perform the test a second time after the unit has been removed from the powerhead to verify the coil is bad.

IGNITION AND ELECTRICAL SYSTEMS 5-17

❊❊ WARNING

Do not rotate the flywheel during these tests or damage to the multimeter will result.

Resistance Tests

♦ See Figure 33

1. Label and disconnect the trigger coil wires.
2. Connect the multimeter leads to the wires specified in the "Ignition Testing Specifications" chart.
3. Note the resistance reading. Resistance should be within specifications listed in the "Ignition Testing Specifications" chart.

➡ Resistance readings are temperature sensitive. The specifications given are for components at 68°F (20°C). If temperature is higher or lower, resistance readings may be off slightly.

4. If resistance is not within specification, remove the trigger coil and retest. If readings are similar, the trigger coil may be defective.
5. Disconnect the multimeter.
6. Connect the coil wires as previously labeled.

Direct Voltage Adapter Tests

♦ See Figures 34, 35, 36, 37 and 38

1. Remove all the spark plugs.
2. Install a spark plug gap tool to each spark plug boot and attach the alligator clips to a good engine ground.
3. Disconnect the powerhead from its fuel supply.
4. Locate the trigger/charge coil.
5. There are two ways to connect a multimeter with peak reading adapter to the trigger/charge coil harness. The first is to use a factory style wiring harness adapter. These adapters can be obtained from your Mercury dealer. The second way is to use piercing probes available from tool manufacturers.

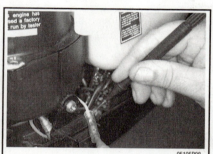

Fig. 33 Use the multimeter to check the trigger coil resistance. The trigger coil does not need to be removed to perform this test

Fig. 34 A CDI Electronics peak reading voltage adapter (#511-9773) plugged into a high quality multimeter

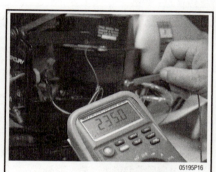

Fig. 35 Using a multimeter and DVA attachment to check the peak voltage on the trigger/charging circuit

Ignition Testing Specifications

Year	Cyl	CDI Part #	Stator Part # ①	Trigger Ohms ②	Charge Coil Ohms High Speed	Charge Coil Ohms Low Speed	Trigger Cranking DVA	Charge Coil Cranking DVA High Speed	Charge Coil Cranking DVA Low Speed	CD Module (Switch Box) Cranking DVA	Trigger Wire Testing Colors	Charge Coil Wire Testing Colors
1990-94	2	339-7452	86617	750-1400	200-250	3250-3650	4V+	20V+	180V+	150V+	Between Trigger Wires	Ground to B/Y Ground to B/W
1990-94	2	114-7452	174-6617	750-1400	200-250	2250-2650	4V+	20V+	180V+	150V+	Br/W to Br/Y	B to B/Y B to B/W
1994-97 (20-25)	2	18495	86617	1200-1400	200-250	3250-3650	4V+	20V+	180V+	150V+	Br/W to Br/Y	B to Bl B to R
1990-94	2	18495	174-6617	750-1400	200-250	2250-2650	4V+	20V+	180V+	150V+	Br/W to Br/Y	B to B/Y B to B/W
1994-99 (30)	2	827509	398-832074	—	—	600-700	4V+	—	180V+	150V+	Br/W to Br/Y	W/G to G/W
1987-93	3	332-7778	398-8778 398-9710	750-1400	75-90	3250-3650	4V+	20V+	180V+	150V+	W/B to P W/B to W W/B to Br	Ground (B) to R Ground (B) to Bl
1987-93	3	114-7778	174-8778K1 174-9710K1	750-1400	28-32	500-600	4V+	20V+	180V+	150V+	W/B to P W/B to W W/B to Br	R to R/W Bl to Bl/W
1991-95 (50,55,60)	3	114-9052	174-8778K1 174-9710K1	1200-1400	28-32	500-600	4V+	20V+	180V+	150V+	W/B to P W/B to W W/B to Br	R to R/W Bl to Bl/W
1994-97 (75-90)	3	18495	398-8778 398-9710	1200-1400	75-90	3250-3650	4V+	20V+	180V+	150V+	W/B to P W/B to W W/B to Br	Ground (B) to R Ground (B) to Bl
1994-97 (75-90)	3	114-4953	174-8778K1 174-9710K1	1200-1400	28-32	500-600	4V+	20V+	180V+	150V+	W/B to P W/B to W W/B to Br	R to R/W Bl to Bl/W
1997-00	3	827509	398-832075	—	—	600-700	1.5V+	—	180V+	—	W/B to B	W/G to G/W
1997-00	3	827509	174-832075	—	—	600-700	1.5V+	—	180V+	—	W/B to B	W/G to G/W
1990-96	4	332-5772	398-8778 398-9710	1200-1400	75-90	3250-3650	4V+	20V+	180V+	150V+	Br to B (B/W) P to W	R to R/W Bl to Bl/W
1990-96	4	114-5772	174-8778K1 174-9710K1	1200-1400	28-32	500-600	4V+	20V+	180V+	150V+	Br to B (B/W) P to W	R to R/W Bl to Bl/W
1995-96 (SportJet)	4	332-826866	398-8778 398-9710	200-1400	75-90	3250-3650	4V+	20V+	180V+	150V+	Br to B B/W to W	B to B/W R to R/W

5-18 IGNITION AND ELECTRICAL SYSTEMS

Ignition Testing Specifications

Year	Cyl	CDI Part #	Stator Part # ①	Trigger Ohms ②	Charge Coil Ohms High Speed	Charge Coil Ohms Low Speed	Trigger Cranking DVA	Charge Coil Cranking DVA High Speed	Charge Coil Cranking DVA Low Speed	CD Module (Switch Box) Cranking DVA	Trigger Wire Testing Colors	Charge Coil Wire Testing Colors
1995-96 (SportJet)	4	332-826866	174-8778K1 174-9710K1	1200-1400	75-90	3250-3650	4V+	20V+	180V+	150V+	Br to B (B/W) P to W	B to B/W R to R/W
1996-99	4	827509	398-832075	—	—	600-700	1.5V+	—	180V+	—	W/B to B	W/G to G/W
1996-99	4	827509	174-832075	—	—	600-700	—	—	180V+	150V+	W/B to B	W/G to G/W
1990-00 (9-15 amp)	6	332-7778	398-5454	750-1400	135-165	5800-7000	4V+	20V+	180V+	150V+	Br (Y sleeve) to P (B sleeve) P (Y sleeve) to W (B sleeve) W (Y sleeve) to Br (B sleeve)	Ground (B) to R and R/W Ground (B) to Bl and B/W
1900-00 (9-15 amp)	6	114-7778	174-5454	750-1400	45-55	2200-2400	4V+	20V+	180V+	150V+	Br (Y sleeve) to P (B sleeve) P (Y sleeve) to W (B sleeve) W (Y sleeve) to Br (B sleeve)	Ground (B) to R and R/W Ground (B) to B and B/W
1900-00 (16 amp)	6	332-7778	398-5454	750-1400	135-165	5800-7000	4V+	20V+	180V+	150V+	Br (Y sleeve) to P (B sleeve) P (Y sleeve) to W (B sleeve) W (Y sleeve) to Br (B sleeve)	Ground (B) to R and R/W Ground (B) to B and B/W
1900-00 (16 amp)	6	114-7778	174-5454	750-1400	45-55	2200-2400	4V+	20V+	180V+	150V+	Br (Y sleeve) to P (B sleeve) P (Y sleeve) to W (B sleeve) W (Y sleeve) to Br (B sleeve)	Ground (B) to R and R/W Ground (B) to B and B/W
1990-00 (40 amp)	6	332-7778	398-9610	1100-1400	90-140	3500-4200	4V+	20V+	180V+	150V+	Br (Y sleeve) to P (B sleeve) P (Y sleeve) to W (B sleeve) W (Y sleeve) to Br (B sleeve)	Ground (B) to R and R/W Ground (B) to B and B/W
1990-00 (40 amp)	6	114-7778	174-9610	750-1400	80-100	2200-2400	4V+	20V+	180V+	150V+	Br (Y sleeve) to P (B sleeve) P (Y sleeve) to W (B sleeve) W (Y sleeve) to Br (B sleeve)	Ground (B) to R and R/W Ground (B) to B and B/W

① Stators 398-8778 and 398-9710 are superceeded by the Mercury Red Stator Kit
② Stator and Trigger readings should be consistent with no trace of varnish leakage visable

B - Black   G - Green   W - White
Br - Brown   P - Purple   Y - Yellow
Bl - Blue   R - Red

Fig. 36 High quality piercing probes (#511-9770), like those available from CDI Electronics (256-772-3829), allow you to perform dynamic ignition testing without the use of factory wiring harness adapters

Fig. 37 The tiny pins in the piercing probes contact the wires to read resistance or voltage . . .

Fig. 38 . . . but do not destroy the insulation on the wire

6. Connect the multimeter to the trigger coil wires listed in the "Ignition Testing Specifications" chart.
7. Use the starter motor or recoil starter to crank the engine.

➡ If it is not possible to obtain a cranking speed high enough to properly test the trigger coil, the engine can be tested in the same manner while running.

8. Note the cranking voltage and compare it to the voltage listed in the "Ignition Testing Specifications" chart.
9. If voltage is below specification, the trigger/charge coil may be faulty.
10. If voltage is above specification, the CDI unit should be tested.
11. Disconnect the multimeter with peak reading adapter.

REMOVAL & INSTALLATION

2.5–3.3 HP

▶ See accompanying illustrations

1. Remove the cowling.
2. Remove the three bolts securing the hand rewind starter to the powerhead. Lift off the hand rewind starter from the powerhead.

Hand rewind starter

Step 2

IGNITION AND ELECTRICAL SYSTEMS

Step 4

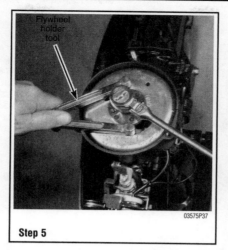
Step 5

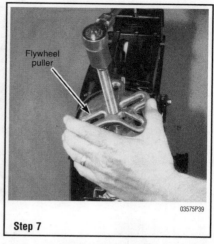
Step 7

Step 8

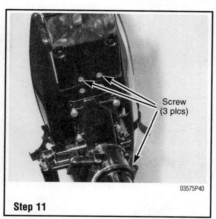
Step 11

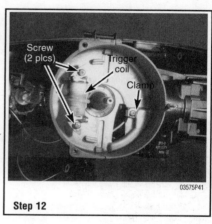
Step 12

3. Loosen the three bolts securing the rope cup to the flywheel. A strap wrench or large pair of channel lock pliers around the outside edge of the rope cup may be needed to hold the flywheel while loosening the rope cup bolts.

4. Remove the three bolts from the rope cup. Lift off the rope cup and flywheel cover (if equipped) free of the flywheel.

5. Install a flywheel holder and hold the flywheel steady while removing the flywheel nut.

6. An alternate method may be used to remove the flywheel nut. Install a grade five or stronger bolt into one of the flywheel puller bolt holes. Insert a pry bar between the bolt and deep socket on the flywheel nut. Using the pry bar as a lever, remove the flywheel nut from the flywheel.

7. Obtain a flywheel puller tool as illustrated.

8. An alternate method of preventing the flywheel from rotating, is to use a bolt and a pry bar as illustrated.

➡ Never attempt to use a puller that pulls on the outside edge of the flywheel. This action will result in a bent, distorted or possibly even broken flywheel, a damaged flywheel must be replaced.

9. Install a flywheel puller. Attach the proper size wrench onto the end of the puller tool threaded shaft. Rotate the wrench in a clockwise direction, continue to tighten on the special tool and at the same time, shock the crankshaft with a gentle to moderate tap with a hammer on the end of the puller. This shock will assist in "breaking" the flywheel loose from the crankshaft.

10. When the flywheel has broken loose from the crankshaft taper fit, remove the flywheel puller tool and lift the flywheel from the powerhead.

11. Remove the four screws from the port side of the lower cowling cover and remove the cover from the powerhead.

12. Disconnect the lead from the capacitor charging/trigger coil at the bullet connector on the port side of the powerhead.

13. Remove the screw and clamp on top of the stator plate securing the lead to the capacitor charging trigger coil. Feed the wire up through the opening below the stator plate. Remove the two screws securing the capacitor charging trigger coil to the stator plate and lift off the free trigger coil.

To install:

14. Feed the wire lead of the capacitor charging/trigger coil down through the opening below the stator plate. Position the charging/trigger coil onto the stator plate and secure with two screws. Connect the lead from the capacitor charging/trigger coil to the lead from the CDI at the bullet connector on the port side of the powerhead.

15. Install the clamp onto the White lead of the capacitor charging trigger coil at the stator plate and secure with the screw.

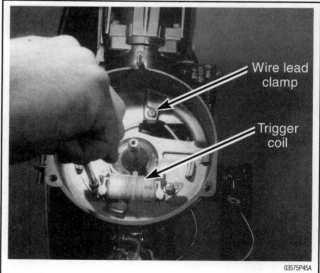
Step 14

5-20　IGNITION AND ELECTRICAL SYSTEMS

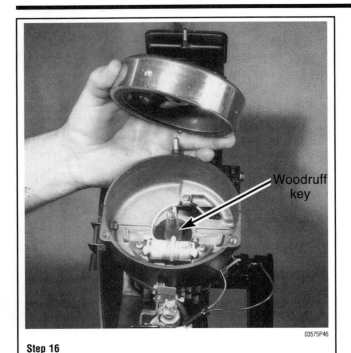

Step 16

20. Position the flywheel cover (if equipped) and rope cup in place over the flywheel with the holes in the cover and cup aligned with the holes in the flywheel. Install and tighten the three bolts to 5.8 ft. lbs. (8 Nm).
21. Lower the hand rewind starter onto the powerhead and align the mounting tabs on the starter with the powerhead. Tighten the three attaching bolts to 5.8 ft. lbs. (8 Nm).
22. Install the proper heat range spark plug into the powerhead. Tighten the spark plug to 20 ft. lbs. (21.7 Nm). Install the top cowling onto the powerhead and secure it with the two snap latches.
23. Place the port side of the lower cowling cover into position and align the holes in the cover with the powerhead. Secure the cover in place with the five mounting screws. Tighten the screws securely.

4–5 HP

♦ **See accompanying illustrations**

1. Remove the cowling to expose the flywheel.
2. Disconnect the start in neutral only linkage. This linkage must be disconnected before "pulling" the flywheel.
3. Remove the flywheel nut and washer (if equipped) from the end of the crankshaft. A flywheel strap wrench may be required to hold the flywheel securely while the nut is loosened.

Step 2

16. Apply just a dab of thick lubricant onto the curved surface of the Woodruff key to hold it in place while the flywheel is being installed. Press the Woodruff key into place in the crankshaft recess. Wipe away any excess lubricant to prevent the flywheel from "walking" during powerhead operation.
17. Check the flywheel magnets to ensure they are free of any metal particles. Double-check the taper in the flywheel hub and the taper on the crankshaft to verify they are clean and contain no oil.
18. Now, slide the flywheel down over the crankshaft with the keyway in the flywheel aligned with the Woodruff key in place on the crankshaft. Rotate the flywheel counterclockwise to be sure it does not contact any part of the stator plate or wiring.
19. Slide the washer onto the crankshaft and then thread the flywheel nut onto the crankshaft. Using a flywheel holder tool tighten the flywheel nut to 30 ft. lbs. (40.6 Nm).

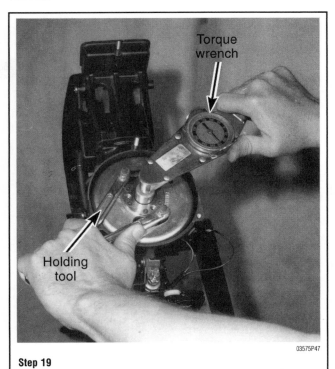

Step 19

Step 3

IGNITION AND ELECTRICAL SYSTEMS 5-21

Step 4

6–25 HP

▶ See Figure 39

1. Remove the cowling.
2. Remove the recoil assembly.
3. Remove the flywheel assembly.
4. Disconnect the stator and trigger coil wires.
5. Remove the retaining screw holding down the stator assembly.
6. Remove the trigger coil assembly from the powerhead.

To install:

7. Install the trigger coil assembly onto the powerhead.
8. Install the stator assembly onto the powerhead above the trigger coil and tighten the retaining screws to 40 inch lbs. (4.5 Nm).
9. Clean and inspect the both the crankshaft and flywheel tapers for grease and damage.
10. Install the flywheel Woodruff key onto the crankshaft.
11. Install the flywheel assembly and tighten the flywheel nut to 50 ft. lbs. (67.8 Nm).
12. Reconnect the stator and trigger coil connectors at the switch box.
13. Install the recoil assembly and tighten the retaining screws to 70 inch lbs. (7.9 Nm).
14. Install the cowling.

4. If a strap wrench is not available, an alternate method is to use a bolt and pry bar to hold the flywheel. First, thread a 5/16 in. bolt into one of the threaded flywheel holes. Position a pry bar between the bolt and the socket on the flywheel nut. Proceed to loosen and remove the flywheel nut.
5. Obtain the proper type flywheel puller. Never attempt to use a puller which pulls on the outside edge of the flywheel or the flywheel may be damaged. After the puller is installed, tighten the center screw onto the end of the crankshaft. Continue tightening the screw until the flywheel is released from the crankshaft. Remove the flywheel.
6. Remove the screw and wire clamp at the stator plate.
7. Disconnect the coil leads from the CDI unit leads.
8. Remove the screws securing the capacitor charging coil and then remove the coil.

To install:

9. Carefully route the charging coil leads along the stator plate and place the coil over the mounting bosses of the crankcase cover.
10. Apply a drop of thread-locking compound to the coil securing screws and torque them to 14 inch lbs. (1.6 Nm).
11. Secure the coil leads/sleeve, using the wire clamp and screw. Tighten the screw securely.
12. Connect the capacitor charging coil leads to the CDI unit leads.
13. Reinstall the flywheel assembly. Using a flywheel holder tool tighten the flywheel nut to 30 ft. lbs. (40.6 Nm).
14. Install the cowling.

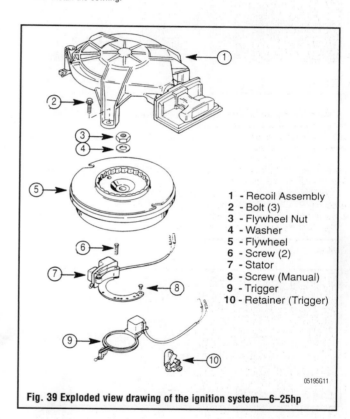

1 - Recoil Assembly
2 - Bolt (3)
3 - Flywheel Nut
4 - Washer
5 - Flywheel
6 - Screw (2)
7 - Stator
8 - Screw (Manual)
9 - Trigger
10 - Retainer (Trigger)

Fig. 39 Exploded view drawing of the ignition system—6–25hp

30 and 40 HP (2-Cylinder)

▶ See Figures 40 and 41

1. Remove the cowling and flywheel cover.
2. Hold the flywheel with a flywheel holder and remove the flywheel nut and washer.
3. Install the crankshaft protector cap (91–24161), then install the flywheel puller on the flywheel.
4. Hold the flywheel tooth with the wrench while tightening the bold down on the protector cap. Tighten the bolt until the flywheel comes free.
5. Remove the flywheel.

➡**Neither heat nor the use of a hammer should be used to aid in the removal of the flywheel. Damage to the electrical components underneath the flywheel may occur.**

6. Disconnect both the stator and trigger coil leads.

Step 5

5-22 IGNITION AND ELECTRICAL SYSTEMS

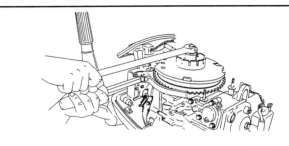

Fig. 40 Hold the flywheel with a flywheel holder and remove the flywheel nut and washer

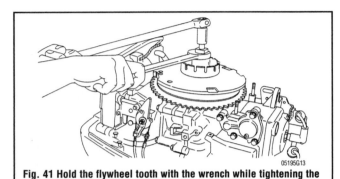

Fig. 41 Hold the flywheel tooth with the wrench while tightening the bold down on the protector cap

7. Remove the stator retaining screws and remove the stator.
8. Remove the trigger coil assembly.

To install:
9. Install the trigger coil assembly onto the powerhead.
10. Next, install the stator assembly onto the powerhead above the trigger coil,

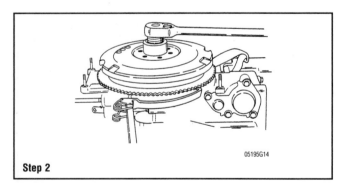

Step 2

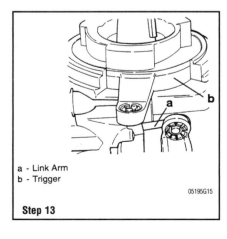

a - Link Arm
b - Trigger

Step 13

apply a small amount of thread-locking compound and tighten the retaining screws to 50 inch lbs. (5.6 Nm).
11. Clean and inspect the both the crankshaft and flywheel tapers for grease and damage.
12. Install the flywheel Woodruff key onto the crankshaft.
13. Install the flywheel assembly and tighten the flywheel nut to 95 ft. lbs. (129 Nm).
14. Reconnect the stator and trigger coil connectors at the switch box.

40–60 HP (3-Cylinder)

♦ See accompanying illustrations

1. Remove the flywheel cover from the engine.

✳✳ WARNING

The engine could possibly start when you are turning the flywheel during removal and installation. Make sure to remove the spark plug leads from the spark plugs to prevent the engine from staring.

2. While holding the flywheel with a flywheel holder , remove the flywheel nut and washer.
3. Install the crankshaft protector cap (91–24161), then install a flywheel puller on the flywheel.
4. Hold the flywheel tooth with the wrench while tightening the bold down on the protector cap. Tighten the bolt until the flywheel comes free.

➡Neither heat nor the use of a hammer should be used to aid in the removal of the flywheel. Damage to the electrical components underneath the flywheel may occur.

5. Remove the flywheel and flywheel key.
6. carefully inspect the flywheel for cracks or defects.
7. Inspect the crankshaft and flywheel tapers and key ways for wear and damage.
8. Check for loose or damaged flywheel magnets (located in the outer rim and center hub). If necessary, replace the flywheel.
9. Remove the Yellow stator leads from the regulator/rectifier leads.
10. Disconnect all stator leads from the CDI wiring harness.

➡Removal of the ignition plate may be may be necessary to gain access to the stator leads.

11. Remove the screws and lift the stator off the bearing cage.
12. Disconnect the trigger coil leads from the CDI wiring harness.
13. Disconnect the link arm and remove the trigger.
14. Lift the trigger coil assembly off the bearing cage.

To install:
15. Install the trigger coil and connect the link arm.
16. Place the trigger on the bearing cage.
17. Route the lead wires under the ignition plate and down to the CDI wiring harness.
18. Connect the corresponding trigger coil leads to the CDI wiring harness.
19. Install the stator on the bearing cage. Apply a small amount of thread-locking compound and tighten the screws to 60 inch lbs. (6.8 Nm).
20. Connect the Yellow stator leads to the Yellow voltage regulator leads (electric start models).
21. Connect the remaining stator lead to the appropriate CDI wiring harness leads.
22. Install the flywheel key into the crankshaft slot with the outer edge of the key parallel to the center-line of the crankshaft.

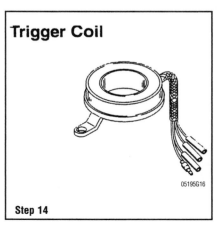

Trigger Coil

Step 14

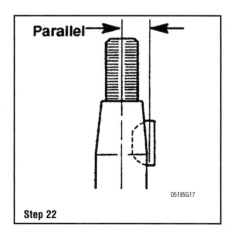

Parallel

Step 22

IGNITION AND ELECTRICAL SYSTEMS 5-23

23. Align the slot in the in the flywheel center bore with the flywheel key and install the flywheel onto the crankshaft.
24. Install the washer and nut.
25. While holding the flywheel with a flywheel holder , torque the flywheel nut to 125 ft. lbs. (169.5 Nm)
26. Replace the flywheel cover.

75–125 HP

♦ See accompanying illustrations

1. Remove the flywheel cover from the engine.

※ WARNING

The engine could possibly start when you are turning the flywheel during removal and installation. Make sure to remove the spark plug leads from the spark plugs to prevent the engine from staring.

2. While holding the flywheel with the flywheel holder (91--52344), remove the flywheel nut and washer.
3. Install the crankshaft protector cap (91-24161), then install a flywheel puller on the flywheel.
4. Hold the flywheel tooth with the wrench while tightening the bold down on the protector cap. Tighten the bolt until the flywheel comes free.

➡ **Neither heat nor the use of a hammer should be used to aid in the removal of the flywheel. Damage to the electrical components underneath the flywheel may occur.**

5. Remove the flywheel and flywheel key.
6. carefully inspect the flywheel for cracks or defects.
7. Inspect the crankshaft and flywheel tapers and key-ways for wear and damage.
8. Check for loose or damaged flywheel magnets (located in the outer rim and center hub). If necessary, replace the flywheel.
9. Remove the stator screw.

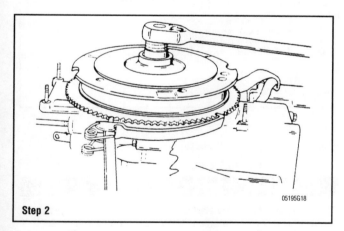

Step 2

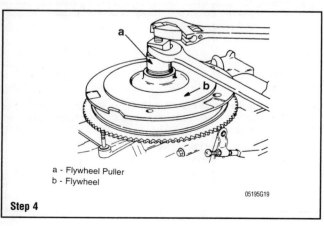

a - Flywheel Puller
b - Flywheel

Step 4

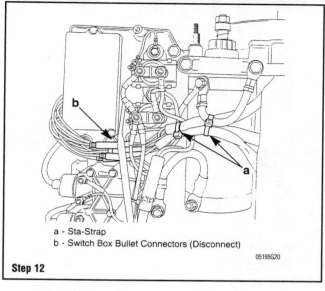

a - Sta-Strap
b - Switch Box Bullet Connectors (Disconnect)

Step 12

10. Remove the starter motor.
11. Remove the wire ties on the wiring harness.
12. Disconnect the stator leads from the switch box on the 3-Cylinder models and remove the stator from the engine.
13. Disconnect the stator leads from the switch box on the 4-Cylinder models and remove the stator from the engine.
14. Disconnect the link arm.
15. Disconnect the trigger coil leads from the switch box.
16. Remove the trigger coil assembly.

To install:

17. Install the trigger coil and link arm.
18. Connect the trigger leads to the switch box.
19. Install the sta-strap.

➡ **There are two types of stators used on the 1994 thru 1996 75 to 125hp engines. These stators can be identified by a large rim or small rim on the under side of the stator where the stator harness exits the stator. These stators must be installed as shown respectively or premature stator failure may occur as a result of stator interference with the engine block.**

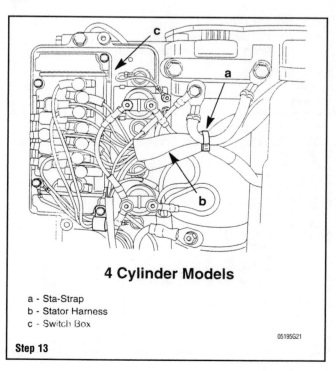

4 Cylinder Models

a - Sta-Strap
b - Stator Harness
c - Switch Box

Step 13

5-24 IGNITION AND ELECTRICAL SYSTEMS

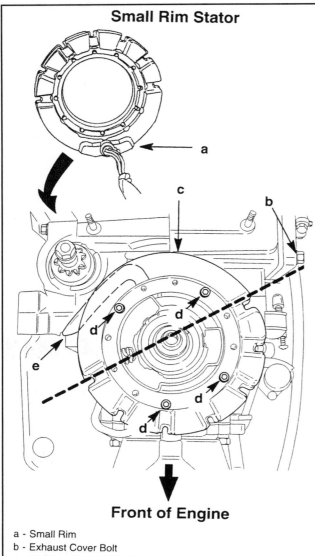

Step 20

a - Small Rim
b - Exhaust Cover Bolt
c - High/Low Speed Winding Module of Stator
d - Stator Screws [Apply Loctite 222 to threads] [Torque screws to 60 lb. in. (6.8 N·m)]
e - Stator Harness

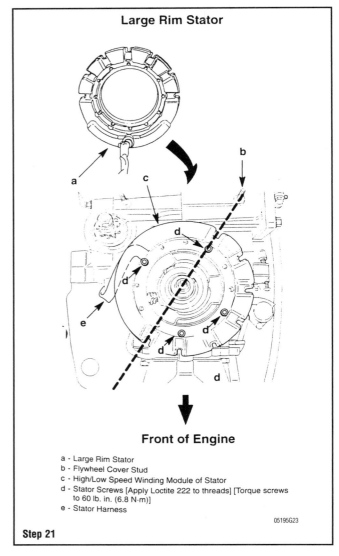

Step 21

a - Large Rim Stator
b - Flywheel Cover Stud
c - High/Low Speed Winding Module of Stator
d - Stator Screws [Apply Loctite 222 to threads] [Torque screws to 60 lb. in. (6.8 N·m)]
e - Stator Harness

20. Install the small rim stator assembly aligned as illustrated to prevent premature stator wear.
21. Install the large rim stator assembly aligned as illustrated to prevent premature stator wear.
22. After installing the correct stator, apply a thread-locking compound and tighten the screws to 60 inch lbs. (6.8 Nm)
23. Connect the stator leads.
24. Install the flywheel key into the crankshaft slot with the outer edge of the key parallel to the center-line of the crankshaft.
25. Align the slot in the in the flywheel center bore with the flywheel key and install the flywheel onto the crankshaft.
26. Install the washer and nut.
27. While holding the flywheel with a flywheel holder , torque the flywheel nut to 100 ft. lbs. (136 Nm)
28. Replace the flywheel cover.

135–225 HP

▶ **See accompanying illustrations**

1. Remove the flywheel cover from the engine.

✱✱ WARNING

The engine could possibly start when you are turning the flywheel during removal and installation. Make sure to remove the spark plug leads from the spark plugs to prevent the engine from staring.

2. While holding the flywheel with a flywheel holder , remove the flywheel nut and washer.
3. Install the crankshaft protector cap (91–24161), then install a flywheel puller on the flywheel.
4. Hold the flywheel tooth with the wrench while tightening the bold down on the protector cap. Tighten the bolt until the flywheel comes free.

➙**Neither heat nor the use of a hammer should be used to aid in the removal of the flywheel. Damage to the electrical components underneath the flywheel may occur.**

5. Remove the flywheel and flywheel key.
6. Carefully inspect the flywheel for cracks or defects.
7. Inspect the crankshaft and flywheel tapers and key-ways for wear and damage.

IGNITION AND ELECTRICAL SYSTEMS

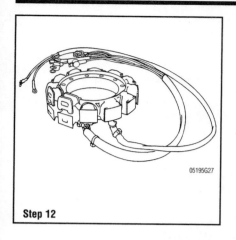
Step 12

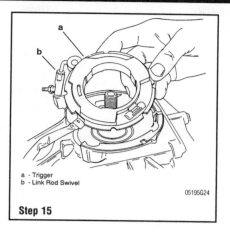
a - Trigger
b - Link Rod Swivel
Step 15

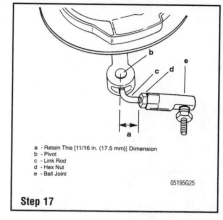
a - Retain This [11/16 in. (17.5 mm)] Dimension
b - Pivot
c - Link Rod
d - Hex Nut
e - Ball Joint
Step 17

8. Check for loose or damaged flywheel magnets (located in the outer rim and center hub). If necessary, replace the flywheel.
9. Remove the stator screws.
10. Remove the wire ties on the wiring harness.
11. Remove the screws, which secures the stator assembly to the upper end cap. Lift off the end cap and move it to the side.
12. Disconnect all the stator leads from the terminals and remove the stator assembly.
13. Remove the lock nut that secures the link rod swivel into the spark advance lever. Pull the link rod out of the lever.
14. Remove the two screws and lift the outer switchbox from the inner switchbox. Make sure not to loose the spacers.
15. Disconnect all the trigger leads from the terminals. Cut the sta-strap and remove the trigger plate assembly from the engine.
16. If the trigger assembly is faulty, remove and retain the link rod swivel from the trigger assembly.

To install:

17. Install the trigger coil and link arm. Make sure to keep a distance of ¹¹⁄₁₆in. (17.5mm) between the pivot and the locknut on the ball joint.
18. Place the trigger plate assembly in the upper end cap. Fasten the link rod swivel to the spark advance lever with the locknut.
19. Carefully route the wiring harness and reconnect the wires to their proper terminals. Wires with a Yellow identification sleeve must be connected to the outer switch box.
20. Clean the stator attaching screws and apply a thread-locking compound. Install the stator in position on the upper end cap and torque the screws to 50 inch lbs. (5.5 Nm).
21. Reconnect the stator leads to the voltage regulator/rectifier and switch boxes. Leads with the Yellow sleeve must be connected to the outer switchbox.
22. Install the switchboxes to the engine with the screws and spacers. Make sure that both boxes are grounded to the engine thru the screws and spacers.

✲✲ CAUTION

Switchboxes must be grounded to the engine before cranking the engine or the switchboxes will be terminally damaged.

23. Install the flywheel key into the crankshaft slot with the outer edge of the key parallel to the center-line of the crankshaft.
24. Align the slot in the in the flywheel center bore with the flywheel key and install the flywheel onto the crankshaft.
25. Install the washer and nut.
26. While holding the flywheel with a flywheel holder , torque the flywheel nut to 120 ft. lbs. (163 Nm)
27. Replace the flywheel cover.

Ignition Coil

TESTING

Spark Test

♦ See Figure 42

A preliminary check of ignition coil function may be made by using a spark tester. Simply connect the spark tester between the ignition high tension lead and a good engine ground. Crank the powerhead and observe the spark tester. Spark should occur inside the glass enclosure. If it does not, the ignition coil, or another ignition system component, may be faulty.

Resistance Test

➥**A multimeter test can only detect certain faults in the ignition coil. Replace the ignition coil if the multimeter reading is not as specified. If the coil tests satisfactory and the coil is still suspected of being faulty, If the resistance is not within specification, the coil may be defective.**

1. Retest the coil using a dedicated ignition coil tester or replace the coil with a known good unit to verify if the coil is bad.

The primary and secondary coil resistance tests may be performed with the components installed on the powerhead. If chaffing or broken wiring is suspected,

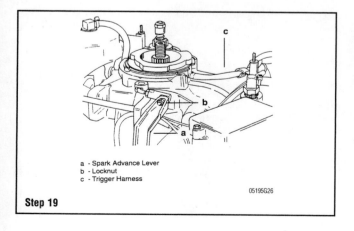
a - Spark Advance Lever
b - Locknut
c - Trigger Harness
Step 19

Fig. 42 Using a spark checker to check the ignition system spark

IGNITION AND ELECTRICAL SYSTEMS

removal of the component is recommended to permit a complete inspection of all wiring leads.

PRIMARY

♦ See Figure 43

1. Disconnect the primary coil electrical lead.
2. Connect a multimeter to the coil primary lead and the other end to a good coil ground.
3. Resistance should be approximately 0.02–0.04 ohms at 68°F (20°C).
4. If the resistance is not within specification, the coil may be defective.
5. Retest the coil using a dedicated ignition coil tester or replace the coil with a known good unit to verify if the coil is bad.

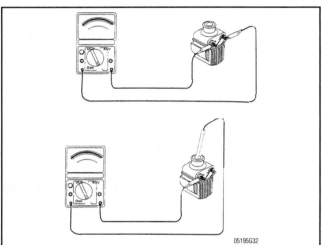

Fig. 43 Connect the positive (+) meter lead to the coil and the negative (-) meter lead to the coil ground to test primary resistance. Connect the (+) meter lead to the spark plug lead terminal and the negative (-) to the coil ground to test secondary resistance

SECONDARY

♦ See Figure 43

1. Disconnect the secondary coil (high tension spark plug) lead.
2. Connect a multimeter to the coil secondary tower and the other end to a good coil ground.
3. Resistance should be approximately 800–1000 ohms at 68°F (20°C).
4. If the resistance is not within specification, the coil may be defective.
5. Retest the coil using a dedicated ignition coil tester or replace the coil with a known good unit to verify if the coil is bad.

Voltage Output Tests

One of the most accurate ways of determining coil condition is to do a dynamic coil output voltage test. This test will show the actual coil output under varying conditions and will not only determine if the coil defective, but will also alert the user to when coil is about to fail. Coil Voltage Output Testers are available from CDI Electronics (256-772-3829).

1. Remove the cowling.
2. Disconnect the spark plug high tension lead and connect the voltage output tester between the spark plug and the high tension lead.
3. Start (crank) the outboard and read the voltage front the tester.
4. Voltage should be 18–20 Kv.
5. If voltage is not as specified, the coil may be defective.
6. Disconnect the tester and connect the spark plug high tension lead.
7. Install the cowling.

REMOVAL & INSTALLATION

2.5–5 HP

1. Remove the engine cover.
2. Disconnect the CDI unit lead from the coil.
3. Disconnect the spark plug wire from the spark plug.
4. Disconnect the ground wires.
5. Remove the bolts and washers securing the ignition coil to the engine block and remove the ignition coil.

To install:

6. Secure the coil to the engine block with the bolts and washers. Make sure to reconnect the ground wires.
7. Reconnect the CDI unit lead to the ignition coil.
8. Connect the spark plug wire to the spark plug.

6–25 HP

♦ See Figure 44

1. Remove the engine cover.
2. Disconnect the spark plug wires from the spark plugs.
3. Disconnect the coil tower boots and remove the spark plug wire from the ignition coil.
4. Disconnect the positive and negative lead wires at the ignition coil. Mark these wires for proper installation during reassembly.
5. Unbolt and remove the coil assembly from the engine block. Make sure to keep track of any spacers (and ground plates if equipped) needed for assembly.

To install:

6. Install the coil assembly onto the engine block. Make sure that all spacers or ground plates are in place.
7. Install the coil tower boots and spark plug wire. Make sure to make a watertight seal between the coil towers and the spark plug leads by using Quicksilver Insulating Compound or equivalent.
8. Attach the positive and negative lead wires onto the correct terminals. Tighten the nuts to 35 inch lbs. (4 Nm), then coat the terminals and nuts with Quicksilver Liquid Neoprene or equivalent.

30 and 40 HP (2-Cylinder)

♦ See Figure 45

1. Disconnect the negative battery cable.
2. Remove the engine cover.
3. Disconnect the high-tension spark plug lead at the spark plug.
4. Unplug the wiring harness and unscrew the ground wire (if equipped).
5. Remove the retaining screws and remove the coil from the electrical plate.

To install:

6. Install the coil onto the electrical plate with the retaining screws.
7. Reconnect the wiring harness and tighten ground wire (if equipped).
8. Connect the high-tension spark plug lead to the spark plug.
9. Replace the engine cover.
10. Connect the battery negative cable.

40–60 HP (3-Cylinder)

♦ See Figure 46

→Always disconnect the battery and spark plug high-tension lead before working on the ignition system.

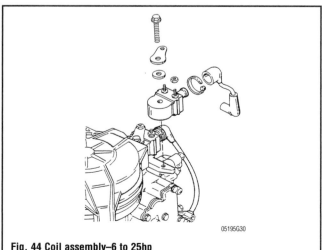

Fig. 44 Coil assembly–6 to 25hp

IGNITION AND ELECTRICAL SYSTEMS

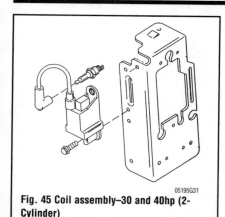

Fig. 45 Coil assembly–30 and 40hp (2-Cylinder)

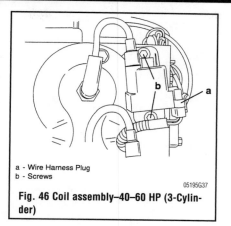

a - Wire Harness Plug
b - Screws

Fig. 46 Coil assembly–40–60 HP (3-Cylinder)

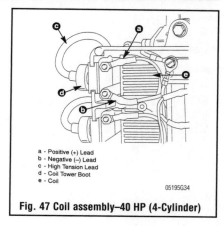

a - Positive (+) Lead
b - Negative (–) Lead
c - High Tension Lead
d - Coil Tower Boot
e - Coil

Fig. 47 Coil assembly–40 HP (4-Cylinder)

1. Remove the engine cover.
2. Disconnect the wiring harness connector.
3. Remove the screws that secure the coil assembly to the ignition plate and remove the coil assembly.

To install:

4. Position the coil on the ignition plate and install the retaining screws. Torque the screws to 60 inch lbs. (6.8 Nm).
5. Connect the wiring harness.
6. Install the engine cover.

40 HP (4-Cylinder)

♦ See Figure 47

1. Remove the engine cover.
2. Remove the bolts securing the electrical box access cover and remove the cover.
3. Disconnect the wires from the positive (+) and negative (-) terminals on the faulty ignition coil.
4. Remove the spark plug boot off the spark plug.
5. Remove the spark plug high-tension lead/coil tower boot assembly from the ignition coil tower.
6. Remove and discard the faulty ignition coil.

To install:

7. Install the spark plug high-tension lead/coil tower boot assembly onto the new ignition coil.
8. Position the ignition coil into the electrical box.
9. Reconnect the positive (+) and negative (-) meter leads to the terminals with two nuts. Torque the nuts to 30 inch lbs. (3.4 Nm).
10. Reconnect the spark plug boot to the spark plug.
11. Reinstall the electrical access cover and tighten the securing bolts.
12. Install the engine cover.

70–115 HP

1. Remove the engine cover.
2. Disconnect the battery cables.
3. Remove the spark plug high-tension lead from the spark plug.
4. Remove the primary wires from the coil terminals. Mark the wires for proper placement during assembly.
5. Remove the retaining bolts and remove the faulty coil from the engine.

To install:

6. Align the coil on the engine and install the retaining bolts. Tighten to 20 inch lbs. (2.3 Nm).
7. Install the primary wires and nuts onto the correct terminal. Tighten the nuts to 30 inch lbs. (3.4 Nm) and coat the terminals and nuts with Quicksilver Liquid Neoprene.
8. Install the spark plug high-tension wire on to the coil tower. Make sure that there is a watertight seal between the wire and coil tower by using Quicksilver Insulating Compound or equivalent.
9. Reconnect the battery cables.
10. Replace the engine cover.

135–225 HP

1. Remove the bolts securing the ignition coil/relay cover and remove the cover.
2. Remove the spark plug high-tension leads from the faulty coil.
3. Disconnect the primary wires from the coil terminals.

4. Remove the screws and nuts and lift the coil cover along with the coils from the engine. Remove the defective coil from the cover.

To install:

5. Install the coil into the coil cover and install it back on the engine with the retaining screws and nuts.
6. Reconnect the switch box wire to the positive (+) coil terminal and the Black ground wire to the negative(-) coil terminal.
7. Pull the boot back and insert the spark plug lead into the coil. Make sure that there is a watertight seal between the wire and coil tower by using Quicksilver Insulating Compound or equivalent.

OptiMax

♦ See Figures 48 and 49

1. Disconnect the coil harness and spark plug leads.
2. Loosen the electrical mounting plate to access the rear locknuts.
3. Remove the module attaching bolts and remove the ignition module.

To install:

4. Attach the module to the electrical mounting plate.
5. Reconnect the spark plug lead and coil wiring harness.

Capacitor Discharge Module (Switch Box)

TESTING

Diode Test (2.5-3.3 HP Only)

♦ See Figures 50 and 51

1. Connect a multimeter between the White and Orange module leads.
2. Continuity should exist with the leads connected in one direction and should not exist with the leads reversed..
3. If the diode does not function as specified, the module may be faulty.

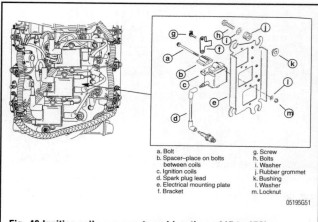

a. Bolt
b. Spacer–place on bolts between coils
c. Ignition coils
d. Spark plug lead
e. Electrical mounting plate
f. Bracket
g. Screw
h. Bolts
i. Washer
j. Rubber grommet
k. Bushing
l. Washer
m. Locknut

Fig. 48 Ignition coil components and location—115 to 150hp

5-28 IGNITION AND ELECTRICAL SYSTEMS

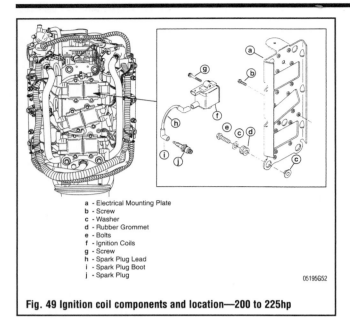

Fig. 49 Ignition coil components and location—200 to 225hp

a - Electrical Mounting Plate
b - Screw
c - Washer
d - Rubber Grommet
e - Bolts
f - Ignition Coils
g - Screw
h - Spark Plug Lead
i - Spark Plug Boot
j - Spark Plug

Direct Voltage Output

1. Remove all the spark plugs.
2. Install a spark plug gap tool to each spark plug boot and attach the alligator clips to a good engine ground.
3. Disconnect the powerhead from its fuel supply.
4. Locate the CD module (switch box).
5. There are two ways to connect a multimeter with peak reading adapter to the module harness. The first is to use a factory style wiring harness adapter. These adapters can be obtained from your Mercury dealer. The second way is to use piercing probes available from aftermarket tool manufacturers.
6. Connect the multimeter at the wires connected to the ignition coil Positive (+) and Negative (-) terminals.
7. Use the starter motor or recoil starter to crank the engine.

➡ If it is not possible to obtain a cranking speed high enough to properly test the switch box, the engine can be tested in the same manner while running.

8. Voltage should be at least 100V +.
9. If voltage is within specification and there is a no spark or weak spark condition, the ignition coil may be faulty.
10. If voltage is below specification, test the ignition coil.
11. If the ignition coil is functioning properly, replace the CD module (switch box) and retest.
12. Disconnect the multimeter with peak reading adapter.

Stop Circuit

1. Remove all the spark plugs.
2. Install a spark plug gap tool to each spark plug boot and attach the alligator clips to a good engine ground.
3. Disconnect the powerhead from its fuel supply.
4. Locate the wire leading from the CD module (switch box) to the stop switch.
5. There are two ways to connect a multimeter with peak reading adapter to the module harness. The first is to use a factory style wiring harness adapter. These adapters can be obtained from your Mercury dealer. The second way is to use piercing probes available from aftermarket tool manufacturers.
6. Connect the multimeter between the White charge/trigger coil wire and ground.
7. Use the starter motor or recoil starter to crank the engine. Note the cranking voltage.
8. If cranking voltage is below 120 volts, disconnect the White wire and retest. If voltage increases above 200 volts, the CD module (switch box) is faulty and should be replaced.
9. If cranking voltage is above 320 volts, and there is no change in voltage when you disconnect the White wire, CD module (switch box) is faulty and should be replaced.
10. Disconnect the multimeter with peak reading adapter.

Switch Box Bias

➡ Switch box bias tests must be performed using a conventional multimeter. If the outboard is equipped with an idle speed stabilizer or spark advance, disconnect these prior to testing.

1. Remove all the spark plugs.
2. Install a spark plug gap tool to each spark plug boot and attach the alligator clips to a good engine ground.
3. Disconnect the powerhead from its fuel supply.
4. Locate the bias wire (usually Black/White).
5. Connect the multimeter between the bias wire and ground.
6. Use the starter motor or recoil starter to crank the engine. Note the cranking voltage.
7. Switch box bias voltage should be between 2–106 volts DC at cranking speed.
8. If cranking voltage is below specification, one or both of the switch boxes may be faulty.
9. If switch box voltage is within specification and the outboard still runs poorly, check the trigger/charge coil. If the trigger/charge coil is functioning properly, replace the switch boxes and retest.

➡ Frequently, switch box bias failure can be attributed to poor grounding between the switch boxes.

10. Disconnect the multimeter.

REMOVAL & INSTALLATION

2.5–3.3 HP

♦ See Figure 52

1. Remove the top cowling.
2. Remove the PORT side lower cowling.

Fig. 50 Multimeter showing continuity . . .

Fig. 51 . . . and zero continuity—CDM (Diode Test) 2.5 and 3.3hp

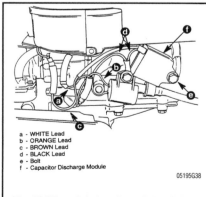

a - WHITE Lead
b - ORANGE Lead
c - BROWN Lead
d - BLACK Lead
e - Bolt
f - Capacitor Discharge Module

Fig. 52 CD module location—2.5 to 3.3hp

IGNITION AND ELECTRICAL SYSTEMS

3. Disconnect the bullet connectors at the CD module (switch box). It is a good idea to label the wiring connectors before disassembly. This will aid in assembly. Also make sure to disconnect the ground lead from the secondary ignition coil retaining bolt.
4. Remove the bolt securing the CD module (switch box) to the engine block.

To install:
5. Secure the CD module (switch box) to the engine block with the retaining bolt.
6. Fasten the ground lead to the block with the secondary ignition coil retaining bolt.
7. Reconnect the bullet connectors in the correct order.
8. Reinstall the PORT side lower engine cowling.
9. Reinstall the top cowling.

4 and 5 HP

1. Remove the top cowling.
2. Disconnect the bullet connectors at the CD module (switch box). It is a good idea to label the wiring connectors before disassembly. This will aid in assembly.
3. Disconnect the ground lead from the secondary ignition coil retaining bolt.
4. Slide the CD module (switch box) out of the rubber brackets and remove it.

To install:
5. Slide the CD module (switch box) back into the rubber brackets.
6. Reconnect the bullet connectors in the correct order.
7. Fasten the ground lead to the block with the secondary ignition coil retaining bolt.
8. Reinstall the top cowling.

6–25 HP

♦ See Figure 53

1. Remove the top cowling.
2. Disconnect the bullet connectors at the CD module (switch box). It is a good idea to label the wiring connectors before disassembly. This will aid in assembly.
3. Disconnect the ground lead wire.
4. Remove the retaining screws and lift off the CD module (switch box) and remove it from the engine.

To install:
5. Align the CD module (switch box) onto the engine block and install the retaining screws, tightening them firmly.
6. Reconnect the bullet connectors in the correct order.
7. Fasten the ground lead wire.
8. Reinstall the top cowling.

30 and 40 HP (2-Cylinder)

♦ See Figure 54

1. Disconnect the negative battery cable.
2. Remove the engine cover.
3. Disconnect the high-tension spark plug lead at the spark plug.
4. Unplug the CD module (switch box) wiring harness and unscrew the ground wire (if equipped).
5. Remove the retaining screws and remove the CD module (switch box) from the electrical plate.

To install:
6. Install the CD module (switch box) onto the electrical plate with the retaining screws.
7. Reconnect the CD module (switch box) wiring harness and tighten ground wire (if equipped).
8. Connect the high-tension spark plug lead to the spark plug.
9. Replace the engine cover.
10. Connect the battery negative cable.

30–125 HP

1. Remove the top cowl.
2. Remove the retaining bolts from the electrical box access cover and remove the cover.
3. Disconnect the leads from the switchbox.

➡ Mark each lead as to its proper location. This greatly aids in reassembly later.

4. Remove the bolts from the switchbox and remove the switchbox.

To install:
5. Secure the switchbox to the electrical component box with the retaining bolts and ground leads.
6. Torque the bolts to 40 inch lbs. (4.5 Nm).
7. Reconnect the leads to the switchbox terminals. Make sure all the leads are installed in their proper places.
8. Torque the nuts to 30 inch lbs. (3.4 Nm).
9. Reinstall the electrical box access cover and secure it with the retaining bolts.
10. Install the top cowl.

135 to 275 HP

♦ See Figure 55

The switchboxes are grounded to the engine through the mounting screws, spacers and ground wires (if so equipped).

1. Make sure the ignition switch is off and disconnect the battery cables.
2. Remove the screws and lift the switchboxes off the engine. Make sure not to loose the metal spacers.
3. Remove the rubber cover on each wire terminal and remove the terminal nuts. Disconnect the wire leads and remove the switchboxes.

➡ Mark each lead as to its proper location. This greatly aids in reassembly later.

To install:
4. Reconnect the leads to the correct terminals in the switchbox. The wires with the Yellow identification sleeves must be connected to the outer switchbox. Connect a ground lead (if equipped) to each switchbox.
5. Apply a thread-locking compound to the switchbox mounting screws. Install the switchboxes onto the powerhead using the screws and spacers.

➡ Be certain the switchboxes are properly grounded to the powerhead or damage to the switchboxes will occur when cranking or running the engine.

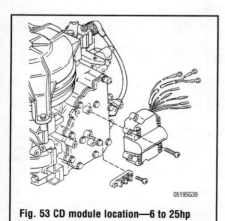

Fig. 53 CD module location—6 to 25hp

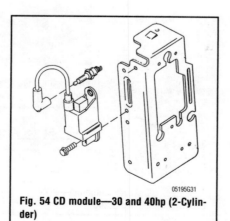

Fig. 54 CD module—30 and 40hp (2-Cylinder)

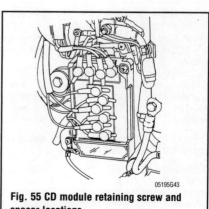

Fig. 55 CD module retaining screw and spacer locations

5-30 IGNITION AND ELECTRICAL SYSTEMS

CHARGING CIRCUIT

Description and Operation

The battery stores electricity and acts as a sponge for the whole system. It mops up generated current until it's fully charged and it releases energy on demand.

The flywheel holds the permanent magnets that create the moving magnetic field. If your engine has good spark, you can take it for granted that the magnets are in working order because the ignition and charging systems share the same magnets.

The stator windings are the stationary coils of wire the flywheel magnets rotate around. They produce the electrical charge. Simply put, the more windings in your stator, the greater the potential output in amps your charging system will have.

The rectifier consists of a series of diodes or one-way electrical valves. It rectifies or corrects the alternating current (AC) produced within the windings to charge the direct current (DC) battery.

the disadvantages of a current-generating system using permanent magnets and stator windings, which is that the current produced within the windings is alternating current (AC). You can't use AC to charge batteries. They accept only direct current (DC). So the rectifier is designed to convert AC current to a usable form of DC current simply called rectified AC.

On midsize and large outboard engines, there may be a voltage regulator, either combined with the rectifier or standing alone. The regulator automatically reduces the output of generated current as the battery becomes fully charged.

The voltage regulator controls the alternators field voltage by grounding one end of the field windings very rapidly. The frequency varies according to current demand the more the field is grounded, the more voltage and current the alternator produces. Voltage is maintained at about 13.5–15 volts. During high engine speeds and low current demands, the regulator will adjust the voltage of the alternator field to lower the alternator output voltage. Conversely, when the engine is idling and the current demands may be high, the regulator will increase field voltage, increasing the output of the alternator.

The rectifier is primarily used for charging the battery circuit on electric start models with the 9-amp stator alternator. The unit is a solid state sealed unit containing a series of diodes. This means that the unit is non-repairable and must be replaced if found to be defective.

Troubleshooting the Charging System

The charging system should be inspected if:
- The charging system warning light is illuminated
- The voltmeter on the instrument panel indicates improper charging (either high or low) voltage
- The battery is overcharged (electrolyte level is low and/or boiling out)
- The battery is undercharged (insufficient power to crank the starter)

The starting point for all charging system problems begins with the inspection of the battery, related wiring and charging system components. The battery must be in good condition and fully charged before system testing.

Do a visual check of the battery, wiring and fuses. Are there any new additions to the wiring? An excellent clue might be, everything was working ok until I added that live well pump. With a comment like this you would know where to check first.

→**Check battery condition thoroughly. It is the #1 culprit in charging system failures.**

The regulator/rectifier assembly is the brains of the charging system. The regulator controls current flow in the charging system. If battery voltage is below about 14.6 volts the regulator sends the available current to the battery. If the battery is fully charged (about 14.5 to 15 volts) the regulator diverts the current/amps to ground.

Do not expect the regulator to send current to a fully charged battery. Check the battery for a possible draw with the key off. This draw may be the cumulative effect of several radio and/or clock memories. If these accessories are wired to the battery then a complaint of charging system failure may really be excessive draw. Draw in excess of 25 milliamps should arouse your suspicions.

→**Do not forget to check through the fuses. It can be embarrassing to overlook a blown fuse.**

You must pull the battery voltage down below 12.5 volts to test charging system output. Running the power trim and tilt will reduce the battery voltage. Once the battery's good condition is verified and it has been reduced to below 12.5 volts you can test further.

Install an ammeter to check actual amperage output. Verify that the system is delivering sufficient amperage. Too much amperage and a battery that goes dry very quickly indicates that the rectifier/regulator should be replaced.

If the system does not put out enough amperage, then test the lighting coil. Isolate the coil and test for correct resistance and short to ground.

During these test procedures the regulator/rectifier has not been bench checked. Usually it is advisable to avoid troubleshooting the regulator/rectifier directly. The procedures listed so far have focused on checking around the rectifier/regulator. If you verify that all other systems stator are good then what is left in the system to cause the verified problem? The process of elimination has declared the rectifier/regulator bad.

Stator (Alternator)

TESTING

Except OptiMax

▶ See Figures 56 and 57

1. The stator may be tested without removing the flywheel, by merely disconnecting the Yellow leads and using a multimeter to check the resistance.
2. Check resistance between the two Yellow stator wires.
3. Resistance should be as specified in the "Stator Alternator Resistance Specification" chart.
4. If the stator is installed on the powerhead, check continuity between each Yellow stator wire and ground.
5. If the stator is not installed on the powerhead, check continuity between each Yellow stator wire and the Black ground wire.

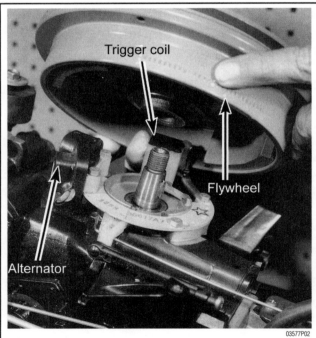

Fig. 56 Once the flywheel is removed, major ignition components are exposed for service

Stator Alternator Resistance Specifications

Model	Stator	Resistance
30	9 Amp Manual Start	.17 - .19
30	14 Amp	.22 - .24
30	18 Amp	.16 - .18
40-60	9-16 Amp	.16 - .19
50-125	9 Amp	.16 - 1.1
50-125	16-18 Amp	.17 - 1.9
V6	16 Amp Regulator/Rectifier	.25 - .45
V6	16 Amp Regulator	0.05 - 0.19

IGNITION AND ELECTRICAL SYSTEMS

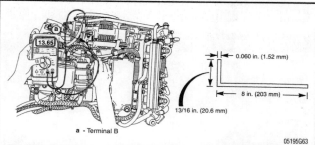

Fig. 58 If the voltage reading is less than normal, this tool will need to be fabricated

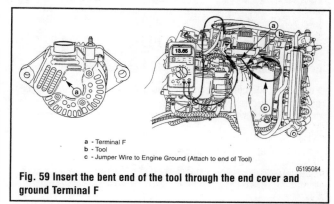

a - Terminal F
b - Tool
c - Jumper Wire to Engine Ground (Attach to end of Tool)

Fig. 59 Insert the bent end of the tool through the end cover and ground Terminal F

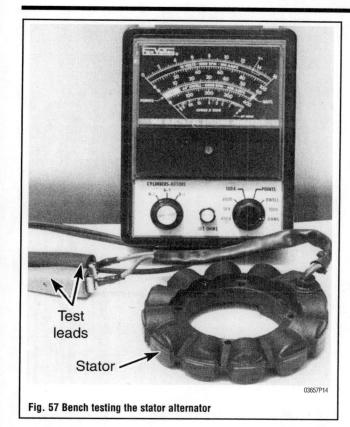

Fig. 57 Bench testing the stator alternator

6. If resistance is not within specification, or continuity exists, the stator may be faulty.

OptiMax

OUTPUT CIRCUIT

1. Connect a multimeter between alternator Terminal B (output Terminal) and case ground on the alternator.
2. Shake the alternator wiring harness.
3. The meter should indicate battery voltage and should not vary.
4. If battery voltage is not obtained or the voltage reading varies while shaking the wires, check for loose or dirty connections or damaged wiring.

SENSING CIRCUIT

1. Unplug the Red and Purple leads from the alternator.
2. Connect the voltmeter between the Red lead and ground.
3. The voltmeter should indicate battery voltage.
4. If the correct voltage is not present, check the sensing circuit (Red lead) for loose or dirty connections or some damaged wiring.

VOLTAGE OUTPUT

♦ See Figures 58 and 59

1. Using a 0–20 volt DC voltmeter, connect the positive (+) lead of the voltmeter to Terminal B of the alternator and the negative (-) meter lead to ground.
2. Start the engine allow it to reach operating temperature. Increase the engine rpm from idle to 2000. Normal voltage output should be 13.5–15.1 volts. If the voltage reading is greater than normal, the voltage regulator is faulty and will need to be replaced.
3. If the reading is less than normal, you will need to fabricate a special tool using a piece of stiff wire.
4. With Terminal F grounded, voltage should rise to within normal range (13.5–15.1 volts). If the voltage rises, the alternator will need to be replaced.
5. If the voltage does not rise to within normal specifications with Terminal F grounded, you will need to perform the "Current Output" test.

CURRENT OUTPUT

1. With the engine shut off, install an ammeter (one that is capable of measuring 60-plus amps) in series between Terminal B on the alternator and the positive terminal of the battery.

2. Start the engine and allow it to reach operating temperature.
3. Advance the engine rpm to 2000 rpm.
4. Insert the fabricated special tool through the end cover and ground Terminal F.
5. Normal output is 52–60 amps @2000 rpm at the alternator.
6. If alternator output is normal, replace the regulator.
7. If the output is low, a disassembly of the alternator is necessary to inspect and test individual alternator components.

REMOVAL & INSTALLATION

➙The stator is located under the flywheel. To gain access to the stator the flywheel must be removed.

6–25 HP

♦ See Figure 60

1. Disconnect the battery leads from the battery terminals. Remove the front cowling cover and the wrap-around cowling, if one is used. Remove the top cowling. Remove the nut on the crankshaft in the center of the flywheel. A flywheel holder will be required to prevent the flywheel from rotating while the nut is loosened.
2. Obtain the proper flywheel puller to "pull" the flywheel. Never use a puller which pulls on the outside edge of the flywheel or the flywheel may be damaged. After the puller is installed, tighten the center screw onto the end of the crankshaft. Continue tightening the screw until the flywheel is released from the crankshaft. Remove the flywheel puller. Lift the flywheel free of the crankshaft. The Woodruff key in the flywheel may remain in place but take care the flywheel is not jarred and the key falls free and is lost.
3. Disconnect the Yellow and Gray wires leading from the alternator to the terminal block on the rectifier. Remove the retaining bolts from the alternator and lift the alternator from the powerhead.

To install:

4. Place the alternator in position on the powerhead. Coat the threads of the alternator attaching bolts with blue Loctite® or equivalent. Secure the alternator with the attaching bolts through the alternator into the powerhead.
5. Insert the Woodruff key into the flywheel recess with just a bit of grease to hold it in place. Check to be sure the inside taper of the flywheel and the taper on the crankshaft are clean of dirt or oil, to prevent the flywheel from attempting to "walk" on the crankshaft during operation. Slide the flywheel down over the crankshaft with the keyway in the flywheel aligned with the Woodruff key in the crankshaft. Rotate the flywheel clockwise and check to be sure the flywheel does not contact any powerhead part or any of the wiring. Thread the flywheel nut onto the end of the crankshaft and then tighten it to specification.

5-32 IGNITION AND ELECTRICAL SYSTEMS

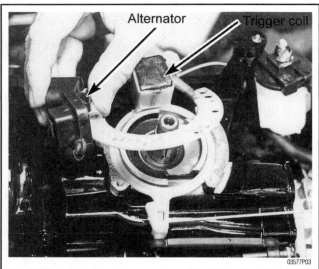

Fig. 60 The alternator can be removed without disturbing the trigger coil

30 to 125 HP

▶ See accompanying illustrations

1. Disconnect the battery leads from the battery terminals. Unlatch and swing open or lift off the powerhead cowling, lift up and remove the cowling halves from the powerhead.
2. Remove the spark plug from all cylinders to prevent the powerhead from starting.
3. Remove the three bolts securing the flywheel cover or hand rewind starter—if so equipped. Remove the nut on the crankshaft in the center of the flywheel. A flywheel holding tool (91–52344) will be required to prevent the flywheel from turning in order to loosen the nut.
4. Install crankshaft protector cap (9124161) over the end of the crankshaft. Thread flywheel puller (91–73687AI) as far into the flywheel as possible. Never use a puller which pulls on the outside edge of the flywheel or the flywheel may be damaged. Hold the outer portion of the flywheel puller tool stationary and turn the center bolt until the flywheel is free of the crankshaft. Lift the flywheel from the crankshaft. Lift the flywheel from the crankshaft and remove the crankshaft protector tool.
5. Disconnect the wire leads from the stator to the switch box, voltage regulator/rectifier and starter solenoid. Remove the cap screws securing the stator to the powerhead and lift the stator from the powerhead.

To install:

6. Place the stator into position on the powerhead. Coat the threads of the cap screws securing the stator with Loctite® Grade "A" or equivalent. Install the cap screws through the stator and into the powerhead. Tighten the screws to 35 inch lbs. (3.9 Nm) on the 40hp model and 60 inch lbs. (6.8 Nm) for all other models Connect the wire leads from the stator to the switch box, voltage regulator/rectifier and starter solenoid using the Wiring Diagram for the model powerhead being serviced.
7. Insert the Woodruff key into the crankshaft keyway. Check to be sure the inside taper of the flywheel and the taper on the crankshaft are absolutely clean of dirt

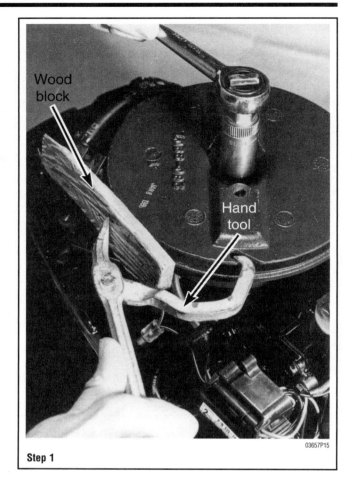

Step 1

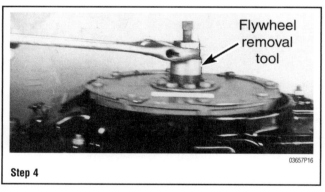

Step 4

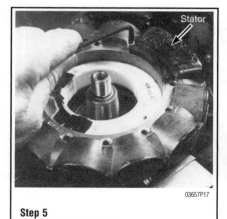

Step 5

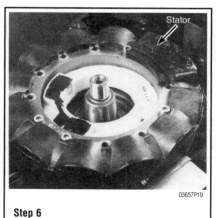

Step 6

Step 7

IGNITION AND ELECTRICAL SYSTEMS

or oil, to prevent the flywheel from "walking" on the crankshaft during operation. Slide the flywheel down over the crankshaft with the keyway in the flywheel aligned with the key on the crankshaft. Rotate the flywheel clockwise a couple turns and check to be sure the flywheel does not contact any powerhead components or wiring. Thread the flywheel nut onto the crankshaft, using holding tool (91–52344), tighten the flywheel nut to 120 ft. lb. (162.7 Nm).

8. Install the spark plugs back into the cylinders. Install the flywheel cover on electric start models or hand rewind starter for manual start models.

9. Before considering the job completed, perform the Stator Alternator Output Test for the amperage size stator once again to verify stator operation.

10. Install the cowling halves by lowering them onto the hinge pins. Swing the halves together and latch closed. Connect the battery leads to the battery terminals.

135–275 HP

♦ See accompanying illustrations

1. Disconnect the battery leads from the battery terminals. Unlatch and lift off the powerhead cowling.
2. Remove the spark plugs from all cylinders to prevent the powerhead from accidentally starting.
3. Remove the three wing nuts securing the flywheel cover and lift the cover off the powerhead.
4. Remove the nut on the crankshaft in the center of the flywheel. A flywheel holding tool, (91–24161) or equivalent, will be required to prevent the flywheel from turning in order to loosen the nut.
5. Both the flywheel removal/installation tool and the tool to protect the threads on the upper end of the crankshaft will be needed.

6. Install crankshaft protector cap (91–24161) or equivalent, over the end of the crankshaft. This cap is used to protect the threads on the end of the crankshaft. Thread flywheel puller (91–73687A1) as far into the flywheel as possible.

※※ CAUTION

Never use a puller which pulls on the outside edge of the flywheel or the flywheel may be damaged.

7. Hold the outer portion of the flywheel puller tool stationary and turn the center bolt until the flywheel is free of the crankshaft and then remove the crankshaft protector tool.
8. Disconnect the wire leads from the stator to the switch box, voltage regulator/rectifier and starter solenoid. Remove the cap screws securing the stator to the powerhead and lift the stator from the powerhead.

To install:

9. Place the stator in position on the powerhead. Coat the threads of the cap screws securing the stator with threadlocking compound.
10. Install the cap screws through the stator and into the powerhead.
11. Tighten the 135 to 200hp screws to 50 inch lbs. (5.6 Nm) and the 275hp screws to 150 inch lbs. (16.9 Nm).
12. Route the stator wire harness down the powerhead exactly as removed.
13. Connect the wire leads from the stator to the voltage regulator/rectifier and switch boxes using the Wiring Diagrams for the model being serviced.
14. Connect all the ground leads to a clean surface. Be sure the switch box leads with Yellow identification sleeves are connected to the outer switch box.
15. Insert the Woodruff key into the crankshaft keyway.

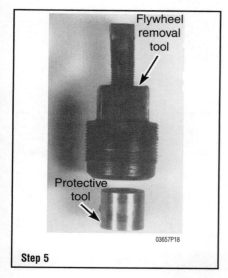

Step 5

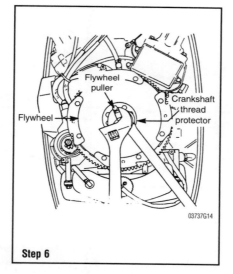

Step 6

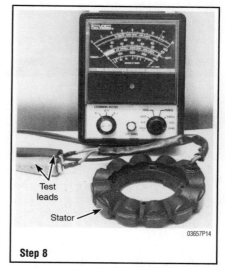

Step 8

Step 9

Step 15

5-34 IGNITION AND ELECTRICAL SYSTEMS

➡Check to be sure the inside taper of the flywheel and the taper on the crankshaft are clean, especially of oil, to prevent the flywheel from "walking" on the crankshaft during operation.

16. Slide the flywheel down over the crankshaft with the keyway in the flywheel aligned with the key on the crankshaft.
17. Rotate the flywheel clockwise a couple turns and check to be sure the flywheel does not contact powerhead components or wiring.
18. Thread the flywheel nut onto the crankshaft, using holding tool (91-52344).
19. Tighten the 135 to 200hp flywheel nut to 120 ft. lbs. (162.7 Nm) and the 275hp nut to 100 ft. lbs. (135.6 Nm)
20. Install the spark plugs and connect the high tension leads to the plugs. Install the flywheel cover over the flywheel and secure it with three wing nuts.
21. Before considering the job completed, perform the stator output test for the amperage size stator once again to verify stator operation.
22. Install the cowling over the powerhead. Connect the battery leads to the battery terminals.

OptiMax

▶ See Figure 61

1. Remove the top cowling.
2. Disconnect the battery cables.
3. Disconnect the wiring harness from the alternator.
4. Remove the pivot bolt and the tension bolt.
5. Remove the alternator from the powerhead.

To install:

6. Secure the alternator to the powerhead with the attaching bolts. Torque the top bolt to 40 ft. lbs. (54 Nm) and the bottom bolt to 35 ft. lbs. (47.5 Nm).
7. Install the alternator belt into the V-groove of the flywheel and alternator pulley.
8. Reconnect the electrical harness to the alternator.

Rectifier

TESTING

▶ See Figure 62

The rectifier can be tested without removal from the powerhead. Disconnect the battery cables from the battery before proceeding with this test.

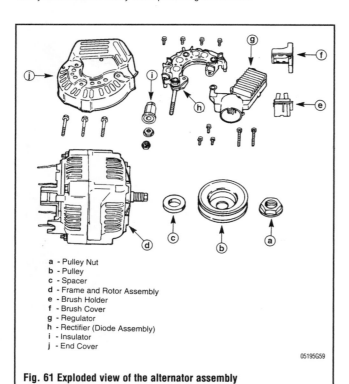

Fig. 61 Exploded view of the alternator assembly

a - Pulley Nut
b - Pulley
c - Spacer
d - Frame and Rotor Assembly
e - Brush Holder
f - Brush Cover
g - Regulator
h - Rectifier (Diode Assembly)
i - Insulator
j - End Cover

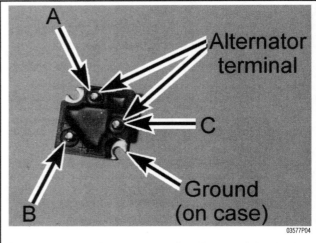

Fig. 62 View of a typical voltage rectifier with terminal identification

➡Depending on the internal polarity of the individual multimeter used during the following test, the results may be the exact opposite of what is described.

1. Disconnect the leads from the rectifier terminals.
2. Connect the positive (+) meter lead to the rectifier ground and the negative (-) meter lead alternating between terminals A and C. No continuity should be indicated.
3. Connect the negative (-) meter lead to terminal B and the positive (+) meter lead alternating between terminals A and C. The meter should indicate continuity.
4. Connect the positive (+) meter lead to terminal B and the negative (-) meter lead alternating between terminals A and C. The meter should indicate no continuity.
5. If the above readings are not shown, replace the rectifier.

REMOVAL & INSTALLATION

1. Disconnect the negative battery cable.
2. Disconnect the rectifier or voltage regulator/rectifier connectors.
3. Remove the fastener attaching the unit to the engine or electrical component and remove the unit.
4. If so equipped, remove the ground lead and disconnect it from the powerhead.

To install:

5. Install the rectifier or voltage regulator/rectifier onto the powerhead or electrical component.
6. Connect all leads and make sure all connections are tight and free of corrosion.
7. Connect the negative battery cable.

Regulator/Rectifier

TESTING

The alternator output and the voltage regulator/rectifier are tested simultaneously. place the powerhead in a test tank or the boat in a body of water.

✱✱ WARNING

Do not operate the powerhead without water for cooling or advance the throttle above idle setting with a flush device attached.

9 Amp Alternator with Rectifier

1. Disconnect the red wire from the rectifier positive (+) terminal going to the starter solenoid.
2. Connect the meter positive (+) lead to the rectifier positive (+) terminal and the meter negative (-) meter lead to the red wire removed from the rectifier terminal.
3. Secure any loose wires on the powerhead so that they are clear of the moving flywheel. Place the powerhead in a test tank or the vessel in a body of water. Do not operate the powerhead without water for cooling or advance the throttle above idle while the flushing device is attached.
4. Start the powerhead and slowly advance the throttle until it reaches 3000 rpm. The ammeter should indicate between 7–9 amps. If the ammeter dies not indicate the correct amps, replace the stator assembly.

IGNITION AND ELECTRICAL SYSTEMS

9 Amp Alternator with Small Regulator/Rectifier

1. Connect the multimeter positive (+) lead to the positive battery terminal and the negative (-) meter lead to the negative battery terminal.
2. Crank the powerhead with the starter motor while monitoring the voltmeter. If the battery voltage below 9.5 volts while cranking the engine, the battery is weak and should be recharged.
3. If the cranking voltage is acceptable, start the powerhead and let it idle. Battery voltage at idle should be 12.5 to 14.5 volts. If the voltage is over 14.5 volts, replace the voltage regulator/rectifier.
4. If the voltage is acceptable, shut the powerhead down and disconnect the battery. Disconnect the two red wires from the voltage regulator at the starter solenoid positive (+) terminal.
5. Connect the small diameter red (sensing lead) from the voltage regulator to the positive (+) terminal of a 9-volt transistor battery. Connect a jumper wire from the negative (-) terminal of the 9-volt battery to a good engine ground.
6. Connect the positive (+) multimeter lead to the large red wire from the voltage regulator. Connect the negative (-) meter lead to the to the positive (+) terminal on the starter motor solenoid. Make sure all loose wires are clear of the turning flywheel.
7. Connect the battery and start the powerhead. Slowly advance the throttle to 3000 rpm. The meter should indicate between 7 and 9 amps. If the meter indicates 7–9 amps, the charging system is functioning properly but the battery is being discharged, because the amp draw on the battery is greater than the amperage output of the alternator.
8. If the meter indicates 7 amps or less, perform a resistance test on the stator, alternator and coil. If the stator tests good, replace the voltage regulator/rectifier.

16 and 18 Amp Alternator with Small Regulator/Rectifier

1. Connect the multimeter positive (+) lead to the battery positive (+) terminal. Connect the meter negative (-) meter lead to the negative (-) battery terminal.
2. Crank the powerhead with the starter motor while monitoring the meter. If the battery voltage drops below 9.5 volts while cranking, the battery is weak and should be recharged.
3. If the cranking voltage is acceptable, start the powerhead and let it idle. Battery voltage at idle speed should be 12.5 to 14.5 volts. If the voltage is over 14.5 volts, replace the voltage regulator/rectifier.
4. If the idle voltage is acceptable, shut off the engine and disconnect the battery. Disconnect the two wires from the voltage regulator at the starter motor solenoid positive (+) terminal.
5. Connect the small diameter red (sensing lead) from the voltage regulator to the positive (+) terminal of a 9-volt transistor battery. Connect a jumper wire from the negative (-) terminal of the transistor battery to a good ground on the powerhead.
6. Connect the positive (+) lead of an ammeter to the large red wire from the voltage regulator. Connect the negative (-) meter lead of the ammeter to the positive (+) terminal on the starter motor solenoid. Secure all loose wires clear of the flywheel.
7. Connect the battery and start the powerhead. Slowly advance the throttle to the rpm indicated below and verify the alternator amps output.

A reading of 18 amps at 3000 RPM indicates the charging system is functioning properly but the battery is being discharged, because the amperage draw on the battery is greater than the amperage output of the alternator.

If the ammeter indicates less than 18 amps, perform a resistance test on the stator, alternator and coil. If the stator test is satisfactory, replace the voltage regulator/rectifier.

16 and 24 Amp Alternator With Large Finned Regulator/Rectifier

♦ See Figures 63 and 64

1. Connect the voltmeter positive (+) lead to the positive terminal on the battery. Connect the negative (-) meter lead to the negative terminal on the battery.
2. Crank the powerhead with the starter motor while monitoring the voltmeter. If the battery voltage drops below 9.5 volts while cranking, the battery is weak and should be recharged.
3. If the cranking voltage is acceptable, start the powerhead and let it idle. Battery voltage at idle speed should be 12.5 to 14.5 volts. If the voltage is over 14.5 volts, replace the voltage regulator/rectifier.
4. If the voltage at idle speed is acceptable, shut the powerhead down and disconnect the battery. Disconnect the red wire from the center terminal on the voltage regulator/rectifier. Secure the remaining lead on the center terminal with the hex nut previously removed.
5. Connect the positive (+) lead of the ammeter to the center terminal of the voltage regulator/rectifier. Connect the negative (-) meter lead of the ammeter to the red wire previously removed from the voltage regulator. Secure all loose wires clear of the flywheel.
6. Connect the battery and start the powerhead. Slowly advance the throttle to the rpm indicated below and verify the alternator amps output.
7. If the ammeter indicates 18 amps for the 16 amp alternator output test or 20

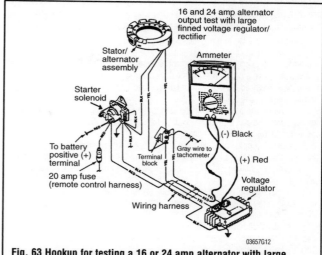

Fig. 63 Hookup for testing a 16 or 24 amp alternator with large finned voltage regulator/rectifier

SMALL VOLTAGE REGULATOR TEST (STATIC)

TEST LEADS	OHMS SCALE	METER READING (Ohms)
DIODE CHECK: Connect NEGATIVE (-) test lead to either YELLOW lead. Connect POSITIVE (+) test lead to thick RED lead.	R x 10	100 - 400
DIODE CHECK: Connect NEGATIVE (-) test lead to thick RED lead. Connect POSITIVE (+) test lead to either YELLOW lead. Note 1	R x 1K	40K to Infinity
SCR CHECKS: Connect NEGATIVE (-) test lead to either YELLOW lead. Connect POSITIVE (+) test lead to case ground.	R x 1K	10K to Infinity
TACHOMETER CIRCUIT CHECK: Connect NEGATIVE (-) test lead to case ground. Connect POSITIVE (+) test lead to GRAY lead.	R x 1K	10K to 30K

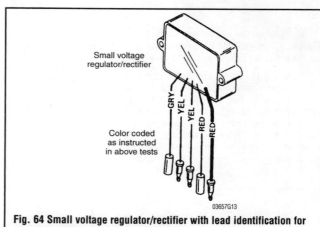

Fig. 64 Small voltage regulator/rectifier with lead identification for testing as outlined in the above listing

amps and greater for the 24 amp alternator test, the charging system is functioning properly but the battery is being discharged, because the amperage draw on the battery is greater than the amperage output of the alternator.

8. If the ammeter indicates less than the specified output above, perform an ohms test on the stator, alternator and coil. If the stator test is satisfactory, replace the voltage regulator/rectifier.

40 Amp Alternator with Large Regulator

♦ See Figure 65

Prior to performing this test, check all charging circuit wiring from the powerhead to the battery. Look for loose terminals, loose bullet connectors and loose or cor-

5-36 IGNITION AND ELECTRICAL SYSTEMS

roded hardware securing the voltage regulator to the powerhead. Check the battery cables for broken insulation and corrosion. Verify the battery posts are clean and the cables are in good serviceable condition. Batteries should have a full charge and must not be connected to any other charging device.

1. Place the outboard in a test tank or move the boat to a body of water. If this is not possible, connect a flush attachment and garden hose to the lower unit. If a flush attachment is used, do not operate the powerhead at a high rpm. Such action could cause the powerhead to runaway from the no load on the propeller, causing severe and expensive damage.
2. Connect the positive (+) meter lead to the positive (+) battery terminal. And the negative (+) meter lead to the negative battery terminal. Observe the meter reading and note this as "battery voltage."
3. Crank the powerhead with the starter motor while monitoring the meter. If the battery voltage drops below 9.5 volts while cranking the powerhead, the battery is weak and should be checked for a defective cell(s) and fully recharged or replaced.
4. If the cranking voltage is acceptable, start and operate the powerhead at 1,000 rpm. Observe the meter reading. Battery voltage should rise to approximately 14.5 volts and then stabilize.
5. If the voltage does not increase above the previously indicated battery voltage, shut down the powerhead and continue the troubleshooting procedures at "Charging Circuit Has No Output".
6. If the battery voltage exceeds 16 volts and/or does not return to approximately 14.5 volts, the regulator may be faulty.

CHARGING CIRCUIT HAS NO OUTPUT

▶ See Figure 66

1. Connect the positive (+) meter lead to the purple wire lead (bullet connector) at the voltage regulator. Do not disconnect the bullet connector. Connect the negative (-) meter lead to a good powerhead ground.
2. Place the key switch in run position. The meter should indicate battery voltage. If battery voltage is not present, troubleshoot for broken or damaged wiring from the key switch to the main powerhead harness.
3. Set the meter switches for AC voltage readings. Connect the positive (+) meter lead to one of the voltage regulator Yellow leads on the terminal block and the (-) negative meter lead to the other voltage regulator Yellow lead on the terminal block. Tape and secure any loose wires clear of the flywheel.
4. Start and operate the powerhead at approximately 1,000 rpm. If the voltage is greater than 16 VAC, the regulator is defective. Perform the Voltage Regulator Static Ohms Test to verify the regulator is defective and must be replaced. If the AC voltage is zero or some value lower than battery voltage, perform the "Stator Alternator Coil Test".

➡The signal for the tachometer originates in the voltage regulator. Therefore, it is possible to have an accurate signal to the tachometer but the regulator itself may still be defective.

5. The voltage regulator is mounted directly to the powerhead casting under the trim and tilt relay panel. If the powerhead is not receiving sufficient cooling water due to a damaged water pump or blocked passage, a thermal overload protection device which is heat sensitive and located inside the regulator is designed to stop current flow to the battery, thereby protecting the voltage regulator from overheating. Therefore, it is possible to have a normal charging circuit when the powerhead is cold but a defective charging circuit when the powerhead heats to or beyond normal operating temperature.

REGULATOR VOLTAGE CHECK

1. Place the outboard in a test tank or move the boat to a body of water. If this is not possible, connect a flush attachment and garden hose to the lower unit. If a flush attachment is used, do not operate the powerhead at a high rpm. Such action could cause the powerhead to runaway from the no load on the propeller, causing severe and expensive damage.
2. Connect the positive (+) meter lead to the positive terminal on the battery. Connect the negative (-) meter lead to the negative battery terminal.
3. Turn all the electrical accessories on the boat to on. Crank the powerhead with the starter motor for approximately 20 seconds with the lanyard switch turned off or the emergency stop switch set on to prevent the powerhead from starting. This cranking will drain some of the charge from the battery.
4. Observe the multimeter reading and note this as battery voltage. Start and operate the powerhead at 1,000 rpm. Observe the multimeter reading. Battery voltage should rise to approximately 14.5 volts and then stabilize.

If the voltage did not increase to about 14.5 volts and stabilize, shut down the powerhead and continue the troubleshooting procedures.

If the battery voltage increases to 16 volts and/or does not return to approximately 14.5 volts, shut down the powerhead and continue the troubleshooting procedures at Charging Circuit Has Constant High Output.

40 Amp Alternator With Dual Voltage Regulators

▶ See Figure 67

Prior to performing this test, check all charging circuit wiring from the powerhead to the battery. Look for loose terminals, loose bullet connectors and loose or corroded hardware securing the voltage regulator to the powerhead. Check the battery cables for broken insulation and corrosion. Check to be sure the battery posts are clean and the cables are in good serviceable condition. Batteries should have a full charge and must not be connected to any other charging device.

1. Place the outboard in a test tank or move the boat to a body of water. If this is not possible, connect a flush attachment and garden hose to the lower unit. If a flush attachment is used, do not operate the powerhead at a high rpm. Such action could cause the powerhead to runaway from the no load on the propeller, causing severe and expensive damage.

※※ CAUTION

Water must circulate through the lower unit to the powerhead any time the powerhead is run to prevent damage to the water pump in the lower unit. Just five seconds without water will damage the water pump.

2. Connect the positive (+) meter lead to the positive battery terminal. Connect the negative (-) meter lead to the negative battery terminal. Observe the meter reading and note this as battery voltage.
3. Crank the powerhead with the starter motor and at the same time monitor the meter. If the battery voltage drops below 9.5 volts while cranking the powerhead, the battery is weak and should be fully charged, checked for defective cell(s) or replaced.
4. If the cranking voltage is acceptable, start and operate the powerhead at 1,000 rpm. Observe the meter reading. Battery voltage should rise to approximately 14.5 volts and then stabilize.

If the voltage did not increase above indicated battery voltage, shut down the powerhead and continue the troubleshooting procedures at Charging Circuit Has No Output-Dual Regulators.

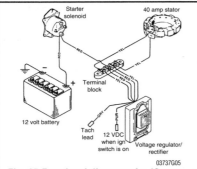

Fig. 65 Functional diagram of a 40-amp charging system. Notice, this system has one large regulator/rectifier

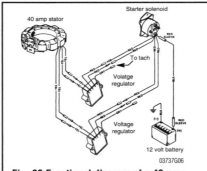

Fig. 66 Functional diagram of a 40-amp charging system with two voltage regulators

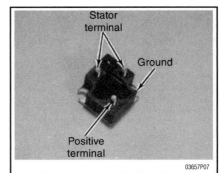

Fig. 67 Up-to-date rectifier installed on the powerhead, with the terminals identified for test purposes

IGNITION AND ELECTRICAL SYSTEMS

If the battery voltage increases to 16 volts and/or does not return to approximately 14.5 volts, continue the troubleshooting procedures at Charging Circuit Has Constant High Output-Dual Regulators.

CHARGING CIRCUIT HAS NO OUTPUT (DUAL REGULATORS)

1. Connect the positive (+) meter lead to either red wires (bullet connector) to the voltage regulators. Do not disconnect the bullet connector. Connect the negative (-) meter lead to a good powerhead ground.
2. Place the key switch in the run position. The meter should indicate battery voltage. If battery voltage is not present, troubleshoot for broken or damaged wiring from the key switch to the main powerhead harness.
3. Set the multimeter on the AC voltage scale. Connect the meter leads to either two of the short Yellow wire leads or the two long Yellow wire leads from the stator to the voltage regulator. Tape and secure any loose wires clear of the flywheel.
4. Start and operate the powerhead at approximately 1,000 rpm. If the voltage is greater than 16 VAC, the regulator is defective and must be replaced. If the AC voltage is zero or lower than battery voltage, perform the "Stator Alternator Coil Test".

CHARGING CIRCUIT HAS CONSTANT HIGH OUTPUT (DUAL VOLTAGE REGULATORS)

1. Remove the flywheel from the powerhead. Visually inspect the stator alternator coils for signs of discoloration, burned or broken wires. If there is any obvious damage to the stator coils, the unit MUST be replaced. If there is no damage to the stator coils or wiring, install the flywheel.
2. Set the multimeter for amps reading. The meter should be capable of carrying up to 40 amps. Disconnect the red wire lead from the terminal block above the voltage regulator. Connect the positive (+) multimeter lead to the terminal block where the red wire was disconnected. Connect the negative (-) multimeter lead to the loose red wire to the starter solenoid.
3. Secure any loose wires on the powerhead clear of the flywheel. Place the outboard in a test tank or move the boat to a body of water. If this is not possible, connect a flush attachment and garden hose to the lower unit. If a flush attachment is used, do not operate the powerhead at a high rpm. Such action could cause the powerhead to runaway from the no load on the propeller, causing severe and expensive damage.

☠ CAUTION

Water must circulate through the lower unit to the powerhead any time the powerhead is run to prevent damage to the water pump in the lower unit. Just five seconds without water will damage the water pump.

4. Disconnect one short and two long Yellow stator leads from the bullet connectors. Tape and secure any loose wires clear of the flywheel. Start and operate the powerhead between 1000 and 2000 rpm while observing the meter for current output. If there is no current output observed, shutdown the powerhead, reconnect the one short and two long Yellow stator leads, previously disconnected. Now, disconnect the two short and the opposite one long Yellow stator leads. Tape and secure any loose wires clear of the flywheel. Start and operate the powerhead between 1000 and 2000 rpm. Observe the meter for current output.
5. If output current was observed during this test, the stator is shorted to ground and must be replaced. If there was no current output indicated during this test with either short or long Yellow leads to the stator disconnected, the voltage regulators are defective. Perform the Voltage Regulator Ohms Test to verify the regulator is defective and must be replaced.
6. The alternator used on the 225hp is a 60-amp alternator with an internal voltage regulator and rectifier. If troubleshooting indicates the regulator is defective, the alternator MUST be removed from the powerhead and then disassembled to gain access to the regulator. If additional damage or defective parts internal to the alternator are suspected, the authors recommend the alternator be taken to a local marine dealer or electrical shop specializing in marine electrical systems for further evaluation and testing. The alternator and its replacement parts are manufactured to comply with U.S. Coast Guard Rules and Regulations to minimize the risk of fire and/or explosion. Always use only authorized and approved replacement parts.

RECTIFIER TESTING

The rectifier is primarily used for charging the battery circuit on electric start models with the 9-Ampere stator alternator. The unit is a solid-state sealed unit containing a series of diodes. This means the unit is non-repairable and must be replaced if found defective.

If the rectifier is suspected of malfunctioning, perform the rectifier troubleshooting test. Prior to testing, disconnect the battery leads from the battery. The rectifier may also be tested while installed on the powerhead provided all wires are disconnected from the rectifier terminals.

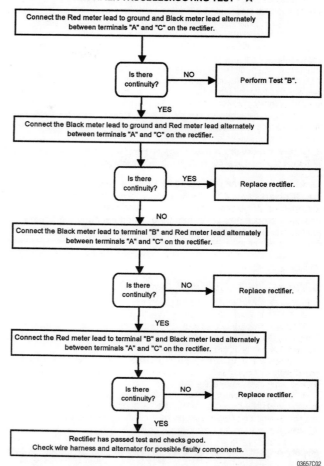

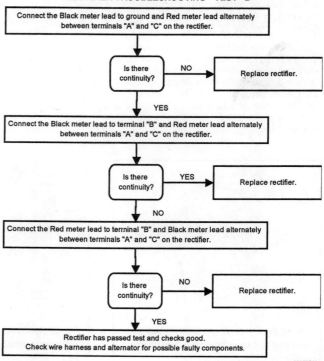

5-38 IGNITION AND ELECTRICAL SYSTEMS

Obtain a multimeter and set the meter switches for "Continuity Testing". Perform the steps as listed in the "Rectifier Troubleshooting Test".

REMOVAL & INSTALLATION

◆ See Figures 68 and 69

In most cases the rectifier is easily removed, because it is housed under the electrical access cover on the starboard side of the powerhead. Before removing the rectifier, be sure to disconnect the battery leads at the battery first. Remove the Yellow stator leads, red positive lead and Gray tachometer lead if equipped. Remove the two mounting screws securing the rectifier to the powerhead and lift off the rectifier.

✱✱ CAUTION

Never attempt to check the polarity of the battery leads by sparking the terminals against the battery posts. Such action will burn out the rectifier. A burned-out rectifier in most cases is caused by improper procedures at the battery or when handling the battery leads as outlined in the following short list.

1. The battery leads of the electrical control harness are connected to the wrong terminals at the battery.
2. The battery leads were disconnected from the battery terminals while the powerhead was running.
3. An open circuit resulting from a broken wire, loose connection, corroded switch contact or a loose harness connector at the powerhead.

To install:

4. Place the rectifier in position on the powerhead and secure it with two screws. Tighten the screws to 30 inch lbs. (3.3Nm).
5. Connect the two Yellow leads from the stator to the alternator terminals on the rectifier. If the powerhead is equipped with a tachometer—connect the Gray wire from the tachometer to one of the Yellow wire terminals on the rectifier.
6. Connect the positive lead from the starter solenoid to the positive (+) terminal on the rectifier. After all connections are completed, coat all terminals with Mercury liquid neoprene (92-25711) or equivalent, to prevent corrosion on the terminals.

Battery

The battery is one of the most important parts of the electrical system. In addition to providing electrical power to start the engine, it also provides power for operation of the running lights, radio and electrical accessories.

Because of its job and the consequences (failure to perform in an emergency), the best advice is to purchase a well-known brand, with an extended warranty period, from a reputable dealer.

The usual warranty covers a pro-rated replacement policy, which means the purchaser is entitled to consideration for the time left on the warranty period if the battery should prove defective before the end of the warranty.

Many manufacturers have specifications on the size and type of battery to use for their engines. If in doubt as to how large the boat requires, make a liberal estimate and then purchase the one with the next higher amp rating. Outboards equipped with an onboard computer, should be equipped with a battery of at least 100 to 105 amp/hour capacity.

BATTERY CONSTRUCTION

◆ See Figure 70

A battery consists of a number of positive and negative plates immersed in a solution of diluted sulfuric acid. The plates contain dissimilar active materials and are kept apart by separators. The plates are grouped into elements. Plate straps on top of each element connect all of the positive plates and all of the negative plates into groups.

The battery is divided into cells holding a number of the elements apart from the others. The entire arrangement is contained within a hard plastic case. The top is a one-piece cover and contains the filler caps for each cell. The terminal posts protrude through the top where the battery connections for the boat are made. Each of the cells is connected to its neighbor in a positive-to-negative manner with a heavy strap called the cell connector.

MARINE BATTERIES

◆ See Figure 71

Because marine batteries are required to perform under much more rigorous conditions than automotive batteries, they are constructed differently than those used in automobiles or trucks. Therefore, a marine battery should always be the No. 1 unit for the boat and other types of batteries used only in an emergency.

Marine batteries have a much heavier exterior case to withstand the violent pounding and shocks imposed on it as the boat moves through rough water and in extremely tight turns. The plates are thicker and each plate is securely anchored within the battery case to ensure extended life. The caps are spill proof to prevent acid from spilling into the bilge when the boat heels to one side in a tight turn or is moving through rough water. Because of these features, the marine battery will recover from a low charge condition and give satisfactory service over a much longer period of time than any type intended for automotive use.

✱✱ WARNING

Never use a Maintenance-free battery with an outboard engine that is not voltage regulated. The charging system will continue to charge as long as the engine is running and it is possible that the electrolyte could boil out if periodic checks of the cell electrolyte level are not done.

BATTERY RATINGS

◆ See Figure 72

Three different methods are used to measure and indicate battery electrical capacity:

- Amp/hour rating
- Cold cranking performance
- Reserve capacity

The amp/hour rating of a battery refers to the battery's ability to provide a set amount of amps for a given amount of time under test conditions at a constant temperature. Therefore, if the battery is capable of supplying 4 amps of current for 20 consecutive hours, the battery is rated as an 80 amp/hour battery. The amp/hour rating is useful for some service operations, such as slow charging or battery testing.

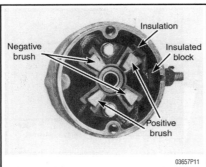

Fig. 68 View of a typical voltage rectifier with terminal identification, as mentioned in the text

Fig. 69 Removing the cover exposing much of the powerhead electrical connections and showing typical location of the rectifier

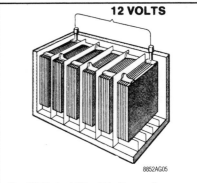

Fig. 70 Typical 12 volt battery cell arrangement

IGNITION AND ELECTRICAL SYSTEMS 5-39

Fig. 71 Cut-away look at a typical marine battery

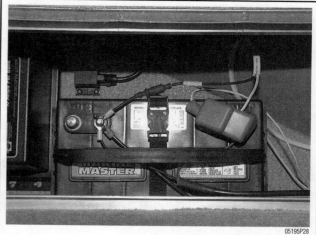

Fig. 73 A good example of a secured, well protected and ventilated battery area

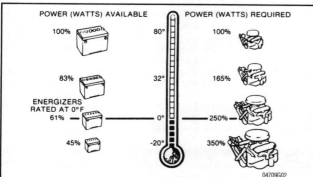

Fig. 72 Comparison of battery efficiency and engine demands at various temperatures

Cold cranking performance is measured by cooling a fully charged battery to 0°F (-17°C) and then testing it for 30 seconds to determine the maximum current flow. In this manner the cold cranking amp rating is the number of amps available to be drawn from the battery before the voltage drops below 7.2 volts.

The illustration depicts the amount of power in watts available from a battery at different temperatures and the amount of power in watts required of the engine at the same temperature. It becomes quite obvious—the colder the climate, the more necessary for the battery to be fully charged.

Reserve capacity of a battery is considered the length of time, in minutes, at 80°F (27°C), a 25 amp current can be maintained before the voltage drops below 10.5 volts. This test is intended to provide an approximation of how long the engine, including electrical accessories, could operate satisfactorily if the stator assembly or lighting coil did not produce sufficient current. A typical rating is 100 minutes.

→If possible, the new battery should have a power rating equal to or higher than the unit it is replacing.

BATTERY LOCATION

♦ See Figure 73

Every battery installed in a boat must be secured in a well protected, ventilated area. If the battery area lacks adequate ventilation, hydrogen gas, which is given off during charging, is very explosive. This is especially true if the gas is concentrated and confined.

BATTERY CHARGERS

♦ See Figure 74

Before using any battery charger, consult the manufacturer's instructions for its use. Battery chargers are electrical devices that change Alternating Current (AC) to a lower voltage of Direct Current (DC) that can be used to charge a marine battery. There are two types of battery chargers—manual and automatic.

A manual battery charger must be physically disconnected when the battery has come to a full charge. If not, the battery can be overcharged and possibly fail. Excess charging current at the end of the charging cycle will heat the electrolyte, resulting in loss of water and active material, substantially reducing battery life.

→As a rule, on manual chargers, when the ammeter on the charger registers half the rated amperage of the charger, the battery is fully charged. This can vary and it is recommended to use a hydrometer to accurately measure state of charge.

Automatic battery chargers have an important advantage—they can be left connected (for instance, overnight) without the possibility of overcharging the battery. Automatic chargers are equipped with a sensing device to allow the battery charge to taper off to near zero as the battery becomes fully charged. When charging a low or completely discharged battery, the meter will read close to full rated output. If only partially discharged, the initial reading may be less than full rated output, as the charger responds to the condition of the battery. As the battery continues to charge,

Fig. 74 Automatic chargers, such as the Battery Tender® from Deltran, are equipped with a sensing device to allow the battery charge to taper off to near zero as the battery becomes fully charged

IGNITION AND ELECTRICAL SYSTEMS

the sensing device monitors the state of charge and reduces the charging rate. As the rate of charge tapers to zero amps, the charger will continue to supply a few milliamps of current—just enough to maintain a charged condition.

BATTERY CABLES

Battery cables don't go bad very often but like anything else, they can wear out. If the cables on your boat are cracked, frayed or broken, they should be replaced.

When working on any electrical component, it is always a good idea to disconnect the negative (-) battery cable. This will prevent potential damage to many sensitive electrical components

Always replace the battery cables with one of the same length or you will increase resistance and possibly cause hard starting. Smear the battery posts with a light film of dielectric grease or a battery terminal protectant spray once you've installed the new cables. If you replace the cables one at a time, you won't mix them up.

➡ **Any time you disconnect the battery cables, it is recommended that you disconnect the negative (-) battery cable first. This will prevent you from accidentally grounding the positive (+) terminal when disconnecting it, thereby preventing damage to the electrical system.**

Before you disconnect the cable(s), first turn the ignition to the **OFF** position. This will prevent a draw on the battery which could cause arcing. When the battery cable(s) are reconnected (negative cable last), be sure to check all electrical accessories are all working correctly.

STARTER CIRCUIT

Description and Operation

▶ See Figure 75

In the early days, all outboard engines were started by simply pulling on a rope wound around the flywheel. As time passed and owners were reluctant to use muscle power, it was necessary to replace the rope starter with some form of power cranking system. Today, many small engines are still started by pulling on a rope but others have a powered starter motor installed.

The system utilized to replace the rope method was an electric starter motor coupled with a mechanical gear mesh between the starter motor and the powerhead flywheel, similar to the method used to crank an automobile engine.

As the name implies, the sole purpose of the starter motor circuit is to control operation of the starter motor to crank the powerhead until the engine is operating. The circuit includes a relay or magnetic switch to connect or disconnect the motor from the battery. The operator controls the switch with a key switch.

A neutral safety switch is installed into the circuit to permit operation of the starter motor only if the shift control lever is in neutral. This switch is a safety device to prevent accidental engine start when the engine is in gear.

The starter motor is a series wound electric motor which draws a heavy current from the battery. It is designed to be used only for short periods of time to crank the engine for starting. To prevent overheating the motor, cranking should not be continued for more than 30-seconds without allowing the motor to cool for at least three minutes. Actually, this time can be spent in making preliminary checks to determine why the engine fails to start.

Power is transmitted from the starter motor to the powerhead flywheel through a Bendix drive. This drive has a pinion gear mounted on screw threads. When the motor is operated, the pinion gear moves upward and meshes with the teeth on the flywheel ring gear.

When the powerhead starts, the pinion gear is driven faster than the shaft and as a result, it screws out of mesh with the flywheel. A rubber cushion is built into the Bendix drive to absorb the shock when the pinion meshes with the flywheel ring gear. The parts of the drive must be properly assembled for efficient operation. If the screw shaft assembly is reversed, it will strike the splines and the rubber cushion will not absorb the shock.

The sound of the motor during cranking is a good indication of whether the starter motor is operating properly or not. Naturally, temperature conditions will affect the speed at which the starter motor is able to crank the engine. The speed of cranking a cold engine will be much slower than when cranking a warm engine. An experienced operator will learn to recognize the favorable sounds of the powerhead cranking under various conditions.

The job of the starter motor relay is to complete the circuit between the battery and starter motor. It does this by closing the starter circuit electromagnetically, when activated by the key switch. This is a completely sealed switch, which meets SAE standards for marine applications. Do not substitute an automotive-type relay for this application. It is not sealed and gasoline fumes can be ignited upon starting the powerhead. The relay consists of a coil winding, plunger, return spring, contact disc and four externally mounted terminals. The relay is installed in series with the positive battery cables mounted to the two larger terminals. The smaller terminals connect to the neutral switch and ground.

To activate the relay, the shift lever is placed in neutral, closing the neutral switch. Electricity coming through the ignition switch goes into the relay coil winding which creates a magnetic field. The electricity then goes on to ground in the powerhead. The magnetic field surrounds the plunger in the relay, which draws the disc contact into the two larger terminals. Upon contact of the terminals, the heavy amperage circuit to the starter motor is closed and activates the starter motor. When the key switch is released, the magnetic field is no longer supported and the magnetic field collapses. The return spring working on the plunger opens the disc contact, opening the circuit to the starter.

When the armature plate is out of position or the shift lever is moved into forward or reverse gear, the neutral switch is placed in the open position and the starter control circuit cannot be activated. This prevents the powerhead from starting while in gear.

Troubleshooting the Starting System

If the starter motor spins but fails to crank the engine, the cause is usually a corroded or gummy Bendix drive. The drive should be removed, cleaned and given an inspection.

1. Before wasting too much time troubleshooting the starter motor circuit, the following checks should be made. Many times, the problem will be corrected.
   - Battery fully charged.
   - Shift control lever in neutral.
   - Main 20-amp fuse located at the base of the fuse cover is good (not blown).
   - All electrical connections clean and tight.
   - Wiring in good condition, insulation not worn or frayed.
2. Starter motor cranks slowly or not at all.
   - Faulty wiring connection
   - Short-circuited lead wire
   - Shift control not engaging neutral (not activating neutral start switch)
   - Defective neutral start switch
   - Starter motor not properly grounded
   - Faulty contact point inside ignition switch
   - Bad connections on negative battery cable to ground (at battery side and engine side)
   - Bad connections on positive battery cable to magnetic switch terminal
   - Open circuit in the coil of the magnetic switch (relay)
   - Bad or run-down battery
   - Excessively worn down starter motor brushes
   - Burnt commutator in starter motor
   - Brush spring tension slack
   - Short circuit in starter motor armature
3. Starter motor keeps running.
   - Melted contact plate inside the magnetic switch
   - Poor ignition switch return action
4. Starter motor picks up speed, put pinion will not mesh with ring gear.

Fig. 75 A typical starting system converts electrical energy into mechanical energy to turn the engine. The components are: battery, to provide electricity to operate the starter, ignition switch, to control the energizing of the starter relay or relay, starter relay or relay, to make and break the circuit between the battery and starter, starter, to convert electrical energy into mechanical energy to rotate the engine, starter drive gear, to transmit the starter rotation to the engine flywheel

IGNITION AND ELECTRICAL SYSTEMS

- Worn down teeth on clutch pinion
- Worn down teeth on flywheel ring gear

5. Two more areas may cause the powerhead to crank slowly even though the starter motor circuit is in excellent condition
- A tight or frozen powerhead
- Water in the lower unit.

Starter Motor

REMOVAL & INSTALLATION

8–25 HP

♦ See Figure 76

1. Before beginning any work on the starter motor, disconnect the positive (+) lead from the battery terminal.
2. Remove the cowling from the powerhead.
3. Disconnect the Yellow cable at the starter motor terminal or at the solenoid.
4. Remove the mounting bolts from the starter motor housing and remove the motor from the powerhead.

To install:

5. Install the starter motor to the powerhead and secure it in place with the two mounting bolts. Tighten the bolts securely. Connect the lead to the solenoid. Connect the leads to the battery terminals.
6. Reconnect the Yellow cable at the starter motor terminal or at the solenoid.
7. Install the cowling onto the powerhead.
8. Connect the positive (+) battery lead.

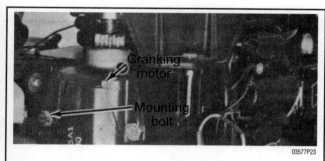

Fig. 76 Starter motor and mounting bolt location

30–275 HP

1. Before beginning any work on the starter motor, disconnect the positive (+) lead from the battery terminal.
2. Unlatch and remove the cowling from the powerhead.
3. Disconnect the cable from the solenoid to the starter motor, at the starter motor terminal.
4. Disconnect the ground cable from the powerhead frame to the starter motor.
5. Remove the clamp bolts, clamps and starter motor from the powerhead.

To install:

6. Align the starter motor onto the powerhead and install the clamps and bolts.
7. Connect the ground cable from the powerhead frame to the starter motor.
8. Connect the cable from the solenoid to the starter motor, at the starter motor terminal.
9. Install the engine cowling.
10. Attach the positive (+) battery cable to the battery.

Starter Solenoid

TESTING

Resistance Test

♦ See Figures 77 and 78

The following test MUST be conducted with the solenoid removed from the powerhead.

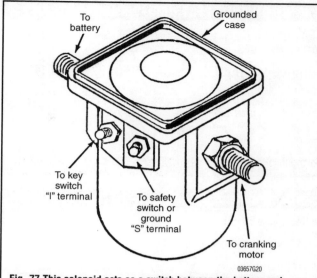

Fig. 77 This solenoid acts as a switch between the battery and starter motor. If the unit is found to be defective, it must be replaced with a marine approved solenoid

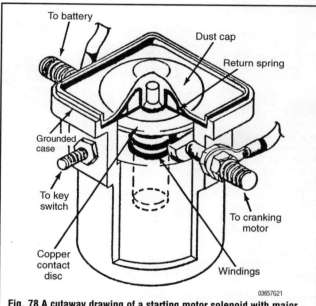

Fig. 78 A cutaway drawing of a starting motor solenoid with major parts identified

1. Connect one test lead of a multimeter to each of the large solenoid terminals.
2. Connect the positive (+) lead from a fully charged 12-volt battery to the small solenoid terminal marked S.
3. Momentarily make contact with the ground lead from the battery to the small solenoid terminal marked I. If a loud "click" sound is heard and the multimeter indicates continuity, the solenoid is in serviceable condition. If, however a "click" sound is not heard and/or the multimeter does not indicate continuity, the solenoid is defective and must be replaced and only with a marine solenoid.

Solenoid Voltage Test

1. Connect the voltmeter between the common powerhead ground and the No. 1 point.
2. Turn the ignition key switch to the start position.
3. Observe the voltmeter. If there is no reading, the starting motor solenoid is defective and must be replaced. If a reading is indicated and a click sound is heard, the solenoid is functioning properly.

5-42 IGNITION AND ELECTRICAL SYSTEMS

IGNITION AND ELECTRICAL WIRING DIAGRAMS

The following diagrams represent the most popular models with the most popular optional equipment.

Engine Wiring Diagram—1990–92 2.5, 3 HP

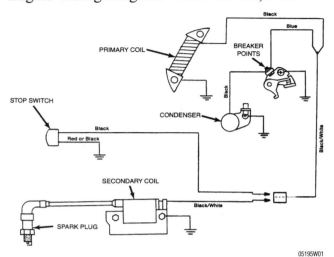

Engine Wiring Diagram—1993 2.5, 3.3 HP

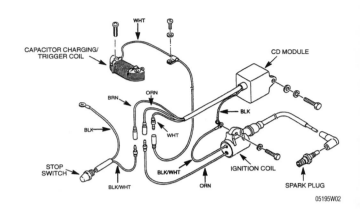

Engine Wiring Diagram—4, 5 HP (Serial #0A809601 and above)

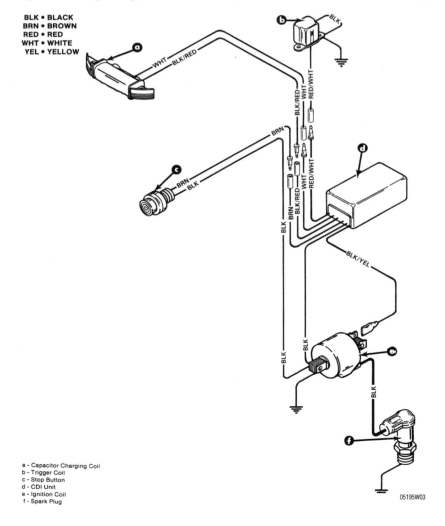

a - Capacitor Charging Coil
b - Trigger Coil
c - Stop Button
d - CDI Unit
e - Ignition Coil
f - Spark Plug

IGNITION AND ELECTRICAL SYSTEMS 5-43

Engine Wiring Diagram—6, 8, 9.9, 15 HP Manual Start (Serial #0A197112–0D280999)

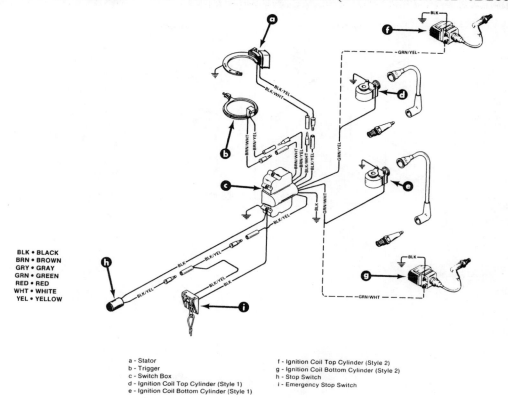

a - Stator
b - Trigger
c - Switch Box
d - Ignition Coil Top Cylinder (Style 1)
e - Ignition Coil Bottom Cylinder (Style 1)
f - Ignition Coil Top Cylinder (Style 2)
g - Ignition Coil Bottom Cylinder (Style 2)
h - Stop Switch
i - Emergency Stop Switch

Engine Wiring Diagram—6, 8, 9.9, 15 HP Electric Start with Tiller Handle (Serial #0A197112–0D280999)

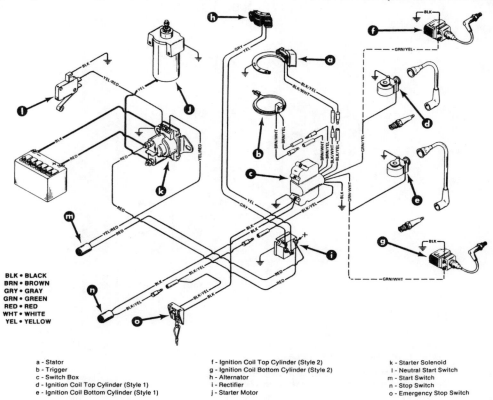

a - Stator
b - Trigger
c - Switch Box
d - Ignition Coil Top Cylinder (Style 1)
e - Ignition Coil Bottom Cylinder (Style 1)
f - Ignition Coil Top Cylinder (Style 2)
g - Ignition Coil Bottom Cylinder (Style 2)
h - Alternator
i - Rectifier
j - Starter Motor
k - Starter Solenoid
l - Neutral Start Switch
m - Start Switch
n - Stop Switch
o - Emergency Stop Switch

5-44 IGNITION AND ELECTRICAL SYSTEMS

Engine Wiring Diagram—6, 8, 9.9, 15 HP Electric Start with Remote Control (Serial #0A197112–0D280999)

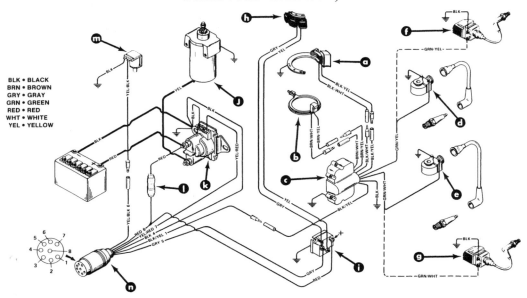

BLK • BLACK
BRN • BROWN
GRY • GRAY
GRN • GREEN
RED • RED
WHT • WHITE
YEL • YELLOW

a - Stator
b - Trigger
c - Switch Box
d - Ignition Coil Top Cylinder (Style 1)
e - Ignition Coil Bottom Cylinder (Style 1)
f - Ignition Coil Top Cylinder (Style 2)
g - Ignition Coil Bottom Cylinder (Style 2)
h - Alternator
i - Rectifier
j - Starter Motor
k - Starter Solenoid
l - Fuse Holder (20 Amp. Fuse)
m - Choke Solenoid
n - Wiring Harness

05195W06

Commander 2000 Side Mount Remote Control with Key/Choke Switch—6, 8, 9.9, 15 HP (Serial #0A197112–0D280999)

BLK • BLACK
PUR • PURPLE
RED • RED
YEL • YELLOW

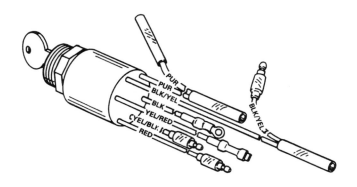

"OFF"   BLK/YEL - BLK
"RUN"   RED - PUR
"START"   RED - PUR - YEL/RED
PUSH (CHOKE)* RED - YEL/BLK

*Key switch must be positioned to "RUN" OR "START" and key pushed in to actuate choke, for this continuity test.

05195W08

IGNITION AND ELECTRICAL SYSTEMS 5-45

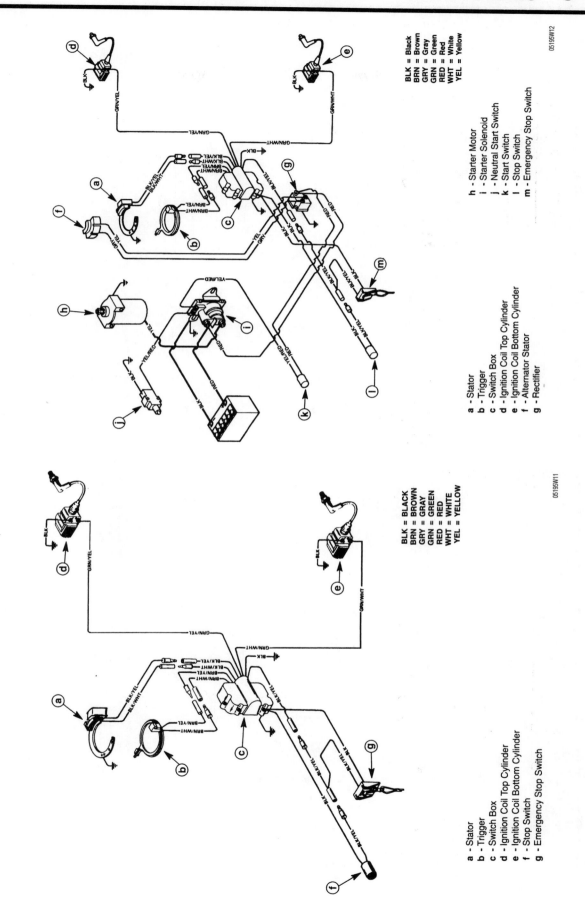

5-46 IGNITION AND ELECTRICAL SYSTEMS

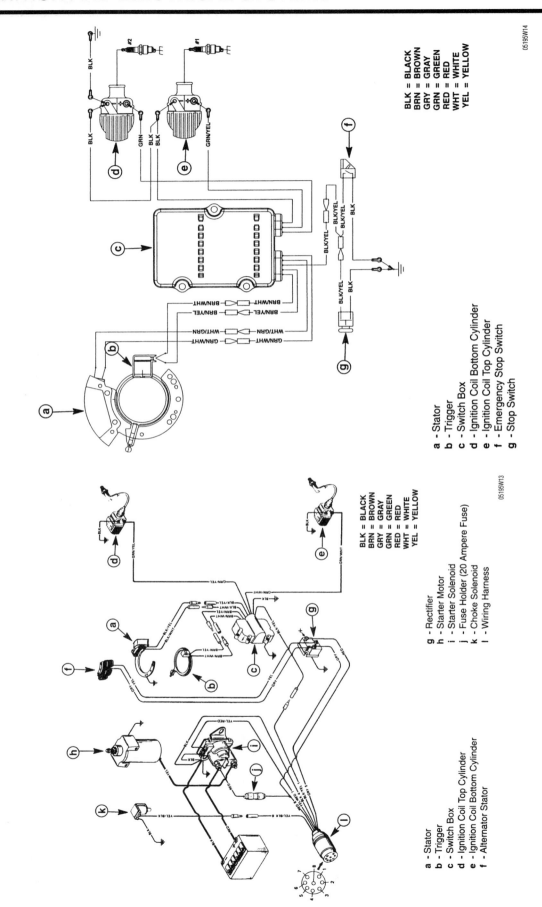

Engine Wiring Diagram—6, 8, 9.9, 10, 15, 20 JET, Manual Start with RED Stator (Serial #0D281000 and above) and 1999–00 20, 25 HP

- a - Stator
- b - Trigger
- c - Switch Box
- d - Ignition Coil Bottom Cylinder
- e - Ignition Coil Top Cylinder
- f - Emergency Stop Switch
- g - Stop Switch

Engine Wiring Diagram—1998 6, 8, 9.9, 10, 15 HP Electric with Remote Control and BLACK Stator (Serial #0D281000 and above) and 1998 20, 25 HP with Mechanical Advance

- a - Stator
- b - Trigger
- c - Switch Box
- d - Ignition Coil Top Cylinder
- e - Ignition Coil Bottom Cylinder
- f - Alternator Stator
- g - Rectifier
- h - Starter Motor
- i - Starter Solenoid
- j - Fuse Holder (20 Ampere Fuse)
- k - Choke Solenoid
- l - Wiring Harness

IGNITION AND ELECTRICAL SYSTEMS 5-47

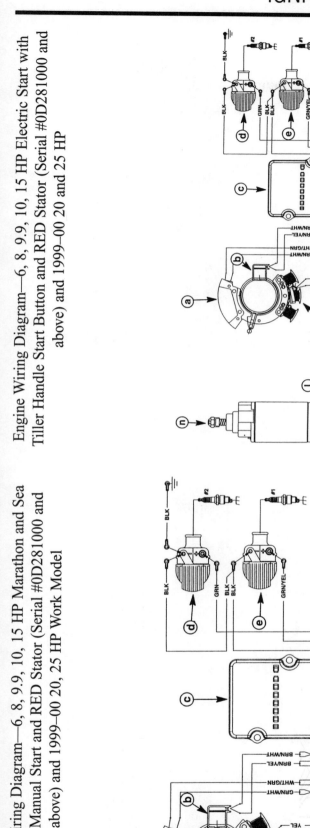

Engine Wiring Diagram—6, 8, 9.9, 10, 15 HP Electric Start with Tiller Handle Start Button and RED Stator (Serial #0D281000 and above) and 1999–00 20 and 25 HP

BLK = BLACK
BRN = BROWN
GRY = GRAY
GRN = GREEN
RED = RED
WHT = WHITE
YEL = YELLOW

a - Stator
b - Trigger
c - Switch Box
d - Ignition Coil Bottom Cylinder
e - Ignition Coil Top Cylinder
f - Emergency Stop Switch
g - Push Button Stop Switch
h - Rectifier
i - Alternator Stator
j - Neutral Start Switch
k - Push Button Start Switch
l - Start Solenoid
m - 12 VDC Battery
n - Starter Motor

Engine Wiring Diagram—6, 8, 9.9, 10, 15 HP Marathon and Sea Pro with Manual Start and RED Stator (Serial #0D281000 and above) and 1999–00 20, 25 HP Work Model

BLK = BLACK
BRN = BROWN
GRY = GRAY
GRN = GREEN
RED = RED
WHT = WHITE
YEL = YELLOW

a - Stator
b - Trigger
c - Switch Box
d - Ignition Coil Bottom Cylinder
e - Ignition Coil Top Cylinder
f - Stop Switch
g - Emergency Stop Switch
h - Alternator Stator

5-48 IGNITION AND ELECTRICAL SYSTEMS

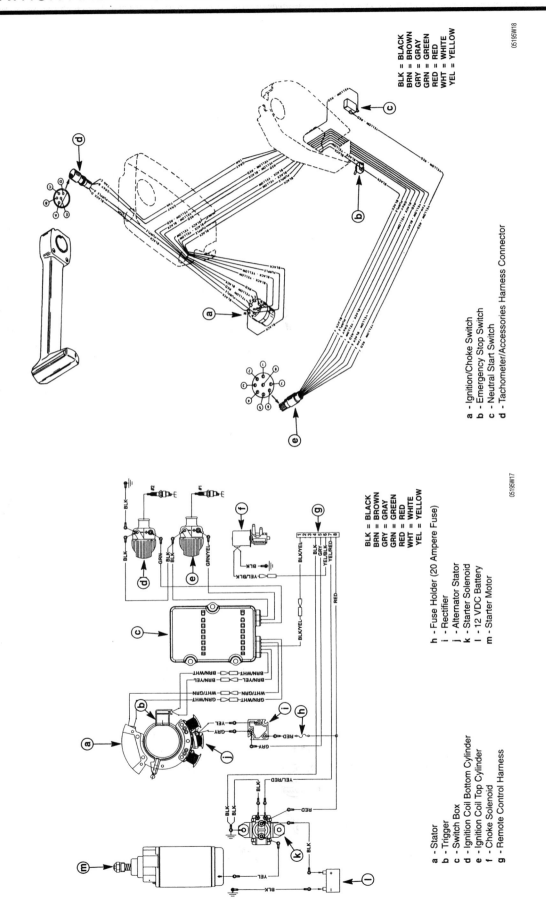

Commander Remote Control Electric Start—6, 8, 9.9, 10, 15 HP (Serial #0D281000 and above)

a - Ignition/Choke Switch
b - Emergency Stop Switch
c - Neutral Start Switch
d - Tachometer/Accessories Harness Connector

Engine Wiring Diagram—6, 8, 9.9, 10, 15 HP Electric with Remote Control and RED Stator and Electric Start with Tiller Handle Start Button and RED Stator (Serial #0D281000 and above) and 1990-00 20, 25 HP

a - Stator
b - Trigger
c - Switch Box
d - Ignition Coil Bottom Cylinder
e - Ignition Coil Top Cylinder
f - Choke Solenoid
g - Remote Control Harness
h - Fuse Holder (20 Ampere Fuse)
i - Rectifier
j - Alternator Stator
k - Starter Solenoid
l - 12 VDC Battery
m - Starter Motor

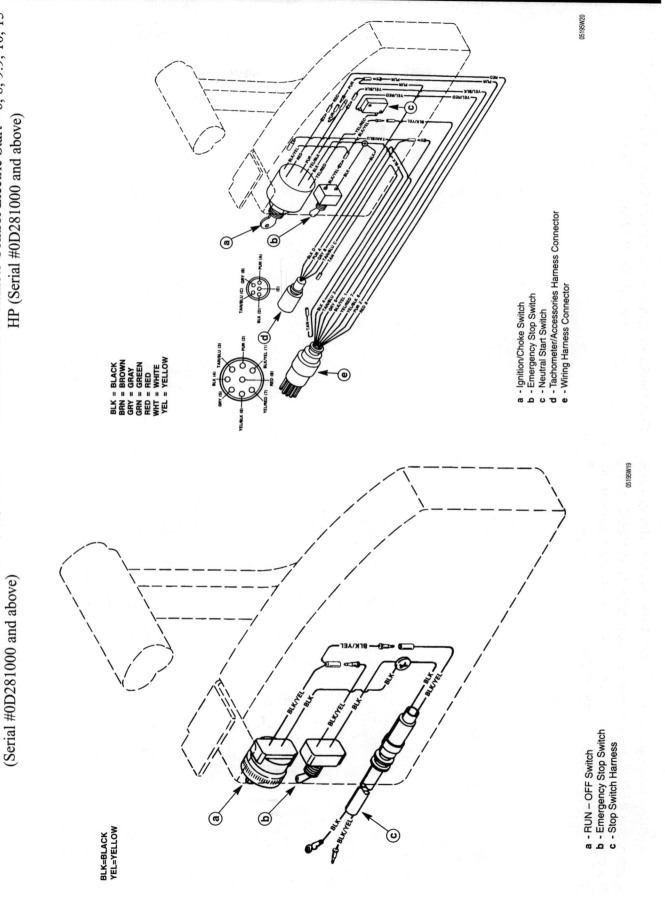

5-50 IGNITION AND ELECTRICAL SYSTEMS

Commander 3000 Panel Mount Control—6, 8, 9.9, 10, 15 HP (Serial #0D281000 and above)

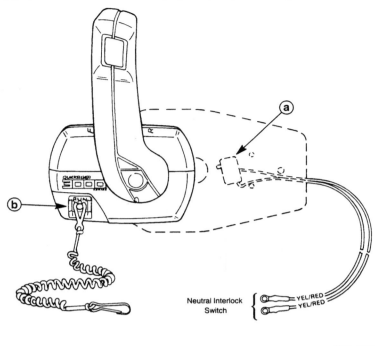

a - Neutral Interlock Switch
b - Emergency Stop Switch

RED=RED
YEL=YELLOW

Engine Wiring Diagram Manual Start—18, 18XD, 20, 25, 25XD, 25 Marathon, 25 Sea Pro HP (Serial #6416713–0G044026)

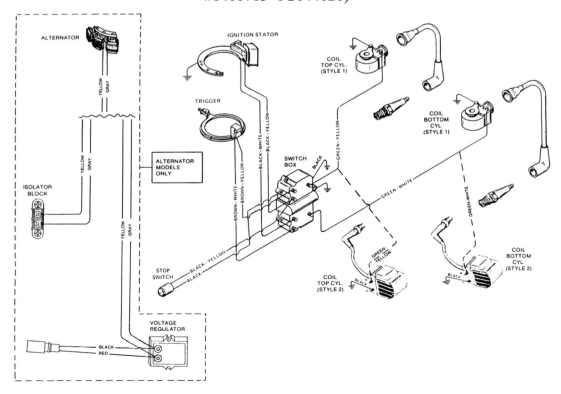

IGNITION AND ELECTRICAL SYSTEMS 5-51

Engine Wiring Diagram Electric Start with Tiller Handle Button—18, 18XD, 20, 25, 25XD, 25 Marathon, 25 Sea Pro HP (Serial #6416713–0G044026)

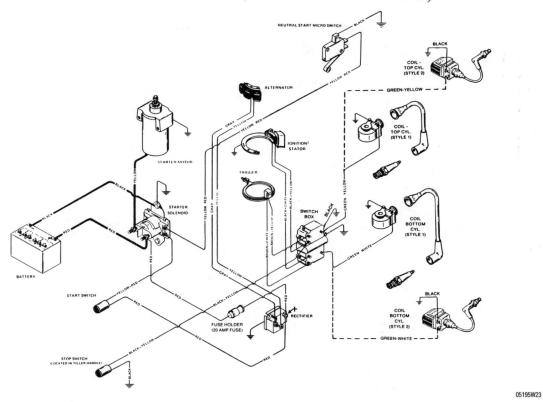

Engine Wiring Diagram—18, 18XD, 20, 25, 25XD, 25 Marathon, 25 Sea Pro HP Electric Start with Remote Control (Serial #6416713–0G044026)

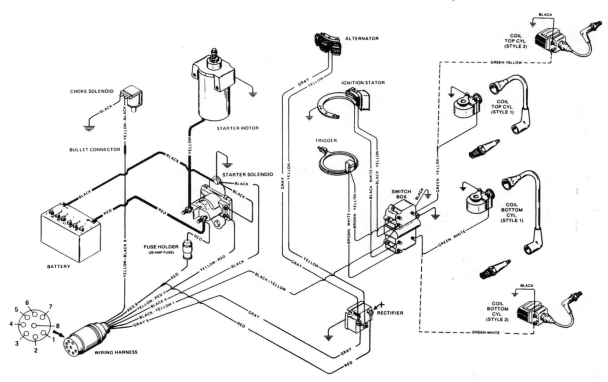

5-52 IGNITION AND ELECTRICAL SYSTEMS

Engine Wiring Diagram—18, 18XD, 20, 25, 25XD, 25 Marathon, 25 Sea Pro HP Electric Start with Tiller Handle and Ignition Key/Choke Panel (Serial #6416713–0G044026)

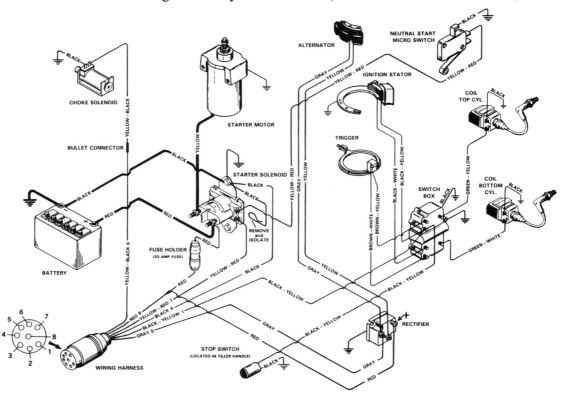

Engine Wiring Diagram—30, 40 HP Manual Start (Serial #0G380075 and above)

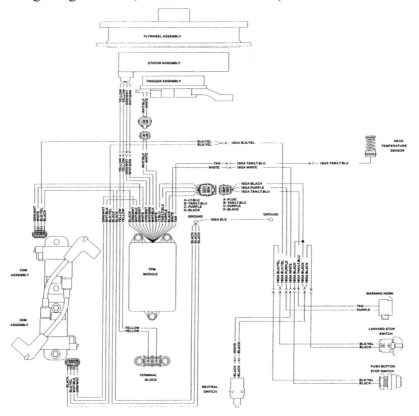

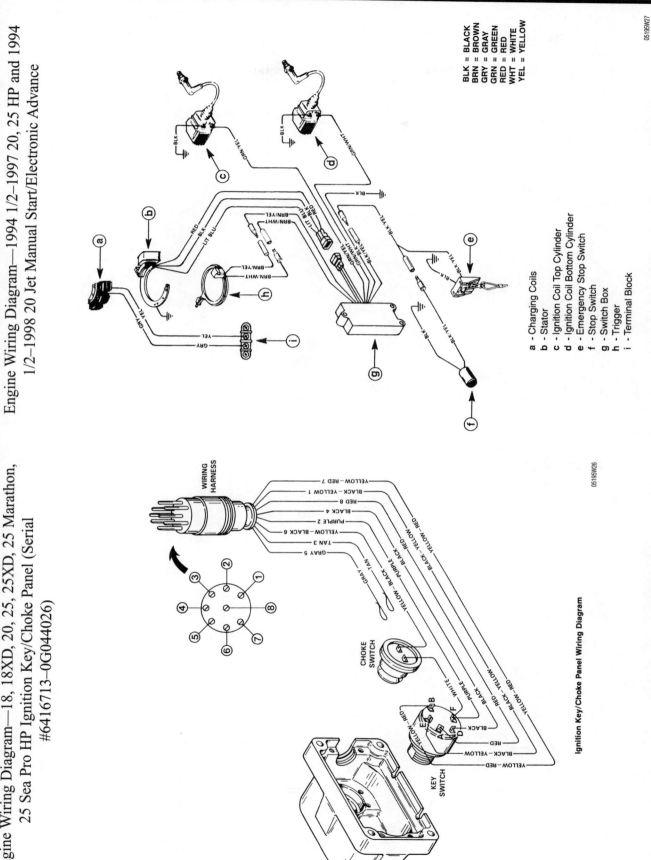

5-54 IGNITION AND ELECTRICAL SYSTEMS

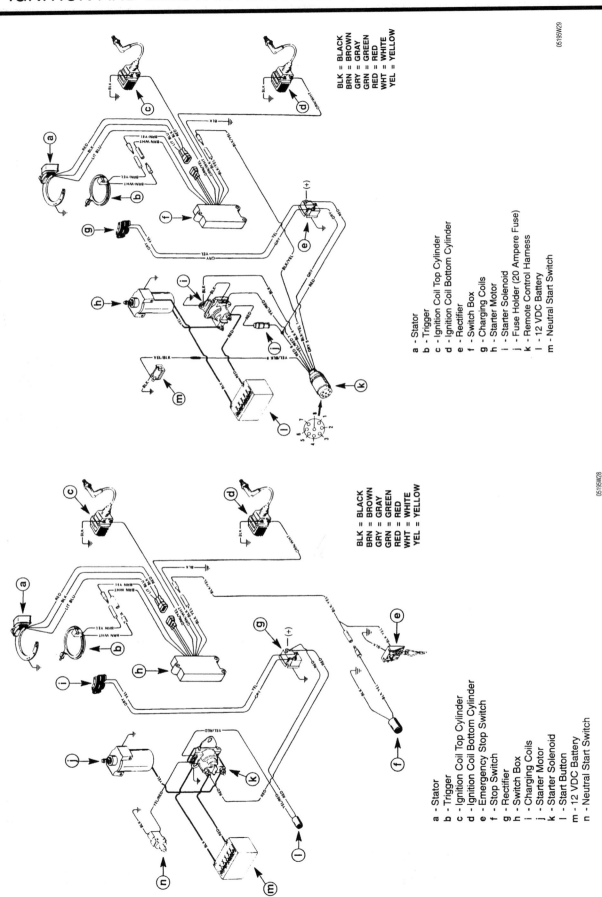

IGNITION AND ELECTRICAL SYSTEMS 5-55

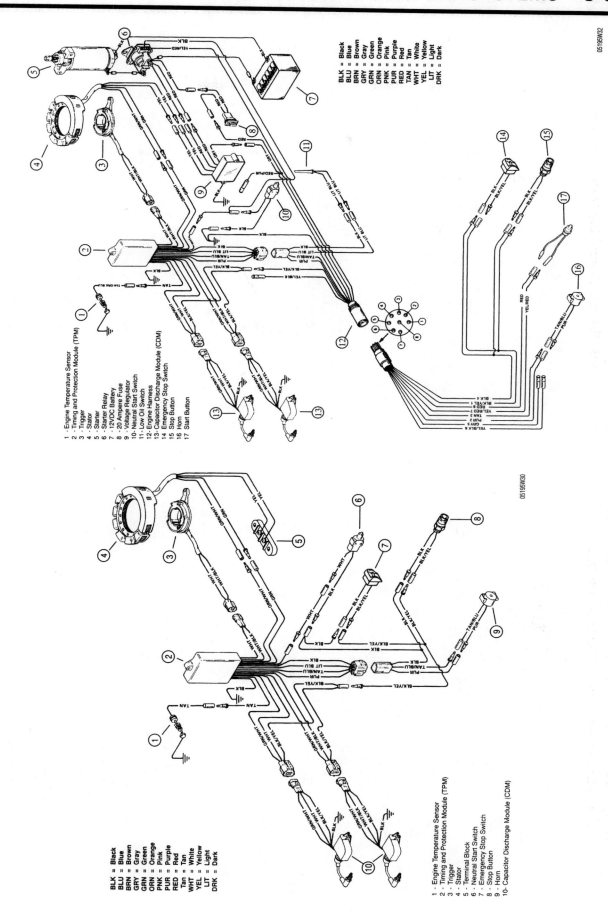

IGNITION AND ELECTRICAL SYSTEMS

Engine Wiring Diagram—30, 40 HP Electric Start with Tiller Handle Ignition (Serial #0G380075 and above)

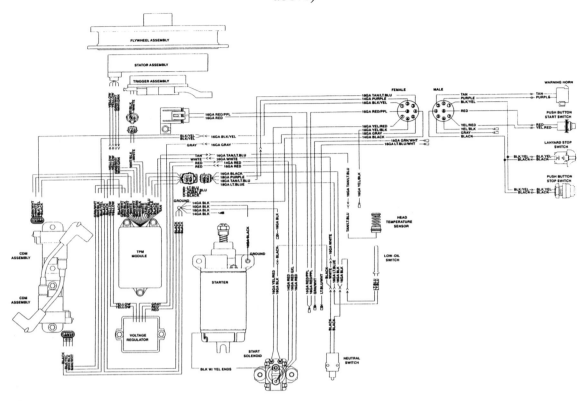

Engine Wiring Diagram Electric Start—30, 40 HP (Serial #0G380075 and above)

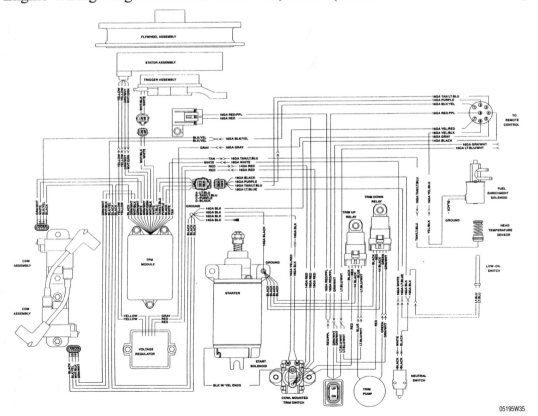

IGNITION AND ELECTRICAL SYSTEMS 5-57

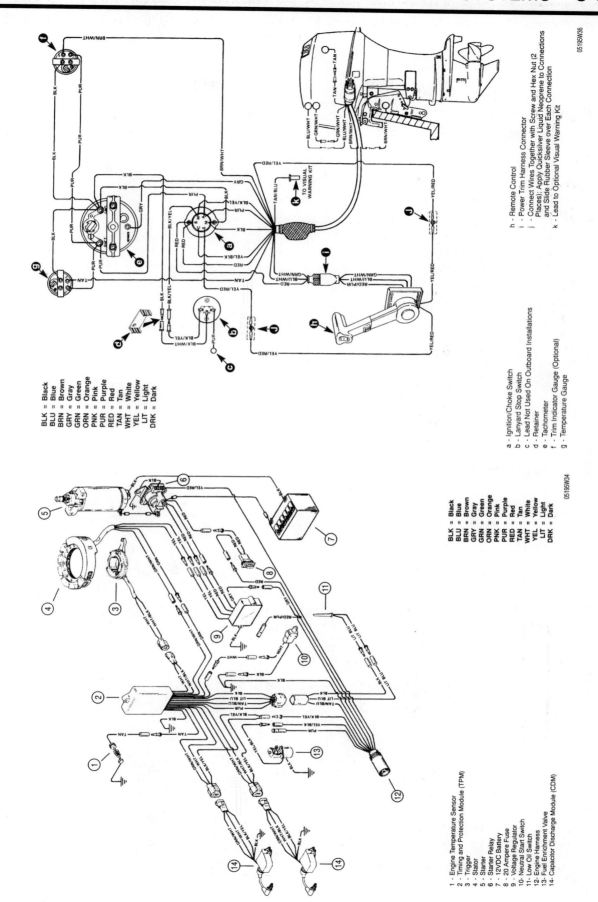

5-58 IGNITION AND ELECTRICAL SYSTEMS

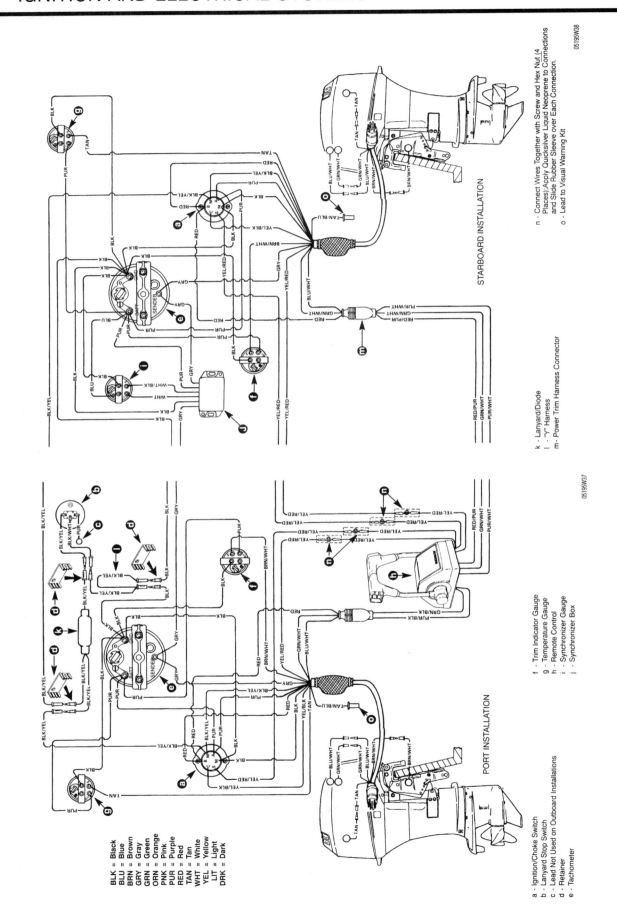

Instrument/Lanyard Stop Switch Wiring Diagram Dual Outboard—30, 40 HP (Serial #0G044027 and above) (Cont.)

IGNITION AND ELECTRICAL SYSTEMS 5-59

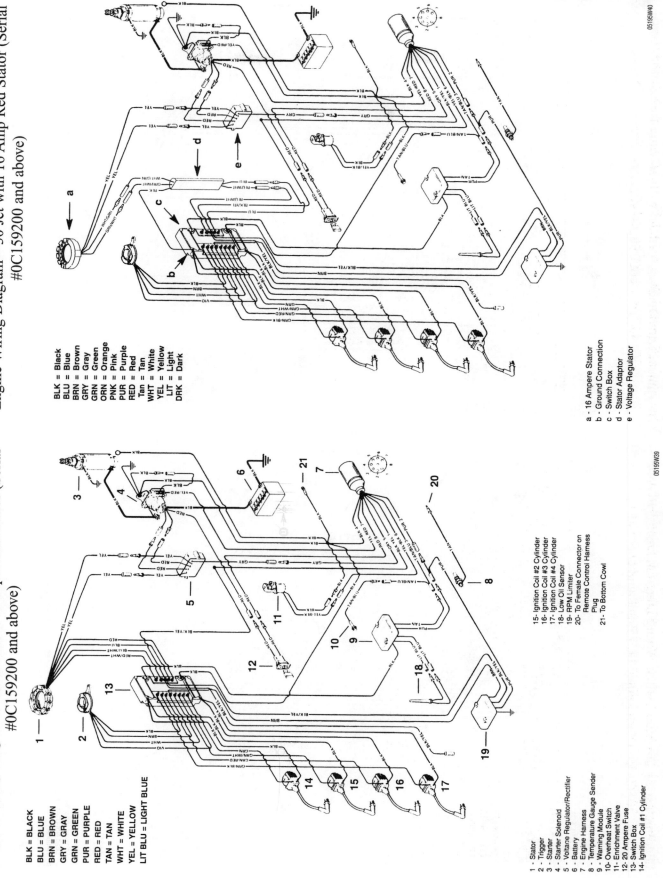

5-60 IGNITION AND ELECTRICAL SYSTEMS

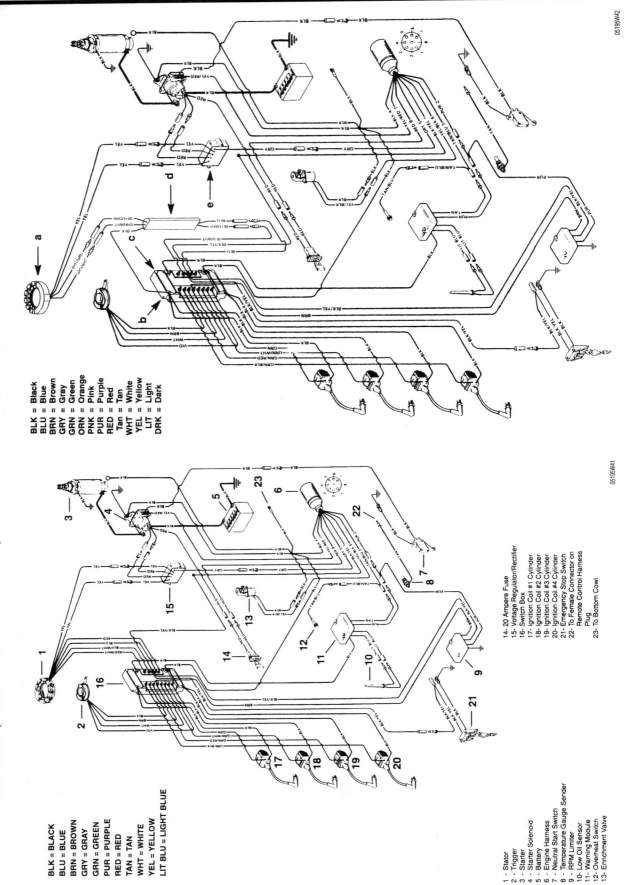

IGNITION AND ELECTRICAL SYSTEMS 5-61

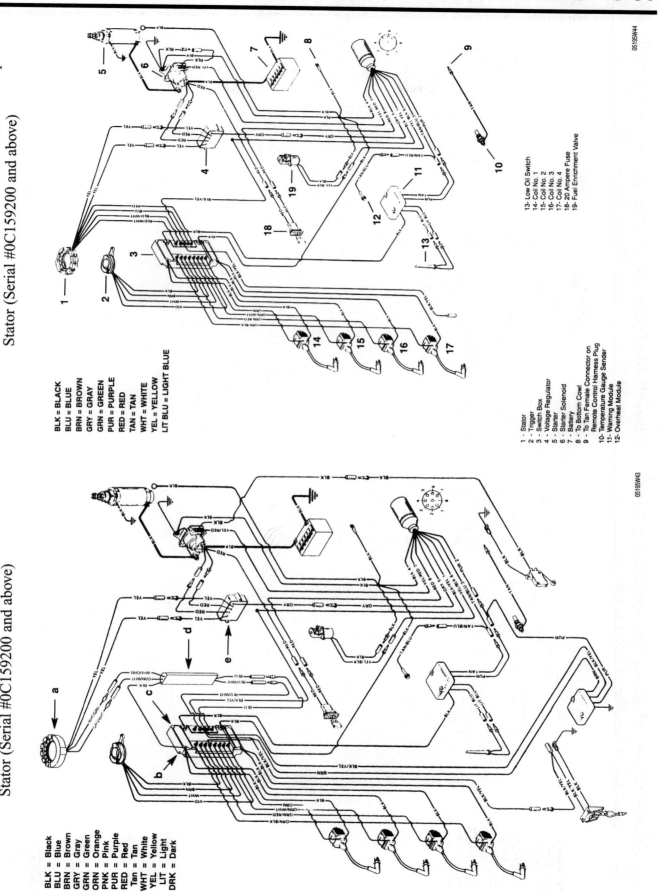

5-62 IGNITION AND ELECTRICAL SYSTEMS

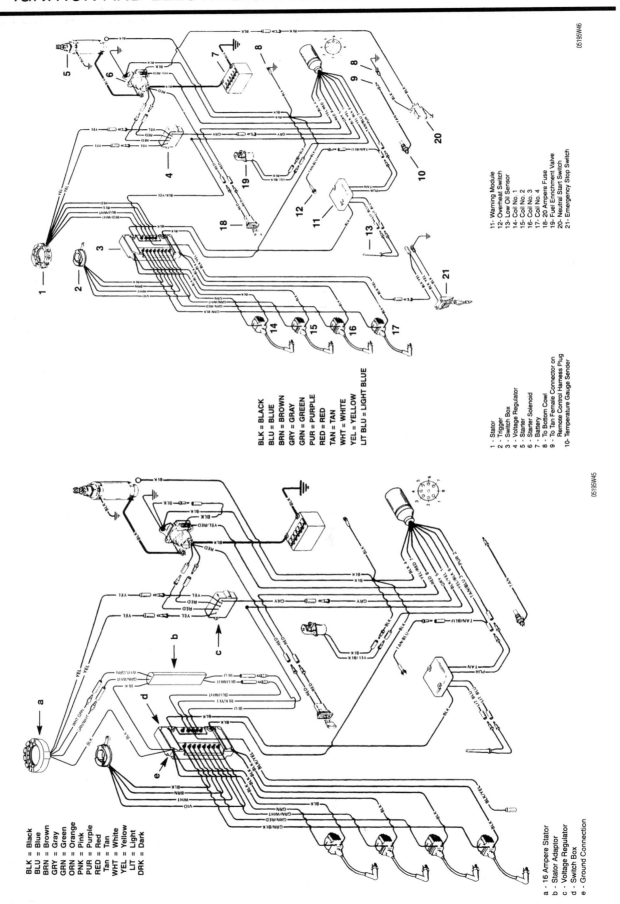

IGNITION AND ELECTRICAL SYSTEMS 5-63

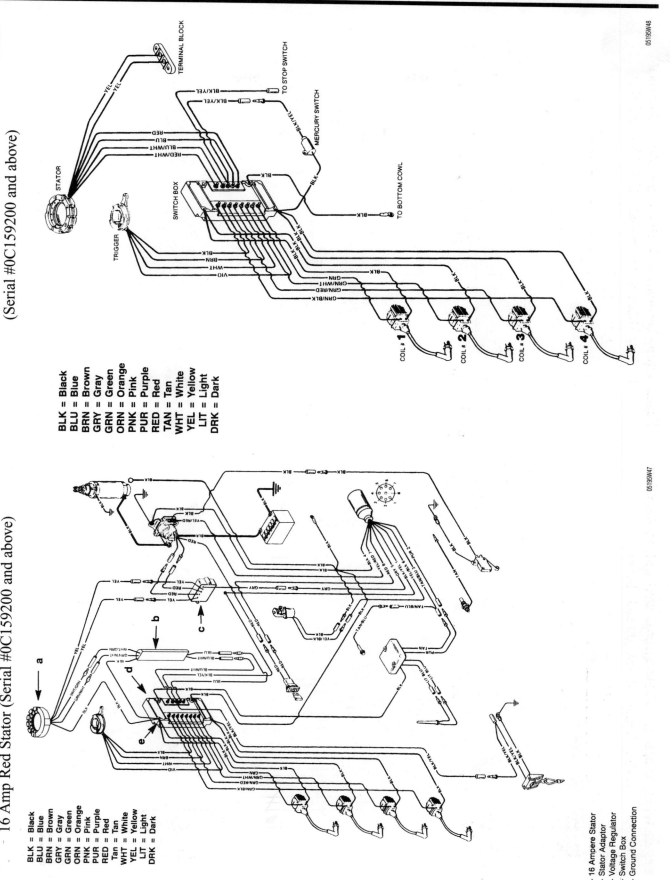

5-64 IGNITION AND ELECTRICAL SYSTEMS

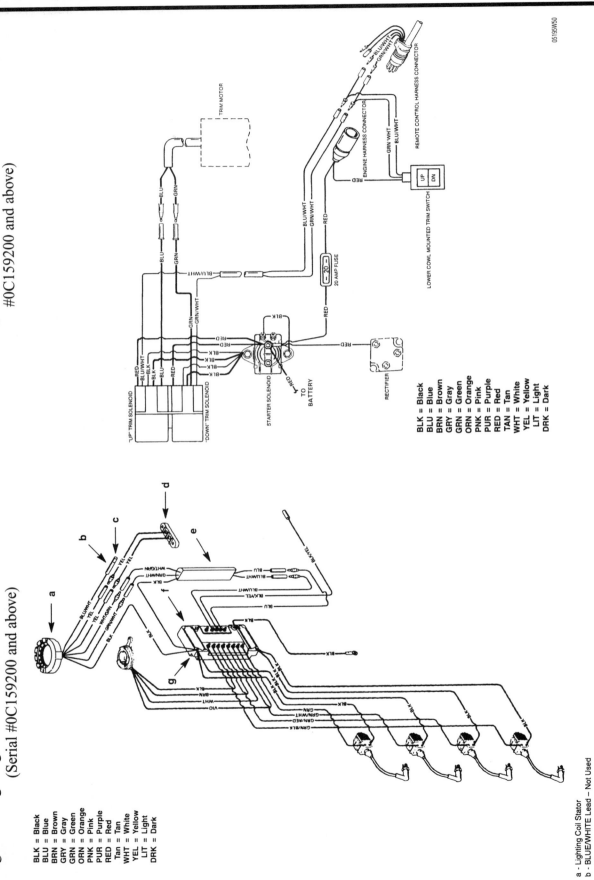

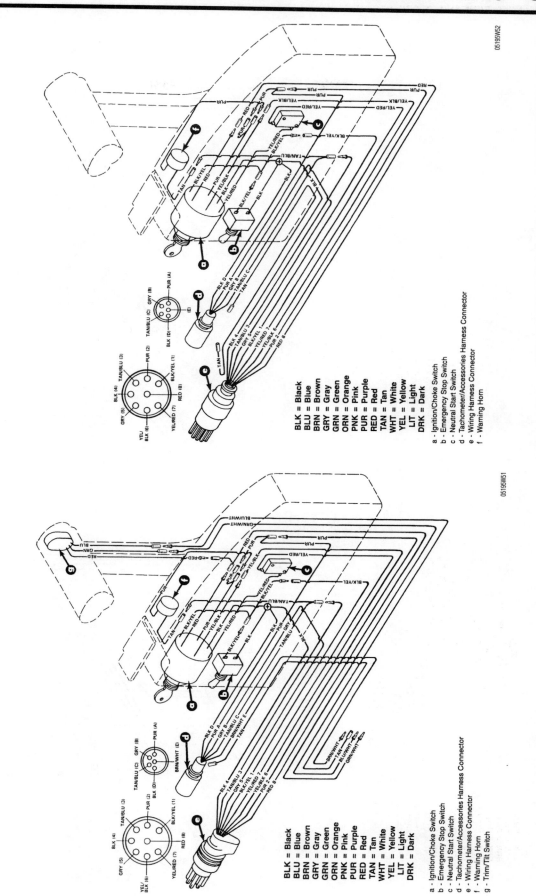

5-66 IGNITION AND ELECTRICAL SYSTEMS

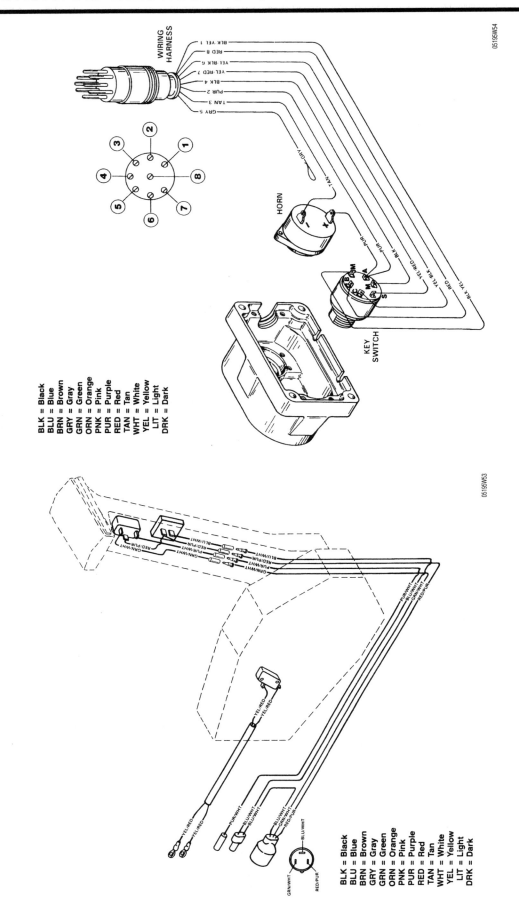

Remote Key Switch and Warning Horn—30 Jet, 40 (4-Cylinder) HP (Serial #0C159200 and above)

Panel Mount Remote Control Wiring Diagram—30 Jet, 40 (4-Cylinder) HP (Serial #0C159200 and above)

IGNITION AND ELECTRICAL SYSTEMS 5-67

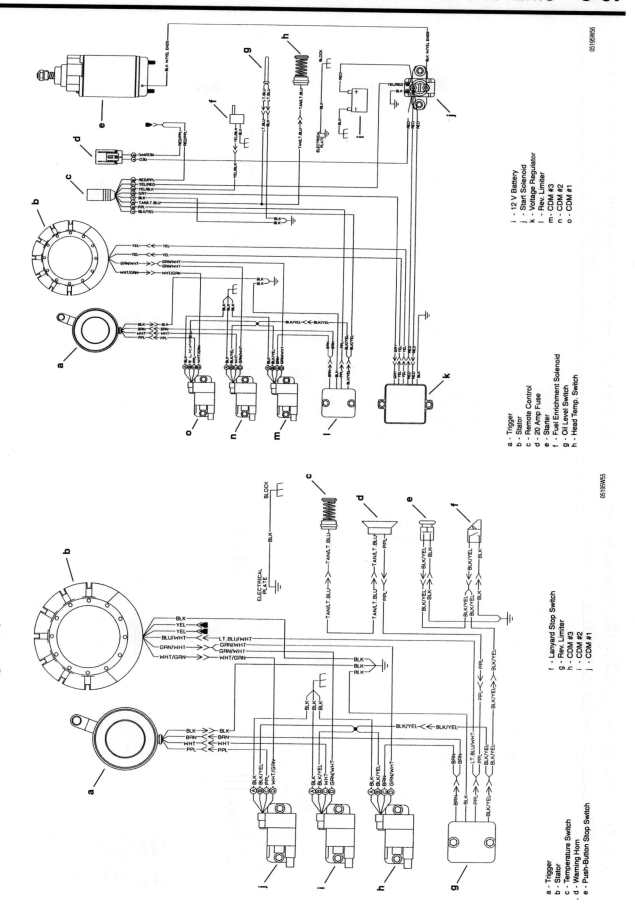

5-68 IGNITION AND ELECTRICAL SYSTEMS

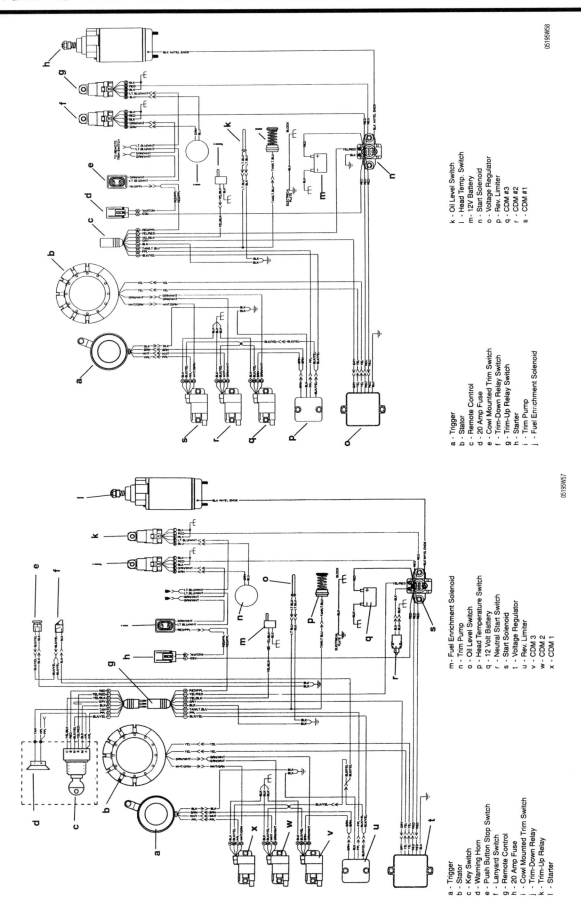

Engine Wiring Diagram—40, 50, 60 HP EPTO (Serial #0G531301 and above)

a - Trigger
b - Stator
c - Remote Control
d - 20 Amp Fuse
e - Cowl Mounted Trim Switch
f - Trim-Down Relay Switch
g - Trim-Up Relay Switch
h - Starter
i - Trim Pump
j - Fuel Enrichment Solenoid
k - Oil Level Switch
l - Head Temp. Switch
m - 12V Battery
n - Start Solenoid
o - Voltage Regulator
p - Rev. Limiter
q - CDM #3
r - CDM #2
s - CDM #1

Engine Wiring Diagram—40, 50 HP EHPTO (Serial #0G531301 and above)

a - Trigger
b - Stator
c - Key Switch
d - Warning Horn
e - Push Button Stop Switch
f - Lanyard Switch
g - Remote Control
h - 20 Amp Fuse
i - Cowl Mounted Trim Switch
j - Trim-Down Relay
k - Trim-Up Relay
l - Starter
m - Fuel Enrichment Solenoid
n - Trim Pump
o - Oil Level Switch
p - Head Temperature Switch
q - 12 Volt Battery
r - Neutral Start Switch
s - Start Solenoid
t - Voltage Regulator
u - Rev. Limiter
v - CDM 3
w - CDM 2
x - CDM 1

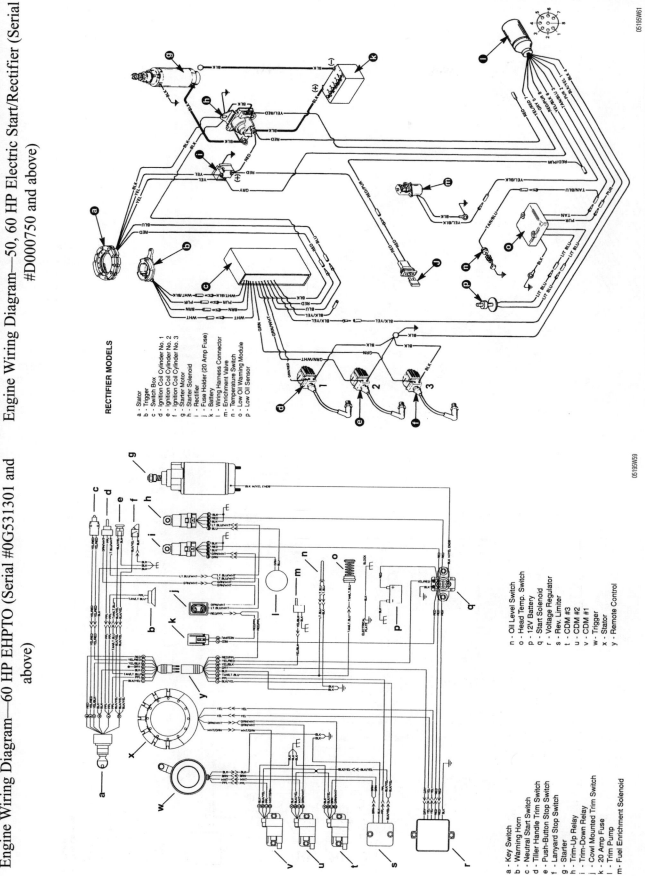

5-70 IGNITION AND ELECTRICAL SYSTEMS

Power Tilt—40, 50, 55, 60 HP Electric Start with Warning Horn (Serial #0G531301 and above)

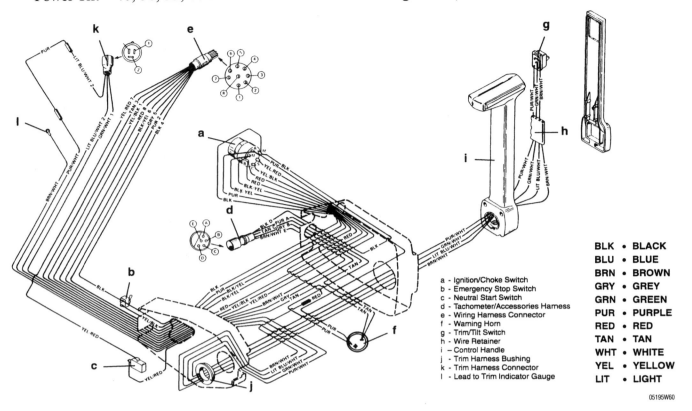

Power Trim Wiring Diagram—50, 60 HP (Serial #D000750 and above)

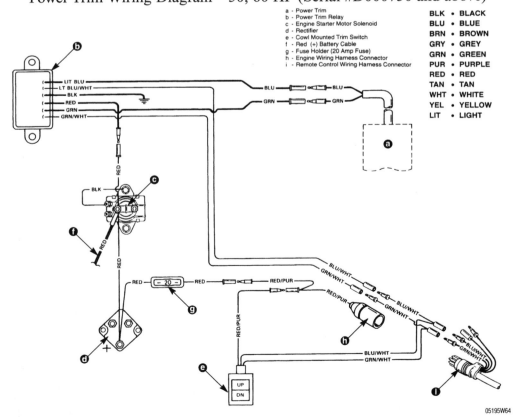

IGNITION AND ELECTRICAL SYSTEMS 5-71

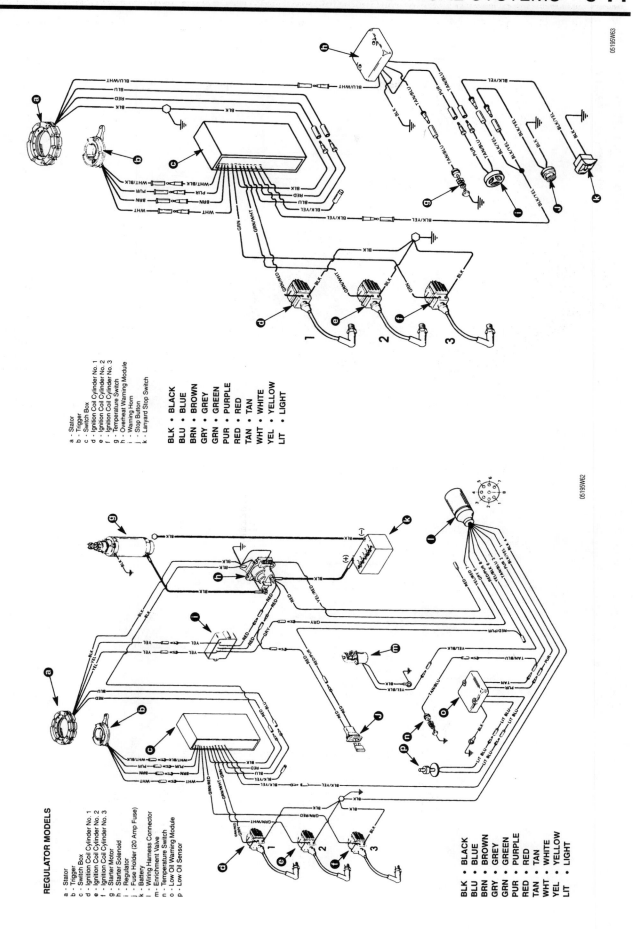

Engine Wiring Diagram—55 HP Manual Start (Serial #D000750 and above)

Engine Wiring Diagram—50, 60 HP Electric Start/Regulator (Serial #D000750 and above)

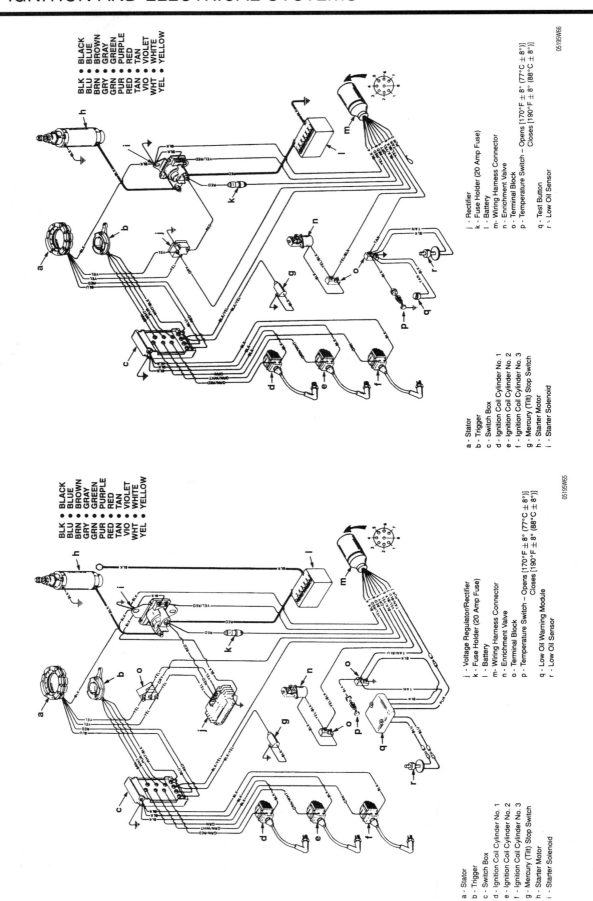

IGNITION AND ELECTRICAL SYSTEMS 5-73

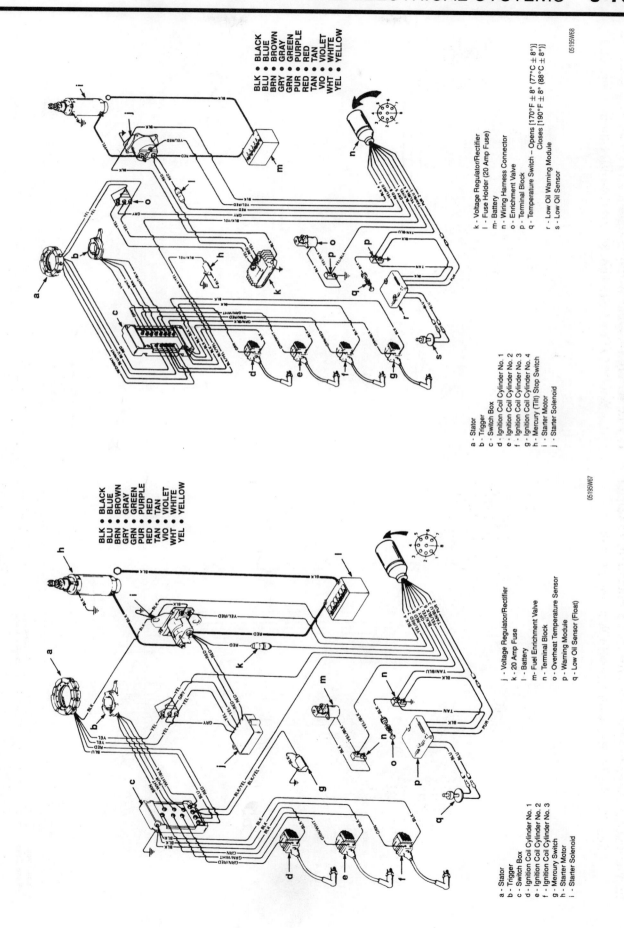

5-74 IGNITION AND ELECTRICAL SYSTEMS

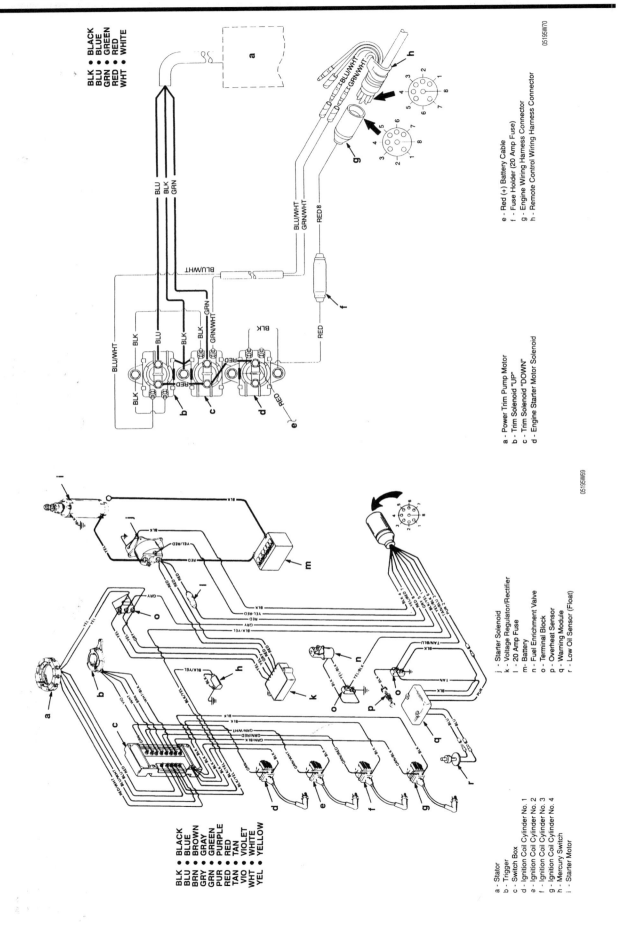

IGNITION AND ELECTRICAL SYSTEMS 5-75

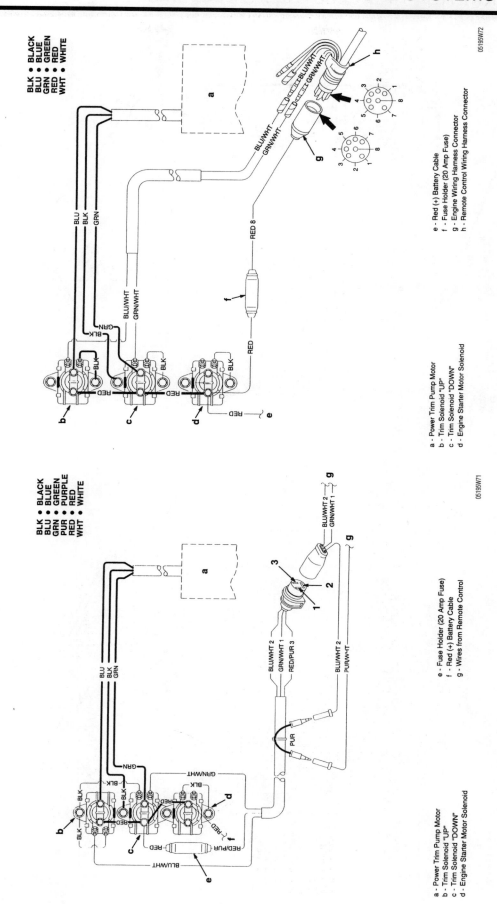

5-76 IGNITION AND ELECTRICAL SYSTEMS

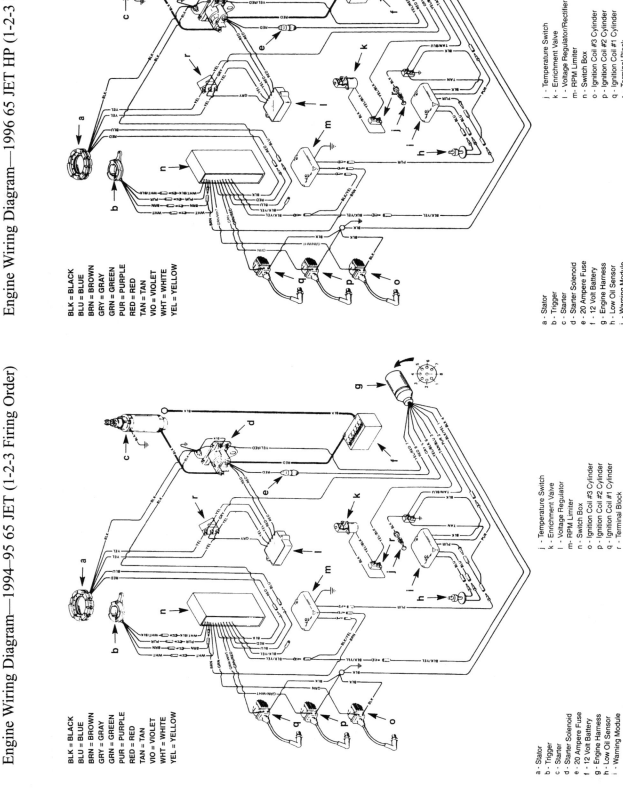

IGNITION AND ELECTRICAL SYSTEMS   5-77

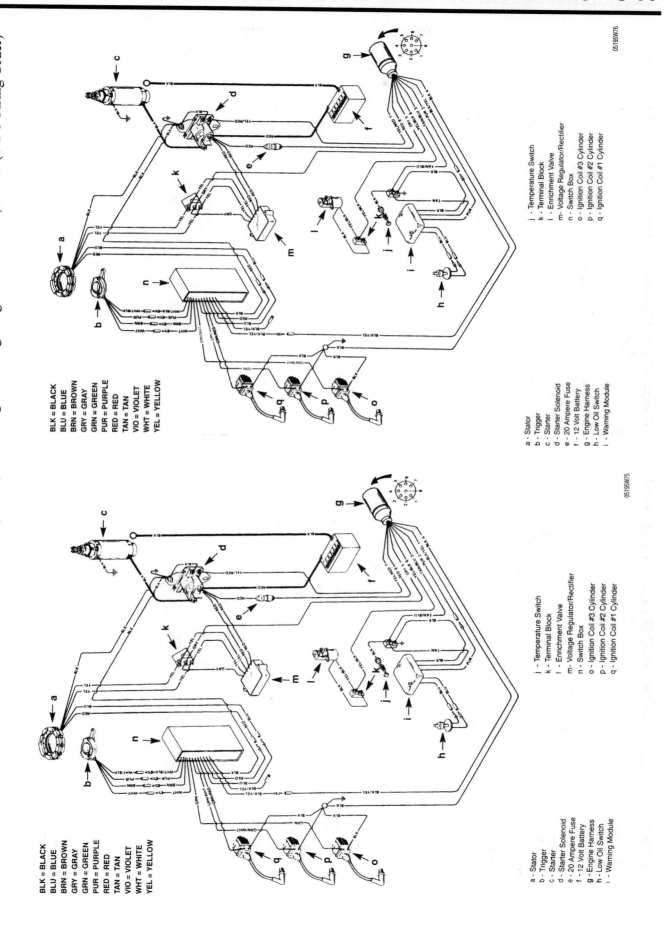

5-78 IGNITION AND ELECTRICAL SYSTEMS

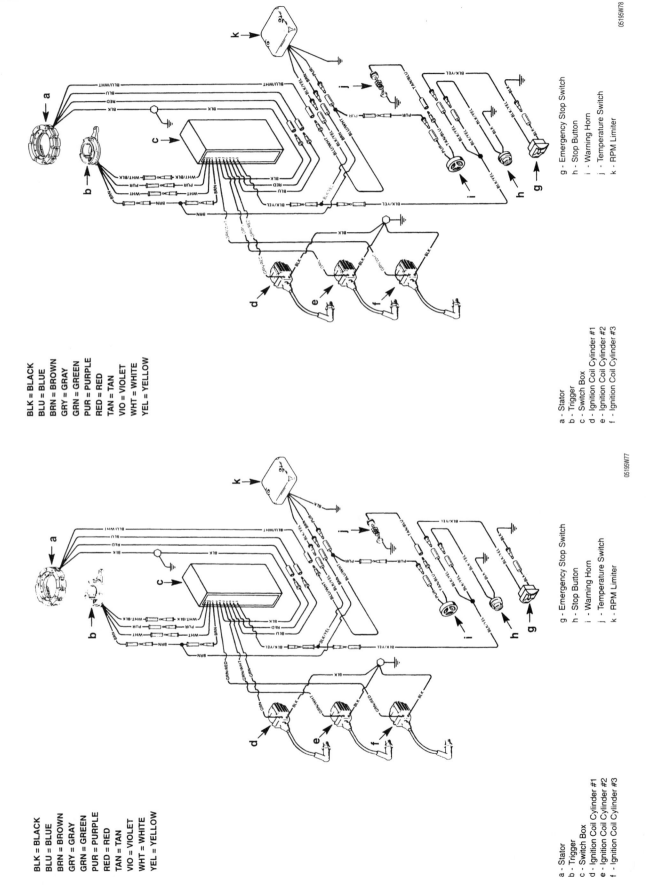

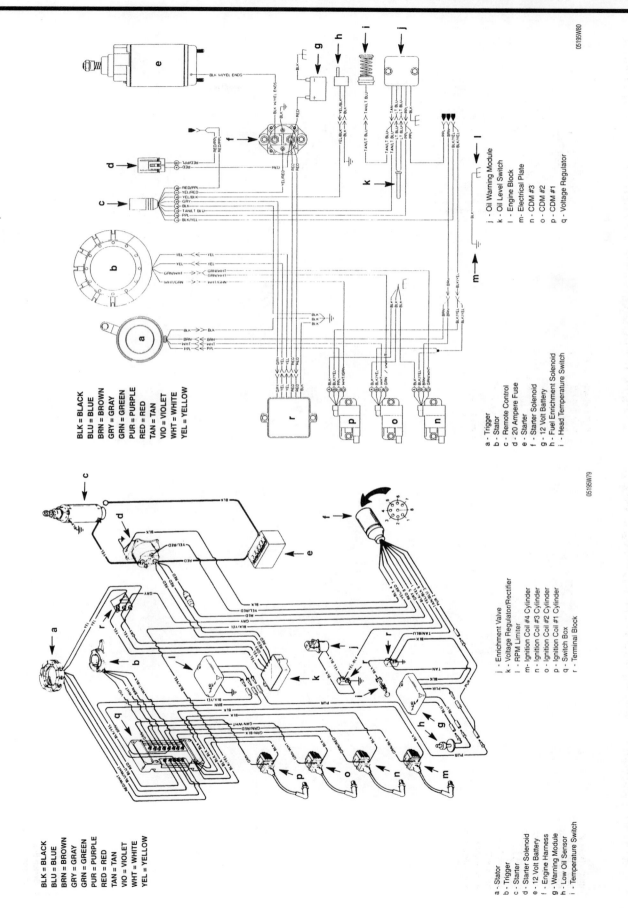

5-80 IGNITION AND ELECTRICAL SYSTEMS

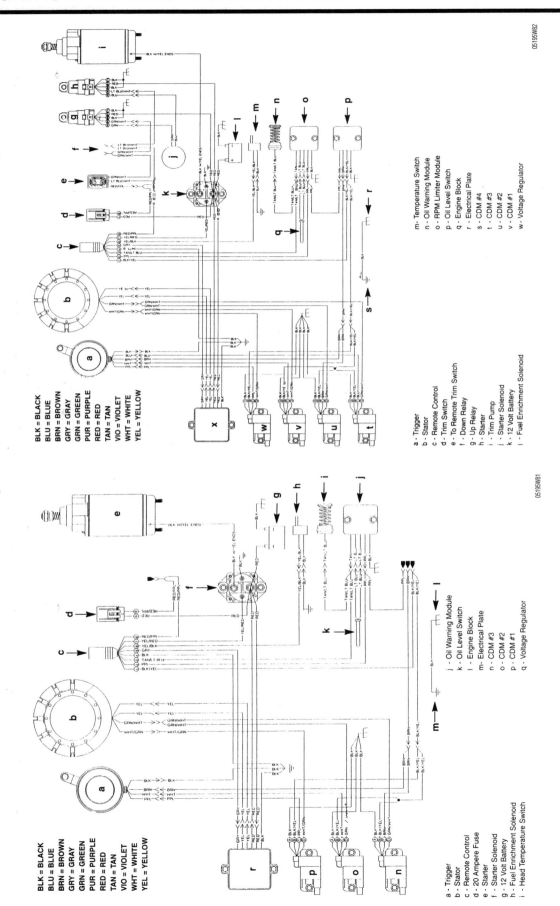

IGNITION AND ELECTRICAL SYSTEMS 5-81

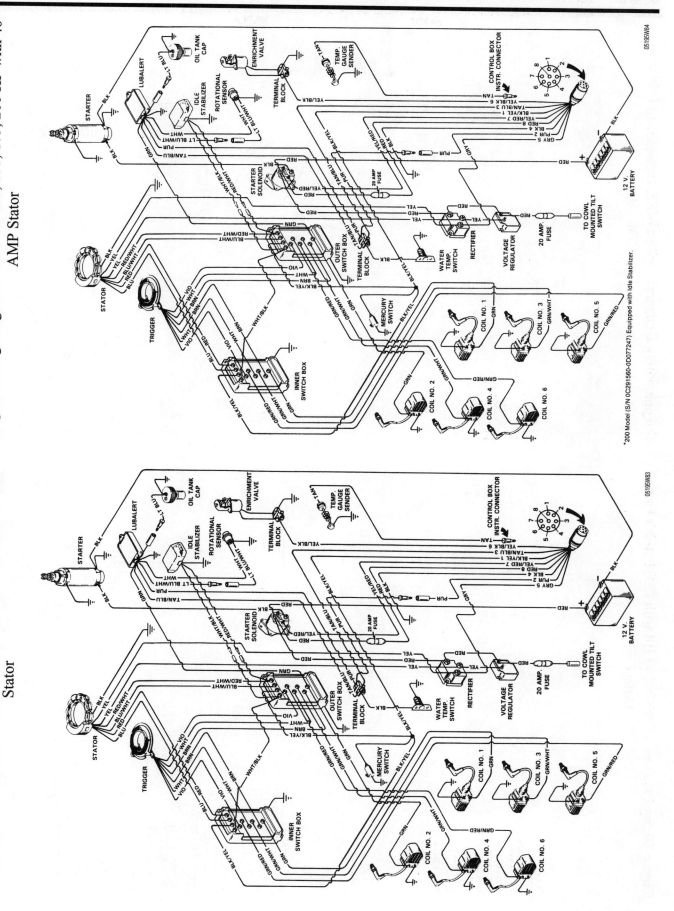

Engine Wiring Diagram—1991-92 135, 150, 175, 200 HP with 40 AMP Stator

Engine Wiring Diagram—1991-92 135, 150, 175 HP with 16 AMP Stator

5-82 IGNITION AND ELECTRICAL SYSTEMS

Engine Wiring Diagram—1991–92 XR4, Magnum II (Serial #C254932 and above)

Engine Wiring Diagram—1991–92 135, 150, 175 HP with 40 AMP Stator

IGNITION AND ELECTRICAL SYSTEMS 5-83

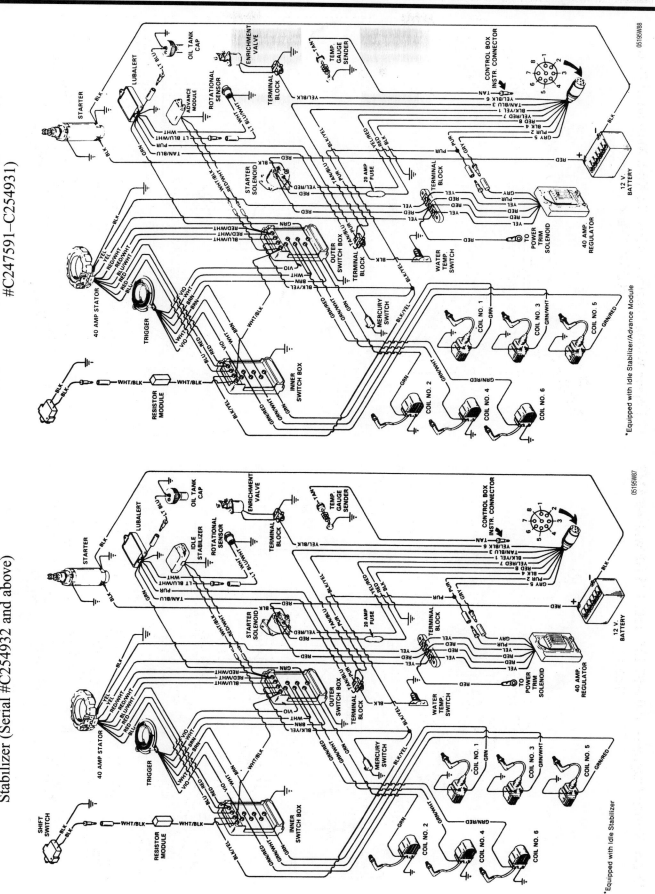

5-84 IGNITION AND ELECTRICAL SYSTEMS

Engine Wiring Diagram—1990–91 175 HP EFI

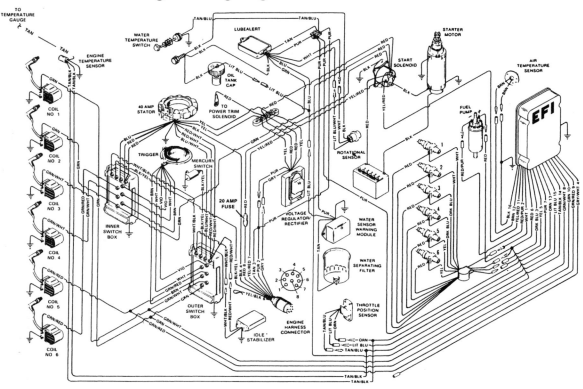

*NOTE: 175 EFI Models (S/N OD007414 on up) are equipped with Spark Advance Module in place of Idle Stabilizer. This will result in an ignition timing change. See Timing/Synchronizing/Adjusting - EFI Models section for specifications.

Engine Wiring Diagram—1990–91 200 HP EFI

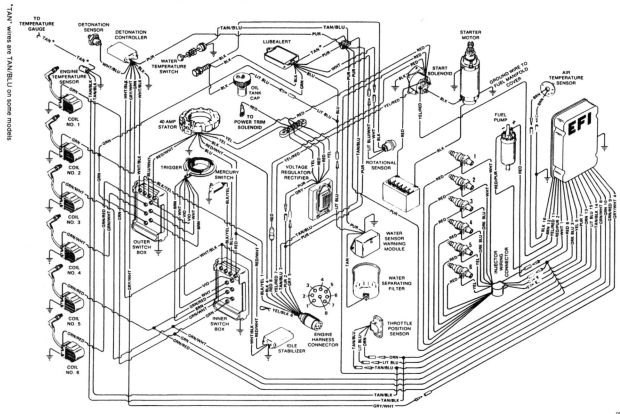

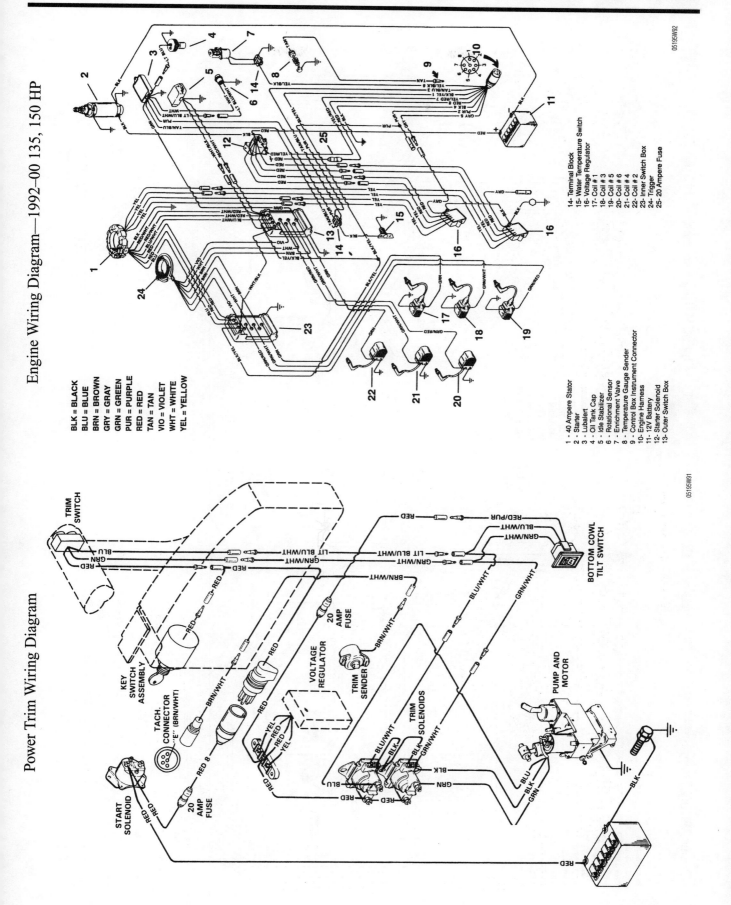

5-86 IGNITION AND ELECTRICAL SYSTEMS

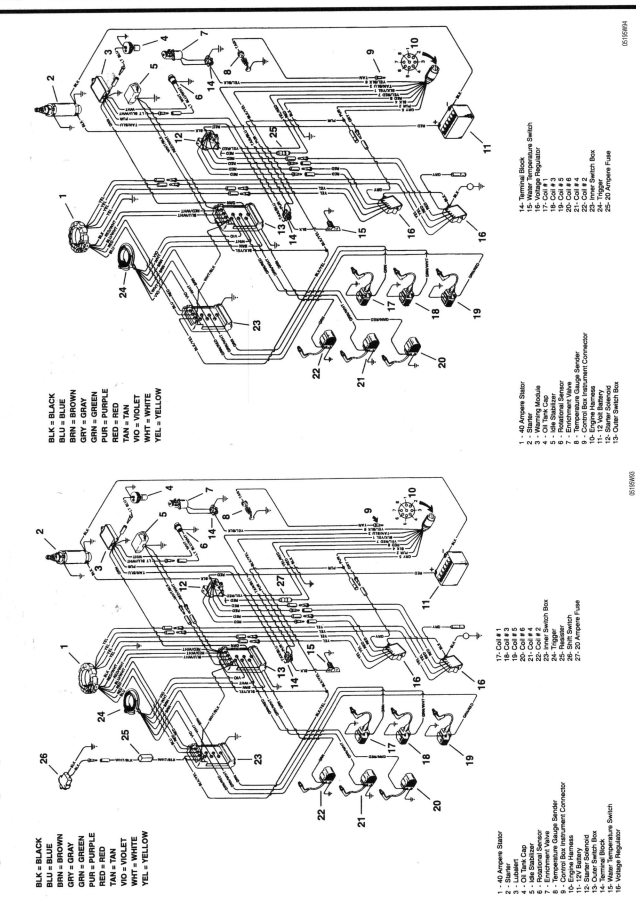

IGNITION AND ELECTRICAL SYSTEMS 5-87

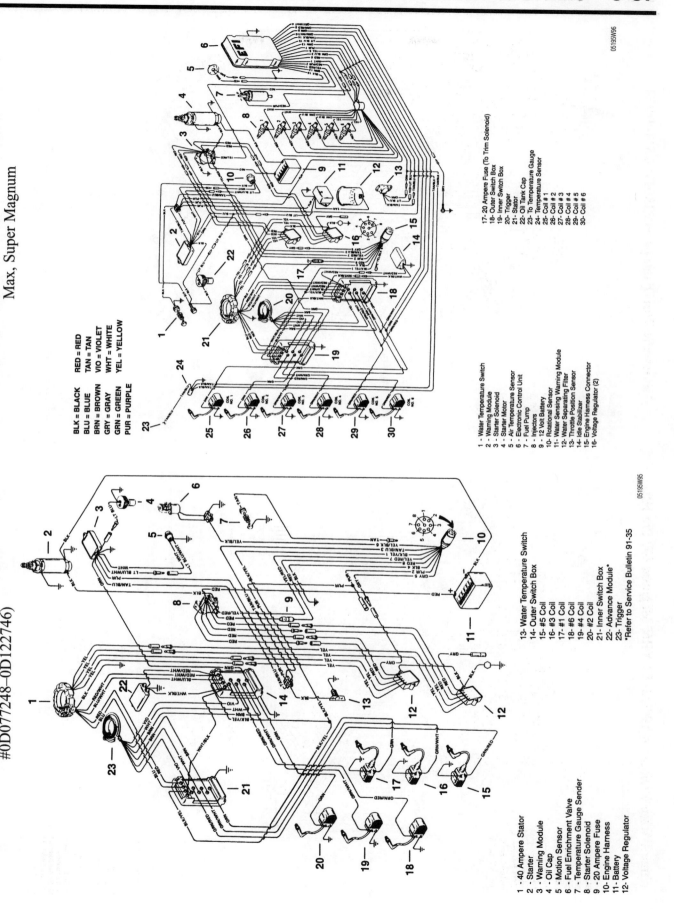

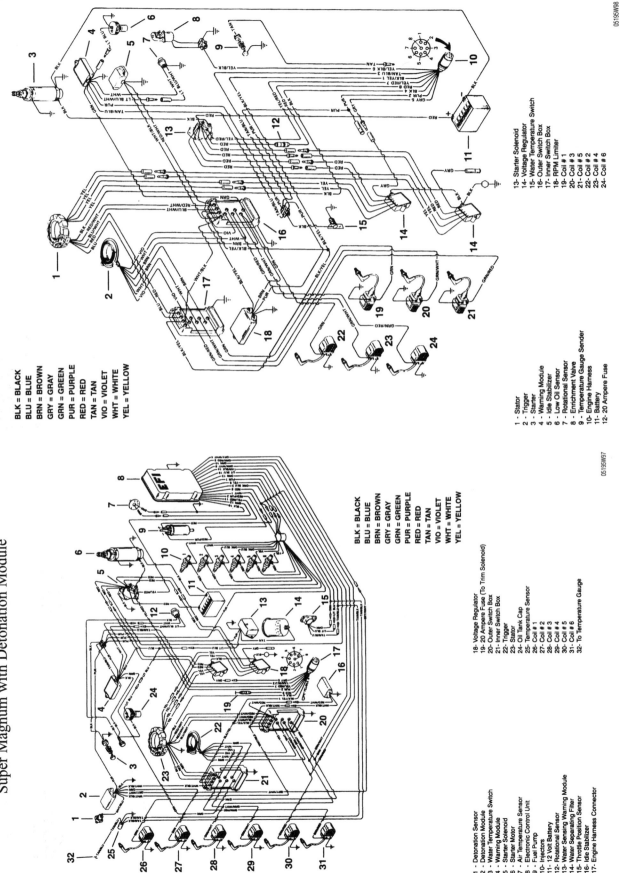

IGNITION AND ELECTRICAL SYSTEMS

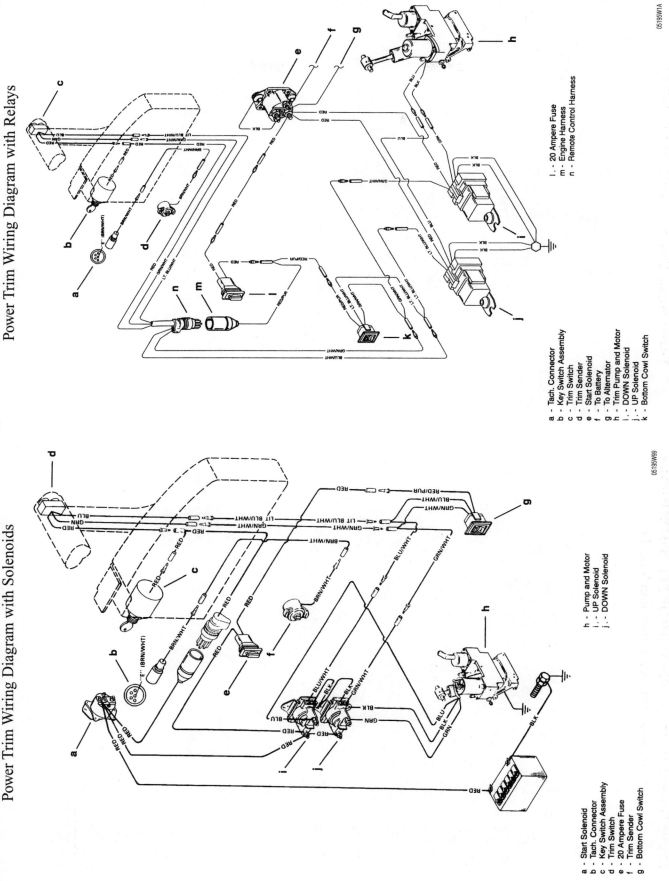

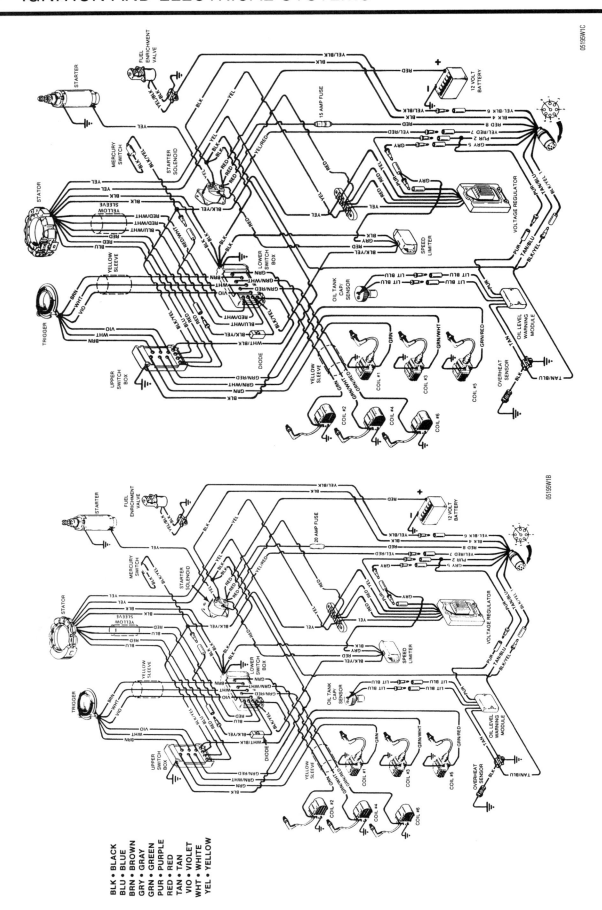

IGNITION AND ELECTRICAL SYSTEMS 5-91

Power Trim Wiring Diagram with Fuse

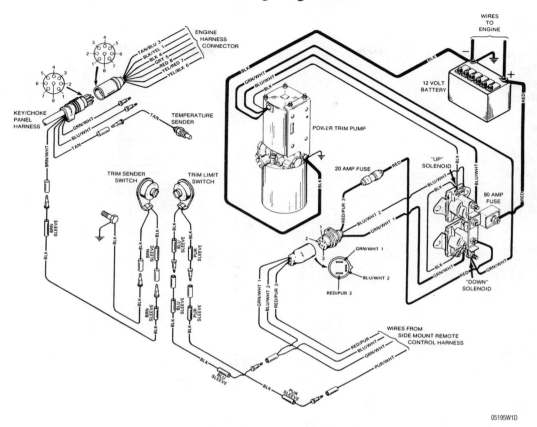

Side Mount Control Wiring Diagram

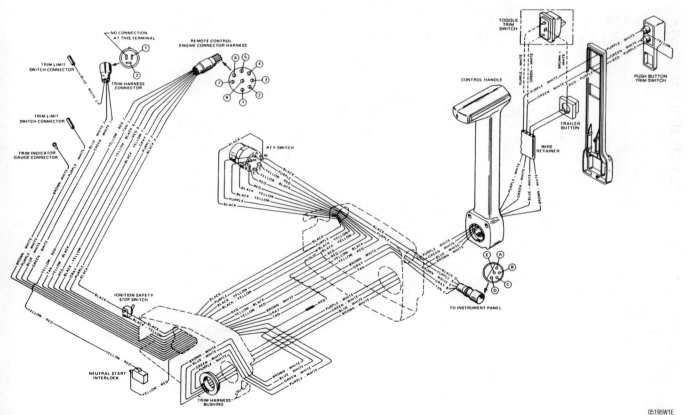

5-92 IGNITION AND ELECTRICAL SYSTEMS

Engine Wiring Diagram—1994 225 HP

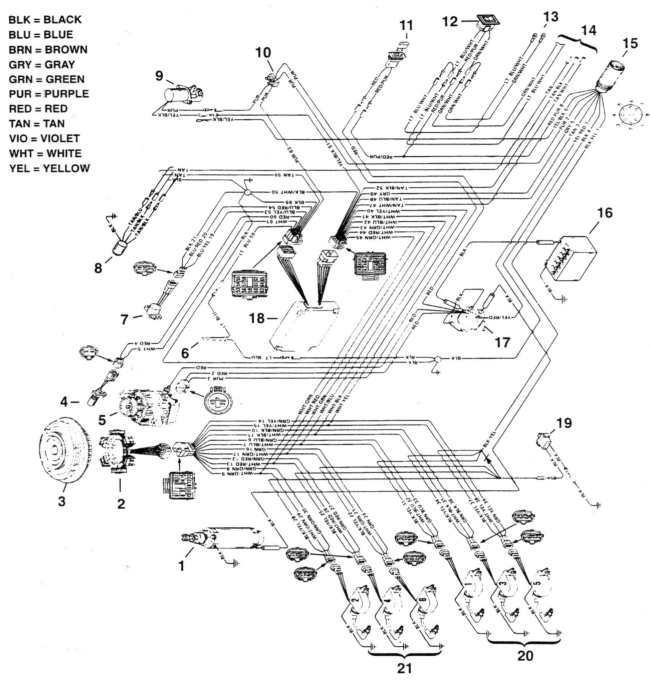

BLK = BLACK
BLU = BLUE
BRN = BROWN
GRY = GRAY
GRN = GREEN
PUR = PURPLE
RED = RED
TAN = TAN
VIO = VIOLET
WHT = WHITE
YEL = YELLOW

1 - Starter
2 - Ignition Stator
3 - Flywheel
4 - Crank Position Sensor
5 - 60 Ampere Alternator
6 - Low Oil Sensor
7 - Throttle Position Sensor
8 - Overheat Sensor
9 - Fuel Enrichment Valve
10 - Terminal Block
11 - 20 Ampere Fuse
12 - Cowl Trim Switch
13 - To Trim Solenoids
14 - To Remote Control Harness
15 - Engine Harness Plug
16 - 12 Volt Battery
17 - Starter Solenoid
18 - Electronic Control Module (ECM)
19 - Shift Interrupt Switch
20 - Starboard Ignition Modules - 1,3,5
21 - Port Ignition Modules - 2,4,6
22 - Ignition Stator

05195W1F

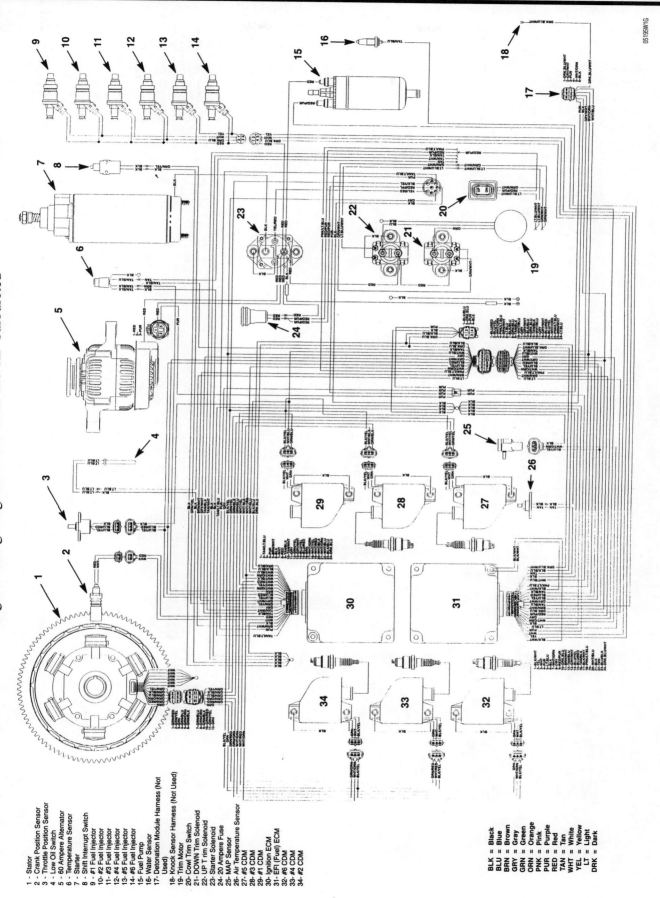

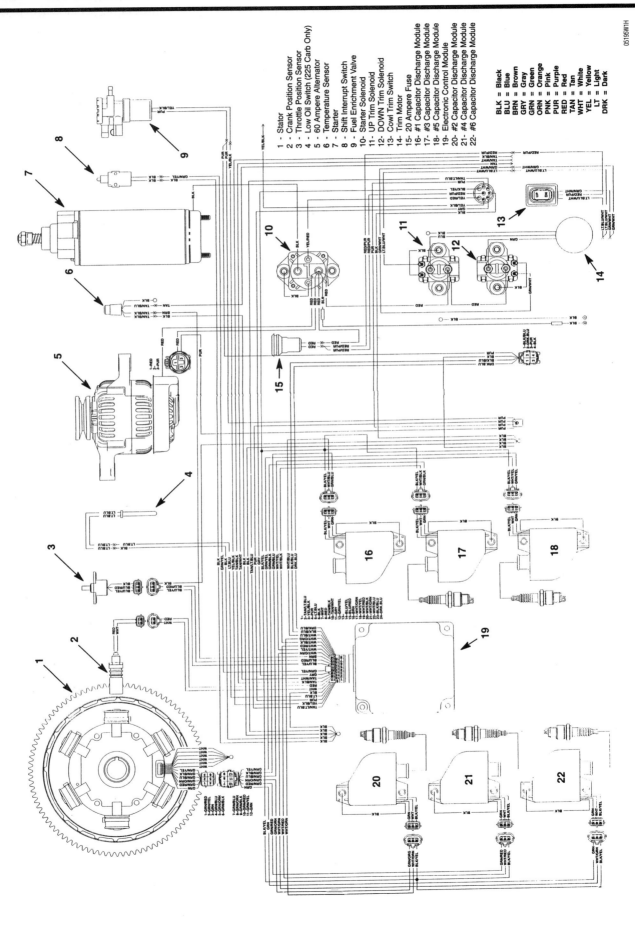

IGNITION AND ELECTRICAL SYSTEMS 5-95

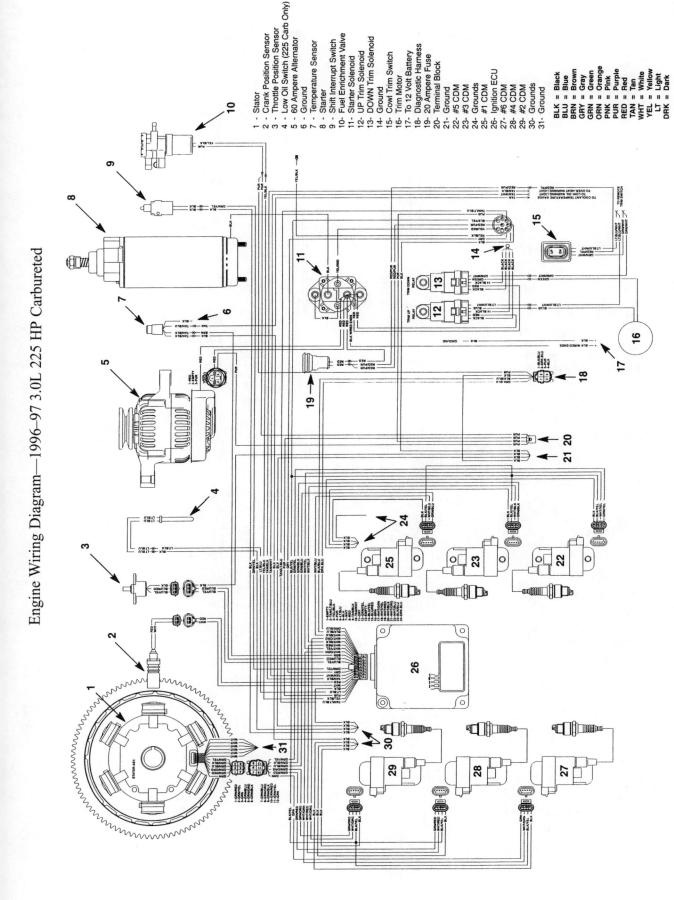

Engine Wiring Diagram—1996–97 3.0L 225 HP Carbureted

1 - Stator
2 - Crank Position Sensor
3 - Throttle Position Sensor
4 - Low Oil Switch (225 Carb Only)
5 - 60 Ampere Alternator
6 - Temperature Sensor
7 - Starter
8 - Shift Interrupt Switch
9 - Fuel Enrichment Valve
10- Starter Solenoid
11- Terminal Block
12- UP Trim Solenoid
13- DOWN Trim Solenoid
14- Ground
15- Cowl Trim Switch
16- Trim Motor
17- To 12 Volt Battery
18- Diagnostic Harness
19- 20 Ampere Fuse
20- Terminal Block
21- Ground
22- #5 CDM
23- #3 CDM
24- Grounds
25- #1 CDM
26- Ignition ECU
27- #6 CDM
28- #4 CDM
29- #2 CDM
30- Grounds
31- Ground

BLK = Black
BLU = Blue
BRN = Brown
GRY = Gray
GRN = Green
ORN = Orange
PNK = Pink
PUR = Purple
RED = Red
TAN = Tan
WHT = White
YEL = Yellow
LT = Light
DRK = Dark

5-96 IGNITION AND ELECTRICAL SYSTEMS

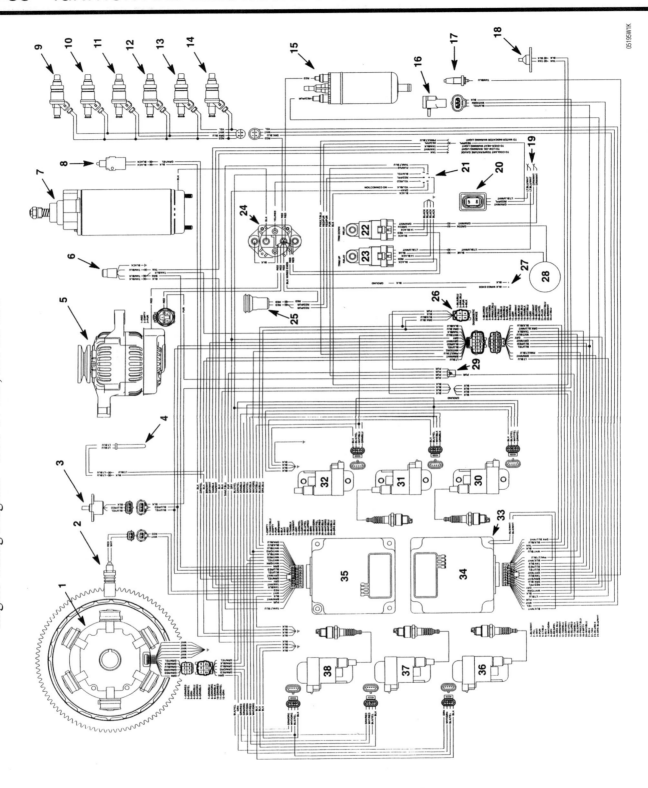

Engine Wiring Diagram—1996-97 225, 250 HP EFI

1 - Stator
2 - Crank Position Sensor
3 - Throttle Position Sensor
4 - Low Oil Switch
5 - 60 Ampere Alternator
6 - Temperature Sensor
7 - Starter
8 - Shift Interrupt Switch
9 - #1 Fuel Injector
10 - #2 Fuel Injector
11 - #3 Fuel Injector
12 - #4 Fuel Injector
13 - #5 Fuel Injector
14 - #6 Fuel Injector
15 - Fuel Pump
16 - Map Sensor
17 - Water Sensor
18 - Air Temperature Sensor
19 - To Remote Trim Switch
20 - Cowl Trim Switch
21 - Remote Control Harness
22 - DOWN Trim Solenoid
23 - UP Trim Solenoid
24 - Starter Solenoid
25 - 20 Ampere Fuse
26 - Diagnostic Harness
27 - To 12 Volt Battery
28 - Trim Motor
29 - Terminal Block
30 - #5 CDM
31 - #3 CDM
32 - #1 CDM
33 - Ground
34 - EFI (Fuel) ECM
35 - Ignition ECM
36 - #2 CDM
37 - #4 CDM
38 - #6 CDM

BLK = Black
BLU = Blue
BRN = Brown
GRY = Gray
GRN = Green
ORN = Orange
PNK = Pink
PUR = Purple
RED = Red
TAN = Tan
WHT = White
YEL = Yellow
LT = Light
DRK = Dark

IGNITION AND ELECTRICAL SYSTEMS 5-97

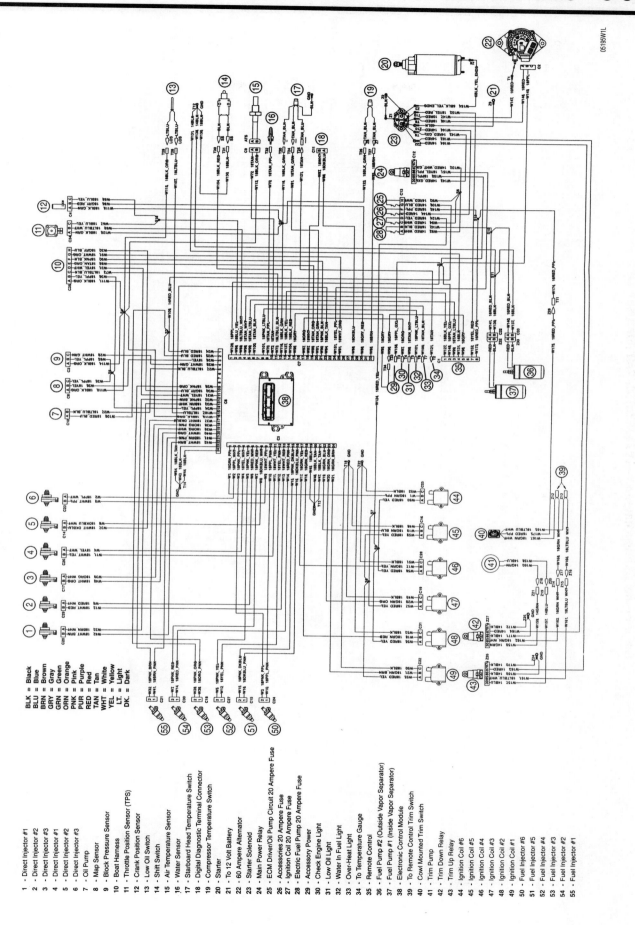

Engine Wiring Diagram—1998 135, 150 HP DFI

5-98 IGNITION AND ELECTRICAL SYSTEMS

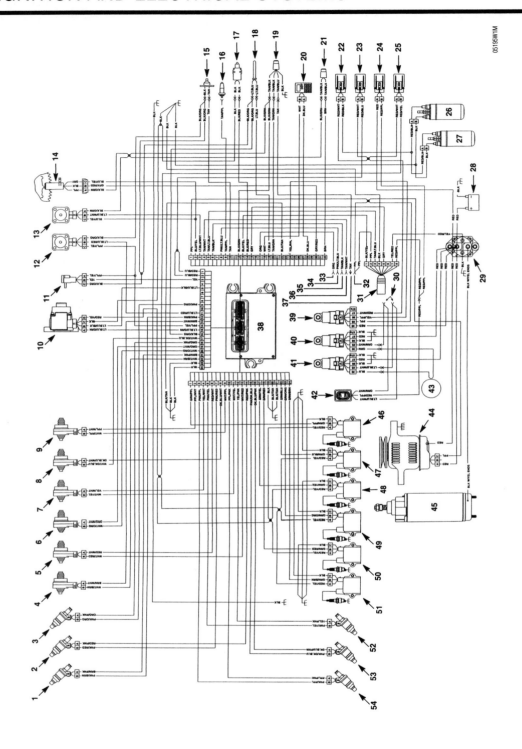

Engine Wiring Diagram—2000 115, 135, 150 HP Optimax DFI

1 - Fuel Injector #1
2 - Fuel Injector #2
3 - Fuel Injector #3
4 - Direct Injector #1
5 - Direct Injector #2
6 - Direct Injector #3
7 - Direct Injector #4
8 - Direct Injector #5
9 - Direct Injector #6
10 - Oil Pump
11 - MAP Sensor
12 - T.P.S. #1 (Inner)
13 - T.P.S. #2 (Outer)
14 - Crank Position Sensor
15 - Air Temperature Sensor
16 - Water Sensor
17 - Shift Switch
18 - Low-Oil Switch
19 - Cylinder Head Temperature Switch
20 - Digital Diagnostic Terminal Connector
21 - Compressor Temperature Sensor
22 - Fuel Pump 20 Ampere Fuse
23 - ECM 20 Ampere Fuse
24 - ACC. and Trim Pump 20 Ampere Fuse
25 - Ignition Coils and Oil Pump 20 Ampere Fuse
26 - Fuel Pump #1 (Outside Vapor Separator)
27 - Fuel Pump #2 (Inside Vapor Separator)
28 - 12 Volt Battery
29 - Starter
30 - To Remote Control
31 - Remote Control Trim Switch
32 - Auxiliary
33 - Check Engine Light
34 - Water In Fuel Light
35 - Over-Heat Light
36 - Low Oil Light
37 - Temperature Gauge
38 - Electronic Control Module
39 - Main Power Relay
40 - Trim Down Relay
41 - Trim Up Relay
42 - Cowl Mounted Trim Switch
43 - Starter Solenoid
44 - 60 Ampere Alternator
45 - Starter
46 - Ignition Coil #6
47 - Ignition Coil #5
48 - Ignition Coil #4
49 - Ignition Coil #3
50 - Ignition Coil #2
51 - Ignition Coil #1
52 - Fuel Injector #4
53 - Fuel Injector #5
54 - Fuel Injector #6

BLK = Black
BLU = Blue
BRN = Brown
GRY = Gray
GRN = Green
ORN = Orange
PNK = Pink
PUR = Purple
RED = Red
TAN = Tan
WHT = White
YEL = Yellow
LT. = Light
DK. = Dark

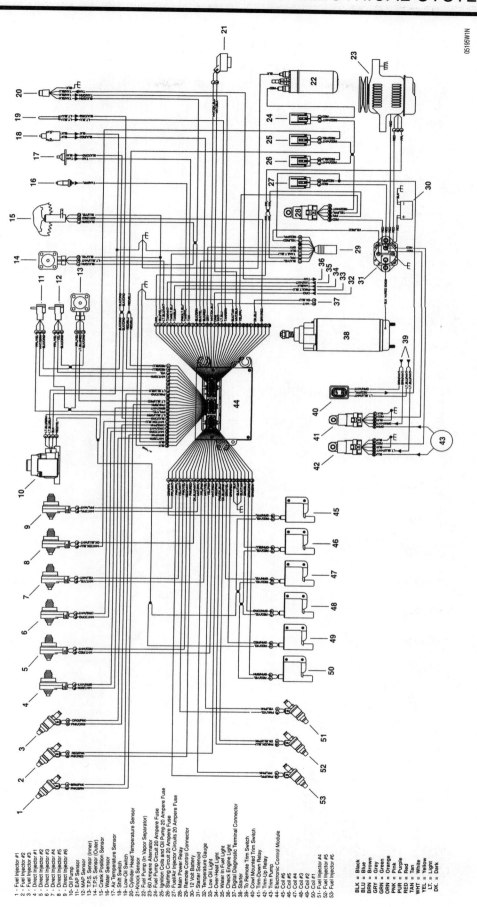

5-100 IGNITION AND ELECTRICAL SYSTEMS

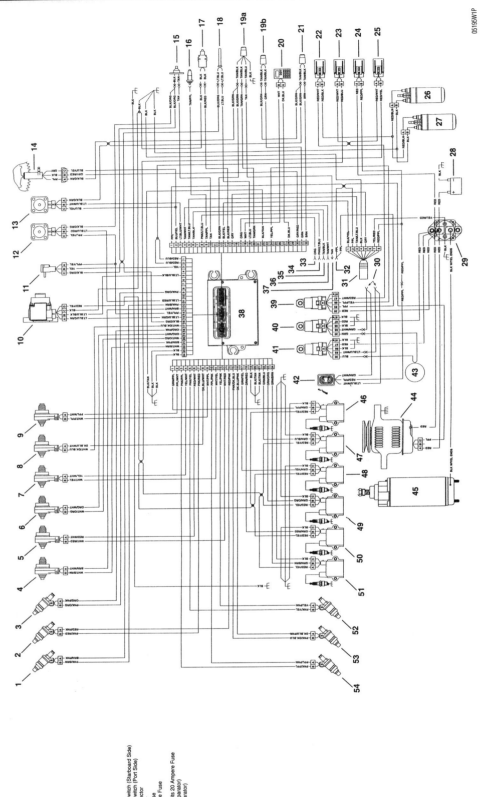

Engine Wiring Diagram—1998 200, 225 HP DFI

1 - Fuel Injector #1
2 - Fuel Injector #2
3 - Fuel Injector #3
4 - Direct Injector #1
5 - Direct Injector #2
6 - Direct Injector #3
7 - Direct Injector #4
8 - Direct Injector #5
9 - Direct Injector #6
10 - Oil Pump
11 - MAP Sensor
12 - T.P.S. #1 (Inner)
13 - T.P.S. #2 (Outer)
14 - Crank Position Sensor
15 - Air Temperature Sensor
16 - Water Sensor
17 - Shift Switch
18 - Low-Oil Switch
19 - a - Cylinder Head Temperature Switch (Starboard Side)
19 - b - Cylinder Head Temperature Switch (Port Side)
20 - Digital Diagnostic Terminal Connector
21 - Compressor Temperature Sensor
22 - Temperature Sensor
23 - Fuel Pump Circuit 20 Ampere Fuse
24 - Starting Circuit 20 Ampere Fuse
25 - Ignition Coils and Oil Pump Circuits 20 Ampere Fuse
26 - Fuel Pump #2 (Outside Vapor Separator)
27 - Fuel Pump #1 (Inside Vapor Separator)
28 - 12 Volt Battery
29 - Starter Solenoid
30 - To Remote Control
31 - Remote Control
32 - Auxiliary
33 - Check Engine Light
34 - Water in Fuel Light
35 - Over-Heat Light
36 - Low Oil Light
37 - Temperature Gauge
38 - Electronic Control Module
39 - Main Power Relay
40 - Trim Down Relay
41 - Trim Up Relay
42 - Cowl Mounted Trim Switch
43 - Trim Pump
44 - 60 Ampere Alternator
45 - Starter
46 - Ignition Coil #6
47 - Ignition Coil #5
48 - Ignition Coil #4
49 - Ignition Coil #3
50 - Ignition Coil #2
51 - Ignition Coil #1
52 - Fuel Injector #4
53 - Fuel Injector #5
54 - Fuel Injector #6

BLK = Black
BLU = Blue
BRN = Brown
GRY = Gray
GRN = Green
ORN = Orange
PNK = Pink
PUR = Purple
RED = Red
TAN = Tan
WHT = White
YEL = Yellow
LT. = Light
DK. = Dark

IGNITION AND ELECTRICAL SYSTEMS

5-101

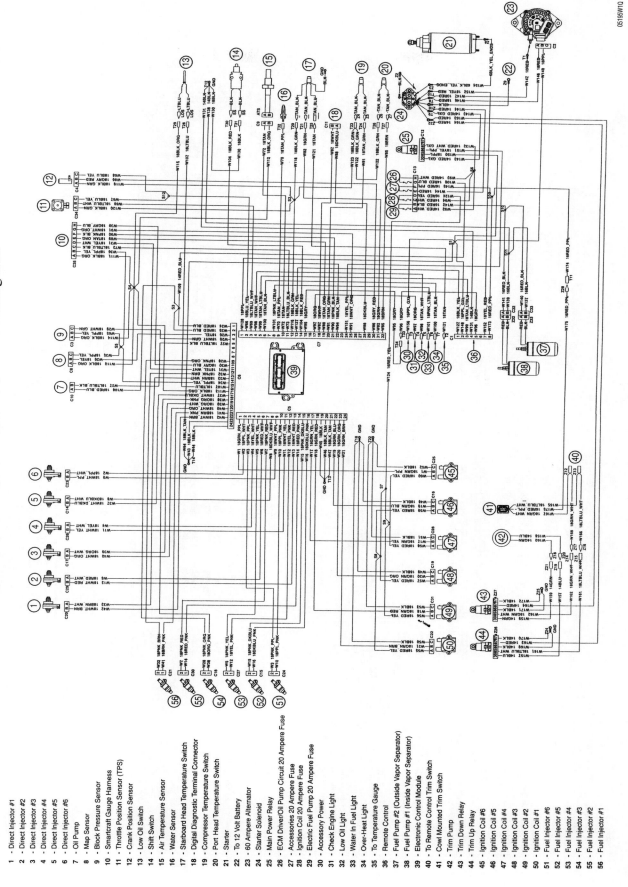

Engine Wiring Diagram—2000 200, 225 HP Analog DFI

1 - Direct Injector #1
2 - Direct Injector #2
3 - Direct Injector #3
4 - Direct Injector #4
5 - Direct Injector #5
6 - Direct Injector #6
7 - Oil Pump
8 - Map Sensor
9 - Block Pressure Sensor
10 - Smartcraft Gauge Harness
11 - Throttle Position Sensor (TPS)
12 - Crank Position Sensor
13 - Low Oil Switch
14 - Shift Switch
15 - Air Temperature Sensor
16 - Water Sensor
17 - Starboard Head Temperature Switch
18 - Digital Diagnostic Terminal Connector
19 - Compressor Temperature Switch
20 - Port Head Temperature Switch
21 - Starter
22 - To 12 Volt Battery
23 - 60 Ampere Alternator
24 - Starter Solenoid
25 - Main Power Relay
26 - ECM Driver/Oil Pump Circuit 20 Ampere Fuse
27 - Accessories 20 Ampere Fuse
28 - Ignition Coil 20 Ampere Fuse
29 - Electric Fuel Pump 20 Ampere Fuse
30 - Accessory Power
31 - Check Engine Light
32 - Low Oil Light
33 - Water In Fuel Light
34 - Over-Heat Light
35 - To Temperature Gauge
36 - Remote Control
37 - Fuel Pump #2 (Outside Vapor Separator)
38 - Fuel Pump #1 (Inside Vapor Separator)
39 - Electronic Control Module
40 - Cowl Mounted Trim Switch
41 - To Remote Control Trim Switch
42 - Trim Pump
43 - Trim Down Relay
44 - Trim Up Relay
45 - Ignition Coil #6
46 - Ignition Coil #5
47 - Ignition Coil #4
48 - Ignition Coil #3
49 - Ignition Coil #2
50 - Ignition Coil #1
51 - Fuel Injector #6
52 - Fuel Injector #5
53 - Fuel Injector #4
54 - Fuel Injector #3
55 - Fuel Injector #2
56 - Fuel Injector #1

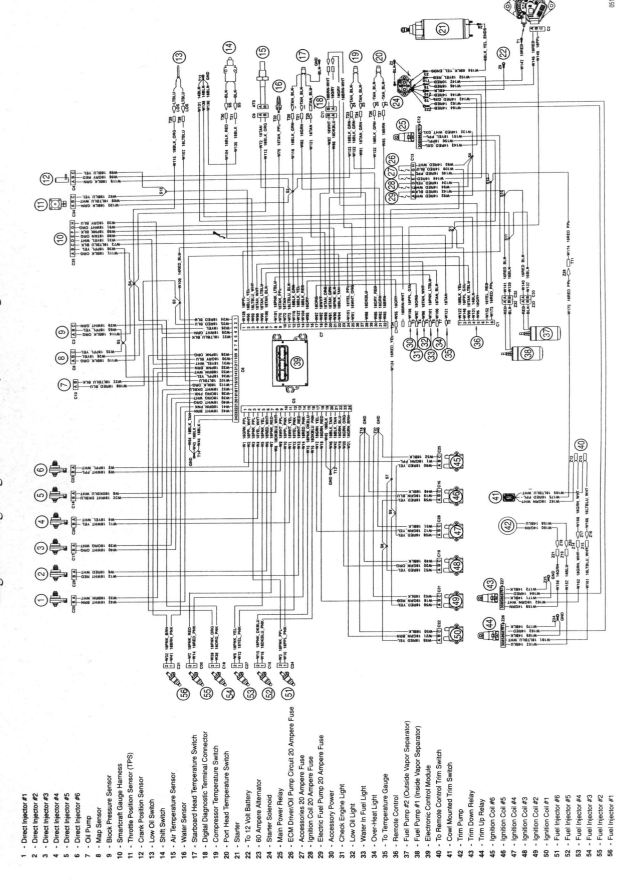

OIL INJECTION SYSTEM 6-2
  DESCRIPTION AND OPERATION 6-2
    MERCURY OIL INJECTION 6-2
    OPTIMAX OIL INJECTION 6-2
  TROUBLESHOOTING THE OIL INJECTION
    SYSTEM 6-3
  BLEEDING THE OIL INJECTION
    SYSTEM 6-3
    PROCEDURE 6-3
  OIL TANK 6-4
    REMOVAL & INSTALLATION 6-4
    CLEANING & INSPECTION 6-6
  OIL PUMP 6-6
    REMOVAL & INSTALLATION 6-6
  OIL LINES 6-7
    OIL LINE CAUTIONS 6-7
  OIL PUMP DISCHARGE RATE 6-7
    TESTING 6-7
  OIL PUMP CONTROL ROD 6-7
    ADJUSTMENT 6-7
COOLING SYSTEM 6-8
  DESCRIPTION AND OPERATION 6-8
    OPTIMAX 6-8
  TROUBLESHOOTING THE COOLING
    SYSTEM 6-9
  WATER PUMP 6-9
    REMOVAL & INSTALLATION 6-9
    CLEANING & INSPECTION 6-20
  THERMOSTAT 6-20
    REMOVAL & INSTALLATION 6-20
    CLEANING & INSPECTION 6-22
    TESTING 6-22
WARNING SYSTEMS 6-22
  DESCRIPTION AND OPERATION 6-22
    OIL QUANTITY INDICATOR 6-22
    OVERHEAT TEMPERATURE
      WARNING 6-23
  TROUBLESHOOTING THE WARNING
    SYSTEMS 6-24
  WATER TEMPERATURE SENSOR 6-24
    REMOVAL & INSTALLATION 6-24
    TESTING 6-26
  OIL GAUGE 6-26
    TESTING 6-26
  OIL LEVEL SWITCH 6-26
    TESTING 6-26
    REMOVAL & INSTALLATION 6-26
OPTIMAX WARNING SYSTEMS 6-27
  DESCRIPTION AND OPERATION 6-27
    1998–99 MODEL YEARS 6-27
    GUARDIAN PROTECTION SYSTEM—
      2000 6-27
  TROUBLESHOOTING THE WARNING
    SYSTEMS 6-28
SPECIFICATIONS CHART
  OIL PUMP DISCHARGE RATE 6-7

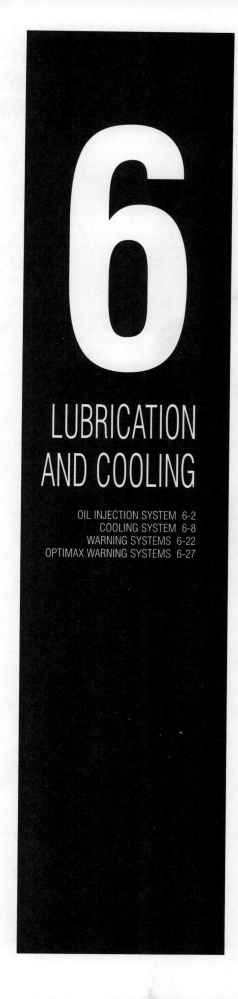

6

LUBRICATION AND COOLING

OIL INJECTION SYSTEM 6-2
COOLING SYSTEM 6-8
WARNING SYSTEMS 6-22
OPTIMAX WARNING SYSTEMS 6-27

6-2 LUBRICATION AND COOLING

OIL INJECTION SYSTEM

Description and Operation

Over the years, manufacturers have tried various ways of mixing oil with the fuel on 2-stroke outboard engines. This was not only to make refueling easier, but to also provide more accurate fuel/oil ratios over the entire engine operational range.

An outboard's requirement for a fuel/oil mix does not stay constant. Mixture requirement varies depending on engine RPM and load. A mixture that is too rich in oil results in excess exhaust smoke.

Pre-mixing the engine lubrication oil with gasoline before use is by far the simplest method. Unfortunately, it's also the messiest and the greatest polluter of the environment. Nothing complicated here, just add the correct amount of oil to a certain volume of fuel in your tank, and the job is done.

One important consideration often overlooked is the grade of 2-stroke oil used. Current standards call for TCW-3, but check with your manufacturer for their recommendation. TCW-3 has a far superior additives that reduce many common outboard-engine problems such as sticking rings and carbon build-up in the combustion chamber.

It's also important to know what ratio of fuel to oil your engine manufacturer recommends. In recent years, many manufacturers have reduced the ratio from 50:1 to 100:1. This means 100 parts of gasoline to 1 part of oil or half as much oil as before.

The order in which you do things counts, too. Remember to put the correct amount of oil in the tank before adding the gasoline, so the gasoline will mix with the new oil as you fill the tank.

The second popular type of lubrication for 2-strokes is a mechanical oil pump systems. Most outboard manufacturers use a mechanically driven oil pump mounted on the engine block. The oil pump is connected to the throttle by way of a linkage arm. The theory of operation here is that the crankshaft drives a gear in the pump, creating oil pressure. As the throttle lever is advanced to increase engine speed, the linkage arm also moves, opening a valve that allows more oil to flow into the oil pump.

There are two schools of thought on the matter of where to inject the oil and this is where the significant difference lies between the mechanical oil-pump systems. Some manufacturers inject the oil into a port in the carburetor-intake system behind the carburetor throttle plate. This means that only pure fuel passes through the carburetor. The oil blends with the incoming fuel/air mixture immediately after it leaves the carburetor throat, just before the mix enters the crankcase through the reed-valve plates.

Other manufacturers inject the oil from the mechanical pump into a port in the engine mounted fuel pump, where it is mixed with the fuel before it enters the carburetor.

All outboard engines since the late 60's are equipped with a system that's designed to recycle excess oil accumulating in the crankcase. Oil that's injected into the engine is not completely consumed as it enters the combustion chamber with the air/fuel mix. This oil puddles inside the crankcase.

To remove this excess oil, manufacturers have devised a simple system of small fittings screwed directly into the side of the powerhead block assembly at key points. These fittings have hoses connected to the car buretor intake system, so the excess oil from the crankcase can be mixed with the incoming air/fuel mixture.

MERCURY OIL INJECTION

The constant ration oil injection pump on the 40–60hp models is driven by a gear on the crankshaft and delivers a constant 50:1 fuel/oil mixture to each cylinder. The pump injects oil into the fuel stream ahead of the fuel pump.

A 2-psi (13.8 kPa) check valve is installed in the fuel line between the fuel line connector and the oil pump discharge line. The check valve is used to prevent gasoline from entering the pump discharge line.

The variable ratio oil pump on the 40–275hp models is mounted on the powerhead and is driven by a gear and shaft arrangement off the crankshaft. Therefore, as soon as the crankshaft begins to rotate, even during the cranking process, the pump also rotates and begins to deliver oil to the fuel/oil mixer. The pump will meter the injection oil for a fuel/oil ratio of approximately 50:1 at wide open throttle, with the mixture ratio changing to 100:1 at idle speed.

OPTIMAX OIL INJECTION

▶ See Figures 1 and 2

Oil in this engine is not mixed with the fuel before entering the combustion chamber. Oil is stored inside the remote oil tank in the boat. Crankcase pressure forces the oil from the remote oil into the engine mounted oil reservoir. The engine oil reservoir feeds oil to the oil pump. The oil pump is ECM driven and controls the oil distribution to the crankcase and air compressor. The oil pump has seven oil discharge ports. Six of the oil discharge ports inject oil into the crankcase through hoses, with one hose for each cylinder. The last oil discharge port sends oil to the air compres-

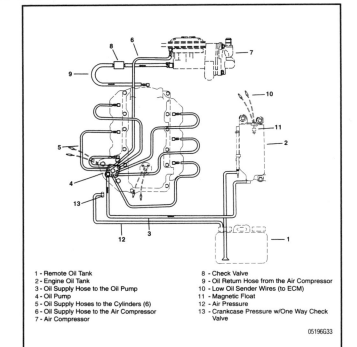

1 - Remote Oil Tank
2 - Engine Oil Tank
3 - Oil Supply Hose to the Oil Pump
4 - Oil Pump
5 - Oil Supply Hoses to the Cylinders (6)
6 - Oil Supply Hose to the Air Compressor
7 - Air Compressor
8 - Check Valve
9 - Oil Return Hose from the Air Compressor
10 - Low Oil Sender Wires (to ECM)
11 - Magnetic Float
12 - Air Pressure
13 - Crankcase Pressure w/One Way Check Valve

Fig. 1 Oil injection system schematic—115 to 150hp

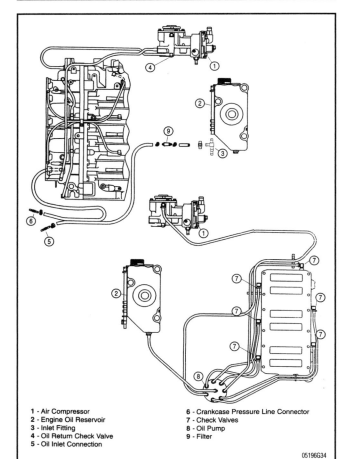

1 - Air Compressor
2 - Engine Oil Reservoir
3 - Inlet Fitting
4 - Oil Return Check Valve
5 - Oil Inlet Connection
6 - Crankcase Pressure Line Connector
7 - Check Valves
8 - Oil Pump
9 - Filter

Fig. 2 Oil injection system schematic—200 to 225hp

LUBRICATION AND COOLING 6-3

sor for lubrication. Unused oil from the air compressor returns to the plenum and is ingested through the crankcase.

The ECM is programmed to automatically increase the oil supply to the engine during the initial engine break-in period. The oil ratio is doubled during the first 120 minutes of operation whenever the engine speed exceeds 2500 rpm and is under load; below 2500 rpm the oil pump provides oil at the normal ratio. After the engine break-in period, the oil ratio will return to normal: 300–400:1 at idle and 60:1 at wide open throttle (depending on throttle load).

➡ On some light boat applications after break-in is completed and the engine is being run at cruising speed (between 4000–5000 rpm) the fuel to oil ratio may be as high as 40:1. This results from a reduced throttle opening with a corresponding reduction in fuel consumption.

The oil pump on the OptiMax models is electrically operated by the ECM and controls oil distribution to the crankcase and air compressor. The oil pump has seven discharge ports. Six of the discharge ports inject oil into the crankcase through hoses, one hose for each cylinder. The last discharge port discharges oil to the air compressor for lubrication.

Troubleshooting the Oil Injection System

Like other systems on a 2-stroke engine, the oil injection system emphasizes simplicity. On pre-mix engines, when enough oil accumulated in the crankcase it passed into the combustion chamber where it burned along with the fuel. Since oil is a fuel just like gasoline, it burns as well. Unlike gasoline, oil doesn't burn as efficiently and may produce blue smoke that is seen coming out of the exhaust. If the engine was cold, such as during initial start-up, oil has an even harder time igniting resulting in excessive amounts of smoke.

Other problems, such oil remaining in the combustion chamber, tend to foul spark plugs and cause the engine to misfire at idle. Most of these problems have been alleviated by the introduction of automatic oil injection systems.

➡ One of the most common problems with oil injection systems is the use of poor quality injection oil. Poor quality oil tends to gel in the system, clogging lines and filters.

It is normal for a 2-stroke engine to emit some blue smoke from the exhaust. The blue color of the smoke comes from the burning 2-stroke oil. An excessive amount of blue smoke indicates too much oil being injected into the engine. On most engines, this is usually caused by an incorrectly adjusted injection control rod.

If the exhaust smoke is white, this is a sign of water entering the combustion chamber. Water may enter as condensation or more seriously may enter through a defective head gasket or cracked head. Usually white smoke from condensation will disappear quickly as the engine warms.

If the exhaust smoke is black, this is a sign of an excessively rich fuel mixture or incorrect spark plugs. The black color of the smoke comes from the fuel burning.

Bleeding the Oil Injection System

PROCEDURE

♦ See Figures 3 and 4

30–275 HP

OIL PUMP INLET HOSE

1. With the engine shut off, hold an absorbent cloth below the oil injection pump and be prepared to catch the oil as it oozes from the bleed screw.
2. Loosen the bleed screw three, maybe four full turns, and allow the air trapped in the system to exit the inlet hose.

➡ This procedure also allows the oil pump to fill with oil.

3. When a steady stream of oil is observed with no sign of air, tighten the bleed screw securely.

OIL PUMP OUTLET HOSE

1. On all engines except the V-6 models, purge air from the outlet hose by running the engine (on a 50:1 fuel/oil mixture) at idle speed until no more air bubbles are present in the outlet hose.
2. On V-6 models, if any air bubbles persist, they can be purged out of the hose by removing the link rod and rotating the pump arm full clockwise while operating the engine at 1000–1500 rpm.

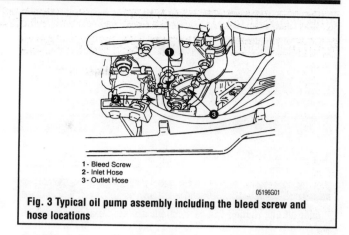

Fig. 3 Typical oil pump assembly including the bleed screw and hose locations
1 - Bleed Screw
2 - Inlet Hose
3 - Outlet Hose

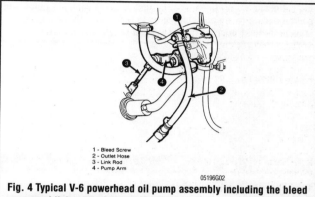

Fig. 4 Typical V-6 powerhead oil pump assembly including the bleed screw and link rod locations
1 - Bleed Screw
2 - Outlet Hose
3 - Link Rod
4 - Pump Arm

3. If necessary, gently pinch the fuel line between the remote fuel line connector and the oil injection pump "T" fitting. This will cause the fuel pump to provide a partial vacuum which will aid in the removal of any trapped air in the system.
4. After this procedure, reinstall the link rod.

OptiMax

➡ If a new powerhead is being installed or the oil hoses or pump have been removed, it is recommended that all air be purged from the oil pump and hoses. This can be accomplished by using a gearcase leakage tester (FT-8950). Connect the leakage tester to the inlet t-fitting on the onboard oil reservoir. While clamping off the inlet hose, manually pressurize the reservoir to 10 psi. Using the DDT, activate the oil pump prime sequence. Maintain the 10 psi pressure throughout the auto prime sequence. when the auto prime is completed, remove the leakage tester and refill the onboard oil reservoir.

Priming the oil pump (filling the pump and hoses) is required on new or rebuilt engines and any time maintenance is performed on the oiling system. There are three methods for priming the oil pump:

SHIFT SWITCH ACTIVATION PRIME

This method does three things:
• Fills the oil pump, oil supply hose feeding pump and oil hoses going to the crankcase and air compressor
• Activates the break-in oil ratio
• Initiates a new 120 minute engine break-in cycle

DIGITAL DIAGNOSTIC TERMINAL (DDT)—RESET BREAK-IN

This method is the same as the above method, except the run history and fault history are erased from the ECM.

DIGITAL DIAGNOSTIC TERMINAL (DDT)—OIL PUMP PRIME

This method fills the oil pump, oil supply hose feeding the pump and the oil hoses going to the crankcase and air compressor.

6-4 LUBRICATION AND COOLING

Use the DDT to activate the Auto Prime function. The oil pump should discharge 3.44–3.99 ounces (102–118 ml) during the auto prime period.
1. To check the oil pump output:
   - Verify the on board oil reservoir is full
   - Release any pressure (loosen cap) from the remote oil tank in the boat
   - With the engine running, use the DDT to activate the Auto Prime
   - Using a graduated container, record the amount of oil discharged
   - Retighten the cap on the remote oil tank

Oil Tank

REMOVAL & INSTALLATION

※※ WARNING

Proper oil line routing and connections are essential for correct oil injection system operation. The line connections to the powerhead and oil pump look the same but may contain check valves of differing calibrations. Oil lines must be installed between the pump and powerhead correctly and connected to the proper fittings on the intake manifold in order for the system to operate properly.

30–60 HP

♦ See Figures 5, 6 and 7

1. Disconnect the battery cables at the battery.
2. Remove the engine cowling.
3. Remove the starter/oil tank bracket (if necessary).
4. Disconnect the oil hose and low oil sensor wires at the bullet connectors.
5. Remove the low oil sensor from the oil tank.

※※ CAUTION

The sensor is fragile. Always handle it with care.

6. Remove the oil tank assembly from the powerhead.

To install:

7. Install the oil sensor into the tank and tighten the screw securely.
8. Connect the oil level sensor wires at the bullet connectors.
9. Connect the oil line to the pump.

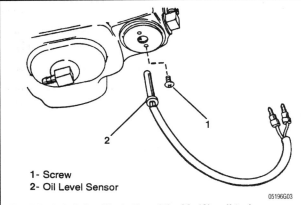

1- Screw
2- Oil Level Sensor

Fig. 5 Exploded view illustration of the 30–40hp oil tank assembly . . .

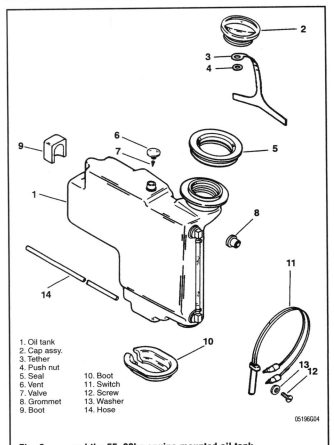

1. Oil tank
2. Cap assy.
3. Tether
4. Push nut
5. Seal
6. Vent
7. Valve
8. Grommet
9. Boot
10. Boot
11. Switch
12. Screw
13. Washer
14. Hose

Fig. 6 . . . and the 55–60hp engine mounted oil tank

※※ WARNING

Any time the oil tank hose is disconnected, the oil injection pump must be purged (bled) of any trapped air. Failure to bleed the system could lead to powerhead seizure due to lack of adequate lubrication.

10. Install the oil tank onto the powerhead.
11. Tighten the oil tank retaining bolts securely to the powerhead.
12. Install the engine cowling.
13. Connect the battery cables.

75–125 HP

♦ See Figure 7

1. Disconnect the battery cables.
2. Remove the engine cowling.
3. Disconnect the oil reservoir outlet hose from the fitting and let the reservoir drain into a suitable container.
4. Disconnect the low oil level sensor at the bullet connectors. These connectors are located at the bottom of the oil reservoir.
5. Remove the starter motor upper bracket and remove the oil reservoir screws from the bracket.
6. Remove the lower screws securing the oil reservoir and remove the reservoir from the powerhead.

LUBRICATION AND COOLING

To install:

7. Install the oil reservoir onto the powerhead and align the mounting holes.
8. Secure the oil reservoir with the mounting screws.
9. Connect the low oil level sensor wires at the bullet connectors.
10. Connect the oil reservoir outlet hose.

> ※※ **WARNING**
>
> Any time the oil tank hose is disconnected, the oil injection pump must be purged (bled) of any trapped air. Failure to bleed the system could lead to powerhead seizure due to lack of adequate lubrication.

11. Refill the oil reservoir. Do not over fill the reservoir. Add only enough oil to bring the oil level up to the bottom of the filler neck.

➡ The oil tank capacity for 3-cylinder models is 3.2 quarts (3.0 liters) and 4-cylinder models is 5.13 quarts (4.9 liters)

135–275 HP

ENGINE MOUNTED OIL RESERVOIR

♦ See Figures 8 and 9

1. Disconnect the reservoir oil hoses at the oil pump. If the reservoir still has some oil in it, plug the hoses to prevent oil from leaking out.
2. Disconnect the low oil level sensor wires or remove the fill cap with the wires still attached.
3. Remove the three bolts securing the oil reservoir to the powerhead and then remove the oil reservoir.

To install:

4. Apply a threadlocking compound, such as Loctite® 222, or equivalent, to the threads of the mounting bolts and then secure the oil reservoir to the powerhead. Tighten the bolts to 25 inch lbs. (2.8 Nm) on the 135–200hp models or 150 inch lbs. (16.9 Nm) on the 250 and 275hp models.
5. Install the input hose to the top of the oil reservoir and secure it with a tie-wrap.

> ※※ **WARNING**
>
> Any time the oil tank hose is disconnected, the oil injection pump must be purged (bled) of any trapped air. Failure to bleed the system could lead to powerhead seizure due to lack of adequate lubrication.

6. Connect the oil level sensor leads to the bullet connectors or install the filler cap.
7. Fill the reservoir tank. Be careful to not overfill the tank.

REMOTE OIL TANK

♦ See Figures 9, 10 and 11

The remote oil tank should be installed in an area in the boat where there is access for refilling the tank. The tank should be restrained to keep it from moving around, causing possible damage.

An acceptable means of restraining the tank would be the use of eye bolts and an elastic restraining strap across the center of the tank. Taking care that any metal hooks do not puncture the tank.

Keep in mind, when installing the tank in tight areas, that this tank will be under pressure while the engine is operating and will expand slightly. Therefore, the restraints used with the tank should be left with a small amount of slack to enable the tank expand.

Fig. 7 The tank is provided with an oil level indicator which is visible through a window in the cowling

Fig. 8 A typical oil reservoir installation on top of the powerhead providing gravity flow to the oil pump

Fig. 9 Remote oil tank for the oil injection system ready for installation in the stern of the boat

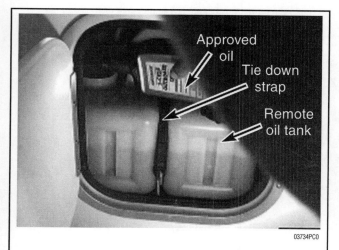

Fig. 10 Adding oil to a typical remote oil tank installation in the stern of a boat under a seat against the transom

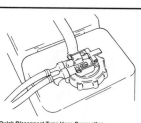
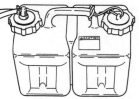

Fig. 11 Two different types of oil injection hose connections, the quick-disconnect type and one-piece type

6-6 LUBRICATION AND COOLING

Oil hoses routed through the engine well must be able to reach the hose fittings on the engine through its full range of movement and not bind, stretch or kink. Any lose of injector oil could cause serious damage to the powerhead.

CLEANING & INSPECTION

One of the most common problems with oil injection systems is the use of poor quality injection oil. This poor quality oil tends to gel in the system, clogging oil lines and filters. If this is found to be the case with your system, or if the powerhead has been sitting in storage for a length of time, it is wise to remove the oil tank and clean it with solvent.

While it is removed, take the opportunity to inspect it for damage and replace it as necessary. The oil tank is the only source of oil for the powerhead. If it should leak, the powerhead will eventually run out of injection oil, with catastrophic and very costly results. Remember, there are no parts stores when you are miles out at sea.

Oil Pump

REMOVAL & INSTALLATION

> ※※ **WARNING**
>
> Proper oil line routing and connections are essential for correct oil injection system operation. The line connections to the powerhead and oil pump look the same but may contain check valves of differing calibrations. Oil lines must be installed between the pump and powerhead correctly and connected to the proper fittings on the intake manifold in order for the system to operate properly.

The only purpose for disassembling oil injection pump is to locate a problem in oil delivery. For example, if the pump is frozen due to debris or rust, the pump can be disassembled and cleaned.

Constant ratio (40–60 HP)

▸ See Figures 12, 13 and 14

1. Disconnect the oil inlet and discharge hose from the pump.
2. Remove the pump mounting screws.
3. Remove the pump assembly from the powerhead.

➡ **If the pump driven gear remains in the engine block, use a pair of needle nose pliers to remove it.**

4. Remove the O-rings and inspect them for cuts, pinching or any damage.

To install:

5. Throughly lubricate the driven gear shaft with needle bearing assembly lube.
6. Insert the driven gear into the bearing assembly. Make sure that the gear is properly engaged on the pump shaft.
7. Coat the O-rings with needle bearing assembly lube and install them onto the pump assembly.
8. Install the pump assembly into the powerhead. Apply a threadlocking compound, such as Loctite® 271, or equivalent, to the threads of the pump mounting screws. Install the screws and tighten to 45 inch lbs.. (5.1 Nm).
9. Connect the oil inlet and discharge hoses to the pump and clamp them securely to the pump using new tie-wraps.

Variable Ratio (40–275 HP)

▸ See accompanying illustrations

1. Pry the oil injection link rod free of the ball joint on the injection pump lever. Take care not to alter the length of this rod.
2. Position a suitable container as far as possible under the oil pump to receive oil drained from the tank
3. Snip the Sta-strap at the inlet fitting. Squeeze the oil supply line from the tank to the pump, to restrict the flow of oil while pulling it free of the fitting.
4. Allow the contents of the tank to drain into the container.
5. Pull the oil line free of the other oil pump fitting.
6. Remove the two bolts securing the pump to the powerhead and lift the pump clear.

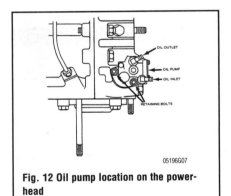

Fig. 12 Oil pump location on the powerhead

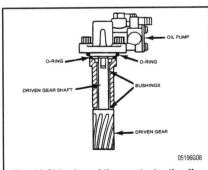

Fig. 13 Side view of the constant ratio oil pump

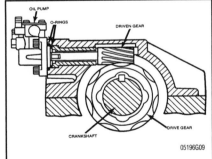

Fig. 14 The oil pump is driven off the crankshaft

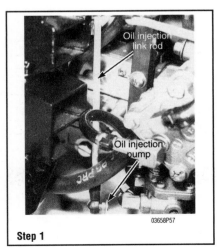

Step 1

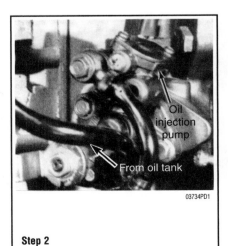

Step 2

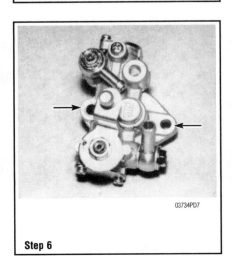

Step 6

LUBRICATION AND COOLING 6-7

Step 7

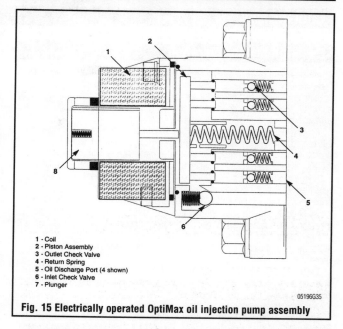

1 - Coil
2 - Piston Assembly
3 - Outlet Check Valve
4 - Return Spring
5 - Oil Discharge Port (4 shown)
6 - Inlet Check Valve
7 - Plunger

Fig. 15 Electrically operated OptiMax oil injection pump assembly

To install:

7. Check to be sure the shaft of the oil pump will index into the slot at the center of the crankshaft driven shaft. If the two are no longer aligned, rotate the pump shaft to match the slot in the driven shaft. Install the oil pump with the pump shaft indexed into the slot on the driven shaft. Secure the pump to the powerhead with the two attaching bolts. Tighten the bolts securely.

8. Connect the line from the tank to the lower pump fitting. Connect the line from the fuel pump to the upper pump fitting.

9. Snap the oil injection link rod back onto the ball joint on the pump lever.

10. Adjust the return spring tension on the lever shaft to permit the correct amount of friction for the lever to stay in place when the rod is moved.

11. Fill and bleed the oil injection system.

OptiMax

♦ See Figure 15

The manufacturer has made no provisions for rebuilding this pump. Spare parts are not available. If any part is found to be defective and no longer fit for service the pump must be replaced.

Save the O-rings, even if they are defective. The old ring will be essential when purchasing a new ring to ensure the proper type and size is obtained.

1. Disconnect the wiring harness from the pump.
2. Mark the hoses for correct location and disconnect the oil hoses.
3. Remove the bolts securing the pump to the powerhead and remove the pump assembly.

To install:

4. Install the pump assembly onto the powerhead tighten the bolts.
5. Reconnect the oil hoses in the correct locations.
6. Connect the wiring harness.
7. Refill the oil system. Refer to the oil pump priming procedures to prime the oil pump.

Oil Lines

OIL LINE CAUTIONS

- Do not bend or twist the oil lines when installing.
- When installing clips, position the tabs toward the inside and make sure they are not in contact with other parts.
- Check the oil lines, when installed in position, do not come in contact with rods and levers during engine operation.
- Secure all valves and sensors using their original fasteners.
- Install hose protectors in their original locations.
- Extreme caution should be taken not to scratch or damage oil lines.
- Do not excessively compress an oil line when installing clamps.
- Always use factory type clamps when installing fuel lines. Never use screw type clamps.
- When installing the oil tank, ensure oil lines will not be pinched between the tank and the powerhead.

Oil Pump Discharge Rate

TESTING

1. Connect a remote fuel source to the powerhead with a 50:1 oil and fuel pre-mix.
2. Install a flush device to the lower unit or place the outboard in a test tank.
3. Remove the cowling and disconnect the oil pump output line (clear hose) from the fuel line "Tee" fitting. Plug the fuel line to prevent any fuel leakage while the outboard is operating.
4. Disconnect the link rod from the oil pump lever. Set the oil pump lever as indicated in the flow specifications below.
5. Place the oil pump output hose into a graduated container, 0 to 250cc or equivalent. Start the powerhead and operate it at the rpm and for the length of time specified.

If the oil injection pump output is less than specified, the pump will need to be replaced.

Oil Pump Discharge Rate

Model	RPM	Time	Min. Flow	Max Flow
75-90 HP	700	15 min	18.7 cc	N/A
100-115 HP	700	15 min	25.5 cc	N/A
135-175 HP	1500	3 min	6.12 - 7.48 cc	15.3 - 18.7 cc
200 HP	1500	3 min	7.38 - 9.02 cc	17.28 - 21.12 cc
225 HP	1500	3 min	6.12 - 7.48 cc	28.35 - 34.65 cc
275 HP	1500	3 min	N/A	55.12 - 67.37 cc

Oil Pump Control Rod

ADJUSTMENT

See the "Maintenance and Tune-Up" section for oil pump linkage adjustment.

6-8 LUBRICATION AND COOLING

COOLING SYSTEM

Description and Operation

Water cooling is the most popular method in use to cool outboard powerheads. A "raw-water" type pump delivers seawater to the powerhead, circulating it through the cylinder head(s), the thermostat, the exhaust housing, and back down through the outboard. The water runs down the exhaust cavity and away, either through an exhaust tube or through the propeller hub.

Routine maintenance of the cooling system is quite important, as expensive damage can occur if it overheats. The cooling system is so important, that many outboards covered in this manual incorporate overheat alarm systems and speed limiters, in case the engine's operating temperature exceeds predetermined limits.

Poor operating habits can play havoc with the cooling system. For instance, running the engine with the water pickup out of water can destroy the water pump impeller in a matter of seconds. Running in shallow water, kicking up debris that is drawn through the pump, can not only damage the pump itself, but send the debris throughout the entire system, causing water restrictions that create overheating.

The water pumps used on most outboards are a displacement type water pump. Water pressure is increased by the change in volume between the impeller and the pump case.

On most outboards, the water pump is mounted on top of the lower unit. A driveshaft key engages a flat on the driveshaft and a notch in the impeller hub. As the driveshaft rotates, the impeller rotates with it.

A thermostat is used to control the flow of engine water, to provide fast engine warm-up and to regulate water temperatures. An element in the thermostat expands when heated and contracts when cooled. The element is connected through a piston to a valve. When the element is heated, pressure is exerted against a rubber diaphragm, which forces the valve to open. As the element is cooled, the contraction allows a spring to close the valve. Thus, the valve remains closed while the water is cold, limiting circulation of water.

As the engine warms, the element expands and the thermostat valve opens, permitting water to flow through the powerhead. This opening and closing of the thermostat permits enough water to enter the powerhead to keep the engine within operating limits.

OPTIMAX

♦ See Figures 16 and 17

Cooling water enters the cooling system through the lower unit water inlets. The water pump assembly forces water through the water tube and exhaust adapter plate passages filling the powerhead central water chamber (located behind the exhaust cavity). Water enters the exhaust cover cavity through 2 holes near the top of the exhaust cover On the 200–225hp models there are 6 holes—3 on each side and 1 slot (top) that connect the central chamber to the exhaust cover cavity.

Water exits the exhaust cover cavity through 4 slots (2 on each side) filling the water passages around the cylinders. Water flows around each bank of cylinders to the top of the cylinder block. On the 200–225hp models, water flow is directed around each cylinder sleeve by 6 water dams.

Water flow exiting the cylinder block is controlled by the thermostats (1 in each cylinder head) and the poppet valve (on the 115–150hp models that is located at the bottom starboard side of the powerhead and on the 200–225hp models is located in the exhaust adapter plate). At low rpm (below 1500 rpm), the thermostats control water flow depending upon engine temperature. When the thermostats are open, water passes through the cylinder heads and exits to the driveshaft housing. At higher rpm (above 1500 rpm), the poppet valve will control the water flow.

On the 115–150hp models, water that passes through the poppet valve enters the water passages in the adapter plates. Water passes through the adapter plates into the driveshaft housing. On the 200–225hp models, water that passes through the poppet valve enters the water passages in the exhaust tube to help cool the exhaust tube. Water will exit the exhaust tube through 2 slots at the top of the exhaust relief holes area (helping keep the holes clear of carbon and salt buildup) and through 2 holes at the lower rear of the exhaust tube into the driveshaft housing.

Water dumped into the driveshaft housing builds up a wall of water around the exhaust tube. This performs two functions:
- It helps to silence the exhaust
- Prevents air from being drawn into the water pump

Water exits the engine in 3 locations:
• On the 115–150hp models, excess water from the wall of water exits around anodes on the gear housing. On the 200–225hp models, the water exits through the bottom aft area of the driveshaft housing
• Water that passes through the air compressor exits through the tell tale hole
• On the 115–150hp models, water exits through two 1/8 in. (3.175 mm) holes in the lower adapter plate into the exhaust. On the 200–225hp models, water exits through a strainer screen in the exhaust adapter plate into the exhaust tube mixing with the exhaust gases.

To allow complete water passage filling and to prevent steam pockets, all of the cooling passages are interconnected. Small passages are incorporated to allow the cooling system to drain.

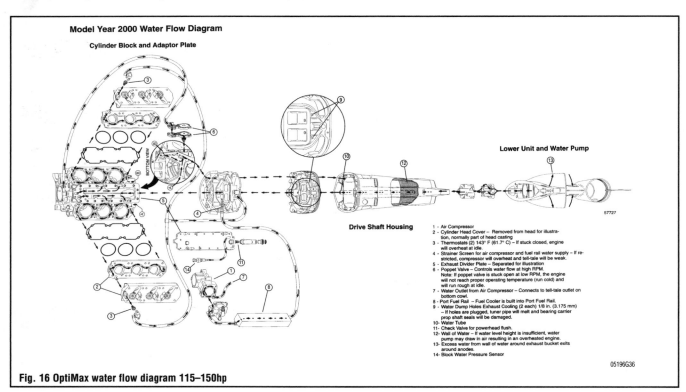

Fig. 16 OptiMax water flow diagram 115–150hp

LUBRICATION AND COOLING   6-9

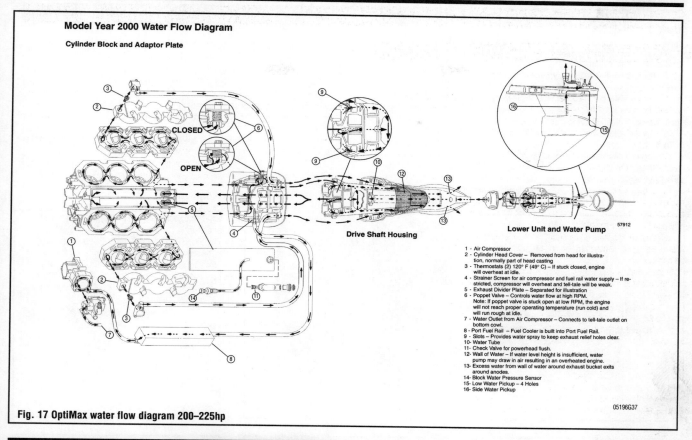

Fig. 17 OptiMax water flow diagram 200–225hp

Troubleshooting the Cooling System

◆ See Figure 18

Poor operating habits can play havoc with the cooling system. For instance, running the engine with the water pickup out of water can destroy the water pump impeller in a matter of seconds. Running in shallow water, kicking up debris that is drawn through the pump, can not only damage the pump itself, but send the debris throughout the entire system, causing water restrictions that create overheating.

Symptoms of overheating are numerous and include:
- A "pinging" noise coming from the engine, commonly known as detonation
- Loss of power
- A burning smell coming from the engine
- Paint discoloration on the powerhead in the area of the spark plugs and cylinder heads

➡ If these symptoms occur, immediately seek and correct the cause. If the engine has overheated to the point where paint has discolored, it may be too late to save the powerhead. Powerheads in this state usually require at least partial overhaul.

So what are major causes of overheating? Well the most prevalent cause is lack of maintenance. Other causes which are directly attributable to lack of maintenance or poor operating habits are:

- Fuel system problems causing lean mixture
- Incorrect oil mixture in fuel or a problem with the oil injection system
- Spark plugs of incorrect heat range
- Faulty thermostat
- Restricted water flow through the powerhead due to sand or silt buildup
- Faulty water pump impeller
- Sticking thermostat

Water Pump

REMOVAL & INSTALLATION

✳✳✳ WARNING

Since proper water pump operation is critical to outboard operation, all seals and gaskets should be replaced whenever the water pump is removed. Also, installation of a new impeller each time the water pump is disassembled is good insurance against overheating.

➡ Never turn a used impeller over and reuse it. The impeller rotates with the driveshaft and the vanes take a set in a clockwise direction. Turning the impeller over will cause the vanes to move in the opposite and result in premature impeller failure.

2.5, 3 and 3.3 HP

◆ See accompanying illustrations

➡ One of the features making this lower unit unique from all others covered in this manual is the location of the water pump. Instead of the pump being installed on the driveshaft, as on most units, the water pump is installed on the propeller shaft.

The pump impeller may be replaced without disassembling the lower unit. In fact the propeller shaft does not have to be disassembled. The only work required is to remove the propeller and a couple other simple tasks to replace an impeller.

1. Pull the cotter pin from the propeller shaft. A spare cotter pin is provided to a

Fig. 18 A new water pump impeller (left), alongside an impeller unfit for further service. Lack of water will render an impeller useless in just a few seconds.

6-10 LUBRICATION AND COOLING

new owner and may be found in the spark plug access cover, courtesy of the manufacturer.

2. Slide the propeller rearward and free of the propeller shaft. Remove the shear pin. A spare shear pin may also be found in the spark plug access cover. Again, compliments of the manufacturer.

3. Remove the two bolts and washers securing the water pump cover to the lower unit.

4. Jar the water pump cover free of the lower unit using a soft head mallet. Remove the cover from the propeller shaft. Check the inside surface of the cover for signs of wear indicating foreign particles had entered the water pump.

5. Pry the water pump impeller from the lower unit recess.

To install:

6. Install a new seal into the water pump housing, using the correct size socket. The accompanying cross-section illustration indicates the proper location for the spring and seal in the water pump housing.

7. If the bearing was removed, press a new bearing into the water pump housing until the bearing race is fully seated in the recess. Position a new O-ring in place around the water pump base. Install the water pump housing into place on the lower gear housing.

8. Rotate the propeller shaft until the hole in the shaft is facing the largest part of the water pump cavity in the gear housing. (This opening is not perfectly round.) Insert the drive pin through the hole in the propeller shaft. Slide a new water pump impeller onto the propeller shaft. As the impeller begins to enter the pump housing, rotate the impeller clockwise to permit the impeller vanes to curl in the proper direction. Continue to move the impeller into the housing with the cutout in the impeller indexed over the drive pin. As the name implies, the drive pin "drives" the impeller.

9. Install the water pump cover onto the water pump housing. Apply Quicksilver Perfect Seal, or equivalent to the threads of the attaching bolts.

10. Secure the cover in place with the two bolts with a washer on each bolt. Tighten the bolts alternately and evenly to a torque value of 25 inch lbs. (2.8 Nm).

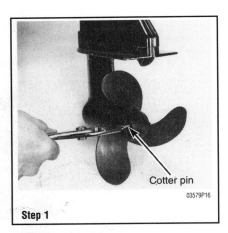
Step 1

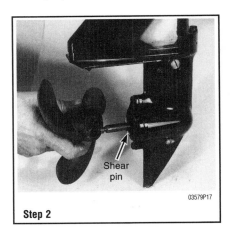
Step 2

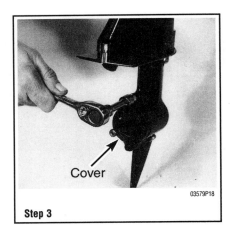
Step 3

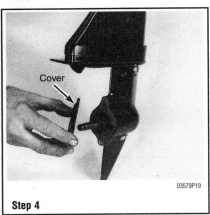
Step 4

Step 5

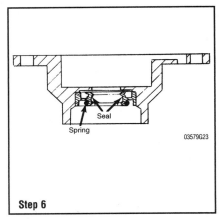
Step 6

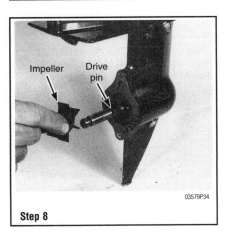
Step 8

Step 9

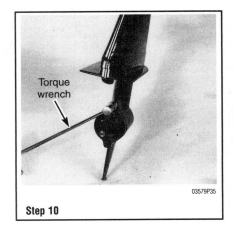
Step 10

4–5 HP

The driveshaft is one continuous shaft extending from the powerhead crankshaft into the torpedo bore of the lower unit. Therefore, the lower unit must be removed from the exhaust housing before the driveshaft assembly or water pump can be removed from the lower unit housing.

1. Rotate and swing the lower end unit in the full UP tilt position. Remove the rubber access plug on the starboard side of the upper gear housing. If servicing a 2.5 or 3.3hp unit place the gear shift in the **NEUTRAL** position. If servicing a model 4 or 5hp unit place the gear shift in the **REVERSE** gear position.
2. Loosen, but do not remove the bolt for the shift rod clamp. Loosen the bolt only enough to allow the shift shaft to slide free of the clamp when the lower unit is removed from the exhaust housing.
3. Remove the two attaching bolts and washers securing the lower unit to the exhaust housing. Pull down on the lower unit gear housing and separate the lower unit from the exhaust housing. Guide the shift rod and driveshaft out of the exhaust housing as the lower unit is lowered.
4. Remove the four bolts, flat washers and lockwashers securing the pump housing to the pump base assembly. Remove the bolt, retainer and washer from the water pump base.
5. Pull up on the pump housing and disengage the water tube from the lower unit housing. Slide the pump housing and gasket up and free of the driveshaft. Discard the gasket.
6. Lift up and slide the water pump cartridge and impeller off the driveshaft. Remove the drive pin from the recess in the driveshaft. Place the small drive pin into a safe place for later use. Remove the guide plate and gasket.

To install:

7. Slide the first gasket down the driveshaft and into place on the water pump base plate. The bolt holes are offset and there are a pair of alignment pins protruding from the pump base plate. Therefore, the gasket holes will only align properly one way.
8. Slide the plate down the driveshaft and into place. Apply just a "dab" of petroleum jelly to the Woodruff key, and then place it in the driveshaft keyway. The jelly will hold the key in place.
9. Slide another gasket down over the driveshaft.
10. Slide the water pump impeller down the driveshaft with the keyway in the pump aligned with the Woodruff key in the driveshaft.

➡ If an old impeller is used, the impeller must be installed in the same position from which it was removed. Never turn the impeller over, thinking it will extend its life. On the contrary, the blades will probably crack and break after just a short time of operation.

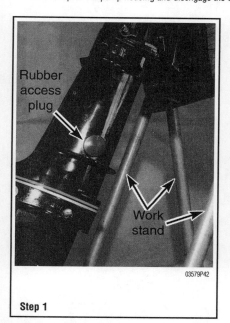
Step 1

Step 2

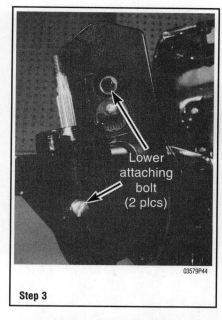
Step 3

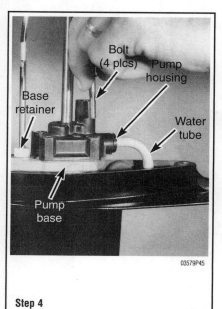
Step 4

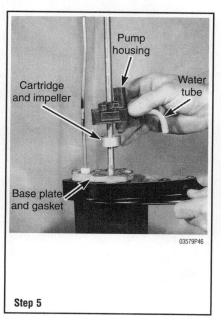
Step 5

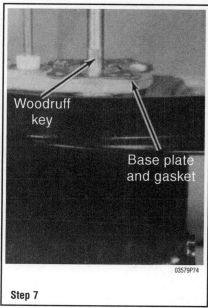
Step 7

6-12 LUBRICATION AND COOLING

11. Install the water pump cartridge into the water pump cover, with the locating tab indexing into the hole in the cover.
12. Lubricate the inside diameter of the two water tube seals with petroleum jelly.
13. Install the seals into the water pump cover. Be sure to index the tabs on the seals with the holes in the water pump cover. Insert the lower aft water pickup tube into the water pump housing.
    Slide the water pump cover down the driveshaft until it makes contact with the impeller. Apply a small amount of downward pressure on the water pump cover and at the same time rotate the driveshaft clockwise until the cover seats on the plate. Rotation of the driveshaft ensures all water pump vanes are bent properly and in the correct direction.
14. Secure the pump in place with the four attaching bolts. Tighten the bolts to a torque value of 70 inch lbs. (8Nm).
15. Push a new water tube seal onto the water pump cover.
16. Push one end of the water pickup tube into the grommet on the water pump cover.
17. Push the other end of the tube into the grommet at the aft end of the lower unit housing.

6–15 HP

▸ See accompanying illustrations

1. Pull upward and separate the water tube guide from the water pump outlet grommet.
2. Loosen and remove the four bolts and washers securing the water pump cover to the lower unit.
3. Slide the water pump cover and gasket up and free of the driveshaft. Discard the old gasket.
4. Slide the impeller and fiber washer up and free of the driveshaft. Remove the Woodruff key from the driveshaft cutout. Place the Woodruff key in a safe place for later use.
5. Slide the second fiber washer, face plate, and gasket up and free of the driveshaft. Discard the old gasket
6. Remove the retaining bolt from the base of the water pump. Place a screwdriver under the pump base and break the seal between the pump base and the lower unit.
7. Slide the pump base, water tube, gasket and shift shaft up and free of the driveshaft. The water tube is retained in the bottom of the pump base by a retainer and screw. The water tube will pull out of the grommet in the lower unit housing.
8. Using a pair of needlenose pliers, remove the "E" clip from the shift shaft. Place the "E" clip in a safe place for later use.
9. Unscrew the shift cam from the end of the shift shaft.
10. Slide the shift shaft out of the water pump base. Remove the O-rings and seal from the water pump base. Discard the seals and O-rings.
11. Turn the base plate over and remove the screw retainer securing the water pump pickup tube to the base.

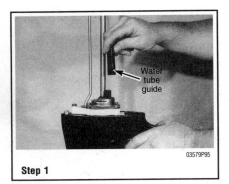
Step 1

Step 2

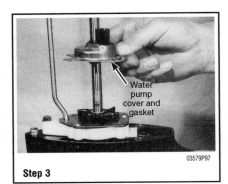
Step 3

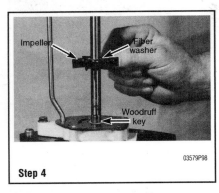
Step 4

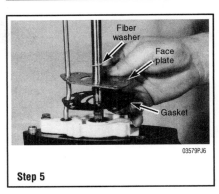
Step 5

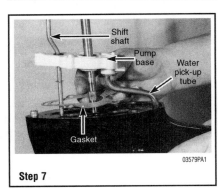
Step 7

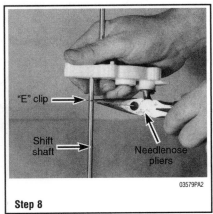
Step 8

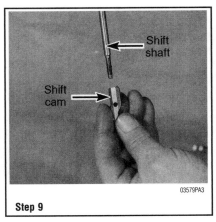
Step 9

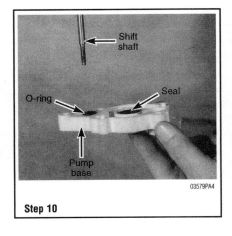
Step 10

LUBRICATION AND COOLING   6-13

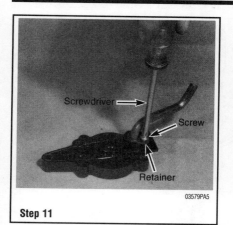
Step 11

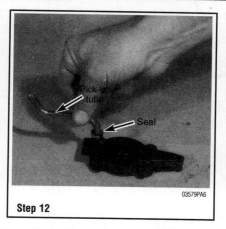
Step 12

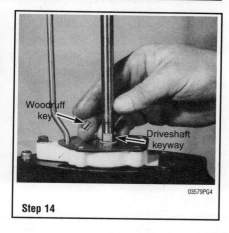
Step 14

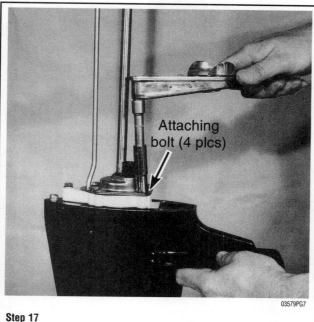
Step 17

12. Pull the water tube out of the base plate and remove the rubber seal on the end of the tube.

To install:

13. Slide a gasket, plate, and fiber washer down the driveshaft and into place on the lower unit. The screw holes are offset. Therefore, the gasket holes and plate will only line up properly one way.
14. Apply a "dab" of petroleum jelly onto the Woodruff key, and then place it in the driveshaft keyway. The petroleum jelly will hold the Woodruff key in place while installing the impeller.
15. Slide the water pump impeller down the driveshaft with the keyway in the impeller aligned with the Woodruff key in the driveshaft. Press the impeller into place with the keyway aligned.

➡ If an old impeller is installed be sure the impeller is installed in the same manner from which it was removed. Never turn the impeller over thinking it will extend its life. On the contrary, the blades would crack and break after just a short time of operation.

16. Install a new O-ring or gasket to the bottom of the pump cover. Slide the pump cover down the driveshaft until it makes contact with the plate. Apply a small amount of downward pressure on the water pump cover, and at the same time, rotate the driveshaft clockwise until the cover seats on the plate. Rotation of the driveshaft ensures all water pump vanes are bent in the proper direction.
17. Secure the pump in place with the four attaching bolts. Tighten the bolts to a torque value of 40 inch lbs. (4.5 Nm).
18. Push a new water tube seal into the water pump cover recess. Insert the water tube guide over the new seal.

20 and 25 HP

♦ See accompanying illustrations

1. Remove the lower unit.
2. Remove the flat washer on the end of the shift shaft and the O-ring on the end of the driveshaft. Discard the O-ring.
3. Slide the rubber slinger on top of the pump cover up and free of the driveshaft.
4. Loosen and remove the four bolts and washers securing the water pump cover to the lower unit.
5. Lift and slide the water pump cover, O-ring, fiber washer and impeller free of the driveshaft. Discard the O-ring inside the water pump cover.

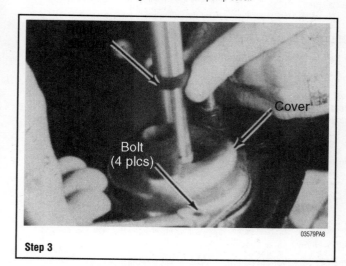
Step 3

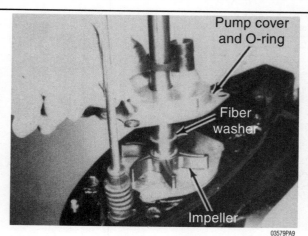
Step 5

6-14 LUBRICATION AND COOLING

6. Remove the impeller Woodruff key from the driveshaft cutout and place the key in a safe place for later use.
7. Slide the fiber washer between the impeller and the base plate up and free of the driveshaft.
8. Lift off the base plate and gasket from the lower unit housing.
9. Slide the base plate and gasket up and free of the driveshaft. Discard the old gasket.

To install:
10. Slide a gasket, plate, and fiber washer down the driveshaft and into place on the lower unit. The screw holes are offset. Therefore, the gasket holes and plate will only line up properly one way.
11. Apply a "dab" of petroleum jelly onto the Woodruff key, and then place it in the driveshaft keyway. The petroleum jelly will hold the Woodruff key in place while installing the impeller.
12. Slide the water pump impeller down the driveshaft with the keyway in the impeller aligned with the Woodruff key in the driveshaft.
13. Press the impeller into place with the keyway aligned.

➡ If an old impeller is installed be sure the impeller is installed in the same manner from which it was removed. Never turn the impeller over thinking it will extend its life. On the contrary, the blades would crack and break after just a short time of operation.

14. Install a new O-ring or gasket to the bottom of the pump cover. Slide the pump cover down the driveshaft until it makes contact with the impeller. Apply a small amount of downward pressure on the water pump cover, and at the same time, rotate the driveshaft clockwise until the cover seats on the plate. Rotation of the driveshaft ensures all water pump vanes are bent in the proper direction.
15. Secure the pump in place with the four attaching bolts.
16. Tighten the bolts to a torque value of 25 inch lbs. (3 Nm).
17. Slide a new rubber slinger down the driveshaft and onto the top of the water pump cover.
18. Push a new water tube seal into the water pump cover recess.

30 HP and 40 HP (2 and 3-Cylinder)

▶ See accompanying illustrations

1. Remove the gear case.

➡ If the water tube seal remained in the drive shaft housing, remove the seal from the housing and reinstall on the water pump cover. Secure the seal to the cover with Loctite® 405, or equivalent.

2. Remove the 4 bolts securing the pump cover.
3. Remove the cover, washer (above impeller), impeller, key and washer (below impeller).
4. Remove the cover gasket, base plate and base gasket.
5. Remove and/or replace the exhaust deflector plate if it is in any way damaged.

To install:
6. Install the water pump seal carrier.
7. Install the exhaust deflector plate if it was removed.
8. Install thew base gasket, base plate, pump cover gasket, nylon washer and impeller key.

➡ Make sure that the neoprene strip on the gasket faces up.

9. Install the impeller and nylon washer. Make sure the keyway in the impeller aligns with the key on the driveshaft.

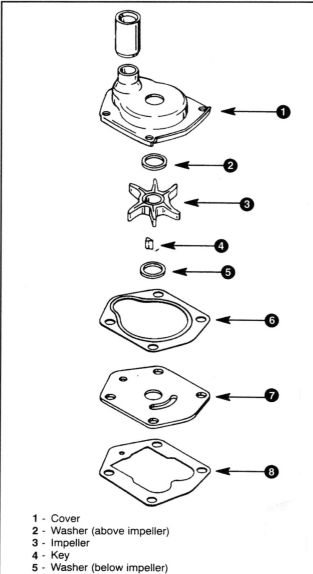

1 - Cover
2 - Washer (above impeller)
3 - Impeller
4 - Key
5 - Washer (below impeller)
6 - Cover Gasket
7 - Base Plate
8 - Base Gasket

Step 3

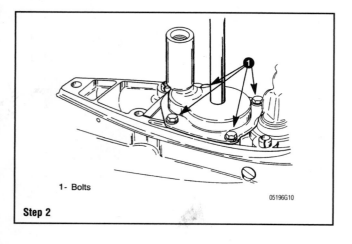

1- Bolts

Step 2

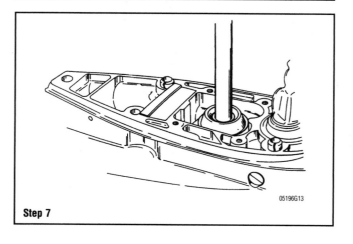

Step 7

LUBRICATION AND COOLING 6-15

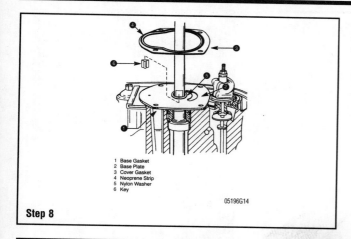

1 Base Gasket
2 Base Plate
3 Cover Gasket
4 Neoprene Strip
5 Nylon Washer
6 Key

Step 8

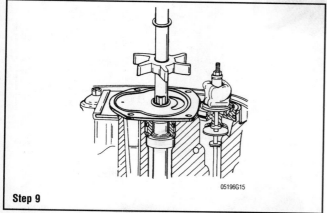

Step 9

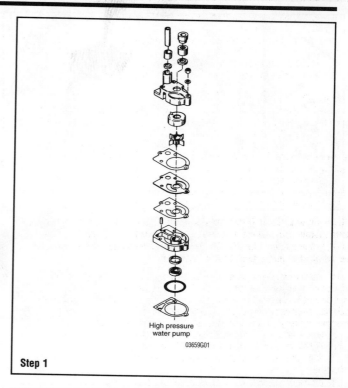

High pressure water pump

Step 1

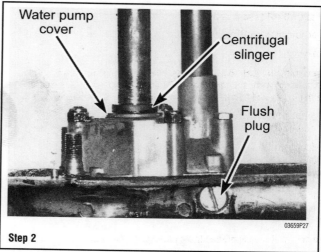

Step 2

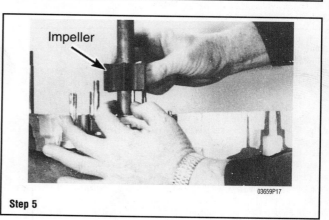

Step 5

➡ Apply a light coat of 2-4-C w/Teflon to the inside of the pump cover to ease the installation of the cover over the impeller.

10. Install the pump cover. Rotate the driveshaft clockwise while pressing down the cover over the impeller.
11. Apply a threadlocking compound to the retaining bolts and torque to 60 inch lbs. (6.8 Nm).

40 HP (4-Cylinder), 50–60 HP (3-Cylinder)

▶ See accompanying illustrations

➡ A high pressure pump is used on the 40hp, (4-Cylinder) and the 50 and 60hp (3-Cylinder) lower units.

1. High pressure pumps can be easily identified by the thick housing and stout impeller blades.
2. Remove the centrifugal slinger from the top of the water pump.
3. Remove the three retaining nuts and washers securing the pump cover to the pump base. Some model units may have three nuts and one bolt. Use two pry bars, one on each side, and pry the pump cover off the studs. If the cover is "frozen" to the pump, it may be necessary to use a chisel to break the cover loose from the studs.

➡ The seal is very difficult to remove. If the seal or insert is unfit for further service, the recommendation is to replace the pump cover. A kit is available from the local dealer and will include a new seal—installed, and a new insert. The new insert will slip right into place.

4. If the cover kit is not available, proceed to remove the seal and insert as follows: Use a punch and hammer and drive the pump insert out of the pump cover. Drive the pump cover seal out of the cover from the insert side.
5. Remove the impeller from the driveshaft. If the impeller is stubborn, it may be necessary to use a punch and hammer to drive the impeller upward and off the driveshaft. If a punch and hammer will not move the impeller, the only answer is to use a chisel and split the impeller. A new impeller should always be installed when

6-16 LUBRICATION AND COOLING

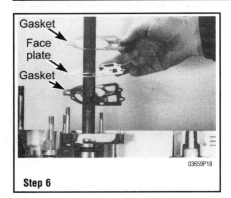
Step 6

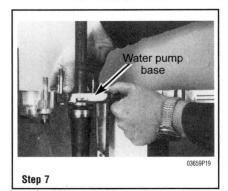
Step 7

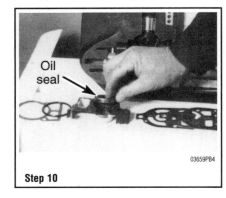
Step 10

Step 11

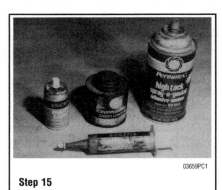
Step 15

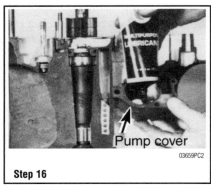
Step 16

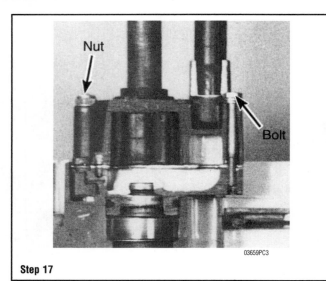
Step 17

the lower unit is opened. Remove and save the impeller drive pin from the flat area of the driveshaft.

6. Lift upward on the face plate and remove it and the gasket from the mounting studs. Clean any gasket material from the face plate and from the pump base.

7. Remove the water pump base by first padding the surface around the pump base.

8. Using a pair of screwdrivers, pry upward and evenly on both ends of the water pump base. Once the water pump base is loose, slide the pump base up and off the end of the driveshaft.

9. Remove the O-ring and oil seals from the base plate assembly.

If the oil seals are unfit for further service, the manufacturer recommends the pump base be replaced. The base is available in kit form and will include new seals. If the kit is not available, proceed to remove the seals as follows: remove the oil seals from the pump base by prying or driving them away from the impeller side of the pump base. Remove the O-ring from the groove in the pump base. Clean any gasket material from the upper and lower surfaces of the pump base.

To install:

10. Position the water pump base on a press. Coat the outer diameter of the oil seals with Blue Loctite®, or equivalent. Use a suitable mandrel and press the smaller diameter of the oil seal into the pump base with the lip of the seal towards the impeller side. Use a suitable mandrel and press the larger diameter oil seal into the pump base with the lip of the seal towards the driveshaft bearing side. Clean any excess Loctite® from the oil seals. Install a new O-ring into the groove in the driveshaft bearing side of the pump base. Lubricate the O-ring and the oil seals with Multi-purpose Lubricant, or equivalent.

11. Remove the lower unit-to-water pump base gasket from the lower unit. Take care not to damage the gasket. Install this gasket over the water pump base O-ring.

12. Install the pump base into the lower unit. Check to be sure the pump base is seated firmly against the lower unit.

13. Install the following items in the order given: The pump base-to-face plate gasket, the face plate, and then the face plate-to-pump cover gasket. The gaskets and the face plate are indexed by dowel pin location and MUST be installed properly.

14. Coat the impeller drive pin area of the driveshaft, the flat area on the shaft just above the face plate, with Multi-purpose Lubricant. Position the impeller drive pin on the driveshaft. Slide a new impeller down over the driveshaft and drive pin.

➡ Use a thin blade screwdriver and check to be sure the impeller drive pin has not moved partially off the flat on the driveshaft. If the drive pin moves off the flat, the impeller would be forced against the pump insert after installation and cause premature impeller wear. The drive pin should be free to move upward approximately 1/16 in.(1.59mm).

15. Obtaining good quality sealing products is well worth the extra cost to prolong satisfactory service of the lower unit.

16. Apply a thin coating of Multi-purpose Lubricant onto the inside diameter of the water pump cover. Slide the assembled water pump cover down over the driveshaft and water pump studs. Rotate the driveshaft clockwise and at the same time push down on the pump cover to ease entry of the impeller into the cover.

17. Install the water pump cover retainer washers, nuts, and bolts. Tighten the bolt and nuts evenly to the torque value as follows:
- Nuts: 1/4-28   30 inch lbs. (3.4Nm)
- Nuts: 5/16-24   40 inch lbs. (4.5Nm)
- Bolt: 1/4-20   20 inch lbs. (2.3Nm)

LUBRICATION AND COOLING 6-17

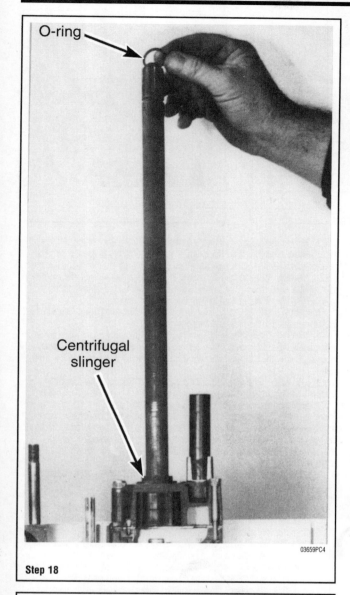

Step 18

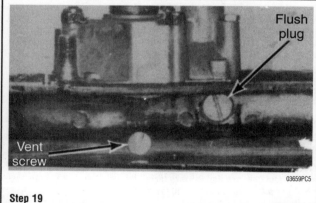

Step 19

➡ **DO NOT overtighten the bolt or nuts on plastic water pump covers. Overtightening could cause the cover to crack during operation.**

18. Install the centrifugal slinger over the driveshaft and down against the pump cover.
19. For units equipped with a flush plug: Position the gasket onto the flush plug, and then install the plug into the lower unit.

50–200 HP
♦ See accompanying illustrations
➡ A high volume pump is used on the 50–200hp lower units.

1. High volume pumps can be easily identified by comparing their appearance to this exploded view illustration.
2. Slide the sealing ring from the top of the water pump. Remove the four bolts,

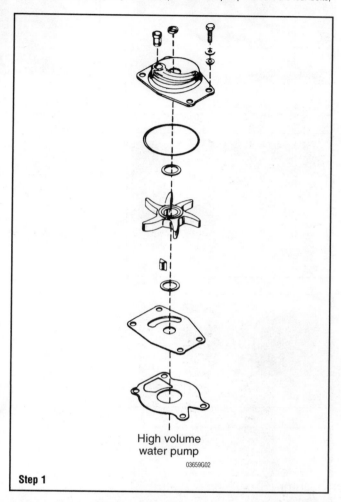

Step 1

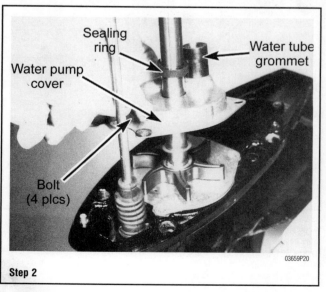

Step 2

6-18 LUBRICATION AND COOLING

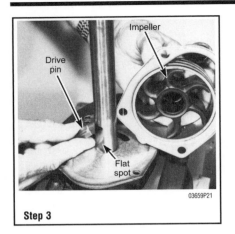
Step 3

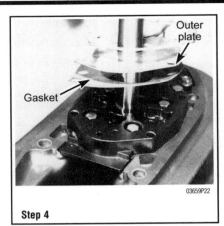
Step 4

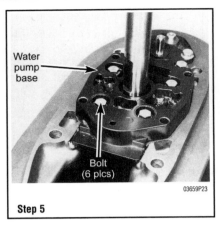
Step 5

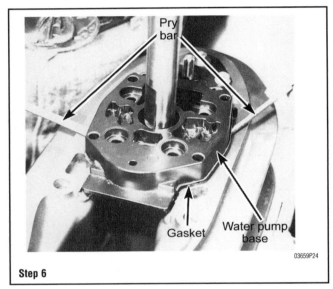
Step 6

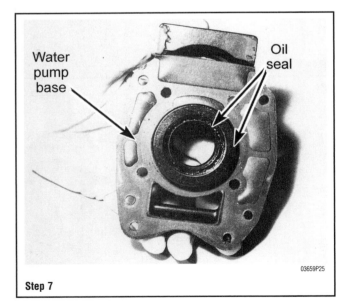
Step 7

washers, and isolators securing the pump cover to the pump base. If necessary, use two screwdrivers, one on each side, and pry the pump up and free of the outer plate. Remove and discard the gasket. The water tube grommet, if still in serviceable condition may remain in place on the water pump cover. The impeller may remain on the driveshaft or may stick in the pump cover.

3. If the impeller remained on the driveshaft, slide the impeller up and free of the driveshaft. If the impeller is stubborn, it may be necessary to use a punch and hammer to drive the impeller upward and off the driveshaft. If a punch and hammer will not move the impeller, the only answer is to use a chisel and split the impeller. A new impeller should always be installed when the lower unit is opened. Remove and save the impeller drive pin from the flat area of the driveshaft.

4. Lift the outer plate and gasket up and free of the pump base. Clean all old gasket material from the plate and the pump base.

5. If working on a 50hp or 60hp model (since 1990), pry up the water pump base from the top of the lower unit—refer to the exploded drawing. For all other models, Remove the bolts and washers securing the water pump base to the lower unit.

6. Use two screwdrivers and pry the water pump base from the lower unit. Remove and discard the gasket under the base.

7. Inspect the condition of the two oil seals housed in the water pump base. Check the oil seals for correct installation. On this type water pump, the seals must be installed back-to-back. One prevents lubricant from escaping the lower unit and the other prevents water from entering. If either seal is no longer fit for service, pry both seals out with a screwdriver, one at a time.

To install:

8. If the two oil seals were not removed during disassembly, skip this step and proceed directly to the next step. If the two oil seals were removed during disassembly, proceed as follows:
   - Obtain Oil Seal Driver (C-91-13949), Loctite ® 271, or equivalent, and Quicksilver Needle Bearing Assembly Lubricant (C-92-42649A-1), or appropriate substitutes.
   - Apply Loctite ® to the outer circumference and pack the lip of the two oil seals to be installed with lubricant. Slide the first seal, with the seal lip facing away from the tool, onto the longer shoulder side of the installation tool. Support the water pump base on in arbor press and press in the seal until the shoulder of the tool bottoms against the pump base. If working without the special tool, carefully and squarely tap in the seal until the seal seats inside the base.

9. Slide the second seal, with the seal lip facing toward the tool, onto the shorter shoulder side of the installation tool. Install this second seal in the same manner as the first.

10. Place a new gasket in place on the lower unit. Slide the water pump base down over the driveshaft. Tap the base down lightly to seat it over the gasket. Apply Loctite ®, or equivalent, to the threads of the six retaining bolts. Install and tighten the bolts and washers to a torque value of 60 inch lbs. (6.8Nm).

11. Slide the gasket and then the outer plate down the driveshaft.

12. Coat the impeller drive pin area of the driveshaft with Water Resistant Multi-purpose lubricant. Position the impeller drive pin on the driveshaft. Slide a new impeller down the driveshaft to index over the drive pin.

➡ **If at all possible, always install a new water pump impeller. If the old impeller must be installed, always install the impeller in the same position to allow the vanes to rotate in the same direction as the original installation. Installing the impeller in such a manner to cause the vanes to rotate in the opposite direction will only result in premature impeller failure.**

13. Apply a thin coating of water resistant multi-purpose lubricant to the inside diameter of the water pump cover. Apply Loctite ® 271 or equivalent to the threads of the four securing bolts. Slide the cover down the driveshaft. Rotate the shaft clockwise and at the same time push down on the pump cover to ease entry of the impeller into the cover.

LUBRICATION AND COOLING

14. Install the water pump cover retaining bolts, washer, and isolators. Tighten the bolts to a torque valve of 60 inch lbs. (6.8Nm).
15. Slide the sealing ring down the driveshaft onto the top of the water pump cover.
16. Apply a coating of water resistant multi-purpose lubricant around the inside diameter of the water tube grommet as an aid in later assembly to the water tube.

225–275 HP

◆ See accompanying illustrations

1. Lift out the water tube guide and seal from the pump cover.
2. Remove the four retaining bolts, washers and nylon bushings securing the pump cover to the lower unit.
3. Use two screwdrivers—one on each side—and pry the pump cover off the studs. If the cover is "frozen" to the pump, it may be necessary to tap on the cover, with a soft head mallet, to break it loose from the plate.
4. If the seal in the cover is damaged, or unfit for further service, the recommendation is to replace the pump cover. A kit is available from the local dealer and will include a new seal—installed. If the cover kit is not available, remove the seal using a hammer and punch. Drive the pump cover seal out of the cover, from the impeller side. Tap the new seal into place using a deep socket the same size as the seal.
5. Slide the impeller up and free of the driveshaft. If the impeller is stubborn, it may be necessary to use a punch and hammer to drive the impeller upward and off the shaft. If corrosion has developed on the driveshaft, use a piece 300 grit paper and remove the corrosion to make the task of removing the impeller much easier.

➡ In an extreme situation, it may be necessary to split the impeller with a chisel and hammer. Splitting the impeller is not really as bad as it sounds, because a new impeller should always be installed when the lower unit is opened. Be extra careful not to damage the driveshaft or key in the impeller.

6. After the impeller has been removed, lift out the impeller drive (Woodruff) key from the driveshaft.
7. Lift up on the impeller plate and slide the plate free of the driveshaft.
8. Clean all gasket material from both sides of the plate.
9. Remove the seal carrier from the driveshaft by inserting two pump cover bolts partially into the seal carrier—one on each side.
10. Using a pair of screwdrivers under the heads of the bolts, pry the seal carrier up and out of the pocket in the housing.
11. Remove the seal and O-ring from the carrier.

To install:

12. If the two oil seals were not removed during disassembling, skip this step and proceed directly to the next step. If the two seals in the carrier for 225hp model or one seal in the carrier for the 275hp model were removed, install the new seal(s).
13. On the 275hp model, Apply a coating of Loctite ® 271, or equivalent, to the outside diameter of the seal.
   b. Place the seal onto the carrier with the lip of the seal facing UP.
   c. Using a deep socket or mandrel the same diameter as the seal, tap the seal flush with the top of the carrier.
   d. Lubricate the O-ring with bearing assembly lubricant and install the O-ring around the seal carrier.
   e. Slide the seal carrier down the driveshaft and into the lower unit housing. Be sure the seal carrier is flush with the driveshaft bearing retaining nut. Apply a small amount of silicone sealant to the ends of the divider plate and install the plate into the housing.
14. On the 225hp model, Place the first seal, with the seal lip facing up into the bore of the carrier.
   a. Press the seal into the carrier with a mandrel or deep socket.
   b. Place the second seal into the carrier with the lip of the seal facing down, (seals should be back-to-back).

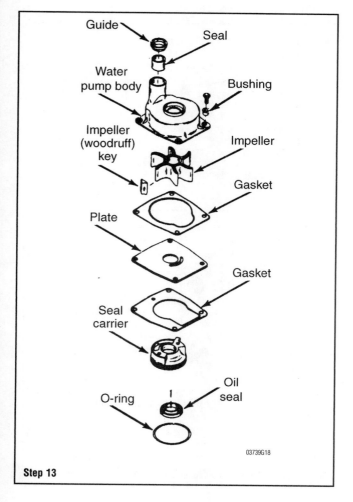

Step 13

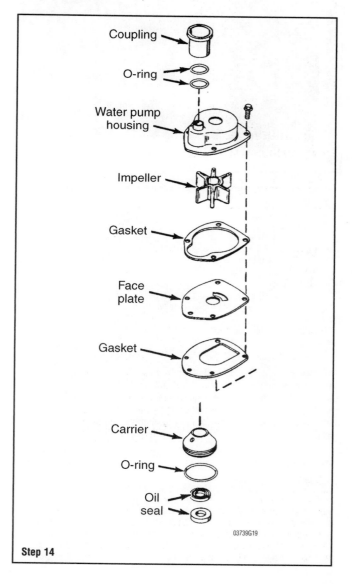

Step 14

6-20 LUBRICATION AND COOLING

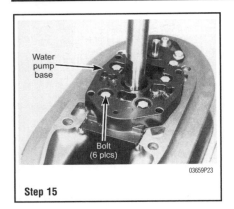

Step 15

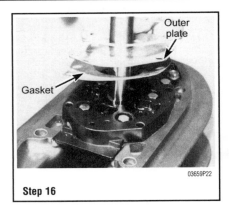

Step 16

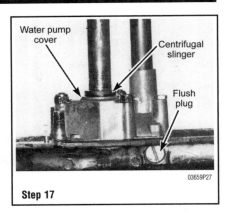

Step 17

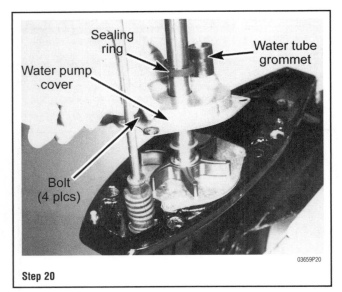

Step 20

CLEANING & INSPECTION

Clean all water pump parts with solvent, and then dry them with compressed air. Inspect the water pump cover and base for cracks and distortion. If possible, always install a new water pump impeller while the lower unit is disassembled. A new impeller will ensure extended satisfactory service and give "peace of mind" to the owner. If the old impeller must be returned to service, never install it in reverse to the original direction of rotation. Installation in reverse will cause premature impeller failure.

Inspect the ends of the impeller blades for cracks, tears, and wear. Check for a glazed or melted appearance, caused from operating without sufficient water. If any question exists, as previously stated, install a new impeller if at all possible.

Thermostat

REMOVAL & INSTALLATION

♦ See Figures 19 thru 27

1. Locate the thermostat cover.
2. Remove the retaining bolts or screws.
3. Carefully remove the thermostat cover.
4. If the cover does not want to come loose, tap it gently with a plastic hammer.
5. Remove the thermostat cover.
6. Remove the thermostat gasket.
7. Remove the thermostat from the housing.

To install:

8. Throughly clean the gasket mating surfaces.
9. After thoroughly inspecting, cleaning and testing the thermostat, install a new gasket and then install the thermostat.
10. Install the housing cover and tighten the bolts snugly.

   c. Press the seal into the carrier with a mandrel or deep socket until it contacts the other seal.
   d. Lubricate the O-ring with bearing assembly lubricant and install the O-ring around the carrier.
   e. Slide the carrier down the driveshaft and into the lower unit housing. Be sure the carrier is flush with the driveshaft bearing retaining nut.
   f. Install the filler block into the housing behind the driveshaft.

15. Place a new gasket in place on the lower unit. Slide the water pump base down over the driveshaft. Tap the base down lightly to seat it over the gasket.
16. Slide the gasket and then the outer plate down the driveshaft.
17. Coat the impeller drive pin area of the driveshaft with water resistant multi-purpose lubricant.
18. Position the impeller drive pin on the flat spot on the driveshaft. The lubricant will hold the pin in place.
19. Slide a new impeller down the driveshaft and over the drive pin.

➡If at all possible, always install a new water pump impeller. If the old impeller must be installed, install the impeller in the same position to allow the vanes to rotate in the same direction as the original installation. If the impeller is installed causing the vanes to rotate in the opposite direction, premature impeller failure of the impeller will surely occur.

20. Apply a thin coating of water resistant multi-purpose lubricant to the inside diameter of the water pump cover. Apply Loctite® 271, or equivalent, to the threads of the four securing bolts.
21. Slide the cover down the driveshaft. Rotate the shaft clockwise and at the same time push down on the pump cover to ease entry of the impeller into the cover.
22. Install the water pump cover retaining bolts, washer, and isolators. Tighten the bolts to a torque value of 60 inch lbs. (6.8Nm).
23. Slide the sealing ring down the driveshaft onto the top of the water pump cover.
24. Apply a coating of water resistant multi-purpose lubricant around the inside diameter of the water tube grommet as an aid in later assembly to the water tube.

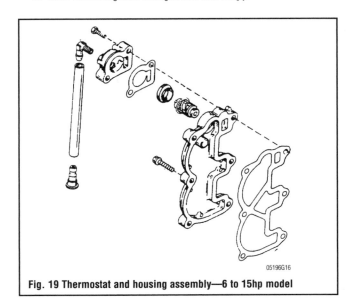

Fig. 19 Thermostat and housing assembly—6 to 15hp model

LUBRICATION AND COOLING 6-21

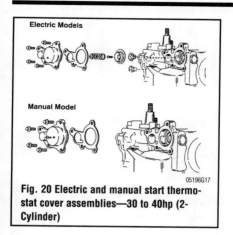

Fig. 20 Electric and manual start thermostat cover assemblies—30 to 40hp (2-Cylinder)

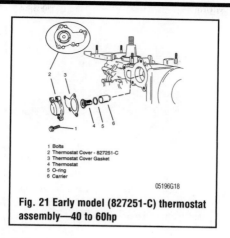

Fig. 21 Early model (827251-C) thermostat assembly—40 to 60hp

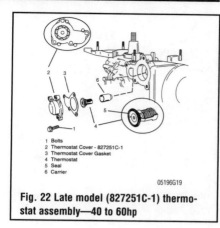

Fig. 22 Late model (827251C-1) thermostat assembly—40 to 60hp

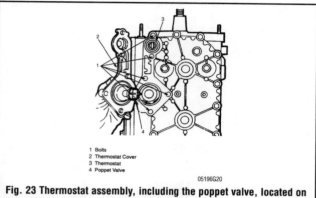

Fig. 23 Thermostat assembly, including the poppet valve, located on the cylinder block—70 to 115hp

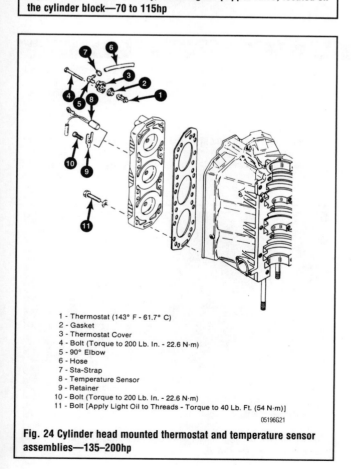

Fig. 24 Cylinder head mounted thermostat and temperature sensor assemblies—135–200hp

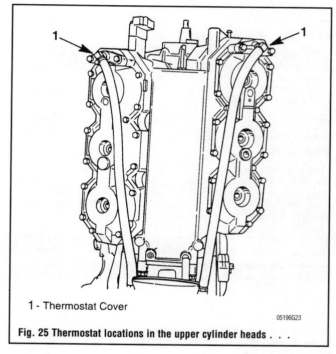

Fig. 25 Thermostat locations in the upper cylinder heads . . .

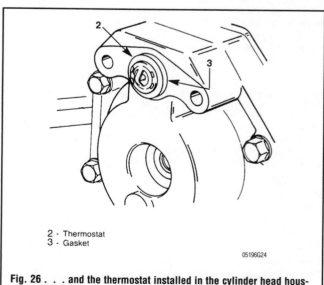

Fig. 26 . . . and the thermostat installed in the cylinder head housing—225–250hp

6-22 LUBRICATION AND COOLING

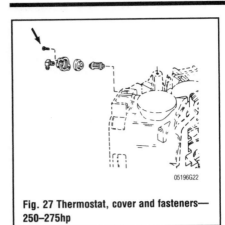

Fig. 27 Thermostat, cover and fasteners—250–275hp

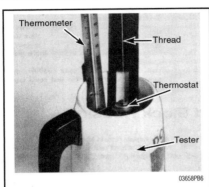

Fig. 28 Corrosion inside the thermostat bore signals a lack of maintenance. Always flush your outboard with fresh water, especially after boating in salt water

Fig. 29 Thermostat ready for testing as described in the text

CLEANING & INSPECTION

♦ See Figure 28

The cause of a malfunctioning thermostat is often foreign matter stuck to the valve seat. Inspect the thermostat to make sure that it is clean and free of foreign matter. If necessary, test the removed thermostat for operation.

Outboards used in salt water should be flushed with fresh water after each use to prevent corrosion from forming.

TESTING

♦ See Figure 29

1. Inspect the thermostat cover and the thermostat opening in the cylinder head cover for cracks and corrosion damage. Such damage could cause leakage. Remove and discard the old thermostat gasket. Wash the thermostat with clean water.
2. Obtain a thermostat tester or similar device, as shown in the accompanying illustration. Test the thermostat as follows:
3. Open the thermostat valve. Insert a length of thread between the valve and the thermostat body. Allow the valve to close against the thread.
4. Suspend the thermostat by the thread inside the tester. Do not allow the thermostat to touch the bottom or sides of the tester.
5. Suspend a thermometer inside the tester, with the bottom of the thermometer even with the bottom of the thermostat. Do not allow the thermometer to touch the bottom or sides of the tester.
6. Fill the tester with water to cover the thermostat. Plug the tester into an electrical outlet.
7. Observe the temperature at which the thermostat begins to open. As soon as the thermostat starts to open, it will drop off the thread. The thermostat must begin to open at 140° to 145°F.
8. Continue to heat the water until the thermostat is completely open. Unplug the tester.
9. Allow the water in the tester to cool before testing the next thermostat.
10. Replace the thermostat, if it fails to open at the specified temperature, or if it does not fully open.

WARNING SYSTEMS

Description and Operation

♦ See Figure 30

OIL QUANTITY INDICATOR

♦ See Figure 31

Oil is drawn from a remote mounted oil tank and mixed with the powerhead fuel system via a mechanically driven oil pump. Two oil tanks, one fairly large, is usually mounted in the stern of the boat close to the transom. The second tank, referred to as the reservoir, is mounted on the powerhead over the carburetors. Both oil tanks are made from a translucent white plastic material so a visual indication of the oil quantity can quickly be determined.

Oil is moved from the main tank to the reservoir by crankcase pressure through a hose with a one way check valve.

An audible alarm system is incorporated in the system to warn the operator when the oil level in the reservoir drops—leaving only about one quart (.95 liter) remaining. Such a condition means oil in the main tank has dropped below the pickup tube. The low oil condition warning is accomplished through a float switch mounted in the

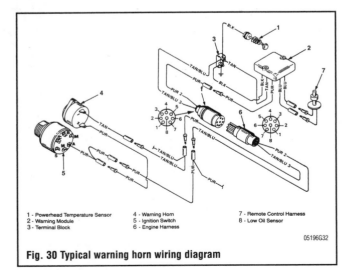

Fig. 30 Typical warning horn wiring diagram

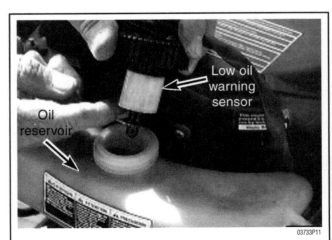

Fig. 31 Typical low oil quantity sending unit in the reservoir mounted on the powerhead.

LUBRICATION AND COOLING 6-23

reservoir oil tank cap. When the float descends to a predetermined level, a signal is sent directly to the Warning Module mounted on the powerhead.

The warning module completes the circuit and the horn in the remote control shift box sounds. In most cases there is sufficient running time—approximately 30 to 35 minutes at full throttle—for the operator to return to shore and service the main oil tank in the boat.

As mentioned earlier, all models equipped with oil injection are also equipped with a Warning Module or ECU warning system.

Test Button Model (If equipped)

When the ignition key is placed in the run position and the Test button, on the oil tank is depressed—the warning horn will sound with a continuous "Beep" tone. This test verifies both the overheat and low oil quantity warning circuits are functioning properly. There is only one tone sound for both the overheat and low oil quantity warnings on these models.

Warning Module

♦ See Figure 32

When the ignition key is placed in the RUN position, the warning module begins a self test mode. Two different tones are used to warn the operator of a powerhead problem. The first warning tone is a brief steady "Beep" verifying the overheat warning system is functioning. The next tone, is a series of short "Beep-Beep" tones verifying the low oil quantity circuit is functioning properly.

An oil tank gauge kit for the remote oil tank mounted in the boat is available from the Mercury/Mariner dealer as an accessory. The kit comes complete with a control panel gauge, remote oil tank sender unit, and necessary wiring. This gauge unit provides the operator a visual indication of oil quantity in the main tank.

Fig. 32 Typical oil pump, fuel pump, and warning module installed on a V6 powerhead.

Oil Quantity Gauge

The oil quantity gauge operates on the variable resistance to ground principle. A moveable float arm on the sending unit rises and falls as the oil quantity changes. A variable resistor or rheostat is connected to the opposite end of the float arm. When the oil level decreases—resistance is increased—allowing less current to flow through the gauge. Since the reduced current passes through the gauge, a bi-metal strip in the gauge cools proportionately and moves the needle toward a higher mark on the gauge face. Resistance is lowest when the oil tank is full. As the oil quantity begins to drop, the resistance increases, causing more current to pass through the gauge. The bi-metal strip begins to heat-up and moves the needle towards a lower gauge indication.

OVERHEAT TEMPERATURE WARNING

The powerheads are equipped with an Overheat Temperature Sensor mounted in the cylinder cover. If the temperature should exceed a set temperature during powerhead operation, the overheat temperature sensor in the water jacket cover will close. When sensor closes, the circuit to the horn or warning module is completed and the horn in the remote shift box will sound—with continuous steady "Beep" tone to warn the operator of the overheat condition.

Water Temperature Sensor

30–40 (2-CYLINDER)

Uses the Timing Protection Module (TPM) to provide control of overheating and low-oil conditions. Warning is provided through the activation of a continuous tone warning horn for either condition.

An over-heat condition occurs when the engine temperature rises above 182–198°F (75–101°C).

The TPM will intermittently interrupt the ignition voltage to the capacitor discharge modules (CDM) to reduce the maximum rpm to approximately 2500. The rpm will be limited and the warning horn will activate until the engine temperature drops below 162–178°F (64–90°C).

40 HP (4-CYLINDER), 50–60 HP (3-CYLINDER)

When the ignition key is initially turned on, the warning module will briefly provide a self-test of its engine overheat warning system. This will be a steady BEEP tone.

If the powerhead temperature should exceed 300°F (65°C), the overheat temperature sensor in the water jacket cover will signal the warning module to activate the warning horn.

70–115 HP

When the ignition key is initially turned on, the warning module will briefly provide a self-test of its engine overheat warning system. This will be a steady BEEP tone.

135–200 HP AND 275 HP

If the temperature exceeds approximately 190°F (87°C) the overheat temperature sensor in the water jacket of the cylinder head will close, completing a circuit to the warning module. The warning module will activate the horn in the remote shift box. This warning horn will sound a continuous steady "Beep" tone to warn the operator of the overheat condition.

225 HP

♦ See Figure 33

The temperature sensor is activated if the powerhead temperature exceeds 200°F (93.3°C). The temperature sensor signal is relayed to the electronic control unit (ECU). If an overheat condition should develop, the ECU will perform two functions. First, it will activate the warning horn in the remote shift box alerting the operator of the condition. Secondly, it will retard powerhead timing, restricting powerhead speed

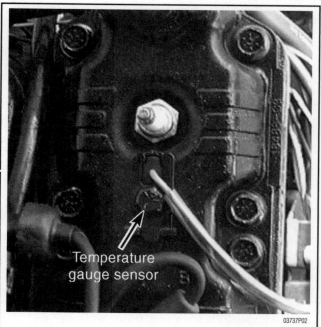

Fig. 33 An optional temperature gauge sending unit may be mounted in the starboard head, below the No. 1 spark plug.

6-24 LUBRICATION AND COOLING

to a maximum of 3,000 rpm. Once the powerhead temperature drops below 190°F (87°C), the ECU will shut off the warning horn and restore full timing functions for normal powerhead operation.

Temperature Gauge

Temperature warning systems utilized on many of today's outboards no longer require the use of a temperature gauge for the operator to monitor. The following is a brief description of a temperature gauge system. The temperature gauge must have a 12-volt power and ground wire connected to the gauge. A sensing wire from the gauge should be connected to the temperature sending unit on the powerhead. When the ignition key switch is set to the run position, the bi-metallic resistor in the sending unit will have little or no resistance. As the powerhead begins to warm to normal operating temperature, the resistance increases causing the needle on the gauge to move. As powerhead temperature increases, the more resistance and consequently more gauge needle movement.

TESTING

1. To test the temperature gauge perform the following:
   - Set the key switch to the RUN position and then disconnect the sensor wire at the sending unit.
   - Make contact with the sensor wire to a good powerhead ground. The gauge should move full scale.
   - If the gauge moves—replace the sending unit, if no movement is observed—check for 12-volts at the gauge unit and the ground connections.
   - Repair defective wire connections or replace the gauge.
2. To test the temperature warning light
   - When the ignition key is turned on, the light assembly is supplied with 12-volts and grounded through the sending unit mounted on the powerhead. When the temperature switch makes contact—because the water temperature has reached an unsafe figure—the circuit to ground is completed and the lamp should light.
   - Set the ignition switch to run. Disconnect the wire at the sending unit, and make contact with the wire to a good powerhead ground. The lamp on the control panel should light. If the lamp does not light, check for a burned-out bulb or a break somewhere in the wiring to the light.

Troubleshooting the Warning Systems

Thermomelt Sticks

♦ See Figure 34

Thermomelt sticks are an easy method of determining if the powerhead is operating at the proper temperature. Thermomelt sticks are not expensive and are available at most local marine dealers.

1. Start the powerhead with the propeller in the water and operate it for about five minutes at roughly 3000 rpm, allowing it to warm to normal operating temperature.

Fig. 34 A thermomelt stick is a quick, simple, inexpensive, and fairly accurate method of determining the operating temperature of the powerhead.

※※ CAUTION

Water must circulate through the lower unit to the powerhead any time the unit is operating to prevent damage to the water pump in the lower unit. Just five seconds without water will damage the water pump.

2. The 140 degree stick should melt when it makes contact with the lower thermostat housing or on the top cylinder. If it does not melt, the thermostat is stuck in the open position and the powerhead temperature is too low.
3. Make contact with the 170 degree stick to the same spot on the lower thermostat housing or on the top cylinder. The stick should NOT melt. If it does melt, the thermostat is stuck in the closed position or the water pump is not operating properly because the powerhead is operating too hot.

If the powerhead is not equipped with a thermostat, the problem may be solved by reverse flushing to clean out the cooling system and/or servicing the water pump. See the service procedures for the thermostat and water pump.

Water Temperature Sensor

REMOVAL & INSTALLATION

30–40 HP (2-Cylinder)

♦ See Figure 35

1. Disconnect the sensor wires at the bullet connectors.
2. Remove the retaining bolts and lift out the sending unit.

To install:

3. Clean the sending unit and all mating surfaces. Make sure that no corrosion or foreign objects are fouling the tip of the sending unit.
4. Install the sending unit in the cylinder head.
5. Tighten the retaining screws snugly. Do not overtighten.

40 HP (4-Cylinder)

♦ See Figure 36

1. Disconnect the temperature switch at the bullet connector.
2. Remove the temperature switch from the cylinder block cover.

To install:

3. Clean the tip of the temperature switch and reinstall it back into the cylinder block cover.
4. Tighten the switch snugly. Do not overtighten the switch.

40–60 HP (3-Cylinder)

♦ See Figures 37 and 38

1. Disconnect the temperature switch at the bullet connector.
2. Remove the retaining screw and remove the temperature switch from the cylinder head.

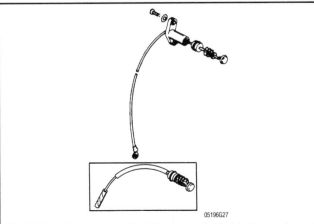

Fig. 35 Location drawing of both the new and old style temperature sending units

LUBRICATION AND COOLING 6-25

To install:

3. Clean the tip of the temperature switch and O-ring seal and reinstall it back into the cylinder block cover.
4. Tighten the switch snugly. Do not overtighten.

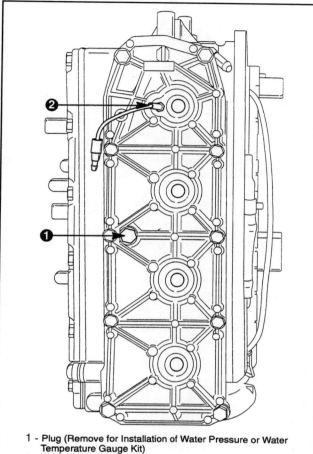

1 - Plug (Remove for Installation of Water Pressure or Water Temperature Gauge Kit)
2 - Temperature Switch

Fig. 36 Location of the water temperature switch and optional temperature or pressure gauge plug

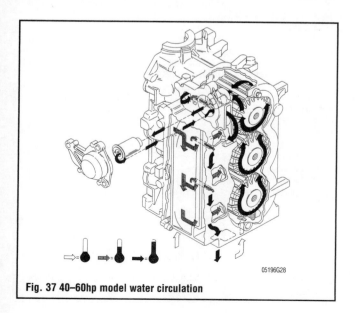

Fig. 37 40–60hp model water circulation

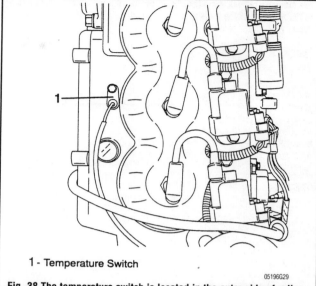

1 - Temperature Switch

Fig. 38 The temperature switch is located in the outer side of cylinder head as shown

70–275 HP

▶ See Figure 39

1. Disconnect the temperature switch at the bullet connector.
2. Remove the screw and retainer and remove the temperature sensor from the cylinder head.

➡ After removing the screw and retainer from the overheat sensor, inserting a small flat-tip screwdriver between the sensor and the cylin-

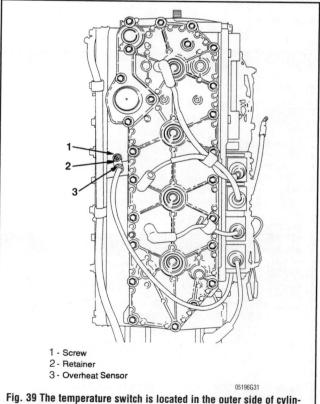

1 - Screw
2 - Retainer
3 - Overheat Sensor

Fig. 39 The temperature switch is located in the outer side of cylinder head as shown

6-26 LUBRICATION AND COOLING

der block will break the adhesion of the O-ring seal to the block and aid in the removal of the sensor.

To install:
3. Clean the tip of the temperature switch and O-ring seal and reinstall it back into the cylinder block cover.
4. Tighten the switch snugly. Do not overtighten.

TESTING

30–60 HP

♦ See Figure 40

1. During normal engine operation, the temperature switch is open. The switch will close when the temperature reaches 182–198°F (84–92°C). The switch will reset back to an open circuit at 162–178°F (73–81°C).
2. Place the switch end into water with a thermometer and heat the water to 182–198°F (84–92°C).
3. Take the switch out and perform an ohms test to check when the switch resets.
4. If the switch does not meet specification, it will need to be replaced.

70–115 HP

1. During normal engine operation, the temperature switch is closed. The switch will open when the temperature reaches 162–178°F (69–85°C). The switch will reset back to an closed circuit at 182–198°F (80–96°C).
2. Place the switch end into water with a thermometer and heat the water to 182–198°F (80–96°C)
3. Take the switch out and as it cools, perform an ohms test to check when the switch resets.
4. If the switch does not meet specification, it will need to be replaced.

135–225 HP

1. During normal engine operation, the temperature switch is open. The switch will close when the temperature reaches 232–248°F (102–129°C).
2. Place the switch end into water with a thermometer and heat the water to 232–248°F (102–129°C).
3. Take the switch out and as it cools, perform an ohms test to check when the switch resets.
4. If the switch does not meet specification, it will need to be replaced.

Oil Gauge

TESTING

1. Set the key switch to the **RUN** position.
2. Disconnect the sensor wire at the remote oil tank sending unit.
3. Make contact with the sensor wire to a good powerhead ground. The gauge should move towards empty.
4. If the gauge moves, replace the sending unit.
5. If no movement is observed, check for 12-volts DC at the gauge terminal embossed "I" and for a good ground connection at terminal "G".
6. Repair the damaged wire connections or replace the defective oil quantity gauge.

Oil Level Switch

TESTING

1. Disconnect both low oil sensor leads from the terminal connections.
2. Connect an ohmmeter between both leads.

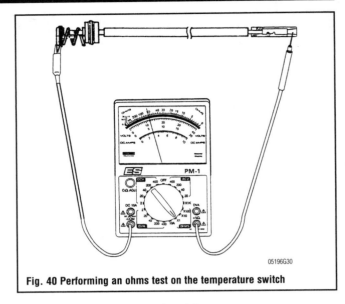

Fig. 40 Performing an ohms test on the temperature switch

3. There should be NO continuity through the sensor.
4. If continuity exists, the sensor is faulty and must be replaced.

REMOVAL & INSTALLATION

1. On the 40hp models with constant-ratio oil injection, tilt the oil reservoir enough to access the bottom of the oil reservoir.
2. Remove the screw securing the low-oil sensor to the bottom of the reservoir and remove the sensor.
3. On the 50–60hp models with constant-ratio oil injection, disconnect the oil sensor wires at their bullet connectors.
4. Disconnect the oil outlet hose and remove the oil reservoir from the engine.
5. Remove the low-oil sensor from the bottom of the reservoir.
6. On the 75–125hp models equipped with variable-ratio oil injection, disconnect the low-oil sensor wires at their bullet connectors located at the bottom of the oil reservoir.
7. Disconnect and plug any oil lines at the oil reservoir fitting.
8. Remove the starter motor upper mounting bracket and the screws securing the reservoir upper support bracket.
9. Remove the reservoir lower support screw and remove the oil reservoir.
10. Remove the low-oil sensor from the bottom of the reservoir.
11. On the 135–275hp V-6 models equipped with variable-ratio oil injection, the engine mounted reservoir is equipped with a float-type low-oil sensor built into the oil reservoir fill cap.
12. Disconnect the oil level sensor wires at their bullet connectors and remove the cap and sensor assembly.

To install:
13. On the 135–275hp models, connect the oil level sensor wires and replace the oil reservoir cap and sensor assembly.
14. On the 75–125hp models, install the oil level sensor into the bottom of the oil reservoir and secure it with the retaining screw.
15. Install the oil reservoir lower support screw and oil reservoir on the powerhead.
16. Install the starter motor upper mounting bracket and secure the oil reservoir to the upper support bracket.
17. Reconnect the oil level sensor wires at their bullet connectors.
18. Connect the oil lines to the oil reservoir fitting..
19. On the 40–60hp models, install the oil level sensor assembly into the bottom of the oil reservoir and secure it.
20. Install the oil reservoir onto the powerhead and secure it in place.
21. Connect the oil lines.
22. Connect the low-oil sensor wires at their bullet connectors.

LUBRICATION AND COOLING 6-27

OPTIMAX WARNING SYSTEMS

Description and Operation

1998–99 MODEL YEARS

The outboard warning system incorporates a warning light gauge and warning horn. The warning horn is located inside the remote control or is part of the ignition key switch wiring harness.

When the key switch is turned to the **ON** position, the warning lights and horn will turn on for a moment as a system test to let the operator know the system is operational.

Low Oil Level

The system is activated when the oil in the engine mounted oil reservoir drops below 50 fl. oz. (1.5 liters). You will still have an oil reserve remaining for 50 minutes of full speed operation.

➙ The engine mounted oil reservoir along with the remote oil tank will need to be refilled.

The OIL light on the gauge will come on and the warning horn sounds a series of four short tones. If you continue to operate the engine, the light will stay on and the horn will sound four short tones every two minutes. The engine needs to be shut down to reset the warning system.

No Oil Flow To The Electric Oil Pump

The system is activated when the flow of oil to the oil pump is blocked. No lubricating oil is being supplied to the engine. Stop the engine as soon as possible or severe engine damage will occur.

The OIL light and the CHECK ENGINE light will come on and the warning horn will begin sounding. The warning system will automatically reduce and limit the engine speed to 3000 rpm. The engine has to shut down to reset the warning system.

Engine Overheat

The system is activated when the engine temperature is too hot. The TEMP light will come on and the warning horn will begin sounding. the warning system will automatically limit the engine speed to 3000 rpm. After the engine has cooled, shift the outboard into neutral to reset the warning system.

Ignition Coil, Sensor or Injector Not Functioning

The system is activated if an ignition coil, sensor or injector is not working properly. The CHECK ENGINE light will come on.

Throttle Sensor Not Functioning

The system is activated if the throttle sensors are not working properly. The CHECK ENGINE light will come on and the warning horn will begin sounding.

Water Separating Fuel Filter Full of Water

The water level detection warning is activated when water in the water separating fuel filter reaches the full level. The water can be drained from the filter.

The WATER DETECTION light will come on and the warning horn will begin sounding a series of four beeps. As the engine continues to be operated, the light will stay on and the horn will sound every two minutes.

Engine Over-Speed Protection

The system is activated when engine speed exceeds the maximum allowable rpm. anytime the engine over-speed system is activated, the warning horn begins to sound continuously. The system will automatically reduce the engine speed to within the allowable limits.

➙ Engine speed should never reach the maximum limit to activate the warning system unless the propeller is ventilating, an incorrect propeller is being used or the propeller itself is faulty.

GUARDIAN PROTECTION SYSTEM—2000

The Guardian Protection system monitors critical engine functions and will reduce engine power accordingly in an attempt to keep the engine running within safe operating parameters.

※ **WARNING**

The Guardian System cannot guarantee that powerhead damage will not occur when adverse operating conditions are encountered. The Guardian System is designed to (1) warn the operator that the engine is operating under adverse conditions and (2) reduce power by limiting maximum rpm in an attempt to avoid or reduce the possibility of engine damage.

Smartcraft Warning System

♦ See Figure 41

The SmartCraft warning system incorporates the display screens and warning horn and the Guardian Protection System. The warning horn is located inside the remote control or is part of the ignition key switch wiring harness.

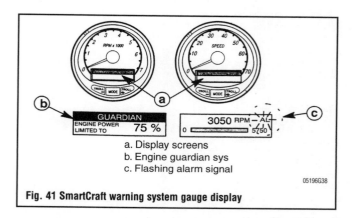

a. Display screens
b. Engine guardian sys
c. Flashing alarm signal

Fig. 41 SmartCraft warning system gauge display

Alarm Warnings

When a problem is detected, the warning horn sounds and the name of the alarm appears in the display on the gauge face. If the problem can cause immediate engine damage, the horn will sound continuously and the Engine Guardian System will respond to the problem by limiting engine power. Immediately reduce throttle speed to idle and refer to the warning messages covered in the chart. If the problem will not cause immediate engine damage, the horn will sound but not continuously.

The alarm message will stay displayed until the mode button is pressed. If there are multiple alarms, these will cycle on the display at five second intervals. If the mode button is pressed to a different screen, the flashing alarm signal "AL" will appear in the upper right hand corner to indicate that there is still a problem

Guardian Protection System

Monitors the critical sensors on the engine for any early indications of problems. The system will respond to a problem by reducing engine power in order to maintain a safe operating condition. The display screen will show the percentage of power loss.

GUARDIAN SYSTEM ACTIVATION

- **Engine overheat**: Engine power level can be reduced to any percentage down to idle speed, if an overheat condition persists
- **Air compressor overheat**: Engine power level can be reduced to any percentage down to idle speed, if an overheat condition persists
- **Block water pressure low**: Engine power level can be reduced to any percentage down to a fast idle speed, if condition persists

6-28 LUBRICATION AND COOLING

- **Throttle position sensor failure**: If the throttle position sensor fails or becomes disconnected, power will be limited to a maximum of approximately 4500 rpm. when the TPS is in the fail mode, the ECM will use the MAP sensor for a reference to determine fuel calibration
- **Temperature sensor (cylinder head and air compressor) failure**: If a temperature sensor should fail or become disconnected, power will be reduced 25 percent.
- **Battery voltage (too high or too low)**: Battery voltage greater than 16.5 volts or less than 10.5 volts will result in engine output power being reduced. The higher or lower the voltage is outside of these parameters, the greater the percentage of power reduction. In an extreme case, power could be reduced to idle speed.
- **Oil pump failure**: If the oil pump fails or an open circuit occurs between the pump and the ECM, engine power will be reduced to idle speed.

Troubleshooting the Warning Systems

The DDT is used to troubleshoot the OptiMax series of engines. Attach the diagnostic cable to the ECM diagnostic connector and plug in the software cartridge. You will then be able to monitor sensors and ECM values including status switches.

The ECM program can help diagnose intermittent engine problems. It will also record the state of the engine sensors and switches for a period of time and then can be played back to review the recorded information.

Troubleshooting without the DDT is limited to checking the resistance on some of the sensors. Typical failures do not usually include the ECM. Connectors, set-up and mechanical wear are usually at fault.

Since this system share sensors with other systems on the outboard, sensor testing and service is covered in both the "Fuel System" and "Ignition and Electrical" chapters.

ENGINE MECHANICAL 7-2
THE TWO-STROKE CYCLE 7-2
POWERHEAD 7-2
    REMOVAL & INSTALLATION 7-2
REED VALVE 7-13
2.5–3.3 HP 7-14
    REMOVAL & INSTALLATION 7-14
4–5 HP 7-14
    REMOVAL & INSTALLATION 7-14
6–15 HP 7-15
    REMOVAL & INSTALLATION 7-15
20–25 HP 7-15
    REMOVAL & INSTALLATION 7-15
30 AND 40 HP (2-CYLINDER) 7-15
    REMOVAL & INSTALLATION 7-15
40 HP (4-CYLINDER) 7-16
    REMOVAL & INSTALLATION 7-16
40 HP (3-CYLINDER), 50 AND 60 HP 7-16
    REMOVAL & INSTALLATION 7-17
75–125 HP 7-17
    REMOVAL & INSTALLATION 7-17
135-250 HP 7-17
    REMOVAL & INSTALLATION 7-17
135–225 DFI OPTIMAX 7-18
    REMOVAL & INSTALLATION 7-18
275 HP 7-18
    REMOVAL & INSTALLATION 7-19
POWERHEAD RECONDITIONING 7-19
DETERMINING POWERHEAD CONDITION 7-19
BUY OR REBUILD? 7-19
POWERHEAD OVERHAUL TIPS 7-20
    TOOLS 7-20
    CAUTIONS 7-20
    CLEANING 7-20
    REPAIRING DAMAGED THREADS 7-21
POWERHEAD PREPARATION 7-21
CYLINDER BLOCK AND HEAD 7-22
    GENERAL INFORMATION 7-22
    INSPECTION 7-23
CYLINDER BORES 7-23
    GENERAL INFORMATION 7-23
    INSPECTION 7-23
    REFINISHING 7-24
PISTONS 7-24
    GENERAL INFORMATION 7-24
    INSPECTION 7-24
PISTON PINS 7-25
    GENERAL INFORMATION 7-25
    INSPECTION 7-25
PISTON RINGS 7-26
    GENERAL INFORMATION 7-26
    INSPECTION 7-27
CONNECTING RODS 7-27
    GENERAL INFORMATION 7-27
    INSPECTION 7-27
CRANKSHAFT 7-28
    GENERAL INFORMATION 7-28
    INSPECTION 7-29
BEARINGS 7-29
    GENERAL INFORMATION 7-29
    INSPECTION 7-30
POWERHEAD EXPLODED VIEWS 7-30
TORQUE SEQUENCE DIAGRAMS 7-36
SPECIFICATIONS CHARTS
    ENGINE TORQUE SPECIFICATIONS 7-39
    ENGINE REBUILDING SPECIFICATIONS 7-41

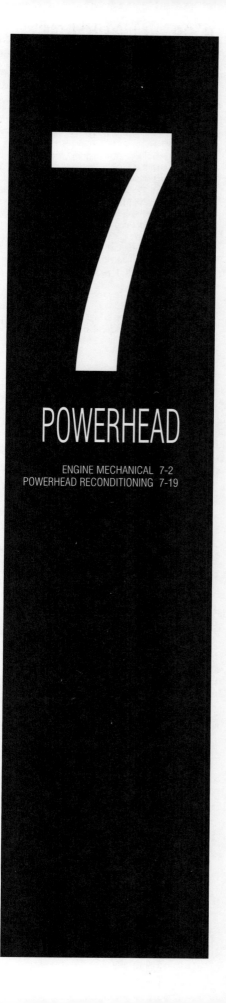

7
POWERHEAD

ENGINE MECHANICAL 7-2
POWERHEAD RECONDITIONING 7-19

7-2 POWERHEAD

ENGINE MECHANICAL

The Two-Stroke Cycle

The two-stroke engine can produce substantial power for its size and weight. But why is a two-stroke so much smaller and lighter than a four-stroke? Well, there is no valvetrain. Camshafts, valves and pushrods can really add weight to an engine. A two-stroke engine doesn't use valves to control the air and fuel mixture entering and exiting the engine. There are holes, called ports, cut into the cylinder which allow for entry and exit of the fuel mixture. The two-stroke engine also fires on every second stroke of the piston, which is the primary reason why so much more power is produced than a four-stroke.

Since two-stroke engines discharge approximately one fourth of their fuel unburned, they have come under close scrutiny by environmentalists. Many states have tightened their grip on two-strokes and most manufacturers are hard at work developing new efficient models that can meet the tough emissions standards. Check out your state's regulations before you buy any two-stroke outboard.

The two-stroke engine is able to function because of two very simple physical laws. The first, gases will flow from an area of high pressure to an area of lower pressure. A tire blowout is an example of this principle. The high-pressure air escapes rapidly if the tube is punctured. Second, if a gas is compressed into a smaller area, the pressure increases and if a gas expands into a larger area, the pressure is decreased. If these two laws are kept in mind, the operation of the two-stroke engine will be easier understood.

Two-stroke engines utilize an arrangement of port openings to admit fuel to the combustion chamber and to purge the exhaust gases after burning has been completed. The ports are located in a precise pattern in order for them to be opened and closed at an exact moment by the piston as it moves up and down in the cylinder. The exhaust port is located slightly higher than the fuel intake port. This arrangement opens the exhaust port first as the piston starts downward and therefore, the exhaust phase begins a fraction of a second before the intake phase.

Actually, the intake and exhaust ports are spaced so closely together that both open almost simultaneously. For this reason, the pistons of most two-stroke engines have a deflector-type top. This design of the piston top serves two purposes very effectively. First, it creates turbulence when the incoming charge of fuel enters the combustion chamber. This turbulence results in more complete burning of the fuel than if the piston top were flat. Second, it forces the exhaust gases from the cylinder more rapidly.

Beginning with the piston approaching top dead center on the compression stroke, the intake and exhaust ports are closed by the piston, the reed valve is open, the spark plug fires, the compressed air/fuel mixture is ignited and the power stroke begins. The reed valve was open because as the piston moved upward, the crankcase volume increased, which reduced the crankcase pressure to less than the outside atmosphere.

As the piston moves downward on the power stroke, the combustion chamber is filled with burning gases. As the exhaust port is uncovered, the gases, which are under great pressure, escape rapidly through the exhaust ports. The piston continues its downward movement. Pressure within the crankcase increases, closing the reed valves against their seats. The crankcase then becomes a sealed chamber. The air/fuel mixture is compressed ready for delivery to the combustion chamber. As the piston continues to move downward, the intake port is uncovered. A fresh air/fuel mixture rushes through the intake port into the combustion chamber striking the top of the piston where it is deflected along the cylinder wall. The reed valve remains closed until the piston moves upward again.

When the piston begins to move upward on the compression stroke, the reed valve opens because the crankcase volume has been increased, reducing crankcase pressure to less than the outside atmosphere. The intake and exhaust ports are closed and the fresh fuel charge is compressed inside the combustion chamber.

Pressure in the crankcase decreases as the piston moves upward and a fresh charge of air flows through the carburetor picking up fuel. As the piston approaches top dead center, the spark plug ignites the air/fuel mixture, the power stroke begins and one full cycle has been completed.

The exact time of spark plug firing depends on engine speed. At low speed the spark is retarded, fires later than when the piston is at or beyond top dead center. Engine timing is built into the unit at the factory.

At high speed, the spark is advanced, fires earlier than when the piston is at top dead center. On all but the smallest horsepower outboards the timing can be changed adjusted to meet advance and retard specifications.

Because of the design of the two-stroke engine, lubrication of the piston and cylinder walls must be delivered by the fuel passing through the engine. Since gasoline doesn't make a good lubricant, oil must be added to the fuel and air mixture. The trick here is to add just enough oil to the fuel to provide lubrication. If too much oil is added to the fuel, the spark plug can become "fouled" because of the excessive oil within the combustion chamber. If there is not enough oil present with the air/fuel mixture, the piston can "seize" within the cylinder. What usually happens in this case is the piston and cylinder become scored and scratched, from lack of lubrication. In extreme cases, the piston will turn to liquid and eventually disintegrate within the cylinder.

Most two-stroke engines require that the fuel and oil be mixed before being poured into the fuel tank. This is known as "pre-mixing" the fuel. This can become a real hassle. You must be certain that the ratio is correct. Too little oil in the fuel could cause the piston to seize to the cylinder, causing major engine damage and completely ruining your weekend. Most modern two-stroke engines have an oil injection system that automatically mixes the proper amount of oil with the fuel as it enters the engine.

Powerhead

REMOVAL & INSTALLATION

When removing any powerhead, it is a good idea to make a sketch or take an instant picture of the location, routing and positioning of electrical harnesses, brackets and component locations for installation reference.

The following procedures assume that the outboard has been removed from the boat and placed on a suitable work stand. If the powerhead is being removed with the outboard still mounted on the boat and the powerhead is equipped with an electric starter, disconnect first the negative, then the positive battery cables to prevent accidental starting.

On some powerheads it will be necessary to remove attached components if the powerhead is to be overhauled. Refer to the specific sections covering these components for removal and installation information.

➡ **Sometimes when attempting to remove the powerhead it won't come loose from the adapter. The gasket may be holding the powerhead. Rock the powerhead back and forth or give it a gentile nudge with a pry bar. If the gasket breaks loose and the powerhead still will not come loose, then the driveshaft is seized to the crankshaft at the splines.**

2.5–5 HP

♦ See accompanying illustrations

1. On the 2.5–3.3 hp models, unsnap and lower the spark plug access cover.
2. Separate the cowling halves free of the powerhead and set them aside.
3. Pull the spark plug lead free of the spark plug. Use a pulling and twisting motion on the molded cap portion. Never pull on the wire or on the connector inside, the cap or boot may be separated or damaged.
4. Remove the spark plug. Take care not to tilt the socket as the plug is removed or the insulator may be cracked.
5. Remove the hand rewind starter.
6. Rotate the fuel shut-off valve to the off position.
7. Remove the clamp securing the fuel line to the carburetor or fuel pump and remove the nuts, washers and spacers securing the fuel tank to the powerhead. Lift the fuel tank free of the powerhead.
8. On the 2.5–3.3 hp models, remove the screws securing the throttle and choke knobs to each lever. Remove the screws securing the cover plate to the front of the carburetor. Slide the front plate forward, off the choke and throttle levers. Set the front plate to one side of the carburetor.
9. On the 4–5 hp models, loosen the screw securing the throttle wire to the throttle arm of the carburetor and then pull the wire through and free of the throttle arm. Disconnect the choke link rod at the carburetor by unsnapping the nylon clip from the link rod and rotating the nylon clip. Pull the link rod free of the carburetor choke shaft.
10. Remove the screws securing the baffle cover to the front of the carburetor.

POWERHEAD 7-3

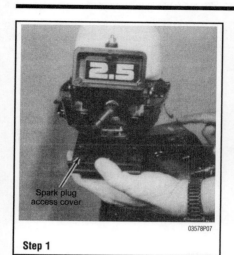
Step 1

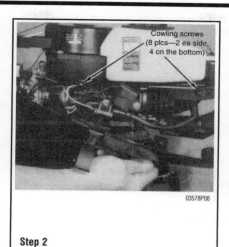
Step 2

Step 5

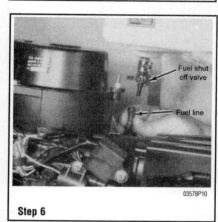
Step 6

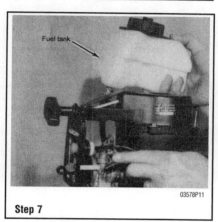
Step 7

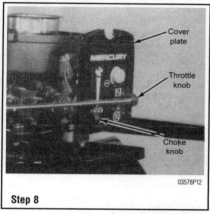
Step 8

Step 10

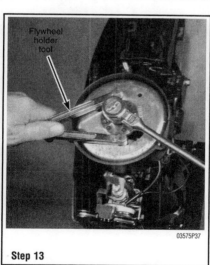
Step 13

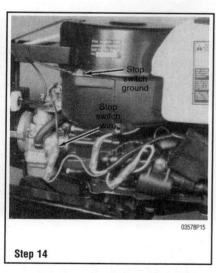
Step 14

11. Remove the through bolts on the front of the carburetor and lift the carburetor free of the powerhead.

12. On the 2.5–3.3 hp models, disconnect the stop switch wire lead from the carburetor front plate and then separate the wire lead going to the CD module at the quick disconnect bullet connector. The front plate of the carburetor is now free. Loosen the screw on the clamp securing the carburetor. Pull the carburetor away from the powerhead.

13. On the 2.5–5 hp models, obtain a strap wrench or equivalent tool. Hold the rope cup steady with the strap wrench and "break" the flywheel nut loose, then remove the nut with the proper size socket. The nut has standard right hand threads.

14. Continue holding the rope cup with the strap wrench and remove the bolts securing the rope cup.

15. Obtain a universal puller and set-up the puller over the crankshaft. Use the holes from which the rope cup bolts were removed to secure the puller. Never attempt to use a puller which pulls on the outside edge of the flywheel or the flywheel

7-4 POWERHEAD

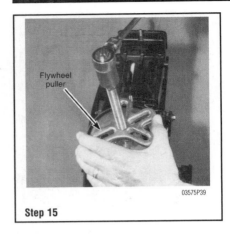
Step 15

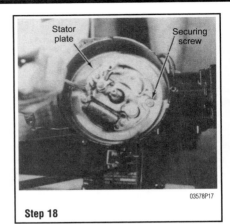
Step 18

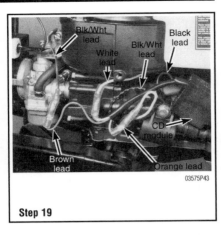
Step 19

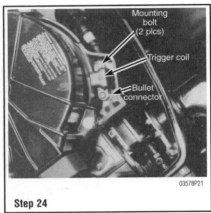
Step 24

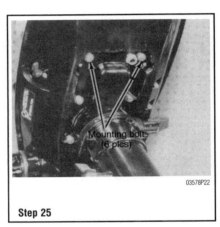
Step 25

Step 26

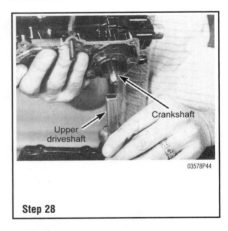
Step 28

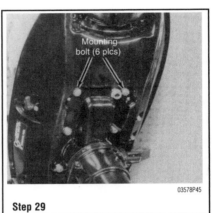
Step 29

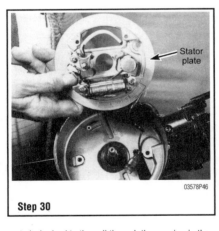
Step 30

may be damaged. After the puller is installed and ready, tighten the center screw onto the end of the crankshaft. Continue tightening the screw until the flywheel is released from the crankshaft.

16. After the flywheel is released, remove the puller and lift the flywheel free of the crankshaft. A "pull" may be felt as the flywheel is lifted due to the permanent magnets installed on the inside rim of the flywheel.

➡ **Handle the flywheel carefully because any sudden shock (such as dropping the flywheel,) will lessen the strength of the magnets. A weak magnet will seriously affect the ignition circuit. Lift out the Woodruff key from the keyway in the crankshaft.**

17. On the 2.5 and 3 hp models equipped with a magneto point ignition, disconnect the wire lead from the stator assembly to the secondary ignition coil on the side of the powerhead. Remove the bolts securing the secondary ignition coil to the powerhead.

18. Remove the screws securing the stator assembly. Lift the stator plate and at the same time feed the disconnected wire lead to the coil through the opening in the cylinder block cover.

19. On 2.5–3.3 hp models equipped with CDI ignitions, disconnect all the wire bullet connectors on the side of the powerhead. Remove the bolts securing the secondary ignition coil and CD module to the powerhead.

20. Lift the CD module and secondary coil free of the powerhead.

21. Remove the screws securing the capacitor charging/trigger coil and the wire lead to the stator. Lift off the capacitor charging/trigger coil and at the same time, feed the wire lead through the opening in the cylinder block cover.

22. On the 4 and 5 hp models, disconnect the blade terminal from the secondary ignition coil. Now, remove the bolts securing the coil to the powerhead and lift off the ignition coil. Disconnect all the wire bullet connectors on the side of the powerhead. Pull the rubber bracket securing the CD module back and then lift the module from the rubber bracket.

23. Remove the screws securing the capacitor charging coil and the wire lead to

POWERHEAD 7-5

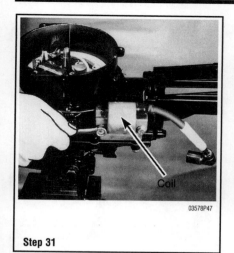
Step 31

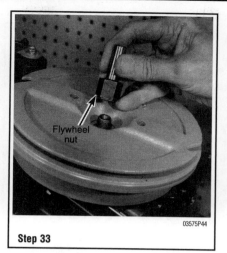
Step 33

Step 34

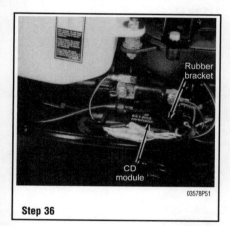
Step 36

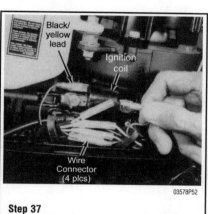
Step 37

Step 39

the stator. Lift off the capacitor charging coil and at the same time, feed the wire lead through the opening in the cylinder block cover.

24. Remove the screws securing the trigger coil and wire lead. Lift out the trigger coil and at the same time, feed the wire lead through the opening in the cylinder block cover.

25. On the 2.5–5 hp models, remove the bolts securing the powerhead to the driveshaft housing.

26. Carefully pry the powerhead free of the driveshaft housing. It may be necessary to tap on the joint with a soft head mallet to break the powerhead loose. Lift the powerhead straight up and clear of the driveshaft.

➡ If the unit is several years old or if it has been operated in salt water or has not had proper maintenance or shelter or any number of other factors, then separating the powerhead from the driveshaft housing may not be a simple task. An air hammer may be required on the studs to shake the corrosion loose; heat may be applied to the casting to expand it slightly; or other devices employed in order to remove the powerhead.

One very serious condition would be the driveshaft "frozen" with the crankshaft. In this case, a circular plug-type hole must be drilled and a torch used to cut the driveshaft. Let's assume the powerhead will come free on the first attempt.

To install:

27. Secure the lower unit in a holding fixture, vise equipped with soft jaws or any other suitable restraining method, as an assist to installing the powerhead and all the supporting components.

28. Slide a new gasket down the driveshaft into position on the driveshaft housing. Now, lower the powerhead onto the driveshaft housing with the external splines on the lower end of the crankshaft indexing with the internal splines of the driveshaft.

29. Install and tighten the bolts securing the powerhead to the driveshaft housing.

30. On the 2.5 and 3.0hp with magneto ignition, hold the stator plate assembly over the powerhead. Feed the wire from the stator plate through the opening in the base of the cylinder block cover. Now, lower the stator plate down over the crankshaft and into place. Secure the stator plate with Phillips head screws.

31. Install the secondary coil, on the port side of the powerhead with the attaching bolts. Connect the Black/White wire leads from the stator and secondary coil together.

32. On 2.5, 3.0 and 3.3hp with CDI, hold the capacitor charging/trigger coil over the powerhead. Feed the wire from the capacitor charging/trigger coil through the opening in the base of the cylinder block cover. Now, lower the capacitor charging/trigger coil down over the crankshaft and into place. Secure the capacitor charging/trigger coil and wire lead clamp with Phillips head screws.

33. Install the secondary ignition coil and CD module onto the side of the powerhead and secure it with the bolts. Be sure to slide the Black/White lead from the secondary ignition coil and the Black lead from the CD module with the large eyelets, onto the mounting bolt for the ignition coil. Connect the Orange lead to the Orange lead at the bullet connector and then the White lead to the White lead at the bullet connector.

34. On 4 and 5hp, insert the trigger coil wire lead through the opening in the crankcase cover. Feed the wire leads through while lowering the trigger coil down into the crankcase cover recess. Secure the trigger coil with Phillips screws, be sure to secure the trigger coil ground lead terminal under one of the Phillips head screws.

35. Lower the capacitor charging coil wire leads down through the opening in the crankcase cover. Guide the wire leads while lowering the capacitor charging coil down into the crankcase cover. Secure the capacitor charging coil and wire lead clamp with Phillips head screws.

36. Insert the CD module into the rubber bracket on the port side of the powerhead.

37. Install the secondary ignition coil onto the port side of the powerhead and secure it with the bolts. Be sure to slide the Black leads, one from the CD module and one from the stop button, onto one of the mounting bolts for the ignition coil.

38. Connect the wire leads as follows: Brown to Brown, Black/Red to Black/Red, White to White, Red/White to Red/White and Black/Yellow to Secondary coil terminal.

39. Place the Woodruff key into the groove of the crankshaft. Slide the flywheel down the crankshaft with the groove in the flywheel indexed over the key on the crankshaft. Rotate the flywheel clockwise and check to be sure the flywheel does not make contact with any part of the ignition and/or components around the flywheel.

7-6 POWERHEAD

Step 40

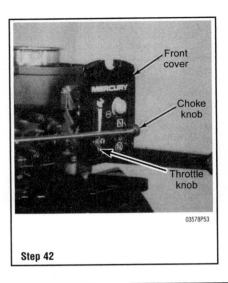
Step 42

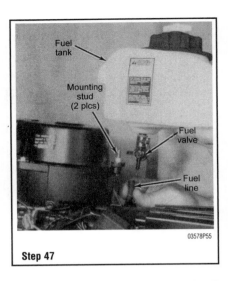
Step 47

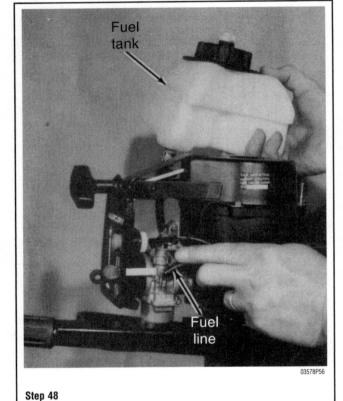
Step 48

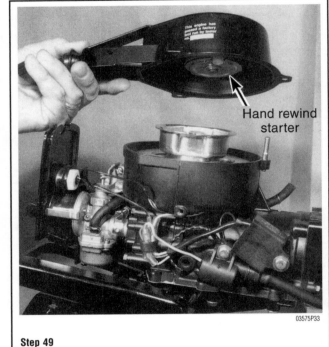
Step 49

40. Install the flywheel nut and tighten the nut to the torque value listed in the Appendix. Use the same method of holding the flywheel as used for disassembling.

41. Install the rope cup and secure it in place with the attaching bolts. Use a strap wrench or similar tool to prevent the rope cup from rotating while the bolts are being tightened.

42. On 2.5, 3.0 and 3.3hp, slide the carburetor onto the intake manifold and secure it in place with the clamp.

43. Position the carburetor front cover plate over the throttle and choke levers and secure the cover plate with screws. Place the throttle knob and choke knob onto the ends of the levers and secure with screws.

44. Connect the Black/White wire lead from the cover plate to the Brown wire lead from the CD module or the Black/White lead from the stator if the model has points.

45. On 4 and 5hp, place the carburetor and a new carburetor gasket against the crankcase cover. Insert the bolts and washers through the baffle bracket, carburetor and into the crankcase cover. Tighten the bolts firmly. Secure the baffle cover to the bracket with screws.

46. Connect the throttle cable and choke linkages.

47. Place the fuel tank in position over the powerhead and then secure it with the spacers, flat washers, locking washers and nuts.

48. Connect the fuel line directly to the carburetor bowl fitting or fuel pump inlet fitting on the side of the carburetor. Turn the fuel shut-off valve to the ON position.

49. Install the hand rewind starter onto the powerhead and secure in place with attaching bolts.

50. Thread a new spark plug into the powerhead and tighten it to the torque value listed in the Appendix.

51. Connect the spark plug high tension lead onto the spark plug.

52. Snap the spark plug access cover into place.

6–25 HP

▶ See accompanying illustrations

1. Remove the engine cover.
2. Disconnect the spark plug leads and remove the spark plugs.

POWERHEAD 7-7

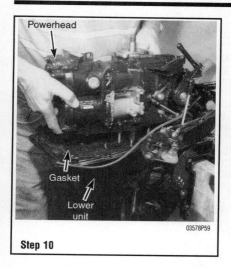
Step 10

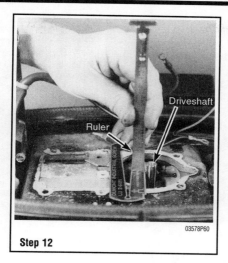
Step 12

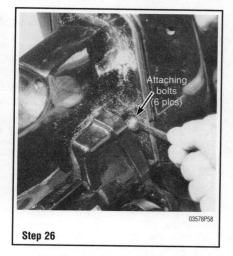
Step 26

   3. Remove the hand rewind starter assembly.
   4. Remove the flywheel and stator assembly.
   5. Remove the ignition coil and switch box.
   6. Remove the carburetor.
   7. Disconnect the tattle-tale hose (if equipped) from the fitting at the bottom of the cowling. Disconnect the shift rods and remove the bolts securing the throttle shift control platform to the powerhead.
   8. Remove the throttle and shift linkage.
   9. On the bottom side of the lower cowling, remove the bolts and nuts securing the powerhead to the driveshaft housing.
   10. Lift the powerhead straight up from the driveshaft housing to prevent tearing the gasket.
   11. Place the powerhead onto a workbench in an upright position for disassembly.

➡ If the crankshaft and driveshaft splines bind together and do not separate smoothly when the powerhead is being lifted, it is possible the driveshaft pulled up/out of the gear housing and has disengaged the water pump impeller key.

   12. Measure the height of the driveshaft to determine if the driveshaft has pulled up/out of the gear housing as follows: Place a straight edge across the driveshaft housing where the powerhead was previously mounted. Measure the distance between the straight edge and the top of the driveshaft, as shown. Next, push down on the driveshaft until the driveshaft bottoms out in the gear housing. If the driveshaft moves downward more than ¼ in. (6.35mm), it will be necessary to remove the gear housing and reinstall the water pump impeller key.
   13. Disconnect the battery cables (if equipped).
   14. Remove the hand rewind starter assembly.
   15. Disconnect the spark plug leads and remove the spark plugs.
   16. Disconnect the fuel hoses and remove the carburetor.
   17. Remove the flywheel, stator assembly, trigger plate, switchbox and the ignition coils.
   18. Place the shift linkage in the neutral position.
   19. Loosen the jam nuts that hold the control cables to the anchor bracket and remove the cables from around the pulley of the primary gear.
   20. Disconnect the lower cable from the throttle/shift gear. Then, rotate the tiller handle twist grip all the way clockwise and disconnect the upper cable from the throttle/shift gear.
   21. Remove the set screw holding the shift shaft coupler to the shift shaft. Slide the coupler off the shaft.
   22. Remove the 6 bolts securing the powerhead to the driveshaft housing and remove the powerhead from the driveshaft housing.

To install:
   23. Observe if the bottom cowling is separated from the driveshaft housing. If it is separated, install a new gasket between the cowling and the housing.
   24. Position a new gasket over the powerhead studs and into place on the powerhead base.
   25. Apply a thin coating of Multipurpose Lubricant or equivalent, to the driveshaft splines.
   26. Install the powerhead to the bottom cowling and driveshaft housing. If necessary, rotate the flywheel slightly to allow the coupler splines to index with the driveshaft splines and allow the powerhead to become fully seated.
   27. Secure the powerhead to the bottom cowling and driveshaft housing with bolts and nuts on studs.

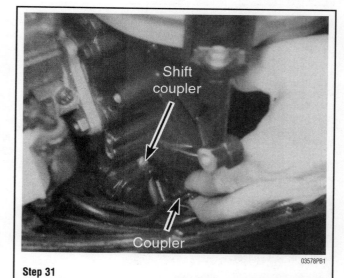
Step 31

   28. Apply Loctite® No. 35 thread sealing compound or equivalent to the threads of the bolts and studs.
   29. Tighten the mounting bolts and nuts evenly, alternately and in stages to 200 inch lbs. (22.6Nm).
   30. Tighten all bolts and nuts to ⅓ the torque value, then repeat the sequence tightening to ⅔ the torque value. Finally, on the third and last sequence, tighten to the full torque value.
   31. Connect the shift shaft coupler to the powerhead.

30 and 40 HP (2-Cylinder)

   1. Disconnect the spark plug leads and remove the spark plugs.
   2. On electric start models, disconnect the battery cables and remove them from the powerhead.
   3. On manual start models, disconnect the safety lanyard switch, stop button assembly and the warning horn at the bullet connectors.
   4. On electric start models, disconnect the remote control harness at the main wiring harness connector. Disconnect any additional remote control harness leads from the bullet connectors. If the unit is equipped with tilt and trim, disconnect the lower cowl switch leads at the bullet connectors.
   5. Disconnect the neutral safety switch leads.
   6. On the manual start models, disconnect the primer bulb lines from the intake manifold and carburetor fittings.
   7. Remove the hand rewind starter assembly (if equipped).
   8. On electric start models, remove the flywheel cover.
   9. Disconnect the tell-tale water discharge hose from the fitting on the powerhead.
   10. On tiller handle equipped models, remove the throttle cables.

7-8 POWERHEAD

11. On remote control equipped models, remove the throttle and shift cables.
12. Disconnect the fuel supply line from the lower cowl.
13. Remove the fasteners securing the trim cover to the driveshaft housing.
14. Remove the 6 screws securing the powerhead to the driveshaft housing.
15. Remove the plastic cap from the center of the flywheel and install the lifting eye (91–90455) into the flywheel a minimum of 5 turns.

➥ **At this point double check for any cables, hoses, wires or linkages that will interfere with the removal of the powerhead.**

16. Move the powerhead back and forth to break the gasket seal between the powerhead and driveshaft housing. When the seal breaks loose, lift the powerhead off the driveshaft and driveshaft housing using a suitable lifting hoist.
17. Remove the powerhead to a clean work area and placed in a suitable holding fixture secured in a vice.

To install:

18. Throughly clean all old gasket material from the driveshaft housing and powerhead mating surfaces.
19. Lubricate the splines on the driveshaft with Quicksilver Special Lubricant No. 101 or equivalent. Wipe off any excess lubricant from the top of the driveshaft.
20. Place a new gasket onto the driveshaft housing.
21. Install the powerhead on the driveshaft housing. You may have to rotate the crankshaft to index the crankshaft and driveshaft splines.
22. Apply Quicksilver Perfect Seal or equivalent to the powerhead mounting screws. Install and tighten evenly to 29 ft. lbs. (39.3Nm).
23. Install the trim cover over the driveshaft housing. Tighten the fasteners securely.
24. Reconnect the tell-tale hose at the powerhead fitting.
25. Install the fuel supply line connector to the lower cowling.
26. On tiller handle equipped models, reinstall and adjust the throttle cable.
27. On remote control equipped models, reinstall the throttle and shaft cables.
28. Install the hand rewind starter (if equipped).
29. Install the flywheel cover.
30. On manual start models, connect the primer bulb lines to the intake manifold and carburetor fittings.
31. Connect the neutral safety switch lead connectors.
32. On manual start models, connect the safety lanyard switch, stop button switch and warning horn to the engine wiring harness.
33. On remote control equipped models, connect the remote control harness to the main wiring harness connector. Make sure to reconnect any remaining remote control harness leads.
34. On electric start equipped models, connect the battery cables.
35. Install the spark plugs and reconnect the spark plug leads.

40–60 HP (3-Cylinder)

♦ **See accompanying illustrations**

1. Disconnect the engine battery cables from the battery terminals.
2. Disconnect the engine fuel line from the fuel tank.
3. Remove the front engine cowling cover.
4. Remove the port and starboard halves of the engine cowling.
5. Separate the electrical extension harness connectors.
6. Disconnect the remote control cables from the powerhead.
7. Stop and carefully observe the wiring and hose connections before proceeding. Because there are so many different powerheads and the arrangement is slightly different on each, it is not possible to illustrate each and every one. Even if they were shown, the reader would not be able to identify the powerhead being serviced. Therefore, Take time to make notes and tag the wire leads and hoses. You may elect to follow the practice of many professional mechanics by taking a series of photographs of the powerhead, one from the top and a couple from the sides showing the wiring and arrangement of parts.
8. On the 50 and 60 hp models, disconnect both leads from the battery.
9. Disconnect the first Black cable from the lower terminal on the cranking motor.
10. Disconnect the second large Black cable from the upper terminal on the cranking motor.
11. Make sure to identify the Black cables connected to the starting motor on the 50 and 60 hp powerheads. One of the bracket securing bolts also secures the oil tank to the powerhead.
12. Remove the bolts securing the cranking motor bracket, the third large Black cable and the oil injection tank to the powerhead.
13. Remove the bracket and cranking motor from the powerhead.
14. Disconnect the high tension leads from the spark plugs. Always use a pulling and twisting motion as a precaution against damaging the connection.
15. Remove the spark plugs.
16. Remove the bolts securing the cover to the electrical box. The ignition switchbox, ignition coil, fuse, starter solenoid and rectifier/regulator are all mounted inside the electrical box. All these electrical components will be removed as an assembly when the box is removed.
17. Disconnect the White/Black, Purple, Brown and White leads between the electrical box and the stator and trigger assembly.
18. Disconnect the wiring harness connector.
19. Locate the low oil warning module secured to the inside of the lower powerhead cover and disconnect the following leads: Tan, Purple, light Blue leads and a Black ground eyelet lead.
20. Remove the bolts securing the module to the cover and remove the module.
21. Remove the securing hardware and lift out the electrical box with all electrical components undisturbed.
22. Disconnect the Blue/White, Green/White and Red/Black leads at the trim switch located on the side of the lower powerhead cover.
23. Remove the screws securing the cover to the air box.
24. Remove the long bolts securing the carburetors to the intake manifold. The carburetors are held together as an assembly by the forward straps, throttle and choke linkage and fuel lines.
25. Disconnect the fuel supply line and the primer line, if equipped.
26. Disconnect the fuel line between the enricher valve and the fitting on the fuel bowl of the top carburetor and the line between the valve and the fitting at the base of the oil pump.
27. Lift off the carburetors, as an assembly, with linkage and fuel lines between the carburetors still intact.

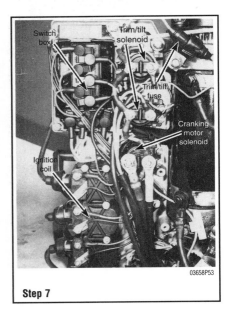

Step 7

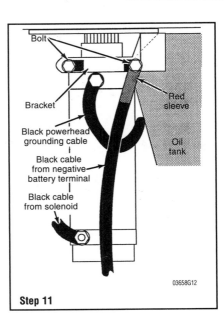

Step 11

Step 27

POWERHEAD 7-9

28. Disconnect the leads from the low oil sensor, located at the base of the oil tank, at their quick disconnect fittings. If the oil tank contains oil, make arrangements to plug the oil line once it is pulled free of the fitting, to prevent oil from spilling into the lower powerhead cover. Snip the tie wrap from the oil supply line, from the tank, at the oil injection pump. Ease the line free of the fitting on the pump and lift out the oil tank.

29. Disconnect the oil outlet line between the pump and the 2 psi check valve next to the fuel pump. Remove the attaching bolts and remove the oil pump from the powerhead.

30. Snip the tie wraps around the inlet, outlet and pulse lines at the fuel pump. Remove the Phillips head screws securing the pump to the powerhead. Lift off the pump.

31. Remove the top bolt securing the barrel retainer over the control cable barrels. Swing the retainer down to clear both barrels.

32. Remove the locknuts and washers securing the throttle and shift cable ends to the throttle lever and shift actuator stud. Slide the cables and barrels away from the barrel receptacles cast into the block and lift both cables clear.

33. Make a final check to make sure no other leads or attachments will impede the removal of the powerhead.

34. Remove the bolts securing the lower powerhead cover to the intermediate housing and remove the cover.

35. Next, remove the bolts securing the powerhead to the intermediate housing.

36. Remove the wing nuts securing the plastic flywheel cover to the powerhead.

37. Remove the plastic cap from the end of the crankshaft.

38. Thread a lifting eye onto the end of the crankshaft as far as it will go.

39. Using a suitable hoist, lift the powerhead assembly clear of the intermediate housing.

40. After lifting off the powerhead, remove all traces of the base gasket.

41. Mount the powerhead onto some type of stand, to facilitate easy access to all parts. Never attempt to mount the powerhead in a stand secured in a vise. Such an attempt will only lead to damage of the powerhead and possible personal injury.

42. Remove the lifting eye from the crankshaft.

To install:

43. Thread a lifting eye onto the end of the crankshaft as far as it will go. For safety, check to be sure the lifting eye is properly installed.

44. Using a suitable hoist, lift the powerhead. Place a new gasket around the powerhead studs and into position on the base of the intermediate housing.

45. Lubricate the driveshaft splines with multipurpose lubricant. Wipe off any excess lubricant from the top of the driveshaft.

46. Slowly lower the powerhead down onto the intermediate housing. It may be necessary to rotate the flywheel slightly to index the crankshaft splines with the driveshaft splines.

47. Once the splines index, lower the powerhead fully into place on the intermediate housing.

48. Secure the powerhead to the intermediate housing with the bolts. Tighten the bolts in stages to 28 ft. lbs. (38Nm).

49. Install the lower cover around the intermediate housing and secure the cover in place with the bolts. Tighten the bolts to 80 inch lbs. (9Nm).

50. Disconnect the hoist from the lifting eye and then remove the eye from the crankshaft. Install the plastic cap onto the end of the crankshaft. Install the flywheel cover and secure it in place with flat washers and wingnuts.

51. Reconnect the tell-tale hose at the powerhead fitting.

52. Install the fuel supply line connector to he lower cowling.

53. Reinstall the throttle and shaft cables.

54. Install the flywheel cover.

55. Connect the remote control harness to the main wiring harness connector. Make sure to reconnect any remaining remote control harness leads.

56. Connect the battery cables.

57. Install the spark plugs and reconnect the spark plug leads.

40 HP (4-Cylinder)

▶ **See accompanying illustrations**

1. Disconnect the battery leads at the battery.
2. Disconnect the fuel line at the powerhead fuel connector.
3. Remove the top cowling from the powerhead.
4. Disconnect the spark plug leads and remove the spark plugs.
5. Disconnect the remote control electrical connector at the powerhead.
6. Remove the positive lead from the cranking motor. Remove negative lead, clamps and bolts securing the cranking motor to the powerhead.
7. Stop and take the time now to carefully observe the wiring and hose connections before proceeding.
8. Stop and carefully observe the wiring and hose connections before proceeding. Because there are so many different powerheads and the arrangement is slightly different on each, it is not possible to illustrate each and every one. Even if they were shown, the reader would not be able to identify the powerhead being serviced. Therefore, Take time to make notes and tag the wire leads and hoses. You may elect to follow the practice of many professional mechanics by taking a series of photographs of the powerhead, one from the top and a couple from the sides showing the wiring and arrangement of parts.
9. Remove the bolt securing the fuel connector to the lower cowling. Disconnect the hose from the tattle-tale fitting at the rear of the powerhead at the lower cowling. Remove the bolt securing the Black lead ground strap next to the tattle-tale hose fitting.

Step 3

Step 7

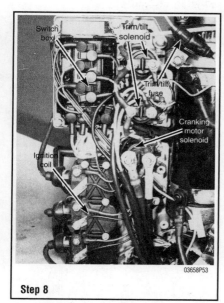

Step 8

7-10 POWERHEAD

10. Disconnect the inlet oil hose at the oil pump coming from the oil tank. Be sure to plug the end of the hose if there is any oil remaining in the tank, otherwise the oil will leak out of the tank.
11. Remove the bolts securing the flywheel cover to the powerhead. Lift off the cover and oil tank as an assembly.
12. Remove the bolts securing the trim cover below the lower cowling. Slide the cover downward and aft until clear of the powerhead.
13. Remove the plastic cap from the flywheel and thread a lifting eye (91-75132) or equivalent into the flywheel. Turn the lifting eye as far as it will go but a minimum of five turns is required for safe lifting.
14. Attach a hoist capable of lifting 500 lbs. or more to the lifting eye.
15. Remove the nuts from the studs securing the powerhead to the driveshaft housing.
16. Using a suitable hoist, lift the powerhead up and free of the driveshaft housing.

➡ If the unit is several years old or if it has been operated in salt water or has not had proper maintenance or shelter or any number of other factors, then separating the powerhead from the driveshaft housing may not be a simple task.

➡ An air hammer may be required on the studs to shake the corrosion loose; heat may have to be applied to the casting to expand it slightly; or other devices employed in order to remove the powerhead.

One very serious condition would be the driveshaft "frozen" with the crankshaft. In this case, a circular plug-type hole must be drilled and a torch used to cut the driveshaft.

To install:
17. Check to be sure old gasket material has been removed from the exhaust housing extension plate and the powerhead. These mating surfaces must be clean. Position a new gasket in place.

18. Thread a lifting eye onto the end of the crankshaft as far as it will go. Use a suitable hoist and lower the powerhead onto the exhaust housing plate with the studs on the powerhead aligned with the holes in the exhaust housing. Use care to prevent the studs from damaging the gasket. As the powerhead is slowly lowered, it will probably be necessary to rotate the flywheel slightly to allow the splines of the driveshaft to index with the crankshaft.
19. Thread the nuts onto the powerhead studs. Tighten the nuts alternately and evenly to 150 inch lbs. (16.9Nm).

➡ The nuts must be tightened to the required torque value in progressive stages. Tighten to ⅓ torque value, then ⅔ torque value and finally to the full torque value.

20. Remove the lifting eye from the crankshaft.

70–125 HP

◆ See Figure 1

1. Disconnect the engine battery cables from the battery terminals.
2. Disconnect the engine fuel line from the fuel tank.
3. Remove the front engine cowling cover.
4. Remove the port and starboard halves of the engine cowling.
5. Separate the electrical extension harness connectors.
6. Disconnect the remote control cables from the powerhead.
7. Stop and carefully observe the wiring and hose connections before proceeding. Because there are so many different powerheads and the arrangement is slightly different on each, it is not possible to illustrate each and every one. Even if they were shown, the reader would not be able to identify the powerhead being serviced. Therefore, Take time to make notes and tag the wire leads and hoses. You may elect to follow the practice of many professional mechanics by taking a series of photographs of the powerhead, one from the top and a couple from the sides showing the wiring and arrangement of parts.
8. Disconnect both leads from the battery.
9. Disconnect the first Black cable from the lower terminal on the cranking motor.
10. Disconnect the second large Black cable from the upper terminal on the cranking motor.
11. Remove the bolts securing the cranking motor bracket, the third large Black cable and the oil injection tank to the powerhead.
12. Remove the bracket and cranking motor from the powerhead.
13. Disconnect the high tension leads from the spark plugs. Always use a pulling and twisting motion as a precaution against damaging the connection.
14. Remove the spark plugs.
15. Remove the bolts securing the cover to the electrical box. The ignition switchbox, ignition coil, fuse, starter solenoid and rectifier/regulator are all mounted inside the electrical box. All these electrical components will be removed as an assembly when the box is removed.
16. Disconnect the White/Black, Purple, Brown and White leads between the electrical box and the stator and trigger assembly.
17. Disconnect the wiring harness connector.
18. Locate the low oil warning module secured to the inside of the lower powerhead cover and disconnect the following leads: Tan, Purple, light Blue leads and a Black ground eyelet lead.
19. Remove the bolts securing the module to the cover and remove the module.
20. Remove the securing hardware and lift out the electrical box with all electrical components undisturbed.

Step 10

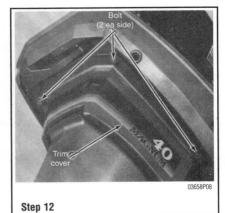

Step 12

Step 13

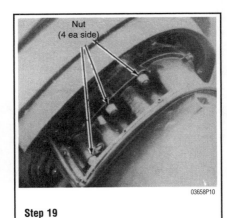

Step 19

POWERHEAD 7-11

Fig. 1 Make sure to identify the fuel system and throttle control components on the portside of the engine—70–125hp

33. Remove the bolts securing the lower powerhead cover to the intermediate housing and remove the cover.
34. Next, remove the bolts securing the powerhead to the intermediate housing.
35. Remove the wing nuts securing the plastic flywheel cover to the powerhead.
36. Remove the plastic cap from the end of the crankshaft.
37. Thread a lifting eye onto the end of the crankshaft as far as it will go.
38. Using a suitable hoist, lift the powerhead assembly clear of the intermediate housing.
39. After lifting off the powerhead, remove all traces of the base gasket.
40. Mount the powerhead onto some type of stand, to facilitate easy access to all parts. Never attempt to mount the powerhead in a stand secured in a vise. Such an attempt will only lead to damage of the powerhead and possible personal injury.
41. Remove the lifting eye from the crankshaft.
42. Snip the Sta-strap from the tattle-tale hose at the aft cowling support bracket.
43. Remove the bolts securing the bracket to the powerhead.
44. Remove the bolts securing the ignition plate cover to the powerhead.
45. Disconnect the large Black lead from the cranking motor and the large Red lead from the cranking motor solenoid.
46. Unplug the power trim fuse mounted on the cranking motor.
47. Loosen but do not remove the Phillips head screws on the clamp securing the main harness to the powerhead. Slide the harness from its retaining clamp.
48. Disconnect the throttle cable from the throttle lever. Disconnect the shift cable and the shift arm from the shift bracket. Remove the bolts and the bracket from the powerhead.
49. Remove the bolts securing the powerhead to the intermediate housing.
50. Disconnect the oil inlet hose from the oil pump and plug the line quickly to prevent oil from draining from the oil reservoir.
51. Remove the attaching hardware and lift the reservoir free of the powerhead.
52. Remove the wing nuts securing the plastic flywheel cover to the powerhead.
53. Remove the plastic cap from the end of the crankshaft.
54. Thread a lifting eye on the end of the crankshaft as far as it will go.
Using a suitable hoist, lift the powerhead assembly clear of the intermediate housing.
55. After lifting off the powerhead, remove all traces of the base gasket.
56. Disconnect the high tension leads from the spark plugs. Always use a pulling and twisting motion as a precaution against damaging the connection.
57. Remove the spark plugs.
58. Mount the powerhead onto some type of stand, to facilitate easy access to all parts. Never attempt to mount the powerhead in a stand secured in a vise. Such an attempt will only lead to damage of the powerhead and possible personal injury. Remove the lifting eye from the crankshaft.

To install:

59. Thread a lifting eye onto the end of the crankshaft as far as it will go. For safety, check to be sure the lifting eye is properly installed.
60. Using a suitable hoist, lift the powerhead. Place a new gasket around the powerhead studs and into position on the base of the intermediate housing.
61. Lubricate the driveshaft splines with multipurpose lubricant.
62. Slowly lower the powerhead down onto the intermediate housing. It may be necessary to rotate the flywheel slightly to index the crankshaft splines with the driveshaft splines.
63. Once the splines index, lower the powerhead fully into place on the intermediate housing.
64. Install the lower cover around the intermediate housing and secure the cover in place with the bolts. Tighten the bolts to 80 inch lbs. (9Nm).
65. Disconnect the hoist from the lifting eye and then remove the eye from the crankshaft. Install the plastic cap onto the end of the crankshaft. Install the flywheel cover and secure it in place with flat washers and wingnuts.
66. Secure the powerhead to the intermediate housing with the flat washers and locknuts. Tighten the locknuts in stages to the specified torque value.
67. On the 75 and 90 hp models, 165 inch lbs. (19Nm). On the 100, 115 & 125 hp models, 44 ft. lbs. (60Nm)
68. Disconnect the hoist from the lifting eye and then remove the eye from the crankshaft.

135–275 HP

♦ See accompanying illustrations

1. Disconnect the battery cables from the battery terminals.
2. Disconnect the fuel supply line from the fuel tank.
3. Unlatch and remove the top cowling or remove the front cover and swing the port and starboard cowling halves open. Lift the cowlings free of the rear support pins and set them aside.

21. Disconnect the Blue/White, Green/White and Red/Black leads at the trim switch located on the side of the lower powerhead cover.
22. Remove the screws securing the cover to the air box.
23. Remove the long bolts securing the carburetors to the intake manifold. The carburetors are held together as an assembly by the forward straps, throttle and choke linkage and fuel lines.
24. Disconnect the fuel supply line and the primer line, if equipped.
25. Disconnect the fuel line between the enrichener valve and the fitting on the fuel bowl of the top carburetor and the line between the valve and the fitting at the base of the oil pump.
26. Lift off all carburetors, as an assembly, with linkage and fuel lines between the carburetors still intact.
27. Disconnect the light Blue leads from the low oil sensor, located at the base of the oil tank, at their quick disconnect fittings. If the oil tank contains oil, make arrangements to plug the oil line once it is pulled free of the fitting, to prevent oil from spilling into the lower powerhead cover. Snip the tie wrap from the oil supply line, from the tank, at the oil injection pump. Ease the line free of the fitting on the pump and lift out the oil tank.
28. Disconnect the oil outlet line between the pump and the 2 psi check valve next to the fuel pump. Remove the attaching bolts and remove the oil pump from the powerhead.
29. Snip the tie wraps around the inlet, outlet and pulse lines at the fuel pump. Remove the Phillips head screws securing the pump to the powerhead. Lift off the pump.
30. Remove the top bolt securing the barrel retainer over the control cable barrels. Swing the retainer down to clear both barrels.
31. Remove the locknuts and washers securing the throttle and shift cable ends to the throttle lever and shift actuator stud. Slide the cables and barrels away from the barrel receptacles cast into the block and lift both cables clear.
32. Make a final check to make sure no other leads or attachments will impede the removal of the powerhead.

7-12 POWERHEAD

4. If the model is equipped with a remote control harness retainer across the lower front cowling, remove the screws and lift out the retainer. Disconnect the remote control cable harness from the side of the powerhead.

➡ **Carefully observe the wiring and hose connections before proceeding. Because there are so many different outboards and the arrangement is slightly different on each, it is not possible to illustrate all of them. Even if they were shown, the reader would not be able to identify the outboard being serviced. Therefore, take time to make notes and tag the wiring and hoses. You may elect to follow the practice of many professional mechanics by taking a series of photographs of the powerhead, one from the top and a couple from the sides showing the wiring and hose arrangements along with other parts.**

5. Disconnect the following color coded wire leads from the plug-in connectors on the starboard side of the powerhead: Trim Indicator lead—Brown/White; Trim Up lead—Blue/White; Trim Down lead—Green/White; Overheat Sensor lead—Tan; and the Trim Up/Down harness connector.

6. Disconnect the Blue, Green and Black trim harness leads from the up solenoid and the down solenoid, located on the aft surface of the powerhead. Release the leads from the J-Clip on the exhaust cover.

7. Disconnect the throttle cable by loosening the screw on the throttle latch.

8. Rotate the latch 45 degrees and lift the throttle cable end free of the anchor pin. Loosen the locknut on the shift cable latch assembly and rotate the latch exposing the end of the shift cable. Lift the shift cable end free of the latch assembly pin.

9. Unlatch the throttle and shift cable barrel retainer, slide both the throttle and shift cable barrels from the retainer.

10. Disconnect the water hose from the tattle-tale nozzle on the lower cowling.

11. Remove the bolts securing the lower cowling halves together and then lift them free of the powerhead.

12. Disconnect the water by-pass hose from the fitting on the exhaust adapter plate below the powerhead.

13. Remove the nuts and lockwashers securing the powerhead to the driveshaft housing.

14. On the 275 hp model, disconnect the throttle cable by loosening the screw on the throttle latch.

15. Rotate the latch 45 degrees and lift the throttle cable end free of the anchor pin. Remove the bolt, flat washer and locknut from the throttle cable barrel nut. Slide the barrel nut free of the throttle cable bracket.

16. Disconnect the shift cable by removing the bolt, flat washer and locknut from both the shift cable barrel nut and the end of the shift cable. Slide the barrel nut free of the anchor bracket.

17. Disconnect the water hose from the tattletale nozzle on the rear of the cowling support bracket. Remove the locknuts and flat washers securing the rear cowling bracket assembly to the exhaust manifold cover and then lift the cowling bracket assembly from the powerhead.

18. Disconnect the remote oil tank hose from the oil reservoir on the powerhead. If the reservoir has oil, cap the fitting to prevent oil leaking out.

19. Disconnect the pulse oil tank hose from the check valve under the powerhead.

20. Remove the pin securing the shift arm to the shift shaft, using a hammer and punch. Tap the pin out and then lift the arm from the shift shaft.

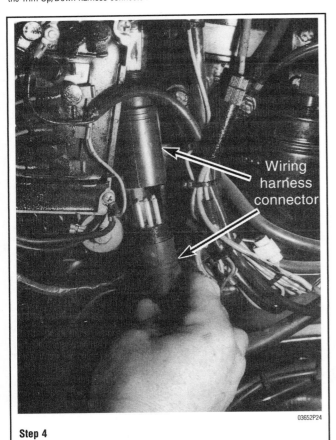

Step 4

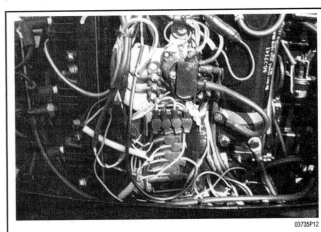

Step 5

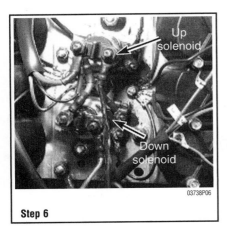

Step 6

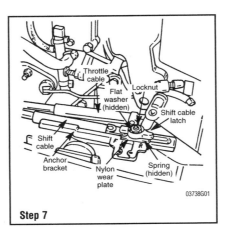

Step 7

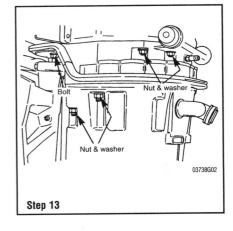

Step 13

POWERHEAD 7-13

21. Remove the wing nuts securing the flywheel cover to the powerhead and lift the cover free.
22. Remove the nuts and lock washers securing the powerhead to the exhaust section.
23. Remove the plastic cap from the center of the flywheel. Thread a lifting eye into the flywheel as far as possible.
24. Using a suitable hoist, lift the powerhead assembly clear of the driveshaft housing. Remove the powerhead base gasket. The gasket may remain on the driveshaft housing.

25. Mount the powerhead onto some type of stand, to facilitate easy access to all parts. Never attempt to mount the powerhead in a stand secured in a vise. Such an attempt will only lead to damage of the powerhead and possibly to personal injury.
26. Remove the lifting eye from the crankshaft.

To install:
27. Thread a lifting eye into the flywheel as far as it will go. Using a suitable hoist, lift the powerhead. Place a new gasket around the powerhead studs and into position on the base of the powerhead.
28. Lubricate the driveshaft splines with multipurpose lubricant.
29. Slowly lower the powerhead down onto the driveshaft housing. It may be necessary to rotate the flywheel slightly to align the crankshaft splines with the driveshaft splines.
30. Once the splines index, lower the powerhead fully into place on the driveshaft housing.
31. On the 135–225 hp (1990 to 1991 and early 1992), secure the powerhead to the driveshaft housing with the 10 flat washers and 10 locknuts.
32. Tighten the locknuts in stages to 30 ft. lbs. (40.7Nm).
33. Disconnect the hoist from the lifting eye and then remove the eye from the flywheel. Install the plastic cap into the center of the flywheel.
34. On the 135–225 hp (1992–2000 models), secure the powerhead to the driveshaft housing with the washers and locknuts, on each side, port and starboard and the bolts, one on each side, port and starboard.
35. Tighten the locknuts and bolts in stages to 30 ft. lbs. (40.7Nm).
36. Disconnect the hoist from the lifting eye and then remove the eye from the flywheel.
37. Install the plastic cap into the center of the flywheel.
38. On the 275 hp model, secure the powerhead to the driveshaft housing with the lockwasher and locknuts plus the one hex bolt in the center aft.
39. Tighten the locknuts and bolt to 30 ft. lbs. (40.7Nm) in the proper sequence.

Step 24

Step 27

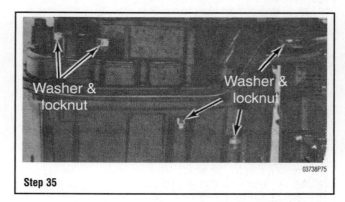
Step 35

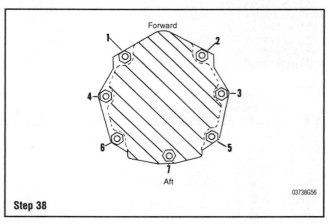
Step 38

Reed Valve

♦ See Figure 2

All Mercury/Mariner 2-Stroke outboard motors are equipped with one set of reed valves per cylinder. The reed valves allow the air/fuel mixture from the carburetor to enter the crankcase, but not exit. They are in essence, one way check valves.

Reed valves are essentially maintenance free and cause very few problems. How-

7-14 POWERHEAD

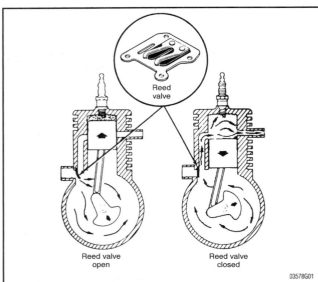

Fig. 2 Reed valves allow the air/fuel mixture from the carburetor to enter the crankcase, but not exit. They are in essence, one way check valves

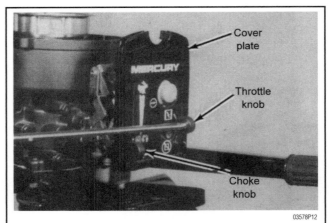

Fig. 3 Check the reed stop opening by measuring from the inside edge of the reed stop to the surface of the closed reed petal—2.5–5hp

ever, if a reed valve does not seal, the air/fuel mixture will escape the crankcase and not be transported to the combustion chamber. Some slight spitting of fuel back out of the carburetor throat at idle can be considered normal, but any substantial discharge of fuel from the carburetor throat indicates reed valve failure.

On most models, the reeds can be inspected with the carburetor removed. A small flashlight and dental mirror can be used to inspect for broken, cracked or chipped reeds. If reed damage is discovered, it is important to attempt to locate the missing pieces of the reed petal. The reed petals are made of stainless steel and will cause internal engine damage if allowed to pass through the crankcase and combustion chamber.

Anytime the reed valves are removed from the powerhead, the reeds should be inspected for excessive stand open. Reed valve specifications are listed in the "Engine Rebuilding Specifications" charts for each powerhead. On models with reed stop specifications, the reed stop opening should also be measured. The easiest way to measure the reed stop opening is to use a drill bit of the specified diameter. The shank (smooth portion) of the drill bit should just fit between the top surface of the closed reed petal and the inner edge of the reed stop. Some reed stops are adjustable, while some require replacement if not within specification. The text will differentiate between such models.

Reed petals must never be turned over and reinstalled. This can lead to a preloaded condition. Preloaded reeds require a higher crankcase vacuum level to open. This causes acceleration and carburetor calibration problems. The reed petals should be flush to nearly flush along the entire length of the reed block mating surface with no preload. Most larger engines have a maximum stand open specification.

Many new models use rubber coated reed blocks. These reed blocks cushion the impact as the reed closes and improves sealing. Reed block assembly part numbers automatically supersede where applicable. Rubber coated reed blocks are serviced as assemblies only.

2.5–3.3 HP

The reed valves are mounted to the crankcase cover. Powerhead disassembly is required to service the reeds.

REMOVAL & INSTALLATION

▶ See Figures 3 and 4

1. Remove the crankcase cover.
2. Remove the screws securing the reed stop and reed petals to the crankcase cover.
3. Remove the reed stop and reed petals. Discard the reed petals.

To install:

4. Clean and inspect the crankcase cover and reed stop. Check for indentation (wear) on the face of the seat area in the crankcase cover. If the reeds have worn indentations in the seat, the cylinder block and crankcase cover must be replaced.
5. Install a new reed petal assembly into the crankcase cover. Place the reed stop over the reed petals.

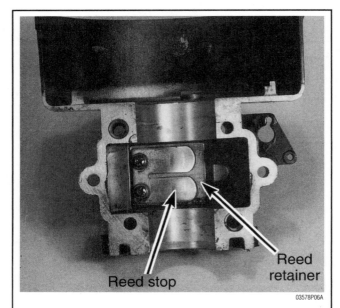

Fig. 4 Reeds mounted on the crankcase cover—2.5–5hp

6. Coat the threads of the reed valve screws with Loctite® 271. Install the screws and tighten securely.
7. The reed stop is not adjustable but can be checked. Replace the reed stop if the measurement is not within the specification listed in the "Engine Rebuilding Specifications" chart.
8. Reinstall the crankcase cover.

4–5 HP

The reed valves are mounted to the crankcase cover. Powerhead disassembly is required to service the reeds.

REMOVAL & INSTALLATION

▶ See Figures 3 and 4

1. Remove the crankcase cover.
2. Remove the screws securing the reed stop and reed petals to the crankcase cover.
3. Remove the reed stop and reed petals. Discard the reed petals.
4. Thoroughly clean the reed block assembly in clean solvent.
5. Check for excessive wear (indentations), cracks or grooves in the seat area of the reed block. Replace the reed block if any damage is noted.

POWERHEAD 7-15

6. Check the reed petals for cracks, chips or evidence of fatigue. Replace the reed petals or reed block assembly if any damage is noted.

To install:

7. Install a new reed petal assembly into the crankcase cover. Place the reed stop over the reed petals.
8. Coat the threads of the reed valve screws with Loctite® 271. Install the screws and tighten securely.
9. Check the reed stop opening by measuring between the reed stop and the top of the closed reed. Carefully bend the reed stop to obtain the dimension specified in the "Engine Rebuilding Specifications" chart.

➡ On 4 hp models, one reed stop is flat and holds that reed petal closed making it inoperative.

10. Reinstall the crankcase cover.

6–15 HP

Two sets of reed petals are mounted to a single reed block assembly. Each set of petals feeds one cylinder. A rubber seal separates the petals.

The reed block is the intake manifold and the carburetor bolts directly to the reed block. The reed block is serviceable and the rubber seal replaceable on models so equipped. The reed block on 1997–00 models is rubber coated and is serviced as an assembly. The rubber seal is not replaceable on 1997–00 models. The rubber coated reed block will retrofit to earlier models.

REMOVAL & INSTALLATION

♦ See Figures 5 and 6

1. Remove the carburetor
2. Remove the bolts securing the reed block to the powerhead.
3. Carefully remove the reed block assembly from the powerhead.

➡ Do not scratch, warp or gouge the reed block or crankcase cover. Damaging these components will result in vacuum leaks.

4. Remove all gasket material from the reed block and crankcase cover.
5. As required, remove and discard the rubber seal from the center of the reed block assembly.
6. Thoroughly clean the reed block assembly in clean solvent.
7. Check for excessive wear (indentations), cracks or grooves in the seat area of the reed block. Replace the reed block if any damage is noted.
8. Check the reed petals for cracks, chips or evidence of fatigue. Replace the reed petals or reed block assembly if any damage is noted.

To install:

9. Installing a new rubber seal in the center of the reed block assembly, as required. Use needle bearing assembly grease to hold the rubber seal in position.
10. Using a new gasket, install the reed block over the carburetor mounting studs and seat it against the crankcase cover. If a rubber seal is used, make sure it does not slip out of position.
11. Install the screws and tighten evenly to 60 inch lbs. (7 Nm).
12. Check the maximum reed opening between the petals and the reed block mating surface. Replace the reeds or reed block assembly if the reeds are stuck to the block or stand open more than the maximum limit specified in the "Engine Rebuilding Specifications" chart.
13. Check the reed stop opening by measuring between the reed stop and the top of the closed reed. Carefully bend the reed stop to obtain the dimension specified in the "Engine Rebuilding Specifications" chart.
14. Install the carburetor.

20–25 HP

Two sets of reed petals are mounted to a single reed block assembly. Each set of petals feeds one cylinder. A rubber seal separates the petals. An intake manifold (carburetor adapter) bolts between the carburetor and reed block.

➡ These models do not have serviceable reed valves. The rubber center seal is not removable. The reed block is replaced as an assembly if any defect is noted.

REMOVAL & INSTALLATION

♦ See Figures 5 and 6

1. Remove the carburetor.
2. Remove the bolts securing the intake manifold and reed block to the powerhead.
3. Carefully remove the intake manifold and reed block assembly from the powerhead. Do not scratch, warp or gouge the reed block, intake manifold or crankcase cover. Separate the reed block from the intake manifold. Remove all gasket material from the reed block, intake manifold and crankcase cover.
4. Thoroughly clean the reed block assembly in clean solvent.
5. Check for excessive wear (indentations), cracks or grooves in the seat area of the reed block. Replace the reed block if any damage is noted.
6. Check the reed petals for cracks, chips or evidence of fatigue. Replace the reed block assembly if any damage is noted.

To install:

7. Using new gaskets, install the reed block and intake manifold over the carburetor mounting studs and seat them against the crankcase cover.
8. Install the screws and tighten evenly to 80 inch lbs. (9 Nm) on 1994–00 models and 70 inch lbs. (8 Nm) on all others.
9. Check the maximum reed opening between the petals and the reed block mating surface. Replace the reed block assembly if the reeds are stuck to the block or stand open more than the maximum limit specified in the "Engine Rebuilding Specifications" chart.
10. Install the carburetor.

30 and 40 HP (2-Cylinder)

Two sets of reed valves are mounted to a reed plate on late models (serial # 0G380075 and later) and directly to the intake manifold on early models. A reed stop is used on 30 hp models only. The reed valves are replaceable on early model powerheads only. On late model powerheads, the reed valves are serviced with the reed plate assembly.

REMOVAL & INSTALLATION

♦ See Figures 7 and 8

1. Remove the carburetor.
2. Remove the bolts securing the intake manifold and/or reed block to the powerhead.

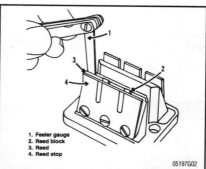

Fig. 5 Checking the maximum reed opening between the petals and the reed block mating surface—6–25hp

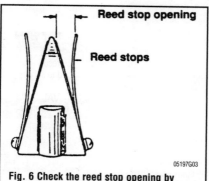

Fig. 6 Check the reed stop opening by measuring between the reed stop and the top of the closed reed—6–25hp

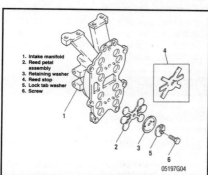

Fig. 7 Early model powerheads use reed petals that are mounted to intake manifold . . .

7-16 POWERHEAD

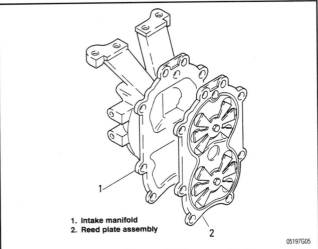

Fig. 8 . . . while late model powerheads use reed petals mounted to a reed block assembly

1. Intake manifold
2. Reed plate assembly

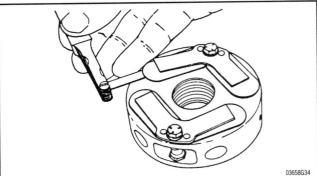

Fig. 9 Check the maximum reed opening between the petals and the reed block mating surface—40hp (4-Cylinder)

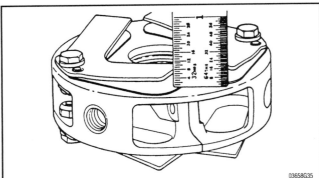

Fig. 10 Check the reed stop opening by measuring between the reed stop and the top of the closed reed —40hp (4-Cylinder)

3. Carefully remove the intake manifold and reed block assembly from the powerhead. Do not scratch, warp or gouge the reed block, intake manifold or crankcase cover. Separate the reed block from the intake manifold. Remove all gasket material from the reed block, intake manifold and crankcase cover.
4. Thoroughly clean the reed block assembly in clean solvent.
5. Check for excessive wear (indentations), cracks or grooves in the seat area of the reed block. Replace the reed block if any damage is noted.
6. Check the reed petals for cracks, chips or evidence of fatigue. Replace the reed block assembly if any damage is noted.

➥Do not remove the reed petals from the intake manifold or reed plate unless they are going to be replaced.

To install:

7. Using new gaskets, install the reed block and/or intake manifold over the carburetor mounting studs and seat them against the crankcase cover.
8. Install the screws and tighten evenly to 17 ft. lbs. (22 Nm) in a circular pattern, starting in the center and working outward.
9. Check the maximum reed opening between the petals and the reed block mating surface. Replace the reed block assembly if the reeds are stuck to the block or stand open more than the maximum limit specified in the "Engine Rebuilding Specifications" chart.
10. On models so equipped, check the reed stop opening by measuring between the reed stop and the top of the closed reed. Carefully bend the reed stop to obtain the dimension specified in the "Engine Rebuilding Specifications" chart.
11. To replace the reed petals, proceed as follows:
    a. On models with replaceable reed petals, bend the lock washer lock tabs away from the screw heads. Remove the screws attaching the reed petals, retaining washer or reed stop to the intake manifold. Discard the lock tab washer.
    b. Apply Loctite® 271 to the threads of the screws. Position the new reed petals, retaining washers or reed stops and new lock tab washers on the intake manifold. Make sure all components are aligned with the alignment pins in the intake manifold.
    c. Tighten the screws to 60 inch lb. (7 Nm) and check the alignment of the screw head to the lock tab washer. If necessary, continue to tighten up to a maximum of 100 inch lbs. (11 Nm) to align the screw head with the lock tab washer. Bend the lock tab of each washer up against each screw head.
12. Install the carburetor.

40 HP (4-Cylinder)

REMOVAL & INSTALLATION

♦ See Figures 9 and 10

1. These models use 2 reed block assemblies mounted to the crankshaft. The reed blocks also serve as crankshaft labyrinth seals. Disassembly of the powerhead is necessary to remove the reed blocks.
2. Clean the reed block assemblies thoroughly in clean solvent.
3. Check the condition of the locating pins in the reed block halves.

➥Locating pins that have been subjected to overheating (melted) will not properly position the reed petals and reed stops. This can adversely affect the starting, idle and overall engine operation. As necessary, replace the reed block.

4. Assemble the reed block halves and secure with the appropriate fasteners.
5. Check for excessive wear (indentations), cracks or grooves in the seat area of the reed block. Replace the reed block if any damage is noted.
6. Check the reed petals for cracks, chips or evidence of fatigue. Replace the reed block assembly if any damage is noted.
7. Check the reed block labyrinth seal area (inside diameter) for excessive wear. Replace the reed blocks as necessary.
8. Check the maximum reed opening between the petals and the reed block mating surface. Replace the reeds or reed block assembly if the reeds are stuck to the block or stand open more than the maximum limit specified in the "Engine Rebuilding Specifications" chart.
9. Check the reed stop opening by measuring between the reed stop and the top of the closed reed. Carefully bend the reed stop to obtain the dimension specified in the "Engine Rebuilding Specifications" chart.
10. To replace the reeds, proceed as follows:
    a. Remove the reed retaining screw from each reed block half, then remove the reed stops and reed petals. Discard the reed petals.
    b. Reassemble the reed valve assembly by installing new reeds and the original reed stops onto the reed block locating pins. Apply Loctite® 271 to the retaining screws and tighten to 30 inch lbs. (3 Nm). Recheck maximum reed opening and reed stop opening.

40 HP (3-Cylinder), 50 and 60 HP

A single reed plate and single intake manifold is used on 40-60 hp. The 40 hp (3-Cylinder) uses a reed stop on each reed plate.

✻✻✻ CAUTION

Do not allow the internal bleed (recirculation) system check valves to fall out or become misplaced while the intake manifold is removed.

POWERHEAD 7-17

REMOVAL & INSTALLATION

♦ See Figure 11

1. Remove the carburetors.
2. Remove the screws securing the intake manifold and reed plate to the powerhead.
3. Carefully remove the intake manifold and reed plate from the powerhead. Do not scratch, warp or gouge the reed plate, intake manifold or crankcase cover.
4. Separate the reed plate from the intake manifold. Remove all gasket material from the reed plate, intake manifold and crankcase cover.
5. Thoroughly clean the reed block assembly in clean solvent.
6. Check for excessive wear (indentations), cracks or grooves in the seat area of the reed block. Replace the reed block if any damage is noted.
7. Check the reed petals for cracks, chips or evidence of fatigue. Replace the reed block assembly if any damage is noted.

➡ Do not remove the reed petals from the reed plate unless replacement is necessary. Always replace reed petals in complete sets. Never turn a reed over for reuse or attempt to straighten a damaged reed.

To install:

8. Check the maximum reed opening between the petals and the reed block mating surface. Replace the reeds or reed block assembly if the reeds are stuck to the block or stand open more than the maximum limit specified in the "Engine Rebuilding Specifications" chart.
9. Check the reed stop opening by measuring between the reed stop and the top of the closed reed. Carefully bend the reed stop to obtain the dimension specified in the "Engine Rebuilding Specifications" chart.
10. To replace the reed petals, proceed as follows:
    a. Bend the lockwasher lock tabs away from the screw heads. Remove the screws attaching the reed petals, reed stop or retaining washer to the reed plate. Discard the lock tab washers.
    b. Apply Loctite® 271 to the threads of the screws. Position the new reed petals, the original reed stops or retaining washers and new lock tab washers on the intake manifold. Make sure all components are aligned with the alignment pins in the intake manifold.
    c. Tighten the screws to 60 inch lbs. (7 Nm) and check the alignment of the screw head to the lock tab washer. If necessary, continue to tighten to a maximum of 100 inch lbs. (11 Nm) to align the screw head with the lock tab washer.
    d. Bend the lock tab washer of each washer against each screw head.

➡ Make sure the gaskets are correctly orientated. The gaskets only fit correctly in one direction.

11. Using new gaskets, install the reed plate and intake manifold to the powerhead. Install the retaining screws finger tight.
12. Tighten the screws evenly to 18 ft. lbs. (24 Nm) in the correct sequence.
13. Install the carburetors.

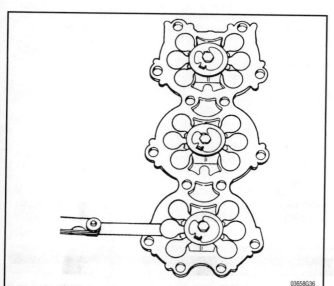

Fig. 11 Check the maximum reed opening between the petals and the reed block mating surface—40hp (3-Cylinder), 50–125hp

75–125 HP

Multiple reed plates and intake manifolds are used on these powerheads.

➡ Do not allow the internal bleed (recirculation) system check valves to fall out or become misplaced while the intake manifolds are removed.

REMOVAL & INSTALLATION

♦ See Figure 11

1. Remove the carburetors.
2. Remove the screws securing the intake manifolds and reed plates to the powerhead.
3. As necessary, disconnect the fuel primer valve line from the balance tubes at the fuel primer valve.
4. Carefully remove the intake manifolds and reed plates from the powerhead. Do not scratch, warp or gouge the reed plate, intake manifold or crankcase cover. Separate the reed plates from the intake manifolds. Remove all gasket material from the reed plates, intake manifolds and crankcase cover.
5. Thoroughly clean the reed block assembly in clean solvent.
6. Check for excessive wear (indentations), cracks or grooves in the seat area of the reed block. Replace the reed block if any damage is noted.
7. Check the reed petals for cracks, chips or evidence of fatigue. Replace the reed plates if any damage is noted.

➡ Do not remove the reed petals from the reed plate unless replacement is necessary. Always replace reed petals in complete sets. Never turn a reed over for reuse or attempt to straighten a damaged reed.

To install:

8. Using new gaskets, install the reed plates and intake manifolds to the powerhead. Install the retaining screws finger tight.
9. Check the maximum reed opening between the petals and the reed block mating surface. Replace the reeds or reed block assembly if the reeds are stuck to the block or stand open more than the maximum limit specified in the "Engine Rebuilding Specifications" chart.
10. To replace the reed petals, proceed as follows:
    a. Bend the lockwasher lock tabs away from the screw heads. Remove the screws attaching the reed petals, reed stop or retaining washer to the reed plate. Discard the lock tab washers.
    b. Apply Loctite® 271 to the threads of the screws. Position the new reed petals, the original reed stops or retaining washers and new lock tab washers on the intake manifold. Make sure all components are aligned with the alignment pins in the intake manifold.
    c. Tighten the screws to 80 inch lbs. (9Nm) and check the alignment of the screw head to the lock tab washer. If necessary, continue to tighten to a maximum of 100 inch lbs. (11 Nm) to align the screw head with the lock tab washer.
    d. Bend the lock tab washer of each washer against each screw head.
11. Tighten the intake manifold screws evenly to 18 ft. lbs. (24 Nm) in a crossing pattern (75–90 hp) or a circular pattern starting with the 2 middle screws in each manifold (100–125 hp). As necessary, reconnect the fuel primer valve hose from the balance tubes to the fuel primer valve. Secure the connection using a new tie-strap clamp.
12. Install the carburetors.

135-250 HP

The 135-250 hp (Carbureted and EFI) use a one-piece intake manifold with 6 reed valve blocks. Up to 1995 serviceable reed blocks with tapered reed petals were used. Since 1996, all models have been equipped with rubber coated reed blocks and square-tipped reeds. Rubber coated reed blocks are only serviced as assemblies.

REMOVAL & INSTALLATION

♦ See Figures 12, 13 and 14

1. On carbureted models, remove the carburetors. On EFI models, remove the intake manifold.

➡ There are 20 screws on the intake manifold. Only 8 screws actually secure the intake manifold to the crankcase cover. If you are unsure which screws to remove, simply remove all 20 screws.

2. Remove the screws securing the intake manifold assembly to the powerhead.
3. Matchmark all bleed lines and note their position and routing for installation reference.

7-18 POWERHEAD

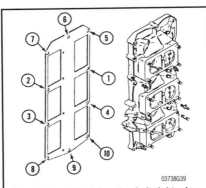

Fig. 12 Reed block housing bolt tightening sequence—135–200hp with vertical reeds

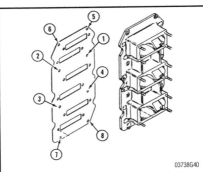

Fig. 13 Reed block housing bolt tightening sequence—135–200hp with horizontal reeds

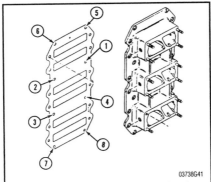

Fig. 14 Reed block housing bolt tightening sequence—225hp with horizontal reeds

4. Disconnect all bleed lines from the intake manifold.
5. Carefully remove the intake manifold assembly from the powerhead. Do not scratch, warp or gouge the intake manifold or crankcase cover.
6. If not already removed, remove the screws securing the reed boxes to the intake manifold. Separate the reed boxes from the intake manifold.

To install:
7. Remove all gasket material from the reed boxes, intake manifold and crankcase cover.
8. Clean the gasket surfaces of the reed boxes thoroughly. Wash the reed boxes, intake manifold, reed mounting blocks and spacer plates (as equipped) in clean solvent.
9. Inspect the intake manifold, reed mounting blocks and spacer plates (as equipped) for distortion, cranks, blocked passages and fittings.
10. Check the reed boxes for distortion, cracks, deep grooves or any other damage that may cause leakage. Replace as necessary.
11. Check for excessive wear (indentations), cracks or grooves in the seat area of the reed boxes. Replace the reed box assembly if any damage is noted. On rubber coated reed blocks, check for rubber delamination from the reed box casting. Replace any reed box showing rubber delamination.
12. Check the reed petals for cracks, chips or evidence of fatigue. Replace the reed petals or reed box assembly if any damage is noted.
13. Check the stand-open gap between the reed petals and the reed plate mating surface. Replace the reed petals or reed box assembly if any are preloaded (stick tightly to the reed plate) or stand open more than specified .
14. To replace the reed petals on models with serviceable reeds, remove the screws attaching the reed petals and reed clamp plates to the reed box. Discard the reed petals.
15. To reinstall, apply Loctite® 271 to the threads of the screws. Position the new reed petals and reed retaining plates
16. Using a new gasket, install the reed boxes to the intake manifold. Install the 12 retaining screws (2 each reed box) finger-tight.
17. Verify that the gasket is properly positioned and that all holes align correctly. Tighten the 12 reed box retaining screws evenly to 80–90 inch lbs. (9–10 Nm).
18. Position the intake manifold against the crankcase cover and install the 8 retaining screws. Tighten the retaining screws evenly to 105 inch lbs. (12 Nm).
19. Reconnect all bleed lines to their original positions.
20. Install the carburetors and/or induction manifold.

135–225 DFI Optimax

The 135–225 DFI Optimax intake manifolds are unique in that the electric oil pump is mounted directly to the intake manifold and an additional spacer (lubrication) plate and gasket are used. The oil pump moves oil through the reed plate's machined channels to each cylinder's reed box area. When removing the intake manifold, remember to disconnect the oil pump electrical connection and external oil supply and air compressor lubrication lines.

REMOVAL & INSTALLATION

1. Remove the induction manifold.
2. Disconnect and cap the air compressor lubrication line the oil supply line and any bleed lines that are in the way.
3. Disconnect the oil pump harness from the oil pump.
4. Remove the screws securing the oil pump bracket to the powerhead.
5. Remove the screws securing the intake manifold assembly to the powerhead.

6. Carefully remove the intake manifold assembly from the powerhead. Do not scratch, warp or gouge the intake manifold or crankcase cover.

→Internal recirculation valves are used on this engine. The valves fit into machined grooves on the mating surface of the crankcase cover to intake manifold. The valves are installed in rubber carriers. Do not lose or misplace the check valves and carriers.

7. Remove the screws securing the oil pump to the intake manifold. Remove the oil pump from the intake manifold.
8. Remove the screws securing the reed boxes to the space plate and intake manifold Separate the reed boxes from the spacer plate. Separate the spacer plate from the intake manifold. Remove all gasket material from the reed boxes, spacer plate, oil pump, intake manifold and crankcase cover.
9. Clean and inspect the reed boxes and intake manifold. The reed boxes are not serviceable. If any defects are noted, replace the suspect reed box. Make sure all machined passages in the intake manifold are clean.
10. Clean the gasket surfaces of the reed boxes thoroughly. Wash the reed boxes, intake manifold, reed mounting blocks and spacer plates (as equipped) in clean solvent.
11. Inspect the intake manifold, reed mounting blocks and spacer plates (as equipped) for distortion, cranks, blocked passages and fittings.
12. Check the reed boxes for distortion, cracks, deep grooves or any other damage that may cause leakage. Replace as necessary.
13. Check for excessive wear (indentations), cracks or grooves in the seat area of the reed boxes. Replace the reed box assembly if any damage is noted. On rubber coated reed blocks, check for rubber delamination from the reed box casting. Replace any reed box showing rubber delamination.
14. Check the reed petals for cracks, chips or evidence of fatigue. Replace the reed petals or reed box assembly if any damage is noted.
15. Check the maximum reed opening gap between the reed petals and the reed plate mating surface. Replace the reed petals or reed box assembly if any are preloaded (stick tightly to the reed plate) or stand open more than specified .
16. Using new gaskets, install the reed boxes to the spacer plate and intake manifold. Coat the threads of the reed box retaining screws with Loctite® 271. Install the retaining screws finger tight.
17. Verify that the gaskets and spacer plate are properly positioned and that all holes are aligned. Tighten the reed box retaining screws evenly to 90 inch lbs. (10 Nm).
18. Using a new gasket, install the oil pump and oil pump bracket to the intake manifold assembly. Install and tighten the screws securely.
19. Position the intake manifold assembly against the crankcase cover and install the retaining screws. Tighten the retaining screws evenly to 100 inch lbs. (11 Nm).
20. Install the screws securing the oil pump bracket to the powerhead. Tighten the screws securely.
21. Reconnect the oil pump electrical connector, oil supply line and air compressor lubrication line. Secure the line connections with new tie straps.
22. Install the induction manifold.
23. Purge the electric oil pump.

275 HP

The 275 hp models are unique in that they use a separate intake manifold for each cylinder. The intake manifolds and gaskets are different from port to starboard. Each intake manifold contains 2 reed valve block assemblies. The reed valve blocks are only serviced as assemblies.

POWERHEAD 7-19

REMOVAL & INSTALLATION

♦ See Figure 15

1. Remove the carburetors.
2. Remove the screws securing the fuel primer valve bracket to the lower intake manifolds. Lay the fuel primer valve and bracket to one side.
3. Note the position and routing of the balance hoses and fuel primer lines to and from each intake manifold. Disconnect and cap the balance hoses, bleed lines and fuel primer lines from the intake manifolds.
4. Remove the screws securing each intake manifold assembly to the powerhead. Note the position of the control cable support bracket mounted to the bottom of the port lower intake manifold.
5. Carefully remove each intake manifold assembly from the powerhead. Do not scratch, warp or gouge any intake manifold or the crankcase cover.
6. Remove the lower screws securing the reed mounting block to each intake manifold.
7. Remove the upper reed box screws securing the reed mounting block and upper reed block to each intake manifold. Separate each manifold assembly's reed mounting block upper reed box and intake manifold from each other.
8. Remove the screws securing the lower reed box to each reed mounting block. Separate each lower reed box from each reed mounting block.
9. Remove all gasket material from the reed boxes, reed mounting blocks, intake manifolds and crankcase cover.

To install:

10. Clean and inspect the reed boxes and intake manifolds. The reed boxes are not serviceable. If any defects are noted, replace the suspect reed box. Make sure all machined passages and fittings in the intake manifold are clean.
11. Clean the gasket surfaces of the reed boxes thoroughly. Wash the reed boxes, intake manifold, reed mounting blocks and spacer plates (as equipped) in clean solvent.
12. Inspect the intake manifold, reed mounting blocks and spacer plates (as equipped) for distortion, cranks, blocked passages and fittings.
13. Check the reed boxes for distortion, cracks, deep grooves or any other damage that may cause leakage. Replace as necessary.
14. Check for excessive wear (indentations), cracks or grooves in the seat area of the reed boxes. Replace the reed box assembly if any damage is noted. On rubber coated reed blocks, check for rubber delamination from the reed box casting. Replace any reed box showing rubber delamination.
15. Check the reed petals for cracks, chips or evidence of fatigue. Replace the reed petals or reed box assembly if any damage is noted.
16. Check the maximum reed open gap between the reed petals and the reed plate mating surface. Replace the reed petals or reed box assembly if any are preloaded (stick tightly to the reed plate) or stand open more than specified .
17. Check the reed stop opening by measuring from the inside edge of the reed stop to the surface of the closed reed petal. A drill bit of the specified dimension should just fit between the widest gap of the reed stop and the reed petal. Replace the reed box assembly if the measurement is not within specification .
18. To replace the reed petals on models with serviceable reeds, proceed as follows:
    a. Remove the screws attaching the reed petals and reed clamp plates to the reed box. Discard the reed petals.

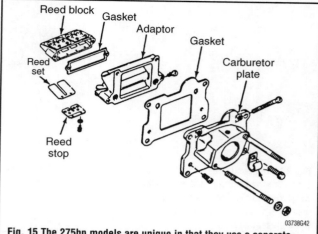

Fig. 15 The 275hp models are unique in that they use a separate intake manifold for each cylinder

   b. Apply Loctite® 271 to the threads of the screws. Position the new reed petals and the reed retaining plates
19. Using new gaskets, install the lower reed boxes to the reed mounting blocks. Coat the threads of the reed box retaining screws with Loctite® 271. Install and evenly tighten the screws to 60 inch lbs. (7 Nm).
20. Using new gaskets install the reed mounting blocks to the intake manifolds. Coat the threads of the reed block lower retaining screws with Loctite® 271. Install the screws finger-tight.
21. Using new gaskets install the upper reed boxes to the reed mounting block and intake manifold assemblies. Coat the threads of the reed box retaining screws with Loctite® 271. Install the screws finger-tight.
22. Check the alignment of the reed boxes, reed mounting blocks and intake manifolds. Make sure all holes are aligned. Tighten the upper screws and lower screws on each intake manifold assembly evenly to 60 inch lbs. (7 Nm).
23. Install the intake manifold assemblies to the crankcase cover. Coat the threads of the retaining screws with Loctite® 271. Install the retaining screws (4 each intake manifold) finger-tight.
24. Verify that the intake manifolds are correctly positioned and that control cable bracket is installed on the bottom of the lower port manifold. Tighten each manifold's 4 retaining screws evenly to 150 inch lbs. (17 Nm).
25. Reconnect the balance hose, bleed lines and fuel primer lines to the intake manifolds.
26. Position the fuel primer valve bracket on the lower port intake manifolds. Coat the threads of the retaining screws with Loctite® 271. Install and tighten the screws to 60 inch lbs. (7 Nm).
27. Install the carburetors.
28. Reconnect the negative battery cable.

POWERHEAD RECONDITIONING

Determining Powerhead Condition

Anything that generates heat and/or friction will eventually burn or wear out (for example, a light bulb generates heat, therefore its life span is limited). With this in mind, a running powerhead generates tremendous amounts of both; friction is encountered by the moving and rotating parts inside the powerhead and heat is created by friction and combustion of the fuel. However, the powerhead has systems designed to help reduce the effects of heat and friction and provide added longevity. The oil injection system combines oil with the fuel to reduce the amount of friction encountered by the moving parts inside the powerhead, while the cooling system reduces heat created by friction and combustion. If either system is not maintained, a break-down will be inevitable. Therefore, you can see how regular maintenance can affect the service life of your powerhead.

There are a number of methods for evaluating the condition of your powerhead. A secondary compression test can reveal the condition of your pistons, piston rings, cylinder bores and head gasket(s). A primary compression test can determine the condition of all engine seals and gaskets. Because the 2-stroke powerhead is a pump, the crankcase must be sealed against pressure created on the down stroke of the piston and vacuum created when the piston moves toward top dead center. If there are air leaks into the crankcase, insufficient fuel will be brought into the crankcase and into the cylinder for normal combustion. Information on compression testing can be found in the "Maintenance and Tune-up" section.

Buy or Rebuild?

♦ See Figures 16 and 17

Now that you have determined that your powerhead is worn out, you must make some decisions. The question of whether or not a powerhead is worth rebuilding is largely a subjective matter and one of personal worth. Is the powerhead a popular one or is it an obsolete model? Are parts available? Is the outboard it's being put into worth keeping? Would it be less expensive to buy a new powerhead, have your powerhead rebuilt by a pro, rebuild it yourself or buy a used powerhead? Or would it be simpler and less expensive to buy another outboard? If you have considered all these matters and more and have still decided to rebuild the powerhead, then it is time to decide how you will rebuild it.

➡**The editors at Seloc® feel that most powerhead machining should be performed by a professional machine shop. Don't think of it as wasting money, rather, as insurance that the job has been done right the first**

7-20 POWERHEAD

Fig. 16 The question of whether or not a powerhead is worth rebuilding is largely a subjective matter and one of personal worth. This powerhead is not worth much in its present condition

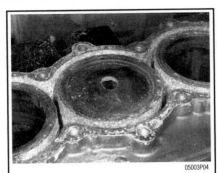

Fig. 17 A burned piston like this one will be replaced during an overhaul. The condition which caused the hole in the top of the piston must be identified and corrected or the same thing will happen again

Fig. 18 Much of the assembly work (crankshaft, bearings, pistons, connecting rods and other components) is well within the scope of the average do-it-yourself mechanic's tools and abilities

time. There are many expensive and specialized tools required to perform such tasks as boring and honing a powerhead. Even inspecting the parts requires expensive micrometers and gauges to properly measure wear and clearances. Also, a machine shop can deliver to you clean and ready to assemble parts, saving you time and aggravation. Your maximum savings will come from performing the removal, disassembly, assembly and installation of the powerhead and purchasing or renting only the tools required to perform the above tasks. Depending on the particular circumstances, you may save 40 to 60 percent of the cost doing these yourself.

A complete rebuild or overhaul of a powerhead involves replacing or reconditioning all of the moving parts (pistons, rods, crankshaft, etc.) with new or remanufactured ones and machining the non-moving wearing surfaces of the block and heads. Unfortunately, this may not be cost effective. For instance, your crankshaft may have been damaged or worn but it can be machined for a minimal fee.

So, as you can see, you can replace everything inside the powerhead but, it is wiser to replace only those parts which are really needed and, if possible, repair the more expensive ones.

Powerhead Overhaul Tips

▶ See Figure 18

Most powerhead overhaul procedures are fairly standard. In addition to specific parts replacement procedures and specifications for your individual powerhead, this section is also a guide to acceptable rebuilding procedures. Examples of standard rebuilding practice are given and should be used along with specific details concerning your particular powerhead.

Competent and accurate machine shop services will ensure maximum performance, reliability and powerhead life. In most instances it is more profitable for the do-it-yourself mechanic to remove, clean and inspect the component, buy the necessary parts and deliver these to a shop for actual machine work.

Much of the assembly work (crankshaft, bearings, pistons, connecting rods and other components) is well within the scope of the do-it-yourself mechanic's tools and abilities. You will have to decide for yourself the depth of involvement you desire in a powerhead repair or rebuild.

TOOLS

The tools required for a powerhead overhaul or parts replacement will depend on the depth of your involvement. With a few exceptions, they will be the tools found in an average do it yourselfer's tool kit. More in-depth work will require some or all of the following:
- A dial indicator (reading in thousandths) mounted on a universal base
- Micrometers and telescope gauges
- Jaw and screw-type pullers
- Scraper
- Ring groove cleaner
- Piston ring expander and compressor
- Ridge reamer
- Cylinder hone or glaze breaker
- Plastigage®

- Powerhead stand

The use of most of these tools is illustrated in this section. Many can be rented for a one-time use from a local parts store or tool supply house.

Occasionally, the use of special tools is necessary. See the information on Special Tools and the Safety Notice in the front of this book before substituting another tool.

CAUTIONS

Aluminum is extremely popular for use in powerheads, due to its low weight. Observe the following precautions when handling aluminum parts:
- Never hot tank aluminum parts, the caustic hot tank solution will eat the aluminum
- Remove all aluminum parts (identification tag, etc.) from powerhead parts prior to hot tanking
- Always coat threads lightly with oil or anti-seize compounds before installation, to prevent seizure
- Never overtighten bolts or spark plugs especially in aluminum threads

When assembling the powerhead, any parts that will be exposed to frictional contact must be prelubed to provide lubrication at initial start-up. Any product specifically formulated for this purpose can be used.

When semi-permanent (locked but removable) installation of bolts or nuts is desired, threads should be cleaned and coated with Loctite® or another similar, commercial non-hardening sealant.

CLEANING

Before the powerhead and its components are inspected, they must be thoroughly cleaned. You will need to remove any varnish, oil sludge and/or carbon deposits from all of the components to insure an accurate inspection. A crack in the block or cylinder head can easily become overlooked if hidden by a layer of sludge or carbon.

Most of the cleaning process can be carried out with common hand tools and readily available solvents or solutions. Carbon deposits can be chipped away using a hammer and a hard wooden chisel. Old gasket material and varnish or sludge can usually be removed using a scraper and/or cleaning solvent. Extremely stubborn deposits may require the use of a power drill with a wire brush. Always follow any safety recommendations given by the manufacturer of the tool and/or solvent. You should always wear eye protection during any cleaning process involving scraping, chipping or spraying of solvents.

➡**If using a wire brush, use extreme care around any critical machined surfaces (such as the gasket surfaces, bearing saddles, cylinder bores, etc.). Use of a wire brush is not recommended on any aluminum components.**

An alternative to the mess and hassle of cleaning the parts yourself is to drop them off at a local machine shop. They will, more than likely, have the necessary equipment to properly clean all of the parts for a nominal fee.

✱✱ CAUTION

Always wear eye protection during any cleaning process involving scraping, chipping or spraying of solvents.

POWERHEAD 7-21

Remove any plugs or pressed-in bearings and carefully wash and degrease all of the powerhead components including the fasteners and bolts. Small parts should be placed in a metal basket and allowed to soak. Use pipe cleaner type brushes and clean all passageways in the components.

Use a ring expander to remove the rings from the pistons. Clean the piston ring grooves with a ring groove cleaner or a piece of broken ring. Scrape the carbon off of the top of the piston. You should never use a wire brush on the pistons. After preparing all of the piston assemblies in this manner, wash and degrease them again.

REPAIRING DAMAGED THREADS

♦ See Figures 19, 20, 21, 22 and 23

Several methods of repairing damaged threads are available. Heli-Coil®, Keenserts® and Microdot® are among the most widely used. All involve basically the same principle, drilling out stripped threads, tapping the hole and installing a prewound insert, making welding, plugging and oversize fasteners unnecessary.

Two types of thread repair inserts are usually supplied: a standard type for most inch coarse, inch fine, metric course and metric fine thread sizes and a spark lug type to fit most spark plug port sizes. Consult the individual tool manufacturer's catalog to determine exact applications. Typical thread repair kits will contain a selection of prewound threaded inserts, a tap (corresponding to the outside diameter threads of the insert) and an installation tool. Spark plug inserts usually differ because they require a tap equipped with pilot threads and a combined reamer/tap section. Most manufacturers also supply blister-packed thread repair inserts separately in addition to a master kit containing a variety of taps and inserts plus installation tools.

Before attempting to repair a threaded hole, remove any snapped, broken or damaged bolts or studs. Penetrating oil can be used to free frozen threads. The offending item can usually be removed with locking pliers or using a screw/stud extractor. After the hole is clear, the thread can be repaired, as shown in the series of accompanying illustrations and in the kit manufacturer's instructions.

Powerhead Preparation

♦ See Figure 24

To properly rebuild a powerhead, you must first remove it from the outboard, then disassemble and inspect it. Ideally you should place your powerhead on a stand. This affords you the best access to the components. Follow the manufacturer's directions for using the stand with your particular powerhead.

Now that you have the powerhead on a stand, it's time to strip it of all but the necessary components. Before you start disassembling the powerhead, you may want to take a moment to draw some pictures, fabricate some labels or get some containers to mark and hold the various components and the bolts and/or studs which fasten them. Modern day powerheads use a lot of little brackets and clips which hold wiring harnesses and such and these holders are often mounted on studs and/or bolts that can be easily mixed up. The manufacturer spent a lot of time and money designing your outboard and they wouldn't have wasted any of it by haphazardly placing brackets, clips or fasteners. If it's present when you disassemble it, put it back when you assemble it, you will regret not remembering that little bracket which holds a wire harness out of the path of a rotating part.

You should begin by unbolting any accessories attached to the powerhead. Remove any covers remaining on the powerhead. The idea is to reduce the powerhead to the bare necessities (cylinder head(s), cylinder block, crankshaft, pistons and connecting rods), plus any other `in block' components.

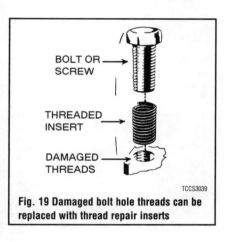

Fig. 19 Damaged bolt hole threads can be replaced with thread repair inserts

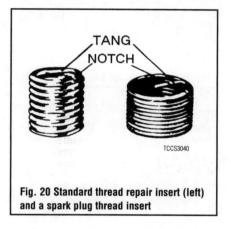

Fig. 20 Standard thread repair insert (left) and a spark plug thread insert

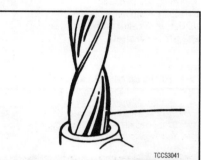

Fig. 21 Drill out the damaged threads with the specified size bit. Be sure to drill completely through the hole or to the bottom of a blind hole

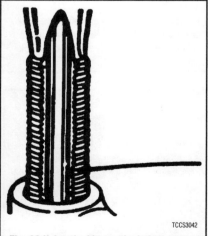

Fig. 22 Using the kit, tap the hole to receive the thread insert. Keep the tap well oiled and back it out frequently to avoid clogging the threads

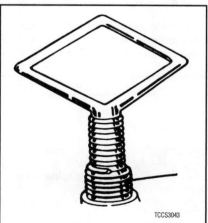

Fig. 23 Screw the insert onto the installer tool until the tang engages the slot. Thread the insert into the hole until it is 1/4–1/2 turn below the top surface, then remove the tool and break of the tang using a punch

Fig. 24 Large powerheads will require the installation of a lifting eye and the use of an engine hoist to remove them from the outboard

7-22 POWERHEAD

Cylinder Block and Head

GENERAL INFORMATION

♦ See Figures 25 thru 35

The cylinder block is made of aluminum and may have cast-in iron cylinder liners. It is the major part of the powerhead and care must be given to this part when service work is performed. Mishandling or improper service procedures performed on this assembly may make scrap out of an otherwise good casting. The cylinder assembly casting and other major castings on the outboard are expensive and need to be cared for accordingly.

There are three parts to the cylinder assembly, the cylinder block, the crankcase half and on some powerheads and separate the cylinder head. The cylinder block and crankcase half are married together and line bored to receive the crankshaft bearings, reed blocks and on some powerheads sealing rings. After this operation they are treated as one casting.

➡Remember that anything done to the mating surfaces during service work will change the inner bore diameter for the main bearings, reed blocks and sealing rings and possibly prevent the block and crankcase mating surfaces from sealing.

The only service work allowed on the mating surface is a lapping operation to remove nicks from the service. Carefully guard this surface when other service work is being performed. The different sealing materials used to seal the mating surfaces are sealing strips, sealing compound and Loctite®.

Since the 2-stroke powerhead operates like a pump with one inlet and one outlet for each cylinder, special sealing features must be designed into the cylinder assembly to seal each individual cylinder in a multi-cylinder powerhead. Each inlet manifold must be completely sealed both for vacuum and pressure. One way of doing this internally is with a labyrinth seal, which is located between two adjacent cylinders next to the crankshaft. It may be of aluminum or brass, formed in the assembly and machined with small circular grooves running very close to a machined area on the crankshaft. The tolerance is so close that fuel residue puddling in the seal effectively

Fig. 25 The cylinder block . . .

Fig. 26 . . . crankcase half . . .

Fig. 27 . . . and cylinder head make up the major components of the cylinder assembly

Fig. 28 In this cylinder, an exhaust port can be seen above the level of the piston. The inlet port is on the opposite side of the cylinder wall, below the piston

Fig. 29 The cylinder block and crankcase half are machined to fit together perfectly. They provide a cradle for the spinning crankshaft

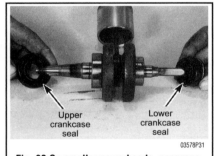
Fig. 30 On smaller powerheads, neoprene seals are installed to seal the ends of the cylinder block assembly around the crankshaft

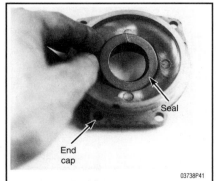
Fig. 31 On larger powerheads, end caps which use seals . . .

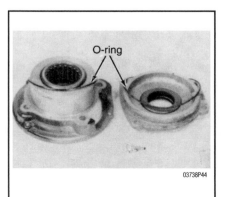
Fig. 32 . . . and O-rings . . .

Fig. 33 . . . are installed at both the upper . . .

POWERHEAD 7-23

Fig. 34 . . . and lower ends of the cylinder block assembly . . .

Fig. 35 Sealing rings are also installed into grooves in the crankshaft. When the crankshaft is installed, the sealing rings mate up to and seal against the web in the cylinder block crankcase halves

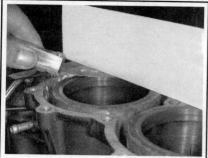

Fig. 36 Every time the cylinder head is removed, the cylinder head and cylinder block deck should be checked for warping using a straight edge and a feeler gauge

completes the seal between the cylinder block and crankcase halves against the crankshaft. Crankcase pressures are therefore retained to each individual cylinder. No repair of the labyrinth seal is made. If damage has occurred to the seal, the main bearings have allowed the crankshaft to run out and rub.

Another method of internal sealing between the crankcases is with seal rings. These rings are installed in grooves in the crankshaft. When the crankshaft is installed, the sealing rings mate up to and seal against the web in the cylinder block crankcase halves and crankshaft. Sealing rings of different thickness are available for service work. The side tolerance is close, so puddled fuel residue will effectively complete the seal between crankcases and crankshaft.

To seal the ends of the cylinder assembly around the crankshaft, O-rings are installed around the end caps and neoprene seals are installed inside the cap and seal against the crankshaft.

INSPECTION

▶ See Figures 36, 37 and 38

Everytime the cylinder head is removed, the cylinder head and cylinder block deck should be checked for warping. Do this with a straight edge or a surface block. If the cylinder head or cylinder block deck are warped, the surface should be machined flat by a competent machine shop. Minor warpage may be cured by using emery paper in a figure eight motion on a surface block until the surface is true.

Inspect the cylinder head and cylinder block for cracks and damage to the bolt holes caused by galvanic corrosion. On models which do not use a cylinder head, check the cylinder dome for holes or cracks caused by overheating and pre-ignition. The spark plug threads may also be damaged by overtorquing the spark plug.

Quite often the small bolts around the cylinder block sealing area are seized by corrosion. If white powder is evident around the bolts, stop. Galvanic corrosion is probably seizing the shank of the bolt and possibly the threads as well. Putting a wrench on them may just twist the head off, creating one big mess. Know the strength of the bolt and stop before it breaks. If it does break, don't reach for an easy out, it won't work.

A good way to service these seized bolts is with localized heat (from a heat gun, not a torch) and a good penetrating oil. Heat the aluminum casting, not the bolt. This releases the bolt from the corrosive grip by creating clearance between the bolt, the corrosion and the aluminum casting. Be careful because too much heat will melt the casting. Many bolts can be released in this way, preventing drilling out the total bolt and heli-Coiling the hole or tapping the hole for an oversize bolt.

To help prevent bolts from seizing due to corrosion, coat threads with a good anti-seize compound.

Cylinder Bores

GENERAL INFORMATION

The purpose of the cylinder bore is to help lock in combustion gases, provide a guided path for the piston to travel within, provide a lubricated surface for the piston rings to seal against and transfer heat to the cooling system. These functions are carried out through all engine speeds. To function properly the cylinder has to have a true machined surface and must have the proper finish installed on it to retain lubricant.

INSPECTION

▶ See Figures 39 and 40

The roundness of the cylinder diameter and the straightness of the cylinder wall should be inspected carefully. Micrometer readings should be taken at several points to determine the cylinder condition. Start at the bottom using an outside micrometer or dial bore gauge. By starting at the bottom, below the area of ring travel, cylinder bore diameter can be determined and a determination can be made if the powerhead is standard or has been bored oversize. Take the second measurement straight up from the first in the area of the ports and note that the cylinder is larger here. This is the area where the rings ride and it has worn slightly. Take the third measurement within a half inch of the top of the cylinder, straight up from where the second measurement was taken. These three measurements should be repeated with the measuring instrument turned 90° clockwise.

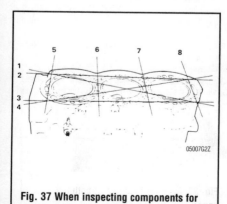

Fig. 37 When inspecting components for warping, check in multiple directions

Fig. 38 To help prevent bolts from seizing due to corrosion, coat threads with a good antiseize compound.

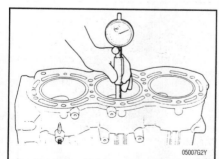

Fig. 39 The roundness of the cylinder diameter and the straightness of the cylinder wall should be inspected using a dial bore gauge

7-24 POWERHEAD

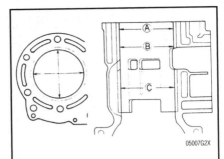

Fig. 40 Readings should be taken at several points to determine the cylinder condition. Start at the bottom and work your way to the top

Fig. 41 A hole placed in the side of the piston, commonly referred to as the piston boss, is used to mount the piston to the piston pin

Fig. 42 The piston has machined grooves in which the rings are installed. They are carried along with the piston as it travels up and down the cylinder wall

After the readings are taken, you will have enough information to access the cylinder condition. This will tell you if the rings can simply be replaced or if the cylinder will need to be overbored. While measuring the cylinder, you should also be noting if there is a cross-hatched pattern on the cylinder walls. Also note any scuffing or deep scratches.

REFINISHING

If the cylinder is out of round, worn beyond specification, scored or deeply scratched, reboring will be necessary. If the cylinder is within specification, it can be deglazed with a flex hone and new rings installed.

➡ Some cylinders are chrome plated and require special service procedures. Consult a qualified machine shop when dealing with chrome plated cylinders.

Almost all engine block refinishing must be performed by a machine shop. If the cylinders are not to be rebored, then the cylinder glaze can be removed with a ball hone. When removing cylinder glaze with a ball hone, use a light or penetrating type oil to lubricate the hone. Do not allow the hone to run dry as this may cause excessive scoring of the cylinder bores and wear on the hone. If new pistons are required, they will need to be installed to the connecting rods. This should be performed by a machine shop as the pistons must be installed in the correct relationship to the rod or engine damage can occur.

When deglazing, it is important to retain the factory surface of the cylinder wall. The cross-hatched patter on the cylinder wall is used to retain oil and seal the rings. As the piston rings move up and down the wall, a glaze develops. The hone is used to remove this glaze and reestablish the basket weave pattern. The pattern and the finish is has a satin look and makes an excellent surface for good retention of 2-stroke oil on the cylinder wall.

There is nothing magic about the crosshatch angle but there should be one similar to what the factory used. (approximately 20–40°). Too steep an angle or too flat a pattern is not acceptable and as it is not good for ring seating. Since the hone reverses as it is being pushed down and pulled up the cylinder wall, many different angles are created. Multiple criss-Crossing angles are the secret for longevity of the cylinder and the rings. The pattern allows 2-stroke oil to flow under the piston ring bearing surface and prevents a metal-to-metal contact between the cylinder wall and piston ring. The satin finish is necessary to prevent early break-in scuffing and to seat the ring correctly.

After the cylinder hone operation has been completed, one very important job remains. The grit that was developed in the machining process must be thoroughly cleaned up. Grit left in the powerhead will find its way into the bearings and piston rings and become embedded into the piston skirts, effectively grinding away at these precision parts. Relate this to emery paper applied to a piece of steel or steel against a grinding stone. The effect is removal of material from the steel. Grit left in the powerhead will damage internal components in a very short time.

Wiping down the cylinder bores with an oil or solvent soaked rag does not remove grit. Cleaning must be thorough so that all abrasive grit material has been removed from the cylinders. It is important to use a scrub brush and plenty of soapy water. Remember that aluminum is not safe with all cleaning compounds, so use a mild dish washing detergent that is designed to remove grease. After the cylinder is thought to be clean, use a White paper towel to test the cylinder. Rub the paper towel up and down on the cylinder and look for the presence of gray color on the towel. The gray color is grit. Re-scrub the cylinder until it is perfectly clean and passes the paper towel test. When the cylinder passes the test, immediately coat it with 2-stroke oil to prevent rust from forming.

➡ Rust forms very quickly on clean, oil free metal. Immediately coat all clean metal with 2-stroke oil to prevent the formation of rust.

Pistons

GENERAL INFORMATION

▶ See Figures 41, 42, 43 and 44

Piston are the moveable end of a cylinder. The cylinder bore provides a guided path for the piston allowing a small clearance between the piston skirt and cylinder wall. This clearance allows for piston expansion and controls piston rock within the cylinder.

Modern piston design is such that the head of the piston directs incoming fuel toward the top of the cylinder and outgoing exhaust to the exhaust port in the cylinder wall. This design is called a deflector type piston head. The deflector dome deflects the incoming fuel upward to the spark plug end of the cylinder, partially cooling the cylinder and spark plug tip. It also purges the spent gases from the cylinder. In essence, the incoming fuel charge is chasing out the exhaust gases from the cylinder.

Not all piston designs are of the deflector head type. Other pistons have a small convex crown on the piston head. In this case, port design aids in directing the incoming fuel upward. The piston head bears the brunt of the combustion force and heat. Most of the heat is transferred from the piston head through the rings to the cylinder wall and then on to the cooling system.

The piston design can be round, cam ground or barrel shaped. The cam ground design allows for expansion of the piston in a controlled manner. As the piston heats up, expansion take place and the piston moves out along the piston pin becoming more round as it warms up. Barrel shaped pistons rock very slightly in the bore which helps to keep the rings free.

The piston has machined ring grooves in which the rings are installed. They are carried along with the piston as it travels up and down the cylinder wall. There is one small pin in each ring groove to prevent the ring from rotating. The piston skirt is the bearing area for thrust and rides on the cylinder wall oil film. The side thrust of the piston is dependent upon piston pin location. If the pin is in the center of the piston, then there will be more thrust. If the pin is offset a few thousandths of an inch from the center of the piston, there will be less thrust. A used piston will have one side of the piston skirt show more signs of wear than the opposite side. The side showing wear is the major thrust side.

Thrust is caused by the pendulum action of the rod following the crankshaft rotation, which pulls the rod out from under the piston. The combustion pressure therefore pushes and thrusts the piston skirt against the cylinder wall. Some heat is also transferred at this point. The other skirt receives only minor pressure. Some pistons have small grooves circling the skirts to retain oil in the critical area between the skirt and the cylinder wall.

INSPECTION

▶ See Figures 45 and 46

The piston needs to be inspected for damage. Check the head for erosion caused by excessive heat, lean mixtures and out of specification timing/synchronization. Examine the ring land area to see if it is flat and not rounded over. Also look for burned through areas caused by preignition. Check the skirt for scoring caused by a

POWERHEAD 7-25

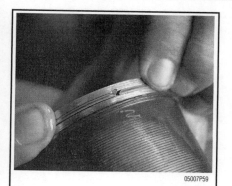
Fig. 43 There is one small pin in each ring groove to prevent the ring from rotating

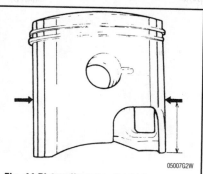
Fig. 44 Piston diameter should be measured at a specific position on the piston which the manufacturer will specify

Fig. 45 This piston is severely scored from lack of lubrication and should not be reused

break through of the oil film, excessive cylinder wall temperatures, incorrect timing/synchronization or inadequate lubrication.

To measure the piston diameter, place an outside micrometer on the piston skirt at the specified location. All pistons in a given powerhead should read the same. Check the specifications for placement of the micrometer when measuring pistons. Generally there is a specific place on the piston. This is especially true of barrel shaped pistons that are larger in the middle than they are at the top and bottom.

If the piston looks reasonably good after cleaning, take a close look at the ring lands. Wear may develop on the bottom of the ring lands. This wear is usually uneven, causing the ring to push on the higher areas and loads the ring unevenly when inertia is the greatest. Such uneven support of the ring will cause ring breakage and the piston will need to be replaced.

When installing a new ring in the groove, measure the ring side clearance against specification. Also check the see if the ring pins are there and that they have not loosened. Measure the skirt to see if the piston is collapsed.

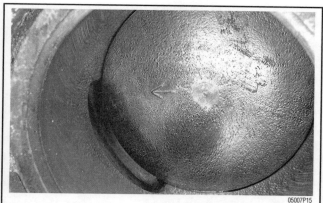
Fig. 46 Pistons should be installed with the arrow facing the exhaust port

Piston Pins

GENERAL INFORMATION

▶ See Figures 47, 48 and 49

A hole placed in the side of the piston, commonly referred to as the piston boss, is used to mount the piston to the piston pin. The combustion pressure is transferred to the piston pin and connecting rod bearing, then on to the crankshaft where it is converted to rotary motion. The pin is fitted to the piston bosses. The piston pin is the inner bearing race for the bearing mounted in the small end of the connecting rod. This transfers the combustion pressures into the connecting rod and allows the rod to swing with a pendulum-like action.

Piston pins are secured into both piston bosses. All have retainers and in addition some use a press fit to secure the pin. There are some models which use a slip fit. These may require special installation techniques.

Another type of pin fitting is loose on one side and tight on the other. This type aids in removal of the pin without collapsing the piston. With this design, always press on the pin from the loose boss side. The piston is marked on the inside of the piston skirt with the word "loose" to identify the loose boss. Always press with the loose side up and press the pin all the way through and out. When installing, press with the loose side up.

In all pressing operations, set the piston in a cradle block to support the piston. Some pistons require heating to expand the piston bosses so the pin can be pressed out without collapsing the piston. Other pistons just have a slip fit.

INSPECTION

▶ See Figures 50 thru 55

Check the piston pin retainer grooves for evidence of the retainers moving as they may have been distorted. Always replace the retainers once they have been removed. If there is evidence of wear in any of these areas, the piston should be replaced.

Inspect piston pin for wear in the bearing area. Rust marks caused by water will leave a needle bearing imprint. Chatter marks on the pin indicate that the piston pin

Fig. 47 The holes in the bottom of this piston pin bore provide oiling to the piston pin . . .

Fig. 48 A similar hole in the connecting rod also oils the pin

Fig. 49 This piston uses a floating pin design. Once the retainers are removed, the pin should slide out easily

7-26 POWERHEAD

Fig. 50 Measuring the piston pin bore inside diameter. This reading will be compared with the piston pin outside diameter to determine pin-to-bore clearance

Fig. 51 Measuring the piston pin outside diameter with an outside micrometer at the point where the pin aligns with the piston pin bore . . .

Fig. 52 . . . and also at the point where the pin aligns with the connecting rod bore

Fig. 53 Some powerheads use a caged bearing design used to support piston pins . . .

Fig. 54 . . . while others use individual needle bearings

Fig. 55 The small end bore in the connecting rod must be perfectly round to prevent bearing troubles

should be replaced. If these marks are not too heavy, they may possibly be cleaned with emery paper for loose needle bearings or crocus cloth for caged bearings.

If the piston pin checks out visually, measure its outside diameter and compare that measurement with the inside diameter of the piston pin bore. Proper clearance is vital to providing enough lubrication.

Piston Rings

GENERAL INFORMATION

♦ See Figure 56

The piston ring seals the piston to the cylinder bore, just as other seals are used on the crankshaft and lower unit. To perform correctly, the rings must conform to the cylinder wall and maintain adequate pressure to insure their sealing action at required operating speeds and temperatures. There are different designs used throughout the outboard industry. A given manufacturer will select a ring design that meets the operating requirements of the powerhead. This may be a standard ring, a pressure back (Keystone) ring or a combination of rings.

The functions of the piston ring include sealing the combustion gases so they cannot pass between the piston and the cylinder wall into the crankcase upsetting the pulse and maintaining an oil film in conjunction with the cylinder wall finish throughout the ring travel area. The rings also transfer heat picked up by the piston during combustion. This heat is transferred into the cylinder wall and thus to the cooling system. There are either two or three rings per piston, which perform these functions.

➥An oil control ring is not used on 2-stroke engines.

All piston rings used are of the compression type. This means that they are for sealing the clearance between piston and cylinder wall. They are not allowed to rotate on the piston as automotive piston rings do. They are prevented from rotating by a

Fig. 56 The piston ring seals the piston to the cylinder bore

pin in the piston ring groove. If the ring was allowed to turn, a ring end could snap into the cylinder port and become broken. The ring ends are specially machined to compensate for the pin. As the rings warm up in a running powerhead they expand, thereby requiring a specific end gap between the ring ends for expansion. This ring gap decreases upon warm-up, effectively limiting blowing gases (from the combustion process) from going into the crankcase. The rings ride in a piston ring groove with minimal side clearance, which gives them support as they move up and down the cylinder wall. With this support, combustion gas pressure and oil effectively seal the piston ring against the ring land and the cylinder wall. As long as the oil mix is correct and temperatures remain where they should, the rings will provide service for many hours of operation.

POWERHEAD

INSPECTION

▶ See Figures 57 thru 62

One of the first indications of ring trouble is the loss of compression and performance. When compression has been lost or lowered because of the ring not sealing, the ring is either broken or stuck with carbon, gum or varnish. Improper oil mixing and stale gasoline provide the carbon, gum and varnish which cause the rings to stick. Low octane fuel, improperly adjusted timing/synchronization and lean fuel mixtures can damage the ring land, causing the ring to stick or break.

➡ **Running the outboard out of the water for even a few seconds can have damaging effects on the rings, pistons, cylinder walls and water pump.**

To determine if the rings fit the cylinder and piston, two measurements are taken; ring gap and ring side clearance. To determine these measurements, the ring is pushed into the cylinder bore using the piston skirt, so it will be square. Position each ring, one at a time, at the bottom of the cylinder (the smallest diameter) and using a feeler gauge, measure the expansion space between the ring ends. This is known as the ring gap measurement. Compare this measurement against the specifications. If the measurement is too small, the ring must be filed to increase the gap. If it is too large, either the bore is too large or the ring is not correct for the powerhead.

After ring end gap has been determined, position each ring in the piston ring groove and using a feeler gauge, measure between the ring and the piston ring land. Compare this measurement against the specifications. If the measurement is too small, the ring groove may be compressed. Inspect the ring groove and ring land condition. If it is too large the ring may not correct for the powerhead.

Connecting Rods

GENERAL INFORMATION

The connecting rod transfers the combustion pressure from the piston pin to the crankshaft, changing the vertical motion into rotary motion. In doing so, the connecting rod swings back and forth on the piston pin like a pendulum while it is traveling up and down. It goes down by combustion pressure and goes up by flywheel momentum and/or other power strokes on a multi-cylinder powerhead. The connecting rod can be of aluminum on smaller horsepower fishing outboards or of steel on larger horsepower models.

Most connecting rod designs use a steel liner with needle bearings in the large end and a pressed-in needle bearing in the small end.

The steel rod is a bearing race at both the large and small ends of the rod. It is hardened to withstand the rolling pressures applied from the loose or caged needle bearings. Unlike many connecting rod designs, these rods do not use two piece caps. The connecting rod big end is one piece. This requires the crankshaft to be pressed together to form a rotating assembly with the connecting rods.

The connecting rods are mist lubricated. Some of the rods have a trough design in the shank area. Oil holes may be drilled into the bearing area at both ends of this trough. Oil mist that falls out of the fuel will settle into the rod trough and collect. As the rod moves in and out, the oil is sloshed back and forth in the trough and out the oil holes into the rod or piston pin bearings. This provides sufficient lubrication for these bearings. When the rod is equipped with oil holes, the oil holes have to be placed in the upward position toward the tapered end of the crankshaft when reassembled.

INSPECTION

▶ See Figures 63, 64, 65 and 66

Damage to the connecting rod can be caused by lack of lubrication and will result in galling of the bearing and eventual seizing to the crankshaft. Over speeding of the powerhead may also cause the upper shank area of the rod to stretch and break near the piston pin.

Steel rods are inspected in the bearing areas, much like you would inspect a roller bearing. Look for scoring, pit marks, chatter marks, rust and color change. A Blue color indicates overheating of the bearing surface. Minor rust marks or scoring may be cleaned up using crocus cloth for caged needle bearings or emery paper for loose needle bearings. A piece of round stock, cut with a slot in one end to accept a small piece of emery paper and mounted in a drill motor, can be used to clean up the rod ends.

Fig. 57 To determine ring gap, use a feeler gauge to measure the expansion space between the ring ends with the ring installed in the cylinder

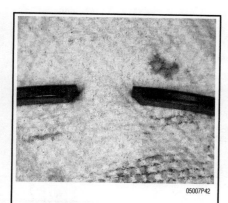

Fig. 58 Some rings are square . . .

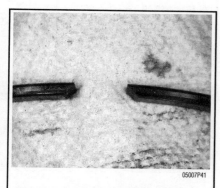

Fig. 59 . . . while other rings have a notched shape

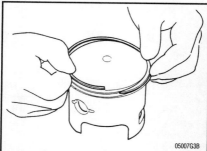

Fig. 60 One way to install piston rings is to place the ring in the groove and work it around the piston using a spiral motion until the ring is properly seated . . .

Fig. 61 . . . but the best way to install piston rings is to use a ring expander

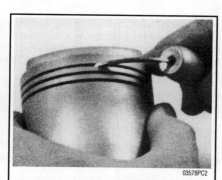

Fig. 62 Decarbon the piston rings using a ring groove cleaner or a broken piece of piston ring

7-28 POWERHEAD

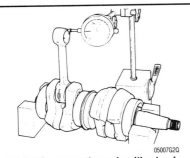

Fig. 63 On connecting rods without rod caps, check for side clearance between the crankshaft journals using a dial indicator as illustrated. For connecting rods with caps, use a feeler gauge between the connecting rod and the crankshaft journal

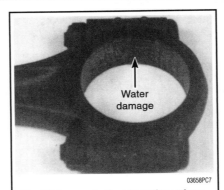

Fig. 64 Damage to the connecting rod can be caused by lack of lubrication and will result in galling of the bearing and eventual seizing to the crankshaft

Fig. 65 Minor rust marks or scoring may be cleaned up using crocus cloth for caged needle bearings or emery paper for loose needle bearings

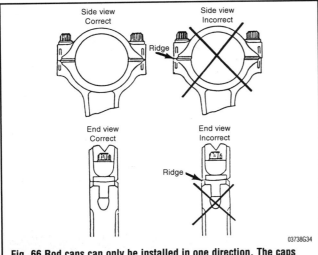

Fig. 66 Rod caps can only be installed in one direction. The caps have been correctly installed if no ridge can be felt

Fig. 67 Crankshaft assembly—V-type powerhead

The rod also needs to be checked to see if it is bent or has a twist in it. To do this, remove the piston and place the rod on a surface plate or a piece of flat glass (automotive widow). Using a flash light behind the rod and looking from in front of the rod, check for any light which can be seen under the rod ends. If light can be seen shining under the rod ends, the rod is bent and it must be replaced. You can also use a .002 feeler gauge. See if it will start under the machined area of the rod. If it will, the rod is bent. Examine the rod bolts and studs for damage and replace the nuts where used. Always reinstall the rod back on the same journal from which it was removed. The needle bearings, rod bearing surface and crankshaft journal are all mated to each other once the powerhead has been run.

➡ When installing the connecting rod, the long sloping side must be installed toward the exhaust side of the cylinder assembly and if there is a hole in the connecting rod, position the oil hole upward. Some piston designs are marked with the word "UP". This side should be placed toward the tapered end of the crankshaft.

Crankshaft

GENERAL INFORMATION

▸ See Figures 67, 68, and 69

The crankshaft is used to convert vertical motion received from the mounted connecting rod into rotary motion, which turns the driveshaft. It mounts the flywheel, which imparts a momentum to smooth out pulses between power strokes. It also provides sealing surfaces for the upper and lower seals and provides a surface for

Fig. 68 Crankshaft assembly—single cylinder powerhead

POWERHEAD 7-29

the labyrinth seal to hold oil against and a groove in which sealing rings are installed to seal pressures into each crankcase. Mounted main bearings control the axial movement of the crankshaft as it accomplishes these functions. The crankshaft bearing journals are case hardened to be able to withstand the stresses applied by the floating needle bearings used for connecting rod and main bearings. In essence, the crankshaft journals are the inner bearing races for the needle bearings.

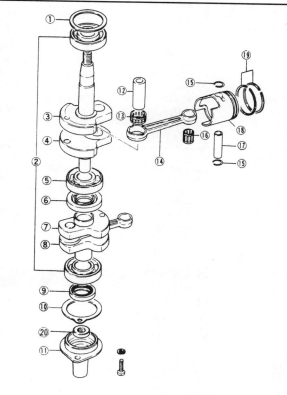

1	Shim	11	Lower oil seal housing
2	Crankshaft assy	12	Crank pin
3	Upper crankshaft	13	Crank pin bearing
4	Upper crankshaft wheel	14	Connecting rod
5	Bearing	15	Circlip
6	Oil seal	16	Piston pin bearing
7	Lower crankshaft wheel	17	Piston pin
8	Lower crankshaft	18	Piston
9	Oil seal	19	Piston ring
10	Lower oil seal gasket	20	Oil seal

Fig. 69 Exploded view of a crankshaft assembly with major components identified

INSPECTION

♦ See Figures 70 and 71

Pressure from the power stoke applied to the crankshaft rod journal by the needle bearings has a tendency to wear the journal on one side. During crankshaft inspection, the journals should also be measured with a micrometer to determine if they are round and straight. They should also be inspected for scoring, pitting, rust marks, chatter marks and discoloration caused by heat.

Check the sealing surfaces for grooves worn in by the upper and lower crankshaft seals. Take a look at the splined area which receives the driveshaft. Inspect the side of the splines for wear. This wear can be caused by lack of lubrication or improper lubricant applied during a seasonal service. An exhaust housing/lower unit that has received a sudden impact can be warped and this can also cause spline damage in the crankshaft.

The crankshaft cannot be repaired because of the case hardening and the possibility of changing the metallurgical properties of the material during the welding and machine operation. Also, there are no oversized bearings available. Repairs are limited to cleaning up the journal surface with 320 emery paper when loose needle bearings are run on the journal. Where caged roller bearings are run, the journal may be polished with crocus cloth.

The tapered end of the crankshaft has a spline or a keyway and key which times the flywheel to the crankshaft. Inspect the spline or key and keyway for damage. The crankshaft taper should be clean and free of scoring, rust and lubrication. The taper must match the flywheel hub. If someone has hit the flywheel with a heavy hammer or has used an improper puller to remove the flywheel, the flywheel and hub may be warped. Place the flywheel on the tapered end of the crankshaft and check the fit. If there is any rocking indication a distorted hub, replace the flywheel. Always use a puller which pulls from the bolt pattern or threaded inner hub of the flywheel. Never use a puller on the outside of the flywheel.

The taper is used to lock the flywheel hub to the crankshaft. When mounting the flywheel to the crankshaft the taper on the crankshaft an in the flywheel hub must be cleaned with a fast evaporating solvent. No lubrication on the crankshaft taper or flywheel hub should be done. The flywheel nut must be torqued to specification to obtain a press fit between the flywheel hub and the crankshaft taper. If the nut is not brought to specifications, the flywheel may spin on the crankshaft causing major damage.

The flywheel key is for alignment purposes and sets the flywheel's relative position to the crankshaft. Check the key for partial shearing on the side. If there is any indication of shearing, replace the key. Also check the keyway in the flywheel and crankshaft for damage. If there is damage which will allow incorrect positioning of the flywheel, the powerhead timing will be off.

Bearings

GENERAL INFORMATION

♦ See Figure 72

Needle bearings are used to carry the load which is applied to the piston and rod. This load is developed in the combustion process and the bearings reduce the friction between the crankshaft and the connecting rod. They roll with little effort and at times have been referred to as anti-friction bearings, as they reduce friction by reducing the surface area that is in contact with the crankshaft and the connecting rod. These needle bearings are of two types, loose and caged. When loose bearings are used, there can be upwards to 32 loose bearings floating between the rod journal of the crankshaft and the connecting rod. These bearings are aided in rolling by the

Fig. 70 Crankshaft seals should always be lubricated prior to installation

Fig. 71 Most crankshafts use two types of seals, an O-ring and sealing ring

Fig. 72 Typical caged bearing assembly

7-30 POWERHEAD

movement of the crank pin journal and the connecting rod pendulum action. The surface installed on the journal and rod encourages needle rotation because of its relative roughness. If the journal and rod surface was polished with crocus cloth, the loose needle bearings would have the tendency to scoot, wearing both surfaces. So, journals and rods which uses the loose needle bearings are cleaned up sing 320 grit emery paper.

Caged needle bearings used a reduced number of needles and the needles are kept separated and are encouraged to roll by the cage. The cage also controls end movement of the bearings. Because of the cage, the journal and rod surfaces can be smoother, so these surfaces are polished with crocus cloth.

Main bearings are used to mount and control the axial movement of the crankshaft. They are either ball, needle or split race needle bearings. The split race needle bearings are held together with a ring and are sandwiched between the crankcase and cylinder assembly. The split race bearings are commonly used as center main bearings, as this is the only type of bearing that can be easily installed in this location. The ball bearings may be mounted as top or bottom mains on the crankshaft.

The bearing is made up of three parts- the inner race, needle and the outer race. In most industrial applications, the outer or inner race of a needle bearing assembly is held in a fixed position by a housing or shaft. The connecting rod needle bearings in the outboard powerhead have the same basic parts but differ in that both inner and outer races are in motion. The outer race-the connecting rod-is swinging like a clock pendulum. The inner race-the crankshaft-is rotating and the needle bearing is floating between the two races.

INSPECTION

When the powerhead is disassembled and inspection of the parts is made, then by necessity along with examining the needle bearings, the crankshaft main bearing journal, rod journal and connecting rod bearing surfaces are also examined. The surfaces of all three of these parts can give a tremendous amount of information and the examination will determine if the parts are reusable.

Surfaces should be examined for scoring, pitting, chatter marks, rust marks, spalling and discoloration from overheating of the bearing surfaces. Minor scoring or pitting and rust marks may be cleaned up and the surfaces brought back to a satisfactory condition. This is done using crocus cloth for caged needle bearings and 320 grit emery paper for loose needle bearings. This is not a metal removing process, rather just a clean up of the surfaces.

Needle bearings are used as main bearings and are inspected for the same conditions as listed above. There are no oversized bearings available for rod or main bearings. Because of the hardness of the crankshaft (a bearing race), it should not be turned or welded up in order to bring it back to standard size. The welding process may stress the metallurgical properties of the crankshaft, developing cracks.

The caged rod bearings and split race main bearings are inspected for the same condition as loose needle bearings, plus the cage is examined for wear, cracks and breaks.

Ball bearings are used for top and bottom main bearings in some powerheads. These may be pressed onto the crankshaft or pressed into the end cap. To examine these bearings, wash, dry, oil and check them on the crankshaft or in the bearing cap. Turn the bearing by hand and feel if there is any roughness or catching. Try to wobble the bearing by grasping the outer race, (inner race) checking for looseness of the bearing. Replace the bearing if any of these conditions are found. If the bearing is pressed off (out) the bearing will probably be damaged and should be replaced.

If new or used bearings are contaminated with grit or dirt particles at the time of installation, abrasion will naturally follow. Many bearing failures are due to the introduction of foreign material into the internal parts of the bearing during assembly. Misalignment of the rod cap,, torque of the rod bolts and lack of proper lubrication also cause failures. Bearing failure is usually detected by a gradual rise in operating noise, excessive looseness (axial) in the bearing and shaft deflection. Keep the work area clean and use needle bearing grease or multipurpose grease to hold the bearings in place. This grease will dissipate quickly as the fuel mixture comes in contact with it. Do not use a wheel bearing or chassis grease as this will cause damage to the bearings. Oil the ball bearings with 2-stroke oil upon installation. Remember to keep them clean.

Powerhead Exploded Views

▶ See Figures 73 thru 93

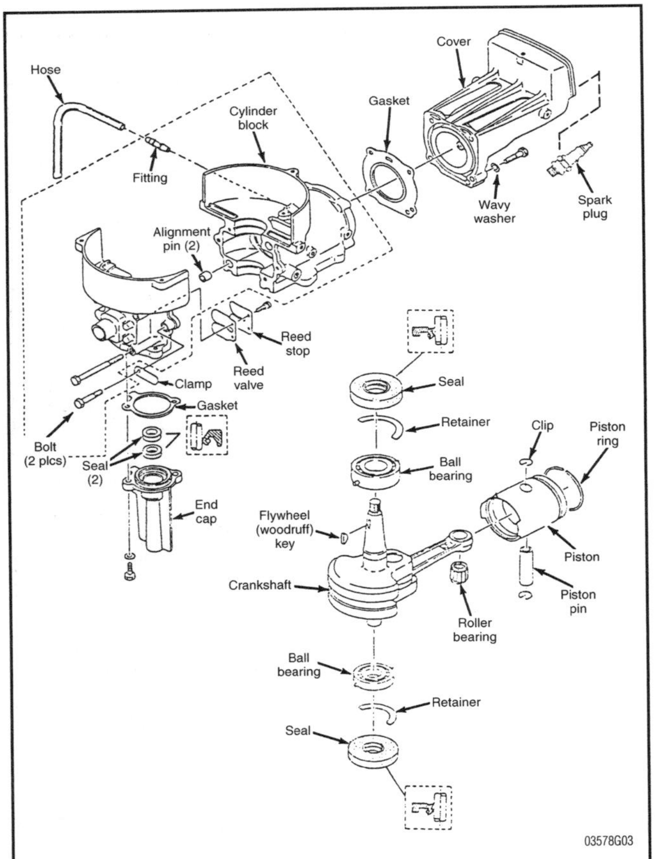

Fig. 73 Exploded view of the cylinder block and crankcase cover with major parts identified—2.5, 3 and 3.3 hp

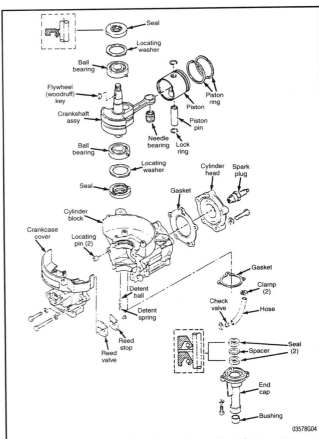

Fig. 74 Exploded view of the cylinder block and crankcase cover with major parts identified—4 and 5 hp

POWERHEAD 7-31

Fig. 75 Exploded view of the cylinder block and crankcase cover with major parts identified—6, 8, 9.9 and 15 hp

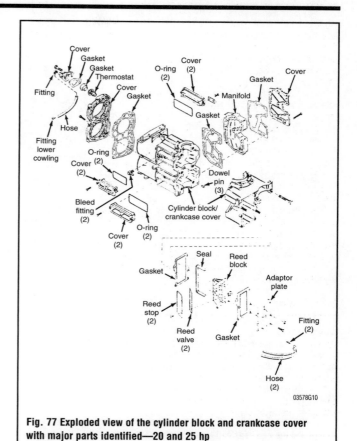

Fig. 77 Exploded view of the cylinder block and crankcase cover with major parts identified—20 and 25 hp

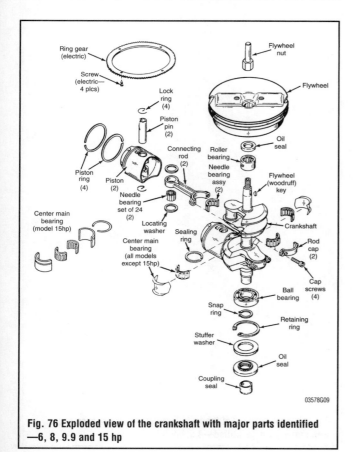

Fig. 76 Exploded view of the crankshaft with major parts identified—6, 8, 9.9 and 15 hp

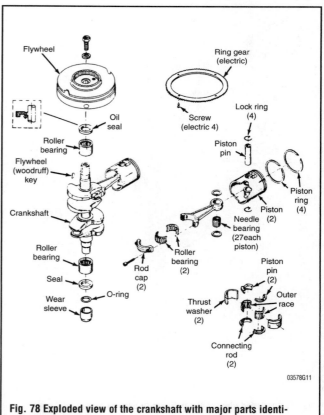

Fig. 78 Exploded view of the crankshaft with major parts identified—20 and 25 hp

7-32 POWERHEAD

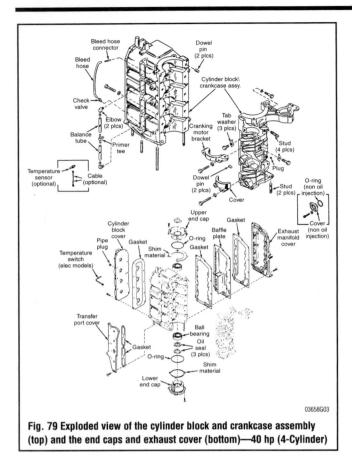

Fig. 79 Exploded view of the cylinder block and crankcase assembly (top) and the end caps and exhaust cover (bottom)—40 hp (4-Cylinder)

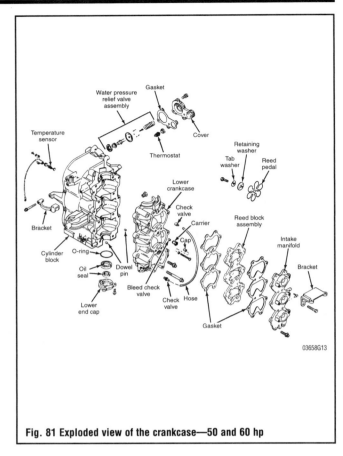

Fig. 81 Exploded view of the crankcase—50 and 60 hp

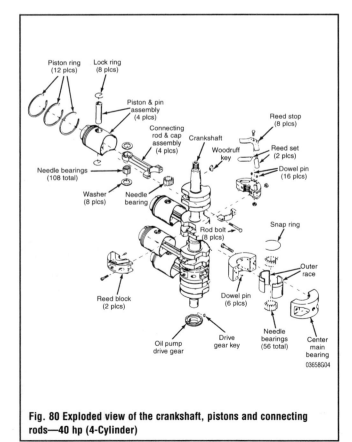

Fig. 80 Exploded view of the crankshaft, pistons and connecting rods—40 hp (4-Cylinder)

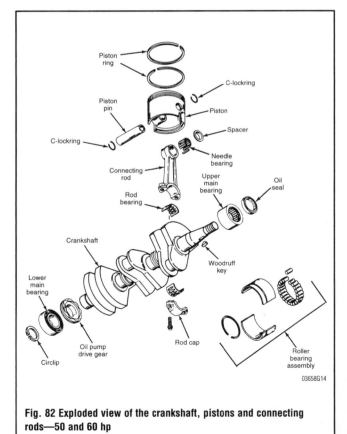

Fig. 82 Exploded view of the crankshaft, pistons and connecting rods—50 and 60 hp

POWERHEAD 7-33

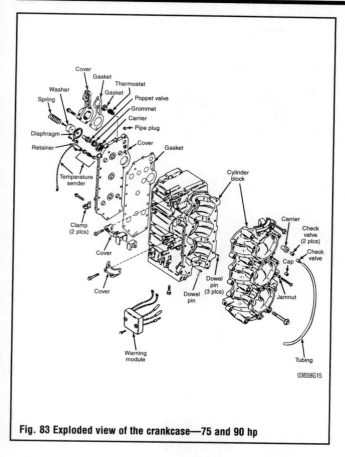

Fig. 83 Exploded view of the crankcase—75 and 90 hp

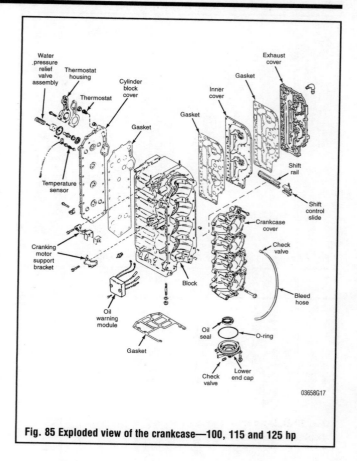

Fig. 85 Exploded view of the crankcase—100, 115 and 125 hp

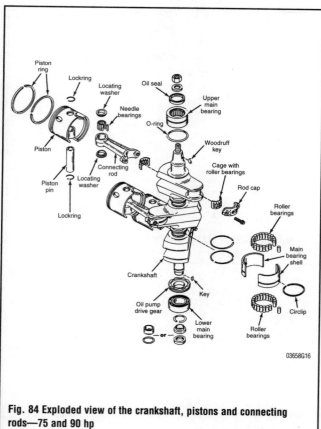

Fig. 84 Exploded view of the crankshaft, pistons and connecting rods—75 and 90 hp

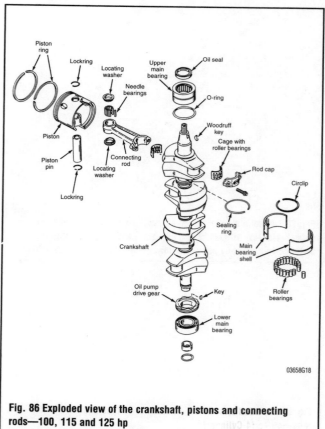

Fig. 86 Exploded view of the crankshaft, pistons and connecting rods—100, 115 and 125 hp

7-34 POWERHEAD

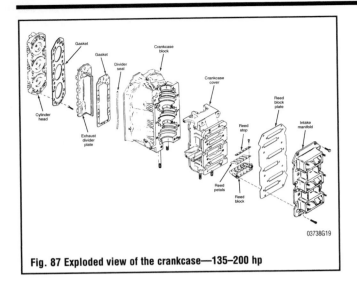

Fig. 87 Exploded view of the crankcase—135–200 hp

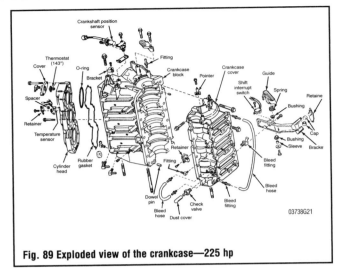

Fig. 89 Exploded view of the crankcase—225 hp

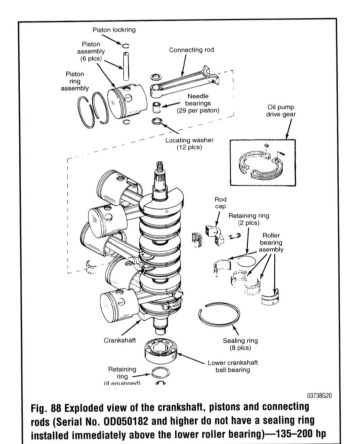

Fig. 88 Exploded view of the crankshaft, pistons and connecting rods (Serial No. OD050182 and higher do not have a sealing ring installed immediately above the lower roller bearing)—135–200 hp

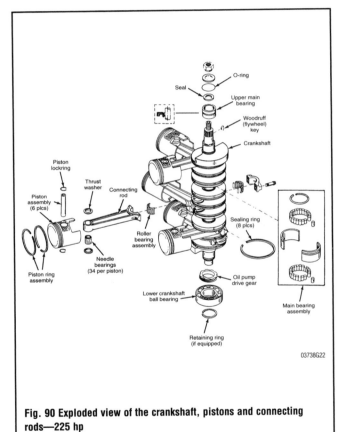

Fig. 90 Exploded view of the crankshaft, pistons and connecting rods—225 hp

POWERHEAD 7-35

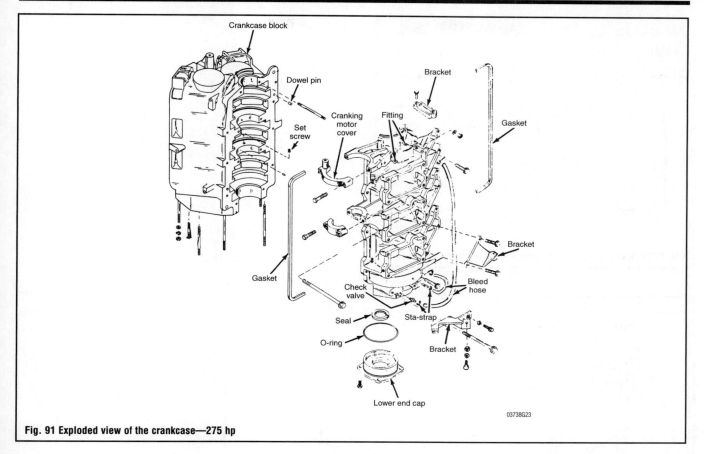

Fig. 91 Exploded view of the crankcase—275 hp

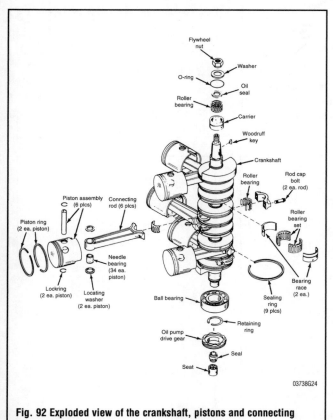

Fig. 92 Exploded view of the crankshaft, pistons and connecting rods—275 hp

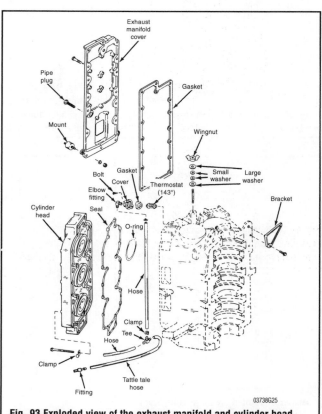

Fig. 93 Exploded view of the exhaust manifold and cylinder head assemblies—275 hp

7-36 POWERHEAD

Torque Sequence Diagrams

♦ See Figures 94 thru 118

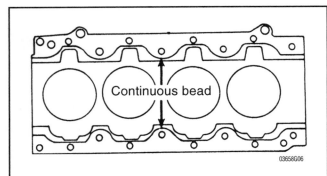

Fig. 94 Apply a 1/16 in. thick continuous bead of Loctite® sealant between the cylinder block and crankcase cover as illustrated—40 hp (4-Cylinder)

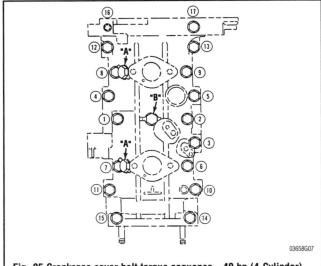

Fig. 95 Crankcase cover bolt torque sequence—40 hp (4-Cylinder)

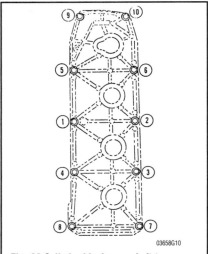

Fig. 96 Cylinder block cover bolt torque sequence—40 hp (4-Cylinder)

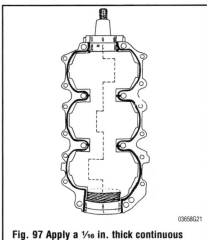

Fig. 97 Apply a 1/16 in. thick continuous bead of Loctite® sealant between the cylinder block and crankcase cover as illustrated—50–90 hp

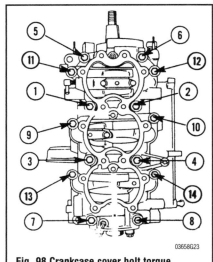

Fig. 98 Crankcase cover bolt torque sequence—50–60 hp

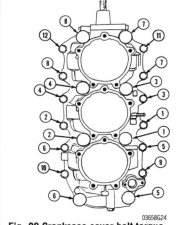

Fig. 99 Crankcase cover bolt torque sequence. Tighten the large bolts first in sequence and then the small bolts in sequence—50–60 hp

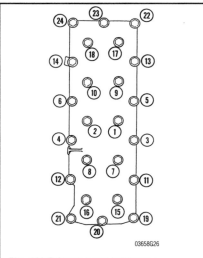

Fig. 100 Exhaust cover bolt torque sequence—50–90 hp

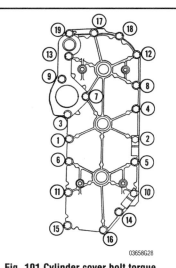

Fig. 101 Cylinder cover bolt torque sequence—50–90 hp

POWERHEAD   7-37

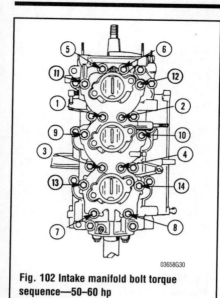

Fig. 102 Intake manifold bolt torque sequence—50–60 hp

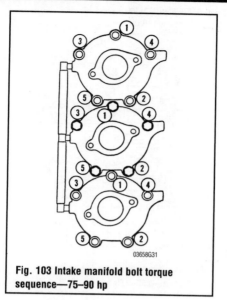

Fig. 103 Intake manifold bolt torque sequence—75–90 hp

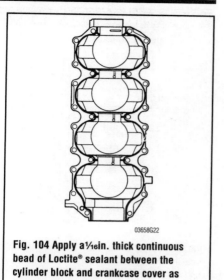

Fig. 104 Apply a 1/16 in. thick continuous bead of Loctite® sealant between the cylinder block and crankcase cover as illustrated—100–125 hp

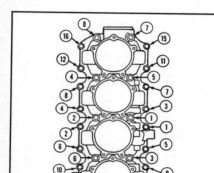

Fig. 105 Crankcase cover bolt torque sequence. Tighten the large bolts first in sequence and then the small bolts in sequence—100–125 hp

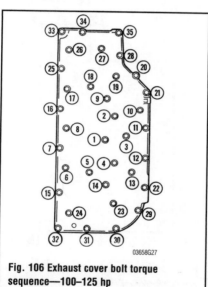

Fig. 106 Exhaust cover bolt torque sequence—100–125 hp

Fig. 107 Cylinder cover bolt torque sequence—100–125 hp

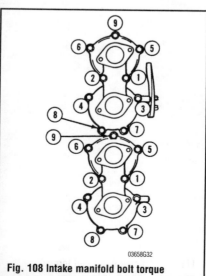

Fig. 108 Intake manifold bolt torque sequence—100–125 hp

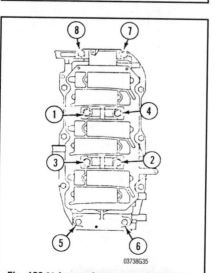

Fig. 109 3/8 in. crankcase cover bolt torque sequence—135–200 hp

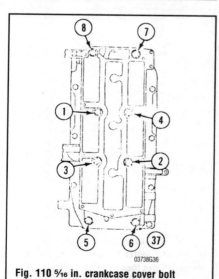

Fig. 110 5/16 in. crankcase cover bolt torque sequence—135–200 hp

7-38 POWERHEAD

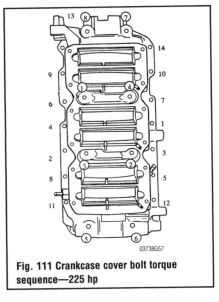

Fig. 111 Crankcase cover bolt torque sequence—225 hp

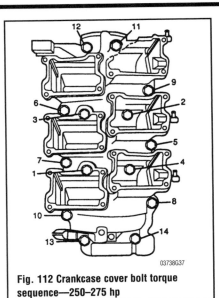

Fig. 112 Crankcase cover bolt torque sequence—250–275 hp

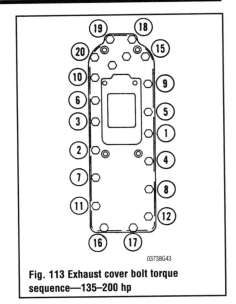

Fig. 113 Exhaust cover bolt torque sequence—135–200 hp

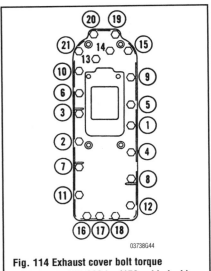

Fig. 114 Exhaust cover bolt torque sequence—150–200 hp (153 cubic inch)

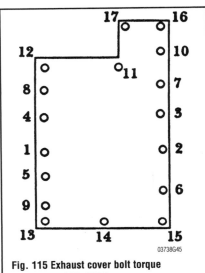

Fig. 115 Exhaust cover bolt torque sequence—225–275 hp

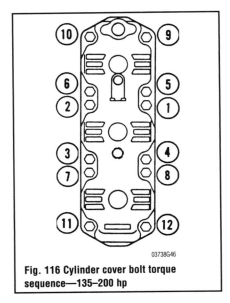

Fig. 116 Cylinder cover bolt torque sequence—135–200 hp

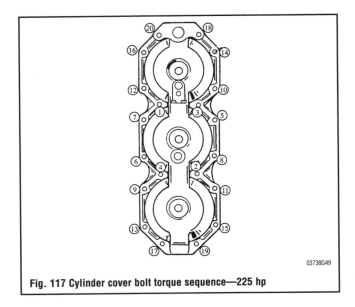

Fig. 117 Cylinder cover bolt torque sequence—225 hp

POWERHEAD 7-39

Engine Torque Specifications

Component/Model	Standard (ft. lbs.)	Metric (Nm)
Conventional bolt/nut		
8mm	3.0	4
10mm	6.0	8
12mm	13	18
14mm	26	36
17mm	31	42
Bearing carrier bolts		
2.5, 3.3 hp	50 in. lbs.	6.0
3, 4, 5 hp	70 in. lbs.	8
6, 8, 9.9, 15 hp	60	81.6
20, 25 hp	80	108.8
50, 60 hp	150 in. lbs	17
Reed mounting screw		
6, 8, 9.9, 15 hp	20 in. lbs.	2.3
20, 25 hp	25 in. lbs.	2.8
Reed block bolts		
40 hp	75 in. lbs.	8.5
135-200 hp	80 in. lbs.	9
Cylinder head to crankcase		
2.2, 3.3 hp	85 in. lbs.	9.6
3, 4, 5 hp	18	24.4
6, 8, 9.9, 15 hp	60	6.8
135-200 hp (1990-93)	40	54.2
135-225 hp (1994-00)	30 (+ 90°)	40.7
Connecting rod bolt		
6, 8, 9.9, 15 hp	8.3	11.3
20, 25 hp	15	19.8
30, 40 hp	16	21.7
50, 55, 60 hp (1/4 in. bolts)	180 in. lbs.	20
50, 55, 60 hp (5/16 in. bolts)	27	37
75, 90, 100, 115, 125 hp	30	40.7
135-275 hp	30	40.7

Engine Torque Specifications

Component/Model	Standard (ft. lbs.)	Metric (Nm)
Crankcase to cylinder block		
2.5, 3.3 hp	50 in. lbs.	5.6
3, 4, 5 hp	90 in. lbs.	10.3
6, 8, 9.9, 15 hp	16.5	22.5
20, 25 hp	30	40.7
30, 40, 50, 60 hp	18	24
75, 90 hp (large bolts)	25	34
75, 90 hp (small bolts)	18.5	25
100, 115, 125 hp	18.5	25
Crankcase cover		
6, 8, 9.9, 15 hp	60 in. lbs.	81.6
20, 25, 30, 40 hp	100 in. lbs.	11.2
75, 90, 100, 115, 125 hp	18.5	25
135-200 hp (3/8 in. bolts)	38	51.5
135-200 hp (5/16 in. bolts)	15	20.3
135, 150, 175, 200 hp (upper end cap bolts)	150 in. lbs.	16.9
135, 150, 175, 200 hp (lower end cap bolts)	80 in. lbs.	6.8
225 hp (8 x 1.25 mm bolts)	28	37.9
225 hp (10 x 1.5mm bolts)	30	40.7
275 hp	30	40.7
Flywheel nut		
2.5, 3.3 hp	30	4.1
3, 4, 5 hp	40	52.2
6, 8, 9.9, 15, 20, 25 hp	50	67.8
30, 40 hp	75	101.6
50-225 hp	120	162.7
250, 275 hp	100	135.6
Exhaust cover		
6, 8, 9.9, 15 hp	60 in. lb	6.8
20, 25 hp	8.5	11.3
30, 40 hp	17	22.5
75-225 hp	18	24.4
Gearcase to mid-section		
2.5, 3.3 hp	50 in. lbs.	5.6
3, 4, 5 hp	70 in. lbs.	8
6, 8, 9.9, 15 hp	10	20
20, 25 hp	25	33.9
30-125 hp	40	54
135-250 hp	55	74.6

Engine Torque Specifications

Component/Model	Standard (ft. lbs.)	Metric (Nm)
Cylinder block cover		
30, 40 hp	100 in. lbs.	11.3
70-125 hp	18.5	24.9
Pinion gear bolt		
20, 25 hp	150 in. lbs.	16.9
30-60 hp	50	67.8
75-125 hp	70	94.9
135-200 hp	75	101.7
225 hp	100	136
275 hp	70	94.9
Powerhead to mid-section		
2.5, 3.3 hp	50 in. lbs.	5.6
3, 4, 5 hp	70 in. lbs.	8
6, 8, 9.9, 15 hp	120 in. lbs.	13.6
20, 25 hp	17	22.6
30, 40 hp	13.5	18
50, 60 hp	28	38
75, 90 hp	14	18.6
100, 115 hp (1990-93)	29	39.5
100, 115 hp (1994-00)	44	59.7
135-225 hp	20	27
250, 275 hp	30	40.7
Propeller nut		
3, 4, 5 hp	12.5	17
6, 8, 9.9, 15 hp	6	8
20, 25 hp	10	13.6
30-275 hp	55	74.6
Starter motor to crankcase		
6-40 hp (as applicable to model)	17	22.6
30, 40 hp	15	20.3
50, 55, 60 hp	18	24
75-115 hp	14.5	19.8
125-225 hp	17.5	23.5
250, 275 hp	12.5	17
Water pump cover		
2.5, 3, 3.3, 4, 5 hp	70 in. lbs.	8
6, 8, 9.9, 15 hp	40 in. lbs.	4.5
20, 25 hp	25 in. lbs.	2.8
50-125 hp	60 in. lbs.	6.8

Engine Torque Specifications

Component/Model	Standard (ft. lbs.)	Metric (Nm)
Water pump bolts		
135-200 hp	35 in. lbs.	3.9
225-275 hp	60 in. lbs.	6.8
Water pump base		
3, 4, 5 hp	70 in. lbs.	8
6, 8, 9.9, 15 hp	40 in. lbs.	4.5
Water pump body		
30, 40 hp	30 in. lbs.	3.4
75-125 hp	60 in. lbs.	6.8
Spark plugs		
2.5, 3.3 hp	20	27.1
3, 4, 5 hp	14	19.7
6-225 hp	20	27.1
275 hp	17	23
Hand rewind starter mount		
2.5-25 hp	70 in. lbs.	8
Transfer port and intake cover bolts		
6, 8, 9.9, 15 hp	60 in. lbs.	6.8
20, 25 hp	30 in. lbs.	3.4
Intake manifold		
50-125 hp	18	24
Carburetor adaptor plate		
135-225 hp	9	12

POWERHEAD

Engine Rebuilding Specifications - 2.5, 3, 3.3 HP

Component	Standard (in.) Maximum	Metric (mm)
Crankshaft runout	0.001	0.05
Connecting rod deflection	0.022-0.056	0.6-1.5
Cylinder bore ①	1.85	47.05
Cylinder bore wear ②	1.852	47.04
Piston to cylinder clearance	0.002-0.005	0.06-0.15
Piston ring end gap	0.006-0.012	0.18-0.33
Piston ring side clearance	0.0003-0.0010	0.01-0.05
Maximum reed stop opening	0.236-0.244	6.0-6.2

① 5 mm oversize: 1.869 in. (47.55 mm)
② Oversize bore: 1.871 in. (47.52 mm)

Engine Rebuilding Specifications - 4 and 5 HP

Component	Standard (in.) Maximum	Metric (mm)
Crankshaft runout	0.002	0.05
Connecting rod deflection	0.005-0.015	0.13-0.37
Cylinder bore ①	2.165	54.991
Cylinder out of round	0.003	0.08
Cylinder taper	0.003	0.08
Cylinder bore wear	0.0024	0.6
Piston diameter ②	2.164	54.965
Piston to cylinder clearance	0.0012-0.0024	0.03-0.06
Piston ring end gap	0.008-0.016	0.2-0.4
Piston ring side clearance (TOP)	0.0012-0.0023	0.03-0.07
Piston ring side clearance (2nd)	0.0008-0.0024	0.02-0.06
Maximum reed stop opening	0.24-0.248	6.0-6.2
Cylinder head warpage	0.002	0.05

① Oversize: 2.185 in. (54.991 mm)
② Oversize: 2.184 in. (55.473 mm)

Engine Rebuilding Specifications - 6, 8, 9.9, 15 HP

Component	Standard (in.) Maximum	Metric (mm)
Crankshaft runout	0.003	0.076
Connecting rod small end I.D.	0.8195	20.789
Connecting rod big end I.D.	1.0635	27.0129
Cylinder bore		
1990-1994: 6/8/9.9 hp	2.125	53.98
1995-2000: 9.9/15 hp	2.375	60.325
Cylinder out of round	0.004	0.1016
Cylinder taper	0.004	0.1016
Cylinder bore wear		
Piston diameter		
1990-1994: 6/8/9.9 hp	2.123	53.92
1995-2000: 9.9/15 hp	2.373	60.27
Piston to cylinder clearance	0.002-0.005	0.05-0.13
Piston ring end gap	0.010-0.018	0.25-0.46
Maximum reed stop opening	0.007	0.178
Reed stop opening	19/64	7.6

Engine Rebuilding Specifications - 20, 25 HP

Component	Standard (in.) Maximum	Metric (mm)
Crankshaft runout	0.004-0.019	0.10-0.64
Connecting rod small end I.D.	0.897	22.78
Connecting rod big end I.D.	1.196	30.38
Connecting rod alignment	0.002	0.051
Cylinder bore	2.562	65.01
Cylinder out of round	0.003	0.08
Bore type:		
① 0G202749 and below	Chrome	
② 0G202750 and above¹	Mercosil	
Cylinder taper	0.003	0.08
Piston diameter	2.5583-2.5593	64.98-65.00
Piston to cylinder clearance	0.003-0.004	0.076-0.101
Piston ring end gap	0.011-0.025	.28-.64
Maximum reed opening	0.007	0.178

① Models 0G202749 and below. The cylinder bores are chrome and cannot be rebored or efficiently honed. Check each cylinder for an out-of-round "egg shaped" cylinder. A maximum of 0.003 in. (0.076 mm) is allowable.
② On models 0G202750 and above: the cylinder block is Mercosil and the cylinders can be rebored to 0.030 in. oversize. check each cylinder bore for an out-of-round "egg shaped" cylinder, a maximum of 0.003 in. (0.076 mm) is allowed.

POWERHEAD

Engine Rebuilding Specifications - 30, 40 (2-cylinder) HP

Component	Standard (in.)	Maximum Metric (mm)
Crankshaft runout	0.003	0.076
Top main bearing jounal	1.375	34.93
Center main bearing journal	1.216	30.89
Bottom ball bearing journal	1.181	29.99
Connecting rod journal	1.181	29.99
Connecting rod small end I.D.	0.957	24.31
Connecting rod big end I.D.	1.499	38.07
Connecting rod alignment	0.002	0.051
Cylinder bore ①	2.993	76
Cylinder out of round	0.003	0.08
Bore type	Cast iron	
Cylinder taper	0.003	0.08
Cylinder finish hone I.D.	2.993	76
Piston diameter at skirt ②	2.988	75.9
Piston to cylinder clearance	0.003-0.004	0.076-0.101
Piston ring end gap	0.010-0.018	.25-.46
Maximum reed opening	0.02	0.508
Reed stop		
30 hp	0.09	2.286
40 hp	Non-Adjustable	

① 0.015 in. oversize: 3.007 in. (76.38 mm)
② O.D. oversize at skirt: 3.003 in. (76.28 mm)

Engine Rebuilding Specifications—40, 50, 55, 60 HP

Component	Standard (in.)	Standard Metric (mm)
Crankshaft runout	0.002	0.05
Cylinder bore		
1990-97	2.955	75.057
.015 in. (0.381 mm) oversize	2.970	75.438
.030 in. (0.762 mm) oversize	2.985	75.819
1998-00	2.993	76.00
.015 in. (0.381 mm) oversize	3.003	76.276
.030 in. (0.762 mm) oversize	3.018	76.657
Cylinder distortion	0.003	0.08
Cylinder bore type	Cast iron	
Cylinder taper	0.003	0.08
Piston diameter		
1990-1997	2.950	74.93
.015 in. (0.381 mm) oversize	2.965	75.31
.030 in. (0.762 mm) oversize	2.980	75.69
1998-2000	2.988	75.895
.015 in. (0.381 mm) oversize	3.003	76.276
.030 in. (0.762 mm) oversize	3.018	76.657
Piston ring end gap	0.010-0.018	.254-.457
Maximum reed opening	0.02	0.5
Reed stop		
40 HP	0.09	2.286
50, 55, 60 HP	Non-Adjustable	
Reed thickness	0.01	0.254

Engine Rebuilding Specifications—40 HP (4-Cylinder)

Component	Standard (in.)	Standard Metric (mm)
Crankshaft runout	0.003	0.08
Crankshaft taper	0.003	0.08
Cylinder bore	2.565	65.15
.015 in. (0.381 mm) oversize	2.573	65.35
.030 in. (0.762 mm) oversize	2.588	65.73
Cylinder distortion	0.004	0.10
Cylinder taper	0.004	0.10
Piston diameter	2.558	64.97
.015 in. (0.381 mm) oversize	2.573	65.35
.030 in. (0.762 mm) oversize	2.588	65.73
Piston to cylinder clearance	0.007-0.011	0.17-0.27
Piston ring end gap	0.0015-0.014	0.038-0.35
Reed stop opening	0.020	0.50

POWERHEAD 7-43

Engine Rebuilding Specifications—75, 90, 100, 115, 125 HP

Component	Standard (in.)	Standard Metric (mm)
Crankshaft runout	0.006	0.152
Cylinder bore		
1990-1993	3.375	85.725
.015 in. (0.381 mm) oversize	3.390	86.106
.030 in. (0.762 mm) oversize	3.405	86.487
1994-2000	3.501	88.925
.015 in. (0.381 mm) oversize	3.516	89.306
.030 in. (0.762 mm) oversize	3.531	89.687
Cylinder bore type	Cast Iron	
Cylinder distortion	0.003	0.076
Cylinder taper	0.003	0.076
Piston diameter		
1990-1993	3.371	85.623
.015 in. (0.381 mm) oversize	3.386	86.004
.030 in. (0.762 mm) oversize	3.401	86.385
1994-2000	3.495	88.773
.015 in. (0.381 mm) oversize	3.510	89.154
.030 in. (0.762 mm) oversize	3.525	89.535
Reed stand opening (max)	0.020	0.508
Reed stop	Non-Adjustable	

Engine Rebuilding Specifications—OptiMax 200 and 225 HP

Component	Standard (in.)	Standard Metric (mm)
Crankshaft runout	0.002	0.0508
Cylinder bore	3.6265	92.1131
.015 in. (0.381 mm) oversize	3.6415	92.4941
.030 in. (0.762 mm) oversize	3.6565	92.6751
Cylinder bore type	Cast Iron	
Cylinder distortion/taper	0.003	0.076
Piston diameter	3.6215-3.6205	91.9861-91.9607
.015 in. (0.381 mm) oversize	3.6365-3.6355	92.3671-92.3417
.030 in. (0.762 mm) oversize	3.6515-3.6505	92.7481-92.7227
Reed stand open	0.020	0.50
Cylinder compression	110-135 psi	753.3-924.5 kPa

① Two measurements 0.700 in. (17.78 mm) from skirt of piston: A: 90 degrees from piston pin; B: in line with piston pin. Dimension A will be 0.001-0.0015 in. less if coating is worn off piston (used)

Engine Rebuilding Specifications - 135, 150, 175, 200, 225 HP

Component	Standard (in.)	Maximum	Metric (mm)
Crankshaft runout	0.006		0.152
Cylinder bore			
135/150 hp ①	3.125		79.375
175/200/225 hp ②	3.501		88.925
Bore type	Cast Iron		
Cylinder out of round	0.006		0.152
Cylinder taper	0.006		0.152
Piston diameter			
135/150 hp ③	3.113-3.117		79.07-79.172
175/200/225 hp ④	3.493-3.495		88.723-88.773
Maximum reed opening	0.02		0.51

① .015" over: 3.140 in. (79.756 mm); .030" over: 3.155 in. (80.137 mm)
② .015" over: 3.516 in. (89.306 mm)
③ .015" over: 3.130 in. (79.502 mm); .030" over: 3.145 in. (79.883 mm)
④ .015" over: 3.509 in. (89.129 mm)

Engine Rebuilding Specifications—OptiMax 115, 135, 150 HP

Component	Standard (in.)	Standard Metric (mm)
Crankshaft runout	0.006	0.152
Cylinder bore	3.501	88.925
.015 in. (0.381 mm) oversize	3.516	89.306
Cylinder bore type	Cast Iron	
Cylinder distortion/taper	0.003	0.076
Piston diameter	3.493-3.492	88.7222-88.6968
.015 in. (0.381 mm) oversize	3.508-3.507	89.1032-89.0778
Reed stand open	0.020	0.50
Cylinder compression	110-135 psi	753.3-924.5 kPa

① Two measurements 0.700 in. (17.78 mm) from skirt of piston: A: 90 degrees from piston pin; B: in line with piston pin. Dimension A will be 0.001-0.0015 in. less if coating is worn off piston (used)

Engine Rebuilding Specifications—250 and 275 HP

Component	Standard (in.)	Standard Metric (mm)
Conrod side clearance	0.003-0.009	0.07-0.22
Cylinder bore	3.74	95.00
Cylinder bore type ①	Chrome	
Cylinder distortion	0.006	0.152
Cylinder taper	0.006	0.152
Piston diameter	3.732	85.6
Piston to cylinder clearance	0.007-0.011	0.17-0.27
Reed stop opening	0.130	3.30

① The cylinder bores are chrome and cannot be rebored or efficiently honed.

LOWER UNIT 8-2
GENERAL INFORMATION 8-2
SHIFTING PRINCIPLES 8-2
    STANDARD ROTATING UNIT 8-2
    COUNTER-ROTATING UNIT 8-2
TROUBLESHOOTING THE LOWER
  UNIT 8-2
PROPELLER 8-3
    REMOVAL & INSTALLATION 8-3
LOWER UNIT 8-5
    REMOVAL & INSTALLATION 8-5
JET DRIVE 8-11
DESCRIPTION AND OPERATION 8-11
MODEL IDENTIFICATION AND SERIAL
  NUMBERS 8-11
JET DRIVE ASSEMBLY 8-12
    REMOVAL & INSTALLATION 8-12
    ADJUSTMENT 8-15
    DISASSEMBLY 8-16
    CLEANING AND INSPECTING 8-17
    ASSEMBLY 8-18
    SHIMMING 8-19
SPECIFICATIONS CHART
  LOWER UNIT SPECIFICATIONS 8-20

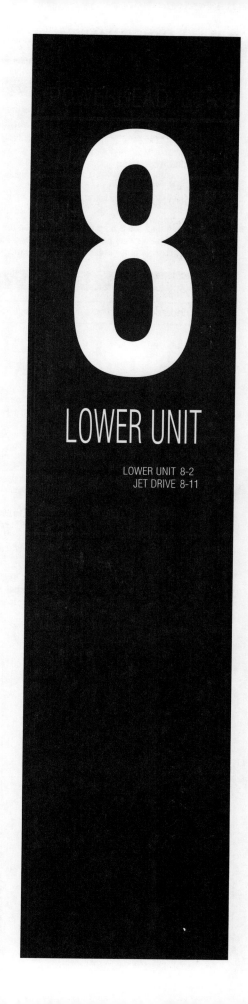

8
LOWER UNIT

LOWER UNIT 8-2
JET DRIVE 8-11

8-2 LOWER UNIT

LOWER UNIT

General Information

The lower unit is considered as that part of the outboard below the exhaust housing. The unit contains the propeller shaft, the driven and pinion gears, the drive shaft from the powerhead and the water pump. The shifting capabilities, including the forward and reverse gears together with the clutch, shift assembly, and related linkage, are all housed within the lower unit.

The lower unit may be removed and serviced without disturbing the remainder of the outboard unit.

➡ The water pump on the 2.5 hp unit is located on the propeller shaft. Therefore, the impeller may be removed and a new impeller installed without disturbing any other areas of the lower unit.

Shifting Principles

STANDARD ROTATING UNIT

Non-Reversing Unit

This type of lower unit has no shifting capability, the unit is always in forward gear anytime the engine is running.

The operator swings the outboard 180° to move the boat in astern.

Some 2.5 and 3.3 hp models have lower units with forward and neutral capability only. The gearcase has only a pinion and forward gear, but the propeller has a sliding clutch to allow forward and neutral capability.

Reverse Type

1. The driveshaft delivers engine torque from the crankshaft to the lower unit gearcase. A pinion gear on the driveshaft is in constant engagement with the forward and reverse gears in the lower unit gearcase turning them in opposite directions.

2. A sliding clutch is splined to the propeller shaft. The clutch is held in the middle or neutral position between the forward and reverse gears. When the shift shaft (rod) is moved, the shift cam (shifter) moves the follower (shift shaft), which in turn moves the clutch dog into the rotating forward or reverse gear. This allows a direct coupling of the drive shaft to the propeller shaft.

➡ Shifting the unit in and out of gear needs to be quick and positive to prevent rounding over the clutch dogs and/or ratchet teeth. Slow engagement will damage the parts. This problem is evident when the unit jumps out of gear.

COUNTER-ROTATING UNIT

Because of the increasing popularity and size of most outboards, especially the large horsepower units, a common practice has evolved—using dual outboards. These dual units are usually installed on larger vessels, such as sport fishing, charter and commercial fishing boats.

In the early years, a torque load problem existed with dual outboard installations. The propellers drastically rocked the vessel to one side as the powerheads were accelerated. To off-set this torque load from the propellers, the manufacturer developed a left-hand drive or counter-rotating lower unit. Therefore, when dual powerheads are accelerated, the torque from the right-hand rotation propeller is off-set by the torque from the left-hand or counter-rotating propeller. This improvement has made dual outboard installations a very popular choice when selecting propulsion for a larger boat.

The major difference between the standard right-hand drive and the left-hand counter-rotating unit is in the gear arrangement within the lower unit gearcase. The forward and reverse gear locations are exchanged, along with the thrust bearings and shims. With the gears in the new locations, the direction of the propeller rotation is reversed, opposite to the right-hand drive lower unit.

The drive shaft, water pump and remainder of the lower unit is identical to the right-hand rotation unit. The procedures and illustrations are valid for both units.

Troubleshooting the Lower Unit

♦ See Figures 1, 2, 3 and 4

Troubleshooting must be done before the unit is removed from the powerhead to permit isolating the problem to one area. Always attempt to proceed with troubleshooting in an orderly manner. The "shot-in-the-dark" approach will only result in wasted time, incorrect diagnosis, frustration, and replacement of unnecessary parts.

Check the propeller and the rubber hub. See if the hub is shredded. If the propeller has been subjected to many strikes against under water objects, it could slip on its hub. If the hub appears to be damaged, replace it with a new hub. Replacement of the hub must be done by a propeller rebuild shop equipped with the proper tools and experienced personnel for such work.

Verify the ignition switch is off and the spark plug has been removed, to prevent possible personal injury, should the powerhead attempt to start during the procedures. Shift the unit into reverse gear (if so equipped), and at the same time have an assistant turn the propeller shaft to ensure the clutch is fully engaged. If the shift handle is hard to move, the trouble may not be in the lower unit, but in the remote control cable, or in the shift box.

Isolate the problem. Disconnect the remote control cable at the powerhead and lift off the remote control shift cable. Operate the shift lever. If shifting is still hard, the

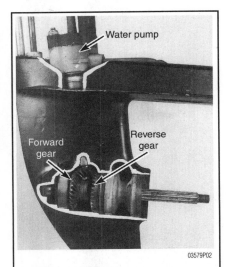

Fig. 1 Lower unit used with the small horsepower engines. This unit has forward, neutral, and reverse gear

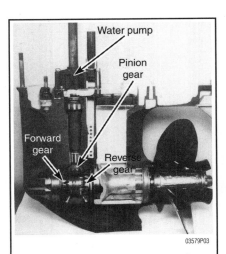

Fig. 2 Classroom type cutaway view of a lower unit with major parts, including the propeller and water pump, installed. Notice how the forward, reverse and pinion gears all are "bevel cut"

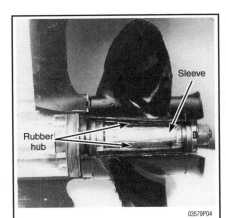

Fig. 3 Cutaway view showing the rubber hub and sleeve. The rubber hub protects the lower unit if the propeller should strike an underwater object. If the rubber hub loses its holding power with the inner hub of the propeller, the propeller hub must be replaced.

LOWER UNIT 8-3

Fig. 4 Check the propeller and the rubber hub

problem is in the shift cable or control box. If the shifting feels normal with the remote control cable disconnected, the problem must be in the lower unit.

To verify the problem is in the lower unit, have an assistant turn the propeller and at the same time move the shift cable back-and-forth. Determine if the clutch engages properly.

Propeller

REMOVAL & INSTALLATION

♦ See accompanying illustrations

1. If the propeller is "frozen" to the shaft, heat must be applied to the shaft to melt out the rubber inside the hub. Using heat will destroy the hub, but there is no other way. As heat is applied, the rubber will expand and the propeller will actually be blown free of the shaft. Therefore, stand clear to avoid personal injury.
2. Use a knife and cut the hub off the inner sleeve.
3. The sleeve can be removed by cutting it with a hacksaw, or it can be removed with a puller. Again, if the sleeve is "frozen" it may be necessary to apply heat. Remove the thrust hub from the propeller shaft.
4. A standard puller can be used to remove the sleeve and rubber hub from the propeller shaft. A puller is required because sealing compound was not used on the shaft during the previous installation.

Step 1

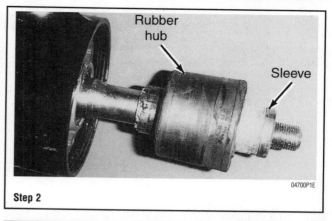

Step 2

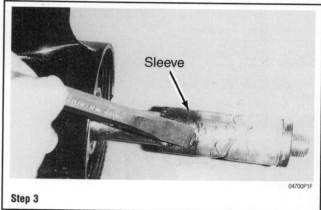

Step 3

Step 4

8-4 LOWER UNIT

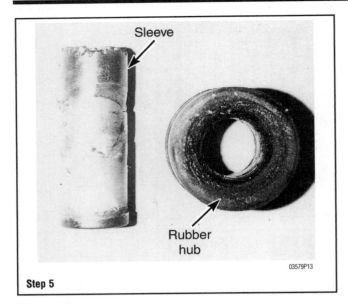

Step 5

5. The sleeve may have to be heated and then cut loose, because of extensive corrosion.

2.5–3.3 HP

Bend the legs of the cotter pin, and then remove the cotter pin from the propeller.
1. Discard the used cotter pin.
2. Slide the propeller aft and free of the propeller shaft.

To install:
3. Inspect the propeller shaft for damage and wear. Replace any components as necessary.
4. Lubricate the propeller shaft with Mercury Special Lubricant 101, 2-4-C multi-purpose lubricant or equivalent.
5. Slide the propeller onto the propeller shaft and over the shear pin.
6. Insert a new cotter pin through the propeller and propeller shaft. Bend the ends of the cotter and secure it in place.

4 and 5 HP

♦ See accompanying illustrations

1. Remove the cotter pin from the castellated nut.
2. Place a block of wood between the anti-cavitation plate and a blade of the propeller.

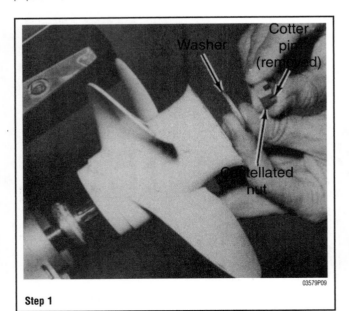

Step 1

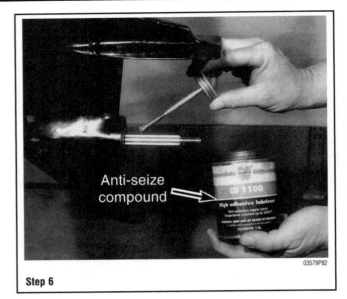

Step 6

3. Using the proper size socket and possibly a "breaker bar", remove the propeller nut and washer.
4. Slide the propeller aft and free of the propeller shaft.
5. Remove the thrust hub from the propeller shaft.

To install:
6. Apply a coating of anti-seize compound or equivalent waterproof lubricant to the propeller shaft.
7. Install the splined thrust hub onto the propeller shaft with the shoulder facing the lower unit.
8. Slide the propeller onto the propeller shaft until it seats against the thrust hub. Install the flat washer.
9. Thread the propeller nut onto the propeller shaft.
10. Place a block of wood between the anti-cavitation plate and a blade of the propeller. Tighten the propeller nut to 150 inch lbs. (17Nm).

6–35 HP

♦ See accompanying illustrations

1. Place a block of wood between the anti-cavitation plate and a blade of the propeller.
2. Use the proper size socket and possibly a "breaker bar" and remove the propeller nut.
3. Slide the thrust hub free of the shaft. The 20 and 25 hp models do not have this thrust hub.
4. Move the propeller aft and free of the shaft. Remove the large thrust hub from the propeller shaft.

To install:
5. Install the splined thrust hub onto the propeller shaft with the shoulder facing toward the lower unit.
6. Lubricate the propeller shaft thoroughly anti-seize compound or an equivalent waterproof grease.

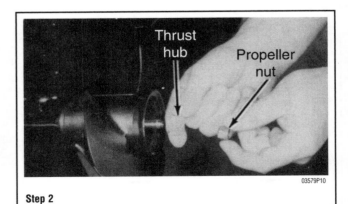

Step 2

LOWER UNIT 8-5

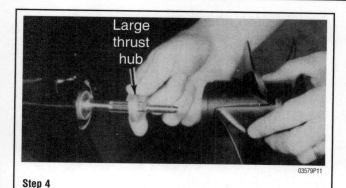
Step 4

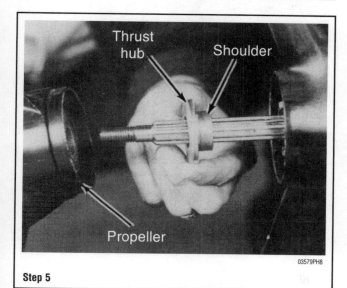
Step 5

7. Slide the propeller onto the propeller shaft until it seats against the thrust hub.
8. Install the flat washer or small thrust hub on the propeller shaft, and then thread the propeller nut onto the propeller shaft.
9. Place a block of wood between the anti-cavitation plate and a blade of the propeller.
10. Tighten the propeller nut on the 6–15 hp models to 70 inch lbs. (8Nm) and on the 20–30 hp models to 120 inch lbs. (13Nm).

40–275 HP

♦ See accompanying illustrations

1. Remove the propeller nut by first placing a block of wood between one of the propeller blades and the anti-cavitation plate to prevent the propeller from turning, and then remove the nut.
2. Remove the splined washer.
3. Remove the outer thrust hub from the propeller shaft. If the thrust hub is stubborn and refuses to budge, use two padded pry bars on opposite sides of the hub and work the hub loose. Take care not to damage the lower unit. Remove the propeller.

To install:

✶✶ WARNING

An outboard powerhead may start very easily. Therefore, anytime the propeller is to be removed or installed check to be sure the key switch is in the OFF position and the spark plug high tension leads are disconnected from the spark plugs and the electrical leads are disconnected at the battery terminals.

4. Position the thrust hub over the propeller with the shoulder side entered into the propeller.

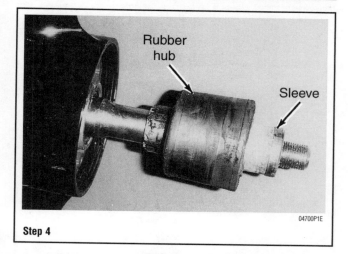
Step 4

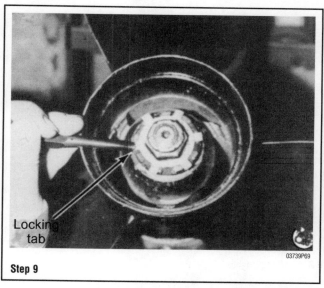
Step 9

5. As an aid to removing the propeller the next time, apply a liberal coating of Perfect Seal or equivalent waterproof lubricant to the propeller shaft splines.
6. Position the propeller on the propeller shaft, and then slide it up against the thrust hub taper. Slide a second splined washer onto the propeller shaft and engage the splines in the washer with the splines of the propeller shaft. Slide the tab washer onto the propeller shaft with the nut recess toward the splined washer.
7. Start the propeller nut onto the propeller shaft threads.
8. Align the tabs on the tab washer with the recessed areas of the splined washer. Place a block of wood between a propeller blade and the anti-cavitation plate to prevent the propeller shaft from rotating while the nut is being tightened. Tighten the propeller nut to a torque value of 55 ft. lbs. (75Nm).
9. Use a punch and hammer to bend the tabs of the tab washer into the recesses of the splined washer.
10. Remove the block of wood.
11. Connect the spark plug high tension leads to the spark plugs and the electrical lead to the battery terminal.

Lower Unit

REMOVAL & INSTALLATION

2.5 HP

♦ See accompanying illustrations

1. Disconnect and ground the spark plug lead to the powerhead to prevent the accidental starting of the powerhead.

8-6 LOWER UNIT

Step 3

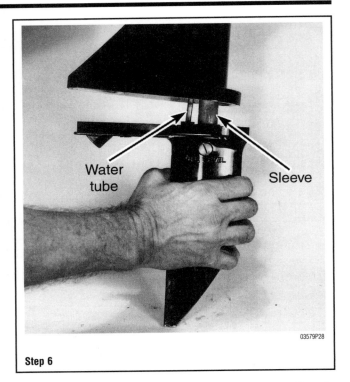

Step 6

Step 5

The upper rectangular driveshaft and sleeve, together with the water tube, will usually remain in the mid-section when the two units are separated. The lower driveshaft will remain in the lower gear housing.

To install:

7. Apply a coating of multi-purpose lubricant onto the outside diameter of the long tube. Insert the long tube into the mid-section housing.

8. Apply a coating of multi-purpose lubricant to the outside diameter of the rectangular driveshaft. Insert the driveshaft into the tube.

9. Apply a thin coating of multi-purpose lubricant to the inside diameter of the seals in the lower gear housing. Position a new gasket in place on the lower gear housing.

10. Begin to bring the midsection housing and the lower gear housing together. As the two units come closer, rotate the propeller shaft slightly to index the upper end of the lower driveshaft with the upper rectangular driveshaft. At the same time, feed the water tube into the water tube seal. Push the lower gear housing and the mid-section together.

11. Secure the lower gear housing and the mid-section together with the two bolts, lockwasher, and regular washer. The lockwasher should be installed between the bolt head and the regular washer.

12. Tighten the bolts alternately and evenly to a torque value of 25 inch lbs. (3 Nm).

3–5 HP

♦ See accompanying illustrations

This section covers the 2.5 and 3.3 hp models which have only a forward gear and neutral position. This section also covers all 4 and 5 hp models with forward, neutral and reverse gear shift capability. Minor differences between these units will be identified as the work progresses throughout the section.

1. Disconnect the high tension spark plug lead and remove the spark plug before working on the lower unit.

2. Position a suitable clean container under the lower unit. Remove the fill screw on the bottom of the lower unit, and then the vent screw. The vent screw must be removed to allow air to enter the lower unit behind the lubricant.

3. Allow the gear lubricant to drain into the container. As the lubricant drains, catch some with your fingers from time-to-time and rub it between your thumb and finger to determine if there are any metal particles present. Examine the fill plug. A small magnet imbedded in the end of the plug will pickup any metal particles. If metal is detected in the lubricant, the unit must be completely disassembled, inspected, and the damaged parts replaced.

4. Check the color of the lubricant as it drains. A whitish or creamy color indicates the presence of water in the lubricant. Check the container for signs of water separation from the lubricant. The presence of any water in the lubricant is bad news.

2. Close the fuel supply shut off valve.

3. Position a suitable container under the lower unit, and then remove the oil screw and the oil level screw. Allow the gear lubricant to drain into the container. As the lubricant drains, catch some with your fingers from time to time, and rub it between your thumb and finger to determine if any metal particles are present. If metal is detected in the lubricant, the unit must be completely disassembled, inspected, and the damaged parts replaced. Check the color of the lubricant as it drains. A whitish or creamy color indicates the presence of water in the lubricant. Check the drain pan for signs of water separation from the lubricant. The presence of any water in the gear lubricant is bad news. The unit must be completely disassembled, inspected, the cause of the problem determined, and then corrected.

4. Position the outboard in the full up position.

5. Remove the two bolts securing the lower unit gear housing to the mid-section. One bolt is located on the front side of the mid-section and the other goes into the mid-section from the underneath side of the lower gear housing, as shown.

6. Separate the lower gear housing from the mid-section. It may be necessary to tap the seam around the lower gear housing with a soft-head mallet to jar it loose.

LOWER UNIT 8-7

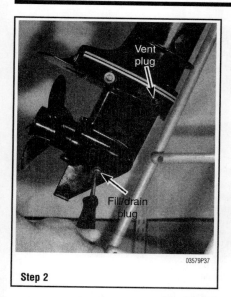
Step 2

Step 3

Step 6

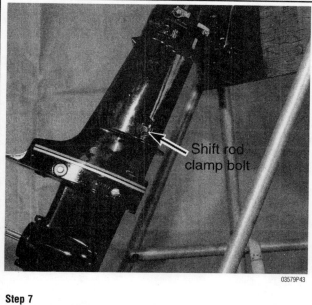
Step 7

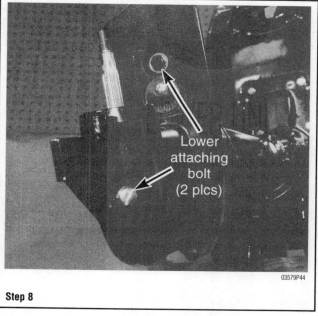
Step 8

The unit must be completely disassembled; inspected; the cause of the problem determined; and then corrected.

※※ WARNING

Check to be sure the spark plug has been removed; the run/stop switch is in the stop position and the battery has been disconnected to prevent any possibility of the powerhead starting during propeller removal.

5. Rotate and swing the lower end unit in the full up tilt position.
6. Remove the rubber access plug on the starboard side of the upper gear housing. If servicing a 2.5 or 3.3 hp unit place the gear shift in the neutral position. If servicing a model 4 or 5hp unit place the gear shift in the reverse gear position.
7. Loosen, but do not remove the bolt for the shift rod clamp. Loosen the bolt only enough to allow the shift shaft to slide free of the clamp when the lower unit is removed from the exhaust housing.
8. Remove the two attaching bolts and washers securing the lower unit to the exhaust housing. Pull down on the lower unit gear housing and separate the lower unit from the exhaust housing. Guide the shift rod and driveshaft out of the exhaust housing as the lower unit is lowered.

To install:

9. Swing the exhaust housing outward until the tilt lock lever can be actuated, and then engage the tilt lock.
10. Both the lower unit and the shift lever must be in matching positions when assembling the lower unit. On the 4 and 5 hp models with forward, neutral & reverse: Move the shift lever to the reverse gear position. On the 2.5 and 3.3 hp models with forward/neutral only: Move the shift lever to the neutral position. This will enable the upper shift shaft to align with the clamp on the lower shift shaft.
11. Rotate the propeller shaft while pushing down on the shift shaft to shift the lower unit. When in reverse, the propeller shaft will not rotate more than a few degrees in either direction. When in neutral position the propeller will move freely in either direction.
12. Coat the driveshaft splines with a light coating of multi-purpose lubricant. Take care not to use an excessive amount of lubricant on the driveshaft splines on any Mercury/Mariner Outboard unit.

➡**An excessive amount of lubricant on top of the driveshaft to crankshaft splines will be trapped in the clearance space. This trapped lubricant will not allow the driveshaft to fully engage with the crankshaft.**

13. Insert the driveshaft into the exhaust housing and at the same time align the water tube with the water pump cover outlet.

8-8 LOWER UNIT

14. Feed the lower shift rod into the coupler on the upper shift rod. Maintain the lower unit mating surface parallel with the exhaust housing mating surface.

15. Push the lower unit toward the exhaust housing and at the same time rotate the flywheel to permit the crankshaft splines to index with the driveshaft splines.

16. Secure the lower unit in place with the two attaching bolts and washers. Tighten the bolts alternately to a torque value of 70 inch lbs. (8Nm).

17. Tighten the shift shaft coupler bolt securely. Install the plug into the intermediate housing to cover the coupler bolt. Check the complete work for proper shifting.

6–25 HP

♦ See accompanying illustrations

1. Disconnect and ground the spark plug leads to the powerhead to prevent accidental engine start.

2. Rotate the outboard unit to the full up position and engage the tilt lock pin.

3. Place the gear shift in the forward gear position.

4. Unsnap the shift shaft retaining clip, inside the cowling next to the carburetor mounting bolts.

Rotate the clip 90°clockwise and pull the clip out of the shift shaft coupling.

5. Loosen the Phillips head screw securing the reverse lock actuator halves to the shift shaft. Separate the two halves of the reverse lock actuator and remove them from the shift shaft.

6. Remove the attachment bolts and washers securing the lower unit to the exhaust housing.

7. Pull down on the lower unit gear housing and separate the lower unit from the exhaust housing.

8. Guide the shift rod and driveshaft out of the exhaust housing as the lower unit is removed.

To install:

9. Swing the exhaust housing outward until the tilt lock lever can be actuated, and then engage the tilt lock.

10. Place the shift lever on the engine in forward position and the lower unit shift lever in neutral position.

11. Measure the length of the shift shaft as shown in the illustration. This distance will vary depending on the driveshaft length. If the distance is not as indicated, turn the shift shaft clockwise to shorten or counterclockwise to increase the shift shaft length to obtain the correct measurement.

12. Rotate the propeller shaft while pulling up on the shift shaft. When the lower unit is in forward gear, the propeller shaft will rotate clockwise.

13. Position the driveshaft and shift shaft into the exhaust housing. Guide the two shafts up into the housing as the lower unit is lifted to meet the exhaust housing. Maintain the lower unit mating surface parallel with the exhaust housing mating surface.

14. As the lower unit approaches closer to the exhaust housing, align the water tube with the water pump tube guide.

15. Rotate the flywheel to permit the crankshaft splines to index with the driveshaft splines.

16. Secure the lower unit in place with the three attaching bolts. Tighten the bolts to a torque value of 180 inch lbs. (20 Nm).

17. Guide the shift shaft yoke up through the lower cowling and into the shifter. Insert the shift shaft retaining clip through the yoke. Rotate the clip 90°counterclockwise and lock the clip to the shifter.

18. Place the shifter into neutral position and install the two halves of the reverse lock actuator onto the shift shaft. Slide the reverse lock actuator up to the reverse lock lever and securely tighten the Phillips head screw.

19. On the 20 and 25 hp models, install a new O-ring into the groove on the driveshaft just below the splines. Lubricate the O-ring with Multi-Purpose lubricant. Insert a flat washer on the end of the shift shaft.

20. Swing the exhaust housing outward until the tilt lock lever can be actuated, and then engage the tilt lock.

21. Place the shift lever on the engine in forward position and the lower unit shift lever in forward position.

22. Check to be sure the water tube inside the driveshaft housing is seated in the water tube seal on the adapter plate.

23. Position the driveshaft and shift shaft into the exhaust housing. Guide the two shafts up into the housing while lifting the lower unit. Maintain the lower unit mating surface parallel with the exhaust housing mating surface.

24. As the lower unit approaches closer to the exhaust housing, guide the shift shaft through the reverse lock loop. Align the water tube with the seal in the water pump.

25. Rotate the flywheel, if needed, to permit the crankshaft splines to index with the driveshaft splines.

Apply a small amount of Loctite®Type "A" or equivalent to the threads of the lower unit attaching bolts. Secure the lower unit in place with the four attaching bolts. Tighten the four bolts to a torque value of 25 ft. lbs. (33.9 Nm).

Insert the shift shaft coupler onto the shift shaft actuator rod and insert the pin through the rod and coupler.

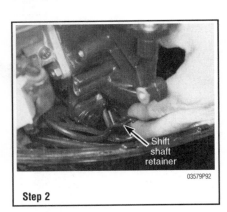

Step 2

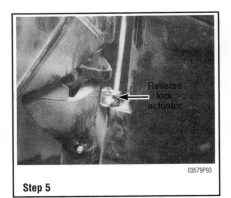

Step 5

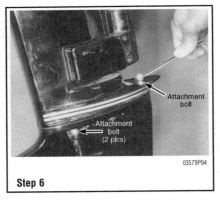

Step 6

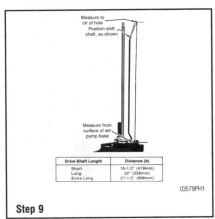

Step 9

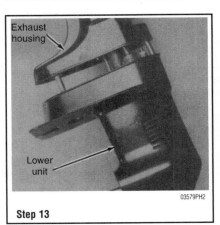

Step 13

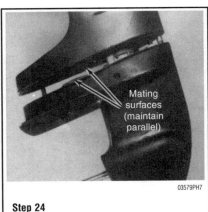

Step 24

30–125 HP

1. Disconnect and ground the spark plug leads to the powerhead to prevent accidental starting.
2. Tilt the outboard to the full up position.
3. Shift the gearcase into the neutral position.
4. Loosen the jamnut above the shift coupler. Then turn the shift rod coupler until the upper and lower shift rods are separated.
5. Mark the trim tab position and remove the trim tab retaining screw.
6. Remove the screws and washers securing the gearcase to the driveshaft housing.
7. Pull the gearcase straight down and away from the driveshaft housing.
8. Place the gearcase in a suitable holding fixture or a clean work area.
9. If the water tube guide and seal remained on the water tube in the driveshaft housing, remove the water tube guide and seal from the water tube. Inspect the guide and seal and for any wear or damage and replace if necessary.

To install:

10. Make sure that the water tube guide and seal are securely attached to the water pump housing. If the guide and seal are loose, secure them to the water pump housing using Loctite®405 sealant or equivalent.

➡ **Do not apply excess lubricant to the top of the driveshaft. Excess lubricant between the top of the driveshaft and the crankshaft can prevent the driveshaft from seating completely in the crankshaft.**

11. Clean and inspect the driveshaft splines and then coat the splines with Quicksilver 2-4-C multi-purpose lubricant or equivalent. Make sure to cat the inside of the water tube seal at the same time.
12. Shift the gearcase in the neutral position.
13. Position the gearcase under the driveshaft housing. Align the water tube into the water pump and the driveshaft splines with the crankshaft splines.

✲✲ WARNING

Do not rotate the flywheel counterclockwise to help align the splines or damage to the water pump impeller can occur.

14. Push the gearcase towards the driveshaft housing, rotating the flywheel clockwise as needed to align the splines on the driveshaft and crankshaft.
15. Make sure the water tube is seated in the water pump guide and seal, then push the gearcase against the driveshaft housing. Install the screws and washers and tighten to 40 ft. lbs. (54.2Nm).
16. Align the upper and lower shift rods. Reconnect the shift rods by turning the shift rod coupler.
17. Adjust the shift linkage.
18. Reinstall the propeller.
19. Install the trim tab and secure it with the bolt and washer. Make sure the previously made alignment marks line up. Tighten the bolt to 15 ft. lbs. (20Nm).
20. Check the lubricant level and refill as needed.
21. Release the tilt lock and return the engine to the normal operating position.
22. Reconnect the spark plug leads.

135–275 HP

♦ See accompanying illustrations

➡ **The E-Z Shift system is installed on 135–200 hp powerheads including the Magnum and XRi series. The Cam Shift-II system is installed on lower units matched with the 150XR4, Magnum II and 150XR6, Magnum III powerheads since about 1989. With the lower unit still attached to the powerhead, and with the unit in reverse gear, rotate the propeller shaft counterclockwise. If the shaft does not ratchet, the unit has a Cam-Shift. If the propeller shaft "ratchets" when turned counterclockwise, the lower unit has the E-Z Shift. With the lower unit separated from the powerhead, and with the unit in neutral gear, rotate the shift shaft clockwise and counterclockwise. If the shift shaft will rotate a full 360°, the unit has the Cam-Shift. If the shift shaft will only rotate about 30° in either direction, the unit has the E-Z Shift.**

1. With the lower unit separated from the powerhead, and with the unit in neutral gear, rotate the shift shaft clockwise and counterclockwise. If the shift shaft will rotate a full 360°, the unit has the Cam-Shift. If the shift shaft will only rotate about 30° in either direction, the unit has the E-Z Shift.
2. On the Cam Shift models only, check to be sure the lower unit is in forward gear. On the E-Z Shift models, check to be sure the lower unit is in the neutral position.
3. Scribe a line between the trim tab and the anti-cavitation plate. This mark will ensure the trim tab will be installed back at the original angle.

4. Remove the plastic cap from the rear edge of the exhaust housing. Insert the proper size wrench into the hole and remove the trim tab adjusting bolt and trim tab from the lower unit.
5. Remove the bolt from the recess where the trim tab was mounted. Remove the two locknuts and washers from the bottom middle section of the anti-cavitation plate. Remove the locknut and washer on the front end of the lower unit.
6. Loosen, but do not remove, the ⅝ in. locknuts on each side of the exhaust housing. Loosen the nuts as far as the stud threads will allow. Separate the lower unit from the exhaust housing as far as the studs and nuts will permit. Hold the lower unit from falling and at the same time remove the two nuts.
7. Remove the lower unit from the intermediate housing.
8. On the 275 hp models, remove the center bolt from the anode, and then the anode. Remove the nut and washer from inside the anode plate cavity. Remove the two bolts and washers from the bottom of the anti-cavitation plate. On the forward end of the lower unit housing, remove the bolt and washer securing the lower unit to the intermediate housing. Some units may have a stud and nut in this same location.

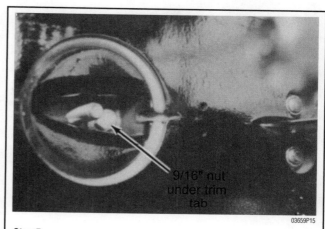

Step 5

Step 8

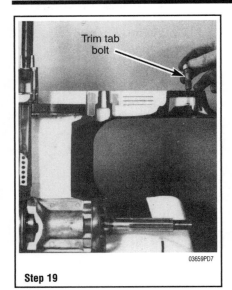
Step 19

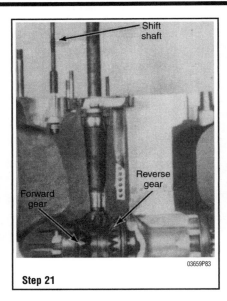
Step 21

Step 24

9. Slowly loosen, but do not remove, the two mounting locknuts from each side of the lower unit. Loosen all four nuts a little at-a-time to prevent damage to the drive shaft housing. As the nuts are loosened, allow the lower unit to separate from the intermediate housing. After the lower unit is moved downward a short distance, disconnect the speedometer hose from the lower unit fitting.

10. Hold the lower unit, and then remove all four nuts. Have an assistant help support and lift the lower unit because these lower units weigh approximately 90 pounds.

11. Separate the lower unit from the intermediate housing.

12. Slowly lower the unit and at the same time, guide the driveshaft out of the intermediate housing.

13. Place the lower unit in a suitable holding fixture or work area.

To install:

14. Always fill the lower unit with lubricant and check for leaks before installing the unit to the driveshaft housing.

15. Take time to remove any old gasket material from the fill and vent recesses and from the screws.

16. Place the lower unit in an upright vertical position. Fill the lower unit with Super-Duty Lubricant, or equivalent, through the fill opening at the bottom of the unit. Never add lubricant to the lower unit without first removing the vent screw and having the unit in its normal operating position—vertical. Failure to remove the vent screw will result in air becoming trapped within the lower unit. Trapped air will not allow the proper amount of lubricant to be added.

17. Continue filling slowly until the lubricant begins to escape from the vent opening with no air bubbles visible.

18. Use a new gasket and install the vent screw. Slide a new gasket onto the fill screw. Remove the lubricant tube and quickly install the fill screw.

19. Check the lower unit for leaks.

20. Install the trim tab bolt into the rear hole in the rear of the lower unit-to-exhaust housing machined surface.

21. Install the water tube guide into the water pump cover above the water tube seal.

➥ **Check the lower unit to be sure it is in the proper gear. Cam-Shift unit in forward gear and the E-Z Shift unit in neutral.**

22. Move the upper shift shaft into the proper gear position also.

23. Align the guide block to position the rear edge of the pin with the front edge of the exhaust cover plate.

24. On Cam-Shift units, the forward end of the shift block must extend 1/8 in. (3.2mm) past the front of the rail.

25. On the 225 hp model, install the splined nylon tube onto the upper end of the driveshaft. Slide the seal onto the driveshaft with the splined portion toward the nylon tube.

26. Tilt the powerhead and intermediate housing outward, and then engage the tilt lever. Check to be sure the water tube is in position in the exhaust extension plate. Check to be sure the reverse lock push rod and the shift shaft guide block are in the correct position. Apply a light coating of multipurpose lubricant to the driveshaft splines and to the shift shaft splines.

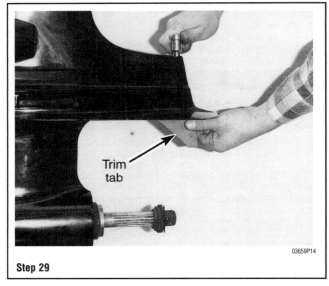
Step 29

✱✱ CAUTION

An excessive amount of lubricant on top of the driveshaft to crankshaft splines will be trapped in the clearance space. This trapped lubricant will not allow the driveshaft to fully engage with the crankshaft. As a result, when the lower unit nuts are tightened, a load will be placed on the driveshaft/crankshaft and will cause damage to either the powerhead or the lower unit or both. Therefore, any lubricant must be cleaned from the top of the driveshaft.

Secure the services of an assistant because of the lower unit weight and because several things must be done—items guided and aligned, all at the same time.

27. Position the lower unit under the exhaust housing with the machined surfaces parallel. With the help of an assistant, raise the lower unit and guide the driveshaft into the exhaust housing. Almost at the same time guide the water tube into the water guide tube with your fingers or with a screwdriver. Route the speedometer hose up through the shift shaft opening.

28. Raise the lower unit further with the housing sliding up over the studs. Raise the unit as far as possible. Maintain pressure on the lower unit against the exhaust housing and at the same time slowly rotate the powerhead flywheel clockwise to per-

LOWER UNIT 8-11

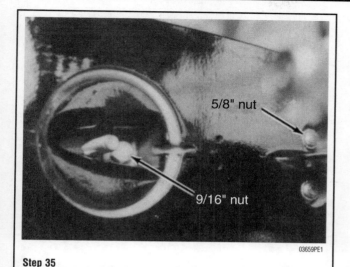

Step 35

32. After the shift shafts are aligned and the driveshaft splines have indexed properly with the crankshaft splines, close the gap between the lower unit and the exhaust housing until the two machined surfaces make contact.
33. Hold the two units in contact with one hand and check the shift operation with the other hand. Shift into forward gear; the propeller shaft should not rotate counterclockwise. Shift into neutral. The propeller shaft should be free to rotate in either direction. Shift into reverse gear. The propeller shaft should be free to turn in either direction only about⅓ turn.
34. If the lower unit fails any of the shift tests just described, the upper and lower shift shafts are not aligned properly. Remove the lower unit and repeat this step.
35. After the shift operations are verified as correct, start the bolt at the rear of the lower unit inside the trim tab recess. Do not tighten the bolt at this time.
36. Alternately and evenly tighten the two nuts started earlier in this step. Slide the washers, and then start the nuts onto the studs at the bottom center of the anti-cavitation plate. Install the washer and nut onto the stud at the leading edge of the intermediate housing. Tighten the bolt started in the trim recess.
37. Tighten the ⅜ in.-16 nuts securing the lower unit to the intermediate housing to a torque value of 55 ft. lbs. (75Nm).
38. Tighten the 7/16 in.-20 nuts or bolts securing the lower unit to the intermediate housing to a torque value of 65 ft. lbs. (88Nm).
39. Tighten the trim tab bolt to a torque value 25 ft. lbs. (34Nm).
40. Install the trim tab in the same position from which it was removed. A scribe mark should have been made on the trim tab and a matching mark on the anti-cavitation plate before the trim tab was removed.
41. On the 225 and 275 hp models, apply a thin coating of Loctite Grade "A" or equivalent to the threads of the anode retaining bolt. Install the anode plate, and then tighten the bolt to a torque value of 15 ft. lbs. (20.3Nm).

➡ **The trim tab should be positioned to enable the operator to handle the boat with equal ease to starboard and port at normal cruising speed. If the boat seems to turn more easily to starboard, loosen the socket head screw and move the trim tab trailing edge to the right. Move the trailing edge of the trim tab to the left if the boat tends to turn more easily to port.**

42. Snap the plastic cap into the trim tab bolt opening at the rear edge of the lower unit. Shift the unit into forward gear, release the tilt lock lever and lower the outboard to the normal operating position.

mit the driveshaft splines index with the crankshaft splines. When the splines are aligned and index with each other, the two units may be moved closer together.

29. The 9/16 in. nut in the recess is covered by the trim tab. A reference mark inscribed on the trim tab and on the anti-cavitation plate will allow the trim tab to be installed back at the same angle from which it was removed.
30. Slide the flat washers onto the studs on both sides of the exhaust housing. Start the nuts onto the studs and bring them up finger tight. Check the upper shift shaft entry into the lower shift shaft by observing the position of the shift shaft at the exhaust extension plate bushing. If the upper and lower shift shafts are not aligned, the upper shaft will be pushed upward as the lower unit is tightened against the exhaust housing.
31. To correct the entry of the upper shift shaft into the lower shift shaft, place a punch against the upper shift shaft, and then strike the punch with a hammer to help align the shaft. Avoid excessive force. If necessary, realign the shafts.

JET DRIVE

Description and Operation

The jet drive unit is designed to permit boating in areas prohibited to a boat equipped with a conventional propeller drive system. The housing of the jet drive barely extends below the hull of the boat allowing passage in ankle deep water, white water rapids and over sand bars or in shoal water which would foul a propeller drive.

The jet drive provides reliable propulsion with a minimum of moving parts. Simply stated, water is drawn into the unit through an intake grille by an impeller driven by a driveshaft off the crankshaft of the powerhead. The water is immediately expelled under pressure through an outlet nozzle directed away from the stern of the boat.

As the speed of the boat increases and reaches planing speed, the jet drive discharges water freely into the air and only the intake grille makes contact with the water.

The jet drive is provided with a gate arrangement and linkage to permit the boat to be operated in reverse. When the gate is moved downward over the exhaust nozzle, the pressure stream is reversed by the gate and the boat moves sternward. Conventional controls are used for powerhead speed, movement of the boat, shifting and power trim and tilt.

Model Identification and Serial Numbers

♦ See Figure 48

A model letter identification is stamped on the rear, port side of the jet drive housing. A serial number for the unit is stamped on the starboard side of the jet drive housing, as indicated in the accompanying illustration.

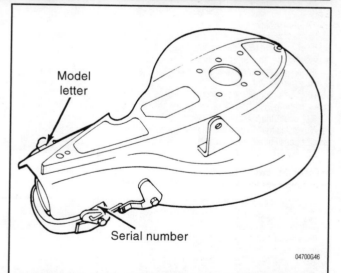

Fig. 48 The model letter designation and the serial numbers are embossed on the jet drive housing

8-12 LOWER UNIT

These numbers reflect the specific size of jet drive unit and to which outboard they are attached. For the most part, jet drive units are identical in design, function and operation. Differences lie in size and securing hardware.

Jet Drive Assembly

REMOVAL & INSTALLATION

♦ See accompanying illustrations

1. Remove the two bolts and retainer securing the shift cable to the shift cable support bracket.
2. Remove the locknut, bolt and washer securing the shift cable to the shift arm. Try not to disturb the length of the cable.
3. Remove the six bolts securing the intake grille to the jet casing.
4. Ease the intake grille from the jet drive housing.
5. Pry the tab or tabs of the tabbed washer away from the nut to allow the nut to be removed.
6. Loosen and then remove the nut.
7. Remove the tabbed washer and spacers. Make a careful count of the spacers behind the washer. If the unit is relatively new, there could be as many as eight spacers stacked together. If less than eight spacers are removed from behind the washer, the others will be found behind the jet impeller, which is removed in the following step. A total of eight spacers will be found.
8. Remove the jet impeller from the shaft. If the impeller is frozen to the shaft, obtain a block of wood and a hammer. Tap the impeller in a clockwise direction to release the shear key.
9. Slide the nylon sleeve and shear key free of the driveshaft and any spacers found behind the impeller. Make a note of the number of spacers at both locations—behind the impeller and on top of the impeller, under the nut and tabbed washer.
10. One external bolt and four internal bolts are used to secure the jet drive to the intermediate housing. The external bolt is located at the aft end of the anti-cavitation plate.

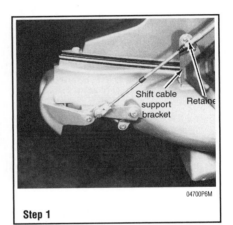
Step 1

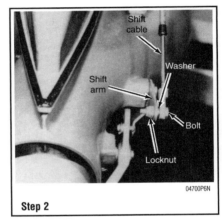
Step 2

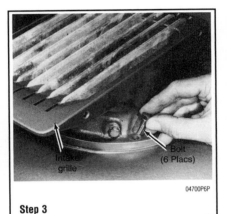
Step 3

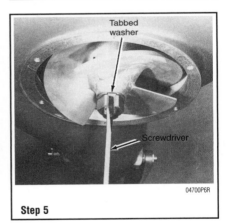
Step 5

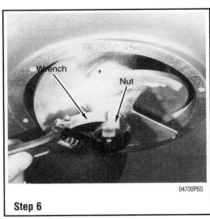
Step 6

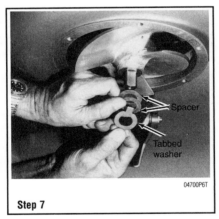
Step 7

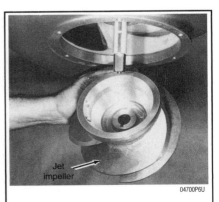
Step 8

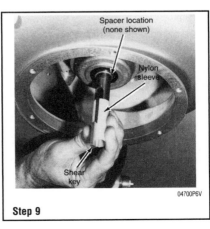
Step 9

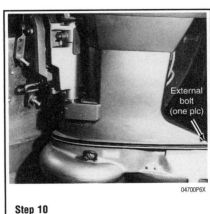
Step 10

LOWER UNIT 8-13

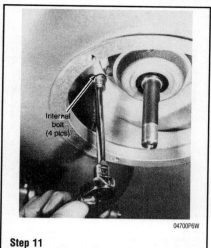
Step 11

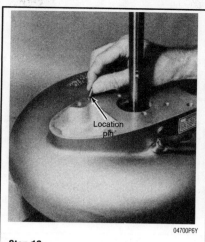
Step 12

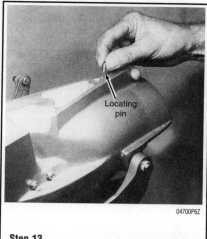
Step 13

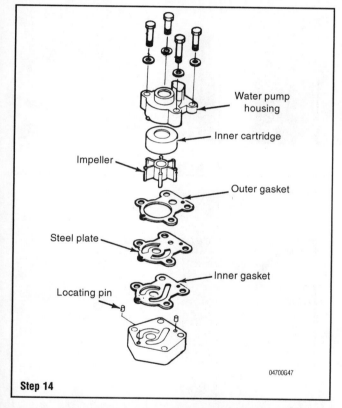
Step 14

11. The four internal bolts are located inside the jet drive housing, as indicated in the accompanying illustration. Remove the five attaching bolts.

12. Lower the jet drive from the intermediate housing. Remove the locating pin from the forward starboard side (or center forward, depending on the model being serviced) of the upper jet housing.

13. Remove the locating pin from the aft end of the housing. This pin and the one removed in the previous step should be of identical size.

14. Remove the four bolts and washers from the water pump housing.

Pull the water pump housing, the inner cartridge and the water pump impeller, up and free of the driveshaft. Next, remove the outer gasket, the steel plate and the inner gasket.

15. Remove the two small locating pins and lift the aluminum spacer up and free of the drive shaft.

Remove the driveshaft and bearing assembly from the housing.

Remove the large thick adapter plate from the intermediate housing. This plate is secured with seven bolts and lock washers. Lower the adapter plate from the intermediate housing and remove the two small locating pins, one on the forward port side and another from the last aft hole in the adapter plate. Both pins are identical in size.

To install:

16. Install the other small locating pin into the forward starboard side (or center forward end, depending on the model being serviced).

17. Raise the jet drive unit up and align it with the intermediate housing, with the small pins indexed into matching holes in the adapter plate. Install the four internal bolts.

18. Install the one external bolt at the aft end of the anti-cavitation plate. Tighten all bolts to a torque value of 11 ft. lbs. (15Nm).

19. Place the required number of spacers up against the bearing housing. Slide the nylon sleeve over the driveshaft and insert the shear key into the slot of the nylon sleeve with the key resting against the flattened portion of the driveshaft.

20. Slide the jet impeller up onto the driveshaft, with the groove in the impeller collar indexing over the shear key.

21. Place the remaining spacers over the driveshaft.

22. Tighten the nut to a torque value of 17 ft. lbs. (23Nm). If neither of the two tabs on the tabbed washer aligns with the sides of the nut, remove the nut and

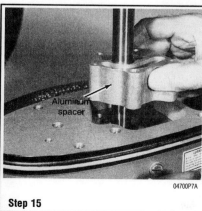
Step 15

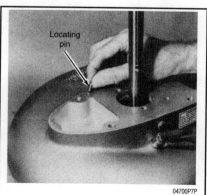
Step 16

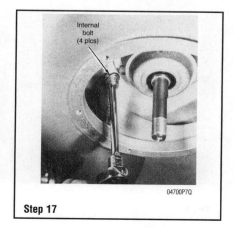
Step 17

8-14 LOWER UNIT

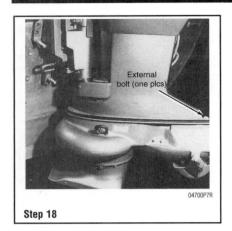
Step 18

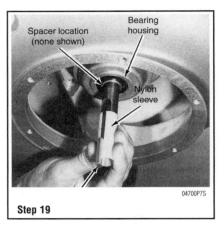
Step 19

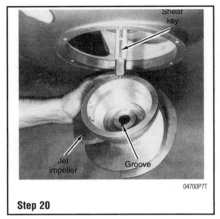
Step 20

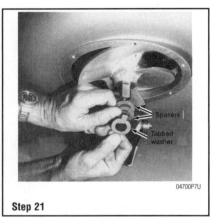
Step 21

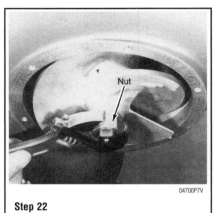
Step 22

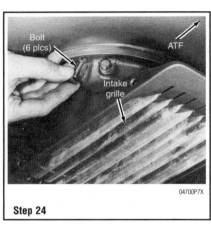
Step 24

washer. Invert the tabbed washer. Turning the washer over will change the tabs by approximately 15°. Install and tighten the nut to the required torque value. The tabbed washer is designed to align with the nut in one of the two positions described.

23. Bend the tabs up against the nut to prevent the nut from backing off and becoming loose.

24. Install the intake grille onto the jet drive housing with the slots facing aft. Install and tighten the six securing bolts. Tighten ¼ in. bolts to a torque value of 5 ft. lbs. (7Nm). Tighten 5/16 in. bolts to 11 ft. lbs. (15Nm).

25. Slide the bolt through the end of the shift cable, washer and into the shift arm. Install the locknut onto the bolt and tighten the bolt securely.

26. Install the shift cable against the shift cable support bracket and secure it in place with the two bolts.

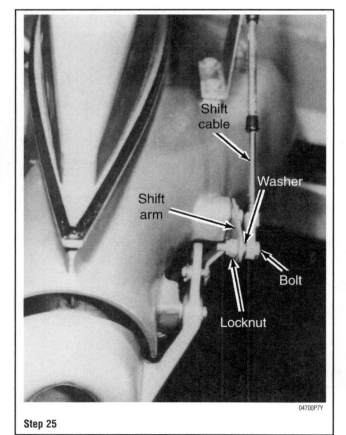
Step 25

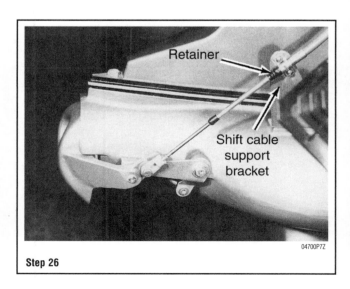
Step 26

LOWER UNIT 8-15

ADJUSTMENT

▶ See accompanying illustrations

1. Move the shift lever downward into the forward position. The leaf spring should snap over on top of the lever to lock it in position.
2. Remove the locknut, washer and bolt from the threaded end of the shift cable
3. Move the shift lever downward into the forward position. The leaf spring should snap over on top of the lever to lock it in position.
4. Remove the locknut, washer and bolt from the threaded end of the shift cable. Push the reverse gate firmly against the rubber pad on the underside of the jet drive housing.
5. Check to be sure the link between the reverse gate and the shift arm is hooked into the LOWER hole on the gate.
6. Hold the shift arm up until the link rod and shift arm axis form an imaginary straight line, as indicated in the accompanying illustration. Adjust the length of the shift cable by rotating the threaded end, until the cable can be installed back onto the shift arm without disturbing the imaginary line. Pass the nut through the cable end, washer and shift arm. Install and tighten the locknut.
7. Loosen, but do not remove the locknut on the neutral stop lever. Check to be sure the lever will slide up and down along the slot in the shift lever bracket
8. Start the powerhead and allow it to operate only at idle speed. With the neutral stop lever in the down position, move the shift lever until the jet stream forces on the gate are balanced. Balanced means the water discharged is divided in both directions and the boat moves neither forward nor sternward. The gate is then in the neutral position with the powerhead at idle speed
9. Move the neutral stop lever up against the shift lever until the stop lever barely makes contact with the shift lever. Tighten the locknut to maintain this new adjusted position. Shut down the powerhead

In the forward position, the reverse gate is neatly tucked underneath and clear of the exhaust jet stream. In the reverse position, the gate swings up and blocks the jet stream deflecting the water in a forward direction under the jet

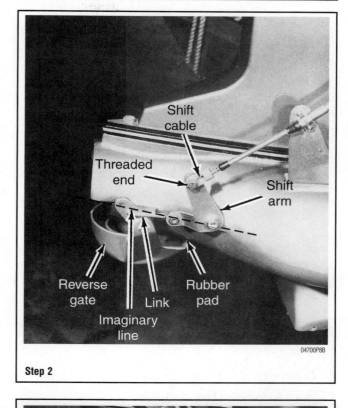
Step 2

Step 1

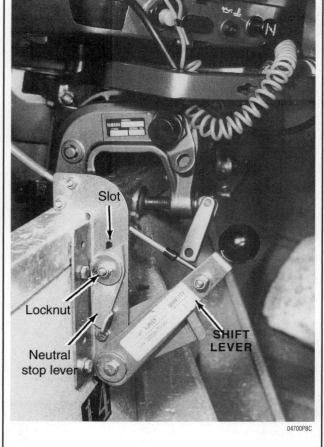
Step 7

8-16 LOWER UNIT

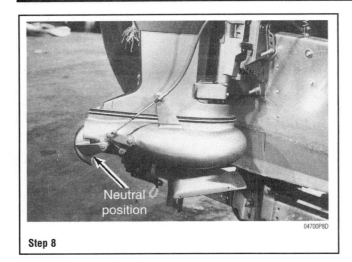
Step 8

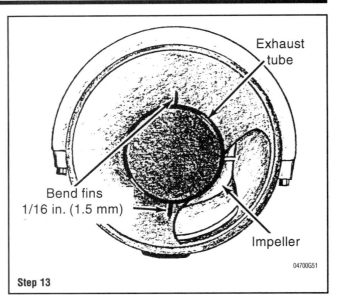
Step 13

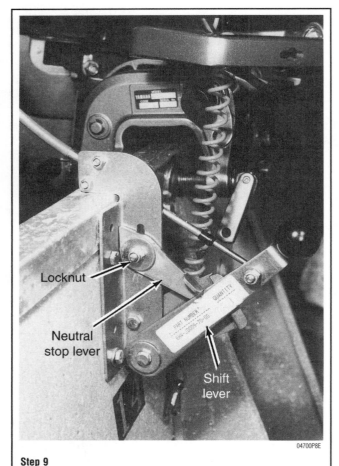
Step 9

housing to move the boat sternward. In the neutral position, the gate assumes a happy medium—a balance between forward and reverse when the powerhead is operating at idle speed. Actually, the gate is deflecting some water to prevent the boat from moving forward, but not enough volume to move the boat sternward.

❈❈ WARNING

The gate must be properly adjusted for safety of boat and passengers. Improper adjustment could cause the gate to swing up to the reverse position while the boat is moving forward causing serious injury to boat or passengers.

10. Loosen, but do not remove the locknut on the neutral stop lever. Check to be sure the lever will slide up and down along the slot in the shift lever bracket.

➡ **The following procedure must be performed with the boat and jet drive in a body of water. Only with the boat in the water can a proper jet stream be applied against the gate for adjustment purposes.**

❈❈ CAUTION

Water must circulate through the lower unit to the powerhead anytime the powerhead is operating to prevent damage to the water pump in the lower unit. Just five seconds without water will damage the water pump impeller.

11. Start the powerhead and allow it to operate only at idle speed. With the neutral stop lever in the down position, move the shift lever until the jet stream forces on the gate are balanced. Balanced means the water discharged is divided in both directions and the boat moves neither forward nor sternward. The gate is then in the neutral position with the powerhead at idle speed.

12. Move the neutral stop lever up against the shift lever until the stop lever barely makes contact with the shift lever. Tighten the locknut to maintain this new adjusted position. Shut down the powerhead.

➡ **The reverse gate may not swing to the full up position in reverse gear after the previous steps have been performed. Do not be concerned. This condition is acceptable, because water pressure in reverse will close the gate fully under normal operation.**

13. During operation, if the boat tends to pull to port or starboard, the flow fins may be adjusted to correct the condition. These fins are located at the top and bottom of the exhaust tube.

14. If the boat tends to pull to starboard, bend the trailing edge of each fin approximately 1/16 in. (1.5mm) toward the starboard side of the jet drive. Naturally, if the boat tends to pull to port, bend the fins toward the port side.

DISASSEMBLY

1. Remove the locating pin from the forward starboard side (or center forward, depending on the model being serviced) of the upper jet housing.

➡ **There will be a total of six locating pins to be removed in the following steps. Make careful note of the size and location of each when they are removed, as an assist during assembling.**

2. Remove the locating pin from the aft end of the housing. This pin and the one removed in the previous step should be of identical size.

3. Remove the four bolts and washers from the water pump housing.

Pull the water pump housing, the inner cartridge and the water pump impeller, up and free of the driveshaft. Remove the Woodruff key from its recess in the driveshaft. Next, remove the outer gasket, the steel plate and the inner gasket.

LOWER UNIT   8-17

4. Remove the two small locating pins and lift the aluminum spacer up and free of the driveshaft.
5. Remove the driveshaft and bearing assembly from the housing.

Remove the large thick adapter plate from the intermediate housing. This plate is secured with seven bolts and lock-washers. Lower the adapter plate from the intermediate housing and remove the two small locating pins, one on the forward port side and another from the last aft hole in the adapter plate. Both pins are identical size.

CLEANING AND INSPECTING

▶ See Figures 49, 50, 51, and 52

Wash all parts, except the driveshaft assembly, in solvent and blow them dry with compressed air. Rotate the bearing assembly on the driveshaft to inspect the bearings for rough spots, binding and signs of corrosion or damage.

1. Saturate a shop towel with solvent and wipe both extensions of the driveshaft.
2. Lightly wipe the exterior of the bearing assembly with the same shop towel. Do not allow solvent to enter the three lubricant passages of the bearing assembly. The best way to clean these passages is not with solvent—because any solvent remaining in the assembly after installation will continue to dissolve good useful lubricant and leave bearings and seals dry. This condition will cause bearings to fail through friction and seals to dry up and shrink—losing their sealing qualities.

The only way to clean and lubricate the bearing assembly is after installation to the jet drive—via the exterior lubrication fitting.

If the old lubricant emerging from the hose coupling is a dark, dirty, gray color, the seals have already broken down and water is attacking the bearings. If such is the case, it is recommended the entire driveshaft bearing assembly be taken to the dealer for service of the bearings and seals.

A complicated procedure must be followed to dismantle the bearing assembly including torching off the bearing housing. Naturally, excessive heat might ruin the seals and bearings. Therefore, the best recommendation is to leave this part of the service work to the experts at your local Honda dealership.

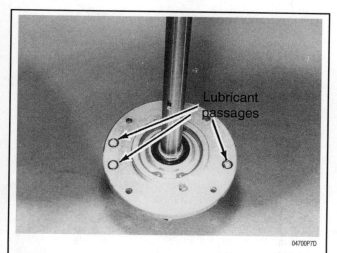

Fig. 49 Take extra precautions to prevent solvent from entering the lubrication passages

Fig. 51 Inspect the slats of the water intake grille for straightness. Straighten any bent slats, if possible. Use the utmost care when prying on any slat, as they tend to break if excessive force is applied. Replace the intake grille if a slat is lost, broken, or bent and cannot be repaired. The slats are spaced evenly and the distance between them is critical, to prevent large objects from passing through and becoming lodged between the jet impeller and the inside wall of the housing.

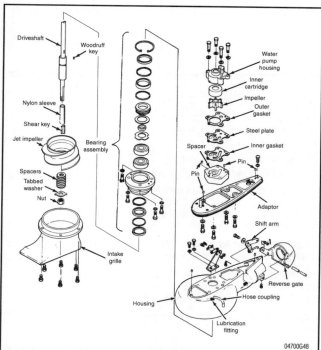

Fig. 50 Exploded view of a typical jet drive lower unit with major parts identified

Fig. 52 Excessive rounding of the jet impeller edges will reduce efficiency and performance. Therefore, the impeller should be inspected at regular intervals

3. Inspect the threads and splines on the driveshaft for wear, rounded edges, corrosion and damage.

4. Carefully check the driveshaft to verify the shaft is straight and true without any sign of damage.

5. Inspect the jet drive housing for nicks, dents, corrosion, or other signs of damage. Nicks may be removed with No. 120 and No. 180 emery cloth.

6. Inspect the gate and its pivot points. Check the swinging action to be sure it moves freely the entire distance of travel without binding.

7. The jet impeller is a precisely machined and dynamically balanced aluminum spiral. Observe the drilled recesses at exact locations to achieve this delicate balancing. Some of these drilled recesses are clearly shown in the accompanying illustration.

8. Excessive vibration of the jet drive may be attributed to an out-of-balance condition caused by the jet impeller being struck excessively by rocks, gravel or cavitation burn.

9. The term cavitation burn is a common expression used throughout the world among people working with pumps, impeller blades and forceful water movement.

10. Burns on the jet impeller blades are caused by cavitation air bubbles exploding with considerable force against the impeller blades. The edges of the blades may develop small dime size areas resembling a porous sponge, as the aluminum is actually eaten by the condition just described.

11. If rounding is detected, the impeller should be placed on a work bench and the edges restored to as sharp a condition as possible, using a file. Draw the file in only one direction. A back-and-forth motion will not produce a smooth edge. Take care not to nick the smooth surface of the jet impeller. Excessive nicking or pitting will create water turbulence and slow the flow of water through the pump.

12. Inspect the shear key. A slightly distorted key may be reused although some difficulty may be encountered in assembling the jet drive. A cracked shear key should be discarded and replaced with a new key.

13. Clean all water pump parts with solvent and then blow them dry with compressed air. Inspect the water pump housing for cracks and distortion, possibly caused from overheating. Inspect the steel plate, the thick aluminum spacer and the water pump cartridge for grooves and/or rough spots. If possible always install a new water pump impeller while the jet drive is disassembled. A new water pump impeller will ensure extended satisfactory service and give peace of mind to the owner. If the old water pump impeller must be returned to service, never install it in reverse of the original direction of rotation. Installation in reverse will cause premature impeller failure.

14. If installation of a new water pump impeller is not possible, check the sealing surfaces and be satisfied they are in good condition. Check the upper, lower and ends of the impeller vanes for grooves, cracking and wear. Check to be sure the indexing notch of the impeller hub is intact and will not allow the impeller to slip.

ASSEMBLY

♦ See accompanying illustrations

1. Place the driveshaft bearing assembly into the jet drive housing. Rotate the bearing assembly until all bolt holes align

2. Identify the two small locating pins used to index the large thick adapter plate

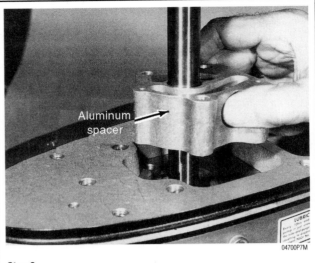

Step 6

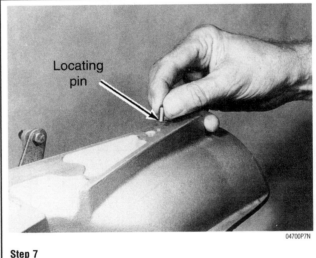

Step 7

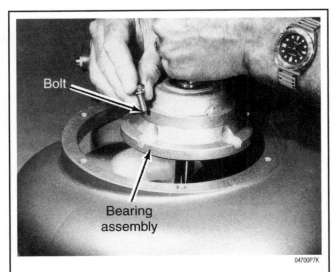

Step 1

to the intermediate housing. Insert one pin into the last hole aft on the topside of the plate. Insert the other pin into the hole forward toward the port side, as shown.

3. Lift the plate into place against the intermediate housing with the locating pins indexing with the holes in the intermediate housing. Secure the plate with the five (or seven) bolts.

➡On the five bolt model, one of the five bolts is shorter than the other four. Install the short bolt in the most aft location.

4. Tighten the long bolts to a torque value of 22 ft. lbs. (30Nm). Tighten the short bolt to a torque value of 11 ft. lbs. (15Nm).

5. Place the driveshaft bearing assembly into the jet drive housing. Rotate the bearing assembly until all bolt holes align. There is only one correct position.

➡If installing a new jet impeller, place all eight spacers at the lower or nut end of the impeller and skip the following step.

6. Place the aluminum spacer over the driveshaft with the two holes for the indexing pins facing upward.

7. Install one of the small locating pins into the aft end of the jet drive housing

➡The manufacturer recommends no sealant be used on either side of the water pump gaskets.

8. Slide the inner water pump gasket (the gasket with two curved openings) over the driveshaft. Position the gasket over the two locating pins. Slide the steel plate down over the driveshaft with the tangs on the plate facing downward and with the holes in the plate indexed over the two locating pins.

Lower Unit

Check to be sure the tangs on the plate fit into the two curved openings of the gasket beneath the plate. Now, slide the outer gasket (the gasket with the large center hole) over the driveshaft. Position the gasket over the two locating pins.

Fit the Woodruff key into the driveshaft. Just a dab of grease on the key will help to hold the key in place. Slide the water pump impeller over the driveshaft with the rubber membrane on the top side and the keyway in the impeller indexed over the Woodruff key. Take care not to damage the membrane. Coat the impeller blades with water resistant lubricant.

Install the insert cartridge, the inner plate and finally the water pump housing over the driveshaft. Rotate the insert cartridge counterclockwise over the impeller to tuck in the impeller vanes. Seat all parts over the two locating pins.

➡ **On some models, two different length bolts are used at this location.**

Tighten the four bolts to a torque value of 11 ft. lbs. (15Nm).

9. Install one of the small locating pins into the aft end of the jet drive housing.

SHIMMING

◆ See Figures 54 and 55

1. The clearance between the outer edge of the jet drive impeller and the water intake housing cone wall should be maintained at approximately 1/32 in. (0.8mm).

This distance can be visually checked by shining a flashlight up through the intake grille and estimating the distance between the impeller and the casing cone, as indicated in the accompanying illustrations. It is not humanly possible to accurately measure this clearance, but by observing closely and estimating the clearance, the results should be fairly accurate.

After continued use, the clearance will increase. The spacers previously removed are used to position the impeller along the driveshaft with a desired clearance of 1/32 in. (0.8mm) between the jet impeller and the housing wall.

2. Spacers are used depending on the model being serviced. When new, all spacers are located at the tapered (or nut) end of the impeller. As the clearance increases, the spacers are transferred from the tapered (nut) end and placed at the wide (intermediate housing) end of the jet impeller.

This procedure is best accomplished while the jet drive is removed from the intermediate housing.

Secure the driveshaft with the attaching hardware. Installation of the shear key and nylon sleeve is not vital to this procedure. Place the unit on a convenient work bench. Shine a flashlight through the intake grille into the housing cone and eyeball the clearance between the jet impeller and the cone wall, as indicated in the accompanying line drawing. Move spacers one-at-a-time from the tapered end to the wide end to obtain a satisfactory clearance. Dismantle the driveshaft and note the exact count of spacers at both ends of the bearing assembly. This count will be recalled later during assembly to properly install the jet impeller.

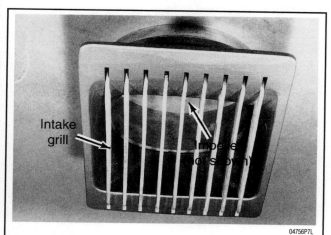

Fig. 54 The clearance between the outer edge of the jet drive impeller and the water intake housing cone wall should be maintained at approximately 1/32 in. (0.8mm)

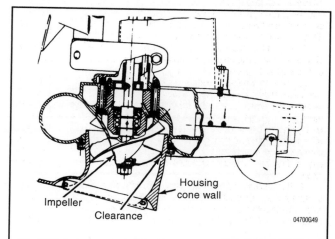

Fig. 55 Spacers are used depending on the model being serviced. When new, all spacers are located at the tapered (or nut) end of the impeller

Lower Unit Specifications

HP(Model)	Cyl.	Lubricant Capacity fl.oz (ml)	Gear Type	Clutch Type	Gear Ratio	Pinion Gear Height in. (mm)	Backlash Forward Gear	Backlash Reverse Gear
2.5 (non-shift)	1	2.5(74)	Spiral Bevel	Sliding Dog	1.85:1	—	①	①
3 (non-shift)	1	2.5(74)	Spiral Bevel	Sliding Dog	2.18:1	—	①	①
2.5-3.3 (shift)	1	2.5(74)	Spiral Bevel	Sliding Dog	2.18:1	—	①	①
4-5	1	6.6(195)	Spiral Bevel	Sliding Dog	2.15:1	—	①	①
6-15	2	6.8(200)	Spiral Bevel	Sliding Dog	2.00:1	Floating	①	①
20-25 (Big Foot)	2	7.8(230)	Spiral Bevel	Sliding Dog	2.42:1	—	①	①
20-25	2	8.8(260)	Spiral Bevel	Sliding Dog	2.25:1	—	①	①
30-40	2	14.9(440)	Spiral Bevel	Sliding Dog	2.00:1	—	①	①
40	4	12.5(370)	Spiral Bevel	Sliding Dog	2.00:1	.025(0.64)	.007-.010 (0.178-0.254)	—
40-50	3	14.9(440)	Spiral Bevel	Sliding Dog	1.83:1	.025(0.64)	—	—
60	3	11.5(340)	Spiral Bevel	Sliding Dog	1.64:1	.025(0.64)	0.013-0.019 (0.33-0.48)	—
60 (Big Foot)	3	22.5(655)	Spiral Bevel	Sliding Dog	2.3:1	.025(0.64)	0.012-0.019 (0.30-0.48)	—
75-90	3	22.5(655)	Spiral Bevel	Sliding Dog	2.3:1	.025(0.64)	0.012-0.019 (0.30-0.48)	—
100-125	4	22.5(655)	Spiral Bevel	Sliding Dog	2.07:1	.025(0.64)	0.015-0.022 (0.38-0.55)	—
135-150	V-6	22.5(655)	Spiral Bevel	Sliding Dog	2.00:1	.025(0.64)	0.015-0.022 (0.38-0.55)	0.030-0.050 (.762-12.7)
XR6/MAGIII/150XRi	V-6	22.5(655)	Spiral Bevel	Sliding Dog	1.87:1	.025(0.64)	0.018-0.027 (0.460-0.686)	0.030-0.050 (.762-12.7)
XR6/MAGIII (4 1/4" Dia.)	V-6	21.0(621)	Spiral Bevel	Sliding Dog	1.78:1	.025(0.64)	0.016-0.019 (0.406-0.482)	0.030-0.050 (.762-12.7)
175-275	V-6	22.5(655)	Spiral Bevel	Sliding Dog	1.87:1	.025(0.64)	0.018-0.027 (0.460-0.686)	0.030-0.050 (.762-12.7)

① The amount of backlash is not critical, but NO backlash is critical.

MANUAL TILT 9-2
DESCRIPTION AND OPERATION 9-2
  SERVICING 9-2
GAS ASSIST TILT SYSTEM 9-5
DESCRIPTION AND OPERATION 9-5
GAS ASSIST DAMPER 9-5
  TESTING 9-5
  REMOVAL & INSTALLATION 9-6
POWER TRIM/TILT 9-6
DESCRIPTION AND OPERATION 9-6
TROUBLESHOOTING THE POWER TRIM/TILT SYSTEM 9-6
SINGLE RAM INTEGRAL POWER TILT/TRIM 9-7
DESCRIPTION AND OPERATION 9-7
TRIM/TILT PUMP 9-7
  REMOVAL & INSTALLATION 9-7
  HYDRAULIC SYSTEM BLEEDING 9-8
TILT/TRIM MOTOR 9-8
  TESTING 9-8
  REMOVAL & INSTALLATION 9-8
  CLEANING & INSPECTION 9-9
TILT/TRIM CYLINDER 9-10
  REMOVAL & INSTALLATION 9-10
  CLEANING & INSPECTION 9-11
TILT/TRIM SWITCH 9-11
  TESTING 9-11
TILT/TRIM RELAY 9-12
  TESTING 9-12
THREE RAM INTEGRAL POWER TILT/TRIM SYSTEM 9-12
DESCRIPTION AND OPERATION 9-12
TILT/TRIM PUMP 9-13
  REMOVAL & INSTALLATION 9-13
  HYDRAULIC SYSTEM BLEEDING 9-15
TILT/TRIM MOTOR 9-16
  TESTING 9-16
  REMOVAL & INSTALLATION 9-16
TILT/TRIM CYLINDER 9-18
  DISASSEMBLY 9-18
  CLEANING & INSPECTION 9-18
TILT/TRIM SWITCH 9-18
  TESTING 9-18
TILT/TRIM SOLENOID 9-18
  TESTING 9-18
TILT/TRIM RELAY 9-20
  TESTING 9-20

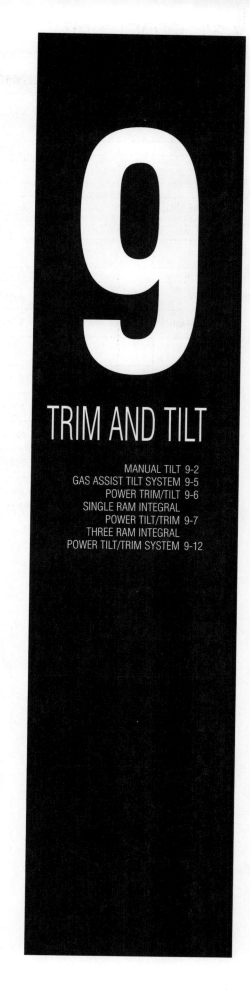

9
TRIM AND TILT

MANUAL TILT 9-2
GAS ASSIST TILT SYSTEM 9-5
POWER TRIM/TILT 9-6
SINGLE RAM INTEGRAL
POWER TILT/TRIM 9-7
THREE RAM INTEGRAL
POWER TILT/TRIM SYSTEM 9-12

9-2 TRIM AND TILT

MANUAL TILT

Description and Operation

◆ See Figures 1, 2 and 3

All outboard installations are equipped with some means of raising or lowering the lower unit for efficient operation under various load, boat design, water conditions and while making the trip to and from a body of water. The most simple form is a mechanical tilt adjustment consisting of a series of holes in the transom mounting bracket through which an adjustment pin passes through to secure the outboard unit at the desired angle. A second and more modern method, especially for the larger horsepower units, is a hydraulically operated system controlled from the helm.

The system employed on the outboard units covered in this section consist of a mechanical tilt pin arrangement.

A mechanical tilt arrangement is found on most outboard units. A change in the tilt angle of the outboard unit is accomplished by inserting the tilt adjustment pin through one of a series of holes in the transom mounting bracket. These holes allow the operator to obtain the desired boat trim under various speeds and loading conditions.

The tilt angle of a lower unit is properly set when the anti-cavitation plate is approximately parallel with the bottom of the boat. The boat trim is corrected by stopping the boat, removing the adjustment pin, tilting the outboard upward or downward, as desired and then installing the pin through the new hole exposed in the transom mounting bracket.

To raise the bow of the boat, the outboard is raised one hole at-a-time until the operator is satisfied with the boat's performance. If the bow is to be lowered, the lower unit is lowered one hole at-a-time.

Performance will generally be improved if the bow is lowered during operation in rough water. The boat should never be operated with the lower unit set at an excessive raised position. Such a tilt angle will cause the boat to "porpoise", which is very dangerous in rough water. Under such conditions the helmsperson does not have complete control at all times.

Instead of making extreme changes in the lower unit angle, it is far better to shift passengers and/or the load to obtain proper performance.

In order to obtain maximum efficiency and safety from the boat and outboard unit, the tilt pin must be installed in the proper position. The wide range of boat designs with their various transom angles, requires a determination be made for each outboard installation.

Actually, the tilt pin is only required if the boat handles improperly in the full trimmed "in" position at WOT (wide open throttle). Usually this occurs when the transom "angle" is too large.

1. Move the outboard unit inward or outward until the anti-cavitation plate is parallel to the boat bottom. With the outboard in this position, notice the position of the swivel bracket in relation to the clamp bracket tilt pin holes. Now, install the tilt pin into the first full pin hole closer to the transom.

2. Lock the tilt pin in position by pushing in on the pin compressing the lock spring. Close the clevis on the end of the pin and release the tilt pin. The tilt pin is secured by the clevis on the end of the tilt pin.

Some earlier models utilize a cotter pin and washer on the end of the tilt pin. After installing the tilt pin through the desired clamp bracket hole, slide the washer over the end of the tilt pin and insert the cotter pin through the hole at the end of the tilt pin. Open the ends of the cotter pin to secure the tilt pin in place.

The angle of the lower unit is properly set when the boat is operating to give maximum performance, including comfort and safety.

SERVICING

◆ See Figures 4 thru 9

Service procedures for the manual tilt system are confined to general lubrication and inspection. If individual components should wear or break, replacement of the defective components is necessary.

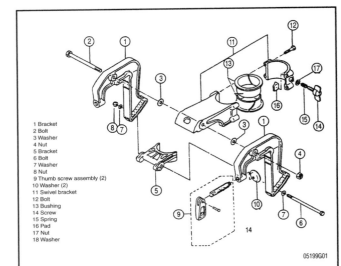

Fig. 4 Engine tilt components and lubrication points—2.5–3.3 hp

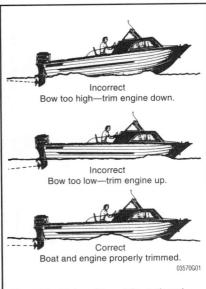

Fig. 1 The trim position of the outboard unit directly affects the bow position and thus the boat performance

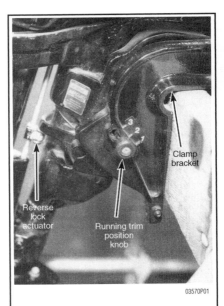

Fig. 2 Typical manual tilt pin arrangement on a smaller outboard unit

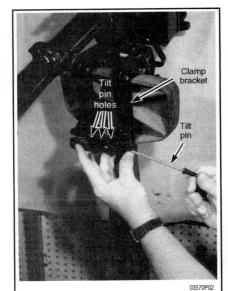

Fig. 3 With the outboard in the proper position, insert the tilt pin into the tilt pin hole

TRIM AND TILT 9-3

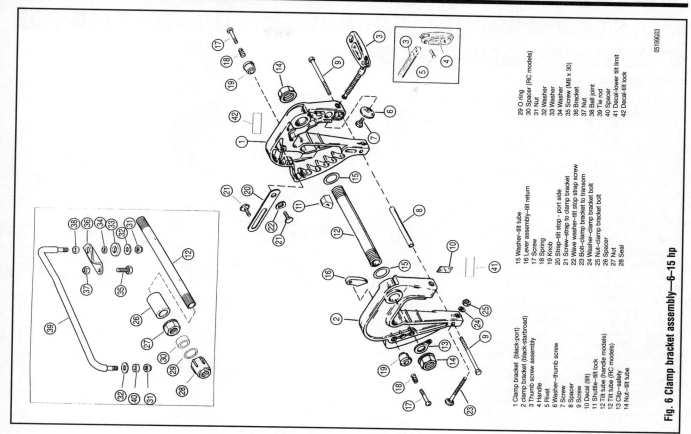

Fig. 6 Clamp bracket assembly—6–15 hp

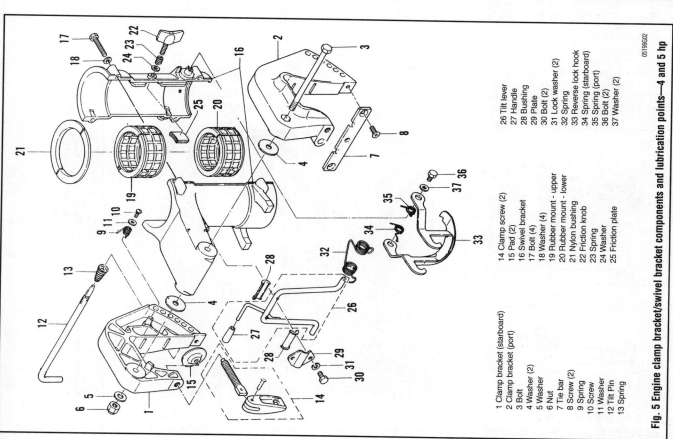

Fig. 5 Engine clamp bracket/swivel bracket components and lubrication points—4 and 5 hp

9-4 TRIM AND TILT

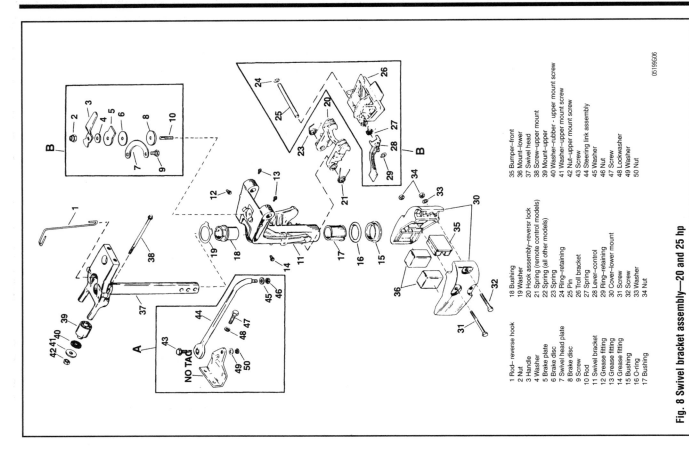

Fig. 8 Swivel bracket assembly—20 and 25 hp

1 Rod – reverse hook
2 Nut
3 Handle
4 Washer
5 Brake plate
6 Brake disc
7 Swivel head plate
8 Brake disc
9 Screw
10 Rod
11 Swivel bracket
12 Grease fitting
13 Grease fitting
14 Grease fitting
15 Bushing
16 O-ring
17 Bushing
18 Bushing
19 Washer
20 Hook assembly–reversr lock
21 Spring (remote control models)
22 Spring (all other models)
23 Spring
24 Ring–retaining
25 Pin
26 Troll bracket
27 Spring
28 Lever–control
29 Ring–retaining
30 Cover–lower mount
31 Screw
32 Screw
33 Screw
34 Nut
35 Bumper–front
36 Mount–lower
37 Swivel head
38 Screw–upper mount
39 Mount–upper
40 Washer–rubber - upper mount screw
41 Washer–upper mount screw
42 Nut–upper mount screw
43 Screw
44 Steering link assembly
45 Washer
46 Nut
47 Screw
48 Lockwasher
49 Washer
50 Nut

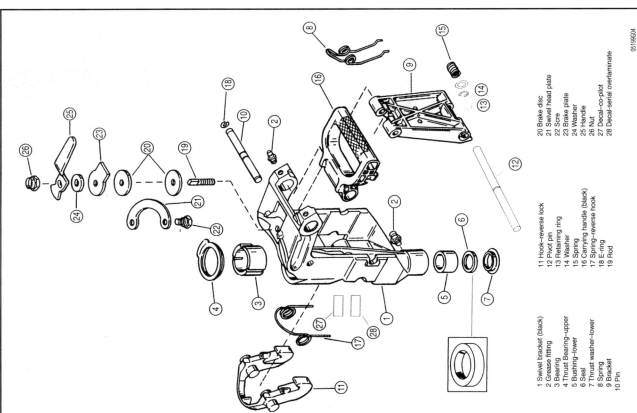

Fig. 7 Swivel bracket and tilt lock pin assembly—6–15 hp

1 Swivel bracket (black)
2 Grease fitting
3 Bearing
4 Thrust Bearing–upper
5 Bushing–lower
6 Seal
7 Thrust washer–lower
8 Spring
9 Bracket
10 Pin
11 Hook–reverse lock
12 Pivot pin
13 Retaining ring
14 Washer
15 Spring
16 Carrying handle (black)
17 Spring–reverse hook
18 E-ring
19 Rod
20 Brake disc
21 Swivel head plate
22 Scre
23 Brake plate
24 Washer
25 Handle
26 Nut
27 Decal–co-pilot
28 Decal–serial overlaminate

TRIM AND TILT

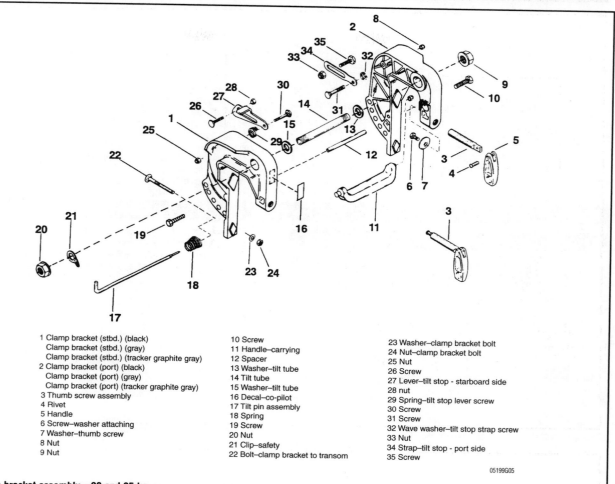

1 Clamp bracket (stbd.) (black)
  Clamp bracket (stbd.) (gray)
  Clamp bracket (stbd.) (tracker graphite gray)
2 Clamp bracket (port) (black)
  Clamp bracket (port) (gray)
  Clamp bracket (port) (tracker graphite gray)
3 Thumb screw assembly
4 Rivet
5 Handle
6 Screw–washer attaching
7 Washer–thumb screw
8 Nut
9 Nut
10 Screw
11 Handle–carrying
12 Spacer
13 Washer–tilt tube
14 Tilt tube
15 Washer–tilt tube
16 Decal–co-pilot
17 Tilt pin assembly
18 Spring
19 Screw
20 Nut
21 Clip–safety
22 Bolt–clamp bracket to transom
23 Washer–clamp bracket bolt
24 Nut–clamp bracket bolt
25 Nut
26 Screw
27 Lever–tilt stop - starboard side
28 nut
29 Spring–tilt stop lever screw
30 Screw
31 Screw
32 Wave washer–tilt stop strap screw
33 Nut
34 Strap–tilt stop - port side
35 Screw

Fig. 9 Clamp bracket assembly—20 and 25 hp

GAS ASSIST TILT SYSTEM

Description and Operation

The gas assist tilt system consists of the shock rod cylinder, valve block, nitrogen filled accumulator and manual release lever.

When the manual release lever is opened, the nitrogen filled accumulator assists the operator while tilting the outboard for shallow water operation or trailering. The entire system is located between the transom brackets.

If the lower unit strikes an underwater object, the force will apply a high pressure on the down side of the cylinder. A shock valve located in the piston will relieve this pressure to the up side of the piston and if necessary, to allow excess oil to return to the reservoir.

If the outboard has trouble staying in the selected tilt position, check for external fluid leaking from the accumulator or shock rod cylinder. Make repairs as necessary. Make sure that the manual release lever and control rod operate freely and make any adjustments if needed.

Gas Assist Damper

TESTING

♦ See Figure 10

1. To check for a discharged accumulator, 30–50 ft. lbs. (47–68Nm) of pulling force must be attained when tilting the outboard from a full down position to a full

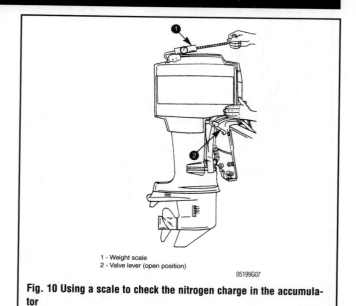

1 - Weight scale
2 - Valve lever (open position)

Fig. 10 Using a scale to check the nitrogen charge in the accumulator

9-6 TRIM AND TILT

up position. If more than 50 ft. lbs. (68Nm) of torque is required, the nitrogen charge in the accumulator is discharged and the accumulator will need to be replaced.

REMOVAL & INSTALLATION

1. Support the outboard in the up position using the tilt lock lever.
2. Remove the link rod.
3. Using a suitable punch, remove (drive down) the upper dowel pin.
4. Use a suitable punch to drive out the upper pivot pin.
5. Use a punch to remove (drive up) the lower dowel pin.
6. Use a punch to drive out the lower pivot pin.
7. Tilt the shock absorber assembly (top first) out from the clamp bracket and remove the assembly.

To install:

8. Apply Quicksilver special lubricant 101 or equivalant to the lower pivot pin hole and pivot pin surface.
9. Start the lower pivot pin into the pivot pin hole and position the lower dowel pin in its hole.
10. Reinstall the manual tilt system. Reconnect the release valve link rod.
11. Using a punch, drive the lower pivot pin into the clamp bracket and trim cylinder assembly until the pivot pin is flush with the outside surface.
12. Using a punch, drive in the lower dowel pin until it is seated.
13. Apply Quicksilver 2-4-C marine lubricant or equivalant to the surface of the upper pivot pin, pivot pin hole and the shock rod hole.
14. Using a mallet, drive the upper pivot pin into the swivel bracket and through the shock rod until the pivot pin is flush with the swivel bracket.
15. Drive the upper dowel pin into its hole until seated.
16. Check the manual release cam adjustment. The cam must open and close freely. Make adjustments to the link rod as necessary.

POWER TRIM/TILT

Description and Operation

The power trim/tilt systems consist of a housing with an electric motor, a gear driven hydraulic pump, hydraulic reservoir and one or more trim/tilt cylinders. The cylinders perform a double function as trim/tilt cylinders and also as a shock absorbers, should the lower unit strike an underwater object while the boat is underway.

The necessary valves, check valves, relief valves and hydraulic passageways are incorporated internally and externally for efficient operation. A manual release valve is provided to permit the outboard unit to be raised or lowered should the battery fail to provide the necessary current to the electric motor or if a malfunction should occur in the hydraulic system.

The gear driven pump operates in much the same manner as an oil circulation pump installed on motor vehicles. The gears rotate in either direction, depending on the desired cylinder movement. One side of the pump is considered the suction side and the other the pressure side, when the gears rotate in a given direction. These sides are reversed, the suction side becomes the pressure side and the pressure side becomes the suction side when gear movement is changed to the opposite direction.

Depending on the model, up to two relays may be used for the electric motor. The relays are usually located at the bottom cowling pan, where they are fairly well protected from moisture.

➡ As a convenience, on some models an auxiliary trim/tilt switch is installed on the exterior cowling.

When the up portion of the trim/tilt switch is depressed, the up circuit, through the relay, is closed and the electric motor rotates in a clockwise direction. Pressurized oil from the pump passes through a series of valves to the lower chamber of the trim cylinders, the pistons are extended and the outboard unit is raised. The fluid in the upper chamber of the pistons is routed back to the reservoir as the piston is extended. When the desired position for trim is obtained, the switch on the control handle is released and the outboard is held stationary.

If the trim cylinder pistons should become fully extended, such as in a tilt up situation, fluid pressure in the lower chamber of the trim cylinders increases. This increase in pressure opens an up relief valve and the fluid is routed to the reservoir. The sound of the electric motor and the pump will have a noticeable change.

When the down portion of the trim/tilt switch is depressed, the down circuit, through the relay, is closed and the electric motor rotates in a counterclockwise direction. The pressure side of the pump now becomes the suction side and the original suction side becomes the pressure side. Pressurized oil from the pump passes through a series of valves to the upper chamber of the trim cylinders, the pistons are retracted and the outboard unit is lowered. The fluid in the lower chamber of the pistons is routed back to the reservoir as the retracted is extended. When the desired position for trim is obtained, the switch on the control handle is released and the outboard is held stationary.

If the trim cylinder pistons should become fully retracted, such as in a tilt down situation, fluid pressure in the upper chamber of the trim cylinders increases. This increase in pressure opens an up relief valve and the fluid is routed to the reservoir. The sound of the electric motor and the pump will have a noticeable change.

In the event the outboard lower unit should strike an underwater object while the boat is underway, the tilt piston would be suddenly and forcibly extended, moved upward. For this reason, the lower end of the tilt piston is capped with a free piston. This free piston normally moves up and down with the tilt piston.

The free piston also moves upward but at a much slower rate than the tilt piston. The action of the tilt piston separating from the free piston causes two actions. First, the hydraulic fluid in the upper chamber above the piston is compressed and pressure builds in this area. Second, a vacuum is formed in the area between the tilt piston and the free piston.

This vacuum in the area between the two pistons sucks fluid from the upper chamber. The fluid fills the area slowly and the shock of the lower unit striking the object is absorbed. After the object has been passed the weight of the outboard unit tends to retract the piston. The fluid between the tilt piston and the free piston is compressed and forced through check valves to the reservoir until the free piston reaches its original neutral position.

A manual relief valve, located on the stern bracket, allows easy manual tilt of the outboard should electric power be lost. The valve opens when the screw is turned counterclockwise, allowing fluid to flow through the manual passage. When the relief valve screw is turn fully clockwise, the manual passage is closed and the outboard lock in position.

A thermal valve is used to protect the trim/tilt motor and allow it to maintain a designated trim angle. Oil in the upper chamber is pressurized when force is applied to the outboard from the rear while cruising. Oil is directed through the right side check valve and activates the thermal valve to release oil pressure and lessen the strain on the motor and pump.

Troubleshooting the Power Trim/Tilt System

Any time a problem develops in the power trim/tilt system the first step is to determine whether it is electrical or hydraulic in nature. After the determination is made, then the appropriate steps can be taken to remedy the problem.

The first step in troubleshooting is to make sure all the connectors are properly plugged in and that all the terminals and wires are free of corrosion. The simple act of disconnecting and connecting a terminal may sometimes loosen corrosion that preventing a proper electrical connection. Inspect each terminal carefully and coat each with dielectric grease to prevent corrosion.

The next step is to make sure the battery is fully charged and in good condition. While checking the battery, perform the same maintenance on the battery cables as you did on the electrical terminals. Disconnect the cables (negative side first), clean and coat them and then reinstall them. If the battery is past its useful life, replace it. If it only requires a charge, charge it in.

Check the power trim/tilt fuse, as appropriate. Many systems will have a fuse to prevent large current draws from damaging the system. If this fuse is blown, the system will cease to function. This is a good indicator that you may have problems elsewhere in the electric system. Fuses don't blow without cause.

After inspecting the electrical side of the system, check the hydraulic fluid level and top it off as necessary. Remember to position the motor properly (full tilt up or down) to get an accurate measurement of fluid level. A slight decrease in the level of hydraulic fluid may cause the system to act sporadically.

Finally, make sure the manual release valve is in the proper position. A slightly open manual release valve may prevent the system from working properly and mimic other more serious problems.

Just remember to check the simple things first. If these simple tests do not diagnose the cause of the problem, then it is time to investigate more deeply. Perform the hydraulic pressure tests in this section to determine if the pump is making adequate pressure. Inspect the entire power trim/tilt electrical harness with a multimeter, checking for excessive resistance and proper voltage.

TRIM AND TILT

SINGLE RAM INTEGRAL POWER TILT/TRIM

Description and Operation

The single cylinder power tilt/trim system consists of an electric motor, pump, pressurized fluid reservoir, one large trim/tilt cylinder and electrical components.

The single cylinder extends very slowly through the first 20° of movement, this is considered the trim range and then accelerates to move the outboard to the desired tilt position for trailering or shallow water operation.

The hydraulically powered single cylinder tilt/trim system permits changing the tilt angle of the outboard unit from the helm position. Controls and indicators for the system are located on or about the remote control handle. A set of switches are also installed on the side of the powerhead on the lower cowling.

Trim/Tilt Pump

REMOVAL & INSTALLATION

30–60 HP

▶ See accompanying illustrations

1. As a safety measure to prevent accidental movement of the outboard while work is being performed, it is strongly recommended a few minutes be used to make a safety support tool. The tool may be made from any metal bar stock or small channel iron of suitable size, with a ⅜ in. (9.53 mm) hole drilled through at each end and 14 in. (35.6 cm) apart, as shown. Cut-off the head of a ⅜ in. bolt about 2-½ in. (6.4 cm) long. Drill a hole through each bolt for a cotter pin. Secure the bolts through the holes made in the bar stock with two nuts, one on each side of the bar. The tool is now ready for installation, one end through the clamp bracket and the other end through the tilt stop bracket. Secure each end of the tool in place with a washer and cotter pin. The trim/tilt system may now be serviced or other work performed with confidence and in safety.

2. Raise the outboard unit to the full up position. If the hydraulic system is inoperative, first rotate the manual release valve counterclockwise about three complete turns and then manually lift the unit to the full up position.

3. Set the tilt lock lever in place. Install the safety support tool as described at the beginning of this section. Secure one end of the tool to the clamp bracket with a washer and cotter pin. Secure the other end to the tilt stop bracket in a similar manner, as shown.

4. Disconnect the trim/tilt motor electrical wires (Blue and green) from the powerhead harness, at the plug connector under the powerhead cowling, as shown.

5. Remove the harness retainer and any other clamps securing the wire harness run to the trim/tilt motor. Remove the trim gauge sending unit from the upper pivot pin by removing two screws and lifting the sender unit free.

6. If the cross pin has a large head, use a suitable punch and remove the cross pin from the upper pivot pin.

7. If the cross pin is headless, use a large blunt nose punch and drive the upper pivot pin out from the clamp bracket and cylinder ram, this will shear-off the cross pin.

8. Use a suitable punch and remove the cross pin from the lower pivot pin. With a large blunt nose punch, drive the lower pivot pin out from the lower clamp brackets and end of the cylinder.

9. Swing the trim/tilt assembly, upper end first, clear of the transom and then lift the unit from the clamp brackets.

To install:

10. Raise the outboard and install safety support tool bar if not previously installed.

11. Position the trim/tilt unit between the clamp brackets, bottom end in first. Slide the pivot pin through the starboard clamp bracket and then through the lower trim cylinder pivot. Using a suitable punch and hammer, drive the pivot pin through until it is flush with the port side transom bracket. Install a new cross pin through the pivot pin.

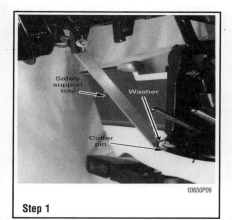
Step 1

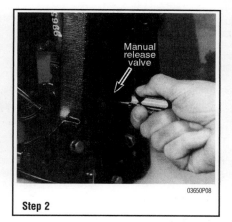
Step 2

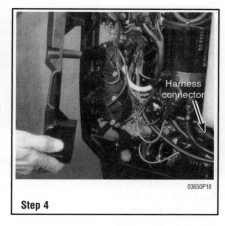
Step 4

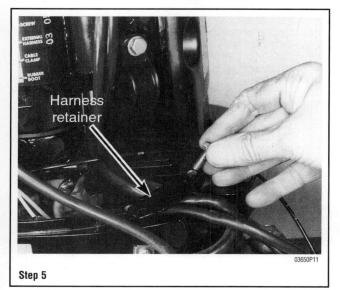
Step 5

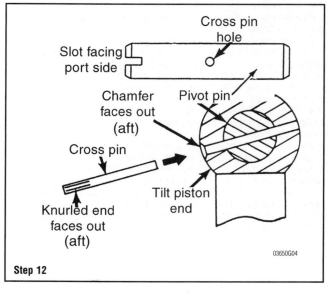
Step 12

9-8 TRIM AND TILT

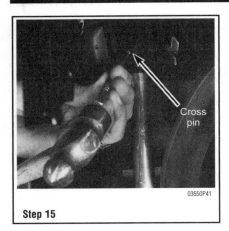

Step 15

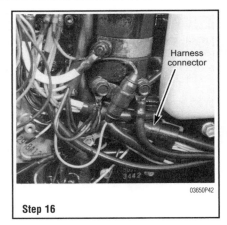

Step 16

Fig. 11 With the trim/tilt unit clamped in a vise, the manual release valve may be opened and closed during the "bleeding" process

12. The pivot pin can be rotated with a large screwdriver until the hole in the pin is aligned with the hole in the end of the tilt piston. The chamfered hole in the end of the piston must face outward. If it does not, remove the pivot pin and rotate the piston in the cylinder ½ turn.

13. Insert the pump motor wiring harness up through the starboard transom bracket and into the powerhead cowling area.

14. Rotate the trim/tilt unit until the upper cylinder piston aligns with the outboard upper pivot point. Align the hole in the piston rod with the pivot and then install the pivot pin. Using a suitable punch and hammer, drive the pivot pin through until it is flush at both ends. Use a flat blade screwdriver and align the pivot pin with the cylinder piston cross pin holes.

15. Insert the cross pin through the pivot pin and cylinder piston rod. Using a punch and hammer, drive the cross pin flush.

16. Route the trim/tilt harness to the starboard side of the powerhead cowling area. Insert the Green and Blue wire connectors into the same color coded connectors from the powerhead electrical harness. Secure the harness and wires with tie straps.

17. Close and secure the powerhead cowlings; remove the safety support tool; and connect the battery. Perform the Hydraulic Bleeding procedures outlined in this chapter.

18. Check to be sure the battery has a full charge and then move the outboard unit through several cycles, full down to full up for trailering. Observe the system for satisfactory operation electrical and hydraulic. Check the system for possible leaks.

HYDRAULIC SYSTEM BLEEDING

This trim/tilt system with one large cylinder has been designed with a "self bleeding" system. Fluid is pumped from one side of the cylinder to the opposite side as the piston extends or retracts. Therefore, any air trapped on one end of the cylinder, is pushed out back to the reservoir, when the piston bottoms out. Usually two or three cycles of the system will remove all entrapped air from the cylinder. The following short sections outline the necessary steps to perform during the cycling sequence just mentioned. Two sets of instructions are included, one with the system installed on the outboard and the other with the system on the workbench.

Installed On Outboard

1. Check to be sure the manual release valve is fully closed prior to activating the pump motor.

2. Press and hold the up trim switch until the powerhead is in the full up position. Release the up switch and engage the uplock lever.

3. Slowly remove the fill plug from the reservoir. Fluid level should be to the edge of the fill plug opening. If necessary add Quicksilver Power Trim and Steering Fluid. If the Quicksilver product is not available, Automatic Transmission Fluid (ATF) Type F, FA or Dexron II, may be used. Tighten the fill plug securely.

4. Press and hold the down trim switch until the powerhead is in the full down position. Release the down switch.

5. Press and hold the up trim switch until the powerhead is in the full up position. Release the up switch. Continue to cycle the system in this manner two or three times and then check the fluid level in the reservoir. Add fluid if necessary and tighten the fill plug.

Removed From Outboard

♦ See Figure 11

1. Check to be sure the manual release valve is fully closed prior to activating the pump motor.

2. Connect a 12-volt power source Red lead to the pump motor Green lead. Connect the power source negative Black lead to the pump Blue lead. Hold the connections until the piston has fully extended.

3. Slowly remove the fill plug from the reservoir. The fluid level should be to the edge of the fill plug opening. If necessary, add Quicksilver Power Trim and Steering Fluid. If the Quicksilver product is not available, Automatic Transmission Fluid (ATF) Type F, FA or Dexron II, may be used. Tighten the fill plug securely.

4. Reverse the 12-volt power source connections, Red lead to the pump Blue lead and Black lead to the pump Green lead. Hold the lead connections until the piston is fully retracted.

5. Continue to cycle the system in this manner two or three times and then check the fluid level in the reservoir. Add fluid if necessary and tighten the fill plug securely.

6. Disconnect the 12-volt power source leads from the pump motor.

Tilt/Trim Motor

TESTING

1. Disconnect the pump motor wire harness from the system wire harness. Connect the pump motor Blue wire, to the battery positive terminal and the Green wire to the battery negative terminal. The motor should operate in one direction.

2. Reverse the two wires connected to the battery and pump. The pump should operate in the opposite direction.

3. If the motor fails to run in one or both directions, disassemble the motor and check for defective components.

4. The trim motor armature and commutator are tested in the same manner as the starter motor armature. Refer to the "Ignition and Electrical" section for the starter motor repair instructions and perform the same checks and test on the pump motor armature.

REMOVAL & INSTALLATION

♦ See accompanying illustrations

1. Loosen and remove the four Phillips head screws from the pump motor end cap.

2. Using two flat blade screwdrivers, insert the blade of the screwdrivers into the slotted space between the end cap on opposite sides of the motor housing.

3. Push down on the screwdrivers and slowly separate the end cap, from the motor housing. Do not insert the screwdriver between the end cap and motor housing except at the slotted space. Damage to the end cap and/or motor housing will result and loosen the waterproof seal to the motor.

4. After the motor end cap has been separated from the motor housing, lift the end cap and wiring harness straight up and free of the motor armature.

5. Reach inside the motor housing and grasp the armature shaft. Due to the magnetic field of the motor frame, both the armature and frame will slide out of the motor housing as one unit. Lift up on the armature shaft, removing both the

TRIM AND TILT

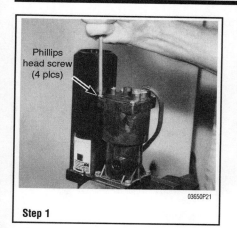
Step 1

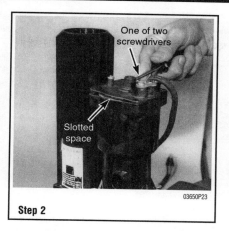
Step 2

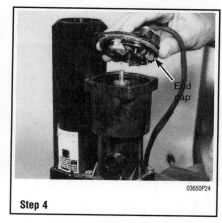
Step 4

Step 5

Fig. 12 Measure the length of the brushes. If less than ⅜ in. remains, the brushes should be replaced

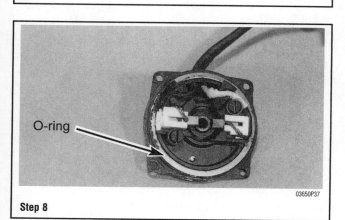
Step 8

armature and motor frame from the motor housing. Take care not to lose the flat washer used on each end of the armature shaft. The washer on the lower end of the armature shaft will usually be in the bottom of the motor housing. Retrieve the washer from the motor housing and place it back on the lower armature shaft.

To install:

6. Insert the armature into the motor frame. Slide one flat washer onto each end of the armature shaft.

7. Insert the pump driveshaft into the lower end of the armature shaft. Lower the motor assembly, with the pump driveshaft going in first, down into the motor housing.

8. Install a new O-ring onto the end cap and lubricate the O-ring with fluid. Slide the brushes into the brush holders. Set the spring tensioner ends outside the holders to prevent the brushes from having any spring force applied.

9. Apply a small amount of marine lubricant or equivalent, to the bushing in the end cap.

10. Lower the end cap down over the armature shaft. As soon as the brushes make contact with the commutator, release the brush spring tensioner against the brushes. The spring tensioner will now keep the brushes against the commutator.

11. Secure the end cap onto the housing with the four Phillips head screws.

CLEANING & INSPECTION

♦ See Figure 12

1. Inspect the brushes in the end cap for wear. If the brushes are worn down to less than ⅜ in. inch, replace the brushes.

2. Inspect the ends of the pump driveshaft for worn or damaged splines. If the shaft is worn or damaged, replace the shaft.

9-10 TRIM AND TILT

Tilt/Trim Cylinder

REMOVAL & INSTALLATION

▶ See accompanying illustrations

1. Clamp the trim/tilt unit in a vise. Take care not to damage the manifold assembly.
2. Obtain a suitable container and be prepared to catch fluid escaping from the unit. Remove the manual release valve from the manifold assembly and the reservoir fill plug from the reservoir.
3. Extend the piston to the full up position by pulling up on the end of the piston. This may not be an easy task, a fair amount of muscle power and some rotation of the piston will probably be necessary. After the piston is fully extended, use a spanner wrench with ¼ in. diameter X ⁵⁄₁₆ in. inch long lugs, loosen and unscrew the cylinder end cap from the cylinder.
4. Slide the trim rod assembly upward until the piston is free of the cylinder.
5. Remove the three screws securing the plate to the trim rod piston.
6. Lift off the plate, five check valve springs, seats and balls from inside the piston. Do not remove the check valve on the trim rod piston, retained with a roll pin.
7. Using a pair of snap ring pliers reach inside the cylinder and grasp the memory piston and pull it up and free of the cylinder.

➡ If the hydraulic system was not completely drained of all fluid, perform the next step over a drain pan to catch the hydraulic fluid behind the memory piston.

8. An alternate method to remove the piston is to first remove the trim/tilt unit from the vise. Turn the unit over and make a couple of quick solid "blows" against a piece of plywood or other hard surface. The memory piston will slide down and out of the cylinder.

To install:

9. Lubricate two new O-rings with fluid and then install them into the recesses of the cylinder.
10. Install two new O-rings onto the pilot valve. Slide the pilot valve and spring into the manifold.
11. Align the pump and manifold assembly with the cylinder. Be sure the two O-rings and pilot valve are aligned.
12. Install the two Allen head cap screws and tighten to a torque value of 100 inch lbs. (11Nm).
13. Install a new O-ring into the groove in the bottom of the reservoir. Lubricate the O-ring with fluid.
14. Slide the reservoir, with the fill plug hole facing aft, down over the pump and flush against the manifold.
15. Install four flat washers and bolts securing reservoir to manifold. Tighten the four bolts to a torque value of 70 inch lbs. (7.7Nm).

Step 1

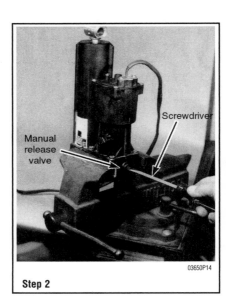

Step 2

Step 3

Step 4

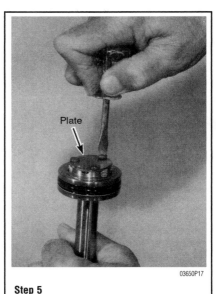

Step 5

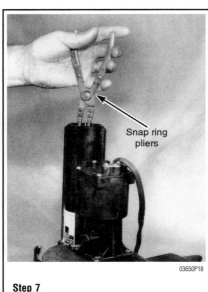

Step 7

TRIM AND TILT 9-11

Step 8

Step 9

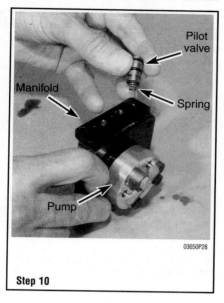
Step 10

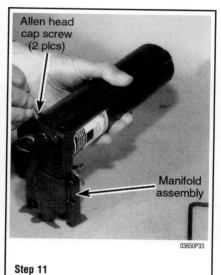
Step 11

Step 13

Step 14

CLEANING & INSPECTION

➡ **Never use an O-ring a second time. Always replace used O-rings with new.**

1. During the assembling work, lubricate O-rings with either Quicksilver Power Trim and Steering Fluid or automatic transmission fluid.
2. Inspect the check valve on the back side of the trim rod piston. Clean the check valve with a parts cleaning solvent but do not remove the valve from the piston. If the valve is suspected to have failed, replace the trim rod piston.
3. Inspect the trim rod cap scraper seal. If the seal does not keep the rod clean, remove the trim rod piston from the end of the trim rod. Slide off the cap and replace the seal inside the cap.
4. Install a new O-ring on the trim rod cylinder piston. Lubricate the O-ring with fluid.
5. Examine the interior of the cylinder for scoring, roughness or corrosion.
6. Inspect the edges of the memory piston and trim rod piston for scoring, roughness and corrosion.
7. Clean all parts with solvent and dry them with compressed air.

Tilt/Trim Switch

TESTING

Both Down and Up Trim Switches Are Inoperative

1. Obtain a digital multimeter and set it for continuity reading. Open or remove the powerhead cowling and check the 20-Amp fuse for continuity. If the fuse is "blown", examine the wiring for breaks or damage which caused a direct short and the fuse to "blow". Repair the damaged wiring. If the fuse is satisfactory, proceed to the next step.
2. Set the digital multimeter for a 12-volt meter reading. Connect the Red meter lead to the starter solenoid positive (battery), terminal and the Black meter lead to a good powerhead ground. If there is no battery voltage indication, check the battery leads for proper connection and the battery charge. If the meter indicates battery voltage, proceed to the next step.
3. Connect the Red meter lead to the Blue/White wire from the remote control wiring harness. Connect the Black wire to a good powerhead ground. Depress the up trim button and check for battery voltage indication. If the meter indicates bat-

9-12 TRIM AND TILT

tery voltage, check the Black ground wires on the Up/Down trim relays for proper connection and continuity. If the continuity check of the relay ground wires is satisfactory, the pump motor is possibly defective. Perform the pump motor test covered in this chapter. If there is no battery voltage indication, proceed to the next step.

4. Connect the Red meter lead to the trim/tilt switch Red wire connection, inside the remote control handle. Connect the Black meter lead to good powerhead ground. If the meter indicates battery voltage, the trim switch has failed or both wires, Blue/White and green/white, from the remote control to the Up/Down relays has been damaged or severed. Inspect the wiring harness and connectors between the remote control and the powerhead. If there is no battery voltage indication, proceed to the next step.

5. Connect the Red meter lead to any instrument positive terminal and the Black meter lead to a good powerhead ground. Set the ignition switch to the on or run position. If battery voltage is indicated, the Red wire in the remote control handle between the ignition switch and the trim/tilt switch is open.

6. Repair the defective wiring. If there is no battery voltage indication, the Red wire between the cranking motor solenoid positive terminal and the ignition switch is open.

7. Repair the defective wiring.

Tilt/Trim Relay

TESTING

The single cylinder trim/tilt system uses a set of relays for the pump motor operation direction and control. Previous trim/tilt systems have used a set of solenoids to perform the same function. The relays are compact, sealed units and are easily changed with their plug-in feature.

If the trim system operates in one direction but fails in the opposite direction, a quick and easy test can be performed to determine if one of the relays has failed and which one has failed.

1. Mark both relays, the top relay is the Up relay, the lower one is the Down relay.
2. Unplug both relays from the plug-in connectors. Plug each relay back into the opposite connector from which it was removed.
3. Depress the trim/tilt switch in the inoperative direction. If the trim/tilt system now operates, the relay is good.
4. Press the trim/tilt switch in the opposite direction. If the trim/tilt system fails to operate in the previously known good direction, the failed relay must be replaced.

Down Trim Inoperative—Up Trim Operates

1. Obtain a digital multimeter and set the meter for 12-volt DC reading. Connect the meter Red lead to the green/White wire from the remote control wiring harness. Connect the Black wire to a good powerhead ground.
2. Depress the Down trim button and observe the battery voltage indication. If the meter indicates battery voltage, then the down relay is defective and must be replaced. If there is no indication, proceed to the next step.
3. Connect the Red meter lead to the green/White wire (trim/tilt switches), inside the remote control handle. Connect the Black meter lead to good powerhead ground.
4. Depress the down trim button and check the meter for battery voltage indication. If the meter indicates battery voltage, the wiring between the remote control and the powerhead trim/tilt down relay is defective and must be repaired or replaced. If there was no voltage indication, proceed to the next step.
5. Connect the Red meter lead to the trim/tilt switch Red wire connection, inside the remote control handle. Connect the Black meter lead to a good powerhead ground.
6. Depress the down trim button and observe the meter for battery voltage indication. If the meter indicates battery voltage, the trim/tilt switch in the remote control handle is defective and must be replaced. If there is no battery voltage indication, check for voltage at the 20-Amp fuse and/or the wiring to the trim system from the main powerhead harness.

Up Trim Inoperative—Down Trim Operates

1. Obtain a digital multimeter and set the meter for 12-volt DC reading. Connect the Red meter lead to the Blue/White wire from the remote control wiring harness. Connect the Black wire to a good powerhead ground.
2. Depress the up trim button and observe for voltage indication. If the meter indicates battery voltage, the up relay is defective and must be replaced. If there was no voltage indication, proceed to the next step.
3. Connect the Red meter lead to the Blue/White wire (trim/tilt switches), connection inside the remote control handle. Connect the Black meter lead to good powerhead ground.
4. Depress the up trim button and check the meter for battery voltage indication. If the meter indicates battery voltage, the wiring between the remote control and the powerhead trim/tilt up relay is defective and must be repaired or replaced. If there was no voltage indication, proceed to the next step.
5. Connect the Red meter lead to the trim/tilt switch Red wire, connection inside the remote control handle. Connect the Black meter lead to a good powerhead ground.
6. Depress the up trim button and check the meter for battery voltage indication. If the meter indicates battery voltage, the trim/tilt switch in the remote control handle is defective and must be replaced. If there is no battery voltage indication, check for voltage at the 20-Amp fuse and/or wiring to the trim/tilt system from the main powerhead harness.

THREE RAM INTEGRAL POWER TILT/TRIM SYSTEM

Description and Operation

▶ See Figure 13

There have been three designs of this Trim/Tilt System used on various outboard units. All three systems work identically but engineering changes and improvements have resulted in some minor component changes along with modification to the disassembly and assembly procedures. The unique differences between these units are as follows:

• Design I Side Fill Reservoir with Round Motor Frame and Internal Hydraulic Pump
• Design II Side Fill Reservoir with Square Motor Frame and Internal Hydraulic Pump
• Design III Rear Fill Reservoir with Square Motor Frame and External Hydraulic Pump

Due to the manufacturers constant improvements from one model year to the next, some outboard models may have any one of the three designs described above. However most of the procedures are identical. Where differences do occur, these differences are clearly indicated. Therefore, before starting work, carefully examine the trim/tilt system on the outboard unit being serviced, then match the physical characteristics with the illustrations provided, to ensure the proper procedures are followed when differences are indicated. Once the model is identified, proceed with the work for that particular model.

The trim/tilt system consists of an electric motor, pump, pressurized fluid reservoir, two small trim cylinders and one large tilt cylinder.

The remote control throttle lever contains switches for adjusting the trim angle, raising the outboard for shallow water operation and tilting the outboard fully up for tilting.

An electronic trim position indicator (optional equipment), may be installed to provide the helmsperson with a visual reference of the outboard trim angle. This indicator system consists of a transducer mounted on one end of the tilt cylinder hinge pin and a trim gauge mounted on the control panel. As the angle of the outboard changes, the transducer senses the change altering the voltage level to the gauge on the control panel.

Trimming Outboard Unit Up

▶ See Figures 14, 15 and 16

1. Depressing the up button will actuate the up solenoid and close the circuit to the electric motor. The electric motor will drive the pump forcing fluid into the up side of the two trim cylinders.

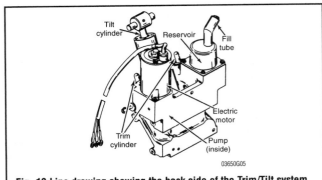

Fig. 13 Line drawing showing the back side of the Trim/Tilt system

TRIM AND TILT 9-13

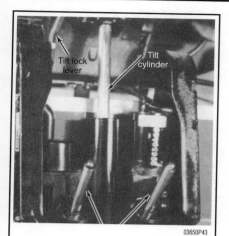

Fig. 14 The outboard unit raised to the full up position by the larger center tilt cylinder. The tilt lock lever has been engaged to take the weight from the cylinder.

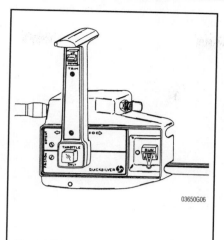

Fig. 15 Controls for this system does not have a "Tilting" button. Holding the up button depressed will activate the center tilt cylinder to raise the outboard

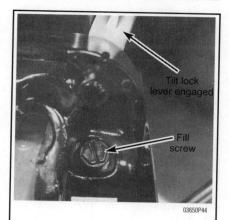

Fig. 16 The system is pressurized! Therefore, heed the warning decal just below the fill screw and never back off the screw unless the outboard unit is in the full up position

As the trim cylinders extend, the outboard unit may be raised to the desired angle. The system is designed to prevent the unit from being raised above 20° if the powerhead is operated above approximately 2000 rpm.

When the port side trim cylinder is fully extended, a check valve opens allowing the system to be regulated at 425 psi. If the powerhead exceeds 2000 rpm, the propeller thrust is sufficient to build up pressure above 425 psi on the up side of the system.

At this point, excess fluid is returned to the reservoir through a check valve, lowering pressure in the system. When the pressure is reduced, the outboard unit will be lowered to the trim limit position. In this manner, the check valve is operating as a tilt limit switch preventing the outboard unit to be operated at a greater angle than the 20° trim limit, if powerhead rpm exceeds 2000.

If powerhead rpm is cut back below 2000 rpm, the maximum angle may be increased above the 20° limit. However, if powerhead rpm exceeds 2000 rpm, the thrust created by the propeller (provided the propeller is deep enough in the water), will cause the trim system to automatically lower the unit back to the 20° maximum trim angle.

Trimming Outboard Unit Down

Depressing the down button will close the down circuit and actuate the down solenoid. The electric motor will operate in the opposite direction (from the up direction), forcing hydraulic fluid into the down side of the tilt cylinder. This action will move the outboard unit downward. When the desired angle of trim is obtained the button is released and movement ceases.

Tilting

Depress the up button. This action will close the up circuit and activate the up solenoid. The two trim cylinders will extend slowly to the full trim up position. Hold the up button depressed and the tilt cylinder will continue to move the outboard unit upward to the full up position for tilting.

➡ As a safety measure to prevent accidental movement of the outboard while tilting to and from the water, it is strongly recommended a tilt bracket always be used to mechanically lock the outboard unit in the up position. Such a bracket may be purchased at modest cost from the local marine store. With the bracket in place, the unit may be tilted in confidence over rough roads without fear of the outboard being jarred suddenly to the down position.

Tilt/Trim Pump

REMOVAL & INSTALLATION

♦ See accompanying illustrations

The hydraulic pump in this system cannot be serviced. Therefore, if troubleshooting leads to a faulty pump the unit must be replaced.

1. Raise the outboard unit to the full up position. If the hydraulic system is inoperative, first rotate the manual release valve counterclockwise about three complete turns and then manually lift the unit to the full up position.
2. Set the tilt lock lever in place. Install the safety support tool. Secure one end of the tool to the clamp bracket with a washer and cotter pin. Secure the other end to the tilt stop bracket in a similar manner, as shown.
3. Disconnect the electrical cables at both battery terminals. Tag and then disconnect the wiring at the upper and lower solenoids under the powerhead cowling.
4. Remove the harness retainer and any other clamps securing wiring running to the electric motor.

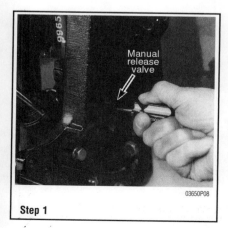

Step 1

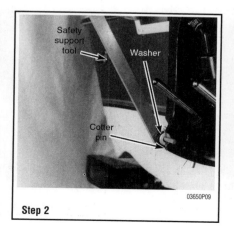

Step 2

Step 4

9-14 TRIM AND TILT

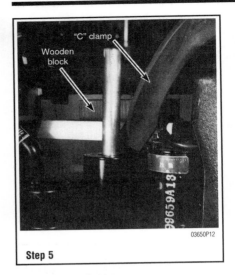
Step 5

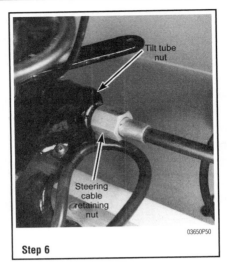
Step 6

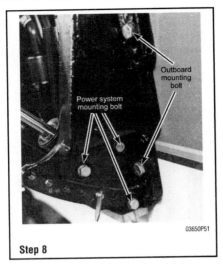
Step 8

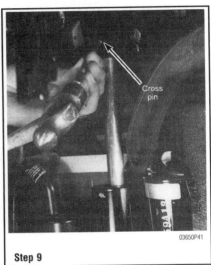
Step 9

Step 10

Step 12

5. Obtain a large C-clamp and a block of wood, approximately 4 in. to 6 in. (10.2 x 21.2 cm) square and about 2 in. (5.1 cm) thick. Support the outboard unit by clamping the block of wood on the inside of the transom up hard against the swivel bracket tube with the large C-clamp, as shown.

6. If the unit being serviced is equipped with thru-the-tilt tube steering, remove the steering cable retaining nut from the end of the tilt tube.

7. Remove the tilt tube nut.

8. Remove the two outboard mounting bolts and the three bolts securing the trim/tilt assembly to the starboard bracket. Carefully move the starboard transom bracket away far enough for the manual release valve to clear when the trim/tilt system is removed.

9. Drive out the cross pin securing the retaining pin for the tilt piston at the upper end. The cross pin should never be used a second time.

10. After the cross pin is free, push out the piston retaining pin. Do not allow the cylinder to fall free.

11. Support the trim/tilt system. Remove the three mounting bolts securing the system to the port side clamp bracket. Lower the system from the outboard unit. Pull the wires out with the system.

12. Clamp the trim/tilt system in a vise with the jaws gripping on the anode installed on the underneath side of the system.

13. The pump will be removed with the motor and the two will be separated after removal.

➡ **The pump cannot be serviced. Therefore, if troubleshooting indicates the pump has failed, it must be replaced.**

➡ **The electric motor and the pump may be removed from the trim/tilt assembly without removing the complete unit from the clamp bracket. To remove the motor and pump without disturbing other parts, move the starboard clamp bracket clear. After the clamp bracket is clear, proceed with the next step.**

14. Remove the two bolts securing the pump to the trim/tilt assembly. Lift the electric motor, with the pump attached, upward and free of the trim/tilt assembly base.

➡ **Before separating the electric motor from the pump, scribe a mark on the motor end cap and a matching mark on the motor case as an aid to assembling.**

15. Remove the wire clamp from one of the thru-bolts. Remove the two thru-bolts passing through the electric motor into the pump and then separate the electric motor from the pump. As the electric motor case is being lifted, exercise care not to drop and damage the armature.

To install:

16. Install new O-rings onto the pump. If the square motor pump is used, install the flat ring into the groove, as shown in the accompanying exploded drawing. Install a new seal into the recess of the upper end of the pump surface. The armature shaft passes through this seal. Coat the O-rings and the seal with Automatic Transmission Fluid.

17. Grasp the motor case with the thumb and forefinger holding the end cap in place and position it over the pump.

18. Push down on the motor case with the lower end of the armature passing through the pump seal. The chamfered part of the pump must be on the same side as the bolt hole opposite the electrical fitting, as shown.

19. Hold the electric motor and end cap with one hand and with the other hand rotate the pump ever so slightly until the hex end of the armature shaft indexes with the recess in the pump. When this occurs, the end surface of the motor case will be firm against the surface of the pump.

TRIM AND TILT 9-15

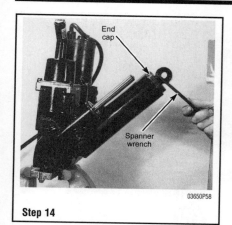
Step 14

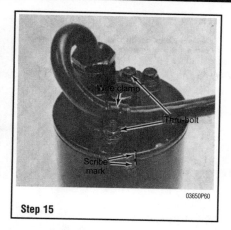
Step 15

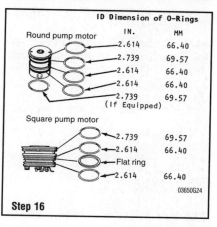
Step 16

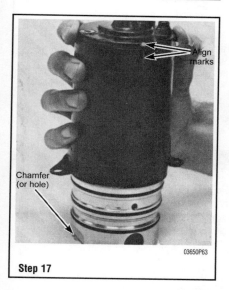
Step 17

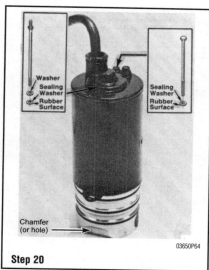
Step 20

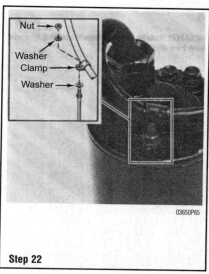
Step 22

20. Slide the sealing washer onto the thru-bolt with the head. The rubber surface of the washer must face down to bear against the surface of the end cap. Start to thread the bolt into the pump but do not tighten it at this time.

21. Slide a regular washer on the other thru-bolt and then the sealing washer with the rubber surface facing down to bear against the surface of the end cap. Start to thread the second bolt into the pump. A slight shifting of the motor case or the pump may be necessary. Tighten thru-bolts securely.

22. Secure the electrical lead to the thru-bolt without the head, using first a washer, then the wire clamp, another washer and finally the nut. Tighten the nut securely. Apply a coating of Quicksilver Liquid Neoprene around the thru-bolts, around the electrical connection and at the seam between the electric motor and the pump, as a waterproof measure.

23. Feed the electrical wire up through the outboard clamp bracket, if the motor and pump were removed with the trim/tilt assembly remaining on the clamp brackets.

24. Move the assembled pump and motor into position on the trim/tilt assembly. Carefully insert the pump into the power trim assembly housing, with the chamfer (or hole) on the lower end of the pump facing toward the tilt cylinder. Apply a light coating of Quicksilver Perfect Seal (or equivalent) to the threads of the attaching bolts.

25. Slide a lockwasher onto each bolt. Secure the pump on the housing with the attaching hardware. Tighten the bolts securely. Apply a coating of Quicksilver Liquid Neoprene on the seam between the pump and the housing as a waterproof measure.

26. If the trim/tilt assembly was not removed from the brackets, carefully move the starboard transom bracket into place with the manual release valve passing through the opening in the bracket.

27. Apply a light coating of Loctite® on the threads of five bolts. Secure the bracket in place with the two outboard mounting bolts and the three bolts through bracket into the power trim assembly. Tighten the bolts to a torque value of 30 ft. lbs. (40.7Nm).

HYDRAULIC SYSTEM BLEEDING

This trim/tilt system, with two small trim cylinders and one large tilt cylinder is almost a "self bleeding" system. Actually, the "bleed" operation is accomplished through the fill screw.

Checking For Air In The System

1. Activate the up circuit and raise the outboard slightly with the trim cylinders.
2. Exert a heavy, steady, downward force on the lower unit. If the trim pistons retract into the trim cylinders more than 1/8 in. (3.2 mm), the system contains air which must be removed.

✸✸ WARNING

The trim/tilt system is pressurized. Do not remove the fill screw unless the outboard unit is in the full up position. Never attempt to move the outboard unless the fill screw is securely tightened.

3. Raise the outboard unit to the full up position. All three pistons will be fully extended. Once the unit is in the full up position, release the button and engage the tilt lock lever.

4. Momentarily depress the up button just a couple times. Depressing the up button, for a second or two, will send current through the up solenoid to the electric motor; the motor will drive the pump; air in the pump will cause it to "squeal"; the pump will draw fluid from the reservoir; the pump will attempt to send fluid to the tilt cylinder; the piston cannot move because it is already extended; the pressure will increase to 425 psi (2930 kPa); the port side pressure relief valve will automatically open; the excess fluid will be returned to the reservoir, purging the system of air. Actually, the air is returned to the reservoir.

9-16 TRIM AND TILT

5. At this time, slowly remove the fill screw and the trapped air will escape. Add fluid, if necessary.

➥A small amount of air will remain on top of the reservoir under the cap. This is caused by the design of the cap and reservoir and as a "cushion" for the system.

6. Remove the tilt lock lever; depress the down button; lower the outboard unit to the full down position.

7. Momentarily depress the down button for just a second or two, a couple of times. Depressing the down button for a second or two will again activate the system; air in the pump will cause it to "squeal"; the pump will draw fluid from the reservoir; the pump will attempt to send fluid to the down side of the tilt cylinder; the cylinder cannot move because it is already retracted; the pressure will increase to 425 psi (2930 kPa); the pressure relief valve in the port side trim piston will automatically open; the excess fluid will be returned to the reservoir, purging the system of air. Actually, the air is returned to the reservoir.

8. Raise the outboard to the full up position again. Engage the tilt lock lever, slowly remove the fill screw again and the trapped air will escape. Add fluid to the system, if necessary.

9. Repeat the above procedure two or three times and any and all excess air will be bled from the system. After the air leaves the system, a noticeable change in the sound of the pump motor laboring with the outboard unit in the full up or down position will be heard.

10. As a further check, make the initial test for air again by first raising the outboard with the trim cylinders just a few degrees and then exerting a steady downward pressure on the lower unit. If the trim rods do not retract into the cylinders more than about 1/8 in. (3.2mm), all excess air has been bled from the system.

Hydraulic System Flushing

♦ See Figures 17 and 18

Only Automatic Transmission Fluid should be used in the power trim system. However, in remote areas or in an emergency, a substitute can be used. SAE 10W-30 or 10W-40 oil rated SE was recommended in the past by the manufacturer for earlier model systems. Therefore, these oils should be a workable substitute until the automatic transmission fluid can be obtained. Once the transmission fluid is available, the system must be flushed of the oil before putting the system into service. This can be accomplished by simply draining all the oil from the system.

1. First, raise the outboard unit to the full up position and engage the tilt lock lever. Place a suitable container under the starboard side trim cylinder.

2. Slowly remove the fill screw and bleed the pressure from the system. Back off the manual release valve about two full turns counterclockwise to release any remaining pressure in the system.

3. Remove the two bolts securing the zinc anode to the underneath side of the transom bracket and then remove the anode.

4. Now, remove the Allen plug from the underneath side of the starboard trim cylinder and drain the oil from the system.

5. After all oil has been drained from the system, install the Allen plug; tighten the manual release valve full clockwise; and then add Automatic Transmission Fluid to the system.

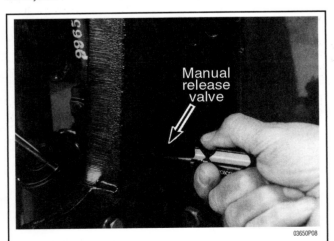

Fig. 17 The manual release valve is used to release pressure in the system to permit raising the outboard manually. Always release the valve slowly

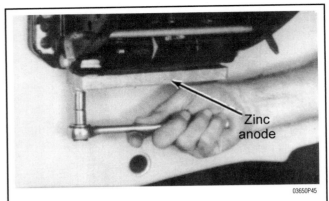

Fig. 18 The zinc anode must be removed to gain access to the Allen screw for draining the system

6. Bleed air from the system.

7. Operate the system several times through the complete cycle from the full down position to the full up position.

8. Drain this first quantity of flushing transmission fluid according to the foregoing procedures. As the fluid is being drained, take special note of the fluid color. If the fluid has a decided crimson tone, the oil has been flushed from the system. However, if the fluid color has a slight brownish tinge, there is still oil in the system. Repeat the flushing procedure until the brownish tinge has disappeared. The system is now thoroughly flushed and ready for service.

Tilt/Trim Motor

TESTING

1. Disconnect the Blue motor wire from the up solenoid. Disconnect the Green motor wire from the down solenoid.

2. Disconnect the Black motor ground wire.

3. Connect a test lead from the positive terminal of a 12-volt battery to the Blue wire just removed from the solenoid.

4. Connect a test lead from the negative (−) terminal of the battery to the Black ground wire disconnected. The motor should run.

5. Leave the ground wire connected and change the positive (+) battery lead to the Green wire removed from the down solenoid. The motor should run.

6. If the motor fails either one or both of these tests, it requires service or replacement.

REMOVAL & INSTALLATION

♦ See accompanying illustrations

1. Raise the outboard unit to the full up position. If the hydraulic system is inoperative, first rotate the manual release valve counterclockwise about three complete turns and then manually lift the unit to the full up position.

2. Set the tilt lock lever in place. Install the safety support tool. Secure one end of the tool to the clamp bracket with a washer and cotter pin. Secure the other end to the tilt stop bracket in a similar manner, as shown.

3. Disconnect the electrical cables at both battery terminals. Tag and then disconnect the wiring at the upper and lower solenoids under the powerhead cowling.

4. Remove the harness retainer and any other clamps securing wiring running to the electric motor.

5. Obtain a large C-clamp and a block of wood, approximately 4 in. to 6 in. (10.2 x 21.2 cm) square and about 2 in. (5.1 cm) thick. Support the outboard unit by clamping the block of wood on the inside of the transom up hard against the swivel bracket tube with the large C-clamp, as shown.

6. If the unit being serviced is equipped with thru-the-tilt tube steering, remove the steering cable retaining nut from the end of the tilt tube. Remove the tilt tube nut.

7. Remove the two outboard mounting bolts and the three bolts securing the trim/tilt assembly to the starboard bracket. Carefully move the starboard transom bracket away far enough for the manual release valve to clear when the trim/tilt system is removed.

8. Drive out the cross pin securing the retaining pin for the tilt piston at the upper end. The cross pin should never be used a second time.

TRIM AND TILT 9-17

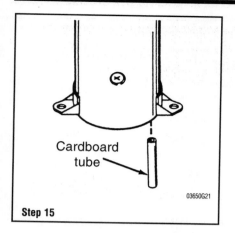
Step 15

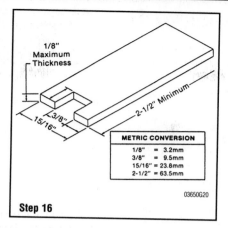
Step 16

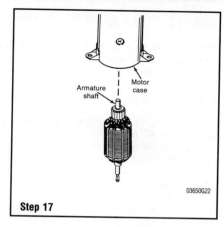
Step 17

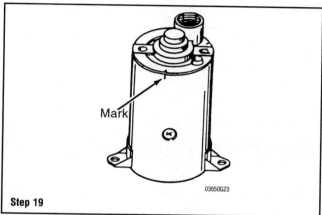
Step 19

mum thickness, the short time involved in cutting out the piece will be time well spent in speeding the assembling process and avoiding frustration in inserting the armature into the case.

17. Apply a light coating of SAE 10W motor oil to the upper end of the armature shaft. Place the tool mentioned in the previous paragraph in place to hold the brushes apart when the armature is installed.

18. Push the armature up into the motor case until the upper end of the armature makes contact with the brush holder tool. Exert just a little upward pressure on the armature and at the same time, slide the brush holder out. The armature will move just a little further into the case and the brushes will bear against the armature in a proper manner.

19. Place the end cap onto the case with the mark, on the mark on the cap aligned with the mark on the case.

20. Install new O-rings onto the pump. If the square motor pump is used, install the flat ring into the groove, as shown in the accompanying exploded drawing. Install a new seal into the recess of the upper end of the pump surface. The armature shaft passes through this seal. Coat the O-rings and the seal with Automatic Transmission Fluid.

21. Grasp the motor case with the thumb and forefinger holding the end cap in place and position it over the pump.

22. Push down on the motor case with the lower end of the armature passing through the pump seal. The chamfered part of the pump must be on the same side as the bolt hole opposite the electrical fitting, as shown.

23. Hold the electric motor and end cap with one hand and with the other hand rotate the pump ever so slightly until the hex end of the armature shaft indexes with the recess in the pump. When this occurs, the end surface of the motor case will be firm against the surface of the pump.

24. Slide the sealing washer onto the thru-bolt with the head. The rubber surface of the washer must face down to bear against the surface of the end cap. Start to thread the bolt into the pump but do not tighten it at this time.

25. Slide a regular washer on the other thru-bolt and then the sealing washer with the rubber surface facing down to bear against the surface of the end cap. Start to thread the second bolt into the pump. A slight shifting of the motor case or the pump may be necessary. Tighten thru-bolts securely.

26. Secure the electrical lead to the thru-bolt without the head, using first a washer, then the wire clamp, another washer and finally the nut. Tighten the nut securely. Apply a coating of Quicksilver Liquid Neoprene around the thru-bolts, around the electrical connection and at the seam between the electric motor and the pump, as a waterproof measure.

27. Feed the electrical wire up through the outboard clamp bracket, if the motor and pump were removed with the trim/tilt assembly remaining on the clamp brackets.

28. Move the assembled pump and motor into position on the trim/tilt assembly. Carefully insert the pump into the power trim assembly housing, with the chamfer (or hole) on the lower end of the pump facing toward the tilt cylinder. Apply a light coating of Quicksilver Perfect Seal (or equivalent) to the threads of the attaching bolts.

29. Slide a lockwasher onto each bolt. Secure the pump on the housing with the attaching hardware. Tighten the bolts securely. Apply a coating of Quicksilver Liquid Neoprene on the seam between the pump and the housing as a waterproof measure.

30. If the trim/tilt assembly was not removed from the brackets, carefully move the starboard transom bracket into place with the manual release valve passing through the opening in the bracket.

31. Apply a light coating of Loctite® on the threads of five bolts. Secure the bracket in place with the two outboard mounting bolts and the three bolts through bracket into the power trim assembly. Tighten the bolts to a torque value of 30 ft. lbs. (40.7Nm).

9. After the cross pin is free, push out the piston retaining pin. Do not allow the cylinder to fall free.

10. Support the trim/tilt system. Remove the three mounting bolts securing the system to the port side clamp bracket. Lower the system from the outboard unit. Pull the wires out with the system.

11. Clamp the trim/tilt system in a vise with the jaws gripping on the anode installed on the underneath side of the system.

12. The pump will be removed with the motor and the two will be separated after removal.

➡ **The pump cannot be serviced. Therefore, if troubleshooting indicates the pump has failed, it must be replaced.**

➡ **The electric motor and the pump may be removed from the trim/tilt assembly without removing the complete unit from the clamp bracket. To remove the motor and pump without disturbing other parts, move the starboard clamp bracket clear. After the clamp bracket is clear, proceed with the next step.**

13. Remove the two bolts securing the pump to the trim/tilt assembly. Lift the electric motor, with the pump attached, upward and free of the trim/tilt assembly base.

➡ **Before separating the electric motor from the pump, scribe a mark on the motor end cap and a matching mark on the motor case as an aid to assembling.**

14. Remove the wire clamp from one of the thru-bolts. Remove the two thru-bolts passing through the electric motor into the pump and then separate the electric motor from the pump. As the electric motor case is being lifted, exercise care not to drop and damage the armature.

To install:

15. Slide the cardboard tube into the case, as shown in the accompanying illustration. Place a new O-ring into place on the end cap. Set the cap aside, ready for installation.

16. As an aid to holding the brushes clear when the armature is inserted into the case, a special tool may be made, as indicated in the accompanying illustration. Because the tool is made of such light weight material, only ⅛ in. (3.2mm) maxi-

9-18 TRIM AND TILT

Tilt/Trim Cylinder

DISASSEMBLY

Trim Cylinders

♦ See accompanying illustrations

→The trim cylinders themselves are integrated into the casting of the tilt/trim assembly and cannot be removed. The trim rod and piston from each cylinder can be removed and serviced without removing the tilt/trim assembly from the boat.

1. Using a spanner wrench with the two tips indexed into the two recesses in the cylinder end cap, remove the end cap by rotating the cap in a counterclockwise direction until it is free.
2. After the end cap is free, carefully withdraw the piston straight up and out of the cylinder.
   To install:
3. After new O-rings have been installed, coat the surface of the piston, O-rings and seal lip with Automatic Transmission Fluid (ATF). Carefully slide the piston into the cylinder.
4. Thread the end cap onto the end of the cylinder and then tighten the cap securely with a spanner wrench. Lubricate the piston rod with anti-corrosion grease.

Tilt Cylinder

→The tilt cylinder can be serviced without being removed from the tilt/trim assembly.

1. Remove the end cap using a spanner wrench with the points indexed into the recesses of the end cap.
2. Rotate the end cap counterclockwise until it is free of the cylinder.
3. Carefully pull the piston and rod assembly straight out of the cylinder, with the end cap on the piston rod.

CLEANING & INSPECTION

♦ See Figure 19

1. Clamp the rod in a vise equipped with soft jaws with the jaws gripping the upper end. Tighten the vise just good and snug to prevent any possible damage to the piston end. Remove the rod end using a spanner wrench and rotating the rod end counterclockwise until it is free. If the rod end refuses to loosen, heat may have to be applied to the upper end of the rod under the piston.

✱✱ WARNING

Do not let the check valve assemblies fall when removing the rod end or the washer.

2. Lift the rod end straight up and out of the piston. Carefully remove the washer and then the check valve assemblies. Slide the piston free of the rod. If necessary, slide the end cap off the rod.
3. If the tilt/trim system is equipped with a design II or design III tilt cylinder, perform the following:
4. Place the rod end into a vise with soft jaws gripping the upper end of the piston. Tighten the vise snug to prevent any possible damage to the upper end. Remove the bolt on (design II) or nut on (design III) rod end. Lift off the washer, O-ring, seven check valve assemblies and the piston. Slide the end cap up and free of the rod. Remove and discard the O-ring from the piston assembly. Clean but do not remove the check valve held in place by the roll pin. If the check valve is defective, the piston assembly must be replaced.
5. Disassemble the end cap by first removing the retaining ring, then the washer. The scraper/seal can now be popped out with a small screwdriver. Remove the O-ring.
6. The check valves on the back side of the piston may be removed, if there is evidence the springs do not function properly. Removal is accomplished by first driving the roll pin out using a fine punch or nail and hammer and then removing the check valve including the spring and ball.
7. Remove the memory piston from the tilt cylinder on Design II and Design III by first draining any fluid remaining in the cylinder. Place a towel or shop cloth on the bench surface and tap the open end of the cylinder (vertical position), onto the cloth, as depicted in the accompanying line drawing. The memory piston will fall free. Remove and discard the O-ring from the memory piston.

Tilt/Trim Switch

TESTING

Complete diagnosis, testing and servicing procedures for the trim/tilt switch are located in the "Remote Control" section of this manual.

Tilt/Trim Solenoid

TESTING

1. Using a multimeter, check for continuity between the "small" terminal on the lower solenoid (the terminal with the Blue/White wire connected) and the tilt button terminal with the Blue/White wire connected. If continuity is not indicated, the wire is open.

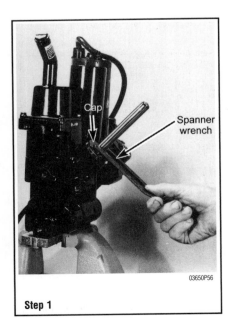

Step 1

Step 2

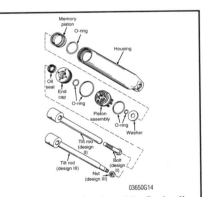

Fig. 19 Exploded drawing of the Design II and Design III tilt piston with major parts identified. O-ring repair kit includes the oil seal. Memory piston for the tilt pistons are different and must be used with the correct tilt rod/cylinder assembly. Memory piston for Design II tilt piston is flat. Design III is dished to clear nut and thread

TRIM AND TILT 9-19

2. If the wire checked OK in the previous step, disconnect all battery leads from the battery. Check for continuity across tilt button terminals with tilt button depressed. If continuity is not indicated, the tilt button is at fault and must be replaced.

Up and Tilt Circuits Both Inoperative—Down Circuit Ok

Before spending time with a multimeter and making other tests, check all connections on the up solenoid and all other wiring for loose or corroded connections, frayed insulation or a break in a wire.

1. Connect the positive multimeter lead to the "small" terminal of the lower solenoid (terminal with the Blue/White wire connected). Connect the negative multimeter lead to the ground terminal of the bottom solenoid (the terminal with the Black lead connected). Depress the tilt button.
2. If no voltage is indicated, there is an open circuit in the Blue/White wire and purple/White (or purple) wires between the up solenoid and the trim buttons.
3. If voltage is indicated, proceed with the next step.
4. Leave the negative meter lead connected to ground. Connect the positive meter lead to the large terminal with two Red leads connected on the lower solenoid.
5. If no voltage is indicated, check for loose or corroded connection at the "large" terminal with the two Red leads connected on the lower solenoid.
6. Also check for loose connection or corrosion at the same terminal on the upper solenoid. Check the condition of the Red lead connecting these two terminals of each solenoid together.
7. If voltage is indicated, proceed with the next step.
8. Leave the negative multimeter lead connected to ground. Connect the positive multimeter lead to the "small" terminal of the lower solenoid (the terminal with the Black wire connected).
9. Depress the tilt button. If no voltage is indicated, the lower solenoid (the up solenoid) is defective and must be replaced.
10. If voltage is indicated proceed with the next step.
11. Inspect the lower solenoid (the up solenoid), for loose or corroded connections. Remove the lower mounting bolt on the lower solenoid and check for paint, grease, dirt, corrosion and the like preventing the connecting wire from making a good ground. Install the bolt.
12. If the up and "tilt" circuits still do not function, disconnect the battery leads at the battery.
13. Label the wires leading to the lower solenoid or draw a quick sketch before disconnecting any of the leads.
14. Remove all connections leading to the solenoid and remove the two attaching bolts.
15. Bench test the solenoid according to the following procedures.

SOLENOID BENCH TESTS

♦ See Figure 20

1. Connect a 12-volt battery across the two smaller terminals. The multimeter should register 0-ohms (zero) indicating continuity and an audible "click" should be heard. The "click" sound will verify the plunger in the solenoid is being drawn up making contact with the inside terminal to close the circuit.

2. If the multimeter indicates a resistance in the solenoid and/or a "click" sound is not heard, the solenoid is defective and must be replaced.
3. If the solenoid checked OK, test the charge condition of the battery and inspect all cables and leads.

Up Circuit Inoperative—Tilt Circuit Ok

♦ See Figures 21 and 22

Before spending time with a multimeter and making other tests, check all connections on the up solenoid and all other wiring for loose or corroded connections, frayed insulation or a break in a wire.

1. Using a multimeter on the ohms scale, check for continuity between the "small" terminal on the lower solenoid (the terminal with the Blue/White wire connected). If continuity is not indicated, there is an open wire in the circuit between these two connections.
2. If continuity is indicated, proceed with the next step.
3. If a push button type switch is being tested connect the positive multimeter lead to the up button on the trim switch (the terminal with the yellow wire connected). If a toggle switch is being tested, connect the positive multimeter lead to the center terminal (the terminal with the brown/White or red/purple wire connected).
4. Connect the negative multimeter lead to a good ground such as the mounting bolt of the lower solenoid.
5. Depress the up trim button. If no voltage is indicated, the trim switch is defective and must be replaced.
6. If voltage is indicated, proceed to the next step.
7. Check for poor or corroded connections at the up button or toggle switch.
8. If the "U" circuit is still inoperative, replace the up button or switch.

Down, Up and Tilt Circuits Inoperative

♦ See Figure 23

Before spending time with a multimeter and making other tests, check all connections on both solenoids and all other wiring for loose or corroded connections, frayed insulation or a break in a wire.

1. Check the two 20 amp in-line fuses.
2. Replace the defective fuse and again test the up, down and tilt operation of the trim/tilt system. If the fuses were not "blown", proceed with the next step.
3. Connect the positive lead of a multimeter to the large terminal of the lower solenoid (the terminal with the two Red wires connected). Connect the negative lead of the multimeter to the "small" terminal of the lower solenoid (the terminal with the Black wire connected).
4. The multimeter should indicate close to 12-volts.
5. If no voltage is indicated, check the battery leads and the Red leads between the starter motor and the large terminal on the upper solenoid (the terminal with 2 Red wires connected) and the same terminal on the lower solenoid (the terminal with the two Red wires connected). Check for poor connections, corrosion or an open circuit.
6. If battery voltage is indicated, proceed with the next step.
7. Leave the negative multimeter lead connected to ground. Connect the posi-

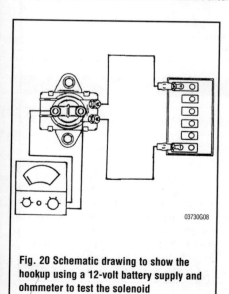

Fig. 20 Schematic drawing to show the hookup using a 12-volt battery supply and ohmmeter to test the solenoid

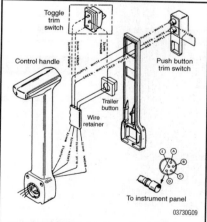

Fig. 21 Exploded drawing of the control handle. This illustration and the wiring diagram in the next column will be helpful during the troubleshooting work.

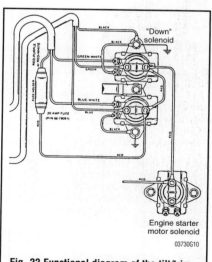

Fig. 22 Functional diagram of the tilt/trim system.

TRIM AND TILT

Fig. 23 Location of the other in-line fuse on powerheads equipped with the Power Trim/Tilt unit.

tive multimeter lead to the "small" terminal on the lower solenoid (the terminal with the Blue/White wire connected.

8. Depress the tilt button and check for close to 12-volts. If battery voltage is indicated, check the Black ground wires at the solenoids (the upper and lower mounting bolts with Black wires attached). Particularly, the Black motor lead from the harness, grounded at the top mounting bolt.

➡ **If this wire is properly grounded and circuits are still inoperative, the pump motor may be faulty. Refer to the "Ignition and Electrical" section for troubleshooting electric motors.**

9. If no voltage is indicated, proceed to the next step.
10. Leave the negative multimeter lead connected to ground as in previous steps. Connect the positive multimeter lead to the "tilt" button (the terminal with the brown/White and red/purple wires connected).
11. If battery voltage is indicated, there is an open circuit in each wire between the trim buttons (or toggle switch) and the trim motor.
12. Check for pinched or severed wires and inspect all trim harness connectors for loose or corroded connections.
13. If no voltage is indicated, proceed with the next step.
14. Verify voltage is being supplied to the controls by performing the following checks:
15. Do not start the engine but turn the ignition switch to the run position. Use a multimeter and check for voltage at any instrument.
16. If close to 12-volts is indicated, there is an open circuit in the wire between the tilt button (the terminal with the brown/White and red/purple wires connected) and the "B" terminal on the back of the ignition switch (the terminal with the gray wire connected), as shown in the illustration.
17. If no voltage is indicated, proceed to the next step.
18. Check for continuity between the large terminal on the starter solenoid (the terminal with the Red wire connected) and the "B" terminal on the back of the ignition switch. Check for open wires, loose or corroded connections.

Down Circuit Inoperative—Up Circuit Ok

Check all connections on the down solenoid and all other wiring for loose or corroded connections, frayed insulation or a break in the wire.

1. Connect the positive multimeter lead to the small terminal of the upper solenoid (the terminal with the green/White wire connected). Connect the negative multimeter lead to the top mounting bolt of the upper solenoid (the ground terminal with the Black wire connected).
2. Depress the down trim button or toggle switch.
3. If no voltage is indicated, leave the negative multimeter lead in place and connect the positive multimeter lead to the down button (the terminal with the green/White wire connected).
4. Depress the down button. If close to 12-volts is indicated, there is an open wire between the down push button (or toggle switch) and the small terminal on the upper solenoid (the terminal with the green/White wire connected).
5. If no voltage is indicated, proceed with the next step.
6. Leave the negative multimeter lead connected. Connect the positive meter lead to the down terminal of the push button (the terminal with the red/purple wire connected). If testing a toggle switch, connect the positive meter lead to the center terminal (the terminal with the brown/White or red/purple wire connected).
7. If close to 12-volts is indicated, the trim switch is faulty and must be replaced.
8. If no voltage is indicated, check for loose or corroded connections at the testing points in this step. Test the down circuit for operation.
9. If still inoperative, proceed with the next step.
10. Leave the negative multimeter lead in place and connect the positive multimeter lead to the "large" terminal on the upper solenoid (the terminal with the two Red wires connected).
11. If no voltage is indicated, there is an open circuit between this terminal and the positive battery terminal. Check the entire length of the wires for damage to insulation, evidence of sparking, loose or corroded connections.
12. If close to 12-volts is indicated, proceed with the next step.
13. Leave the negative multimeter lead connected and connect the positive multimeter lead to the "small" terminal on the upper solenoid (the terminal with the green/White wire connected). Depress the down trim button or toggle switch.
14. If no voltage is indicated the down solenoid is defective and must be replaced.
15. If close to 12-volts is indicated, proceed with the next step.
16. Tag the wires or make a quick sketch and then remove the down solenoid.
17. Test the solenoid according to the following procedures.

SOLENOID BENCH TESTS

♦ See Figure 24

1. The following simple quick test can be performed to check the integrity of the solenoid:
   a. Connect a 12-volt battery across the two smaller terminals. The ohmmeter should register 0-ohms (zero) indicating continuity and an audible "click" should be heard. The "click" will verify the plunger in the solenoid is being drawn up making contact with the inside terminal to close the circuit.
   b. If the ohmmeter indicates a resistance in the solenoid and/or a "click" is not heard, the solenoid is defective and must be replaced.
   c. If the solenoid checked "OK", test the charge condition of the battery and inspect all cables and leads.

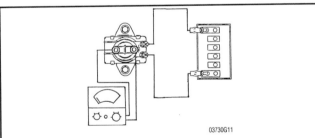

Fig. 24 Functional diagram depicting a battery hooked up directly to one of the solenoids for testing.

Tilt/Trim Relay

Some systems are equipped with relays rather than solenoids to operate the pump motor. The relays are compact, sealed units that are easily changed due to their plug-in design.

TESTING

If the trim system operates in one direction but fails in the opposite direction, a quick and easy test can be performed to determine if one of the relays has failed. And which one has failed.

1. Mark both relays. The top relay is the up relay and the lower one is the down relay.
2. Unplug both relays from the plug-in connectors. Plug each relay back into the opposite connector from which it was removed.
3. Depress the trim/tilt switch. If the system operates, the relay is good.
4. Press the tilt/trim switch in the opposite direction. If the system fails to operate in the correct manner, in the previously good direction, the relay is faulty and needs to be replaced.

REMOTE CONTROL BOX 10-2
DESCRIPTION AND OPERATION 10-2
TROUBLESHOOTING THE REMOTE
  CONTROLS 10-2
REMOTE CONTROL BOX 10-2
    REMOVAL & INSTALLATION 10-2
REMOTE CONTROL CABLES 10-3
    ADJUSTMENT 10-3
NEUTRAL START SWITCH 10-4
    TESTING 10-4
    REMOVAL & INSTALLATION 10-4
ENGINE STOP SWITCH 10-5
    TESTING 10-5
    REMOVAL & INSTALLATION 10-5
EMERGENCY STOP SWITCH
  (LANYARD) 10-5
    TESTING 10-5
    REMOVAL & INSTALLATION 10-5
IGNITION SWITCH 10-5
    TESTING 10-5
    REMOVAL & INSTALLATION 10-5
TRIM/TILT SWITCH 10-6
    TESTING 10-6
    REMOVAL & INSTALLATION 10-6
TILLER HANDLE 10-6
DESCRIPTION AND OPERATION 10-6
TROUBLESHOOTING THE TILLER
  HANDLE 10-6
TILLER HANDLE 10-6
    REMOVAL & INSTALLATION 10-6
CONTROL CABLES 10-10
    REMOVAL & INSTALLATION 10-10
ENGINE STOP SWITCH 10-12
    TESTING 10-12
    REMOVAL & INSTALLATION 10-12
EMERGENCY STOP SWITCH 10-12
    TESTING 10-12
    REMOVAL & INSTALLATION 10-12
ENGINE START SWITCH 10-12
    TESTING 10-12
    REMOVAL & INSTALLATION 10-12

10

REMOTE CONTROL

REMOTE CONTROL BOX 10-2
TILLER HANDLE 10-6

10-2 REMOTE CONTROL

REMOTE CONTROL BOX

Description and Operation

▶ See Figure 1

The remote control box allows the person operating the boat to control the throttle operation and shift movements from a location other than where the outboard is mounted. In most cases, the remote control box is mounted approximately halfway forward (midship) on the starboard side of the boat.

Several control box types are used, but all usually house a key switch, engine stop switch, choke switch, neutral safety switch, warning buzzer and the necessary wiring and cable hardware to connect the control box to the outboard unit.

➡There are many different types of remote control assemblies that can be installed on your boat. The remote control assemblies covered in this manual are the ones most commonly used by the manufacturer.

Troubleshooting the Remote Controls

One of the things taken for granted on most boats is the engine controls and cables. Depending on how they are originally routed, they will either last the life of the vessel or can be easily damaged. These cables should be routinely maintained by careful inspection for kinks or other damage and lubrication with a marine grade grease.

If the cables do not operate properly, have a helper operate the controls at the helm while you observe the cable and linkage operation at the powerhead. Make sure nothing is binding, bent or kinked. Check the hardware that secures the cables to the boat and powerhead to make sure they are tight. Inspect the clevis and cotter pins in the ends of the cables and also give some attention to the cable release hardware.

Another area of concern is the neutral start switch that is sometimes located in the control box and sometimes on the powerhead. This switch may fall out of adjustment and prevent the engine from being started. It is easily inspected using a multimeter by performing a continuity check.

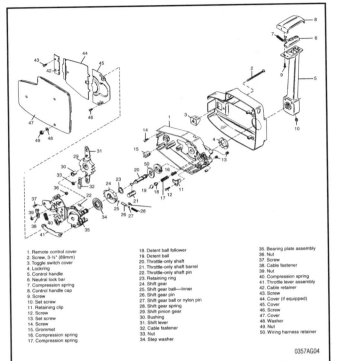

Fig. 1 Exploded drawing of a typical Mercury control box with major parts identified.

Remote Control Box

REMOVAL & INSTALLATION

▶ See accompanying illustrations

1. Turn the ignition key to the off position. Disconnect the high tension leads from the spark plugs, with a twisting motion.
2. Disconnect the remote control wiring harness plug from the outboard trim/tilt motor and pump assembly.
3. Disconnect the tachometer wiring plug from the forward end of the control housing.
4. Remove the three locknuts, flat washers, and bolts securing the control housing to the mounting panel. One is located next to the run button (the ignition safety stop switch), and the second is beneath the control handle on the lower portion of the plastic case. The third is located behind the control handle when the handle is in the neutral position. Shift the handle into forward or reverse position to remove the bolt, then shift it back into the neutral position for the following steps.
5. Pull the remote control housing away and free of the mounting panel.

Remove the plastic cover from the back of the housing. Lift off the access cover from the housing. (Some "Commander" remote control units do not have an access cover.)

6. Remove the two screws securing the cable retainer over the throttle cable, wiring harness, and shift cable. Unscrew the two Phillips-head screws securing the back cover to the control module, and then lift off the cover.
7. Loosen the cable retaining nut and raise the cable fastener enough to free the throttle cable from the pin. Lift the cable from the anchor barrel recess. Remove the grommet.
8. Shift the outboard unit into reverse gear by depressing the neutral lock bar on the control handle and moving the control handle into the reverse position.
9. Loosen, but do not remove, the shift cable retainer nut with a ⅜ in. deep socket as far as it will go without removing it. Raise the shift cable fastener enough to free the shift cable from the pin.

➡Do not attempt to shift into reverse while the cable fastener is loose. An attempt to shift may cause the cable fastener to strike the neutral safety micro-switch and cause it damage.

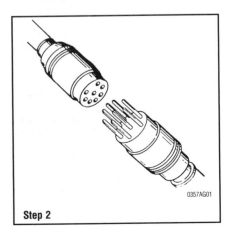

Step 2

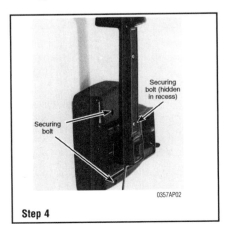

Step 4

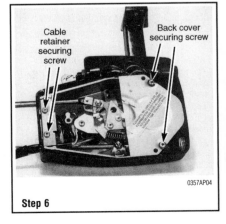

Step 6

REMOTE CONTROL 10-3

Step 7

Step 9

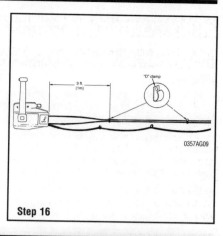
Step 16

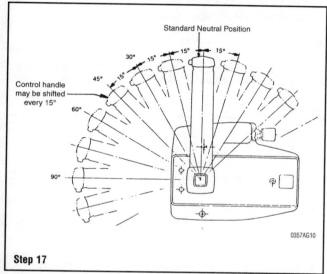
Step 17

10. Lift the wiring harness out of the cable anchor barrel recess and remove the shift cable from the control housing.

To install:

11. Position the control housing in place on the mounting panel and secure it with the three long (3-½ in.) bolts, flat washers, and locknuts. One is located next to the run button (the ignition safety stop switch). The second is beneath the control handle on the power portion of the plastic case. The third bolt goes in behind the control handle when the handle is in the neutral position.

12. In order to install this bolt, shift the handle into the forward or reverse position, and then install the bolt. After the bolt is secure, shift the handle back to the neutral position for the next few steps.

13. Connect the tachometer wiring plug to the forward end of the control housing.

➡**Clean the prongs of the connector with crocus cloth to ensure the best connection possible. Exercise care while cleaning to prevent bending the prongs.**

14. Connect the remote control wiring harness plug from the outboard trim/tilt motor and pump assembly.

15. Install the high-tension leads to their respective spark plugs.

16. Route the wiring harness alongside the boat and fasten with the "Sta-Straps". Check to be sure the wiring will not be pinched or chafe on any moving part and will not come in contact with water in the bilge. Route the shift and throttle cables the best possible way to make large bends and as few as possible. Secure the cables approximately every three feet (one meter).

17. The neutral position of the remote control handle may be changed to any one of a number of convenient angles to meet the owner's preference. The change is accomplished by shifting the handle one spline on the shaft at a time. Each spline equals 15° of arc, as shown.

Remote Control Cables

ADJUSTMENT

Shift Cable

25–125 HP

1. On the remote control units equipped with a neutral lock bar, depress the neutral lock bar and secure it in this position with a strong rubber band or a piece of tape. This is necessary to ensure the correct location of the true neutral detent while installing the shift and throttle cables to the powerhead.

2. On models equipped with a small neutral warm-up lever on the side of the control box, push the lever to the full down position.

3. Before starting the adjustments, check to be sure the cable and guide setscrews have been tightened to a torque value of 25 inch lbs. (2.8Nm).

4. Slowly move the remote control handle toward the forward gear position, and at the same time observe the cable end guides for movement. The shift cable end guide should be the first to move. Return the control handle to the neutral position.

5. Feed the shift cable end guide and brass barrel through the rubber grommet in the bottom cowling.

6. Adjust the brass barrel until the length between the brass barrel and the hole in the cable end guide is equal to the amount between the barrel retainer and the shift linkage peg.

7. Secure the cable end guide in place with the retainer.

8. Move the remote control lever to the forward gear position. With the lever in this position, it should not be possible to rotate the propeller counterclockwise.

➡**If it is possible to rotate the propeller, adjust the brass barrel closer to the cable end guide, repeat the previous steps until the propeller cannot be turned in a counterclockwise direction.**

9. Shift the remote control lever to the neutral position, without going past the neutral detent. The propeller shaft should now be free to rotate without any drag. Adjust the brass barrel away from the cable end guide, if necessary.

10. Rotate the propeller and at the same time shift the remote control lever to the reverse position. The propeller shaft must not be free to rotate more than 120° in either direction. Adjust the brass barrel away from the cable end guide, if necessary.

11. Move the control lever to the neutral position without moving past the neutral detent. The propeller should now be free to rotate in either direction without any drag. Adjust the brass barrel closer to the cable end guide, if necessary.

135–275 HP

1. Disconnect and remove the shift cable. Manually shift the outboard into the neutral position.

2. Slide the guide block forward until some resistance is felt. Mark the location of the guide pin. Slide the guide block aft until resistance is felt again. Mark the location of the guide pin.

3. Slide the guide block forward until the guide pin is centered between the two marks previously made.

4. Position the remote control handle in the neutral position. Feed the shift cable and brass barrel through the rubber grommet in the bottom cowling.

10-4 REMOTE CONTROL

5. Insert the end of the shift cable onto the guide block pin. Center the guide block pin halfway between the two marks made previously.

6. Rotate the cable retainer over the guide block pin and tighten the nut in the center of the guide block securely.

7. Adjust the brass barrel to align and slip into the barrel receptacle without any preload.

➡ **On the XR4/Magnum II or XR6/Magnum III models, adjust the cable barrel so that a slight preload in the reverse direction is required for the barrel to slip into the barrel retainer.**

8. Move the remote control lever to the forward gear position. With the lever in this position, it should not be possible to rotate the propeller counterclockwise. If it is possible to rotate the propeller, adjust the brass barrel closer to the cable end guide and repeat the previous steps needed to complete the cable adjustment.

9. Shift the remote control lever to the neutral position, without going past the neutral detent. The propeller shaft should now be free to rotate without any drag. Adjust the brass barrel away from the cable end guide, if necessary, and repeat the previous steps needed to complete the cable adjustment.

10. Rotate the propeller and at the same time, shift the remote control lever into reverse. The propeller shaft must not turn clockwise. Adjust the brass barrel away from the cable end guide, if necessary, and repeat the previous steps needed to complete the cable adjustment.

11. Move the remote control lever to the neutral position without moving it past the neutral detent. The propeller should now be free to rotate in either direction without any evidence of drag. Adjust the brass barrel closer to the cable end guide, if necessary, and repeat the previous steps needed to complete the cable adjustment.

Throttle Cable

25–125 HP

1. Feed the throttle cable end guide and the brass barrel through the rubber grommet in the bottom cowling.

2. Move the remote control handle to the neutral position. Move the warm-up lever to the full down position.

3. For early models equipped with a distributor, move the throttle lever to a position where the distributor is held lightly against the idle stop screw mounted on the stop plate. For later models without a distributor, move the throttle lever towards the rear of the powerhead until it rests against the idle stop screw mounted against the powerhead. Install the cable end guide over the peg in the throttle lever swivel.

4. Adjust the brass barrel in the necessary direction until the barrel can be installed in the barrel retainer.

5. Move the remote control handle to open the throttle. Hold a thin sheet of paper between the idle stop screw and the idle stop on the engine. Continue to hold the paper and close the throttle by moving the remote control handle to the neutral position. The neutral warm-up lever must be in the full down position. The neutral warm-up lever must be in the full down position.

   a. If the paper cannot be removed at this time, the brass barrel is preloaded against the idle stop screw.

   b. If the paper can be removed with no drag, the barrel is not close enough to the idle stop screw. Adjust the brass barrel and repeat the check to be sure it has not been moved too far.

6. Move the remote control to the forward gear position and continue to move the handle to the end of its travel (the wide-open throttle position). Check to be sure that the throttle plates on the carburetors are at the full open position, but not jammed against the stop.

7. On models so equipped, place the battery lead in the recess and secure the lead with the retainer and strap.

➡ **Excessive preload on the throttle cable may cause difficult shifting from the forward gear position to the neutral position. If this condition should occur, adjust the throttle cable barrel to reduce the preload.**

135–275 HP

1. Feed the end of the throttle cable through the rubber grommet in the bottom cowling.

2. Place the remote control handle to the neutral position. Move the warm-up lever to the full down position.

3. Connect a flushing device to the lower unit and start the powerhead.

4. Adjust the powerhead speed, with the lower unit in forward gear. After the idle speed has been set, shut down the powerhead.

5. Slip the end of the throttle cable over the pin on the throttle lever. Secure the throttle cable in place by closing the latch. While holding the throttle lever against the idle stop, adjust the throttle cable barrel to slip into the barrel receptacle with very light preload on the throttle lever against the idle stop. Secure the throttle and shift cable barrels in place by closing the barrel retainer.

6. Move the remote control handle to open the throttle. Hold a thin sheet of paper between the idle speed stop screw and the idle stop on the powerhead. Continue to hold the paper and close the throttle by moving the remote control handle to the neutral position.

➡ **The neutral warm-up lever must be in the full down position.**

7. If the paper cannot be removed at this time, the brass barrel is preloaded against the idle stop screw.

8. If the paper can be removed with no drag, the barrel is not close enough to the idle stop screw. Adjust the brass barrel and repeat the check to be sure it wasn't moved too far.

9. Move the remote control to the forward gear position and continue to move the handle to the end of its travel (wide-open throttle).

10. Check to be sure that the throttle plates on the carburetors are at the full throttle position, but not jammed against the stop.

✱✱ CAUTION

Excess preload on the throttle cable may cause difficulties when shifting from forward to neutral. If this condition should occur, adjust the throttle cable barrel to reduce the preload.

Neutral Start Switch

TESTING

♦ See Figure 2

1. Disconnect the neutral start switch wiring harness.
2. Connect a multimeter between the switch harness leads.
3. With the remote control lever in the neutral position, continuity should exist. With the remote control lever in the forward or reverse position, continuity should not exist.
4. If the switch does not function as specified, there is a short in either the switch or switch wiring harness. The switch will need to be replaced.
5. If the switch functions properly, there may be a problem with the powerhead wiring harness.

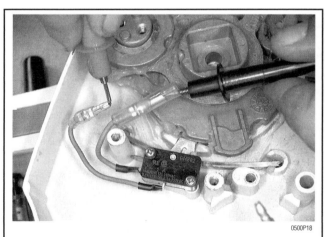

Fig. 2 Testing a typical neutral start microswitch. With the remote control lever in the neutral position, continuity should exist. With the remote control lever in the forward or reverse position, continuity should not exist

REMOVAL & INSTALLATION

1. Remove the control box from the boat and open the side covers to access the internal components.
2. Disconnect the neutral start switch wiring harness.
3. Remove any wire straps that connect the switch to the control box.
4. Remove any retaining nuts/screws that secure the switch to the control box.

REMOTE CONTROL 10-5

5. Remove the switch from the control box.

To install:
6. Install the switch into the control box.
7. As required, install the switch retaining screw/nut and tighten securely.
8. Install any wire straps that connect the switch to the control box or bracket.
9. Test the switch for proper operation.
10. Install the control box side covers and mount the control box in the boat.

Engine Stop Switch

TESTING

1. Disconnect the engine stop switch wiring harness.
2. Connect a multimeter between the switch harness leads.
3. With the switch engaged (button pushed), continuity should exist. With the switch released (button not pushed), continuity should not exist.
4. If the switch does not function as specified there is either a fault in the switch or switch wiring harness and the switch will need to be replaced.
5. If the switch functions properly, there may be a problem with the powerhead wiring harness.

REMOVAL & INSTALLATION

1. Remove the control box from the boat and open the side covers to access the internal components.
2. Disconnect the stop switch wiring harness.
3. Remove any wire straps that connect the switch to the control box.
4. Remove any retaining nuts/screws that secure the switch to the control box.
5. Remove the switch from the control box.

To install:
6. Install the switch into the control box.
7. As required, install the switch retaining screw/nut and tighten securely.
8. Install any wire straps that connect the switch to the control box or bracket.
9. Test the switch for proper operation.
10. Install the control box side covers and mount the control box in the boat.

Emergency Stop Switch (Lanyard)

TESTING

♦ See Figure 3

1. Disconnect the engine stop switch wiring harness.
2. Connect a multimeter between the switch harness leads.
3. With the switch engaged (stop switch lanyard in pulled), continuity should exist. With the switch released (stop switch lanyard in position), continuity should not exist.

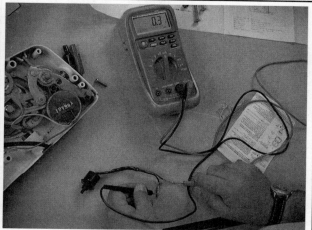

Fig. 3 Connect a multimeter between the switch harness leads. With the switch released (as shown), continuity should not exist

4. If the switch does not function as specified, there is a fault with either the switch or switch wiring harness and the switch will need to be replaced.
5. If the switch functions properly, there may be a problem with the powerhead wiring harness.

REMOVAL & INSTALLATION

1. Remove the control box from the boat and open the side covers to access the internal components.
2. Disconnect the emergency stop switch wiring harness.
3. Remove any wire straps that connect the switch to the control box.
4. Remove any retaining nuts/screws that secure the switch to the control box.
5. Remove the switch from the control box.

To install:
6. Install the switch into the control box.
7. As required, install the switch retaining screw/nut and tighten securely.
8. Install any wire straps that connect the switch to the control box or bracket.
9. Connect the emergency stop switch wiring harness.
10. Test the switch for proper operation.
11. Install the control box side covers and mount the control box in the boat.

Ignition Switch

TESTING

♦ See Figure 4

1. Disconnect the ignition switch wiring harness.
2. Connect a multimeter between the switch harness leads.
3. Consult the wiring diagrams for proper test positions and wiring colors.

➡ **Continuity in the wiring diagram is indicated by two dots with a line drawn connecting them.**

4. With the switch in the stated positions, check for continuity between the various terminals.
5. If the switch functions properly, there may be a problem with the powerhead wiring harness.
6. If the switch does not function as specified, the switch may be faulty and need to be replaced.

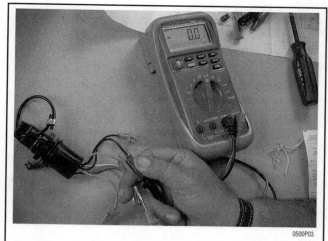

Fig. 4 Connect a multimeter between the switch harness leads and test for continuity

REMOVAL & INSTALLATION

1. Remove the control box from the boat and open the side covers to access the internal components.
2. Disconnect the ignition switch wiring harness.
3. Remove any wire straps that connect the switch to the control box.
4. Remove the switch attaching nut.
5. Remove the switch from the control box.

10-6 REMOTE CONTROL

To install:
6. Install the switch in the control box.
7. Install the switch attaching nut and tighten it securely.
8. Install any wire straps that connect the switch to the control box.
9. Connect the ignition switch wiring harness.
10. Test the switch for proper operation.
11. Install the control box side covers and mount the control box in the boat.

Trim/Tilt Switch

TESTING

1. Disconnect the tilt/trim switch wiring harness.
2. Connect a multimeter between the switch harness terminals as illustrated.
3. With the switch in the stated positions, check for continuity between the leads.
4. If the switch functions properly, there may be a problem with the powerhead wiring harness.
5. If the switch does not function properly as specified, the switch or switch wiring harness may be faulty and needs to be replaced.

REMOVAL & INSTALLATION

1. Remove the control box from the boat and open the side covers to access the internal components.
2. Remove the control lever from the control box assembly. Be careful when removing the handle to not pull the wiring harness.
3. Label and disconnect the tilt/trim wiring harness.
4. Remove the screw securing the neutral lock bar.
5. Remove the neutral lock bar and control lever grip.
6. Remove the tilt/trim switch from the lever.

To install:
7. Install the tilt/trim switch into the control lever.
8. Thread the tilt/trim wiring harness through the control lever and install the neutral lock bar and control lever grip.
9. Connect the tilt/trim wiring harness
10. Install the control lever onto the control box assembly and tighten the retaining fasteners snugly.
11. Install the control box on the boat

TILLER HANDLE

Description and Operation

Steering control for most outboards begins at the tiller handle and ends at the propeller. Tiller steering is the most simple form of small outboard control. All components are mounted directly to the engine and are easily serviceable.

Throttle control is performed via a throttle grip mounted to the tiller arm. As the grip is rotated a cable opens and closes the throttle lever on the engine. An adjustment thumbscrew is usually located near the throttle grip to allow adjustment of the turning resistance. In this way, the operator does not have to keep constant pressure on the grip to maintain engine speed.

An emergency engine stop switch is used on most outboards to prevent the engine from continuing to run without the operator in control. This switch is controlled by a small clip which keeps the switch open during normal engine operation. When the clip is removed, a spring inside the switch closes it and completes a ground connection to stop the engine. The clip is connected to a lanyard that is worn around the helmsman's wrist.

Some tiller systems utilize a throttle stopper system which limits throttle opening when the shift lever is in neutral and reverse. This prevents overrevving the engine under no-load conditions and also limits the engine speed when in reverse.

Troubleshooting the Tiller Handle

If the tiller steering system seems loose, first check the engine for proper mounting. Ensure the engine is fastened to the transom securely. Next, check the tiller hinge point where it attaches to the engine and tighten the hinge pivot bolt as necessary.

Excessively tight steering that cannot be adjusted using the tension adjustment is usually due to a lack of lubrication. Once the swivel case bushings run dry, the steering shaft will get progressively tighter and eventually seize. This condition is generally caused by a lack of periodic maintenance.

Correct this condition by lubricating the swivel case bushings and working the outboard back and forth to spread the lubricant. However, this may only be a temporary fix. In severe cases, the swivel case bushings may need to be replaced.

Tiller Handle

REMOVAL & INSTALLATION

2.5–3.3 HP

1. Remove the engine cover.
2. Remove the tiller handle pivot bolt and carefully remove the tiller handle.
3. Remove the mounting collar, mounting rubber, distance collar and special washer from the tiller handle.

To install:
4. Install the mounting collar, mounting rubber, distance collar and special washer on the tiller handle.
5. Position the tiller handle on the outboard and install the tiller handle pivot bolt. Tighten the pivot bolt securely.
6. Check the handle for smooth operation.
7. Install the engine cover.

4 and 5 HP

♦ See Figure 5

1. Loosen the throttle cable set screw and pull the throttle wire from the throttle arm of the carburetor.

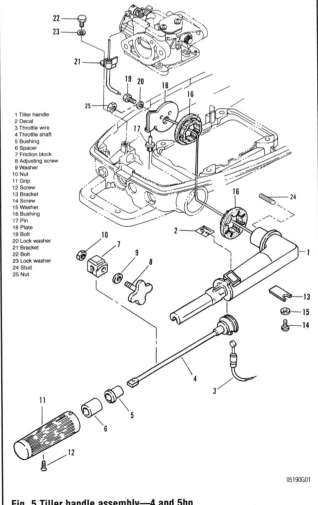

1 Tiller handle
2 Decal
3 Throttle wire
4 Throttle shaft
5 Bushing
6 Spacer
7 Friction block
8 Adjusting screw
9 Washer
10 Nut
11 Grip
12 Screw
13 Bracket
14 Screw
15 Washer
16 Bushing
17 Pin
18 Plate
19 Bolt
20 Lock washer
21 Bracket
22 Bolt
23 Lock washer
24 Stud
25 Nut

Fig. 5 Tiller handle assembly—4 and 5hp

REMOTE CONTROL 10-7

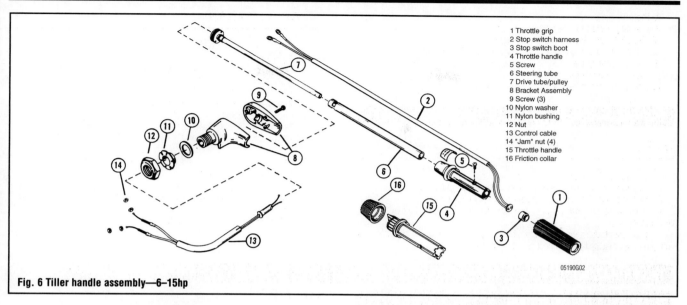

Fig. 6 Tiller handle assembly—6–15hp

2. Remove the cable jacket, bolts and lock washer.
3. Remove the plate.
4. Pull the tiller handle from the bottom cowling. Then remove the inner bushing.

To install:
5. Install the inner bushing into the bottom cowling. Lubricate the bushing with marine grade lubricant.
6. Install the tiller handle, routing the throttle wire into the bottom cowling.
7. Install the plate over the throttle wire.
8. Secure the tiller handle to the bottom cowling using the plate, lockwashers and bolts. Torque the bolt to 70 inch lbs. (8Nm).

6–15 HP

♦ See Figure 6

1. Remove the control cables from the secondary gear pulley.
2. Disconnect the stop switch leads from the switch box.
3. Remove the screws and cap.
4. Pull the handle assembly away from the anchor bracket and unwrap the cables.
5. Pull the cable assembly out, from the back side of the anchor bracket.

To install:
6. Route the switch harness and control cables through the tiller handle mounting bracket.

➡ Do not push the guide into the mounting bracket at this time.

7. Wrap the control cables around the drive pulley.
8. Being careful to not damage the stop switch wiring harness, place the assembled tiller handle into the mounting bracket.
9. Pull the cable ends to remove any slack and secure the tiller handle to the mounting bracket using the cap and screws.
10. Secure the control cables to the secondary gear.
11. Connect the stop switch leads to the switch box.

20 and 25 HP

♦ See Figure 7

1. Remove the bolts securing the tiller handle to the anchor bracket and remove the tiller handle assembly from the powerhead.

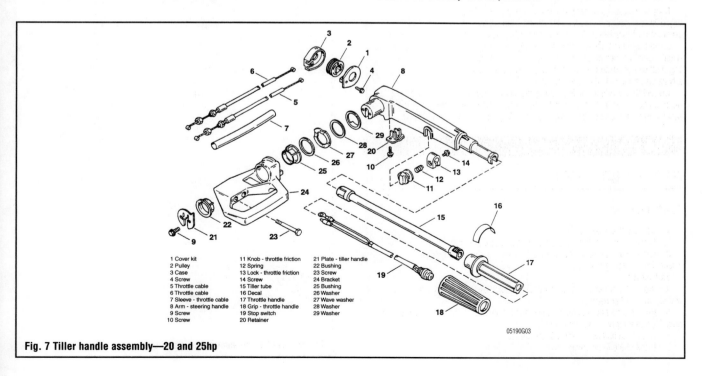

Fig. 7 Tiller handle assembly—20 and 25hp

10-8 REMOTE CONTROL

2. Remove the bushing, flat washers, wave washer and tiller handle washer. Remove the retainer and bolt.
3. Slide the tiller tube out of the pulley case.
4. Remove the pulley case from the tiller handle and remove the cover bolt.
5. Remove the cover and lift the pulley assembly from the case. Replace the control cables if necessary.

To install:
6. Wrap the control cables around the pulley. The top cable wraps and locks in the top groove. The bottom cable wraps and locks in the bottom groove.
7. Place the pulley and cable assembly into the pulley case.
8. Install the pulley cover and secure it with the retaining bolt.
9. Install the pulley assembly into the tiller handle and slide the tiller tube into the pulley.
10. Secure the tiller tube in the handle with the retainer and bolt. Torque the bolt to 50 inch lbs. (5.6Nm).
11. Install the tiller washer (the tab aligns with the slot in the handle), plain washer, wave washer and flanged bushing over the cable/harness assembly.
12. Slide the tiller handle assembly into the anchor bracket.
13. Route the stop button harness through the fuel connector opening in the bottom cowling.
14. Route the control cables through the opening in the bottom cowling.
15. Align the tabs of the inner and outer flanged bushings with the slots in the anchor bracket.
16. Pull on the cable ends to remove any slack and secure the tiller handle to the anchor bracket with thew plate and bolts. Torque the bolts to 80 inch lbs. (9.0Nm).

30 and 40 HP

♦ See Figure 8

1. Disconnect the throttle cable from the engine.
2. Disconnect the lanyard stop switch and neutral start switch wiring harnesses at the bullet connectors.
3. Remove the shift rod link through the opening in the bottom cowling.
4. Remove the tiller handle from the bracket.

To install:
5. Insert the shift rod link through the opening in the bottom cowling.
6. Route the shift link rod around the port side of the powerhead and position it through the lower opening in the rubber grommet.
7. Secure the tiller handle assembly to the studs on the powerhead steering arm using the tab washers and locknuts. Torque the locknuts to 33 ft. lbs. (45Nm) and bend the tab washers against the locknuts.

50 and 60 HP

♦ See Figure 9

1. Remove the nut securing the throttle cable to the engine.
2. Release the latch and remove the throttle cable with grommet from the bottom cowl.
3. Remove the cotter key from the shift rod.
4. Remove the cover.
5. Disconnect the stop switch wire at the bullet connector and screw securing the ground wire.
6. Remove the plug from the tiller handle bracket.
7. Remove the bolt and nut from the tiller handle bracket.
8. Remove the tiller handle, nylon bushings, stainless steel spacer and flat washers from the bracket.
9. Bend the tab washer away from the bolt securing the shift lever and remove the bolt and lever from the bracket.

To install:
10. Install the shift lever with detent assembly to the bracket.
11. Install the tiller handle with bushings to the bracket. Secure in place with the washers, bolt and nut. Tighten the bolt so that it allows tiller handle movement.
12. Torque the nut to 40 ft. lbs. (54Nm).
13. Rinstall the plug.
14. Connect the stop switch wire at the bullet connector and secure the ground wire to the handle with the screw.
15. Install the cover.

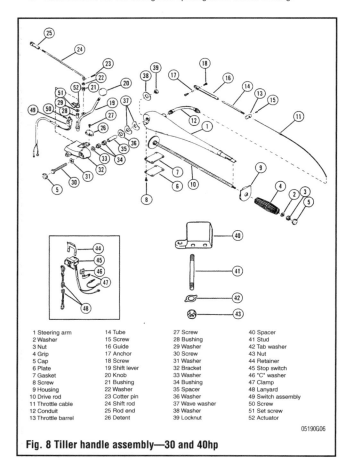

1 Steering arm	14 Tube	27 Screw	40 Spacer
2 Washer	15 Screw	28 Bushing	41 Stud
3 Nut	16 Guide	29 Washer	42 Tab washer
4 Grip	17 Anchor	30 Screw	43 Nut
5 Cap	18 Screw	31 Nut	44 Retainer
6 Plate	19 Shift lever	32 Bracket	45 Stop switch
7 Gasket	20 Knob	33 Washer	46 "C" washer
8 Screw	21 Bushing	34 Bushing	47 Clamp
9 Housing	22 Washer	35 Spacer	48 Lanyard
10 Drive rod	23 Cotter pin	36 Washer	49 Switch assembly
11 Throttle cable	24 Shift rod	37 Wave washer	50 Screw
12 Conduit	25 Rod end	38 Washer	51 Set screw
13 Throttle barrel	26 Detent	39 Locknut	52 Actuator

Fig. 8 Tiller handle assembly—30 and 40hp

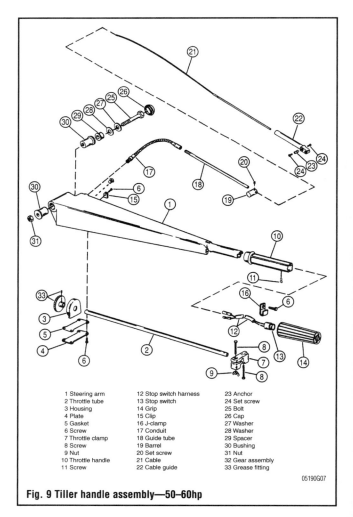

1 Steering arm	12 Stop switch harness	23 Anchor
2 Throttle tube	13 Stop switch	24 Set screw
3 Housing	14 Grip	25 Bolt
4 Plate	15 Clip	26 Cap
5 Gasket	16 J-clamp	27 Washer
6 Screw	17 Conduit	28 Washer
7 Throttle clamp	18 Guide tube	29 Spacer
8 Screw	19 Barrel	30 Bushing
9 Nut	20 Set screw	31 Nut
10 Throttle handle	21 Cable	32 Gear assembly
11 Screw	22 Cable guide	33 Grease fitting

Fig. 9 Tiller handle assembly—50–60hp

REMOTE CONTROL 10-9

16. Install the cotter key and components to the shift rod.
17. Install the throttle cable with grommet.

75–100 HP

♦ See Figure 10

1. Remove the nuts securing the shift link rod and throttle cable to the engine. Release the latch and remove the shift link rod and throttle cable from the anchor bracket.
2. Remove the cotter key, washer, bushing and shift link rod from the shift lever.
3. Remove the access cover from underneath the tiller handle bracket.
4. Disconnect the key switch and tiller stop switch ground leads at the bullet connectors.
5. Remove the nuts securing the key switch and trim switch to the tiller handle bracket.
6. Remove the clip securing the lanyard stop switch and remove the switch from the bracket.
7. Remove the harness retainer from the engine.
8. Remove the grommet from the tiller bracket.
9. Remove the electrical panel access cover.
10. Disconnect the switch harness from the engine harness plug.
11. Disconnect the lead from the trim solenoids.
12. Remove the key switch, trim switch and their wiring harnesses from the tiller bracket.
13. Remove the plug from the tiller handle bracket.
14. Remove the nut and bolt from the tiller handle bracket.
15. Remove the tiller handle, nylon bushings, stainless bushing and washers from the bracket.
16. Bend the tab washer away from the bolt securing the shift lever and remove the bolt and lever from the bracket.
17. Remove the spring and detent pin from the bracket.

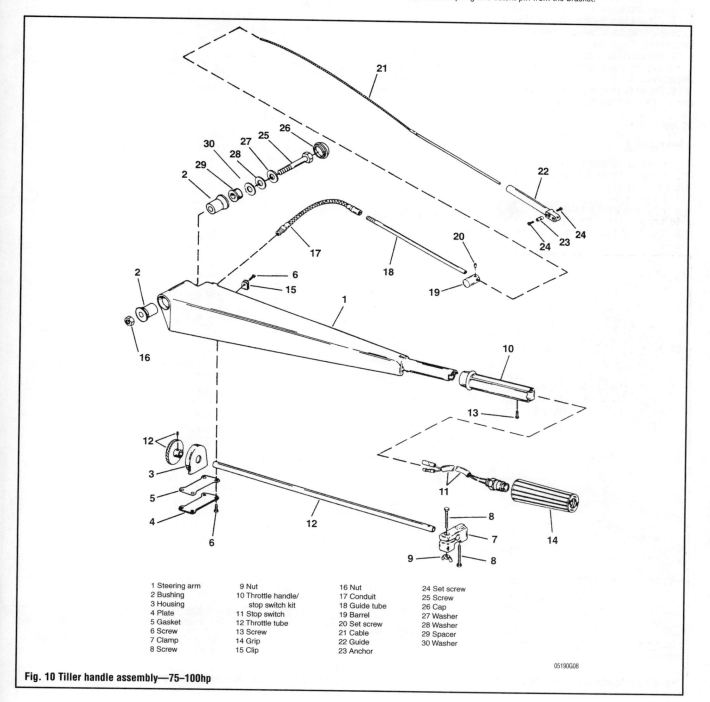

1 Steering arm	9 Nut	16 Nut	24 Set screw
2 Bushing	10 Throttle handle/	17 Conduit	25 Screw
3 Housing	stop switch kit	18 Guide tube	26 Cap
4 Plate	11 Stop switch	19 Barrel	27 Washer
5 Gasket	12 Throttle tube	20 Set screw	28 Washer
6 Screw	13 Screw	21 Cable	29 Spacer
7 Clamp	14 Grip	22 Guide	30 Washer
8 Screw	15 Clip	23 Anchor	

Fig. 10 Tiller handle assembly—75–100hp

10-10 REMOTE CONTROL

18. Bend the tab washers away from the nuts securing the bracket to the steering arm. Remove the nuts, tab washers and bracket from the steering arm.

To install:
19. Slide the shift bracket over the steering arm studs. Secure the bracket to the tiller arm with the nuts and new tab washers. Torque the nuts to 40 ft. lbs. (54 Nm). Bend the tabs against the nut flats.
20. Install the retained spring and detent pin into the tiller handle bracket.
21. Insert the bushings and thrust washer into the shift lever. Install the shift lever onto the bracket.
22. Secure the shift lever to the bracket with the bolt and new tab washer. Align the tab washer with the slot in the bracket. Torque the bolt to 110 inch lbs. (12.5Nm).
23. Install the 2 nylon bushings into the tiller handle. Install the stainless bushing into the bracket.
24. Install the tiller handle to the bracket using the bolt and 2 washers. Secure the bolt in place with the nut and torque to 40 ft. lbs. (54Nm). Reinstall the rubber plug into the bracket.
25. Route the key switch and trim switch wiring harness through the tiller bracket. Secure both switches to the bracket with the nuts.
26. Install the grommet in the wiring harness access hole.
27. Secure the harness with the harness retainer.
28. Reinstall the lanyard stop switch in the bracket and secure it with the clip.
29. Reconnect the key switch, trim switch and remote stop switch leads at the bullet connectors. Make sure the black ground and remote stop switch ground leads are secured to the bracket.
30. Secure the access cover to the tiller bracket. Do not overtighten the screws.
31. Connect the switch harness plug to the engine harness plug.
32. Reconnect the trim solenoid leads.
33. Install the electrical panel access cover.
34. Connect the shift link rod to the shift lever using the bushing, washer and cotter key.

Control Cables

REMOVAL & INSTALLATION

4 and 5 HP

1. Disconnect the cable end from the pulley on the throttle shaft.

To install:
2. Lubricate the throttle shaft pulley with a marine grade lubricant. Then connect the cable end of the throttle wire to the throttle pulley.
3. Thread the jacket over the throttle wire and into the recess of the plate.
4. Route the throttle wire to the carburetor.
5. Rotate the tiller handle twist grip fully clockwise (idle position).
6. Place the throttle wire jacket into the retaining bracket and then thread the throttle wire through the hole in the cable retainer.
7. With the throttle arm of the carburetor against the idle speed screw, pull slack from the throttle cable and secure the cable in the retainer by tightening the set-screw.

6–15 HP

1. Place the tiller handle twist grip in the neutral position.
2. Loosen the jam nuts securing the control cables to the anchor bracket.
3. Unwrap and remove the control cables from the secondary gear pulley.
4. Remove the screws and cap on the tiller handle.
5. Pull the handle assembly away from the anchor bracket and unwrap the cables.
6. Pull the cable assembly out from the back side of the anchor bracket.

To install:
7. Route the control cable assembly out from the inside of the bottom cowl through the tiller handle mounting bracket.

➡ **Do not push the guide into the mounting bracket at this time.**

8. Wrap the control cables around the drive pulley.
9. Being careful not to damage the stop switch wiring harness, place the assembled tiller handle into the mounting bracket.
10. Pull on the cable ends to remove the slack and secure the tiller handle to the mounting bracket.

20 and 25 HP

1. Place the tiller handle twist grip in the neutral position.
2. Remove the throttle link rod from the throttle cam and primary throttle lever.
3. Loosen the jam nuts which secure the control cables to the anchor bracket.
4. Unwrap and remove the control cables from the primary gear pulley.

➡ **If not replacing the cables, mark the top cable to aid in reassembly.**

5. Reomve the nut securing the access cover and remove the cover.
6. Disconnect the stop button bullet connectors at the switch box.

To install:
7. Rotate the tiller handle twist grip to the reverse position.
8. Route the extended cable over the top of the primary gear pulley and secure the cable into the inner groove of the pulley. Place the cable jacket into the top notch of the cable anchor bracket.
9. Rotate the tiller handle twist grip to the forward gear position.
10. Route the remaining cable below the primary gear pulley and secure the cable to the outer groove of the pulley. Place the cable jacket into the lower notch of the cable anchor bracket.
11. After adjusting the cables, reinstall the access cover. Secure the cover with the nut and bolt. Torque the nut to 50 inch lbs. (5.6Nm).

30 and 40 HP

MANUAL START

1. Remove the throttle twist grip.
2. Loosen the screws securing the cable guide to the throttle cable.
3. Remove the conduit and throttle cable from the tiller handle.

To install:
4. Position the drive rod flat surface parallel to the work surface.
5. Rotate the drive rod (in either direction) 1/8 of a turn to attain a drive rod flat surface of 45°.
6. Install the throttle cable into the tiller handle port until movement is felt in the drive rod.
7. Rotate the drive rod counterclockwise 1/8 turn (45°) until the flat surface becomes parallel with the work surface.
8. Rotate the drive rod clockwise 1/4 turn (90°) until the flat surface becomes parallel with the work surface.
9. Measure the distance between the tiller handle port and the end of the throttle cable. A measurement of 18 inches (457mm) must be reached for the carburetor throttle plates to open and close properly.
10. Install the throttle handle to the tiller handle. Align the "IDLE" on the throttle handle with the arrow on the tiller handle without moving the drive rod. Recheck the throttle cable length.

➡ **If 18 inches (457mm) is not measured following the throttle cable installation, repeat the previous steps until it meets specification.**

11. Install the washer, nut and end cap to the throttle grip. Tighten the nut snugly, allowing the grip to turn freely.
12. Install the conduit to the tiller handle. Turn the conduit in until it bottoms out on the tiller handle, then back off one turn.
13. Reinstall the cable guide to the throttle cable.

ELECTRIC START

1. Rotate the throttle twist grip fully clockwise to the idle position.
2. Back out the set screw from the throttle cable barrel until 2 or 3 threads of the set screw are exposed.
3. Place the barrel receptacle onto the throttle barrel. Route the cable around the port side of the powerhead and position the throttle cable through the center opening in the rubber grommet.
4. Position the barrel receptacle into the receptacle guide.
5. Place the throttle cable on the peg of the throttle lever.
6. Secure the throttle cable with the latch.
7. Apply a small amount of Loctite®271 onto the exposed threads of the throttle cable barrel set screw. (Do not tighten the screw at this time).

✳✳ CAUTION

Do not exceed 1/4-turn on the set screw after it has bottomed out.

REMOTE CONTROL 10-11

8. With the throttle lever held lightly against the stop and the twist grip at idle, turn the set screw of the throttle cable barrel until it bottoms out on the tube, then tighten the screw an additional 1/8 turn.
9. Secure the barrel receptacle using the barrel retainer.
10. Check the preload on the throttle cable by placing a thin piece of paper between the idle stop screw and idle stop. Preload is correctly set when the paper can be remove without tearing, but having some drag. If the paper is too loose or too tight, readjust the throttle cable barrel if necessary.

50 and 60 HP

1. Remove the nut securing the throttle cable to the engine.
2. Release the latch and remove the throttle cable with grommet from the bottom cowling.
3. After removing the tiller handle, use a flat tip screw driver to gently pry/push the rubber grip off the tiller handle.
4. Remove the screw from the twist grip.
5. Cut the strap securing the stop switch harness and remove the screw from the harness J-clip.
6. Remove the stop switch and twist grip from the tiller handle.
7. Remove the throttle cable anchor screws and remove the cable guide.
8. Remove the Allen screws from the brass barrel and remove the barrel.
9. Unscrew the stainless conduit from the tiller handle.
10. Pull the throttle cable from the tiller handle.

To install:
11. Insert the throttle cable (curved end facing up) into the tiller handle gear assembly while rotating the tiller arm counterclockwise.
12. Retract the throttle cable into the gear assembly until approximately 17 inches (43Cm) extends from the tiller arm.
13. Slide the stainless steel conduit over the throttle cable and thread it into the tiller arm until it is lightly seated. Rotate the conduit counterclockwise one full turn from a lightly seated position.
14. Slide the brass barrel over the throttle cable tube. Secure the barrel to the tube with the Allen screw approximately 3.5 inches (89mm) from the stainless conduit.

✱✱ CAUTION

Do not overtighten the screw as it may crush the tube and bind the throttle cable.

15. Position the barrel to face towards the tiller handle.
16. Install the throttle cable guide onto the throttle cable with the anchor and two screws. The guide hold should face up.
17. Position the throttle arm slot to face the stop switch harness exit hole in the tiller handle.
18. Route the stop switch harness through the twist grip, into the throttle arm and out through the side of the tiller handle.
19. Secure the stop switch harness to the throttle arm with sta-straps.

➡ **Allow enough slack in the harness (rotate the throttle grip in both directions) before securing the harness to the handle assembly.**

20. Attach the harness to the tiller arm with the J-clip, allowing enough slack in the harness for full throttle operation.
21. Install the twist grip by aligning the ridges on the plastic twist grip with the grooves inside the rubber grip. Applying soapy water to the rubber grip will ease installation.
22. Secure the twist grip to the throttle arm with the attaching screw
23. Rotate the throttle twist grip fully clockwise to the idle position.
24. Back out the set screw from the throttle cable barrel until 2 or 3 threads of the set screw are exposed.
25. Place the barrel receptacle onto the throttle barrel. Route the cable around the port side of the powerhead and position the throttle cable through the center opening in the rubber grommet.
26. Position the barrel receptacle into the receptacle guide.
27. Place the throttle cable on the peg of the throttle lever.
28. Secure the throttle cable with the latch.
29. Apply a small amount of Loctite®271 onto the exposed threads of the throttle cable barrel set screw. (Do not tighten the screw at this time).

✱✱ CAUTION

Do not exceed 1/4-turn on the set screw after it has bottomed out.

30. With the throttle lever held lightly against the stop and the twist grip at idle, turn the set screw of the throttle cable barrel until it bottoms out on the tube, then tighten the screw an additional 1/8 turn.
31. Secure the barrel receptacle using the barrel retainer.
32. Check the preload on the throttle cable by placing a thin piece of paper between the idle stop screw and idle stop. Preload is correctly set when the paper can be remove without tearing, but having some drag. If the paper is too loose or too tight, readjust the throttle cable barrel if necessary.

75–100 HP

1. Remove the nut securing the throttle cable to the engine.
2. Release the latch and remove the throttle cable with grommet from the bottom cowling.
3. After removing the tiller handle, use a flat tip screw driver to gently pry/push the rubber grip off the tiller handle.
4. Remove the screw from the twist grip.
5. Cut the strap securing the stop switch harness and remove the screw from the harness J-clip.
6. Remove the stop switch and twist grip from the tiller handle.
7. Remove the throttle cable anchor screws and remove the cable guide.
8. Remove the Allen screws from the brass barrel and remove the barrel.
9. Unscrew the stainless conduit from the tiller handle.
10. Pull the throttle cable from the tiller handle.

To install:
11. Insert the throttle cable (curved end facing up) into the tiller handle gear assembly while rotating the tiller arm counterclockwise.
12. Retract the throttle cable into the gear assembly until approximately 17 inches (43Cm) extends from the tiller arm.
13. Slide the stainless steel conduit over the throttle cable and thread it into the tiller arm until it is lightly seated. Rotate the conduit counterclockwise one full turn from a lightly seated position.
14. Slidethe brass barrel over the throttle cable tube. Secure the barrel to the tube with the Allen screw approximately 3.5 inches (89mm) from the stainless conduit.

✱✱ CAUTION

Do not overtighten the screw as it may crush the tube and bind the throttle cable.

15. Position the barrel to face towards the tiller handle.
16. Install the throttle cable guide onto the throttle cable with the anchor and two screws. The guide hold should face up.
17. Position the throttle arm slot to face the stop switch harness exit hole in the tiller handle.
18. Route the stop switch harness through the twist grip, into the throttle arm and out through the side of the tiller handle.
19. Secure the stop switch harness to the throttle arm with sta-straps.

➡ **Allow enough slack in the harness (rotate the throttle grip in both directions) before securing the harness to the handle assembly.**

20. Attach the harness to the tiller arm with the J-clip, allowing enough slack in the harness for full throttle operation.
21. Install the twist grip by aligning the ridges on the plastic twist grip with the grooves inside the rubber grip. Applying soapy water to the rubber grip will ease installation.
22. Secure the twist grip to the throttle arm with the attaching screw
23. Rotate the throttle twist grip fully clockwise to the idle position.
24. Back out the set screw from the throttle cable barrel until 2 or 3 threads of the set screw are exposed.
25. Place the end of the throttle cable guide over the peg on the throttle lever and secure it with the locknut and washer. Tighten until snug, then back off 1/4 turn.

✱✱ CAUTION

Do not exceed 1/4-turn on the set screw after it has bottomed out.

26. Hold the engine throttle lever against the idle stop, adjust the throttle cable barrel to slip into the upper hole of the barrel receptacle, with very light preload on the throttle lever against the idle stop.
27. Apply a small amount of Loctite®271 or equivalant to the threads of the Allen screw and tighten until snug, then turn an additional $\frac{1}{8}$ turn.
28. Lock the barrel in place with the barrel retainer.

10-12 REMOTE CONTROL

29. Check the preload on the throttle cable by placing a thin piece of paper between the idle stop screw and idle stop. Preload is correctly set when the paper can be remove without tearing, but having some drag. If the paper is too loose or too tight, readjust the throttle cable barrel if necessary.

Engine Stop Switch

TESTING

1. Disconnect the engine stop switch wiring harness.
2. Connect a multimeter between the switch harness leads.
3. With the switch engaged (button pushed), continuity should exist. With the switch released (button not pushed), continuity should not exist.
4. If the switch does not function as specified there is a short in either the switch or harness and the switch should be replaced.
5. If the switch functions properly, there may be a problem in the powerhead wiring harness.

REMOVAL & INSTALLATION

1. Disconnect the engine stop switch wiring harness.
2. Remove any wire straps that connect the switch to the tiller handle or bracket.
3. Remove any retaining nuts/screws that secure the switch to the tiller handle.

➡ **Some switches are screwed into the end of the tiller handle and simply unscrew from the handle.**

4. Remove the switch from the tiller handle.

To install:
5. Install the switch on the tiller handle.
6. As required, install the switch retaining nut/screw and tighten securely.
7. Install any wire straps that connect the switch to the tiller handle or bracket.
8. Connect the engine stop switch wiring harness.
9. Test the switch for proper operation.

Emergency Stop Switch

TESTING

1. Disconnect the engine stop switch wiring harness.
2. Connect a multimeter between the switch harness leads.
3. With the switch engaged (stop switch lanyard pulled), continuity should exist. With the switch released (stop switch lanyard in position), continuity should not exist.
4. If the switch does not function as specified there is a short in either the switch or harness and the switch should be replaced.
5. If the switch functions properly, there may be a problem in the powerhead wiring harness.

REMOVAL & INSTALLATION

1. Disconnect the emergency stop switch wiring harness.
2. Remove any wire straps that connect the switch to the tiller handle or bracket.
3. Remove any retaining nuts/screws that secure the switch to the tiller handle.

➡ **Some switches are screwed into the end of the tiller handle and simply unscrew from the handle.**

4. Remove the switch from the tiller handle.

To install:
5. Install the switch on the tiller handle.
6. As required, install the switch retaining nut/screw and tighten securely.
7. Install any wire straps that connect the switch to the tiller handle or bracket.
8. Connect the emergency stop switch wiring harness.
9. Test the switch for proper operation.

Engine Start Switch

TESTING

1. Disconnect the engine stop switch wiring harness.
2. Connect a multimeter between the switch harness leads.
3. With the switch engaged (button pressed), continuity should exist. With the switch released (button not depressed), continuity should not exist.
4. If the switch does not function as specified there is a short in either the switch or harness and the switch should be replaced.
5. If the switch functions properly, there may be a problem in the powerhead wiring harness.

REMOVAL & INSTALLATION

1. Disconnect the engine start switch wiring harness.
2. Remove any wire straps that connect the switch to the tiller handle or bracket.
3. Remove any retaining nuts/screws that secure the switch to the engine case.
4. Remove the switch from the engine case.

To install:
5. Install the switch on the engine case.
6. As required, install the switch retaining nut/screw and tighten securely.
7. Install any wire straps that connect the switch wiring harness.
8. Connect the engine start switch wiring harness.
9. Test the switch for proper operation.

HAND REWIND STARTER 11-2
DESCRIPTION AND OPERATION 11-2
TROUBLESHOOTING THE REWIND
 STARTER 11-2
OVERHEAD TYPE STARTER 11-2
2.5, 3 AND 3.3 HP 11-2
 REMOVAL & INSTALLATION 11-2
 DISASSEMBLY 11-3
 CLEANING & INSPECTION 11-4
 ASSEMBLY 11-4
4 AND 5 HP 11-6
 REMOVAL & INSTALLATION 11-6
 DISASSEMBLY 11-7
 CLEANING & INSPECTION 11-8
 ASSEMBLY 11-8
6–15 HP 11-11
 REMOVAL & INSTALLATION 11-11
 DISASSEMBLY 11-12
 CLEANING & INSPECTION 11-13
 ASSEMBLY 11-14
20 AND 25 HP 11-15
 REMOVAL & INSTALLATION 11-15
 DISASSEMBLY 11-16
 CLEANING & INSPECTION 11-17
 ASSEMBLY 11-18
30 AND 40 HP 11-19
 REMOVAL & INSTALLATION 11-19
 DISASSEMBLY 11-20
 CLEANING & INSPECTION 11-22
 ASSEMBLY 11-22

11
HAND REWIND STARTER

HAND REWIND STARTER 11-2
OVERHEAD TYPE STARTER 11-2

11-2 HAND REWIND STARTER

HAND REWIND STARTER

Description and Operation

The main components of a hand rewind starter (recoil starter) are the cover, rewind spring and pawl arrangement. Pulling the rope rotates the pulley, winds the spring and activates the pawl into engagement with the starter hub at the top of the flywheel. Once the pawl engages the hub, the powerhead is spun as the rope unwinds from the pulley.

Releasing the rope on rewind starter moves the pawl out of mesh with the hub. The powerful clock-type spring recoils the pulley in the reverse direction to rewind the rope to the original position.

Newer design manual starters employ the same principles of inertia found in electric starter motors. A nylon pinion gear slides upward and engages the flywheel ring gear as the starter rope is pulled. At the same time, the rewind spring is tightened so there starter rope can be recoiled.

A predetermined ratio designed into the starter provides maximum cranking speed for easy starts and easy pulls. Some models have a cam follower which prevents activation of the manual starter if the shift lever is not in the neutral position. This system is commonly referred to as a Neutral Start Interlock system.

Troubleshooting The Hand Rewind Starter

Repair on hand rewind starter units is generally confined to rope, pawl, nylon pinion gear (on inertia starters) and occasionally spring replacement.

※※ CAUTION

When replacing the recoil starter spring extreme caution must be used. The spring is under tension and can be dangerous if not released properly.

Starters which use friction springs to assist pawl action may suffer from bent springs. This will cause the amount of friction exerted to not be correct and the pawl will not be moved into engagement.

Models equipped with a neutral start interlock system may experience a no-start condition due to a misadjusted interlock cable. The hand rewind starter should only function when the shift handle is in the **NEUTRAL** position.

OVERHEAD TYPE STARTER

2.5, 3, 3.3 HP

♦ See Figures 1 and 2

Fig. 1 Top view of the 2.5–3.3 hp hand rewind starter

Fig. 2 Bottom view of the 2.5–3.3 hp hand rewind starter

REMOVAL & INSTALLATION

♦See accompanying illustrations

1. Remove the powerhead cowling by simply releasing the two snaps—one on each side—and then lifting the cowling free.

Step 1

HAND REWIND STARTER 11-3

Step 2

Step 4

Step 5

2. Remove the three bolts securing the rewind starter assembly to the powerhead, then remove the starter assembly.

To install:

3. Install the rewind hand starter assembly onto the powerhead and secure it in place with the three attaching bolts. Tighten the bolts securely.

4. Connect the high tension lead to the spark plug, if it was removed. Crank the powerhead with the hand rewind starter and check the completed work.

5. Install the cowling down over the powerhead and secure it in place with the two snap retainers.

DISASSEMBLY

♦See accompanying illustrations

1. Turn the starter housing over with the sheave facing up. Pull on the handle to gain some slack in the rope; hold the sheave from rewinding; and then remove the handle. Two types of handle arrangement are used.

2. After the handle is removed, ease the grip on the sheave and allow the sheave to completely rewind with the rope inside the sheave.

3. Snap the circlip out of the groove in the center shaft.

4. Lift the spacer off the friction plate and free of the center shaft.

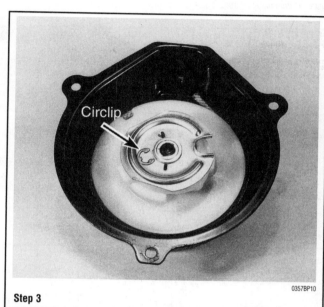
Step 3

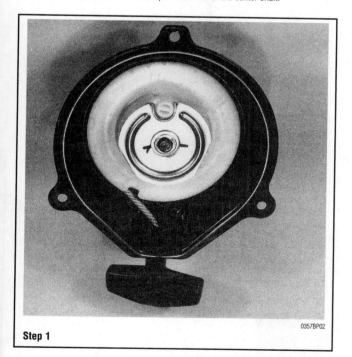
Step 1

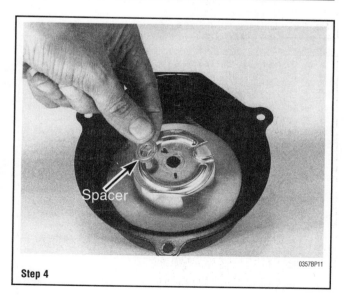
Step 4

11-4 HAND REWIND STARTER

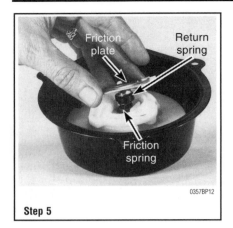
Step 5

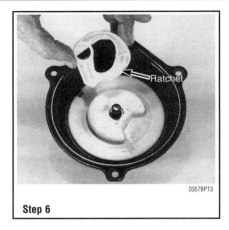
Step 6

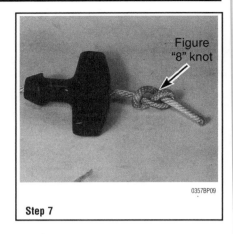
Step 7

Step 8

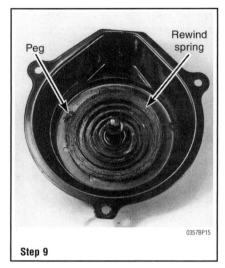
Step 9

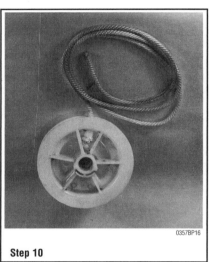
Step 10

5. Lift the friction plate slightly and snap the friction spring out of the hole in the center shaft. Remove the friction plate. The return spring will come with the plate. Remove the spring cover, and then the friction spring.

6. Remove the ratchet from the top of the sheave.

7. The pull rope is secured in the handle recess with a figure "8" knot tied close to the end. Excess rope beyond the knot can be cut off. Burn the end of a non-fiber rope with a match to prevent it from unraveling.

✷✷ CAUTION

The rewind spring is a potential hazard. The spring is under tremendous tension when it is wound. If the spring should accidentally be released, severe personal injury could result from being struck by the spring with force. Therefore, the following two steps must be performed with care to prevent personal injury to self and others in the area.

8. Very carefully "rock" the sheave and at the same time lift the sheave about ½ in. (13mm). The spring will disengage from the sheave and remain in the housing.

9. Now, slowly turn the housing over and gently place it on the floor with the spring facing the floor. Tap the top of the housing with a mallet and the spring will fall free of the housing and partially unwind almost instantly and with considerable force, but it will be contained within the housing. Tilt the container with the opening away from you. The spring will be released from the housing and unwind rapidly. Turn the housing over and unhook the end of the spring from the peg in the housing.

10. Untie the knot in the end of the old starter rope and pull the rope free of the sheave.

CLEANING & INSPECTION

Wash all parts except the rope and the handle in solvent, and then blow them dry with compressed air.

Remove any trace of corrosion and wipe all metal parts with an oil dampened cloth.

Inspect the rope. Replace the rope if it appears to be weak or frayed. If the rope is frayed, check the holes through which the rope passes for rough edges or burrs. Remove the rough edges or burrs with a file and polish the surface until it is smooth.

Inspect the starter spring end hooks. Replace the spring if it is weak, corroded or cracked. Inspect the tab on the spring retainer plate. This tab is inserted into the inner loop of the spring. Therefore, be sure it is straight and solid. Inspect the inside surface of the sheave rewind recess for grooves or roughness. Grooves may cause erratic rewinding of the starter rope.

ASSEMBLY

▶ See accompanying illustrations

➡**Wear a good pair of gloves while winding and installing the spring. The spring will develop tension and the edges of the spring steel are extremely sharp. The gloves will prevent cuts to the hands and fingers.**

It is strongly recommended a pair of safety goggles or a face shield be worn while the spring is being installed. As the work progresses over 14 ft. (4.3m) of spring steel wound into about 4 in. (10.2cm) circumference. If the spring is accidentally released, it will lash out with tremendous ferocity and very likely could cause personal injury to the installer or other persons nearby.

HAND REWIND STARTER 11-5

Step 8

Step 2

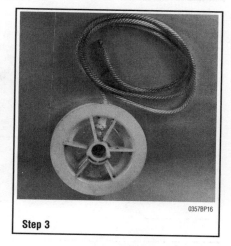
Step 3

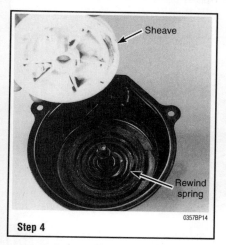
Step 4

Step 5

Step 6

1. Hold the spring in a coil in one gloved hand, as shown.
2. With the other hand, hook the looped end of the spring over the peg in the housing. Feed the spring clockwise into the housing. The easiest way to accomplish this task is to hold the spring in one gloved hand and with the other gloved hand, rotate the housing counterclockwise. Continue working the spring until it is all confined within the housing.
3. Feed one end of a new rope through the hole in the sheave, and then tie a figure "8" knot in the end of the rope. Pull on the rope until the knot is confined inside the sheave recess. Wind the entire pull rope clockwise around the sheave.
4. Lower the sheave down over the center shaft of the housing. As the sheave goes into the housing the hook on the lower face of the sheave must index into the loop end of the spring. In the illustration, the sheave is turned over to expose the hook to view. Once the hook is indexed into the spring, the sheave may be fully seated in the housing.
5. Position the ratchet in place on the sheave, with the flat side of the ratchet against the sheave and the round hole indexed over the post.
6. Slide the friction spring down the center shaft. This spring serves as a spacer and exerts an upward pressure on the friction plate. Install the spring cover on top of the spring.

➡ **Two holes are located in the sheave on the same side of the center shaft. With the holes on the side of the shaft facing you, one end of the friction spring must index into the right hole.**

7. Hook one end of the return spring into the slot of the friction plate. Now, lower the friction plate onto the center shaft and index the free end of the spring into the right hole in the sheave.
8. Place the spacer over the hole on the friction plate.

Step 8

11-6 HAND REWIND STARTER

9. Push down on the friction plate, and then snap the circlip into the groove on the center shaft to secure the plate and associated parts in place.
10. Place the pull rope into the notch in the sheave. With the rope in the notch, rotate the sheave three complete turns counter-clockwise. Hold tension on the rope and at the same time feed the free end of the rope through the rope guide in the starter housing. Continue to hold tension on the sheave for the next step.
11. Feed the free end of the rope through the handle and tie a figure "8" knot in the end.
12. Pull the rope back into the handle recess. Relax the tension on the sheave and allow the sheave to rewind until the handle is against the starter housing.

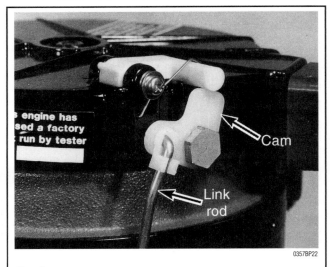
Step 1

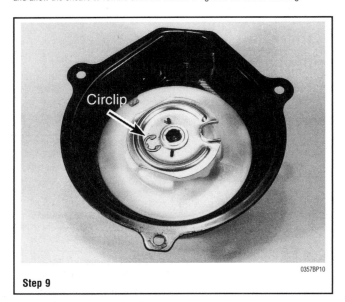
Step 9

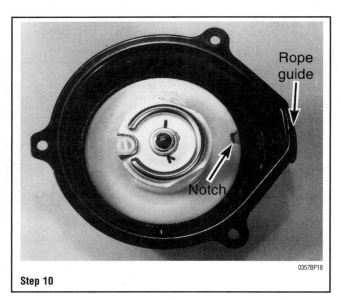
Step 10

Step 2

4 and 5 HP

REMOVAL & INSTALLATION

▶ See accompanying illustrations

1. Pry the shift interlock link rod from the plastic cam on the starter.
2. Remove the three bolts securing the starter legs to the powerhead.
3. Remove the starter and place it upside down on a suitable work surface.

To install:

4. Position the rewind starter in place on the powerhead.

Step 3

HAND REWIND STARTER 11-7

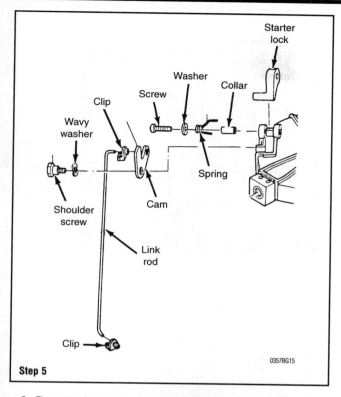
Step 5

6. Apply Loctite® or equivalent to the threads of the three attaching bolts. Secure the starter legs to the powerhead with the bolts, and tighten them to a torque value of 5.8 ft. lbs. (8Nm).
7. Snap the shift interlock link rod into the plastic cam on the starter housing.

➡ **If the link rod adjustment at the shift lever and inside the starter housing was undisturbed, the no-start-in-gear protection system should perform satisfactorily. When the unit is not in neutral, the starter lock should drop down to block the sheave and prevent it from rotating. This means an attempt to pull on the rope with the lower unit in any gear except neutral should fail.**

If the no-start-in-gear protection system fails to function properly, first remove the rewind hand starter from the powerhead. Inspect the blocking surface of the starter lock, the condition of the cam and the return action of the spring. Because no adjustment of the length of the link rod is possible, make sure the lower end of the rod is indeed connected to the shift mechanism below the bottom cowling and the rod itself is not binding or bent.

8. Install the starter on the powerhead and again check the no-start-in-gear system.

DISASSEMBLY

▶**See accompanying illustrations**

1. Pry the circlip from the pawl post using a narrow slotted screwdriver.
2. Lift the pawl, with the spring attached free of the sheave.
3. Rotate the sheave to align the slot in the sheave with the starter handle, as shown. Lift out a portion of rope, feed the rope into the slot and with a controlled motion allow the sheave to rotate in a clockwise direction until the tension on the rewind spring is completely released. Do not allow the sheave to spin without control.
4. Pry the seal from the handle and push out the knot in the end of the rope. Untie the knot and pull the handle free of the rope.
5. Remove the bolt and washer from the center of the sheave.
6. Remove the sheave bushing and starter housing shaft from the sheave.

➡ **If the only work to be performed on the hand rewind starter is to replace the rope, take extra care not to disturb the sheave and spring beneath the sheave.**

5. The no-start-in-gear protection system cannot be adjusted. If the system fails to prevent the sheave from rotating in any shift position except neutral, each component should be thoroughly inspected. The part should be replaced if there is any evidence of excessive wear or distortion.

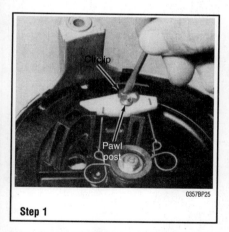
Step 1

Step 2

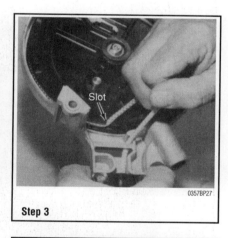
Step 3

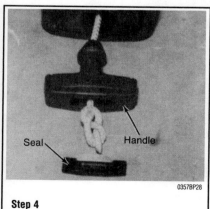
Step 4

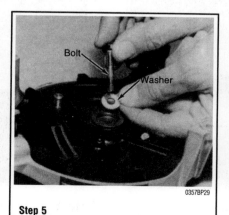
Step 5

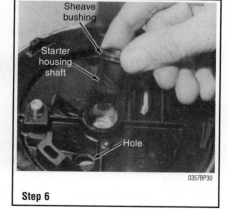
Step 6

11-8 HAND REWIND STARTER

7. Hold the sheave against the starter housing to prevent the spring from disengaging from the sheave and carefully rotate the sheave to allow the rope hole to align with the starter handle. Pull the knotted end out of the sheave until all of the rope is free.

If either the sheave or the starter rewind spring is to be replaced the rope may be left in place until the sheave is removed from the starter housing.

✱✱ CAUTION

Wear a good pair of heavy gloves and safety glasses while performing the following tasks.

✱✱ WARNING

The rewind spring is a potential hazard. The spring is under tremendous tension when it is wound. If the spring should accidentally be released, severe personal injury could result from being struck by the spring with force. Therefore, the following steps must be performed with care to prevent personal injury to self and others in the area. Do not attempt to remove the spring unless it is unfit for service and a new spring is to be installed.

➡ There are two design variations in the housing of the spring. For ease of identification and to ensure the proper procedures are performed, the designs are designated "A" and "B". One design starter has the spring encased in a removable housing and is identified here as Design "A". The other starter has the spring encased inside a recess of the sheave housing and is identified as Design "B".

When necessary, separate instructions are presented for both starter designs.

Design "A"

1. Pry out the two retainer plugs from the sheave. Lift out the tabbed guide plate from the center of the spring housing. Rotate the sheave in a clockwise direction, while holding the spring housing stationary until the two locking tabs disengage.
2. Very carefully lift out the spring housing containing the coiled spring from the sheave and remove the backing plate.

Design "B"

▶See accompanying illustrations

1. Insert a screwdriver into the hole in the sheave, push down on the section of spring visible through the hole. At the same time gently lift up on the sheave and hold the spring down to confine it in the housing and prevent it from escaping uncontrolled. If the rope has not been removed from the sheave, remove it at this time.

➡ The accompanying illustration shows the spring being released from the same type but different model rewind spring. The principle is exactly the same.

2. Obtain two pieces of wood, a short 2 in. x 4 in. (5cm x 10cm) will work fine. Place the two pieces of wood approximately 8 in. (20 cm) apart on the floor. Center the housing on top of the wood with the spring side facing down. Check to be sure the wood is not touching the spring.

3. Stand behind the wood, keeping away from the openings as the spring unwinds with considerable force. Tap the sheave with a soft mallet. The spring retainer plate will drop down releasing the spring. The spring will fall and unwind almost instantly and with force.

CLEANING & INSPECTION

▶ See Figure 3

1. Wash all parts except the rope and the handle in solvent, and then blow them dry with compressed air.
2. Remove any trace of corrosion and wipe all metal parts with an oil dampened cloth.
3. Inspect the rope. Replace the rope if it appears to be weak or frayed. If the rope is frayed, check the holes through which the rope passes for rough edges or burrs. Remove the rough edges or burrs with a file and polish the surface until it is smooth.
4. Inspect the starter spring end hooks. Replace the spring if it is weak, corroded or cracked.
5. Inspect the inside surface of the sheave rewind recess for grooves or roughness. Grooves may cause erratic rewinding of the starter rope.
6. Coat the entire length of the used rewind spring (a new spring will be coated with lubricant from the package), with low-temperature lubricant.

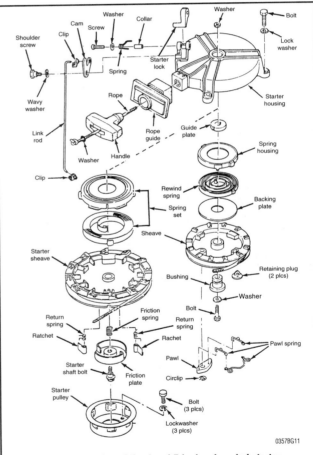

Fig. 3 Exploded drawing of the 4 and 5 hp hand rewind starter

ASSEMBLY

▶See accompanying illustrations

✱✱ CAUTION

Wear a good pair of gloves while winding and installing the spring. The spring will develop tension and the edges of the spring steel

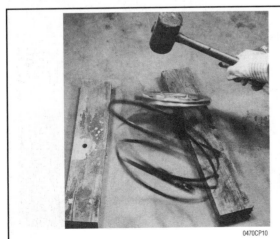

Step 2

HAND REWIND STARTER 11-9

are extremely sharp. The gloves will prevent cuts to the hands and fingers.

1. When installing a new spring, apply a light coating of Quicksilver multi-purpose lubricant or equivalent anti-seize lubricant to the inside surface of the starter housing.

2. A new spring will be wound and held in a steel hoop. Hook the outer end of the new spring onto the starter housing post, and then place the spring inside the housing.

3. Carefully remove the steel hoop. The spring should unwind slightly and seat itself in the housing.

4. When installing an old spring, apply a light coating of Quicksilver multi-purpose lubricant or equivalent anti-seize lubricant to the inside surface of the starter housing. Wind the old spring loosely in one hand in a clockwise direction.

5. Hook the outer end of the spring onto the slot in the removable housing for Design "A" starters, or the starter housing post for Design "B" starters. Rotate the sheave clockwise and at the same time feed the spring into the housing in a counter-clockwise direction. Continue working the spring into the housing until the entire length has been confined.

6. Install the backing plate over the sheave hub. Position the spring housing over the sheave, with the two tabs on the housing aligned with the two grooves in the sheave.

7. Rotate spring housing counterclockwise to engage the two locking tabs against the grooves. Insert the guide plate into the sheave hub with the tab of the plate entering the inner loop of the spring. Insert the retainers into the slots next to the locking tabs of the spring housing.

8. Turn the sheave over and using a pair of needle nose pliers, pull the two ends of the retainers through until seated.

9. Melt the tip of the rope to prevent fraying. Insert the melted end of the rope through the hole in the starter sheave. Tie a figure "8" knot in the end of the rope leaving about one inch (2.5 cm) beyond the knot. Tuck the end of the rope beyond the knot into the groove, if so equipped, next to the knot.

10. Wind the rope in a clockwise direction two turns around the sheave, ending at the slot in the sheave. Lower the sheave into the starter housing. At the same time, for design "A" starters, make sure the tab on the guide plate slides into the hole of the starter housing. For design "B" starters: Use a small screwdriver through the hole to guide the inner loop of the spring onto the post on the underneath side of the sheave.

11. On units where the spring and sheave have been undisturbed, align the hole in the edge of the sheave with the starter handle. Thread the rope through the hole and up through the top side. Tie a figure "8" knot in the end which was just brought through, leaving about one inch (25 cm). Tuck the short free end into the groove next to the hole.

Step 1

Step 4

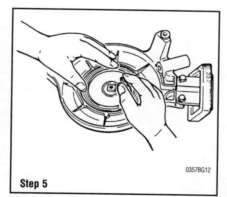

Step 5

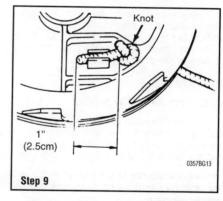

Step 9

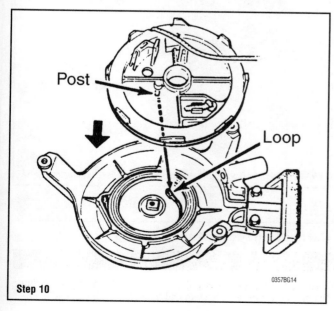

Step 10

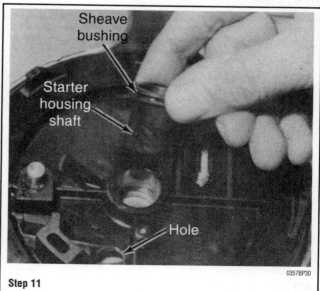

Step 11

11-10 HAND REWIND STARTER

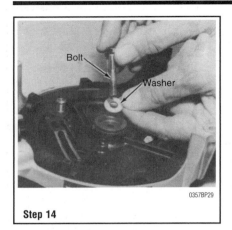
Step 14

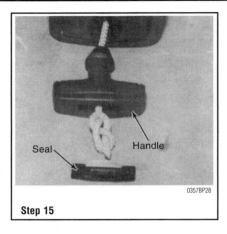
Step 15

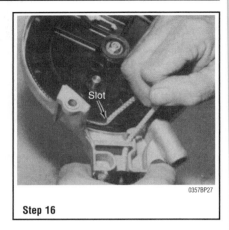
Step 16

12. Without rotating the sheave, feed the rope between the sheave and the edge of the starter housing in a clockwise direction. Push the rope into place with a narrow screwdriver. Continue feeding and tucking the rope for two turns, ending with the rope at the slot of the sheave.

13. On all other units, slide the sheave bushing into the starter housing shaft. Insert the shaft and bushing into the center of the sheave.

14. Coat the threads of the center bolt with Loctite® or equivalent. Install the washer and bolt. Tighten the bolt to a torque value of 5.8 ft. lbs. (8Nm).

15. Thread the rope through the starter handle housing and through the handle. Tie a figure "8" knot in the rope as close to the end as practical. Pull the knot back into the handle recess, and then install the seal in the handle to hide the rope knot.

16. Lift up a portion of rope, and then hook it into the slot of the sheave. Hold the handle tightly and at the same time rotate the sheave counterclockwise until the spring beneath is wound tight. This will take about three complete turns of the sheave. Slowly release the tension on the sheave and allow it to rewind clockwise while the rope is taken up as it feeds around the sheave.

17. With the beveled end of the pawl facing to the left, hook each end of the pawl spring into the two small holes in the pawl, from the underneath side of the pawl, and with the pattern of the spring, as shown. The short ends of the spring will then be on the upper surface of the pawl. Move the spring up against the center of the sheave shaft, and then slide the center of the pawl onto the pawl post, as indicated in the accompanying illustration.

18. Snap the circlip into place over the pawl post to secure the pawl in place.

19. Check the action of the rewind starter before further installation work proceeds. Pull out the starter rope with the handle, then allow the spring to slowly rewind the rope. The starter should rewind smoothly and take up all the rope to lightly seat the handle against the starter housing.

Step 17

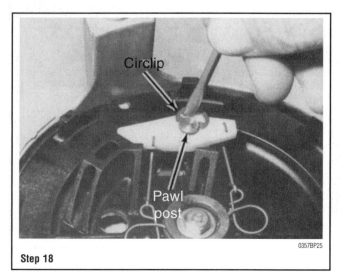
Step 18

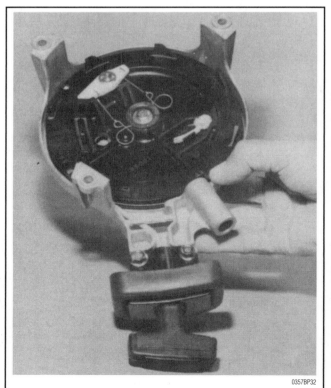
Step 19

HAND REWIND STARTER 11-11

6–15 HP

REMOVAL & INSTALLATION

♦See accompanying illustrations

1. Remove the cowling. Insert a narrow blade screwdriver between the fuel filter and the lower edge of the hand starter housing. Exert a slight downward pressure on the screwdriver and at the same time pull downward on the filter with the other hand. The fuel filter should "pop" free. Move the filter and connecting hoses to one side out of the way.
2. Unsnap the interlock link rod free of the lower lock lever.
3. Remove the three attaching bolts securing the hand rewind starter to the powerhead.
4. Lift the starter housing up and free of the powerhead.

To install:

5. Slide the starter assembly down over the crankshaft and into position on the powerhead.
6. Secure the starter with the attaching hardware. Tighten the bolts alternately and evenly.
7. Snap the interlock link rod into the lower lock lever.

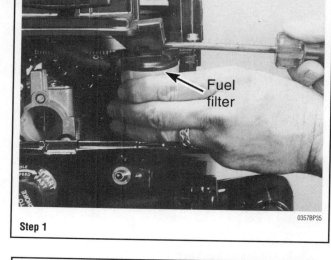
Step 1

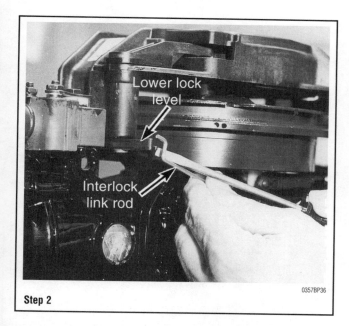
Step 2

Step 3

Step 4

Step 6

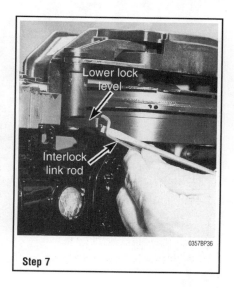
Step 7

11-12 HAND REWIND STARTER

2. Remove the handle retainer; feed the rope back through the handle; untie the knot; then pull the handle free of the rope. Two type handles are used, as shown.

➡ **If the only service on the hand starter is to replace a broken rope, the disassembling procedures may be stopped after the next step. However, if the rewind spring is to be replaced or other service work performed, it is best to leave the rope on the sheave.**

3. Rotate the sheave until the knot in the end of the rope is aligned with the hole in the sheave. Hold the tension on the sheave with one hand, and with the other hand, untie the loose knot in the rope. Pull the rope free of the sheave and housing.

4. Ease your grasp on the sheave and allow the sheave to rotate until all tension on the spring is released. If the rope has not been removed, it will simply wind around the sheave.

Step 8

8. Snap the clear plastic fuel filter into place on the bracket on the starter housing.
9. This type of hand rewind starter has a window on the top of the housing. This window permits a broken pull rope to be replaced without removing the starter from the powerhead.

DISASSEMBLY

▶**See accompanying illustrations**

1. Pull about 1 inch (30cm) of rope out of the starter housing and tie a loose knot in the rope to prevent the rope from rewinding back into the housing. Do not tighten the knot because it will be necessary to untie the knot with one hand later.

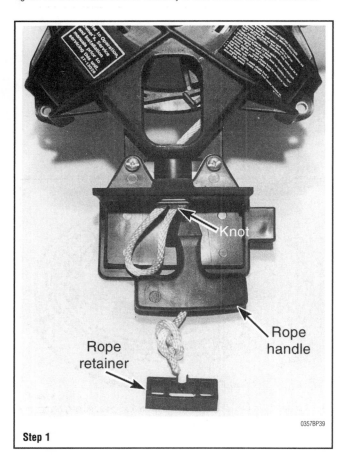
Step 1

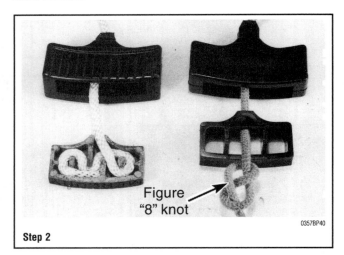
Step 2

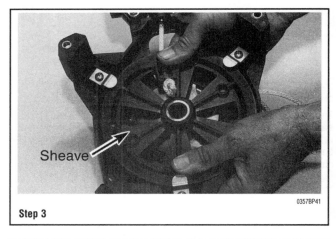
Step 3

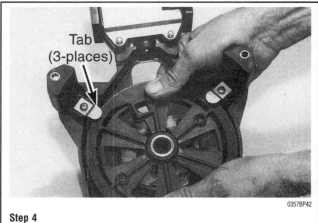
Step 4

HAND REWIND STARTER 11-13

Step 5

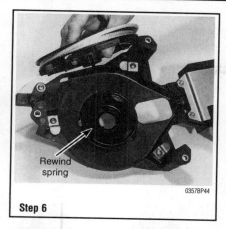
Rewind spring
Step 6

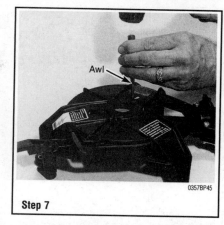
Awl
Step 7

Step 8

Punch
Step 9

Oil seal
Step 10

5. Loosen the three screws securing the tabs holding the sheave in place. It is not necessary to remove the tabs. Rotate the tabs to permit the sheave to clear the housing.

⁂⁂ WARNING

The rewind spring is a potential hazard. The spring is under tremendous tension when it is wound. If the spring should accidentally be released from its container, severe personal injury could result from being struck by the spring with force. Therefore, the following two steps must be performed with care to prevent personal injury to self and others in the area.

6. Very carefully lift and "rock" the sheave. The spring will disengage from the sheave and remain in its container in the housing, as shown.
7. Place the starter housing on a flat surface in the upright position resting on its three legs. Using a hammer and awl, "pop" the rewind spring container out of the starter housing. There is no reason whatsoever to remove the spring from the container. If the spring is in satisfactory condition, it will be lubricated prior to installation. If the spring is broken, a new spring will come in a container, lubricated, and ready for installation.
8. If the rope was not removed, untie the knot and remove it from the sheave.
9. Drive the plastic cap free of the sheave using a punch and hammer.
10. Pry the oil seal out of the sheave, with a long thin blade screwdriver.

CLEANING & INSPECTION

♦ See Figures 4, 5 and 6

1. Wash all parts except the rope and the handle in solvent, and then blow them dry with compressed air.
2. Remove any trace of corrosion and wipe all metal parts with an oil dampened cloth.
3. Inspect the rope. Replace the rope if it appears to be weak or frayed. If the rope is frayed, check the holes through which the rope passes for rough edges or burrs. Remove the rough edges or burrs with a file and polish the surface until it is smooth.
4. Inspect the starter spring end hooks. Replace the spring if it is weak, corroded or cracked. Inspect the tab on the spring retainer plate. This tab is inserted into the inner loop of the spring. Therefore, be sure it is straight and solid. Inspect the inside surface of the sheave rewind recess for grooves or roughness. Grooves may cause erratic rewinding of the starter rope.

➡**If starter operation was erratic or excessively noisy, check the starter clutch for damage from lack of lubrication. If necessary, replace the complete sheave assembly with a pre-lubricated starter clutch installed.**

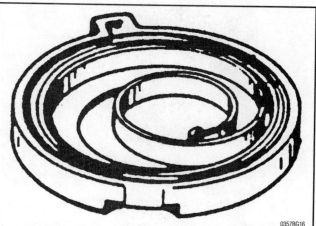

Fig. 4 A new rewind spring will arrive packed in a container, lubricated, and ready for installation.

11-14 HAND REWIND STARTER

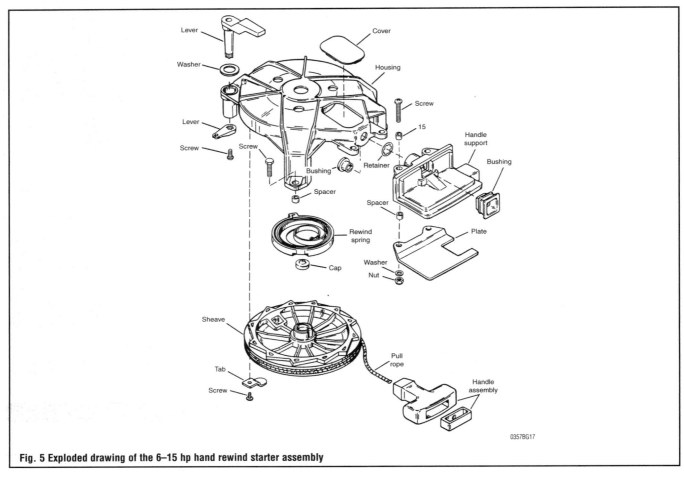

Fig. 5 Exploded drawing of the 6–15 hp hand rewind starter assembly

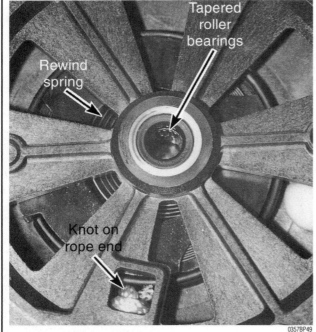

Fig. 6 The sheave of the 6–15 hp hand rewind starter relies on a clutch arrangement containing tapered roller bearings located in the center of the sheave. If the clutch is defective, the sheave assembly must be replaced.

ASSEMBLY

♦See accompanying illustrations

1. Install a new oil seal into the center of the sheave with metal part facing UP. This task may be accomplished using the proper size socket and a hammer.
2. Tap the plastic cap into the center of the sheave using a soft head mallet. The cap secures the seal in place.

Step 2

HAND REWIND STARTER   11-15

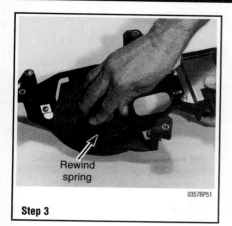
Step 3

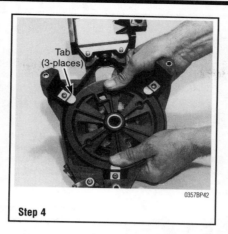
Step 4

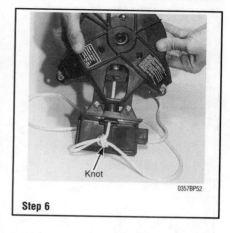
Step 6

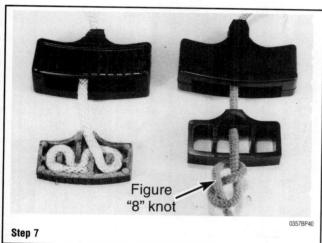
Step 7

the starter housing. Hold tension on the sheave with one hand and tie a figure "8" knot close to the end of the rope. Pull the rope through until the knot is up tight in the rope recess of the sheave. Continue holding tension on the sheave for the next step.

6. Still holding tension on the sheave and at the same time holding the sheave in place within the starter housing, tie a knot in the rope as close as possible to the housing to prevent the sheave from rewinding. Now, release a bit of pressure on the sheave. The spring will rewind just a bit and pull the knot in the rope up tight against the housing.

7. Feed the free end of the rope through the handle and secure it with the retainer, or with a figure "8" knot and the retainer. Two type handles are used, as shown. After the handle is secured, untie the knot next to the housing and allow the sheave to rewind pulling the rope around the sheave.

20 and 25 HP

REMOVAL & INSTALLATION

♦See accompanying illustrations

1. Remove the cowling. Disconnect the high tension leads to the spark plugs. Insert a screwdriver between the rewind starter housing and the fuel filter. A slight downward pressure on the screwdriver will "pop" the fuel filter free. Move the filter and associated fuel lines to one side.

2. Remove the three attaching bolts securing the rewind starter to the powerhead. Lift off the hand starter.

To install:

3. Install the rewind starter assembly onto the powerhead and secure it in place with the three attaching bolts. Check to be sure the upper and lower lock levers operate properly. Tighten the bolts to 70 inch lbs. (8Nm).

3. Snap the rewind spring container, with the spring inside, into the housing. A new spring will come with the container ready for installation.

4. Insert the sheave into the starter housing. Swing the three tabs around to lock the sheave in place. Tighten the tab screws securely. Hold the sheave to maintain tension and at the same time rotate the sheave counterclockwise as far as possible to wind the spring. Continue holding pressure on the sheave, but let it slip a little until the rope knot recess in the sheave is aligned with the rope hole in the starter housing. Hold tension on the sheave for the next step.

5. Feed the rope through the rope knot recess of the sheave and the hole in

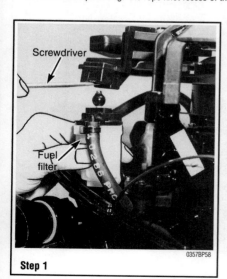
Step 1

Step 2

Step 3

11-16 HAND REWIND STARTER

Step 4

of the knotted rope aligns with the rope guide in the starter housing. Exert a good hold on the sheave and at the same time remove the knot from its recess in the sheave. Hold the sheave securely with one hand and untie the knot you made in the other end of the rope earlier in this step. Now, pull the rope back through the rope guide and the sheave until it is free.

3. Carefully allow the spring to unwind by allowing the sheave to rotate in your grasp, until all the tension on the spring is gone.

➡ **The center bolt on this type rewind starter has standard right-hand threads.**

4. Remove the center Torx bolt from the sheave. If a Torx socket is not available, use the proper size screwdriver. Exercise care not to damage the bolt head. Using a screwdriver provides very little surface for the blade to grasp, therefore, the head may be easily damaged.

5. Remove the center bolt, four arm cam and small spring from the sheave.

Step 3

4. Snap the fuel filter into place on the starter housing bracket. Install the cowling.

DISASSEMBLY

▸**See accompanying illustrations**

1. Hold the shift interlock lever back with the thumb on one hand, and at the same time, pull approximately 1 foot (25cm) of the rope out and tie a loose overhand knot in the rope. Do not tighten the knot, because it will be necessary to untie it later with one hand. Allow the spring to rewind. This knot in the rope will permit removing the handle from the end of the rope.

2. After the handle has been removed, pull the rope further out of the housing. Continue to pull the rope out until the little square window containing the other end

Step 1

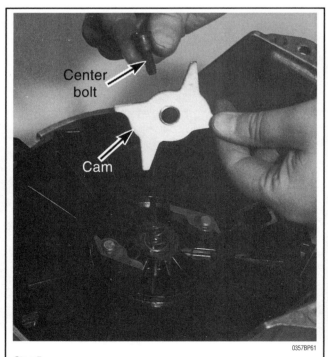
Step 5

HAND REWIND STARTER 11-17

Step 6

Step 7

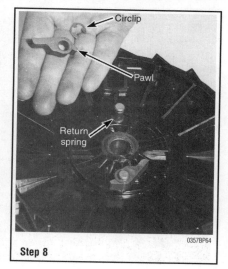
Step 8

⁂ CAUTION

The rewind spring is a potential hazard. The spring is under tremendous tension when it is wound. If the spring should accidentally be released, it will lash out with tremendous ferocity and very likely cause severe personal injury to the service worker or other persons nearby.

6. Carefully Lift the sheave out of the housing. Remove the felt pad from around the shaft in the housing.
7. Turn the sheave over and very carefully lift the container with the rewind spring out of the sheave.

➡ There is no reason whatsoever to remove the spring from the container. If the spring is broken or no longer fit for service for any reason, a new spring must be purchased. The new spring will come in a container, lubricated and ready for installation.

If the old spring is used it will be lubricated, still in the container, prior to installation.

8. Turn the sheave over again. Snap the Circlip off each post. Lift each pawl from its post and then stop. Notice the small spring on each post and the position of the springs. One end of the spring snaps over the pawl and the other end indexes into a recess in the sheave. Lift each spring and observe and remember exactly how it is installed, as an aid during installation.

CLEANING & INSPECTION

▶ See Figure 7

1. Wash all parts except the rope and the handle in solvent, and then blow them dry with compressed air.
2. Remove any trace of corrosion and wipe all metal parts with an oil dampened cloth.
3. Inspect the rope. Replace the rope if it appears to be weak or frayed. If the rope is frayed, check the holes through which the rope passes for rough edges or burrs. Remove the rough edges or burrs with a file and polish the surface until it is smooth.
4. Inspect the starter spring end hooks. Replace the spring if it is weak, corroded or cracked.

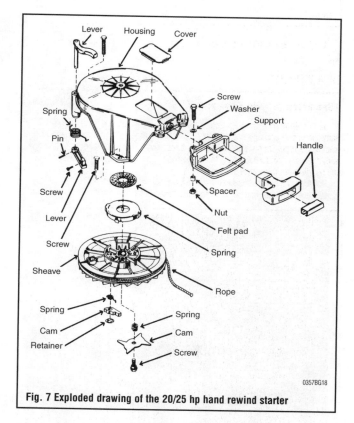

Fig. 7 Exploded drawing of the 20/25 hp hand rewind starter

5. Inspect the inside surface of the sheave rewind recess for grooves or roughness. Grooves may cause erratic rewinding of the starter rope.
6. Coat the entire length of the used rewind spring (a new spring will be coated with lubricant from the package), with low-temperature lubricant.

HAND REWIND STARTER

ASSEMBLY

See accompanying illustrations

1. Slide each small spring down over the pawl posts with one end of the spring indexed into the recess in the sheave as observed during disassembly. Slide each pawl onto the posts with the long end facing inward, as shown. Slip each spring end over the side of the pawl. The illustration shows one pawl correctly installed and the other pawl ready. After installation, the pawls will face in opposite directions with the long flat surface facing outward. Snap a circlip onto each post to restrain the pawls in place.

2. Very carefully place the rewind spring container into the sheave with the three tabs indexed into the notches of the sheave.

3. Apply a couple drops of lubricant to one side of the felt pad. Now, place the felt pad over the center shaft with the "lubricated" side going down first to hold the pad in place. Raise and insert the assembled sheave up into the housing from the underneath side, as shown. Rotate the sheave to index the spring loop into the notch on the shaft. Hold the sheave and housing together, and then turn the complete unit over.

4. Slide the spring down the shaft. This spring serves as a spacer between the cam and the sheave and exerts an upward pressure against the cam to keep it indexed over the shoulder on the center bolt. Install the four arm cam onto the shaft on top of the spring.

5. Shift the cam until the arms are positioned against the two pawls, as shown.

➡ **The center bolt on this type rewind starter has standard right-hand threads.**

6. Secure the cam in place with the Torx bolt. This bolt has a very special shoulder. Tighten the Torx bolt just good and "snug" with a Torx socket or the proper size screwdriver. Exercise care not to damage the bolt head. Using a screwdriver provides very little surface for the blade to grasp. Therefore, the head may be easily damaged. Again, tighten just good and "snug". Some hand rewind starters may have a hex-head bolt with a special shoulder—then no problem.

7. Place tension on the rewind spring by holding the housing and at the same time rotating the sheave counterclockwise. Rotate the sheave as far as possible, thus placing maximum tension on the spring. Hold the tension for the next step.

8. Hold the shift interlock lever clear and allow the sheave to unwind very slowly until the rope knot window in the sheave is aligned with the rope guide in the housing. Hold the sheave and housing in this position and at the same time feed a new rope through the window and out the rope guide. Tie an overhand knot in the rope with one hand, then feed the knot into the window. Continue to hold the sheave and housing and tie a knot about 1 foot (25cm) from the other end of the rope. Check to be sure the knot is imbedded in the window, and then slowly release your grip on the sheave and allow the spring to rewind. After the knot in the other end of the rope is against the rope guide, release your grip on the sheave.

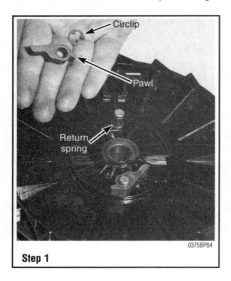
Step 1

Step 2

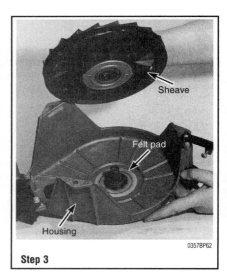
Step 3

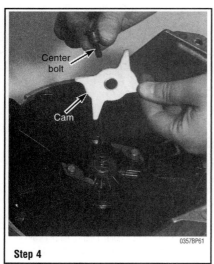
Step 4

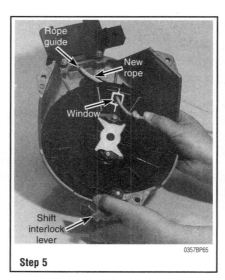
Step 5

Step 7

HAND REWIND STARTER 11-19

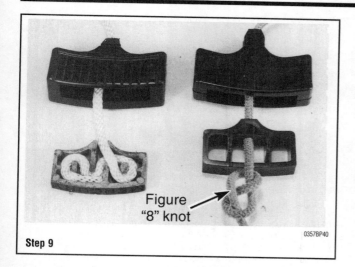

Step 9

9. Install the handle onto the end of the rope. One of two type of handles may be used. After the handle has been installed, untie the knot made about 1 ft (25cm) from the end and allow the rope to completely rewind into the housing, with the rope guide up against the rope guide of the housing.

30 and 40 hp

♦ See Figures 8, 9 and 10

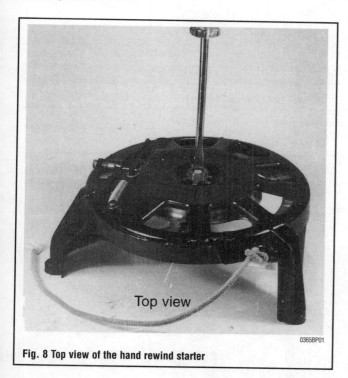

Fig. 8 Top view of the hand rewind starter

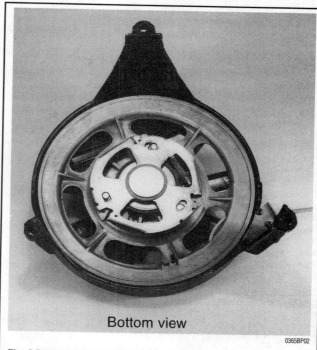

Fig. 9 Bottom view of the hand rewind starter

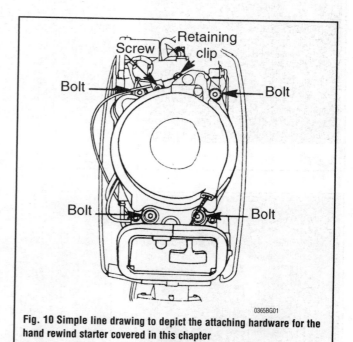

Fig. 10 Simple line drawing to depict the attaching hardware for the hand rewind starter covered in this chapter

REMOVAL & INSTALLATION

♦See accompanying illustrations

1. Remove the wrap around cowling.
2. Remove the rope retainer (handle) from the rope. This is accomplished by simply pushing the inner portion through the outer shell of the handle. Untie the "stop" knot in the rope and allow the rewind spring to unwind within the recess of the sheave.
3. If the choke rod passes down through the top of the cowling, disconnect the upper portion of the shaft from the lower portion by first pulling the cotter pin securing the two together, and then pulling the upper portion of the choke shaft up through the cowling.

11-20 HAND REWIND STARTER

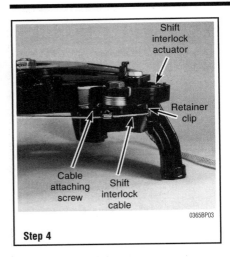

Step 4

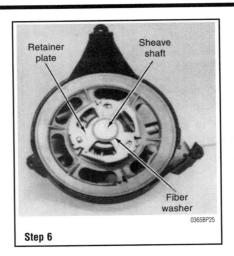

Step 6

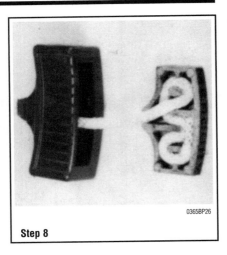

Step 8

4. All powerheads covered in this chapter have a shift interlock system. The system includes a shift interlock actuator (cable). Remove the cable attaching screw and the retainer clip, as shown in the accompanying illustration. Later model rewind starters are equipped with a shift interlock cable mounted on the side of the starter housing. Early models have the cable mounted on the top of the housing. The interlock cable of both models is secured in the same manner.

Now, remove the cowling. Remove the screws securing the three legs of the hand rewind starter mechanism to the powerhead.

To install:

5. Install the assembled rewind starter onto the powerhead and secure the legs with the screws and lockwashers.

Attach the cable securing screw and the retainer clip, as shown in the accompanying illustration.

6. The sheave shaft must be flush with the retainer plate for correct alignment during installation.

7. If the choke handle shaft passes through the top of the cowling, feed it through and make the connection to the lower portion of the shaft with the cotter pin.

8. Feed the rope through the top cowling and install the rope retainer.

9. To adjust the shift interlock, secure the cable attaching screw to the side of the rewind housing, but do not tighten it at this time. Hook the interlock cable over the peg on the interlock actuator with a flat washer and cotter pin. Be sure the control handle is in the neutral position. Adjust the cable until the inter lock actuator is posi-

tioned on the rise of the interlock cam. Tighten the cable attaching screw to hold this position.

10. Shift the control handle into the forward position and attempt to pull the handle outward. The attempt should fail.

11. Shift the handle back into the neutral position and again attempt to pull on the handle. The attempt should be successful and the rope should rewind normally.

12. Install the cowling.

DISASSEMBLY

♦See accompanying illustrations

1. Place the hand starter on a suitable work surface. Pry out the one tab bent into the recess in the rewind housing. Use a hammer and flat punch to push aside the two tabs of the tab washer. This washer secures the sheave shaft retaining nut.

※※ WARNING

The sheave shaft retaining nut has left-hand threads.

2. Remove the nut with a ¾ in. end wrench by rotating the nut clockwise because it is a left-hand nut.

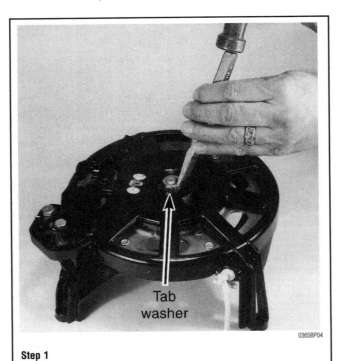

Step 1

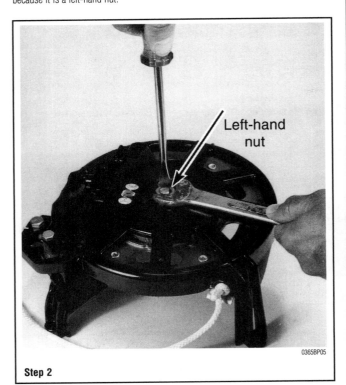

Step 2

HAND REWIND STARTER 11-21

Step 3

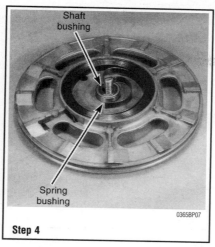
Step 4

Step 5

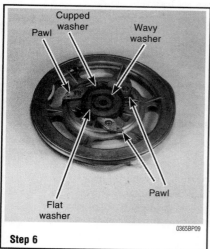
Step 6

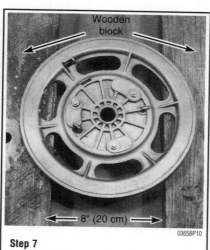
Step 7

Step 8

3. Carefully lift the starter housing from the sheave assembly so as not to disturb the rewind spring encased in the sheave. Place the sheave face down on a flat surface.
4. Lift off the spring retainer plate, and then remove the shaft bushing and the spring bushing from the sheave.
5. Remove the sheave shaft, the fiber washer, and the pawl retainer plate from the sheave.

➡ The return spring will come away with the pawl retainer plate.

6. Remove and discard the wave washer. It cannot be used a second time. Remove the flat washer and the cupped washer. Lift off the three pawls from their pins. Remove and discard the wave washer under each pawl.
7. Obtain two pieces of wood, a short 2 in. x 4 in. (5cm x 10cm) will work fine. Place the two pieces of wood approximately 8 in. (20cm) apart on the floor. Center the sheave on top of the wood with the spring side facing down. Check to be sure the wood is not touching the spring.

✱✱ WARNING

The rewind spring is a potential hazard. The spring is under tremendous tension when it is wound. If the spring should accidentally be released, severe personal injury could result from being struck by the spring with force. Therefore, the following step must be performed with care to prevent personal injury to self and others in the area.

8. Stand behind the wood, keeping away from the openings as the spring unwinds with considerable force. Tap the sheave with a soft mallet. The spring retainer plate will drop down releasing the spring. The spring will fall and unwind almost instantly and with force.
9. Unwind the rope from around the sheave and feed it through the holes. Turn the anchor pin 90° in either direction to release the rope.

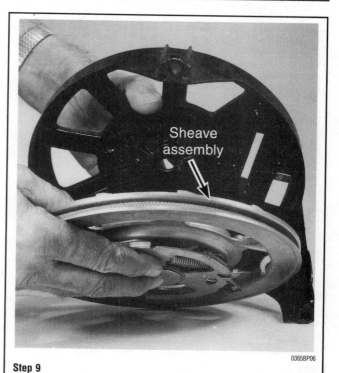
Step 9

11-22 HAND REWIND STARTER

CLEANING & INSPECTION

♦ See Figures 11 and 12

1. Wash all parts except the rope and the handle in solvent, and then blow them dry with compressed air.
2. Remove any trace of corrosion and wipe all metal parts with an oil dampened cloth.
3. Inspect the rope. Replace the rope if it appears to be weak or frayed. If the rope is frayed, check the holes through which the rope passes for rough edges or burrs. Remove the rough edges or burrs with a file and polish the surface until it is smooth.
4. Inspect the starter spring end hooks. Replace the spring if it is weak, corroded or cracked. Inspect the tab on the spring retainer plate. This tab is inserted into the inner loop of the spring. Therefore, be sure it is straight and solid.
5. Inspect the inside surface of the sheave rewind recess for grooves or roughness. Grooves may cause erratic rewinding of the starter rope.
6. Check the condition of the pawl pivot pins for excessive wear. These pins are a part of the sheave casting. They cannot be serviced. Therefore, if they are not acceptable, a new sheave must be purchased and installed.
7. Inspect the pawls for wear around the pivot holes and for rounded outer edges. Replace the pawls in sets of three only. Check the pawl retainer plate for wear, especially the center hole and the areas where the sides of the pawls contact the retainer plate.
8. Inspect the pawl retainer plate return spring (if equipped). The manufacturer recommends this spring be replaced each time the rewind starter is disassembled for service. A weak spring could allow the pawls to contact the flywheel while the powerhead is operating and cause a very unpleasant noise and even damage to the edge of the flywheel. A weak spring will also cause excessive wear to the pawls.
9. It is strongly recommended the wavy washer under the pawl retainer plate, and the three wave washers under the pawls be replaced. New wave washers will ensure smooth operation of the rewind mechanism.
10. Coat the following parts with low-temperature lubricant: the entire length of the used rewind spring (a new spring will be coated with lubricant from the package), the pawl pins, both sides of the cupped washer, the flat washer, the shoulder on the sheave shaft, and the tab on the spring retainer plate.

ASSEMBLY

♦ See accompanying illustrations

→The rewind starter may be assembled with a new rewind spring or a used one. Procedures for assembling are not the same because the new spring will arrive in a shipping container already wound and ready for installation. The used spring must be manually wound into its recess.

The situation may arise when it is only necessary to replace a broken spring. The following few procedures outline the tasks required to replace the spring. A new spring is already properly wound and will arrive in a special shipping container. This container is designed to be used as an aid to installing the new spring.

1. Remove the retainers from the shipping container used to keep the spring from accidentally falling out of the container. Place the shipping container over the spring recess with the tabs resting on the outer edge of the recess.
2. A new rewind spring is installed into the sheave using the shipping container as a helpful tool. Two screwdrivers are used to tap the coiled spring out of the container and into the sheave.
3. Align the hook on the outer end of the spring with the sheave anchor or notch.
4. Place two large blade screwdrivers into the holes of the shipping container over the tensioned spring. Push on both screwdrivers at the same time to press the spring out of the shipping container and into the spring recess of the sheave.

A used spring naturally will not be wound. Therefore, special instructions are necessary for installation.

→Wear a good pair of gloves while winding and installing the spring. The spring will develop tension and the edges of the spring steel are extremely sharp. The gloves will prevent cuts to the hands and fingers.

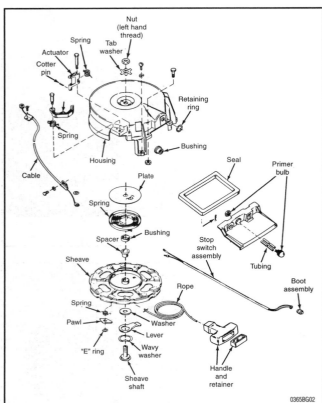

Fig. 11 Exploded drawing of the hand rewind starter covered in this manual. Major parts have been identified.

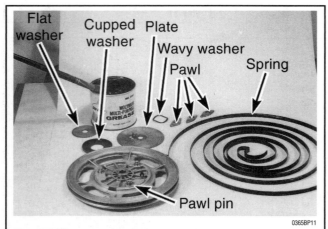

Fig. 12 After the parts shown have been cleaned and are ready for assembling, apply a coating of low-temperature lubricant to ensure long time service.

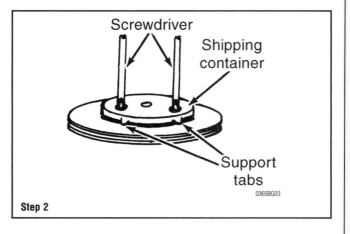

Step 2

HAND REWIND STARTER   11-23

5. Insert the sheave shaft through the sheave with the spring recess facing upward. Obtain two large washers and place them over the sheave shaft. Hold the washers in place and at the same time clamp the sheave shaft in a vise equipped with soft jaws.

6. Shift the shaft in the vise to allow for "up and down" clearance and permit the shaft to rotate freely while the rewind spring is being installed.

➡ **It is STRONGLY recommended a pair of safety goggles or a face shield be worn while the spring is being installed. If the spring is accidentally released, it will lash out with tremendous ferocity and very likely could cause personal injury to the installer or other persons nearby.**

7. Loop the spring loosely into a coil to enable it to be fed into its recess.
8. Insert the hook on the end of the spring into the notch of the recess.
9. Feed the spring around the inner edge of the recess and at the same time rotate the sheave clockwise. Proceed with great care. Guide the spring into place. The spring will be slippery with lubrication. Do not lose control of the spring. Carefully remove the sheave from the vise and the sheave shaft and washers from the sheave.

10. Thread the starter rope thru the holes in the sheave and secure it with the anchor pin. Wind the rope counterclockwise around the outer rim of the sheave (when viewed from the pawl side of the sheave).

11. Place one new wave washer on each pawl pin. Slide the pawls over the pins and wavy washers. Check to be sure the "dot" or depression on the pawl is facing upward and angled outward.

12. Place the cupped washer, with the cupped side facing UPWARD over the center hole of the sheave. Place the flat washer over the cupped washer with the chamfered side of the flat washer facing outward (away from the sheave). Install another new wave washer onto the flat washer.

13. Install a new return spring between the sheave and the pawl retainer plate, and then position the plate over the pawls. Place the fiber washer over the pawl retainer plate.

14. Insert the sheave shaft into the pawl retainer plate. The large diameter of the

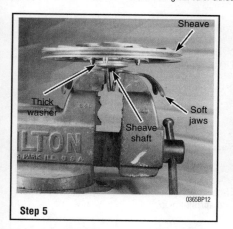
Step 5

Step 7

Step 8

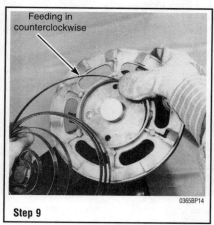
Step 9

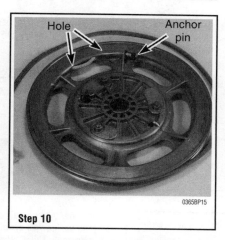
Step 10

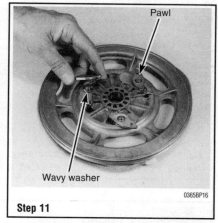
Step 11

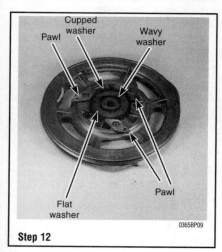
Step 12

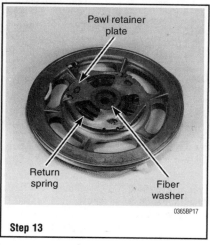
Step 13

Step 14

Step 15

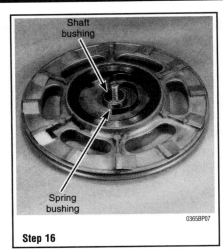
Step 16

Step 17

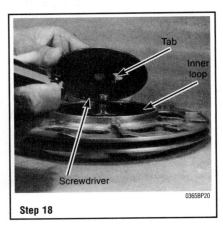
Step 18

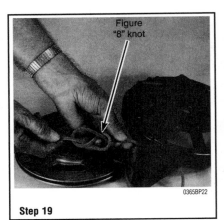
Step 19

Step 20

shoulder on the sheave shaft must be flush against the pawl retainer plate. If the wave washer is not centered beneath the pawl retainer plate, the sheave shaft will appear misaligned. To correct this condition, remove the shaft and the retainer plate, then center the wave washer and again install the plate and shaft. Again, check the alignment of the sheave shaft.

15. Lift the sheave assembly off the work bench, grasp the threaded end of the sheave shaft and turn the assembly over. Place the unit on the work bench with the spring side upward.

16. Slide the shaft bushing and the spring bushing over the sheave shaft.

17. The sheave assembly, with the spring retainer plate installed, is now ready for installation into the rewind starter housing.

18. Lower the spring retainer plate over the shaft. Insert the tab on the plate with the inner loop of the rewind spring. It may be necessary to hook a thin long screwdriver into the loop as an assist to aligning the loop and the tab.

19. Feed the rope through the rope guide. Tie a figure "8" knot in the rope about 1 foot (30 cm) from the end. Place the sheave assembly into the rewind housing.

20. Support the sheave assembly with one hand. Slide a new tab washer over the threaded portion of the sheave shaft with the cupped side of the washer facing down. Thread the left-hand sheave shaft retaining nut onto the shaft in a left-hand (counterclockwise) direction until it is just finger-tight, at this time.

21. To adjust the rewind spring tension, insert a large blade screwdriver into the slot of the sheave shaft. Hold the retaining nut with the proper size wrench and at the same time rotate the shaft with the screwdriver counterclockwise until the figure "8" knot in the rope rests against the rope guide. Continue to rotate the shaft through two full turns after the knot is against the rope guide. The correct tension has now been placed on the rewind spring.

22. Hold the tension on the spring with the screwdriver in the shaft slot and at the same time tighten the retaining nut securely. Remember—left-hand threads—tighten counterclockwise.

➡ **Do not bend the tabs on the washer up against the retaining nut until after the rewind operation has been checked.**

To check the rewind operation, slowly pull the starter rope outward. The pawls must move to the engaged position as the pawl retainer plate begins to turn. If the

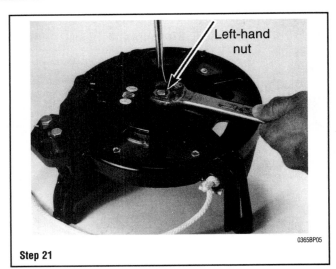
Step 21

pawls fail to engage, check the alignment or replace the wavy washers between the pawls. Again, extend the rope to its full length and allow it to rewind. The rope should rewind smoothly without catching. Never release the rope from the fully extended position.

➡ **If the rewind mechanism catches, but fails to rewind, the sheave shaft and its wave washer are not correctly aligned. The unit must be disassembled to correct the condition.**

Once the rewind operation is satisfactory, proceed to the next step.

23. Bend two tabs of the tab washer up against the flats of the retaining nut. Use tabs opposite each other. Bend one other tab down into the recess of the rewind housing.

GLOSSARY

Understanding your marine mechanic is as important as understanding your outboard. Most boaters know about their boats, but many boaters have difficulty understanding engine terminology. Talking the language of outboards makes it easier to effectively communicate with professional mechanics. It isn't necessary (or recommended) that you diagnose the problem for them, but it will save them time, and you money, if you can accurately describe what is happening. It will also help you to know why your boat does what it is doing, and what repairs were made.

AIR/FUEL RATIO: The ratio of air-to-fuel, by weight, drawn into the engine.

ALTERNATING CURRENT (AC): Electric current that flows first in one direction, then in the opposite direction, continually reversing flow.

AMP/HR. RATING (BATTERY): Measurement of the ability of a battery to deliver a stated amount of current for a stated period of time. The higher the amp/hr. rating, the better the battery.

ATOMIZATION: The breaking down of a liquid into a fine mist that can be suspended in air.

AXIAL PLAY: Movement parallel to a shaft or bearing bore.

BACKFIRE: The sudden combustion of gases in the intake or exhaust system that results in a loud explosion.

BACKLASH: The clearance or play between two parts, such as meshed gears.

BATTERY: A direct current electrical storage unit, consisting of the basic active materials of lead and sulfuric acid, which converts chemical energy into electrical energy. Used to provide current for the operation of the starter as well as other equipment, such as the radio, lighting, etc.

BEARING: A friction reducing, supportive device usually located between a stationary part and a moving part.

BEFORE TOP DEAD CENTER (BTDC): The point just before the piston reaches the top of its travel on the compression stroke.

BORE: Diameter of a cylinder.

BTDC: Before Top Dead Center.

BUSHING: A liner, usually removable, for a bearing; an anti-friction liner used in place of a bearing.

CARBON MONOXIDE (CO): A colorless, odorless gas given off as a normal byproduct of combustion. It is poisonous and extremely dangerous in confined areas, building up slowly to toxic levels without warning if adequate ventilation is not available.

CHECK VALVE: Any one-way valve installed to permit the flow of air, fuel or vacuum in one direction only.

CIRCLIP: A split steel snapring that fits into a groove to hold various parts in place.

CIRCUIT: Any unbroken path through which an electrical current can flow. Also used to describe fuel flow in some instances.

COMPRESSION CHECK: A test involving cranking the engine with a special high pressure gauge connected to an individual cylinder. Individual cylinder pressure as well as pressure variance across cylinders is used to determine general operating condition of the engine.

COMPRESSION RATIO: The ratio of the volume between the piston and cylinder head when the piston is at the bottom of its stroke (bottom dead center) and when the piston is at the top of its stroke (top dead center).

CONDUCTOR: Any material through which an electrical current can be transmitted easily.

CONNECTING ROD: The connecting link between the crankshaft and piston.

CONTINUITY: Continuous or complete circuit. Can be checked with an ohmmeter.

CRANKCASE: The lower part of an engine in which the crankshaft and related parts operate.

CRANKSHAFT: Engine component (connected to pistons by connecting rods) which converts the reciprocating (up and down) motion of pistons to rotary motion used to turn the driveshaft.

CYLINDER: In an engine, the round hole in the engine block in which the piston(s) ride.

DETONATION: An unwanted explosion of the air/fuel mixture in the combustion chamber caused by excess heat and compression, advanced timing, or an overly lean mixture. Also referred to as "ping".

DIAPHRAGM: A thin, flexible wall separating two cavities, such as in a vacuum advance unit.

DIODE: An electrical device that will allow current to flow in one direction only.

DISPLACEMENT: The total volume of air that is displaced by all pistons as the engine turns through one complete revolution.

DVOM: Digital volt ohmmeter

END-PLAY: The measured amount of axial movement in a shaft.

FEELER GAUGE: A blade, usually metal, of precisely predetermined thickness, used to measure the clearance between two parts.

FIRING ORDER: The order in which combustion occurs in the cylinders of an engine.

FLYWHEEL: A heavy disc of metal attached to the rear of the crankshaft. It smoothes the firing impulses of the engine and keeps the crankshaft turning during periods when no firing takes place. The starter also engages the flywheel to start the engine.

FOOT POUND (ft. lbs. or sometimes, ft. lb.): The amount of energy or work needed to raise an item weighing one pound, a distance of one foot.

FUEL FILTER: A component of the fuel system containing a porous paper element used to prevent any impurities from entering the engine through the fuel system. It usually takes the form of a canister-like housing, mounted in-line with the fuel hose, located anywhere on a vessel between the fuel tank and engine.

FUEL INJECTION: A system that sprays fuel into the cylinder through nozzles. The amount of fuel can be more precisely controlled with fuel injection.

FUSE: A protective device in a circuit which prevents circuit overload by breaking the circuit when a specific amperage is present. The device is constructed around a strip or wire of a lower amperage rating than the circuit it is designed to protect. When an amperage higher than that stamped on the fuse is present in the circuit, the strip or wire melts, opening the circuit.

HORSEPOWER: A measurement of the amount of work; one horsepower is the amount of work necessary to lift 33,000 lbs. one foot in one minute. Brake horsepower (bhp) is the horsepower delivered by an engine on a dynamometer. Net horsepower is the power remaining (measured at the flywheel of the engine) that can be used to power the vessel after power is consumed through friction and running the engine accessories (water pump, alternator, fan etc.)

HYDROMETER: An instrument used to measure the specific gravity of a solution.

IMPELLER: The portion of the water pump which provides the propulsion for the coolant to circulate it through the system

INCH POUND (inch lbs.; sometimes in. lb. or in. lbs.): One twelfth of a foot pound.

INJECTOR: A device which receives metered fuel under relatively low pressure and is activated to inject the fuel into the engine under relatively high pressure at a predetermined time.

JOURNAL: The bearing surface within which a shaft operates.

KNOCK: Noise which results from the spontaneous ignition of a portion of the air-fuel mixture in the engine cylinder.

MISFIRE: Condition occurring when the fuel mixture in a cylinder fails to ignite, causing the engine to run roughly.

MULTI-WEIGHT: Type of oil that provides adequate lubrication at both high and low temperatures.

NEEDLE BEARING: A bearing which consists of a number (usually a large number) of long, thin rollers.

OEM: Original Equipment Manufactured. OEM equipment is that furnished standard by the manufacturer.

PING: A metallic rattling sound produced by the engine during acceleration. It is usually due to incorrect timing or a poor grade of fuel.

POLARITY: Indication (positive or negative) of the two poles of a battery.

PRELOAD: A predetermined load placed on a bearing during assembly or by adjustment.

PRESS FIT: The mating of two parts under pressure, due to the inner diameter of one being smaller than the outer diameter of the other, or vice versa; an interference fit.

PSI: Pounds per square inch; a measurement of pressure.

RECTIFIER: A device (used primarily in alternators) that permits electrical current to flow in one direction only.

REGULATOR: A device which maintains the amperage and/or voltage levels of a circuit at predetermined values.

RESISTOR: A device, usually made of wire, which offers a preset amount of resistance in an electrical circuit.

RPM: Revolutions per minute (usually indicates engine speed).

SENSOR: Any device designed to measure engine operating conditions or ambient pressures and temperatures. Usually electronic in nature and designed to send a voltage signal to an on-board computer, some sensors may operate as a simple on/off switch or they may provide a variable voltage signal (like a potentiometer) as conditions or measured parameters change.

SHIM: Spacers of precise, predetermined thickness used between parts to establish a proper working relationship.

SOLENOID: An electrically operated, magnetic switching device.

SPLINES: Ridges machined or cast onto the outer diameter of a shaft or inner diameter of a bore to enable parts to mate without rotation.

STARTER: A high-torque electric motor used for the purpose of starting the engine, typically through a high ratio geared drive connected to the flywheel ring gear.

STROKE: The distance the piston travels from bottom dead center to top dead center.

TACHOMETER: A device used to measure the rotary speed of an engine, shaft, gear, etc., usually in rotations per minute.

THERMOSTAT: A valve, located in the cooling system of an engine, which is closed when cold and opens gradually in response to engine heating, controlling the temperature of the coolant and rate of coolant flow.

TOP DEAD CENTER (TDC): The point at which the piston reaches the top of its travel on the compression stroke.

TORQUE: Measurement of turning or twisting force, expressed as foot-pounds or inch-pounds.

TUNE-UP: A regular maintenance function, usually associated with the replacement and adjustment of parts and components in the electrical and fuel systems of a engine for the purpose of attaining optimum performance.

VOLTAGE REGULATOR: A device that controls the current output of the alternator or generator.

AIR COMPRESSOR 4-70
    REMOVAL & INSTALLATION 4-70
    TESTING 4-70
AIR PRESSURE REGULATOR 4-72
    REMOVAL & INSTALLATION 4-72
AIR TEMPERATURE SENSOR (ELECTRONIC FUEL INJECTION) 4-60
    REMOVAL & INSTALLATION 4-60
    TESTING 4-60
AIR TEMPERATURE SENSOR (OPTIMAX DIRECT FUEL INJECTION) 4-72
    REMOVAL & INSTALLATION 4-72
    TESTING 4-72
ANODES (ZINCS) 3-6
    INSPECTION 3-7
    SERVICING 3-7
AVOIDING THE MOST COMMON MISTAKES 1-3
AVOIDING TROUBLE 1-2
BASIC ELECTRICAL THEORY 5-2
    HOW ELECTRICITY WORKS: THE WATER ANALOGY 5-2
    OHM'S LAW 5-2
BATTERIES (BOAT MAINTENANCE) 3-8
    CLEANING 3-8
    MAINTENANCE 3-8
    STORAGE 3-9
    TESTING 3-8
BATTERY (CHARGING CIRCUIT) 5-38
    BATTERY CABLES 5-40
    BATTERY CHARGERS 5-39
    BATTERY CONSTRUCTION 5-38
    BATTERY LOCATION 5-39
    BATTERY RATINGS 5-38
    MARINE BATTERIES 5-38
BEARINGS 7-29
    GENERAL INFORMATION 7-29
    INSPECTION 7-30
BLEEDING THE OIL INJECTION SYSTEM 6-3
    PROCEDURE 6-3
BOAT MAINTENANCE 3-8
BOATING SAFETY 1-3
BOLTS, NUTS AND OTHER THREADED RETAINERS 2-11
BREAKER POINTS 5-9
    POINT GAP CHECK 5-9
    REMOVAL & INSTALLATION 5-11
BREAKER POINTS IGNITION (MAGNETO IGNITION) 5-8
BUY OR REBUILD? 7-19
CAN YOU DO IT? 1-2
CAPACITIES 3-42
CAPACITOR DISCHARGE IGNITION (CDI) SYSTEM 5-14
CAPACITOR DISCHARGE MODULE (SWITCH BOX) 5-27
    REMOVAL & INSTALLATION 5-28
    TESTING 5-27
CARBURETED FUEL SYSTEM 4-3
  2.5, 3 AND 3.3 HP 4-8
    ASSEMBLY 4-10
    CLEANING & INSPECTION 4-10
    DISASSEMBLY 4-8
    REMOVAL & INSTALLATION 4-8
  4 AND 5 HP 4-11
    ASSEMBLY 4-13
    CLEANING & INSPECTION 4-13
    DISASSEMBLY 4-12
    REMOVAL & INSTALLATION 4-11
  6, 8, 9.9, 10 AND 15 HP 4-14
    ASSEMBLY 4-19
    CLEANING & INSPECTION 4-17
    DISASSEMBLY 4-15
    REMOVAL & INSTALLATION 4-14
  20 AND 25 HP 4-20
    ASSEMBLY 4-22
    CLEANING & INSPECTION 4-21
    DISASSEMBLY 4-20
    REMOVAL & INSTALLATION 4-20

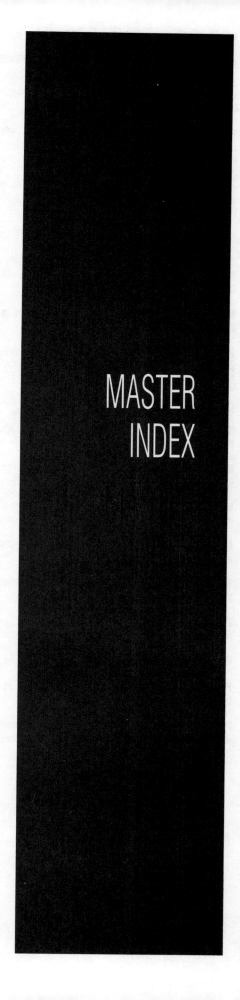

MASTER INDEX

MASTER INDEX

30 AND 40 HP 4-23
  ASSEMBLY 4-25
  CLEANING & INSPECTION 4-25
  DISASSEMBLY 4-24
  REMOVAL & INSTALLATION 4-23
40–125 HP 4-26
  ASSEMBLY 4-31
  CLEANING & INSPECTION 4-30
  DISASSEMBLY 4-28
  REMOVAL & INSTALLATION 4-27
1990 135–200 HP 4-33
  ASSEMBLY 4-36
  CLEANING & INSPECTION 4-35
  DISASSEMBLY 4-35
  REMOVAL & INSTALLATION 4-33
1991–00 135–200 HP 4-37
  ASSEMBLY 4-40
  CLEANING & INSPECTION 4-40
  DISASSEMBLY 4-38
  REMOVAL & INSTALLATION 4-37
1991–00 275 HP 4-41
  ASSEMBLY 4-44
  CLEANING & INSPECTION 4-44
  DISASSEMBLY 4-43
  REMOVAL & INSTALLATION 4-42
CARBURETION 4-3
  BASIC FUNCTIONS 4-3
  CARBURETOR CIRCUITS 4-4
  FUEL & AIR METERING 4-4
  FUEL PUMP 4-5
CHARGING CIRCUIT 5-30
CHEMICALS 2-2
  CLEANERS 2-3
  LUBRICANTS & PENETRANTS 2-2
  SEALANTS 2-3
COMBUSTION 4-2
COMPRESSION TEST 3-10
  PRIMARY COMPRESSION TEST 3-10
  SECONDARY COMPRESSION TEST 3-10
CONDENSER 5-12
  REMOVAL & INSTALLATION 5-12
  TESTING 5-12
CONNECTING RODS 7-27
  GENERAL INFORMATION 7-27
  INSPECTION 7-27
CONTROL CABLES 10-10
  REMOVAL & INSTALLATION 10-10
CONVERSION FACTORS 2-12
COOLANT TEMPERATURE SENSOR 4-60
  REMOVAL & INSTALLATION 4-60
  TESTING 4-60
COOLING SYSTEM 6-8
COURTESY MARINE EXAMINATIONS 1-10
CRANKSHAFT 7-28
  GENERAL INFORMATION 7-28
  INSPECTION 7-29
CRANKSHAFT POSITION SENSOR (CPS) 4-72
  REMOVAL & INSTALLATION 4-72
  TESTING 4-72
CYLINDER BLOCK AND HEAD 7-22
  GENERAL INFORMATION 7-22
  INSPECTION 7-23
CYLINDER BORES 7-23
  GENERAL INFORMATION 7-23
  INSPECTION 7-23
  REFINISHING 7-24
DESCRIPTION & OPERATION (BREAKER POINTS IGNITION) 5-8
  SYSTEM COMPONENTS 5-8

DESCRIPTION AND OPERATION (CAPACITOR DISCHARGE IGNITION SYSTEM) 5-14
  2.5-3.3 HP 5-15
  4-275 HP 5-15
DESCRIPTION AND OPERATION (CHARGING CIRCUIT) 5-30
DESCRIPTION AND OPERATION (COOLING SYSTEM) 6-8
  OPTIMAX 6-8
DESCRIPTION AND OPERATION (ELECTRONIC FUEL INJECTION) 4-50
  FUEL INJECTION BASICS 4-50
  MERCURY ELECTRONIC FUEL INJECTION 4-50
DESCRIPTION AND OPERATION (GAS ASSIST TILT SYSTEM) 9-5
DESCRIPTION AND OPERATION (HAND REWIND STARTER) 11-2
DESCRIPTION AND OPERATION (JET DRIVE) 8-11
DESCRIPTION AND OPERATION (MANUAL TILT) 9-2
  SERVICING 9-2
DESCRIPTION AND OPERATION (OIL INJECTION SYSTEM) 6-2
  MERCURY OIL INJECTION 6-2
  OPTIMAX OIL INJECTION 6-2
DESCRIPTION AND OPERATION (OPTIMAX DIRECT FUEL INJECTION) 4-66
  OPTIMAX COMPONENTS 4-68
DESCRIPTION AND OPERATION (OPTIMAX WARNING SYSTEMS) 6-27
  1998-99 MODEL YEARS 6-27
  GUARDIAN PROTECTION SYSTEM—2000 6-27
DESCRIPTION AND OPERATION (POWER TRIM/TILT) 9-6
DESCRIPTION AND OPERATION (REMOTE CONTROL BOX) 10-2
DESCRIPTION AND OPERATION (SINGLE RAM INTEGRAL POWER TILT/TRIM) 9-7
DESCRIPTION AND OPERATION (STARTER CIRCUIT) 5-40
DESCRIPTION AND OPERATION (THREE RAM INTEGRAL POWER TILT/TRIM SYSTEM) 9-12
DESCRIPTION AND OPERATION (TILLER HANDLE) 10-6
DESCRIPTION AND OPERATION (WARNING SYSTEMS) 6-22
  OIL QUANTITY INDICATOR 6-22
  OVERHEAT TEMPERATURE WARNING 6-23
DETERMINING POWERHEAD CONDITION 7-19
DETONATION SENSOR AND MODULE 4-66
  TESTING 4-66
DIRECT INJECTOR 4-73
  CLEANING & INSPECTION 4-74
  REMOVAL & INSTALLATION 4-73
  TESTING 4-73
DIRECTIONS AND LOCATIONS 1-2
DO'S (SAFTEY IN SERVICE) 1-10
DON'TS (SAFTEY IN SERVICE) 1-10
ELECTRIC FUEL PUMP 4-61
  REMOVAL & INSTALLATION 4-62
  TESTING 4-61
ELECTRICAL COMPONENTS 5-2
  CONNECTORS 5-4
  GROUND 5-3
  LOAD 5-4
  POWER SOURCE 5-2
  PROTECTIVE DEVICES 5-3
  SWITCHES & RELAYS 5-3
  WIRING & HARNESSES 5-4
ELECTRONIC CONTROL MODULE (ECM) 4-74
  REMOVAL & INSTALLATION 4-74
ELECTRONIC FUEL INJECTION (EFI) 4-50
ELECTRONIC TOOLS 2-8
EMERGENCY STOP SWITCH (TILLER HANDLE) 10-12
  REMOVAL & INSTALLATION 10-12
  TESTING 10-12
EMERGENCY STOP SWITCH, LANYARD (REMOTE CONTROL BOX) 10-5
  REMOVAL & INSTALLATION 10-5
  TESTING 10-5
ENGINE MAINTENANCE 3-2

MASTER INDEX 11-29

ENGINE MECHANICAL 7-2
   2.5–3.3 HP 7-14
     REMOVAL & INSTALLATION 7-14
   4–5 HP 7-14
     REMOVAL & INSTALLATION 7-14
   6–15 HP 7-15
     REMOVAL & INSTALLATION 7-15
   20–25 HP 7-15
     REMOVAL & INSTALLATION 7-15
   30 AND 40 HP 7-15
     REMOVAL & INSTALLATION 7-15
   40 HP (3-CYLINDER), 50 AND 60 HP 7-16
     REMOVAL & INSTALLATION 7-17
   40 HP (4-CYLINDER) 7-16
     REMOVAL & INSTALLATION 7-16
   75–125 HP 7-17
     REMOVAL & INSTALLATION 7-17
   135–225 DFI OPTIMAX 7-18
     REMOVAL & INSTALLATION 7-18
   135–250 HP 7-17
     REMOVAL & INSTALLATION 7-17
   275 HP 7-18
     REMOVAL & INSTALLATION 7-19
ENGINE REBUILDING SPECIFICATIONS 7-41
ENGINE START SWITCH 10-12
   REMOVAL & INSTALLATION 10-12
   TESTING 10-12
ENGINE STOP SWITCH (REMOTE CONTROL BOX) 10-5
   REMOVAL & INSTALLATION 10-5
   TESTING 10-5
ENGINE STOP SWITCH (TILLER HANDLE) 10-12
   REMOVAL & INSTALLATION 10-12
   TESTING 10-12
ENGINE TORQUE SPECIFICATIONS 7-39
EQUIPMENT NOT REQUIRED BUT RECOMMENDED 1-9
   ANCHORS 1-9
   BAILING DEVICES 1-9
   FIRST AID KIT 1-9
   SECOND MEANS OF PROPULSION 1-9
   TOOLS AND SPARE PARTS 1-10
   VHF-FM RADIO 1-10
FASTENERS, MEASUREMENTS AND CONVERSIONS 2-11
FIBERGLASS HULL 3-9
FUEL 4-2
   ALCOHOL-BLENDED FUELS 4-2
   HIGH ALTITUDE OPERATION 4-2
   OCTANE RATING 4-2
   RECOMMENDATIONS 4-2
   VAPOR PRESSURE 4-2
FUEL AND COMBUSTION 4-2
FUEL FILTER 3-4
   CLEANING & INSPECTION 3-5
   RELIEVING FUEL SYSTEM PRESSURE 3-4
   REMOVAL & INSTALLATION 3-5
FUEL INJECTOR (OPTIMAX DIRECT FUEL INJECTION) 4-74
   CLEANING & INSPECTION 4-74
   REMOVAL & INSTALLATION 4-74
   TESTING 4-74
FUEL INJECTORS (ELECTRONIC FUEL INJECTION) 4-54
   CLEANING & INSPECTION 4-60
   REMOVAL & INSTALLATION 4-56
   TESTING 4-54
FUEL LINES 4-49
FUEL PRESSURE REGULATOR (ELECTRONIC FUEL INJECTION) 4-64
   REMOVAL & INSTALLATION 4-64
FUEL PRESSURE REGULATOR (OPTIMAX DIRECT FUEL INJECTION) 4-75
   CLEANING & INSPECTION 4-75
   REMOVAL & INSTALLATION 4-75

FUEL PUMP (CARBURETED FUEL SYSTEM) 4-45
   OVERHAUL 4-48
   REMOVAL & INSTALLATION 4-47
   TESTING 4-45
FUEL PUMP (OPTIMAX DIRECT FUEL INJECTION) 4-75
   TESTING 4-75
FUEL RAIL 4-76
   CLEANING & INSPECTION 4-76
   REMOVAL & INSTALLATION 4-76
FUEL/WATER SEPARATOR 3-5
   SERVICE 3-5
GAS ASSIST DAMPER 9-5
   REMOVAL & INSTALLATION 9-6
   TESTING 9-5
GAS ASSIST TILT SYSTEM 9-5
GAUGES 2-9
GENERAL ENGINE SPECIFICATIONS 3-40
GENERAL INFORMATION (LOWER UNIT) 8-2
HAND REWIND STARTER 11-2
HAND TOOLS 2-4
   HAMMERS 2-7
   PLIERS 2-7
   SCREWDRIVERS 2-7
   SOCKET SETS 2-4
   WRENCHES 2-6
HOW TO USE THIS MANUAL 1-2
IGNITION AND ELECTRICAL WIRING DIAGRAMS 5-42
IGNITION COIL (BREAKER POINTS IGNITION) 5-13
   REMOVAL & INSTALLATION 5-14
   TESTING 5-13
IGNITION COIL (CAPACITOR DISCHARGE IGNITION SYSTEM) 5-25
   REMOVAL & INSTALLATION 5-26
   TESTING 5-25
IGNITION SWITCH 10-5
   REMOVAL & INSTALLATION 10-5
   TESTING 10-5
IGNITION SYSTEM 3-14
IGNITION TESTING SPECIFICATIONS 5-17
INTRODUCTION (TUNE-UP) 3-10
JET DRIVE 8-11
JET DRIVE ASSEMBLY 8-12
   ADJUSTMENT 8-15
   ASSEMBLY 8-18
   CLEANING AND INSPECTING 8-17
   DISASSEMBLY 8-16
   REMOVAL & INSTALLATION 8-12
   SHIMMING 8-19
LOWER UNIT 8-5
   REMOVAL & INSTALLATION 8-5
LOWER UNIT 8-2
LOWER UNIT (ENGINE MAINTENANCE) 3-3
   DRAINING AND FILLING 3-3
   OIL RECOMMENDATIONS 3-3
LOWER UNIT SPECIFICATIONS 8-20
LUBRICATION POINTS 3-6
   INSPECTION & LUBRICATION 3-6
MAINTENANCE INTERVAL CHART 3-42
MAINTENANCE OR REPAIR? 1-2
MANIFOLD ABSOLUTE PRESSURE (MAP) SENSOR 4-76
   REMOVAL & INSTALLATION 4-76
   TESTING 4-76
MANUAL TILT 9-2
MEASURING TOOLS 2-9
   DEPTH GAUGES 2-10
   DIAL INDICATORS 2-10
   MICROMETERS & CALIPERS 2-9
   TELESCOPING GAUGES 2-10

MASTER INDEX

MECHANICAL FUEL PUMP 4-61
    REMOVAL & INSTALLATION 4-61
    TESTING 4-61
MODEL IDENTIFICATION AND SERIAL NUMBERS 8-11
NEUTRAL START SWITCH 10-4
    REMOVAL & INSTALLATION 10-4
    TESTING 10-4
OIL GAUGE 6-26
    TESTING 6-26
OIL INJECTION SYSTEM 6-2
OIL LEVEL SWITCH 6-26
    REMOVAL & INSTALLATION 6-26
    TESTING 6-26
OIL LINES 6-7
    OIL LINE CAUTIONS 6-7
OIL PUMP 6-6
    REMOVAL & INSTALLATION 6-6
OIL PUMP CONTROL ROD 6-7
    ADJUSTMENT 6-7
OIL PUMP DISCHARGE RATE 6-7
    TESTING 6-7
OIL TANK 6-4
    CLEANING & INSPECTION 6-6
    REMOVAL & INSTALLATION 6-4
OPTIMAX DIRECT FUEL INJECTION (DFI) 4-66
OPTIMAX WARNING SYSTEMS 6-27
OTHER COMMON TOOLS 2-7
OVERHEAD TYPE STARTER 11-2
    2.5, 3 AND 3.3 HP 11-2
        ASSEMBLY 11-4
        CLEANING & INSPECTION 11-4
        DISASSEMBLY 11-3
        REMOVAL & INSTALLATION 11-2
    4 AND 5 HP 11-6
        ASSEMBLY 11-8
        CLEANING & INSPECTION 11-8
        DISASSEMBLY 11-7
        REMOVAL & INSTALLATION 11-6
    6–15 HP 11-11
        ASSEMBLY 11-14
        CLEANING & INSPECTION 11-13
        DISASSEMBLY 11-12
        REMOVAL & INSTALLATION 11-11
    20 AND 25 HP 11-15
        ASSEMBLY 11-18
        CLEANING & INSPECTION 11-17
        DISASSEMBLY 11-16
        REMOVAL & INSTALLATION 11-15
    30 AND 40 HP 11-19
        ASSEMBLY 11-22
        CLEANING & INSPECTION 11-22
        DISASSEMBLY 11-20
        REMOVAL & INSTALLATION 11-19
PISTON PINS 7-25
    GENERAL INFORMATION 7-25
    INSPECTION 7-25
PISTON RINGS 7-26
    GENERAL INFORMATION 7-26
    INSPECTION 7-27
PISTONS 7-24
    GENERAL INFORMATION 7-24
    INSPECTION 7-24
POWER TRIM/TILT 9-6
POWERHEAD 7-2
    REMOVAL & INSTALLATION 7-2
POWERHEAD EXPLODED VIEWS 7-30
POWERHEAD OVERHAUL TIPS 7-20
    CAUTIONS 7-20

    CLEANING 7-20
    REPAIRING DAMAGED THREADS 7-21
    TOOLS 7-20
POWERHEAD PREPARATION 7-21
POWERHEAD RECONDITIONING 7-19
PRIMARY COIL 5-13
    REMOVAL & INSTALLATION 5-13
    TESTING 5-13
PROFESSIONAL HELP 1-2
PROPELLER (ENGINE MAINTENANCE) 3-6
PROPELLER (LOWER UNIT) 8-3
    REMOVAL & INSTALLATION 8-3
PURCHASING PARTS 1-3
RECTIFIER 5-34
    REMOVAL & INSTALLATION 5-34
    TESTING 5-34
RECTIFIER TROUBLESHOOTING - TEST "A" 5-37
RECTIFIER TROUBLESHOOTING - TEST "B" 5-37
REED VALVE 7-13
REGULATIONS FOR YOUR BOAT 1-3
    CAPACITY INFORMATION 1-4
    CERTIFICATE OF COMPLIANCE 1-4
    DOCUMENTING OF VESSELS 1-4
    HULL IDENTIFICATION NUMBER 1-4
    LENGTH OF BOATS 1-4
    NUMBERING OF VESSELS 1-4
    REGISTRATION OF BOATS 1-4
    SALES AND TRANSFERS 1-4
    VENTILATION 1-4
    VENTILATION SYSTEMS 1-5
REGULATOR/RECTIFIER 5-34
    REMOVAL & INSTALLATION 5-38
    TESTING 5-34
REMOTE CONTROL BOX 10-2
REMOTE CONTROL BOX 10-2
    REMOVAL & INSTALLATION 10-2
REMOTE CONTROL CABLES 10-3
    ADJUSTMENT 10-3
REQUIRED SAFETY EQUIPMENT 1-5
    FIRE EXTINGUISHERS 1-5
    PERSONAL FLOTATION DEVICES 1-6
    SOUND PRODUCING DEVICES 1-8
    TYPES OF FIRES 1-5
    VISUAL DISTRESS SIGNALS 1-8
    WARNING SYSTEM 1-6
SAFETY IN SERVICE 1-10
SAFETY TOOLS 2-2
    EYE & EAR PROTECTION 2-2
    WORK CLOTHES 2-2
    WORK GLOVES 2-2
SERIAL NUMBER IDENTIFICATION 3-2
SHIFT INTERRUPT SWITCH 4-76
    REMOVAL & INSTALLATION 4-76
    TESTING 4-76
SHIFTING PRINCIPLES 8-2
    COUNTER-ROTATING UNIT 8-2
    STANDARD ROTATING UNIT 8-2
SINGLE RAM INTEGRAL POWER TILT/TRIM 9-7
SMALL VOLTAGE REGULATOR TEST (STATIC) 5-35
SPARK PLUG WIRES 3-14
    REMOVAL & INSTALLATION 3-14
    TESTING 3-14
SPARK PLUGS 3-11
    INSPECTION & GAPPING 3-13
    READING SPARK PLUGS 3-12
    REMOVAL & INSTALLATION 3-11
    SPARK PLUG HEAT RANGE 3-11
    SPARK PLUG SERVICE 3-11

SPECIAL TOOLS 2-8
SPECIFICATION CHARTS
  CAPACITIES 3-42
  CONVERSION FACTORS 2-12
  ENGINE REBUILDING SPECIFICATIONS 7-41
  ENGINE TORQUE SPECIFICATIONS 7-39
  GENERAL ENGINE SPECIFICATIONS 3-40
  IGNITION TESTING SPECIFICATIONS 5-17
  LOWER UNIT SPECIFICATIONS 8-20
  MAINTENANCE INTERVAL CHART 3-42
  OIL PUMP DISCHARGE RATE 6-7
  SMALL VOLTAGE REGULATOR TEST (STATIC) 5-35
  STATOR ALTERNATOR RESISTANCE SPECIFICATIONS 5-30
  TUNEUP SPECIFICATIONS CHART 3-41
  WME CARBURETOR SPECIFICATIONS 4-26
SPRING COMMISSIONING CHECKLIST 3-39
STANDARD AND METRIC MEASUREMENTS 2-11
STARTER CIRCUIT 5-40
STARTER MOTOR 5-41
  REMOVAL & INSTALLATION 5-41
STARTER SOLENOID 5-41
  TESTING 5-41
STATOR (ALTERNATOR) 5-30
  REMOVAL & INSTALLATION 5-31
  TESTING 5-30
STATOR ALTERNATOR RESISTANCE SPECIFICATIONS 5-30
TEST EQUIPMENT 5-4
  JUMPER WIRES 5-4
  MULTIMETERS 5-5
  TEST LIGHTS 5-5
THE TWO-STROKE CYCLE 7-2
THERMOSTAT 6-20
  CLEANING & INSPECTION 6-22
  REMOVAL & INSTALLATION 6-20
  TESTING 6-22
THREE RAM INTEGRAL POWER TILT/TRIM SYSTEM 9-12
THROTTLE PLATE ASSEMBLY 4-77
  REMOVAL & INSTALLATION 4-77
THROTTLE POSITION SENSOR (ELECTRONIC FUEL INJECTION) 4-65
  ADJUSTMENT 4-65
  REMOVAL & INSTALLATION 4-65
  TESTING 4-65
THROTTLE POSITION SENSOR (OPTIMAX DIRECT FUEL INJECTION) 4-73
  REMOVAL & INSTALLATION 4-73
  TESTING 4-73
TILLER HANDLE 10-6
TILLER HANDLE 10-6
  REMOVAL & INSTALLATION 10-6
TILT/TRIM CYLINDER (SINGLE RAM INTEGRAL POWER TILT/TRIM SYSTEM) 9-10
  CLEANING & INSPECTION 9-11
  REMOVAL & INSTALLATION 9-10
TILT/TRIM CYLINDER (THREE RAM INTEGRAL POWER TILT/TRIM SYSTEM) 9-18
  CLEANING & INSPECTION 9-18
  DISASSEMBLY 9-18
TILT/TRIM MOTOR (SINGLE RAM INTEGRAL POWER TILT/TRIM SYSTEM) 9-8
  CLEANING & INSPECTION 9-9
  REMOVAL & INSTALLATION 9-8
  TESTING 9-8
TILT/TRIM MOTOR (THREE RAM INTEGRAL POWER TILT/TRIM SYSTEM) 9-16
  REMOVAL & INSTALLATION 9-16
  TESTING 9-16
TILT/TRIM PUMP 9-13
  HYDRAULIC SYSTEM BLEEDING 9-15
  REMOVAL & INSTALLATION 9-13
TILT/TRIM RELAY (SINGLE RAM INTEGRAL POWER TILT/TRIM SYSTEM) 9-12
  TESTING 9-12
TILT/TRIM RELAY (THREE RAM INTEGRAL POWER TILT/TRIM SYSTEM) 9-20
  TESTING 9-20
TILT/TRIM SOLENOID 9-18
  TESTING 9-18
TILT/TRIM SWITCH (SINGLE RAM INTEGRAL POWER TILT/TRIM SYSTEM) 9-11
  TESTING 9-11
TILT/TRIM SWITCH (THREE RAM INTEGRAL POWER TILT/TRIM SYSTEM) 9-18
  TESTING 9-18
TIMING AND SYNCHRONIZATION 3-14
TOOLS 2-4
TOOLS AND EQUIPMENT 2-2
TORQUE 2-11
TORQUE SEQUENCE DIAGRAMS 7-36
TRACKER VALVE 4-77
  REMOVAL & INSTALLATION 4-77
TRIGGER (CHARGE) COIL 5-16
  REMOVAL & INSTALLATION 5-18
  TESTING 5-16
TRIM/TILT PUMP 9-7
  HYDRAULIC SYSTEM BLEEDING 9-8
  REMOVAL & INSTALLATION 9-7
TRIM/TILT SWITCH 10-6
  REMOVAL & INSTALLATION 10-6
  TESTING 10-6
TROUBLESHOOTING CHARTS
  RECTIFIER TROUBLESHOOTING - TEST "A" 5-37
  RECTIFIER TROUBLESHOOTING - TEST "B" 5-37
TROUBLESHOOTING ELECTRONIC FUEL INJECTION 4-53
TROUBLESHOOTING THE BREAKER POINTS IGNITION SYSTEM 5-8
  COMPRESSION TEST 5-9
  SPARK PLUGS 5-9
TROUBLESHOOTING THE CDI SYSTEM 5-15
  TROUBLESHOOTING ALTERNATOR DRIVEN IGNITIONS 5-16
  TROUBLESHOOTING BATTERY CD IGNITIONS 5-16
  TROUBLESHOOTING WITH MINIMAL TEST EQUIPMENT 5-15
TROUBLESHOOTING THE CHARGING SYSTEM 5-30
TROUBLESHOOTING THE COOLING SYSTEM 6-9
TROUBLESHOOTING THE ELECTRICAL SYSTEM 5-6
  OPEN CIRCUITS 5-6
  RESISTANCE 5-6
  SHORT CIRCUITS 5-7
  VOLTAGE 5-6
  VOLTAGE DROP 5-6
TROUBLESHOOTING THE FUEL SYSTEM 4-6
  COMBUSTION RELATED PISTON FAILURES 4-7
  COMMON PROBLEMS 4-6
TROUBLESHOOTING THE LOWER UNIT 8-2
TROUBLESHOOTING THE OIL INJECTION SYSTEM 6-3
TROUBLESHOOTING THE OPTIMAX FUEL INJECTION SYSTEM 4-69
  WITH THE DIGITAL DIAGNOSTIC TERMINAL (DDT) 4-70
  WITHOUT THE DIGITAL DIAGNOSTIC TERMINAL (DDT) 4-69
TROUBLESHOOTING THE POWER TRIM/TILT SYSTEM 9-6
TROUBLESHOOTING THE REMOTE CONTROLS 10-2
TROUBLESHOOTING THE REWIND STARTER 11-2
TROUBLESHOOTING THE STARTING SYSTEM 5-40
TROUBLESHOOTING THE TILLER HANDLE 10-6
TROUBLESHOOTING THE WARNING SYSTEMS (OPTIMAX WARNING SYSTEMS) 6-28
TROUBLESHOOTING THE WARNING SYSTEMS (WARNING SYSTEMS) 6-24

MASTER INDEX

TUNE-UP 3-10
  2.5, 3 AND 3.3 HP   3-15
    IDLE MIXTURE   3-15
    IDLE SPEED   3-15
    IGNITION TIMING   3-15
    THROTTLE JET NEEDLE   3-16
  4 AND 5 HP   3-16
    IDLE SPEED & MIXTURE   3-17
    IGNITION TIMING   3-16
    PRELIMINARY ADJUSTMENTS   3-16
  6, 8, 9.9 AND 15 HP   3-17
    FAST IDLE SPEED   3-18
    IDLE SPEED & MIXTURE   3-18
    IGNITION TIMING   3-18
    PRELIMINARY ADJUSTMENTS   3-17
  20 AND 25 HP   3-19
    CARBURETOR THROTTLE CAM   3-21
    DASHPOT ADJUSTMENT   3-21
    FAST IDLE SPEED   3-21
    FULL THROTTLE STOP   3-19
    IDLE SPEED & MIXTURE   3-20
    IGNITION TIMING   3-19
    PRELIMINARY ADJUSTMENTS   3-19
    STARTER INTERLOCK ADJUSTMENT   3-21
  30 AND 40 HP   3-22
    CAM FOLLOWER   3-22
    IDLE SPEED & MIXTURE   3-22
    IGNITION TIMING   3-22
    OIL PUMP   3-23
    PRELIMINARY ADJUSTMENTS   3-22
  40 (3-CYLINDER), 1994-00 50 AND 60 HP   3-26
    CARBURETOR SYNCHRONIZATION   3-26
    IDLE SPEED & MIXTURE   3-26
    IGNITION TIMING   3-26
    OIL PUMP   3-27
    PRELIMINARY ADJUSTMENTS   3-26
    THROTTLE CAM   3-26
  40 HP (4-CYLINDER)   3-23
    CARBURETOR SYNCHRONIZATION   3-23
    FULL THROTTLE STOP   3-24
    IDLE SPEED & MIXTURE   3-24
    IGNITION TIMING   3-23
    PRELIMINARY ADJUSTMENTS   3-23
  1990-93 50-60 HP   3-25
    CARBURETOR SYNCHRONIZATION   3-25
    IDLE SPEED & MIXTURE   3-25
    IGNITION TIMING   3-25
    PRELIMINARY ADJUSTMENTS   3-25
  75 AND 90 HP   3-27
    CARBURETOR SYNCHRONIZATION   3-27
    IDLE SPEED & MIXTURE   3-28
    IGNITION TIMING   3-27
    OIL PUMP   3-28
    PRELIMINARY ADJUSTMENTS   3-27
    THROTTLE CAM   3-28
  100-125 HP   3-28
    ACCELERATOR PUMP   3-29
    CARBURETOR SYNCHRONIZATION   3-28
    IDLE SPEED & MIXTURE   3-29
    IGNITION TIMING   3-29
    OIL PUMP   3-29
    PRELIMINARY ADJUSTMENTS   3-28
    THROTTLE CABLE PRELOAD   3-29
    THROTTLE CAM   3-28
  115-225 DIRECT FUEL INJECTION (DFI) OPTIMAX   3-37
    CRANK POSITION SENSOR   3-37
    FULL THROTTLE STOP   3-37
    THROTTLE CAM   3-37
    THROTTLE PLATE SCREW   3-38
  135-225 HP CARBURETED   3-30
    CARBURETOR SYNCHRONIZATION   3-30
    FULL THROTTLE STOP   3-30
    IDLE SPEED & MIXTURE   3-31
    IGNITION TIMING   3-30
    OIL PUMP   3-30
    PRELIMINARY ADJUSTMENTS   3-30
    THROTTLE CABLE PRELOAD   3-31
    THROTTLE CAM   3-30
  150-200 HP ELECTRONIC FUEL INJECTION   3-35
    FULL THROTTLE STOP   3-36
    IDLE SPEED   3-36
    IGNITION TIMING   3-35
    OIL PUMP   3-36
    THROTTLE CAM   3-36
    THROTTLE POSITION SENSOR (TPS)   3-37
  225 HP (3.0L) CARBURETED   3-31
    CARBURETOR SYNCHRONIZATION   3-32
    CRANKSHAFT POSITION SENSOR   3-33
    IDLE SPEED   3-33
    IGNITION TIMING   3-32
    OIL PUMP   3-32
    THROTTLE CAM   3-32
    THROTTLE POSITION SENSOR (TPS)   3-33
  250 AND 275 HP   3-33
    CARBURETOR SYNCHRONIZATION   3-34
    FULL THROTTLE STOP   3-34
    IDLE SPEED   3-34
    IGNITION TIMING   3-33
    OIL PUMP   3-34
TUNEUP SPECIFICATIONS CHART   3-41
2-STROKE OIL   3-2
  FILLING   3-2
  OIL RECOMMENDATIONS   3-2

UNDERSTANDING AND TROUBLESHOOTING ELECTRICAL SYSTEMS 5-2
VAPOR SEPARATOR (ELECTRONIC FUEL INJECTION)   4-62
  CLEANING & INSPECTION   4-63
  REMOVAL & INSTALLATION   4-62
VAPOR SEPARATOR (OPTIMAX DIRECT FUEL INJECTION)   4-77
  ASSEMBLY   4-78
  CLEANING & INSPECTION   4-77
  DISASSEMBLY   4-77
  REMOVAL & INSTALLATION   4-77

WARNING SYSTEMS 6-22
WATER PUMP   6-9
  CLEANING & INSPECTION   6-20
  REMOVAL & INSTALLATION   6-9
WATER TEMPERATURE SENSOR (OPTIMAX DIRECT FUEL INJECTION)   4-78
  REMOVAL & INSTALLATION   4-78
  TESTING   4-78
WATER TEMPERATURE SENSOR (WARNING SYSTEMS)   6-24
  REMOVAL & INSTALLATION   6-24
  TESTING   6-26
WHERE TO BEGIN   1-2
WINTER STORAGE CHECKLIST 3-38
WIRE AND CONNECTOR REPAIR   5-7
WME CARBURETOR SPECIFICATIONS   4-26

SELOC PUBLISHING'S FULL-LINE MASTER LIST

ISBN	PART NO		TITLE/DESCRIPTION	YEARS

OUTBOARDS
ISBN	Part	No	Title/Description	Years
089330018-7	018-7	1000	Chrysler Outboards, All Engines	1962-84
089330055-1	055-1	1100	Force Outboards, All Engines	1984-99
089330048-9	048-9	1200	Honda Outboards, All Engines	1978-99
089330007-1	007-1	1300	Johnson/Evinrude Outboards, 1-2 Cyl	1956-70
089330008-X	008-X	1302	Johnson/Evinrude Outboards, 1-2 Cyl	1971-89
089330026-8	026-8	1304	Johnson/Evinrude Outboards, 1-2 Cyl	1990-95
089330009-8	009-8	1306	Johnson/Evinrude Outboards, 3-4 Cyl	1958-72
089330010-1	010-1	1308	Johnson/Evinrude Outboards, 3, 4 & 6 Cyl	1973-91
089330040-3	040-3	1310	Johnson/Evinrude Outboards - All V Engines	1992-96
089330063-2		1310	Johnson/Evinrude Outboards - All V Engines	1992-01
089330052-7		1312	Johnson/Evinrude Outboards, All In-line engines/2 & 4 Stroke	1996-01
089330015-2	015-2	1400	Mariner Outboards, 1-2 Cyl	1977-89
089330016-0	016-0	1402	Mariner Outboards, 3, 4 & 6 Cyl	1977-89
089330012-8	012-8	1404	Mercury Outboards, 1-2 Cyl	1965-91
089330013-6	013-6	1406	Mercury Outboards, 3-4 Cyl	1965-89
089330014-4	014-4	1408	Mercury Outboards, 6 Cyl	1965-89
089330051-9	051-9	1416	Mercury/Mariner Outboards, All Engines	1990-00
089330050-0	050-0	1600	Suzuki Outboards, All Engines	1988-99
089330021-7	021-7	1700	Yamaha Outboards, 1-2 Cyl	1984-91
089330022-5	022-5	1702	Yamaha Outboards, 3 Cyl	1984-91
089330023-3	023-3	1704	Yamaha Outboards, 4 & 6 Cyl	1984-91
089330047-0	047-0	1706	Yamaha Outboards, All Engines	1992-98

STERN DRIVES
ISBN	Part	No	Title/Description	Years
089330029-2	029-2	3000	Marine Jet Drive	1961-96
089330005-5	005-5	3200	Mercruiser Stern Drive	1964-91
089330053-5		3206	Mercruiser Stern Drive - All	1992-01
089330004-7	004-8	3400	OMC Stern Drive	1964-86
089330025-X	025-X	3402	OMC Cobra Stern Drive	1985-95
089330056-X	025-X	3402	OMC Cobra Stern Drive	1985-99
089330011-X	011-X	3600	Volvo/Penta Stern Drives	1968-91
089330038-1	038-1	3602	Volvo/Penta Stern Drives	1992-93
089330041-1	041-1	3604	Volvo/Penta Stern Drives	1992-95
089330057-8	041-1	3604	Volvo/Penta Stern Drives	1992-02

INBOARDS
ISBN	Part	No	Title/Description	Years
089330049-7	049-7	7400	Yanmar Inboards	1975-98

PERSONAL WATERCRAFT
ISBN	Part	No	Title/Description	Years
089330032-2	032-2	9200	Kawasaki	1973-91
089330042-X	042-X	9202	Kawasaki	1992-97
089330045-4	045-4	9400	Polaris	1992-97
089330033-0	033-0	9000	Sea-Doo/Bombardier	1988-91
089330043-8	043-8	9002	Sea-Doo/Bombardier	1992-97
089330034-9	034-9	9600	Yamaha	1987-91
089330044-6	044-6	9602	Yamaha	1992-97

Seloc-On-Line (Internet Access)
ISBN	Part	No	Title/Description	Years
089330075-6		5000	One Mfg/Model - Subscription per user 3 years	1990-01
		5002	Master Pack -Case of 6 CD's including POP Display	

PROFESSIONAL TECHNICIANS MANUALS
ISBN	Part	No	Title/Description	Years
089330060-8		4500	Labor Guide - Johnson and Evinrude	1980-00
089330061-6		4550	Labor Guide - Yamaha	1980-00
089330062-4		4600	Labor Guide - Mercury	1980-00

BRIGGS AND STRATTON
ISBN	Part	No	Title/Description	Years
096376610-4	610-4	610	Briggs & Stratton 4 Stroke Horizontal Crankshaft	1950+
096376611-2	611-2	611	Briggs & Stratton 4 Stroke Vertical Crankshaft	1950+
096376612-0	612-0	612	Briggs & Stratton 4 Stroke Overhead Crankshaft	1950+

ENGINE FINDER

The following listings contain all engines covered in this manual

Model/Engine	Year
2.5 hp, 1 cyl, 2-stroke	1993 - 2000
3 hp, 1 cyl, 2-stroke	1990 - 1992
3.3 hp, 1 cyl, 2-stroke	1993 - 2000
4 hp, 1 cyl, 2-stroke	1990 - 2000
5 hp, 1 cyl, 2-stroke	1990 - 2000
6 hp, 2 cyl, 2-stroke	1993 - 2000
8 hp, 2 cyl, 2 stroke	1990 - 2000
9.9 hp, 2 cyl, 2-stroke	1990 - 2000
10 hp, 2 cyl, 2-stroke	1994 - 2000
15 hp, 2 cyl, 2-stroke	1990 - 2000
20 hp, 2 cyl, 2-stroke	1990 - 2000
25 hp, 2 cyl, 2-stroke	1990 - 2000
30 hp, 2 cyl, 2-stroke	1992 - 2000
30 hp, 2 cyl, 2-stroke	1993 - 2000
40 hp, 2 cyl, 2-stroke	1995
40 hp, 2 cyl, 2-stroke	1999 - 2000
40 hp, 3 cyl, 2-stroke	1998 - 2000
40 hp, 4 cyl, 2-stroke	1990 - 1997
50 hp, 3 cyl, 2-stroke	1990 - 2000
60 hp, 3 cyl, 2-stroke	1990 - 2000
75 hp, 3 cyl, 2-stroke	1990 - 2000
90 hp, 3 cyl, 2-stroke	1990 - 2000
100 hp, 4 cyl, 2-stroke	1990 - 2000
115 hp, 4 cyl, 2-stroke	1990 - 2000
125 hp, 4 cyl, 2-stroke	1993 - 2000
135 hp, V6, 2-stroke	1990 - 2000
135 Optimax/135 hp, V6, 2-stroke	1997 - 2000
150 hp, V6, 2-stroke	1990 - 1996
150 EFI/150 hp, V6, 2-stroke	1993 - 2000
150 Optimax/150 hp, V6, 2-stroke	1997 - 2000
175 hp, V6, 2-stroke	1990 - 2000
175 EFI/175 hp, V6, 2-stroke	1993 - 2000
200 hp, V6, 2-stroke	1990 - 2000
200 EFI/200 hp, V6, 2-stroke	1990 - 2000
200 Optimax/200 hp, V6, 2-stroke	1997 - 2000
225 hp, V6, 2-stroke	1997 - 2000
225 EFI/225 hp, V6, 2-stroke	1995 - 2000
225 Optimax/225 hp, V6, 2-stroke	1997 - 2000
250 EFI/250 hp, V6, 2-stroke	1995 - 2000
275 hp, V6, 2-stroke	1990 - 1994
3.4L/275 hp, V6, 2-stroke	1984 - 1986
Jet Drive	1990 - 2000
XR6 Classic, V6, 2-stroke	1999 - 2000